Transdisciplinary Ecohydrology

Ecohydrology from Catchment to Coast

Transdisciplinary Ecohydrology

Advanced Nature Based Solutions for the Enhancement of Catchment Sustainability Potential

Maciej Zalewski with Invited Co-authors
European Regional Centre for Ecohydrology of the Polish Academy of Sciences
under the auspices of UNESCO, Łódź, Poland

Series Editors

Eric Wolanski, Michael Elliott, and Robert H. Richmond

ELSEVIER

Elsevier
Radarweg 29, PO Box 211, 1000 AE Amsterdam, Netherlands
125 London Wall, London EC2Y 5AS, United Kingdom
50 Hampshire Street, 5th Floor, Cambridge, MA 02139, United States

ISBN: 978-0-443-41596-8

For information on all Elsevier publications visit our website at
https://www.elsevier.com/books-and-journals

Publisher: Peter Llewellyn
Acquisitions Editor: Maria Elekidou
Editorial Project Manager: Surbhi Tyagi
Production Project Manager: M. Balakrishnan
Cover Designer: S. Bakyalakshmi

Typeset by TNQ Tech

Contents

Contributors

Marco Albarracín, UNESCO Scientific Advisory Committee (SAC) on Ecohydrology, Quito, Ecuador; WasiLab - Center for Sustainability Sciences, Pontificia Universidad Católica del Ecuador (PUCE), Quito, Ecuador

Apip, Asia Pacific Centre for Ecohydrology (APCE) – UNESCO C2C, Bogor Regency – West Java, Indonesia; Research Centre for Limnology and Water Resources, National Research and Innovation Agency, Bogor Regency – West Java, Indonesia

Agnieszka Bednarek, University of Lodz, Faculty of Biology and Environmental Protection, UNESCO Chair on Ecohydrology and Applied Ecology, Łódź, Poland

Kamila Belka, European Regional Centre for Ecohydrology of the Polish Academy of Sciences under the auspices of UNESCO, Łódź, Poland

Mulugeta Dadi Belete, Hawassa University, Faculty of Biosystems and Water Resources Engineering, Hawassa, Sidama Region, Ethiopia

Mohammad Mahdi Dorafshan, Department of Civil Engineering, Isfahan University of Technology, Isfahan, Iran

Saeid Eslamian, Department of Water Science and Engineering, College of Agriculture, Isfahan University of Technology, Isfahan, Iran

M. Fakhrudin, Asia Pacific Centre for Ecohydrology (APCE) – UNESCO C2C, Bogor Regency – West Java, Indonesia; Research Centre for Limnology and Water Resources, National Research and Innovation Agency, Bogor Regency – West Java, Indonesia

Arnoldo Font-Nájera, European Regional Centre for Ecohydrology of the Polish Academy of Sciences under the auspices of UNESCO, Łódź, Poland

Piotr Frankiewicz, University of Lodz, Faculty of Biology and Environmental Protection, UNESCO Chair on Ecohydrology and Applied Ecology, Łódź, Poland

Wojciech Frątczak, European Regional Centre for Ecohydrology of the Polish Academy of Sciences under the auspices of UNESCO, Łódź, Poland

Ilona Gagala-Borowska, European Regional Centre for Ecohydrology of the Polish Academy of Sciences under the auspices of UNESCO, Łódź, Poland

Tundisi Jose Galizia, International Institute of Ecology, São Calos, São Paulo, Brazil

Malgorzata Godlewska, European Regional Centre for Ecohydrology of the Polish Academy of Sciences under the auspices of UNESCO, Łódź, Poland

Mohammad Hossein Golmohammadi, Department of Civil Engineering, Isfahan University of Technology, Isfahan, Iran

David M. Harper, University of Leicester, School of Geography, Geology and the Environment, Leicester, United Kingdom; Freshwater Biological Association, Milnthorpe, Cumbria, United Kingdom

Katarzyna Izydorczyk, European Regional Centre for Ecohydrology of the Polish Academy of Sciences under the auspices of UNESCO, Łódź, Poland

Paweł Jarosiewicz, European Regional Centre for Ecohydrology of the Polish Academy of Sciences under the auspices of UNESCO, Łódź, Poland; University of Lodz, Faculty of Biology and Environmental Protection, UNESCO Chair on Ecohydrology and Applied Ecology, Łódź, Poland

Tomasz Jurczak, University of Lodz, Faculty of Biology and Environmental Protection, UNESCO Chair on Ecohydrology and Applied Ecology, Łódź, Poland

Andrzej Kędziora, Institute for Agricultural and Forest Environment of the Polish Academy of Sciences, Poznań, Poland

Edyta Kiedrzyńska, European Regional Centre for Ecohydrology of the Polish Academy of Sciences under the auspices of UNESCO, Łódź, Poland; University of Lodz, Faculty of Biology and Environmental Protection, UNESCO Chair on Ecohydrology and Applied Ecology, Łódź, Poland

Marcin Kiedrzyński, University of Lodz, Department of Biogeography, Paleoecology and Nature Conservation, Łódź, Poland

Kinga Krauze, European Regional Centre for Ecohydrology of the Polish Academy of Sciences under the auspices of UNESCO, Łódź, Poland

Izabela A. Levy, European Regional Centre for Ecohydrology of the Polish Academy of Sciences under the auspices of UNESCO, Łódź, Poland

Joanna Mankiewicz-Boczek, European Regional Centre for Ecohydrology of the Polish Academy of Sciences under the auspices of UNESCO, Łódź, Poland; University of Lodz, Faculty of Biology and Environmental Protection, UNESCO Chair on Ecohydrology and Applied Ecology, Łódź, Poland

Takako Matsumura-Tundisi, International Institute of Ecology, São Calos, São Paulo, Brazil

Elżbieta Mierzejewska-Sinner, University of Lodz, Faculty of Biology and Environmental Protection, UNESCO Chair on Ecohydrology and Applied Ecology, Łódź, Poland

Timothy Mwinami, National Museums of Kenya, Department of Ornithology, Nairobi, Kenya

Arnoldo Font Nájera, European Regional Centre for Ecohydrology of the Polish Academy of Sciences under the auspices of UNESCO, Łódź, Poland

Yohannes Zerihun Negussie, Ministry of Water and Energy, Integrated Water Resources Management, Basin Scale Resilience Initiative for Ethiopia (BaSRINET) Project, Addis Ababa, Ethiopia

Reynard P.S., Industrial Engineering Study Program, Faculty of Engineering, University of Indonesia, Depok City – West Java, Indonesia

Nic Pacini, University of Leicester, School of Geography, Geology and the Environment, Leicester, United Kingdom; Department of Environmental Engineering, University of Calabria, Arcavacata di Rende, Italy

Stéphane Rodrigues, Cités, Territoires, Environnement et Sociétés (CITERES, CNRS UMR 7324) and Polytechnic School of Engineering, University of Tours, Tours, France

Laurent Schmitt, Laboratoire Image, Ville, Environment (LIVE, UMR 7362), University of Strasbourg, CNRS, ENGEES, France

Liliana Serwecińska, European Regional Centre for Ecohydrology of the Polish Academy of Sciences under the auspices of UNESCO, Łódź, Poland

Ali Shokri, School of Engineering, Waikato University, Hamilton, New Zealand

Ignasius D.A. Sutapa, Asia Pacific Centre for Ecohydrology (APCE) – UNESCO C2C, Bogor Regency – West Java, Indonesia; Research Centre for Limnology and Water Resources, National Research and Innovation Agency, Bogor Regency – West Java, Indonesia

Sebastian Szklarek, European Regional Centre for Ecohydrology of the Polish Academy of Sciences under the auspices of UNESCO, Łódź, Poland

Magdalena Urbaniak, University of Lodz, Faculty of Biology and Environmental Protection, UNESCO Chair on Ecohydrology and Applied Ecology, Łódź, Poland

Iwona Wagner, University of Lodz, Faculty of Biology and Environmental Protection, UNESCO Chair on Ecohydrology and Applied Ecology, Łódź, Poland

Silas Wanjala, Lake Naivasha Riparian Association, Naivasha, Kenya

Karl Matthias Wantzen, EUCOR Excellence Chair "Water and Sustainability", Institut Terre et Environnement de Strasbourg (ITES, CNRS/ENGEES UMR 7063), University of Strasbourg, Strasbourg, France; Karlsruhe Institut fur Technologie (KIT), Institute for Water and Environment, Karlsruhe, Germany; UNESCO Chair "Rivers and Heritage", University of Tours, Tours, France; Cités, Territoires, Environnement et Sociétés (CITERES, CNRS UMR 7324) and Polytechnic School of Engineering, University of Tours, Tours, France

Renata Włodarczyk-Marciniak, European Regional Centre for Ecohydrology of the Polish Academy of Sciences under the auspices of UNESCO, Łódź, Poland

Adrianna Wojtal-Frankiewicz, University of Lodz, Faculty of Biology and Environmental Protection, UNESCO Chair on Ecohydrology and Applied Ecology, Łódź, Poland

Maria Yustiningsih, Biology Education, Faculty of Teacher Training and Education (FKIP), University of Timor, Kefamenanu – TTU Regency, East Nusa Tenggara, Indonesia

Maciej Zalewski, European Regional Centre for Ecohydrology of the Polish Academy of Sciences under the auspices of UNESCO, Łodź, Poland; Vice-Chairperson of the Intergovernmental Hydrological Programme (IHP) Council, UNESCO

Preface

Water has become universally recognized as a key factor in achieving the Sustainable Development Goals (SDGs). The increasing anthropogenic impacts have altered the hydrological cycle and accelerated the degradation of water resources. The growing complexity of these pressures requires innovative tools and systemic approaches grounded in integrative science. Over the past 3 decades, Ecohydrology has emerged as an innovative scientific framework that combines hydrology and ecology to strengthen ecosystem resilience and improve water management.

This year marks the 30th anniversary of Ecohydrology at UNESCO and 20 years of UNESCO Ecohydrology Demonstration Sites. Since its inception, Ecohydrology has contributed significantly to global water and sustainability science, knowledge integration and transfer, and supporting scientific insights to decision makers. By combining hydrology and ecology, Ecohydrology promotes nature-based and science-driven solutions that support Member States in reconciling environmental protection with socio-economic development.

The earliest conceptual foundations of Ecohydrology were shaped in the mid-1990s through collaboration between the Man and the Biosphere Programme (MAB) and the Intergovernmental Hydrological Programme (IHP), with early work focused on the role of land–water ecotones in landscape management and restoration. This perspective helped bring the catchment-scale approach into mainstream water resources management. As IHP evolved through successive phases, its focus expanded from water resources management to societal dimensions and ultimately toward a holistic, integrated, and transdisciplinary approach, including the Ecohydrology.

A major theoretical milestone was the formulation of Transdisciplinary Ecohydrology, emphasizing the interplay between the hydrological cycle and biotic processes from the molecular to the catchment scale. This scientific foundation introduced the concept of "dual regulation", demonstrating how ecological processes can be harnessed as management tools. In 2000, the first formal definition of Ecohydrology as "the use of ecosystem properties as management tools" was articulated through extensive international collaboration. This definition later became a cornerstone of what is now widely known as Nature-based Solutions (NbS).

Over the past 2 decades, Ecohydrological NbS has been designed, tested, and implemented across diverse climatic and socio-economic contexts in different countries and regions, forming a Global Network of UNESCO Ecohydrology Demonstration Sites. These sites serve as living laboratories for innovation, enabling the transfer of knowledge and practical wisdom between countries. They demonstrate how low-cost, high-efficiency solutions can bridge the gap between conventional technologies such as wastewater treatment plants and the natural resilience of ecosystems, for example by reducing diffuse agricultural pollution or mitigating urban stormwater impacts.

This book is structured in three parts. The first part of this book introduces the importance of water in addressing today's accelerating sustainability challenges. It argues that in an era of declining global carrying capacity, society must shift from restoration towards the Enhancement of Catchment Sustainability Potential, aiming to increase water, biodiversity, ecosystem services, and resilience (WBSR).

The second part presents empirical research that has advanced understanding of water–biota interactions and enabled the development of Advanced Nature-based Solutions (ANbS) and systemic interventions. This section also encourages ecohydrologists to continue advancing fundamental scientific research alongside applied innovation.

A subsequent section showcases global case studies illustrating how Ecohydrology theory has been translated into practice, including examples that draw on indigenous knowledge, such as the Subak system in Indonesia. These cases highlight the power of co-creation between science, policy, and local communities.

Finally, this book acknowledges the remarkable international cooperation that has shaped Ecohydrology within the IHP framework. Scientists and practitioners from Argentina, Australia, Austria, Belgium, Brazil, China, Costa Rica, Ecuador, Ethiopia, France, Germany, Great Britain, Greece, Hungary, India, Indonesia, Iran, Italy, Japan, Malaysia, the Netherlands, Nigeria, Poland, Portugal, Spain, Tanzania, Tunisia, Ukraine, and the United States have contributed to the development of this field.

Interest in Ecohydrology continues to grow because it enhances the efficiency of water management while reducing costs. Its increasing importance of Ecohydrology for water security and sustainability is reflected in collaborations with scientific associations such as the UNESCO IHP–IAHR through the development of policy brief on integrating Ecohydrology and Ecohydraulics (hydroengineering approach), and its inclusion in the European Partnership "Water4All" Water Security for the Planet initiatives.

Emerging developments in Big Data, artificial intelligence, molecular biology, phytotechnology, and hybrid engineering approaches will further accelerate the evolution of ANbS and enable their adaptation to diverse catchments conditions. Ecohydrology is thus poised to remain a cornerstone of global sustainability science offering a pathway toward resilient ecosystems, thriving societies, and a more water-secure future.

Rahmah Elfithri
Chief of Section for Capacity Development and Water Family Coordination
Division of Water Sciences, Intergovernmental Hydrological Programme (IHP)
UNESCO, Paris, France

From the author

The enhancement of biosphere sustainability

This book is the result of the joint long-term synergistic efforts of scientists from different continents who share their knowledge and wisdom in the framework of the Intergovernmental Hydrological Programme (IHP) of UNESCO towards accelerating the Sustainable Development Goals. Considering the challenges facing humanity, the science of the 21st century has to become integrative, problem-solving, and policy-oriented. The merit of Transdisciplinary Ecohydrology (TEH) lies in the integration of different branches of science, for discovering new system properties and converting these discoveries into Advanced Nature-based Solutions (ANbS) and using them in a systemic way in the context of the evolutionary-ecosystemic perspective proposed in this book. Major reference points in the formulation of the above ideas include strategic documents such as the *Limits to Growth* report of the Club of Rome, the United Nations Sustainable Development Goals, and the conclusions of the 2023 United Nations Water Conference in New York, which identified water as a key driver for accelerating the SDGs, as well as the Water Framework Directive, the Nature Restoration Law, and the Resilience Strategy of the European Commission.

The important conclusion of the 2023 United Nations Water Conference in New York, which identified water as a key driver for accelerating the United Nations Sustainable Development Goals, was the following: to improve the efficiency of water management in the face of climate change there is an urgent need to: (1) "reduce the fragmentation of efforts ", (2) make broader use of Nature-based Solutions in the framework of ecohydrology, and (3) engage in dialogue with society. There is little doubt that this must be preceded by greater integration among scientific disciplines, with Transdisciplinary Ecohydrology providing an important example of such an approach. This, in turn, implies the need for a holistic framework for action, embodied in the three principles of ecohydrology.

The question is: what is the most important novelty of Transdisciplinary Ecohydrology, compared with fields such as restoration ecology and regenerative ecology? While these approaches propose valuable directions for action, they do not fully address what the critical goals are or how they can be achieved.

What? TEH proposes the ambitious yet realistic goal of "WBSR enhancement". Since "water is a key driver of the SDGs," this implicitly requires an extension of the holistic perspective of Integrated Water Resources Management (IWRM), from a predominantly sociocentric and mechanistic approach toward an evolutionary ecosystem-based perspective. To focus the implementation actions, it has been proposed as an integrative goal—the parallel improvement of four factors driven by water, whose synergy enhances catchment sustainability—Water, Bioproductivity/Biodiversity, Services for Society, and Resilience. Achieving this goal is possible only through dialogue among a broad range of specialists, practitioners, decision-makers, and society, and therefore provides an important pathway for reducing the fragmentation of efforts.

How? This is another important novelty of **TEH.** The development of innovative solutions for different catchments, by first identifying the broad scope of catchment-specific Dual Regulation (the water–biota interplay), can be translated into innovative Advanced Nature-based Solutions and hybrid solutions (ANbS + hydroengineering), implemented and harmonized with local societal priorities.

The reason why we are focusing on the catchment rather than a single ecosystem is that water plays a fundamental role in all ecosystem processes and in regulating the interaction between water and biota. To accurately measure the water budget, we need comprehensive hydrological sciences based on well-proven methodologies and, in most cases, long-term monitoring datasets. Water is the common denominator and regulator of most ecological processes, such as nutrient circulation and energy flow. Biota significantly influence and alter the hydrological cycle. Ecosystems are retaining, purifying, recirculating, and recharging groundwater. Both types of interaction act as the background for the intelligent use of ecosystem processes as innovative management tools to enhance the sustainability potential of the catchment. Transdisciplinary Ecohydrology is both the theoretical framework and the practical approach for implementing a new scientific paradigm and addressing sustainability challenges. By integrating deductive reasoning with complex empirical testing, it enables the development of holistic models that drive innovative, systemic, and hybrid solutions.

Why do we use the term "enhancement" rather than "restoration" or "regeneration"?

It is due to the recognition that the biosphere is in a state of continuous change in land cover, climate, societal priorities, and many related parameters, making restoration to pristine conditions neither possible nor realistic. *According to the Merriam-Webster Dictionary, when the word "enhance" was borrowed into English from Latin, it originally meant "to raise something higher." Over time, its definition evolved to mean "to increase or improve value, quality, desirability, or attractiveness." The idea of enhancement is deeply embedded in human culture. One might wonder whether "natural" has ever truly been enough for humans.*

Scientific, technological, and medical discoveries have long been directed toward enhancement, improving movement, appearance, and performance. In medicine, enhancement often refers to achieving a healthier state than that existing prior to disease. This book extends the concept of enhancement to Earth's regulatory processes. It proposes strategies to increase the planet's carrying capacity and presents a framework for addressing today's pressing global challenges through **Transdisciplinary Ecohydrology.**

However, there is both a necessity and an opportunity to integrate knowledge from diverse scientific disciplines to enhance the multidimensional sustainability potential of catchments and the biosphere, helping to ensure the resilience of Earth's life-support systems.

This book is intended for a broad audience, encompassing hydrologists, hydroengineers, students, Non-Governmental Organizations (NGOs) and specialists from related disciplines. Its primary objective is to serve both as a foundational resource and as a catalyst for meaningful discourse among ecohydrologists, hydrologists, hydro-engineers, and strategic water management decision-makers. Furthermore, it seeks to foster interactions between experts from diverse disciplines such as ecologists, hydrologists, landscape architects, and the general public, encouraging the transformation of information and knowledge into collective wisdom. Rather than offering a comprehensive review of the discipline, the book aims to inspire creative scientific approaches that drive positive and lasting environmental change. It examines the evolution of the Transdisciplinary Ecohydrology paradigm, the testing of emerging system properties, and the translation of these properties into applied hybrid solutions to help shape a more sustainable future.

We are confident that providing an accessible summary of the theoretical background of ecohydrology, along with a set of examples and case studies illustrating how ecosystem properties (shaping the structure of ecosystems and regulation of processes) can be used as management tools, will help pave the way for developing a broad range of Advanced Nature-based Solutions, hybrid systems, and systemic solutions for the enhancement of drainage basins' sustainability potential.

UNESCO's Intergovernmental Hydrological Programme (IHP-IX) is clear: the development of the theoretical background for the transdisciplinary, holistic approach is necessary to solve water sustainability problems in different geographic regions, where the hierarchy of challenges and driving factors are different. This approach also has societal priorities related to cultural heritage, which are important reference points. The main goal of this book is to foster dialogue, encourage collaboration among professionals in the built environment, and inspire action from citizens.

The ideas presented in this book were inspired by the work on international teams of scientists during various phases of UNESCO's Intergovernmental Hydrological Programme (IHP) with important contributions from Luis Chicharo, Georg Janauer, David Harper, Jun Xia, Ignasius Sutapa, Jin-song Guo, Eric Wolanski, Karl M. Wantzen, Paweł Jarosiewicz, Zhe Li, Richard Robarts, and William J. Mitsch. Coordination, support, and encouragement by UNESCO Water Division, including Giuseppe Arduino, Andras Szollosi-Nagy, Elfithri Rahmah, Shahbaz Khan, Giovanni Bidoglio, Abou Amani, Frederic Furnier, and Philippe Pypaert, were invaluable. This work has contributed to the development of the holistic approach emphasized in the IHP, and participation in international and national projects funded by the UNESCO Member Countries, the European Commission, and national Polish sources. Many sections and chapters, particularly the case studies and applications discussed in Part II, were authored or coauthored by a diverse group of scholars, each recognized as a leading expert in their respective field. Many contributors are affiliated with the European Regional Centre for Ecohydrology of the Polish Academy of Sciences (ERCE PAS) under the auspices of UNESCO in Łódź, Poland, where I have served as the Founding Director for over 20 years (2000–2021), and I also served as the Founding Chairman and later the Chairholder of Ecohydrology and Applied Ecology at the University of Łódź (1999–2020). During this time, I was supported and inspired by my associates, with special thanks to Piotr Frankiewicz, Bogusława Brewinska Zaraś, Iwona Wagner, Katarzyna Izydorczyk, Edyta Kiedrzyńska, Paweł Jarosiewicz, Małgorzata Tarczyńska, and many students, several of whom contributed across different fields and later developed important directions of their own. A significant contribution to the *Transdisciplinary Ecohydrology* book also emerged through in-depth discussions and essential conceptual clarifications shaped during ongoing conversations with Izabela Levy. The result is an essential introduction to the theory and applications of the rapidly evolving field of ecohydrology. It effectively conveys the importance and immense potential of Transdisciplinary Ecohydrology to contribute to a sustainable future. It also facilitates and fosters broader dialogue and collaboration.

Prof. Maciej Zalewski
European Regional Centre for Ecohydrology of the Polish Academy of Sciences under the auspices of UNESCO, Łodź, Poland
Vice-Chairperson of the Intergovernmental Hydrological Programme (IHP) Council, UNESCO

Acknowledgments

First and foremost, I would like to thank Izabela Levy, whose artistic and large-scale urban landscape planning studies provided an excellent platform for clarifying ideas and making them accessible to a broader audience. I am also grateful to Małgorzata Połatyńska and Małgorzata Łapińska, whose contributions were invaluable during the early stages of the book's preparation. The development of the transdisciplinary field of ecohydrology was made possible thanks to the strong support and stimulating discussions with IHP coordinators Giuseppe Arduino—whose input was of great importance to the development of the transdisciplinary approach in ecohydrology; Shahbaz Khan—whose broad hydrological perspective was crucial in reaffirming the importance of ecohydrology within the IHP; and Rahmah Elfithri—whose networking and communication skills have been instrumental in developing the network of IHP Ecohydrology Demonstration Sites–for the global implementation of Transdisciplinary Ecohydrology for the acceleration of the SDGs. A particularly fundamental milestone was the decision of András Szöllösi-Nagy, who, in 1996, launched ecohydrology in the Vth phase of UNESCO IHP. I would also like to highlight the crucial role of Frédéric Fournier, the first IHP coordinator, whose enthusiasm and support were invaluable in formulating first publication on Ecohydrology concept and working hypothesis UNESCO IHP Document 7 (1997). The scientific foundations of ecohydrology were enriched through numerous joint publications, cooperation, conferences, symposia, and meetings with distinguished scientists worldwide, including Luís Chícharo, David M. Harper, Eric Wolanski, Nicholas Pacini, Richard D. Robarts, Georg Janauer, John E. Thorpe, Fritz Schiemer, William J. Mitsch, Stefano Fazi, Giovanni Bidoglio, Pascal Breil, Yohannes Zerihun Negussie, Pawel Rowinski, Mulugeta Dadi Belete, José G. Tundisi, Johannes Cullmann, Karl M. Wantzen, Marco Albarracín, Jin-song Guo, Artur Magnuszewski, Jun Xia, Yaya, Demin Zhou, Krystian Obolewski, Tadanobu Nakayama, Zhe Li, P.E. Hahanusa, Mario Mendiondo, Wolfgang J. Junk, Vicente Santiago-Fandino, Tomasz Okruszko, Azime Tezer, Vladimir Timchenko, Saeid Eslamian, Angelo A. Agostino, and Jean-Pierre Berton. I would also like to extend my deep gratitude to my team, Piotr Frankiewicz, Bogusława Brewińska-Zaras, Iwona Wagner, Katarzyna Izydorczyk, Edyta Kiedrzyńska, Paweł Jarosiewicz, Małgorzata Godlewska, Tomasz Jurczak, and Małgorzata Tarczyńska, for their tireless work, dedication, and collaboration throughout this journey. I am also especially grateful to those who supported me in the earlier stages and further steps of this endeavour: Włodzimierz Romaniszyn, Tadeusz Pęczak, Anna Hilbricht-Ilkowska, Zdzisław Kajak, Ewa Pieczyńska, Zdzisław Kaczmarek, Leszek Starkel, Zbigniew Kundzewicz. Also, near the final stage of this book the discussion on restoration and regenerative ecology with John Rowan and Delphis Levia were important.

I wish to express my heartfelt appreciation to my wife, Anna, and my daughter, Monika, for their continuous support. They not only shared my enthusiasm, often being the first to offer both inspiring and critical reflections on my ideas but also showed patience and understanding as they endured my frequent absences from home.

Part I

It all starts with water: The origin, evolution, and theoretical foundations of transdisciplinary ecohydrology

Chapter 1

Introduction: Water as a key driver of sustainability

Maciej Zalewski
European Regional Centre for Ecohydrology of the Polish Academy of Sciences under the auspices of UNESCO, Łódź, Poland

1. SDGs too slow: Planetary boundaries at risk

The 1972 report The Limits to Growth, commissioned by the Club of Rome, highlighted the growing complexity of humanity's relationship with the biosphere. More than four decades later, in 2015, the United Nations General Assembly adopted the sustainable development goals (SDGs), a set of 17 interconnected objectives aimed at fostering global peace, prosperity, and environmental stewardship. These goals address health, education, equality, economic development, climate action, and ecosystem protection Fig. 1.1. Designed to be achieved by 2030, their success depends on international cooperation to ensure a sustainable future.

FIGURE 1.1 UN 17 Sustainable Development Goals (2015). The 17 Sustainable Development Goals (SDGs) represent an urgent global call to action for all nations into a global partnership, aiming to end poverty and other deprivations through integrated strategies that promote health, education, equality, and economic growth, while addressing climate change and protecting oceans and forests (United Nations, 2015b). *Source: United Nations (2015). Sustainable Development Goals (SDGs).*

Transdisciplinary Ecohydrology. https://doi.org/10.1016/B978-0-443-41596-8.00010-8

According to a report by the Stockholm Environment Institute (Nilsson and Dalkmann, 2011), based on scenario-based sustainable planning techniques, development is not possible without the universal acceptance of environmentally friendly ecological values. Development focused solely on the progress of civilization, market forces, and politics has proven insufficient, leading instead to social disintegration, increased violence, and poverty (Lempert and Nguyen, 2011).

Only the harmonization of civilizational, economic, and social progress with the natural environment can lead to sustainable development (Zalewski, 2006). For this reason, within a transdisciplinary approach, the basic unit considered in the context of sustainable development is the socioecological system (Redman et al., 2004), also referred to as the coupled human and natural system (Jianguo et al., 2007), which consists of three interconnected components: ecological, economic, and social.

In 2019, the European Commission's Green Deal Strategy reinforced the urgency of sustainability transitions, acknowledging that progress had been insufficient. The Global Carbon Project (Friedlingstein et al., 2025) reports that global CO_2 emissions are expected to reach a record 37.4 gigatons in 2024, driven by fossil fuel use, land-use changes, and deforestation. In addition, forest degradation and climate volatility, intensified by the 2023–24 El Niño phenomenon, are exacerbating environmental instability.

2. Defining a key component of the holistic goal UNESCO IHP–WBSR

UNESCO's Intergovernmental Hydrological Program IHP-IX phase places special emphasis on the development not only the theoretical background for the transdisciplinary holistic approach Box 1.1—necessary to solve water sustainability problems in different geographic regions where the hierarchy of challenges and driving factors are different, but also the translation of the processes into water sustainability science solutions with the consideration of societal priorities related to cultural heritage, which are important reference points. The main goal of this book is to foster dialog, encourage collaboration among professionals in the built environment, and inspire action from citizens. The diagram Fig. 1.2 represents the collaborative efforts of the various thematic groups operating under the objectives of IHP-IX, where at the center of the diagram is a circle labeled *"open-ended thematic working groups"*, symbolizing the core platform for interdisciplinary collaboration. Frame 1 is based on content from UNESCO's *Intergovernmental Hydrological Program IX (IHP-IX) Strategic Plan for 2022–29*, outlining the shift toward a holistic, transdisciplinary approach to water sustainability, integrating ecological, technological, and cultural dimensions in response to escalating global water challenges.

3. European Commission guidelines for achieving good water status

The Water Framework Directive (WFD) (2000/60/EC) imposes an obligation on all member states of the European Union to achieve good status, both qualitative and quantitative, of all waters (rivers, lakes, and groundwater), including marine areas up to one nautical mile from the shore, by the year 2015. It is referred to as a "framework" because it outlines the individual steps required to achieve this goal, rather than simply setting fixed targets in the traditional hierarchical manner.

The aims of the directive go beyond the protection of the aquatic environment; they also include the sustainable socioeconomic development of regions, considering the water demands of local communities and industries. The directive also addresses the prevention of catastrophic events such as droughts and floods (Maciejewski and Walczykiewicz, 2006).

BOX 1.1 Intergovernmental hydrological program IX 2022–29

"To date water management and water engineering were focused on water supply to agriculture, industry, navigation, and domestic use. However, intensive water exploitation, catchment modification and climate change have amplified the stochastic character of the hydrological process and both the intensity and frequency of floods and droughts. This will in turn further negatively impact the water resources per capita ratio.

In a broader context, attaining the SDGs implies recognizing the complex relation between humans and the biosphere, and in particular the role of water as a key driver of bioproductivity, biodiversity, and nutrient cycles, all fundamental life supporting processes, thus the urgent need to harmonize the demand with enhanced water resources. Consequently, the most important challenge for water management is how to increase water resources quantity and quality, and in parallel to increase biodiversity, ecosystem services for society, and resilience to impacts (WBSR). The answer is through a holistic approach which incorporates innovative nature-based and ecohydrological solutions methods (NBS) and catchment-scale systemic solutions, based on understanding the water ecosystems' interplay, as both recipients and producers of water, as well as promoting society involvement through culture and education of water and sustainability (CE). Culture is deeply embodied in societal behavior patterns, defining perception and actions, efficiency and effectiveness."

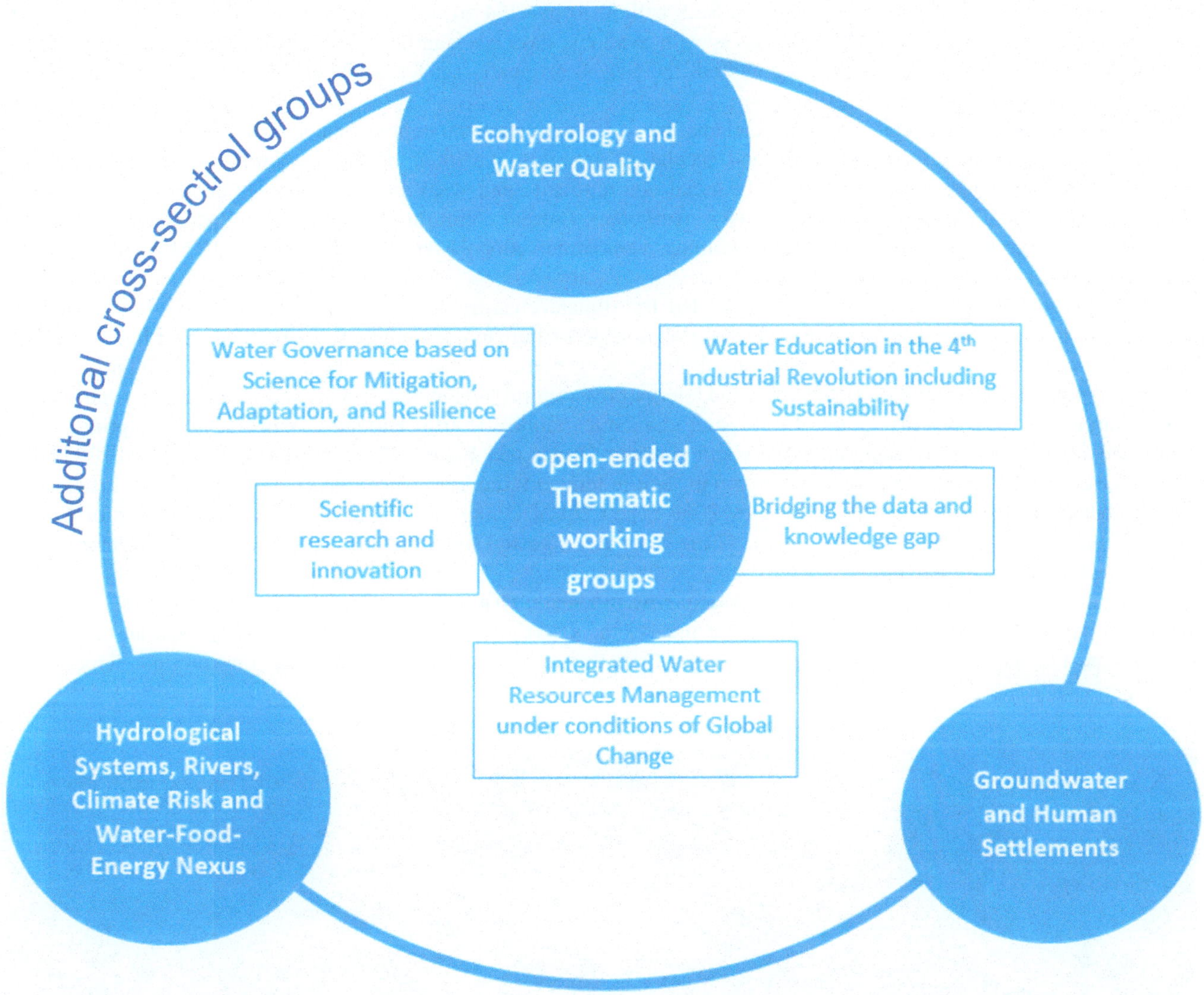

FIGURE 1.2 **UN IHP 2022–29.** "Science for a water secure world in a changing environment." *Source: Operational Framework IHP.*

Importantly, the WFD introduces a division of Europe not based on previous administrative or political borders, but rather on the natural layout of catchment areas, according to geographical and hydrological units. Since rivers often cross-national borders, the WFD requires cooperation between member states within the same catchment area to manage transboundary waters. The main tool for this is the River Basin Management Plan, which is designed to manage catchments at both national and cross-border levels. These plans must include clear management procedures, strategic objectives, and timelines, and must be updated every 6 years.

As a legal act of the European Union, the WFD had to be transposed into national legislation by each member state. However, several factors complicated this process, including political and economic challenges (e.g., lack of financial resources), difficulties in interpreting EU regulations, and issues with implementation at the local level. For these reasons, the timing and effectiveness of transposing the WFD into national law varied between member states, depending on their administrative capacities and level of development (Maciejewski and Walczykiewicz, 2006).

Unfortunately, the goal of good water status as defined in the directive was achieved in only approximately 47% of all waters in the European Union. As a result, the objectives of the WFD remain partially unmet.

4. Financing of SDGs: Gap between needs and reality

The 2024 Report of the Inter-agency Task Force on Financing for Development (United Nations) underscores a pronounced widening gap in financing and investment for the SDGs, particularly in developing nations. Since 2015, the annual investment shortfall has escalated from $2.5 trillion to over $4 trillion, impacting all SDG sectors. The report cautions that nearly half of the 140 SDG targets with available data are lagging behind the required trajectory. Unless current "business-as-usual" approaches are altered, the SDGs, originally set for achievement by 2030, will remain unattainable even by 2050.

Despite their greater financial capacity for adaptation and experience with unprecedented climate extremes, such as the 2024 floods in Central Europe and Spain and escalating droughts, wealthier northern countries have not effectively addressed the issues. However, insufficient financial commitments and weak policy responses continue to undermine global efforts, leaving climate action dangerously insufficient. Current commitments have already resulted in a deviation from the Paris Agreement's objective of maintaining a maximum global warming of 1.5°C or 2°C by the end of the century.

The depletion of natural resources, compounded by climate change, is undermining the regenerative capacity of planetary systems. As illustrated in Fig. 1.3, the global scale of human consumption, primarily driven by fossil fuels, reveals the extent of our impact on Earth, highlighting how human expansion is disrupting fundamental ecological processes, including water and nutrient cycles and energy flows within the biosphere. This accelerating degradation poses serious threats to food security, public health, and economic stability (Earth: Human Expansion is Disrupting Fundamental Ecological Processes, Including Water and Nutrient Cycles and Energy Flows Within the Biosphere, 2024). By 2030, an estimated 150 million people per year will require humanitarian assistance due to environmental disasters, with this number projected to rise to 200 million by 2050 (The Cost of Doing Nothing: The Humanitarian Price of Climate Change and How it Can Be Avoided, 2019). These risks are expected to compound over time, becoming increasingly complex and difficult to manage.

While technological advancements, circular economy models, conservation strategies, and impact mitigation efforts play essential roles, these measures alone cannot resolve the crises of water scarcity, biodiversity loss, and ecosystem degradation. The biosphere undergoes rapid transformation, driven by nonrenewable resource consumption, demographic

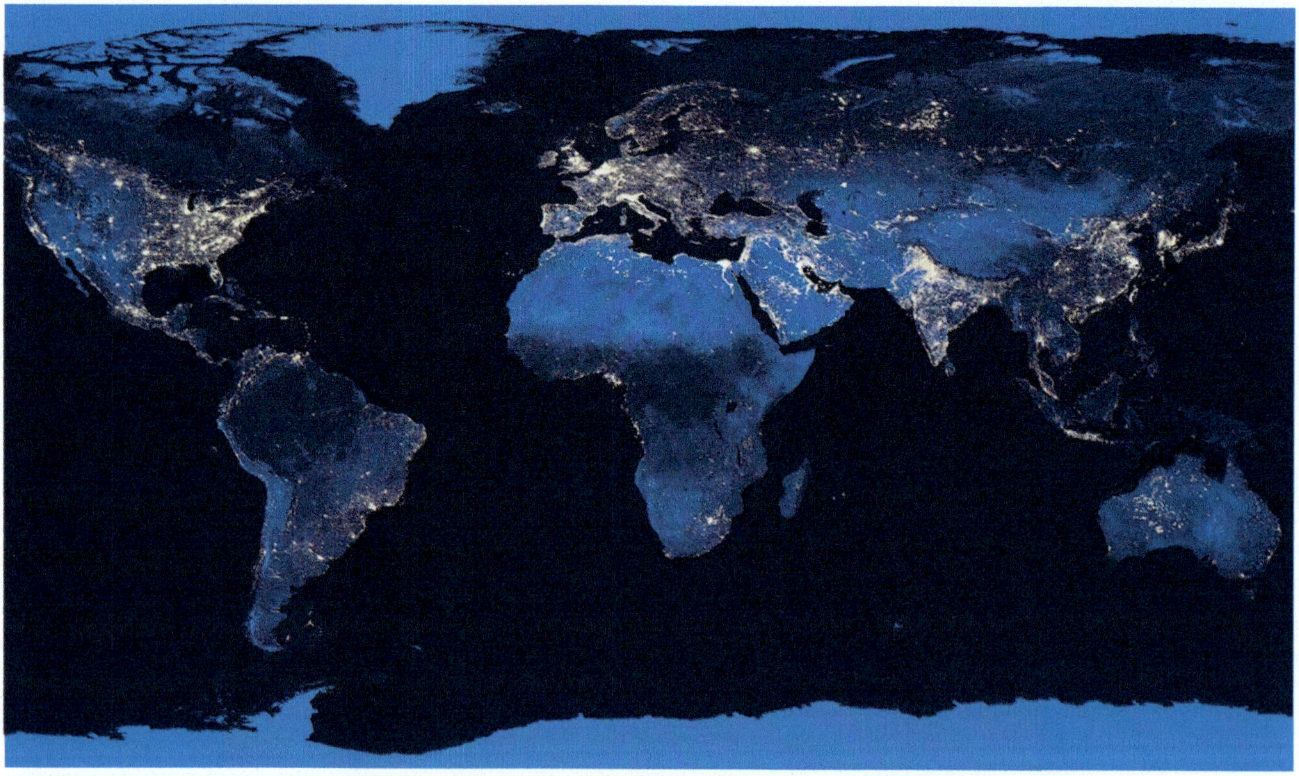

FIGURE 1.3 Earth at night. A view from space of the whole earth at nighttime, plainly showing how many of the planet's areas are currently being used by humankind, and gives some insight into how much energy is utilized daily. The photograph illustrates not only what huge amounts of energy, mostly fossil fuels, we make use of every day, emitting greenhouse gases, but also to what extent and with what intensity we are altering fundamental ecological processes such as water circulation, biogenic circulation, and energy flow in areas that are characterized by intensive human activity. *Photo credit: C. Mayhew and R. Simmon, Source: NASA https://apod.nasa.gov/apod/ap001127.html.*

expansion, land-use changes, and climate variability. These forces are pushing Earth's systems toward critical planetary thresholds (Richardson, 2025; Rockström et al., 2009; Steffen et al., 2018), see Fig. 1.4, Historically, efforts to enhance ecosystem services, such as agricultural intensification, have often led to unintended environmental consequences, including the degradation of water retention capacity, biodiversity, and soil health (Levia et al., 2020). While agriculture has increased food supply, industrial farming has also caused declines in water quality, soil fertility, and ecosystem diversity Fig. 1.5.

Homogenization of agriculture landscape and overgrazing cause:

1. Reduction of water retention capacity in the catchment
2. Loss of organic matter and nutrients from the soil
3. Increase of nonpoint-source pollution
4. Degradation of water quality and eutrophication
5. Decrease in ecosystem resilience to climate change
6. Reduction of groundwater recharge
7. Reservoir siltation

Ethiopia Box 1.2 and the Lake Tanganyika region Box 1.3 exemplify the urgent water-related challenges faced in sub-Saharan Africa, where land degradation, erosion, and population pressures undermine ecosystem services and human well-being. Through transdisciplinary ecohydrology (TEH) approaches supported by UNESCO and international agencies, these regions have piloted integrated approaches to restore catchment function, protect aquatic biodiversity, and enhance water sustainability. In turn, this provided the fundamentals for the recent development of the bioeconomy and circular economy.

Despite previous ecosystem overexploitation, natural recovery processes can be reinitiated when key enabling conditions—such as the reduction of groundwater extraction—are restored, as illustrated in Box 1.3. In the Doñana region of Spain, following severe groundwater depletion in 2000, ecological regeneration occurred between 2001 and 2003, demonstrating the inherent resilience of ecosystems once hydrological stress is alleviated.

5. Key factors of sustainable future

A major constraint remains the scarcity of high-quality soil, as land-use intensification has resulted in landscape homogenization, significantly affecting water resources and hydrological cycles Fig. 1.8. Urbanization, agricultural expansion, and climate change have intensified these pressures, leading to increased pollutant emissions and nutrient runoff (nitrogen, phosphorus) into aquatic ecosystems. These processes heighten pollution, soil erosion rates, cause

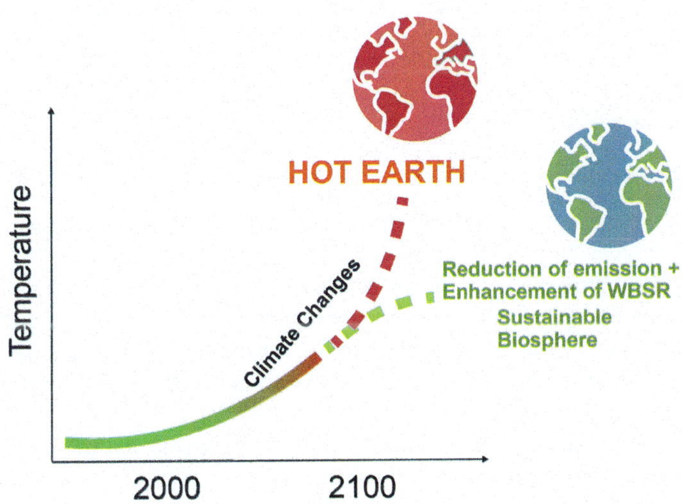

FIGURE 1.4 **"Hothouse Earth" in the Anthropocene.** This diagram illustrates the pathways of the earth system under different scenarios. The dashed pathway represents the potential outcomes: Progressing toward either a "Stabilized Earth" or a "Hothouse Earth," depending on human-driven greenhouse gas emissions and biosphere degradation (~2°C). *Inspired by concepts Steffen, W., Rockström, J., Richardson, K., Lenton, T.M., Folke, C., Liverman, D., et al., 2018. Trajectories of the earth system in the anthropocene. Proc. Natl. Acad. Sci. U. S. A. 115 (33), 8252–8259. https://doi.org/10.1073/pnas.1810141115.*

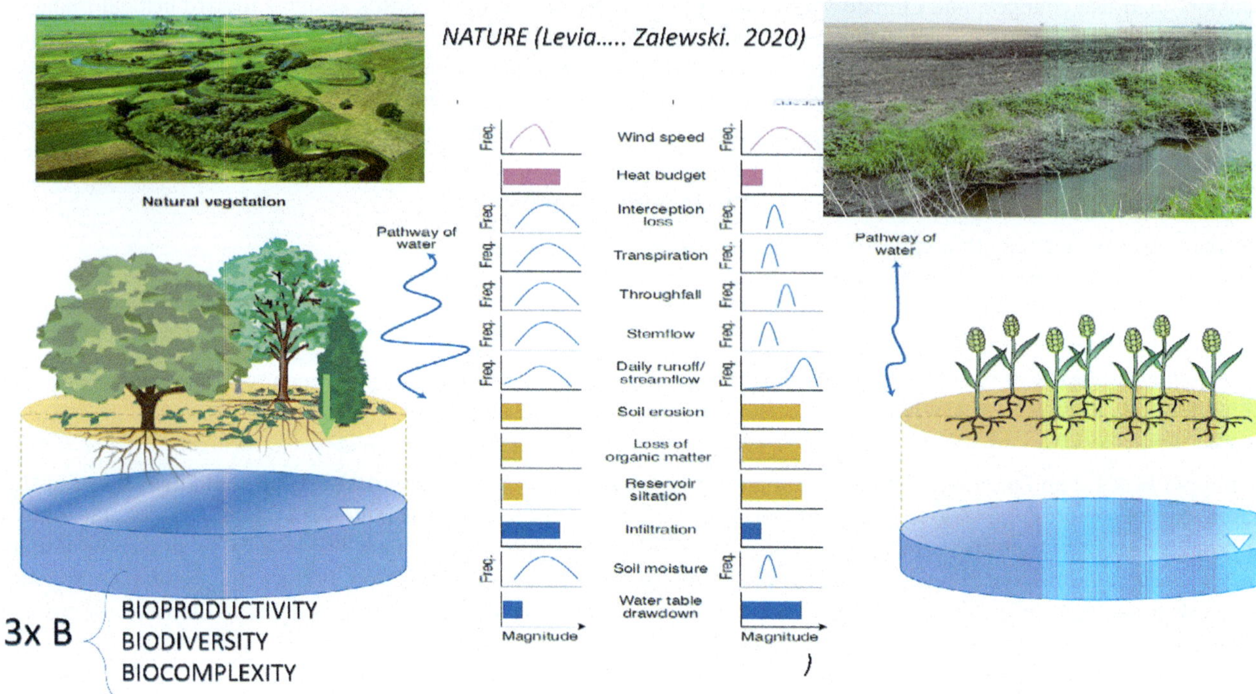

FIGURE 1.5 Fragmentation of knowledge, values, and efforts "Green Deal" versus Industrial Agriculture. Conceptual diagrams showing the differences in the water cycle of natural vegetation and monocultures. Plots show the frequency distribution (lines) and relative magnitude (bars) of the hydrological fluxes (blue), atmospheric variables (pink), and soil processes (yellow). No specific scales are shown; magnitudes depend on region and specific vegetation. The vertical dotted yellow lines indicate the distance to the water table (white triangle). This figure illustrates how important the enhancement of landscape complexity is. *Source: Levia, D. F., Creed, I. F., Hannah, D. M., Nanko, K., Boyer, E. W., Carlyle-Moses, D. E., van de Giesen, N., Grasso, D., Guswa, A. J., Hudson, J. E., Hudson, S. A., Iida, S., Jackson, R. B., Katul, G. G., Kumagai, T., Llorens, P., Lopes Ribeiro, F., Pataki, D. E., Peters, C. A., Sanchez Carretero, D., Selker, J. S., Tetzlaff, D., Zalewski, M., Bruen, M., 2020. Homogenization of the terrestrial water cycle. Nat. Geosci. 13 (10), 656–658. https://doi.org/10.1038/s41561-020-0641-y.*

BOX 1.2 The key challenges of water resources and sustainability in Ethiopia

The key challenges of water resources and sustainability in Ethiopia are overpopulation, deforestation, overgrazing, and reservoir siltation Fig. 1.6.

The Polish Aid Programme for Ethiopia on Ecohydrology for the Enhancement of Water Resources and Sustainability, led by ERCE PAN under the auspices of UNESCO (2010–2015) and financed by the Ministry of Foreign Affairs of Poland, resulted in the establishment of the African Centre for Ecohydrology under UNESCO at the Ministry of Water and Energy of Ethiopia. Reproduced from Zerihun Negussie, Y., Urbaniak, M., Szklarek, S. Lont, K., Gągała, I., Zalewski, M., 2012. Efficiency analysis of two sequential biofiltration systems in Poland and Ethiopia — the pilot study. Ecohydrol. Hydrobiol. *12 (4), 271–285. doi: 10.2478/ v10104-012-0028-9.*

organic matter loss, and destabilize hydrological systems, creating a self-reinforcing feedback loop that accelerates ecosystem degradation.

Human populations operate within evolutionary constraints that regulate the balance between demographic growth and ecosystem productivity. The rapid expansion of human populations could trigger significant social, economic, and geopolitical disruptions. Historical patterns reinforce this concern; civilizations in the Mediterranean Basin, for example, experienced environmental deterioration, leading to social unrest, territorial conflicts, and eventual collapse (Wantzen et al., 2016).

As soils become increasingly nonrenewable, global per capita availability of high-quality soil is projected to shrink from 0.42 ha in 1960 to just 0.22 ha by 2044 (Ye and Van Ranst, 2009). Intensive agriculture and climate change have led to accelerated soil erosion, loss of organic matter, and increased biogenic nutrient loads in rivers, lakes, and coastal waters, fueling harmful cyanobacterial blooms Fig. 1.9.

Until the 1970s, Earth's ecological footprint (EF) remained below 1, indicating that the planet's regenerative capacity could meet human demand. However, in the 21st century, the global EF has surpassed 1.71, with more than 80% of the

FIGURE 1.6 **Challenges in Ethiopia**. (A) This photograph shows the effects of overpopulation and deforestation. (B) This photograph shows the effects of pastures overgrazing. (C) This photograph shows the effects of reservoir siltation due to deforestation and pastures overgreezing. *Photo Credit: M. Zalewski.*

BOX 1.3 Resilience: The importance of groundwater on ecosystem

Effect of upstream ground water overexploitation (2000) in subsequent years of ecosystem recovery (2001–03) due to reduced exploitation of groundwater (Spain, Donana) Fig. 1.7.

Catchment (drainage basin) definition

Catchment (or drainage basin) is the most basic concept of hydrologic science—the area of land contributing water to a nominated point on the earth's surface. This concept works well where there is a well-defined stream and ridge system. Equivalent terms are watersheds (US) and drainage basins (US and UK) (Bren, 2015).

In the context of transdisciplinary ecohydrology, **the catchment serves as the fundamental unit of analysis and intervention**. This is because the hydrological cycle and associated ecological processes are analyzed and quantified at the catchment scale. Furthermore, the catchment provides an appropriate framework for assessing and managing the impacts of dynamic changes, particularly those driven by climate change.

world's population living in regions experiencing ecological deficits (Earth Overshoot Day 2024 Falls on August 1st. Press Release, 2024).

Over the past 6 decades, the human population has grown from 2.5 billion to over eight billion, while global GDP, a key measure of environmental strain, has expanded tenfold, rising from \$5.3 trillion to \$55 trillion. Despite these

FIGURE 1.7 Groundwater overexploitation and ecosystem resilience. From overexploitation to ecosystem recovery as result of reduction of groundwater exploitation. *Courtesy of professor S. Llamas.*

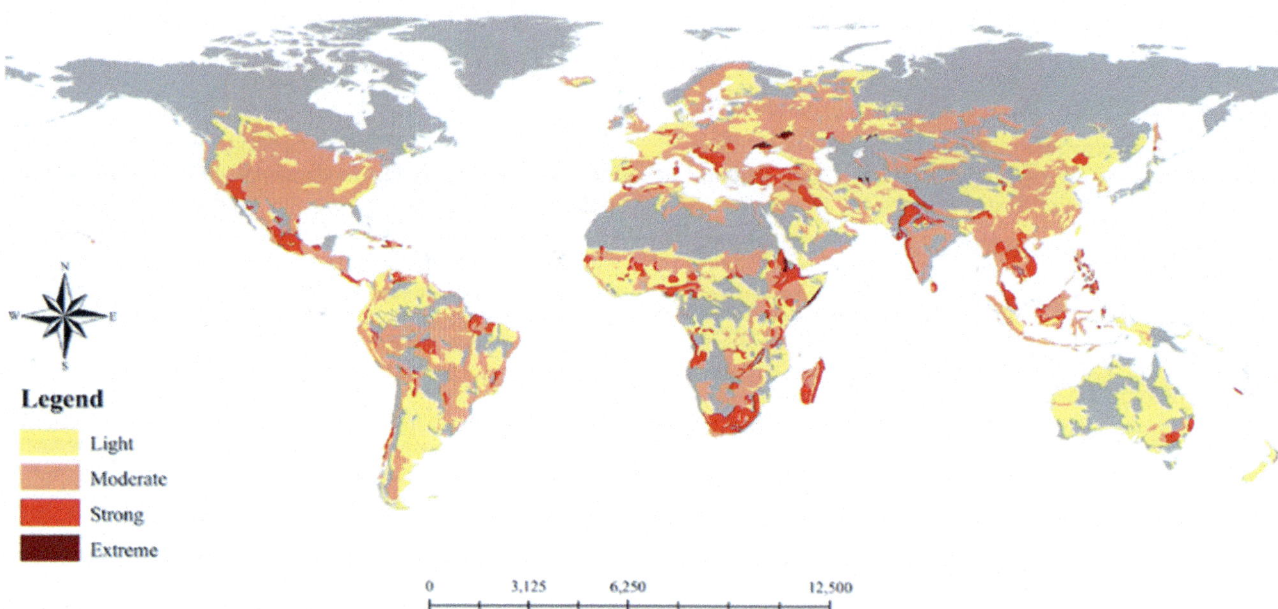

FIGURE 1.8 Land degradation. The global extent of land degradation. *Reproduced from Edrisi, S. A., Sarkar, P., 2022. Assessing the realization of global land restoration: A meta-analysis. Anthropocene Sci. 1 (1), 1–16.*

FIGURE 1.9 The result of the impact of nutrients (N, P). Toxic algal blooms resulting from the transfer of nutrients (N, P) from catchments to aquatic systems. (A) The bay in the Sulejowski reservoir, where a stream flowing from intensively manure-fertilized meadows delivers large amounts of nitrogen and phosphorus compounds, leading to algal blooms. (B) The emerging cyanobacterial bloom near the lake shore results from the infiltration of shallow groundwater originating from the recreational area. Groundwater from the densely built summer cottages, with leaky septic tanks, significantly contributes to this occurrence. The photo illustrates the rate at which nutrient-rich groundwater reaches the lake within 10 h after a storm. (C) Baltic sea cyanobacterial (algal) bloom moving with the wind alongside the Hel Peninsula caused by a failure that occurred in the sewage pumping station in Gdansk due to an overload after a heavy rainfall. This was additionally amplified by Vistula River's floodwaters reaching the Gulf of Gdansk, both causing a massive bloom which moved along the coastline. *Jurata, July 25, 2019, 11:57 a.m. Photo credit: Zalewski.*

challenges, GDP remains the dominant economic metric, reinforcing a consumption-driven economy that shifts responsibility for reducing environmental impact onto others.

This scenario mirrors the tragedy of the commons, a concept originally introduced by William Forster Lloyd (1833) and later formalized by Garrett Hardin (1968), which describes how unregulated resource access leads to depletion (Purvis, 1970). A failure to act endangers humanity's long-term survival, deepens socioeconomic disparities, and threatens the biosphere's capacity to sustain life. Vulnerable populations, who contribute least to climate change, are projected to bear the greatest burdens (Earth: Human Expansion is Disrupting Fundamental Ecological Processes, Including Water and Nutrient Cycles and Energy Flows Within the Biosphere, 2024). According to the UN World Water

Development Report (2022), in 2022, 2.2 billion people lacked access to safely managed drinking water, whereas more than three billion resided in agricultural regions experiencing severe water scarcity (FAO, 2020).

6. The evolving "quantum entanglement" between humans and nature

Water is a fundamental driver of bioproductivity and biodiversity in the biosphere, making its management critical to sustainability. However, traditional approaches to water resource management and hydroengineering have primarily focused on mitigating floods and droughts, securing water supplies for agriculture, supporting industry and navigation, and meeting household water needs. While these objectives remain essential, climate change, ecosystem degradation, and pollution have intensified the unpredictability of hydrological processes. As a result, declining water resources are reducing bioproductivity, damaging infrastructure, threatening food security, and weakening water-related ecosystem services—all of which are crucial to the resilience of drainage basins.

At a planetary scale, ecosystems are increasingly unable to regenerate naturally as human pressures exceed ecological limits. Since historical estimates, natural ecosystems have declined by 47%, with 75% of the Earth's surface, 85% of wetlands, and 66% of oceans significantly altered by human activity (Intergovernmental Science-Policy Platform on Biodiversity and Ecosystem Services (IPBES), 2019). Forests now cover only about 30% of the continents (The State of Food and Agriculture: Climate Change, Agriculture and Food Security, 2016). The degradation of land and ecosystems has led to a 23% decline in terrestrial productivity, which has further impacted pollinators, who play a crucial role in food production. 75% loss of pollination services in global crop yields, amounting to economic losses of $235–577 billion annually based on 2009 market values (Potts et al., 2016; Summary for Policymakers of the Methodological Assessment of Scenarios and Models of Biodiversity and Ecosystem Services, 2016). As ecosystems' capacity to provide essential services is diminished, the financial benefits of resource exploitation are undermined, increasing vulnerabilities and exacerbating biodiversity loss and food insecurity.

Like "quantum entanglement", humankind and the biosphere are deeply interconnected, each dependent on the other's well-being. Yet, managing this complex relationship has become increasingly challenging. The biosphere's responses to anthropogenic pressures and the resulting climate crisis have led to its classification as a "wicked problem", one that is extraordinarily difficult to solve due to its complexity and interconnected nature. Addressing this challenge requires a holistic approach that integrates an understanding of how evolutionary processes shape feedback mechanisms between water and biota across diverse climatic and geomorphological conditions. Such a profound understanding provides the foundation for "using ecosystem properties as management tools" (Zalewski, 2000) and for translating these insights into Advanced Nature-Based Solutions (ANbS) and broader systemic interventions.

7. Transdisciplinary ecohydrology as a pathway to sustainability

Water and temperature are fundamental drivers of all ecosystem processes. To regulate biotic interactions effectively, it is essential to precisely quantify the water and nutrient budgets. This is crucial for understanding ecosystem metabolism at the catchment scale, as it underpins the processes and feedback between hydrological and ecological dynamics. Such understanding is, in turn, vital for reducing environmental impacts and for developing innovative, cost-effective methods of mitigation.

For this purpose, there exist sound hydrological science-based methodologies and, in most cases, long-term monitoring data. This, combined with an understanding of ecohydrological processes, examining the interplay between water and biota from the molecular to the catchment scale, will enable the regulation and enhancement of life-supporting systems.

It is important to understand and regulate processes, and to recognize the hierarchy of drivers (see Chapter 4 for details). Dobzhansky explains that the environment does not impose evolutionary changes on its inhabitants, as postulated by the now-abandoned neo-Lamarckian theories. The most accurate way to conceptualize this situation is as follows: the environment poses challenges to living species, which may respond by undergoing adaptive genetic changes. This significant statement should become broadly accepted as a strong reference point for the achievement of SDGs and Green Deal strategy goals, as well as the inspiration for reflections on and the change of the way of thinking, considering each of us as an element of a dynamic system which emerges as a result of biological evolution. It should be the stimulus for the fast change of mechanistic paradigm generated in the Industrial Era, according to which nature is the provider of endless resources, with nonlimited regenerative and pollutants absorption potential, resulting in exploitative resource use.

Evolution never moves backward, and in today's rapidly changing world, therefore restoration is not a sufficient option. The accelerating pace of development has led to the elimination of numerous ecosystems or has degraded them to the point where they have lost their fundamental characteristics and become "novel ecosystems" (Hobbs et al., 2006). The global rise in temperature has altered ecosystem dynamics and bioproductivity, while shifting precipitation patterns have contributed to water redistribution. As the biosphere undergoes these changes, many ecosystems have already adapted to new conditions (Rocha et al., 2015). However, restoring these systems to their previous state is impractical in evolving ecosystems, particularly in highly impacted areas such as urban environments (Krauze and Wagner, 2019). The challenge of restoration is further intensified by space constraints, limiting efforts to ecosystem-specific initiatives that fall short of addressing large-scale systems such as entire drainage basins. The need for restoration remains crucial for biodiversity conservation, especially as river valleys serve as refuges from human impacts on catchment. However, as illustrated in the diagram Fig. 1.10, ecosystem recovery often follows a nonlinear trajectory, where degradation past a certain threshold makes full restoration unattainable, necessitating structural reshaping of the entire drainage basin with simultaneous modifications of the water–biota interactions (Zalewski, 2014). The extensive human alteration of drainage basins is evident in the presence of over one million dams across Europe, disrupting river continuum processes and altering the flow of water, energy, and biodiversity (Belletti et al., 2020).

Ecohydrology builds an important bridge between ecology and environmental management.
— Sven Erik Jørgensen, Ecohydrology as an Important Concept and Tool in Environmental Management (2016).

The water and carbon cycles have been identified as essential elements for achieving the "sustainable and desirable future we want" (Costanza et al., 2014). Water and biodiversity are intricately linked through ecological cycles, underscoring the interconnected nature of environmental issues. The degradation of terrestrial ecosystems, characterized by a decline in plant cover, adversely impacts the water cycle, self-purification processes, and water resources. Water plays an essential role in bioproductivity, biomass, and biodiversity, which are vital for the assimilation and retention of carbon. To achieve this, it is crucial to limit emissions from resource consumption while simultaneously preventing the loss of ecosystems' absorptive capacity, which is threatened by biomass overexploitation, deforestation, and the depletion of organic matter in soils because of industrial agriculture.

The processes across a river basin (or catchment area) contribute to nutrient retention, self-purification, and ecosystem health Fig. 1.11. They include key ecohydrological processes functioning throughout a river catchment, extending from the watershed to the floodplain and aquatic ecosystems. It outlines mechanisms such as biocomplexity-driven retention, biomass trapping, sedimentation, biofiltration, transformation within land–water ecotones, denitrification under anaerobic

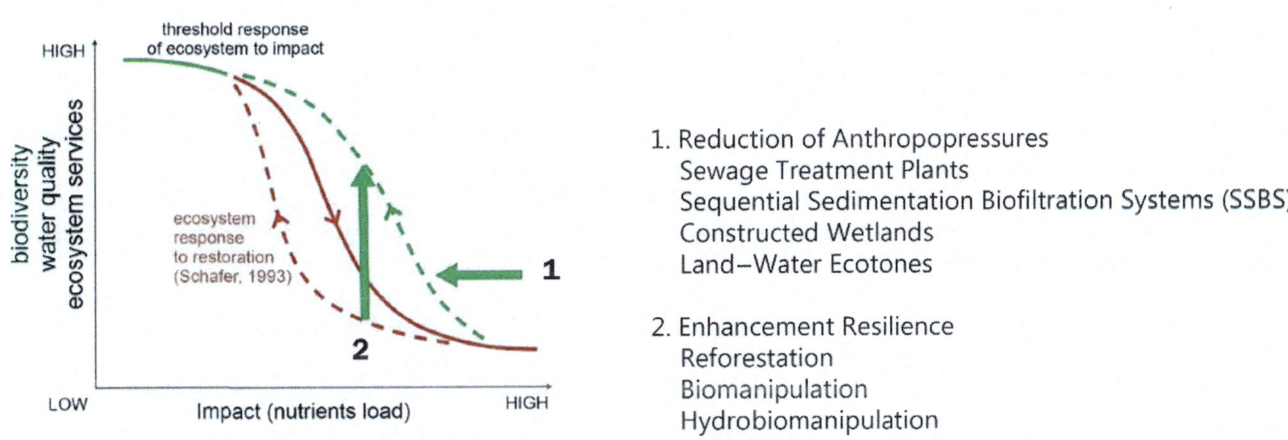

FIGURE 1.10 Ecosystems degradations and pathways to reverse it. The diagram shows the nonlinear response of ecosystems to increasing nutrient load, with a threshold beyond which degradation accelerates. The red line illustrates the limited effectiveness of traditional restoration (Scheffer et al., 1993), the green line represents the application of low-cost advanced nature-based solutions (ANbS) generated on the basis of transdisciplinary ecohydrology that will be discussed in the following chapters. Two key strategies are highlighted: (1) reducing anthropogenic pressures (e.g., sewage treatment, constructed wetlands, land–water ecotones, sequential sedimentation biofiltration systems — SSBS), and (2) enhancing resilience (e.g., reforestation, biomanipulation, and hydrobiomanipulation), emphasizing the need for systemic interventions beyond traditional restoration. Transdisciplinary ecohydrology, described in this book, proposes enhancement of ecosystem sustainability by reducing anthropogenic pressures and strengthening resilience, increasing efficiency while minimizing costs. *Mod. Zalewski, M., 2020. Ekohydrologia. Wydawnictwo Naukowe PWN, Warszawa, ISBN:978-83-01-20804-2.*

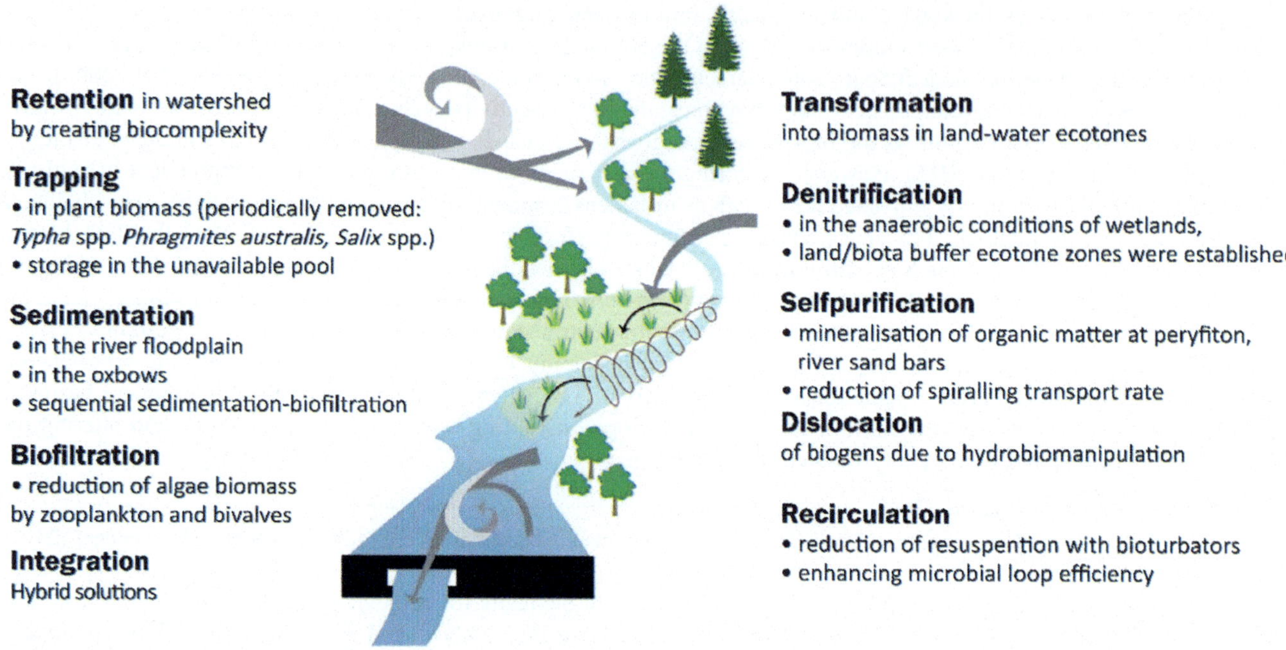

FIGURE 1.11 **Ecohydrological nature-based solutions.** An illustration of the catchment-based, systemic ecohydrological approach, integrating various ecohydrological nature-based solutions to create synergy, mitigate nutrient and pollution inflows (N, P, C) from both point and nonpoint sources. This approach also controls erosion, enhances ecosystem self-purification, and strengthens resilience against climate change, pollution, and eutrophication, and in consequence, significant improvements in water quality and quantity and all related services for society. *Mod. Zalewski, M., 2000. Ecohydrology — the scientific background to use ecosystem properties as management tools toward sustainability of water resources.* Ecol. Eng. *16 (1), 1–8. https://doi.org/10.1016/s0925-8574(00)00071-9.*

conditions, self-purification through organic matter mineralization and, reduced spiraling, biogen dislocation, and microbial loop enhancement via recirculation. Collectively, these interconnected processes in anthropogenically transformed catchments establish a hybrid system that integrates natural dynamics with engineered measures, aiming to improve water quality, strengthen biotic resilience, and enhance ecosystem services at the catchment scale. This approach supports the transformation of catchments into self-regulating systems, where water is retained and recirculated, improving carbon assimilation by plants while reducing its release (e.g., deforestation, wetland degradation).

According to a recent UNESCO publication, ecosystems serve as natural water-retaining, pumping, recirculating, and purification systems on a global scale. They regulate more than 60% of hydrological mesocycles, which are major determinants of drainage basin sustainability. Therefore, the main challenge for water management is this: How can the decline of water resources, in terms of both quantity and quality, be reversed? The proper response, as illustrated throughout this book, involves a holistic approach that combines and incorporates Advanced Nature-based Solutions (ANbS), grounded in ecohydrology.

The hydrological cycle consists of two interrelated phases. The first, terrestrial, involves the critical interactions between water, plants, and soil; a sound understanding of these processes serves as the foundation for enhancing watershed retention capacity and natural purification functions. The second phase is aquatic, where the metabolism of ecosystems and the cycling of nutrients become central to understanding the dynamic interplay between water and ecosystem processes.

This approach offers a dual benefit in that it mitigates the cumulative effects of increasing human impact while improving the four most critical elements of sustainability: water, biodiversity, ecosystem services, and resilience. Achieving these improvements requires the development of a sustainability culture, enhanced education, and the establishment of robust, proactive legal, policy, and governance frameworks. A practical example of this is the use of lateral reservoirs, which demonstrate how enhanced water resources within a catchment area can lead to increased bioproductivity, particularly in diverse habitats (See Chapter 30). This, in turn, stimulates biodiversity, stabilizes the hydrological mesocycle, and enhances nutrient circulation and energy flow within ecosystems. Enhanced water resources, bioproductivity, and biodiversity enable ecosystems to optimize the delivery of ecosystem services based on societal priorities. Ultimately, integrating water, biodiversity, and ecosystem services allows systemic solutions to be developed to enhance the resilience of entire drainage basins. These solutions not only address climate change but also mitigate the cumulative impacts of anthropogenic pressures, fostering a more sustainable and resilient future. To effectively tackle these challenges, ecohydrology must integrate insights from a wide range of disciplines. By nature, it is interdisciplinary, combining hydrology and ecology. However, its influence extends even further when it integrates related disciplines, emphasizing its transdisciplinary approach. The term TEH was first introduced by Zalewski in 2015 and 2023 (Zalewski, 2014; Zalewski and Wantzen, 2023) to emphasize its integrative nature, aiming to address complex challenges at the confluence of multiple disciplines within ecohydrology and facilitating the development of hybrid solutions combining Advanced Nature-based Solutions (ANbS) with hydroengineering. By synthesizing hydrological and ecological processes with ecotechnological innovations, these solutions enhance system resilience, optimize water management, and support ecosystem sustainability in response to evolving environmental pressures.

Transdisciplinary ecohydrology expands the understanding of life-sustaining processes (LSPs) in freshwater and coastal ecosystems by emphasizing the dynamic interplay and feedback between the hydrological cycle and ecosystem functioning (dual regulation). It integrates ecosystem metabolism as a key factor for the development of ANbS for enhancing the sustainability potential (WBSR) of catchments and the biosphere. Using the water–biodiversity–ecosystem services–resilience (WBSR) holistic approach, TEH translates these insights into hybrid, systemic solutions, which enforce Integrated Water Resources Management (IWRM) and WFD.

The challenges of TEH require integrated, catchment-scale strategies that balance ecological functions with human needs. These include enhancing water quantity by reducing atmospheric losses through the evaporation and slowing runoff to increase retention and recharge, as well as improving water quality by minimizing pollution and strengthening natural self-purification processes Box 1.4.

BOX 1.4 Key challenges of transdisciplinary ecohydrology

Water quantity

- Reduce water loss from the catchment to the atmosphere (e.g., through evaporation or evapotranspiration management), e.g., wind speed reduction by increasing complexity of agricultural landscape.
- Reduced outflow of water from the catchment to the sea by increasing retention time by wetlands restoration, stream renaturalization, buffer zones creation, all increases soil water retention and groundwater recharge.

Water quality

- Reduce pollutants: minimize both point-source and nonpoint-source pollution
- Increase resilience, by improvment ecosystem complexity, especially potential for self-purification

This book emphasizes that ecosystem services are not solely determined by the water budget—commonly described in terms of precipitation, flow, and runoff, and pollution—but also by ecohydrological processes, including the hydrological mesocycle (drainage basin), longitudinal patterns (along the river continuum), and seasonal ecological dynamics. This understanding enables:

1. Precise identification and quantification of threats, serve as the foundation for effectively leveraging ecosystem properties as management tools
2. The development of integrated systemic solutions that combine advanced nature-based approaches with biotechnologies, phytotechnology, and hydroengineering, forming ecohydrological hybrid systems designed for catchment and biosphere management.

How and why is transdisciplinary ecohydrology important for developing advanced nature-based and systemic solutions that enhance sustainability?

Concepts and models function as frameworks for the present comprehension of the intricacies inherent in the processes and interactions of human beings, the water cycle, and the biosphere. According to Paul Romer, a renowned American economist, the absence of novel technologies engenders slower growth in gross domestic product (GDP), and the lack of innovative concepts results in technologies not being invented and applied in the most efficacious manner. To expedite the implementation of the sustainable development goals, it is imperative that these are founded on concepts (50%) and technologies (50%). The absence of innovative technologies engenders a decline in GDP growth, whereas the absence of innovative ideas results in resource loss, human capital depreciation, and degradation of the biosphere.

The changes that need to be made consist of two main groups:

1. Tangible change—changing technology, transport, and products
2. Intangible change—changing how we think

TRADITIONAL VIEW ON HYDROLOGY AND BIOLOGY IN ENVIRONMENTAL SCIENCES.

Transdisciplinary ecohydrology is an theoretical and applied science, a subdiscipline of hydrology that seeks to understand ecological processes controlled by the hydrological cycle (Zalewski, 2000). It offers solution-oriented methods to reduce anthropogenic impacts, restore aquatic ecosystems, enhance ecosystem services, and strengthen human–nature connections for sustainable management Fig. 1.12.

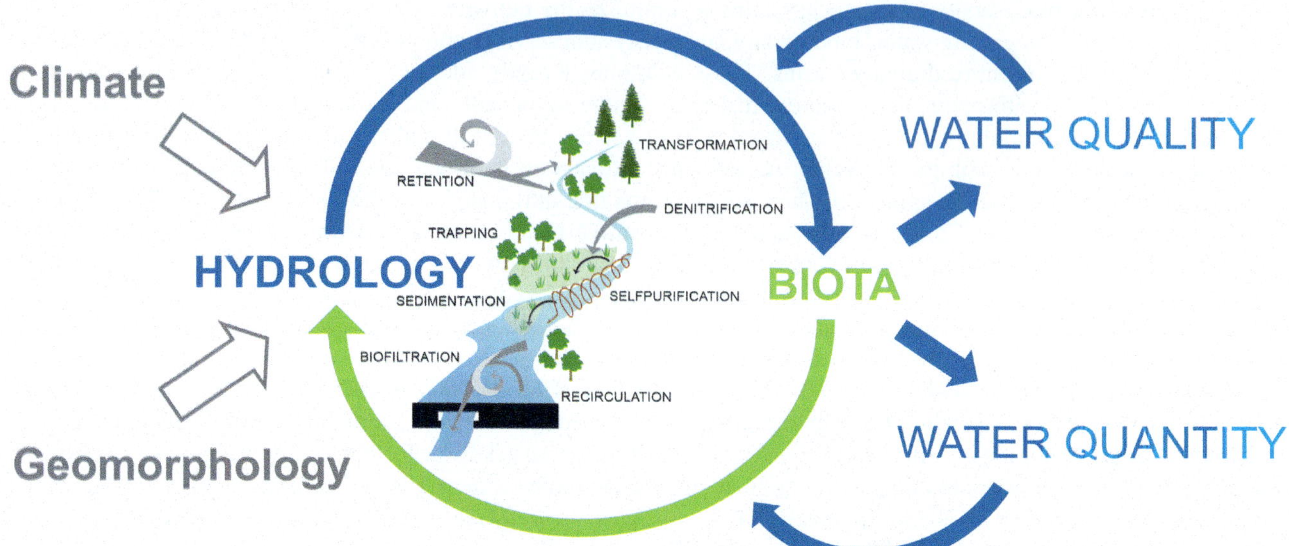

FIGURE 1.12 Dual regulation. Traditional view on hydrology and biology in environmental sciences versus dual regulation concept. *Inspired by concepts Zalewski, M., 2006. Ecohydrology — an interdisciplinary tool for integrated protection and management of water bodies. River Syst. 16 (4), 613–622. https://doi.org/10.1127/lr/16/2006/613.*

8. Dual Regulation: Enhancing ecosystem resilience through hydrological-ecological feedback

Dual Regulation—This term was first introduced to describe the dynamic interplay between water and biota, highlighting their mutual influence on ecosystem processes (Zalewski, 2006), based on earlier research (Zalewski and Naiman, 1985; Zalewski et al., 1990b). It explains how the structure and dynamics of freshwater ecosystems are regulated by the seasonality, timing, and intensity of hydrological mesocycles. In turn, the hydrological cycle itself is influenced by the drainage basin's structural components (such as forests and land–water ecotones) and ecological processes (including phytotechnology, food chain interactions, biomanipulation, and hydrobiomanipulation). This principle describes the reciprocal relationships between the hydrological cycle and ecosystem performance, driven by feedback mechanisms that shape the catchments' capacity to adapt to changes in climate and anthropopressure and further enhance sustainability, captured in the framework of WBSR. Understanding these mechanisms is fundamental to TEH, which promotes the use of ecosystem properties as management tools (Zalewski, 2000) for IWRM Fig. 1.13, Box 1.5.

FIGURE 1.13 Dual regulation. Dual regulation mechanisms encompass all life-sustaining processes within river basins on a global scale, making it applicable to the biosphere's entire planetary life-support system. The concept of dual regulation is not limited to the catchment scale; it can also be effectively applied across planetary ecosystems. By integrating the management of hydrological and ecological processes, it enhances water retention, nutrient cycling, and ecosystem resilience, making it a powerful and scalable approach for addressing sustainability challenges at the biosphere level.

BOX 1.5 Dual regulation

Biomanipulation (top-down approach) involves shaping the biological structure of aquatic ecosystems by stocking predatory fish to reduce populations of planktivorous fish, thereby increasing zooplankton abundance and enhancing their filtration capacity, ultimately leading to the reduction of algal blooms.

Hydromanipulation (feedback, top-down, and bottom-up) refers to the regulation of water levels in reservoirs as an alternative to predatory fish stocking, aiming to reduce the density of fish fry, increase filtering zooplankton thereby reducing algal blooms.

Sulejów Reservoir example: Complex interactions between hydrological status and the trophic cascade influencing water quality and ecosystem services (WBSR) (hydrological processes → biotic interactions → improvement of water quality, biodiversity, ecosystem services and resilience (Zalewski et al., 1990b) Figs. 1.14 and 1.15. See Chapter 4 for more details.

A pathway toward a more resilient future

Although sustainability is a widely recognized concept among both scientists and the general public, deeply embedded in our contemporary culture, its full realization still remains unattainable. The ongoing depletion of nonrenewable resources, for both humanity and the planet's well-being.

Key directions that offer a pathway toward a more resilient future:

To address poverty and social justice

Efforts to eliminate poverty must avoid scenarios where unrestricted access to limited common resources leads to their depletion, as described in Hardin's "tragedy of the commons" (Hardin, 1968). Examples include destruction of pastoral landscapes, overgrazing, deforestation, overfishing, and other issues, which often result in severe consequences—for example, soil erosion. In recent times, this concept has expanded to encompass the broader relationship between humanity and the biosphere, highlighting the need and the way for balanced resource management on a global scale (Ostrom, 2009).

To integrate principles of circular economy

The circular economy is a model of production and consumption, which involves sharing, leasing, reusing, repairing, refurbishing, and recycling existing materials and products as long as possible (Closing the Loop – An EU Action Plan for the Circular Economy, 2015).

To transition toward bioeconomy

The shift from a linear economic model, in which natural resources are extracted, used, and discarded, to a bioeconomy is essential. Bioeconomy relies on renewable biological resources from land and sea, such as crops, forests, fish, animals, and microorganisms, to produce food, materials, and energy. It also integrates circular economy principles, it prioritizes product longevity, resource efficiency, and waste minimization, ensuring that materials are continuously reused and repurposed. This shift not only reduces pressure on natural ecosystems but also fosters resilience, innovation, and a more regenerative economic model.

To develop advanced nature-based solutions

Grounded in the principle of dual regulation, the advancement and implementation of cost-effective green technologies, including ANbS integrated with hydrological engineering—are essential for supporting circular and bioeconomy strategies. These innovations aim to address "quantum entanglement"—the complex and deeply interconnected relationship between humanity and the biosphere. To achieve this, systemic and hybrid strategies must be applied at the catchment scale, coordinating interventions across multiple spatial and temporal scales to reduce anthropogenic pressures and enhance the resilience and long-term sustainability of ecosystems.

9. Factors driving sustainability potential: Water, biodiversity, ecosystem services, and resilience (WBSR)

TEH, as a holistic, integrative, and transdisciplinary science, provides not only strategies for mitigating anthropogenic impacts through supportive technologies (such as hybrid systems) but also a comprehensive framework for maximizing the multidimensional sustainability potential of catchment and the biosphere. A central principle of TEH is that every IWRM intervention within a catchment should simultaneously address and enhance four key elements: water resources (W); biocomplexity, bioproductivity and biodiversity (B); ecosystem services for society (S); and resilience to impacts and climate change (R), together, these form the acronym WBSR Fig. 1.16. The effective implementation of this comprehensive approach relies on the active engagement of society across the entire catchment, achieved through fostering a culture of sustainability (C) and promoting education (E). Moreover, it is crucial to secure the involvement of decision-makers and to embed these principles within proactive law (L), policy (P), and governance (G) structures to ensure the durability and institutionalization of sustainable practices. Proactive, it means based on catchment Foresight, see chapter 9.

The sequence WBSR + CE + LPG (Zalewski, 2014) reflects a deliberate prioritization: beginning with water, the hydrocentric perspective acknowledges hydrology as the principal driver of sustainable transformations. Hydrological processes are fundamental to the regulation of biocomplexity, bioproductivity, and biodiversity, while, in turn, ecological processes play a critical role in shaping hydrological dynamics — together forming the necessary conditions for the

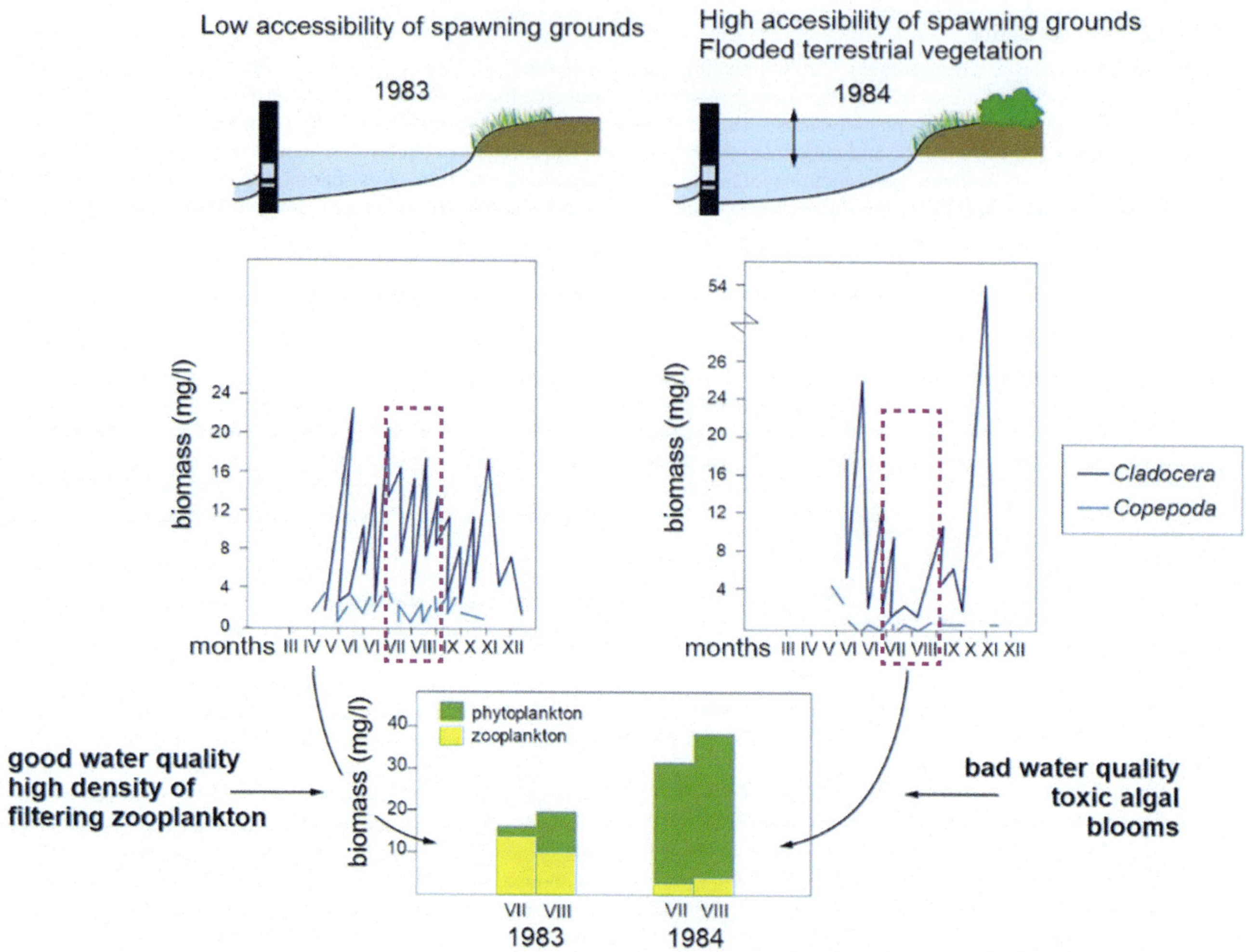

FIGURE 1.14 **Hydrobiomanipulation.** These figures present the empirical data on the relationship between water levels and ecological processes. High accessibility of the spawning grounds stimulate high density of fish frys which reduce drastically filtering zooplankton (Cladocera), thus increase toxic algal blooms (1984). *Mod. Zalewski, M., Brewińska-Zaraś, B., Frankiewicz, P., 1990a. Fry communities as a biomanipulating tool in a temperate lowland reservoir. Arch. Hydrobiol. Beih., Ergebn. Limnol. 33, 763–774.*

FIGURE 1.15 **Sulejów reservoir.** Collection of empirical data at Sulejów reservoir. *Photo credit: M. Zalewski.*

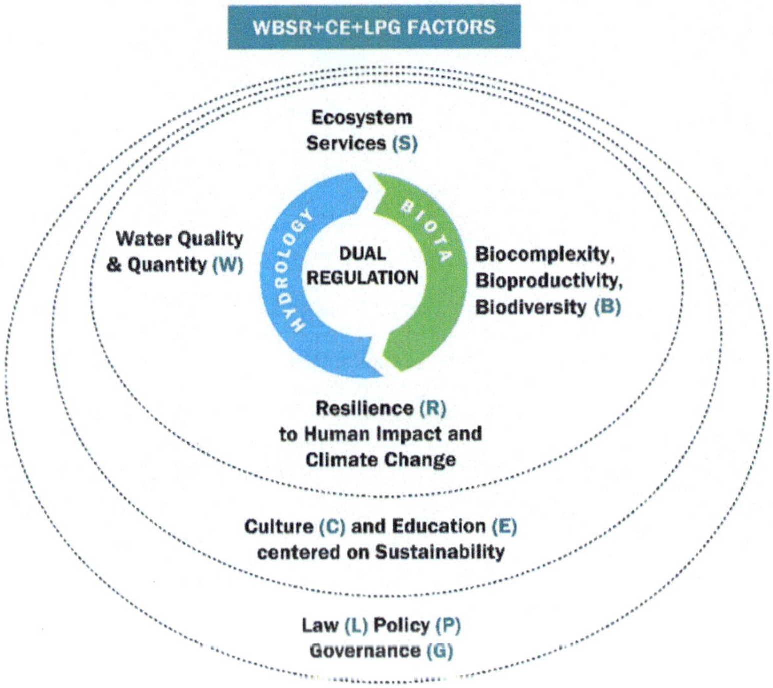

FIGURE 1.16 Conceptual framework. Transdisciplinary ecohydrology to accelerate the SDGs and simultaneously enhance water quality and quantity (W) with biocomplexity, bioproductivity, and biodiversity (B), ecosystem services (S) for society, and resilience (R) to human impact and climate change. These processes are further supported by culture (C) and education (E) centered on sustainability, as well as proactive law (L), policy (P), and governance (G), which reinforce the systemic application of transdisciplinary ecohydrological principles. Together, these factors form the WBSR + CE + LPG acronym.

sustained provision of ecosystem services. The success of this approach ultimately depends on society's capacity to understand, value, and promote these interconnections through a deeply rooted culture of sustainability and education. In the longer term, embedding this understanding within robust, proactive law, policy, and governance frameworks will be essential for maintaining the resilience and functionality of catchment systems in the face of ongoing societal, environmental and climatic challenges.

10. Setting an ambitious goal: Enhancement

Nothing in Biology Makes sense Except in the Light of Evolution. The diversity and the unity of life are equally striking and meaningful aspects of the living world

Dobzhansky (1973)

There are two contrasting future scenarios based on humanity's approach to environmental management and sustainability Fig. 1.17. The red trajectory, labeled "business-as-usual," represents a scenario where human activities continue to exploit natural resources without effective intervention, leading to an overshoot of the Earth's carrying capacity. In this scenario, environmental degradation accelerates, triggering ecosystem collapse sometime after 2050, as planetary boundaries are exceeded, and LSPs deteriorate. In contrast, the blue trajectory represents a sustainability-enhancement approach, where proactive measures are taken to increase the Earth system's capacity for sustainable functioning through the extension of sociocentric-mechanistic paradigm with the evolutionary–ecosystemic, integrating environmental management, hybrid, systemic solutions. This diagram serves as a visual representation of the choices outlined for humanity. It underscores the critical urgency of shifting toward an enhancement-based approach, demonstrating that without immediate and strategic intervention, the stability of ecosystems and human society is at risk.

According to the Merriam-Webster Dictionary, the term "enhancement" refers to "an increase or improvement in value, quality, desirability, or attractiveness." In the medical field, it describes the process of achieving health status that surpasses the preexisting condition before the onset of disease. **In the context of TEH, enhancement extends beyond conservation or restoration, and regeneration reflects the necessity of expanding the biosphere's carrying capacity to support life in response to escalating human and environmental pressures.** As the global population continues to grow, human consumption intensifies, and conflicts over limited natural resources increase, enhancement emerges as a crucial strategy for ensuring long-term sustainability Fig. 1.18.

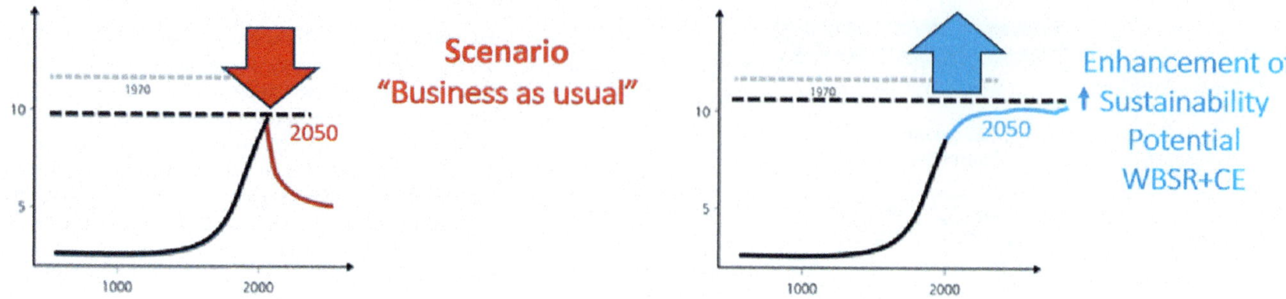

FIGURE 1.17 **"Business-as-usual" versus enhancing earth's carrying capacity.** According to empirical tests of the r-K selection theory, there are two possible scenarios when the population number is advancing to the environment's carrying capacity. The scenario where exponential growth and decline of carrying capacity clashing it, resulting in a decline (red line) of population up to 30% of the clash value (e.g., 10 B to 3 B). The positive K scenario (blue line) shows an exponential growth of the population accompanied by the enhancement of carrying capacity (e.g., from 9.5 to 12 B). The population does not clash with the carrying capacity and has time to adjust, which ensures a sustainable future. Thus fulfilling the SDGs is imperative for humanity's very survival—our "to be or not to be." *On the basis of the original r-K curve by Cohen, J. E., 1995. Ecological Community: The Trophic Dimension. Princeton University Press., modified by Zalewski, M., 2002. Ecohydrology — integrative science for sustainable water, environment and society.* Ecohydrol. Hydrobiol. *2 (1–4), 3–10.*

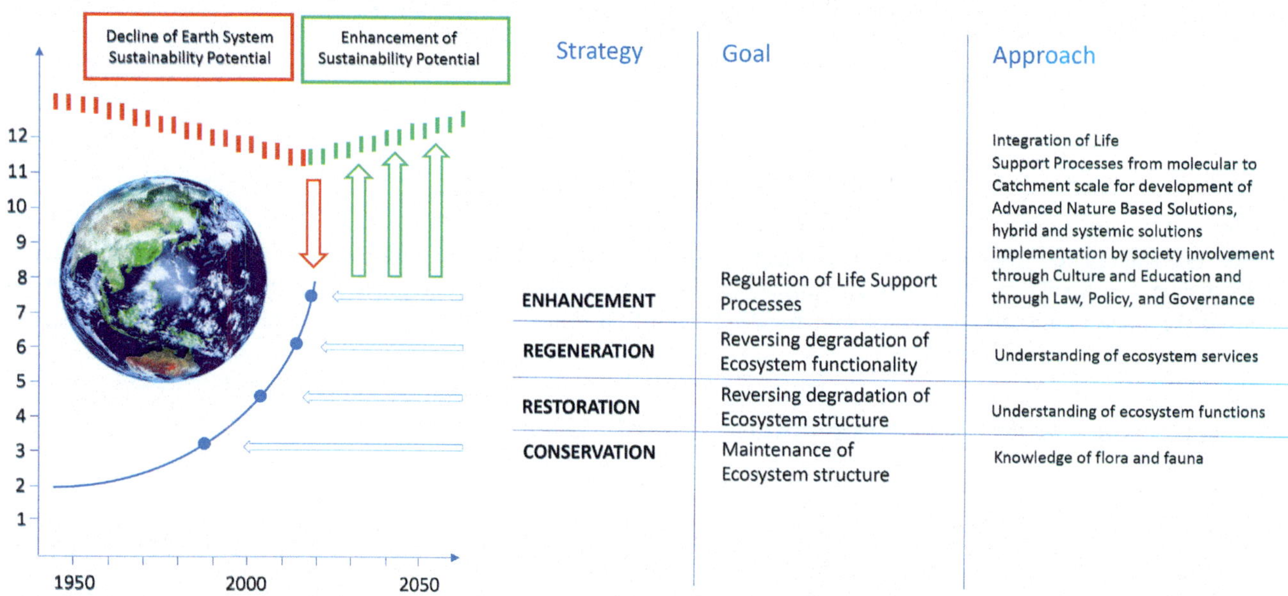

FIGURE 1.18 **The evolution of key paradigms.** The evolution of key paradigms for the integrity of humankind and the biosphere has shifted from conservation and restoration toward enhancement, as a means of compensating for increasing anthropogenic impact (human population in billion) by expanding planetary boundaries.

This diagram presents a strategic framework for environmental management, categorizing approaches into conservation, restoration, and enhancement. It illustrates how the enhancement of Earth System Sustainability Potential (ESSP) can counteract declining carrying capacity by actively improving ecosystem functionality and resilience. The framework also defines key goals, such as simultaneous enhancement of sustainability potential (Water-Biodiversity-Ecosystem Services-Resilience—WBSR + CE + LPG), while proposing an integrative approach that combines hydrological and ecological processes with governance, policy, and education. This TEH approach aims to foster sustainable solutions that go beyond traditional conservation efforts, ensuring that natural and human systems evolve together in a way that strengthens planetary resilience.

TEH primarily emphasizes "enhancement" rather than "restoration" or "regeneration" as a key concept. Why is this? Because it recognizes that the biosphere is in a constant state of change, influenced by shifts in land cover, climate, societal priorities, and numerous other factors that make true restoration to pristine conditions unrealistic. Given the growing demographic pressures, preserving nature at 50% of the biosphere is not a feasible goal. Therefore, we must move beyond restoration to focus on enhancing ecosystems to ensure their resilience and functionality in an ever-evolving environmental landscape to ensure a positive future for generations to come.

We refer to catchments rather than ecosystems because water is the fundamental driver of all ecosystem processes. Regulating the interplay between water and biota requires precise quantification of the water budget, a task best achieved through established hydrological sciences, which are grounded in validated methodologies and, in most cases, supported by long-term monitoring data (Montanari et al., 2015). As water is recognized as a key driver for accelerating progress toward the SDGs, catchments must also be viewed as fundamental units for improving biosphere sustainability as TEH become compatible with strategic development of IWRM and WFD and defines the holistic approach. They provide a structured framework for quantifying ecosystem structures, processes, and impacts. This, in turn, makes it possible to assess the extent of degradation of LSPs and to identify the actions needed to improve them.

TEH focuses on the hydrological mesocycle (approximately 10^1–10^3 km^2), as it operates at an intermediate scale, bridging small-scale processes with larger hydrological events. This approach provides meaningful solutions while remaining practical and adaptable, enabling the development of a step-by-step methodology for sustainable catchments. "In mesoscale catchments, processes from smaller scales combine in complex ways to produce an integrated response" and the intermediate scale is critical from a water resource management perspective (Uhlenbrook et al., 2004).

TEH enhances the sustainability potential of catchments and the biosphere by harnessing their inherent capacity to adapt and improve under specific conditions (Zalewski, 2002). This potential can be managed and transformed through an ecohydrological approach that generates multiple benefits across environmental, hydrological, social, and economic dimensions.

The concept of enhancing biosphere sustainability through TEH goes beyond regenerative ecology, because it is hydrocentric. Regenerative ecology is a sustainability approach focused on actively restoring and revitalizing ecosystems. TEH emphasizes the dynamic interplay between water, ecosystems, and human activity Box 1.6. This approach determines the goal, aiming to transform the biosphere's carrying capacity and strengthen Earth's life-support systems, by addressing sustainability beyond structural regeneration, such as large-scale reforestation, ecohydrological strategies prioritize the restoration of functional and process-based systems, ensuring long-term resilience. TEH defines the holistic WBSR approach, where all four elements require simultaneous enhancement. On the other hand, regenerative sustainability is a design philosophy that aims to create systems and practices that restore, renew, and revitalize their sources of

BOX 1.6 Transdisciplinary ecohydrology: A step beyond regenerative ecology

1. Ecological restoration is the process of assisting the recovery of an ecosystem that has been degraded, damaged, or destroyed (Clewell and Aronson, 2013).
2. Regenerative ecology builds upon ecological restoration by actively assisting the recovery, resilience, and long-term sustainability of ecosystems degraded by human actions, aiming to restore not only structure and function but also the capacity for future adaptation (Clewell and Aronson, 2013).
3. Transdisciplinary ecohydrology: Hydrocentric approach, is defining how to regulate life supporting processes (dual regulation) to increase four key parameters of sustainable future (WBSR). In consequence improve catchment carrying capacity towards harmonisation Man and Biosphere and prevent "tragedy of commons."

The understanding of Water-Biota Interplay is fundamental for the enhancement of catchment-scale sustainability potential WBSR, as water cycling drives LSPs, including:

- Water cycle
- Energy flow
- Nutrient cycling
- Habitat structure and dynamics (e.g., seasonality, succession)

How to implement it?

The key implementation strategy is **Dual Regulation** concept, which forms the foundation for the development of ANbS by integrating hydrological engineering, molecular biology, biotechnology, and phytotechnology into hybrid, systemic solutions aimed at enhancing the carrying capacity of catchments and the biosphere.

Collective intelligence or societal involvement

Collective intelligence is a sociological concept describing how a group's intellect emerges when individuals work together. Through interaction and even competition, groups share information, solve problems collectively, and improve their chances of identifying correct solutions, often by reaching consensus and discarding incorrect or deviant ideas. Sometimes referred to as "swarm intelligence," collective intelligence has a long history and continues to be extensively studied. Prominent definitions include "groups of individuals acting collectively in ways that seem intelligent" and "the general ability of a group to perform a wide range of tasks" (Centola et al., 2022).

energy and materials, going beyond simply reducing harm and aiming for a net positive impact on their environment and community (Mang and Haggard, 2016).

To effectively apply TEH, a deep understanding of its processes is essential, alongside innovative methods grounded in dal regulation, the management of water systems and their ecological interactions. This approach incorporates ANbS, hydroengineering, indigenous ecological knowledge, and biotechnological advancements, as demonstrated in UNESCO IHP Ecohydrology Demonstration Sites (Ecohydrology Web Platform). Additionally, the three TEH principles, hydrological, ecological, and ecological engineering (principles of ecohydrology, see Chapter 8 for details), offer a structured framework for dynamic water management, guiding the enhancement of biotic interactions and the design of engineering interventions that align with natural processes. Identifying and prioritizing key drivers within this framework enables the development of sustainable drainage basis dynamics and innovative broadening IWRM by providing innovative approach. It not only enhances the potential for biosphere sustainability but also advocates for a reduction in anthropogenic pressures.

Beyond its ecological and hydrological contributions, TEH also enhances human well-being by restoring cultural landscapes (Zalewski and Wantzen, 2023), promoting sustainable water governance, and ensuring equitable access to water resources. Human settlements have long been shaped by their proximity to water, not only for drinking and agriculture but also for cultural and spiritual connections. This approach fosters an evolving relationship between humans and the broader natural world, where people actively participate as an integral part of nature, supporting the coevolution of the biosphere (Buckton et al., 2023). By integrating TEH with regenerative ecological principles, biosphere sustainability becomes a structured and adaptive framework for managing environmental change.

The Age of Discovery, which refers to the centuries during which European nations explored the globe, left behind a paradigm that is deeply ingrained, and which assumes that nature has boundless regeneration potential. This paradigm is one of conquest and harnessing of nature, and it led to the intensive exploitation of natural resources in the Industrial Era. While this led to the degradation of many ecosystems, it also created an awareness of the need to protect nature to counteract overexploitation of the natural environment. The advent of biological sciences, marked by seminal contributions such as Linnaeus' Systema Naturae (Linnaeus, 1758), Cuvier's Researches into the Natural History of Living Animals (Cuvier et al., 1817), Buffon's Histoire Naturelle (Buffon et al., 1749), and Darwin's On the Origin of Species (Darwin and Darwin, 1859), along with a burgeoning comprehension of the principles governing ecological succession and its dynamics, has laid the foundation for the contemporary discipline of restoration ecology. Such an approach, predicated on ecological restoration, is of particular importance in the revitalization of areas such as postmining sites, urban areas, and degraded former farmland.

The new evolutionary–ecosystemic approach outlined in this book is predicated on the premise of a perpetual exchange of knowledge, expertise, and opinions between academia, industry, business, and public figures. The approach systematizes and integrates human awareness in major areas of science and economics, whilst also fostering the accumulation of social capital, which the OECD has identified as the key factor stimulating sustainable development. Conventionally, environmental research has been a curiosity-driven science, shaped by the interests of individual scholars who have focused on specific issues, factors, and components of the environment, such as certain species. However, the natural environment is best viewed as an intricately complex superorganism. To address the challenges posed by the Anthropocene era, and despite the increasing specialization required to enhance our understanding of the biosphere and the subsequent development of effective strategies to enhance its potential (WBSR), twentieth-century science must also assume the responsibility to integrate and synthesize knowledge from multiple disciplines into transcapillary problem-solving science.

An examination of the evolution of human–nature relationships reveals a significant transformation from an exploitative and mechanistic perspective to a more ecologically conscious and holistic approach Fig. 1.19. In the early stages, human interaction with nature was largely characterized by conquest and unregulated exploitation, driven by the assumption of nature's infinite abundance. This perspective was particularly pronounced during the Industrial Era, where natural resource exploitation was undertaken with minimal regard for long-term ecological consequences. However, as scientific understanding of biodiversity and ecological structures expanded, started by Linnaean classification, a shift toward protection and conservation emerged. This phase was initiated by a growing recognition of biodiversity loss and the necessity of preserving natural ecosystems. Subsequently, this heightened awareness facilitated the transition to ecological restoration, where the principles of ecological succession and the cultural and esthetic values of nature became increasingly acknowledged and integrated into progressive environmental management strategies.

The shift toward sustainability represents a critical turning point, signifying a broader recognition of the need to preserve natural resources. The following stage, regeneration, is characterized by an increasing awareness of ecosystem degradation and the loss of ecosystem services, prompting proactive measures for ecological recovery. At the highest level

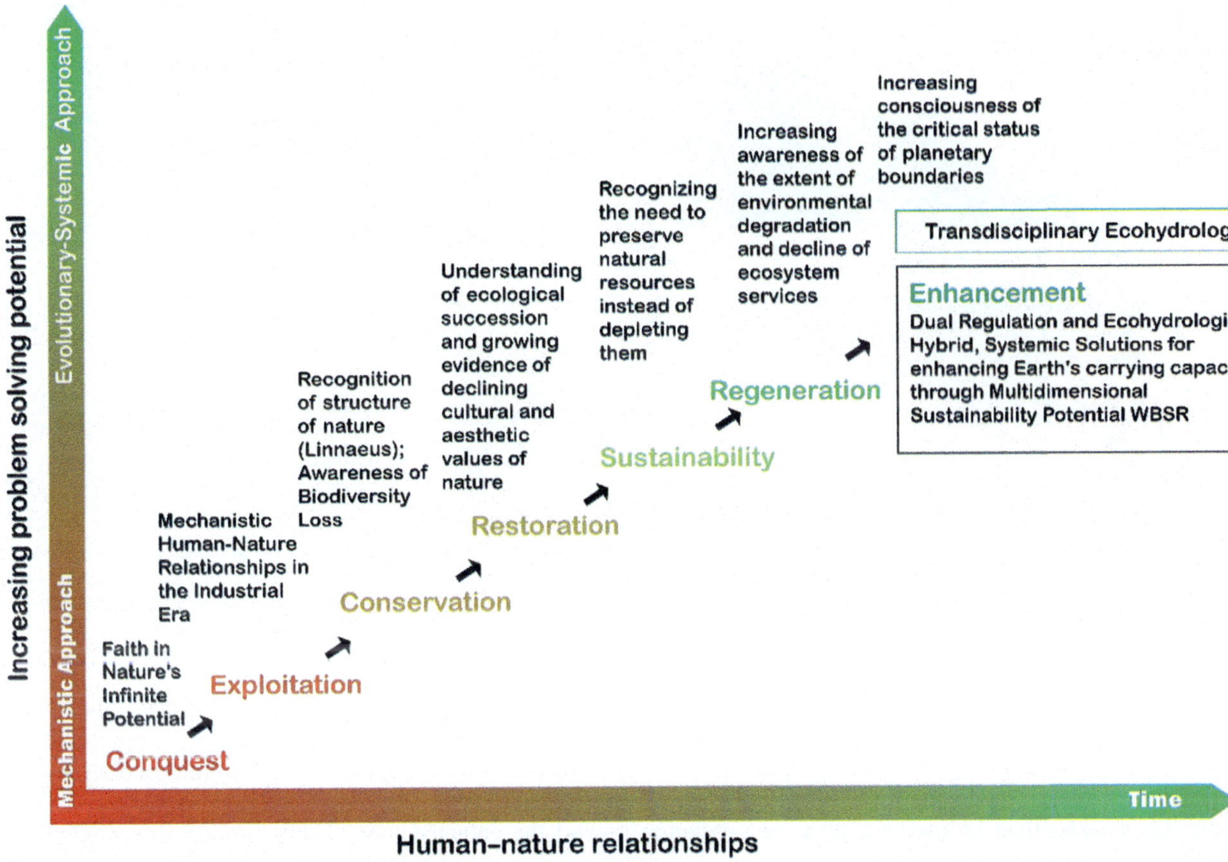

FIGURE 1.19 The evolution of human–nature relationships over time. Conceptual framework outlining the evolution of man and biosphere relationships over time, illustrating a transition from a mechanistic to an evolutionary–ecosystemic approach, accompanied by an increasing problem-solving capacity. The x-axis delineates the progressive stages of human interaction with nature, transitioning from conquest and exploitation to protection, restoration, sustainability, regeneration, and enhancement. The y-axis reflects an expanding capability to address environmental challenges, as the dominant perspective shifts from a mechanistic view, which perceives nature as an unlimited resource, to a systemic approach TEH, that alters fundamental water–ecosystem interplay for the enhancement of WBSR factors. *Zalewski, M., 2020. Ekohydrologia. Wydawnictwo Naukowe PWN, Warszawa, ISBN:978-83-01-20804-2.*

of complexity, the enhancement phase introduces dal regulation and ecohydrological, hybrid, systemic solutions to increase Earth's carrying capacity through the multidimensional sustainability potential (WBSR) framework. By integrating ecohydrological principles, this approach fosters resilience, biodiversity conservation, and sustainable water resource management, ensuring that human activities align with natural processes.

11. New paradigm: Toward achieving the ambitious goal of enhancement

An understanding of ecological succession provides a basis for resolving man's conflict with nature

Odum and Barrett (2004).

It is important to note that every strategy for success is composed of two components, the elimination of threats and the amplification of opportunities Fig. 1.20. The latter component must be clearly defined and strengthen by a new paradigm—the enhancement of the Earth system's sustainability potential.

This can be done by:

- Understanding the complexity of LSPs.
- Understanding how the hierarchy of key drivers, such as water and temperature regulate structure of populations, biocenosis, and ecosystem performance (see Chapter 4 for more details).
- Understanding the water–biota feedback.

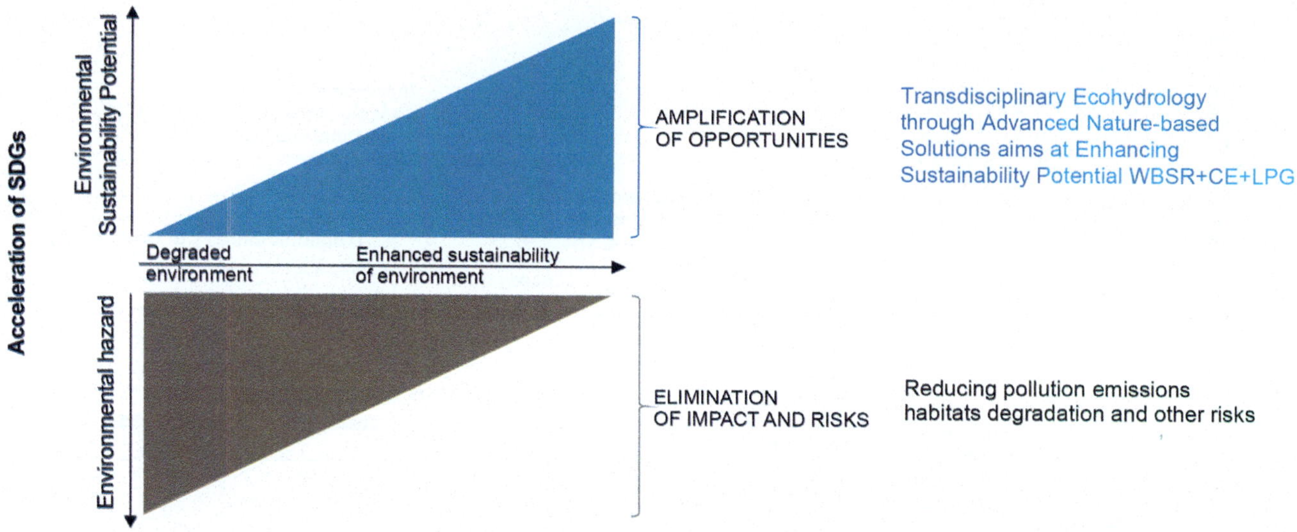

FIGURE 1.20 Application of transdisciplinary ecohydrology. Applying transdisciplinary ecohydrology as a holistic approach to sustainable water management by maximizing opportunities and eliminating threats. *Mod. Zalewski, M., Janauer, G.A., Jolánkai, G., 1997. Ecohydrology: A new paradigm for the sustainable use of aquatic resources. Conceptual Background, Working Hypothesis, Rationale and Scientific Guidelines for the Implementation of the IHP-V Projects 2.3/2.4. UNESCO, Paris.*

All of these are necessary to use ecosystem properties (structure shaping and process regulation) as management tools. To steer the system toward new directions and to avoid fragmentation of efforts (United Nations 2023), it is necessary to precisely define goals and the scope of action. Quantification of processes is fundamental to this, and the water mesocycle in the catchment must serve as a framework for precise identification of threats, process dynamics, and regulatory feedback. These must then be used for efficient regulation toward the enhancement of the sustainability potential of catchment and the biosphere.

To achieve these objectives, it is necessary to harmonize the fundamental bottom-up approach based on hydrology-hydroengineering with the ecohydrological top-down approach. This can be achieved through a profound understanding of how evolution has shaped water-nutrient cycles, water–biota interactions, and how ecosystems are stabilizing hydrological mesocycle.

These, in turn, are key reference points for:

1. Enhancing sustainability potential of catchments and the biosphere
2. Harmonization of society's needs with the SDGs and the IWRM has been expanded by TEH complexity, background, and methodology. Only by integrating a bottom-up and top-down approach, both used as a reference point for foresight (wisely crafting a desirable vision of the future), can we truly achieve sustainability.

In the words of Albert Einstein: "We must not continue to use the same methods that created these problems."

Therefore a new paradigm must be adopted, requiring a more profound transformation in the relationship between humans and their environment Fig. 1.21. The sociocentric-mechanistic approach views the biosphere as a "black box," a system that provides unlimited resources and absorbs all generated pollution and waste. Energy and matter are gained through renewable resources, such as fossil fuels, which are easy to exploit but limited in the future. In contrast, the new evolutionary–ecosystemic approach emphasizes the sustainable usage of renewable resources and a circular economy. The term "evolutionary" underscores that the biosphere is in a state of continuous evolution, driven by ongoing biogeochemical transformations within ecosystems (Bianchi, 2021). Additionally, it reflects the ongoing evolution of the relationship between humans and the biosphere. In contrast, "ecosystemic" highlights the crucial role of ecosystems in maintaining and stabilizing the LSPs essential to human and planetary well-being. In the Anthropocene, human activities significantly alter LSPs, making it imperative for individuals and societies to choose their course of action. We face a critical decision: either to exploit Earth's resources extensively, assuming that the planet can endlessly absorb pollutants—an approach that will inevitably lead to environmental collapse and conflicts—or to manage Earth's finite resources with discipline and foresight, much like operating a self-sustaining spaceship. Achieving the SDGs requires shifting beyond the sociocentric-mechanistic paradigm, which emerged during the Industrial Revolution, toward an

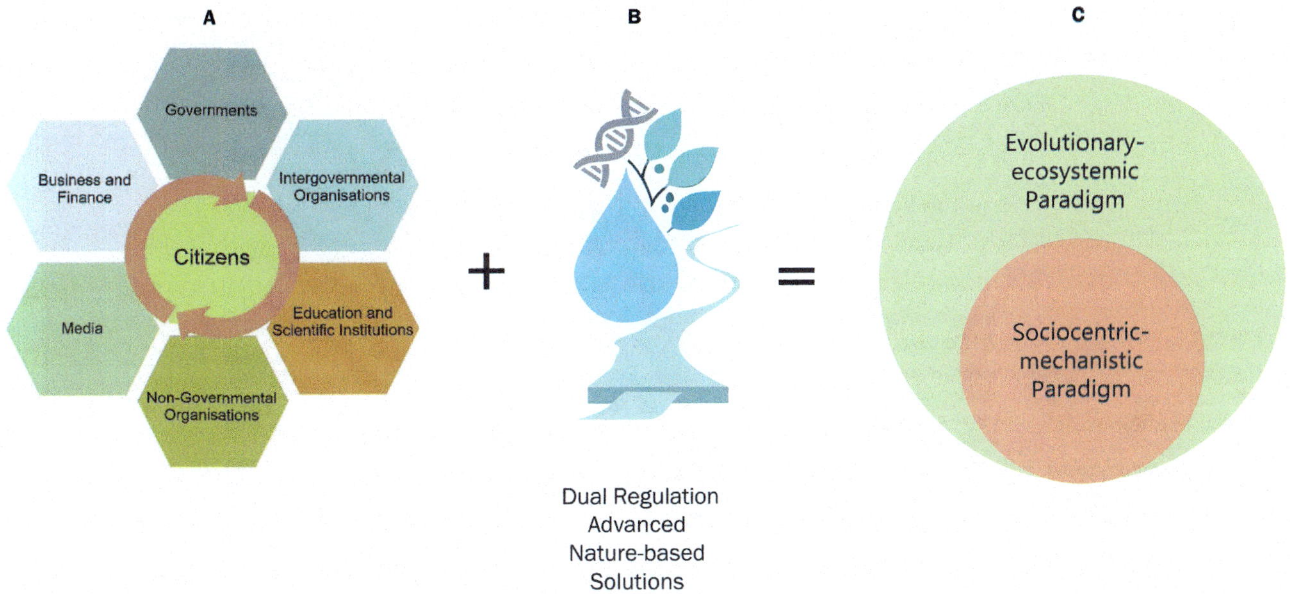

FIGURE 1.21 Evolution of Water Management Paradigms: From the Sociocentric–Mechanistic Model to the Evolutionary–Ecosystemic Framework. (A) The sociocentric–mechanistic paradigm (adapted from Watson et al., 2022), which assumes that technological progress and societal responses can continually resolve environmental problems while largely overlooking the life supporting functions of ecosystems. (B) Integration of Advanced Nature-based Solutions (ANbS) as a transitional approach toward restoring ecosystem processes (Zalewski, 2000). (C) The evolutionary–ecosystemic framework proposed in this book, in which a comprehensive understanding of the water cycle and its interactions with ecosystems provides the foundation for engineering harmony between society and the biosphere. The framework integrates Water–Bioproductivity/Biodiversity–Ecosystem Services–Resilience (WBSR), Culture and Education (CE), and Law, Policy and Governance (LPG) to guide sustainable catchment management.

evolutionary–ecosystemic approach that recognizes the dynamic interplay between biosphere and social systems. This transition is essential for fostering sustainable interactions between humankind and the natural world.

The evolutionary–ecosystemic approach builds on the sociocentric-mechanistic paradigm by integrating hydrological, ecological, and biochemical processes through a TEH methodology. This approach is based on an understanding of dual regulation, which is turn, is fundamental for the development of advanced nature-based solutions (ANbS) and systemic solutions. It employs a range of disciplines, including hydrological engineering, molecular biology, biotechnology, and phytotechnology, to enhance the multidimensional sustainability potential of catchments, encompassing WBSR.

The new paradigm requires new methodology, rooted in TEH; the hybrid, systemic solutions promise a more positive way forward. This approach encourages the creation of hybrid systems that are resilient, sustainable, and capable of providing ecosystem services, while also supporting LSPs. Achieving harmony between society and the environment requires the implementation of integrative sustainability science, with TEH at its core. Insights from TEH are transformed into practical solutions, establishing a clear methodology that integrates ecology, hydrology, molecular biology, biotechnology, phytotechnology, with hydrological engineering, see Chapter 8 for more details, Fig. 1.22. Progressive enhancement of nature-based solutions through transdisciplinary integration through TEH, integrating ecohydraulics, molecular biology, biotechnology, phytotechnology, ultimately leading to the development of ecohydrological hybrid solutions that are systemic, scalable, and highly effective in addressing complex environmental challenges. The concept of dual regulation is central to this progression, emphasizing the necessity of integrating ecological and hydrological processes to achieve sustainable water and ecosystem management.

12. A way to inspire and encourage society for implementation (CE)

In an era of rapid change, particularly with the escalating impacts of climate change, it is imperative to educate society to cope with uncertainty. This education must focus on reshaping societal attitudes and fostering a comprehensive understanding of sustainability. At its core, sustainability recognizes the biosphere as a dynamic system driven by interconnected cycles of water, carbon, phosphorus, and nitrogen. These cycles are influenced by individual and collective human actions, which in turn determine the quantity and quality of ecosystem services available to us.

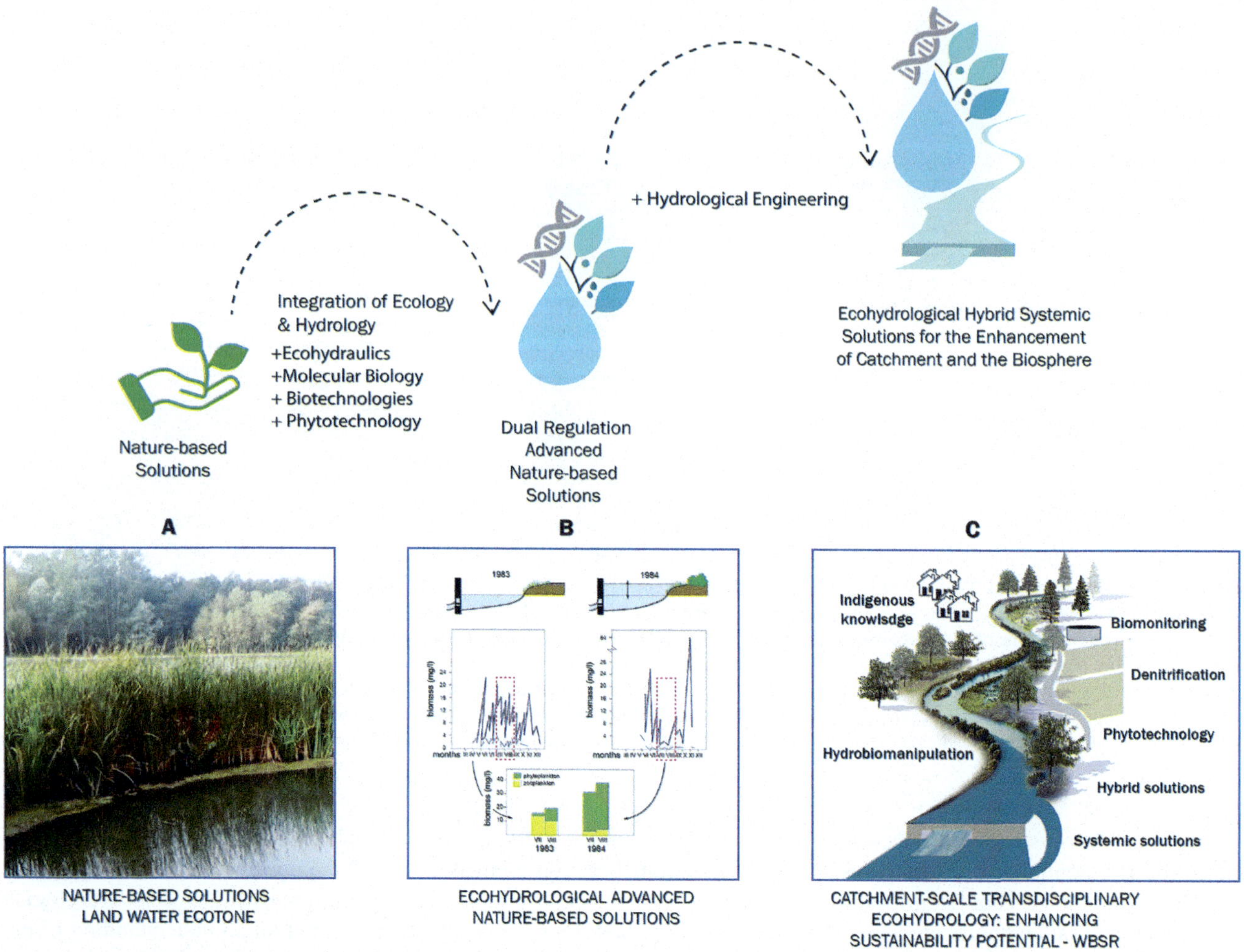

FIGURE 1.22 Integrative hybrid and systemic solutions. (A) nature-based solutions (NbS) have been developed to optimize the functionality of land–water ecotones, contributing to the reduction of non-point source pollution (Naiman et al., 1989). (B) Advanced nature-based solutions (ANbS) extend this approach by incorporating innovations in ecohydraulics, molecular biology, biotechnology, and phytotechnology within a transdisciplinary ecohydrology framework, enhancing their adaptability and enabling a more precise response to specific environmental conditions. Central to ANbS is the concept of dual regulation, in which ecohydrological principles are applied to regulate hydrological and biotic interactions simultaneously, promoting dynamic, adaptive water resource management and increasing ecosystem resilience. In a subsequent phase, hydrological engineering is integrated to further optimize water flow, retention, and purification processes. This addition significantly enhances the scalability and operational effectiveness of ANbS, enabling its application to complex and large-scale environmental challenges. The progression culminates in the development of hybrid and systemic ecohydrological solutions at the catchment scale (C). Grounded in transdisciplinary ecohydrology, these integrated solutions combine NbS, ANbS, and hydroengineering, remote sensing, satellite images, hydroacoustic method, offering a comprehensive approach for monitoring and planning that merges ecological science, hydrology, biotechnology, and engineering for sustainable and adaptive water and ecosystem management.

Engagement of society and culture is essential to the implementation of the paradigm shift from a sociocentric-mechanistic worldview to an evolutionary–ecosystemic one. This shift has to be grounded in three fundamental principles:

1. The unity of humanity and nature: The understanding that humans and the natural world are deeply interconnected and interdependent.
2. Redefining happiness: The recognition that personal happiness is not tied to levels of consumption but is rooted in maintaining fair and meaningful relationships with others and living in a healthy environment.
3. The environment's role in mental well-being: The acknowledgment that our mental and emotional well-being is significantly influenced by the quality of our surrounding environment.

Adopting these principles can stimulate social capital by fostering trust, cooperation, and collective action, and through such an educational framework, society can build resilience and adaptability, ensuring a sustainable and equitable future for

FIGURE 1.23 **Demonstration site—Xian sponge city.** (A) Development of prototype solutions integrated within broader systemic frameworks. (B) An explanation of the fundamentals and general principles, using the educational exhibition as a way to deepen understanding of the demonstration site. *Photo Credit: M. Zalewski.*

all (Zalewski, 2021). This approach is grounded in the concept of collective intelligence, which refers to the enhanced capacity for problem-solving, innovation, and coordination that emerges from collaboration among individuals or groups.

Social capital is an important factor for translating knowledge into wisdom and for developing innovations in watershed management. The philosophy of open idea exchange and receptivity to diverse, even controversial, perspectives, echoing the principles of Kant's Critique of Pure Reason, has expanded the holistic understanding of complex environmental challenges and the pathways to effective innovation. It is worth underlining that encouraging both a "team spirit" and respect for competence, leadership, and seniority toward the achievement of strategic goals is critical for reversing the degradation of the biosphere and generating synergies, amplifying the translation of science into technology. Some implementation approaches have achieved exceptional success, particularly when practical examples Fig. 1.23 are supported by scientific findings and clearly communicated to the public by exhibition boards, Fig. 1.23. Xian is one of the cities in China that has actively embraced the sponge concept, implementing various projects and initiatives to increase water infiltration and reduce runoff.

13. Defining a new legal framework and policy

The precautionary principle serves as a risk management tool in environmental policy, ensuring that scientific uncertainty does not justify inaction when serious harm is possible (Kriebel et al., 2001). By prioritizing prevention over remediation, it provides a framework for safeguarding ecosystems, human health, and biodiversity, reinforcing the idea that preventing environmental harm is inherently more sustainable than attempting to repair it after degradation has occurred. This assumption is particularly relevant in the context of water management and catchment sustainability, where ecological and socioeconomic stability depends on the proactive regulation of water resources and their interactions with biotic systems.

Enhancing catchment resilience and ensuring the long-term sustainability of the biosphere requires recognizing water's central role in shaping ecosystems, biodiversity, and human societies. Strengthening water and biota stability not only improves ecosystem services but also enhances the adaptive capacity of catchments, including the socioeconomic systems that depend on them. However, for such approaches to succeed, it is essential to engage societies by fostering a culture of water stewardship and sustainability through dialog with educational institutions, legal systems, and governmental authorities. A truly comprehensive strategy must integrate all dimensions, ranging from public awareness and institutional education to supportive policies, legal frameworks, and effective governance structures.

Cultural traditions and indigenous knowledge systems offer innovative, place-specific solutions for sustainable water management. Traditional indigenous water management technologies, such as the Hehanusa water-sharing system—Subak—developed at a UNESCO Ecohydrology Demonstration Site in Indonesia (Hehanussa, 2002), integrate traditional principles with contemporary water governance to enhance sustainability and community resilience. Subak is most often related to the term, incorporating these approaches into modern water governance fosters resilient and sustainable management practices, ensuring that local knowledge complements contemporary scientific and policy-driven solutions, and the lack of conflicts.

As outlined in the WBSR framework (Zalewski, 2014), these interconnected dimensions highlight the urgent need for innovative thinking in water management. Engaging stakeholders, particularly decision-makers, is crucial in broadening water management objectives and shifting toward a sustainability-driven perspective. A holistic approach—grounded in science, precaution, proactive management, education, governance, and cultural heritage—is key to ensuring the long-term resilience of water systems and their role in sustaining both natural and human communities.

14. The awakening of human consciousness

Now is the time for every individual to take on the role of a responsible steward of our shared planet—an intricate and fragile system often compared to a magnificent "Spaceship Earth." This responsibility calls for reducing resource consumption, minimizing pollutant emissions, and fostering a circular economy that prioritizes regeneration while unlocking the multidimensional sustainability potential of drainage basins. By integrating a WBSR framework with policy and governance innovations driven by circular economy principles, and broad use of bioeconomy, we can catalyze systemic change—bridging ecological sustainability with cultural awareness, legislative adaptation, and institutional transformation. This shift necessitates both bottom-up and top-down approaches, ensuring that local initiatives align with broader policy reforms to create resilient and adaptable environmental governance.

Achieving this objective necessitates the regulation of interactions between water and biota through ANbS. These solutions encompass innovative, ecosystem-driven approaches that integrate natural processes into infrastructure and resource management. The implementation of these solutions should occur across all scales, with particular emphasis on local action. The adoption of these solutions on a large scale has the potential to create cascading positive impacts, enhance regional resilience, and foster a more sustainable relationship between humanity and the natural world. By transitioning toward enhanced practices for the environment, it is possible to redefine development not as a process of uncontrollable growth, but as a system that prioritizes ecological integrity and long-term sustainability, thereby ensuring a livable future for generations to come.

15. Is transdisciplinary ecohydrology the new rocket science?

In the context of the current critical stage of planetary boundaries—which define the safe operating limits for nine key processes regulating Earth's stability and resilience (Rockström et al., 2009)—there is an urgent need for a paradigm shift. The traditional sociocentric–mechanistic (SM) approach must be expanded by an evolutionary–ecosystemic (EE) approach to address the escalating challenges facing sustainability.

Why is this paradigm shift necessary?

Humanity is approaching the carrying capacity of the biosphere due to increasing consumption, population growth, and declining sustainability. This growing pressure significantly raises the likelihood of conflicts over resources. Furthermore, the relationship between humans and the biosphere has reached a stage that can be metaphorically described using a term from physics: "quantum entanglement", which means two dynamic, interacting subsystems constantly changing. The interdependence between human actions and ecological processes demands a new, integrated approach to managing LSP and enhancing WBSR interactions.

Key Considerations for the evolutionary–ecosystemic paradigm:

1. Catchment processes—Evolution continuously shapes catchment ecosystems, creating intricate water–biota interaction patterns that adapt ecosystems to dynamic environmental changes. Recognizing these self-organizing processes is crucial for sustainable water management.
2. Understanding the evolution of human–nature relationships (Fig. 1.19) for change from reactive to proactive management.
3. Ecosystem functionality—The structure of ecosystems and their ecological succession play a fundamental role in maintaining the integrity of LSP in a flexible and dynamic state. The lack of understanding of these processes can lead to long-term, irreversible environmental degradation.
4. Transdisciplinary science—Effective environmental management requires integrating knowledge across disciplines to understand the complex hierarchies of driving forces, feedback mechanisms, and interdependencies. This holistic approach is essential for cost-effective impact reduction and the enhancement of a catchment's sustainability potential (WBSR).

Why is an integrated scientific approach essential?

The spatial diversity of catchments and ecosystems is shaped by multiple factors, both abiotic (e.g., geology and hydrology) and biotic (e.g., biodiversity, ecological interactions). Some ecosystems remain nearly pristine, while others suffer severe degradation because of human activity.

Over time, the water budget, biological structure, and water quality of a catchment are influenced by long-term ecological processes—such as postglacial colonization in temperate zones—and subsequently modified to varying degrees by human interventions. Understanding these processes is fundamental for developing realistic future scenarios that balance ecological integrity with human needs.

However, human interventions frequently disrupt or unbalance catchment sustainability potential by focusing on isolated parameters (e.g., prioritizing food production without considering impacts on water resources, biocomplexity, biodiversity, and resilience). When these foundational ecological factors decline, the entire WBSR system deteriorates.

Steps for enhancing catchment WBSR:

1. Identify and protect key ecosystems—Priority should be given to ecosystems with unique natural and cultural value, ensuring their conservation.
2. Reduce human pressure on vulnerable areas—Efforts must focus on mitigating pollution and habitat degradation.
3. Restore and enhance degraded landscapes—In areas such as intensive agricultural zones, biological complexity must be increased (e.g., restoring land–water ecotones, tree rows, and biodiversity corridors).
4. Assess and integrate human modifications—All human-made alterations to ecosystem processes—especially hydroengineering solutions—must be harmonized with ANbS at a catchment scale to achieve synergistic WBSR benefits.
5. Align sustainability goals with cultural and social contexts—Societal priorities must be harmonized with ecological sustainability, with consideration of historical and cultural factors. This requires interdisciplinary dialog among experts and translation into integrative sustainability science, where TEH is a starting point for systemic solutions.

16. The future of ecohydrology: A transdisciplinary approach supported by AI

Given the complexity of these challenges, it is essential to integrate geology, ecology, hydrology, and broader environmental sciences. As a result, TEH is poised to become as intellectually advanced as rocket science, or potentially even surpass it. With the integration of AI-driven analytics, future environmental decision-making will be supported by a deep, predictive understanding of ecosystem dynamics integrated with the increasingly stochastic hydrological cycle under climate change. This will significantly enhance our ability to harmonize human development with the biosphere's suitability potential, in alignment with the framework for action and parallel enhancement of WBSR.

AI disclosure

During the preparation of this work the author(s) used OpenAI's ChatGPT. This document includes content generated and/or edited with the assistance of OpenAI's ChatGPT, used to support language refinement and structural improvements. All final content was reviewed and verified by the author. After using this tool/service, the author(s) reviewed and edited the content as needed and take(s) full responsibility for the content of the publication.

References

Belletti, B., Garcia de Leaniz, C., Jones, J., Bizzi, S., Börger, L., Segura, G., Castelletti, A., van de Bund, W., Aarestrup, K., Barry, J., Belka, K., Berkhuysen, A., Birnie-Gauvin, K., Bussettini, M., Carolli, M., Consuegra, S., Dopico, E., Feierfeil, T., Fernández, S., Fernandez Garrido, P., Garcia-Vazquez, E., Garrido, S., Giannico, G., Gough, P., Jepsen, N., Jones, P.E., Kemp, P., Kerr, J., King, J., Łapińska, M., Lázaro, G., Lucas, M. C., Marcello, L., Martin, P., McGinnity, P., O'Hanley, J., Olivo del Amo, R., Parasiewicz, P., Pusch, M., Rincon, G., Rodriguez, C., Royte, J., Schneider, C.T., Tummers, J.S., Vallesi, S., Vowles, A., Verspoor, E., Wanningen, H., Wantzen, K.M., Wildman, L., Zalewski, M., 2020. More than one million barriers fragment Europe's rivers. Nature 588 (7838), 436–441. https://doi.org/10.1038/s41586-020-3005-2.

Bianchi, A., 2021. Impact of river hydrology on hydraulic engineering and hydropower. Water 13 (22), 3165. https://doi.org/10.3390/w13223165.

Bren, L., 2015. Forest Hydrology and Catchment Management: An Australian Perspective. Springer Netherlands, Australia, pp. 1–268. https://doi.org/10.1007/978-94-017-9337-7.

Buckton, S.J., Fazey, I., Sharpe, B., Om, E.S., Doherty, B., Ball, P., Denby, K., Bryant, M., Lait, R., Bridle, S., Cain, M., Carmen, E., Collins, L., Nixon, N., Yap, C., Connolly, A., Fletcher, B., Frankowska, A., Gardner, G., James, A., Kendrick, I., Kluczkovski, A., Mair, S., Morris, B., Sinclair, M., 2023. The regenerative lens: a conceptual framework for regenerative social-ecological systems. One Earth 6 (7), 824–842. https://doi.org/10.1016/j.oneear.2023.06.006.

Buffon, G.L.L., Daubenton, M., Montbéliard, P.G.de, La Cépède, M., 1749. Histoire naturelle, générale et particulière : avec la description du cabinet de roi. De l'Imprimerie royale. https://doi.org/10.5962/bhl.title.105737.

Centola, D., Ticchi, D., Huyck, 2022. Beyond collective intelligence: collective adaptation. J. R. Soc. Interface 19, 20220736.

Clewell, A.F., Aronson, J., 2013. Ecological Restoration: Principles, Values, and Structure of an Emerging Profession, second ed. Island Press-Center for Resource Economics, pp. 1–303. https://doi.org/10.5822/978-1-59726-323-8.

Closing the Loop – An EU Action Plan for the Circular Economy, 2015. European Commission.

Cohen, J.E., 1995. Ecological Community: The Trophic Dimension. Princeton University Press.

Costanza, R., Cumberland, J.H., Daly, H., Goodland, R., Norgaard, R.B., Kubiszewski, I., Franco, C., 2014. An Introduction to Ecological Economics, second ed. CRC Press, pp. 1–328.

Cuvier, G., Louvet, D., Latreille, P.A., 1817. Le règne animal distribué d'après son organisation : pour servir de base a l'histoire naturelle des animaux et d'introduction a l'anatomie comparée. Chez Déterville. https://doi.org/10.5962/bhl.title.41460.

Darwin, C., Darwin, C., 1859. On the Origin of Species by Means of Natural Selection, or, the Preservation of Favoured Races in the Struggle for Life. John Murray. https://doi.org/10.5962/bhl.title.59991.

Dobzhansky, T., 1973. Nothing in biology makes sense except in the light of evolution. Am. Biol. Teach. 35 (3), 125–129. https://doi.org/10.2307/4444260.

Earth: Human Expansion Is Disrupting Fundamental Ecological Processes, Including Water and Nutrient Cycles and Energy Flows within the Biosphere. This Accelerating Degradation Poses Serious Threats to Food Security, Public Health, and Economic Stability Our Common Earth Report, 2024. UNESCO.

Earth Overshoot Day 2024 Falls on August 1st. Press Release, 2024. Global Footprint Network.

Edrisi, S.A., Sarkar, P., 2022. Assessing the realization of global land restoration: A meta-analysis. Anthropocene Sci. 1 (1), 1–16.

FAO, 2020. In: The State of Food and Agriculture 2020: Overcoming Water Challenges in Agriculture. FAO.

Friedlingstein, P., O'Sullivan, M., Jones, M.W., Andrew, R.M., Hauck, J., Landschützer, P., Le Quéré, C., Li, H., Luijkx, I.T., Olsen, A., Peters, G.P., Peters, W., Pongratz, J., Schwingshackl, C., Sitch, S., Canadell, J.G., Ciais, P., Jackson, R.B., Alin, S.R., Arneth, A., Arora, V., Bates, N.R., Becker, M., Bellouin, N., Berghoff, C.F., Bittig, H.C., Bopp, L., Cadule, P., Campbell, K., Chamberlain, M.A., Chandra, N., Chevallier, F., Chini, L. P., Colligan, T., Decayeux, J., Djeutchouang, L.M., Dou, X., Rojas, C.D., Enyo, K., Evans, W., Fay, A.R., Feely, R.A., Ford, D.J., Foster, A., Gasser, T., Gehlen, M., Gkritzalis, T., Grassi, G., Gregor, L., Gruber, N., Gürses, Ö., Harris, I., Hefner, M., Heinke, J., Hurtt, G.C., Iida, Y., Ilyina, T., Jacobson, A.R., Jain, A.K., Jarníková, T., Jersild, A., Jiang, F., Jin, Z., Kato, E., Keeling, R.F., Goldewijk, K.K., Knauer, J., Korsbakken, J.I., Lan, X., Lauvset, S.K., Lefèvre, N., Liu, Z., Liu, J., Ma, L., Maksyutov, S., Marland, G., Mayot, N., McGuire, P.C., Metzl, N., Monacci, N.M., Morgan, E.J., Nakaoka, S.-I., Neill, C., Niwa, Y., Nützel, T., Olivier, L., Ono, T., Palmer, P.I., Pierrot, D., Qin, Z., Resplandy, L., Roobaert, A., Rosan, T.M., Rödenbeck, C., Schwinger, J., Smallman, T.L., Smith, S.M., Sospedra-Alfonso, R., Steinhoff, T., Sun, Q., Sutton, A.J., Séférian, R., Takao, S., Tatebe, H., Tian, H., Tilbrook, B., Torres, O., Tourigny, E., Tsujino, H., Tubiello, F., Werf, G.van der, Wanninkhof, R., Wang, X., Yang, D., Yang, X., Yu, Z., Yuan, W., Yue, X., Zaehle, S., Zeng, N., Zeng, J., 2025. Global carbon budget 2024. Earth Syst. Sci. Data 17 (3), 965–1039. https://doi.org/10.5194/essd-17-965-2025.

Hardin, G., 1968. The tragedy of the commons. Science 162 (3859), 1243–1248. https://doi.org/10.1126/science.162.3859.1243.

Hehanussa, P.E., 2002. Ecohydrology: A Low Cost–High Opportunity Technology for Asian Countries. Conference presentation/paper, UNESCO International Hydrological Programme (IHP), Asia-Pacific region.

Hobbs, R.J., Arico, S., Aronson, J., Baron, J.S., Bridgewater, P., Cramer, V.A., Epstein, P.R., Ewel, J.J., Klink, C.A., Lugo, A.E., Norton, D., Ojima, D., Richardson, D.M., Sanderson, E.W., Valladares, F., Vilà, M., Zamora, R., Zobel, M., 2006. Novel ecosystems: theoretical and management aspects of the new ecological world order. Global Ecol. Biogeogr. 15 (1), 1–7. https://doi.org/10.1111/j.1466-822X.2006.00212.x.

Intergovernmental Science-Policy Platform on Biodiversity and Ecosystem Services (IPBES), 2019.

Jianguo, L., Dietz, T., Carpenter, S.R., Folke, C., Alberti, M., Redman, C.L., Schneider, S.H., Ostrom, E., Pell, A.N., Lubchenco, J., Taylor, W.W., Ouyang, Z., Deadman, P., Kratz, T., Provencher, W., 2007. Coupled human and natural systems. AMBIO 36 (8), 639–649. https://doi.org/10.1579/0044-7447.

Jørgensen, S.E., 2016. Ecohydrology as an important concept and tool in environmental management. Ecohydrol. Hydrobiol. 16 (1), 4–6. https://doi.org/10.1016/j.ecohyd.2015.04.005.

Jørgensen, S.E., Xu, L., Costanza, R., 2016. Handbook of Ecological Indicators for Assessment of Ecosystem Health, 2nd ed. CRC Press.

Krauze, K., Wagner, I., 2019. From classical water-ecosystem theories to nature-based solutions — contextualizing nature-based solutions for sustainable city. Sci. Total Environ. 655, 697–706. https://doi.org/10.1016/j.scitotenv.2018.11.187.

Kriebel, D., Tickner, J., Epstein, P., Lemons, J., Levins, R., Loechler, E.L., Quinn, M., Rudel, R., Schettler, T., Stoto, M., 2001. The precautionary principle in environmental science. Environ. Health Perspect. 109 (9), 871–876. https://doi.org/10.1289/ehp.01109871.

Lempert, D., Nguyen, H., 2011. The global prisoners' dilemma of unsustainability: why sustainable development cannot be achieved without resource security and eliminating the legacies of colonialism. Sustain. Sci. Pract. Policy 7 (1), 16–30. https://doi.org/10.1080/15487733.2011.11908062.

Levia, D.F., Creed, I.F., Hannah, D.M., Nanko, K., Boyer, E.W., Carlyle-Moses, D.E., van de Giesen, N., Grasso, D., Guswa, A.J., Hudson, J.E., Hudson, S.A., Iida, S., Jackson, R.B., Katul, G.G., Kumagai, T., Llorens, P., Lopes Ribeiro, F., Pataki, D.E., Peters, C.A., Sanchez Carretero, D., Selker, J.S., Tetzlaff, D., Zalewski, M., Bruen, M., 2020. Homogenization of the terrestrial water cycle. Nat. Geosci. 13 (10), 656–658. https://doi.org/10.1038/s41561-020-0641-y.

Linnaeus, C., 1758. Systema Naturae per Regna Tria Naturae, Secundum Classes, Ordines, Genera, Species, cum Characteribus, Differentiis, Synonymis, Locis, 1. Typis Ioannis Thomae.

Lloyd, W.F., 1833. Two Lectures on the Checks to Population, Delivered before the University of Oxford in Michaelmas Term 1832. S. Collingwood.

Maciejewski, M., Walczykiewicz, T., 2006. Integrated river basin management—principles and experience of Polish implementation. Ecohydrol. Hydrobiol. 6, 41–50.

Mang, P., Haggard, B., 2016. Regenerative Development and Design: A Framework for Evolving. John Wiley and Sons Inc, pp. 1–235.

Mang, P., Reed, B., 2016. Regenerative Development and Design: A Framework for Evolving Sustainability. Wiley & Sons.

Montanari, Bahr, J., Blöschl, G., Cai, X., Mackay, D.S., Michalak, A.M., Rajaram, H., Sander, G., 2015. Fifty years of water resources research: legacy and perspectives for the science of hydrology. Water Resour. Res. 51 (9), 6797–6803. https://doi.org/10.1002/2015WR017998.

Naiman, R.J., Decamps, H., Fournier, F. (Eds.), 1989. The Role of Land/Inland Water Ecotones in Landscape Management and Restoration: Proposals for Collaborative Research. MAB Digest, 4. UNESCO, Paris.

Nilsson, M., Dalkmann, H., 2011. Decision making and strategic environmental assessment. J. Environ. Assess. Pol. Manag. 3 (3), 305–327. https://doi.org/10.1142/s1464333201000728.

Odum, E.P., Barrett, G.W., 2004. Fundamentals of Ecology, fifth ed. Brooks/Cole-Thomson Learning, p. 624.

Ostrom, E., 2009. A general framework for analyzing sustainability of social-ecological systems. Science 325 (5939), 419–422. https://doi.org/10.1126/science.1172133.

Potts, S.G., Imperatriz-Fonseca, V., Ngo, H.T., Aizen, M.A., Biesmeijer, J.C., Breeze, T.D., Dicks, L.V., Garibaldi, L.A., Hill, R., Settele, J., Vanbergen, A.J., 2016. Safeguarding pollinators and their values to human well-being. Nature 540 (7632), 220–229. https://doi.org/10.1038/nature20588.

Purvis, J.L., 1970. Weekend of Terror. Powell.

Redman, C.L., Grove, J.M., Kuby, L.H., 2004. Integrating social science into the Long-Term Ecological Research (LTER) network: social dimensions of ecological change and ecological dimensions of social change. Ecosystems 7 (2), 161–171. https://doi.org/10.1007/s10021-003-0215-z.

Richardson, K., 2025. Planetary boundaries, polycrisis and politics in the anthropocene. In: Towards Rethinking Politics, Policy and Polity in the Anthropocene. Springer, pp. 339–444.

Rocha, J.C., Peterson, G.D., Biggs, R., Unsworth, R.K.F., 2015. Regime shifts in the Anthropocene: drivers, risks, and resilience. PLoS One 10 (8), e0134639. https://doi.org/10.1371/journal.pone.0134639.

Rockström, J., Steffen, W., Noone, K., Persson, A., Chapin, F.S., Lambin, E.F., Lenton, T.M., Scheffer, M., Folke, C., Schellnhuber, H.J., Nykvist, B., De Wit, C.A., Hughes, T., Van Der Leeuw, S., Rodhe, H., Sörlin, S., Snyder, P.K., Costanza, R., Svedin, U., Falkenmark, M., Karlberg, L., Corell, R.W., Fabry, V.J., Hansen, J., Walker, B., Liverman, D., Richardson, K., Crutzen, P., Foley, J.A., 2009. A safe operating space for humanity. Nature 461 (7263), 472–475. https://doi.org/10.1038/461472a.

Scheffer, M., Hosper, S.H., Meijer, M.-L., Moss, B., Jeppesen, E., 1993. Alternative equilibria in shallow lakes. Trends Ecol. Evol. 8 (8), 275–279.

Steffen, W., Rockström, J., Richardson, K., Lenton, T.M., Folke, C., Liverman, D., Summerhayes, C.P., Barnosky, A.D., Cornell, S.E., Crucifix, M., Donges, J.F., Fetzer, I., Lade, S.J., Scheffer, M., Winkelmann, R., Schellnhuber, H.J., 2018. Trajectories of the earth system in the Anthropocene. Proc. Natl. Acad. Sci. U. S. A 115 (33), 8252–8259. https://doi.org/10.1073/pnas.1810141115.

Summary for Policymakers of the Methodological Assessment of Scenarios and Models of Biodiversity and Ecosystem Services, 2016. IPBES Secretariat, Bonn.

The Cost of Doing Nothing: The Humanitarian Price of Climate Change and How it Can Be Avoided, 2019. IFRC.

The State of Food and Agriculture: Climate Change, Agriculture and Food Security, 2016. FAO.

Uhlenbrook, S., Roser, S., Tilch, N., 2004. Hydrological process representation at the meso-scale: the potential of a distributed, conceptual catchment model. J. Hydrol. 291 (3–4), 278–296. https://doi.org/10.1016/j.jhydrol.2003.12.038.

UN World Water Development Report, 2022. UNESCO World Water Assessment Programme for UN-Water.

United Nations 2023 Water Conference: Midterm Comprehensive Review of the Implementation of the International Decade for Action "Water for Sustainable Development", 2018. United Nations.

United Nations, 2015a. The 17 Sustainable Development Goals. https://sdgs.un.org/goals.

United Nations, 2015b. Transforming Our World: The 2030 Agenda for Sustainable Development (A/RES/70/1). United Nations. https://sdgs.un.org/2030agenda.

Wantzen, K.M., Aziz, B., Longuet, I., Bao, I., Bocoum, H., Cissé, L., Chauhan, M., Girard, P., Gopal, B., Kane, A., Marchese, M.R., Prakash, N., Teixeira, P., Zalewski, M., 2016. River culture: an eco-social approach to mitigate the biological and cultural diversity crisis in riverscapes. Ecohydrol. Hydrobiol. 16 (1), 7–18. https://doi.org/10.1016/j.ecohyd.2015.12.003.

Watson, R., Kundzewicz, Z.W., Borrell- Damián, L., 2022. Covid-19, and the climate change and biodiversity emergencies. Sci. Total Environ. 844, 157188. https://doi.org/10.1016/j.scitotenv.2022.157188.

Ye, L., Van Ranst, E., 2009. Production scenarios and the effect of soil degradation on long-term food security in China. Glob. Environ. Change 19 (4), 464–481. https://doi.org/10.1016/j.gloenvcha.2009.06.002.

Zalewski, M., 2000. Ecohydrology — the scientific background to use ecosystem properties as management tools toward sustainability of water resources. Ecol. Eng. 16 (1), 1–8. https://doi.org/10.1016/s0925-8574(00)00071-9.

Zalewski, M., 2002. Ecohydrology — integrative science for sustainable water, environment and society. Ecohydrol. Hydrobiol. 2 (1–4), 3–10.

Zalewski, M., 2006. Ecohydrology — an interdisciplinary tool for integrated protection and management of water bodies. River Syst. 16 (4), 613–622. https://doi.org/10.1127/lr/16/2006/613.

Zalewski, M., 2014. Ecohydrology and hydrologic engineering: regulation of hydrology-biota interactions for sustainability. J. Hydrol. Eng. 20 (1), A4014012. https://doi.org/10.1061/(ASCE)HE.1943-5584.0000999.

Zalewski, M., 2020. Ekohydrologia. Wydawnictwo Naukowe PWN, Warszawa. ISBN:978-83-01-20804-2.

Zalewski, M., 2021. Ecohydrology: An Integrative Sustainability Science. IntechOpen. https://doi.org/10.5772/intechopen.94169.
Zalewski, M., Naiman, R.J., 1985. The regulation of riverine fish communities by a continuum of abiotic-biotic factors. In: Habitat Modification and Freshwater Fisheries. Butterworths.
Zalewski, M., Wantzen, K.M., 2023. Ecohydrology and river culture: synergies for the enhancement of river and catchment management. In: Ecohydrology and River Culture. UNESCO.
Zalewski, M., Brewińska-Zaraś, B., Frankiewicz, P., 1990. Fry communities as a biomanipulating tool in a temperate lowland reservoir. Archiv für Hydrobiologie–BeiheftErgebnisse der Limnologie 33, 763–774.
Zalewski, M., Cowx, I., Naiman, R.J., 1990. Regulation of hydrology–biota interactions for sustainability. Ecol. Eng. 16 (1), 1–8.
Zalewski, M., Janauer, G.A., Jolánkai, G., 1997. Ecohydrology: A new paradigm for the sustainable use of aquatic resources. Conceptual Background, Working Hypothesis, Rationale and Scientific Guidelines for the Implementation of the IHP-V Projects 2.3/2.4. UNESCO, Paris.
Zalewski, M., Puchalski, W., Frankiewicz, P., Bis, B., 1994. Riparian ecotones and fish communities in rivers – intermediate complexity hypothesis. In: Cowx, L.G. (Ed.), Rehabilitation of Freshwater Fisheries. Fishing News Books.

Further reading

Mang, P., Reed, B., 2012. Regenerative development and design. In: Loftness, V. (Ed.), Sustainable Built Environments. Springer, pp. 303–318. https://doi.org/10.1007/978-1-4614-5828-9_303.
Zalewski, M., 1997. Fish diversity and ecotonal habitat. In: Lachavanne, J.-B., Raphaëlle, J. (Eds.), Biodiversity in Land-Inland Water Ecotones. UNESCO, Paris, pp. 183–203.

Chapter 2

Origins of transdisciplinary ecohydrology: Theoretical framework

Maciej Zalewski
European Regional Centre for Ecohydrology of the Polish Academy of Sciences under the auspices of UNESCO, Łódź, Poland

PHOTO SOURCE: M. ZALEWSKI.

Transdisciplinary Ecohydrology. https://doi.org/10.1016/B978-0-443-41596-8.00009-1

1. Origins of ecohydrology: Theoretical framework

Inspiration is most important driver in science

Andrea Ghez (Nobel Prize Winner).

1.1 The scientific origins of ecohydrology

The notion of ecohydrology, as a scientific discipline concerned with the sustainable management of water resources, has its origins in the Club of Rome Report, entitled "Limits to Growth" Fig. 2.1 (Meadows et al., 1972). However, this report does not combine water resource as the fundamental factor for sustainable development.

2. The role of fundamental physics in ecohydrology

In the current stage of the Anthropocene, ecology must evolve from predicting ecosystem dynamics to shaping the evolution of "novel ecosystems." Rather than focusing on conservation in human-modified areas, there is a pressing need to enhance sustainability by shaping structure and regulating key ecological processes, such as the water and nutrient cycles and energy flow through ecosystems, which have evolved over time. Achieving this ambitious regulatory objective and consideration that water is a key driver of SDGs necessitates a profound comprehension of these processes, predicated on the reduction of observed phenomena to fundamental physics (Grodziński et al., 1975). This theoretical framework facilitates the integration of hydrological and ecological processes, using the common denominator of fundamental physics. This assumption is further substantiated by the observed paradox that decision-makers and society prioritize intensive exploitation and modification of agricultural landscapes, and even conservation of the structure of natural and human-modified ecosystems, without acknowledging the necessity of comprehending the intricacies of fundamental ecological processes, such as water and nutrient cycles, and sustainability dependent on them. Moreover, restoration ecology predominantly concentrates on the restoration of ecosystem structure.

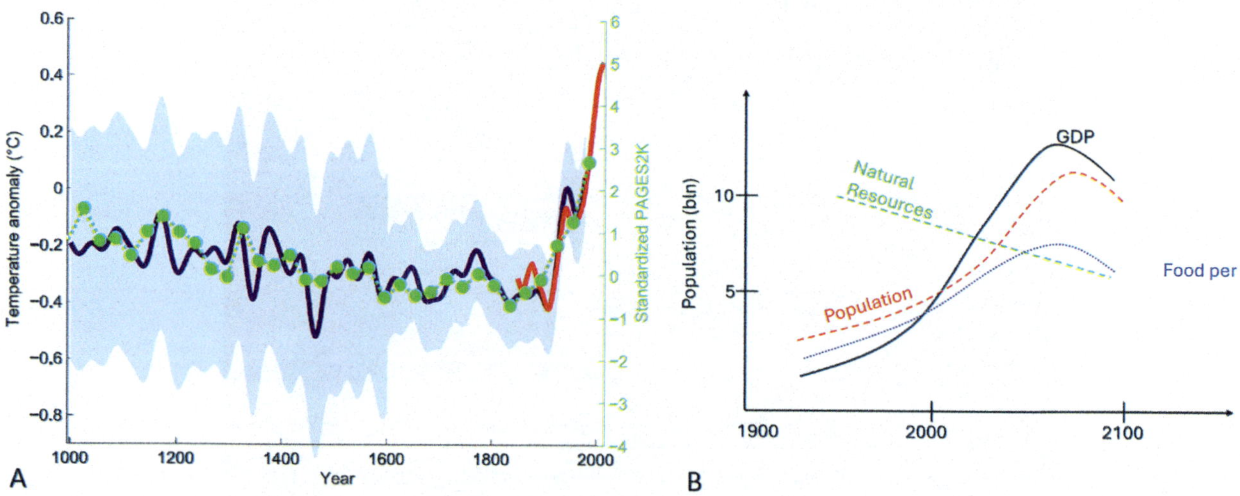

FIGURE 2.1 A. "Hockey Stick Graph". B. Scenarios from the limits to growth report from the club of Rome. (A) The original "hockey stick" temperature reconstruction for the Northern Hemisphere by Mann et al. (1999) is presented as a blue smoothed curve, with its uncertainty range depicted in light blue. This is overlaid with green dots representing the 30-year global average from the PAGES 2k Consortium reconstruction PAGES 2k Consortium, 2013. The red line illustrates the observed global mean temperature from 1850 to 2013, based on HadCRUT4 data. (B) This diagram presents five major variables (population, industrial output, food per capita, non-renewable resources, pollution). (A) *The original northern hemisphere hockey stick graph of Mann, M. E., Bradley, R. S., Hughes, M. K. (1999). Northern hemisphere temperatures during the past millennium: Inferences, uncertainties, and limitations. Geophys. Res. Lett. 26(6), 759–762. https://doi.org/10.1029/1999GL900070. Wikimedia Commons. Licensed under CC BY-SA 4.0. (B) Inspired by concepts from Meadows, D. H., Meadows, D. L., Randers, J., Behrens, W. W. (1972). The limits to growth. Universe Books.*

3. From structure to process: Rethinking ecosystem regulation in ecohydrology

In the past scientific decade (2013–2022), the International Association of Hydrological Sciences (IAHS), a prominent scientific hydrological association, concentrated its development strategy on hydrology, with the objective of enhancing the interpretation of the processes governing the water cycle. This endeavor entailed a focus on the evolving dynamics of these processes in relation to the rapidly evolving human systems, as articulated in the principle "Panta Rhei—Everything flows" (Montanari et al., 2013). This signifies that human interactions have been identified as the most promising reference point for hydrological development, with the objective being the elimination of water-related threats (floods, droughts), the exploitation of water resources, and pollution. Subsequently, there was an appreciation and necessity for interdisciplinary development, as evidenced by "increased efforts to connect with the socioeconomic sciences and geosciences in general." A significant advancement in this regard was the publication of "Twenty-three unsolved problems in hydrology" (Blöschl et al., 2019), which implicitly addressed the need for a holistic perspective. This publication highlighted the necessity for a process-based understanding of hydrological variability and causality, an increased focus on environmental change, and a new set of questions related to human interactions with nature and water cycle feedback in the context of complex water management problems. These statements reflect the growing recognition among hydrologists of the need to advance hydrology with greater consideration for the role of ecosystems in regulating the water cycle. This mirrors earlier calls from landscape ecologists and stream ecologists to bridge research and conservation across spatial and disciplinary boundaries, as exemplified by Fausch et al. (2002), who advocated for connecting local-scale ecological understanding with catchment-scale management. The most optimistic message to be drawn from this publication is that the 23 questions require an explicit linkage of hydrological subdisciplines, as opposed to the "current canalization of subdisciplines." The second important conclusion is that "the first step for organizing the community (hydrologists; IAHS) in a more holistic way, international cooperation is at the core of it." Furthermore, Kundzewicz et al. (2018) emphasized the necessity of increasing the scale of research, particularly in the context of climate change processes and interdisciplinary collaboration. In particular, Blöschl et al. (2019) argued that "questions remain focused on the process-based understanding of hydrological variability and causality at all space and time scales (Habersack and Piégay, 2007). Increased attention to environmental change drives a new emphasis on understanding how change propagates across interfaces within the hydrological system and across disciplinary boundaries (Harms et al., 2021). In particular, the expansion of the human footprint raises a new set of questions related to human interactions with nature and water cycle feedback in the context of complex water management problems."

To affect a transition from a structure-based to a process-oriented mode of thinking, it is necessary to consider the question of how populations and ecosystems are regulated. This issue has been the subject of debate among ecologists with regard to the density-dependent and density-independent regulation of ecosystems (Krebs, 1972). Two reference points for the deductive model have been proposed to explain this controversy: The first of these is the Abiotic-Biotic Regulatory Continuum (Zalewski and Naiman, 1985) which builds upon the River Continuum Concept (RCC) (Vannote et al., 1980). The general assumption is that changes in the hierarchy of drivers, which shape the structure and functions of the ecosystem, are moving from abiotic (density independent) to biotic (density dependent) as ecological processes are reduced to fundamental laws of physics in hydrology, hydraulics, and ecology. In consideration of the positive balance between energy gain and energy expenditure for the persistence of organisms and populations in a given fluvial ecosystem or its part, the Bernoulli law on distribution in stream turbulency and laminar flow was considered as one of the major factors of energy expenditure along the RCC and temperature. The Van't Hoff law governed chemical reactions and, consequently, metabolic processes, biological productivity, and the potential energy gain available to organisms. For example, in mountain streams, high water velocity determined high energy expenditure even by well-adapted (torpedo-shaped) trout. It is evident that riffle–pool sequences play a crucial role in ecosystem function Fig. 2.2. Riffles act as primary zones of productivity by supporting periphyton growth and sustaining diverse invertebrate communities. Pools, together with complex riparian ecotones such as tree roots and woody debris, form a mosaic of rich habitats that supply both allochthonous and autochthonous food sources. This spatial diversity optimizes the balance between energy gain and expenditure for organisms. Moreover, the variety of habitats and flow conditions not only improve energy efficiency but also reduce competition and predation pressures, thereby promoting a more diverse and resilient ecosystem. If mountain streams become channelized and lose their riffle-pool sequence, they will become fishless and lowland streams will have a lot less biodiversity (Fig. 2.2). Understanding how diverse habitats mitigate feeding stress and influence changes in the hierarchy of ecological drivers is essential for regulating ecological processes and enhancing the sustainability of the catchment. Therefore, adopting a transdisciplinary ecohydrological perspective is crucial for fully understanding these interdependencies.

FIGURE 2.2 **(A) Diverse habitat in the mountain stream.** A stream featuring a natural morphology of pools, riffles, and unaltered stream banks. **(B) Channelized mountain stream.** A stream characterized by a straightened, unnatural channel, lacking the natural morphology. *Photo credit: M. Zalewski.*

The aforementioned point constitutes a significant step in the development of coevolution of understanding of eco-hydrological and hydrological processes (Blöschl et al., 2013). Nevertheless, the important step in formulating a holistic perspective on the role of biota in regulating the hydrological mesocycle should be the study of ecohydrology in terrestrial ecosystems, specifically, the interactions among water, plants, and soil (Eagleson, 1982). This foundation was further advanced by Rodriguez-Iturbe, Eagleson (Rodriguez-Iturbe, 2000; Eagleson, 2002), but also specific processes, such as disturbance-driven infiltration dynamics (Beven and Germann, 1982) and the role of soil macrofauna in enhancing water infiltration, also contribute important insights. The land/water ecotones (Naiman et al., 1989) and floodplains (Zalewski, 2006; Kiedrzyńska et al., 2008) are of extreme importance in terms of mitigating the impact of the agricultural and urbanized landscape on freshwater ecosystems, as well as functioning of the freshwater ecosystems (Zalewski et al., 2001) especially river metabolism, which determines bioproductivity, biodiversity, self-purification, and resilience. The quality of water and, consequently, the ecological status of large rivers, reservoirs, lakes, coastal zones, and the ecosystem services they provide are determined by catchment structure and the use of land/water ecotones along small, natural, meandering streams. Consequently, in the era where water is increasingly recognized as a key factor for sustainability, hydrology should focus on understanding the role of catchments and biocenosis in slowing down the hydrological mesocycle, reducing the rate of water and nutrient transfer from the catchment into the ocean. Enhancing river self-purification through renaturalization and converting excessive nutrients in the catchment into biomass by increasing landscape complexity—processes that are critical for achieving sustainability.

The traditional hydrological approach concentrated on the lower zones of the river continuum, thus underscoring the reasons for the steady decline in water resources despite increased expenditure on water. There is growing evidence that biota in the catchment and in freshwater ecosystems are important, particularly in warm climates where a high rate of ecosystem metabolism is the dominant factor shaping ecosystem services and resilience. However, intensive catchment modification and water exploitation due to demographic processes, urbanization, industrialization (industrial agriculture development), and deforestation accelerate the hydrological cycle and reduce catchment retentiveness of water (residence time), and nutrients (e.g., overfertilization, soil erosion), thus increasing water deficit and pollution (Fig. 2.3). The negative effects of these changes

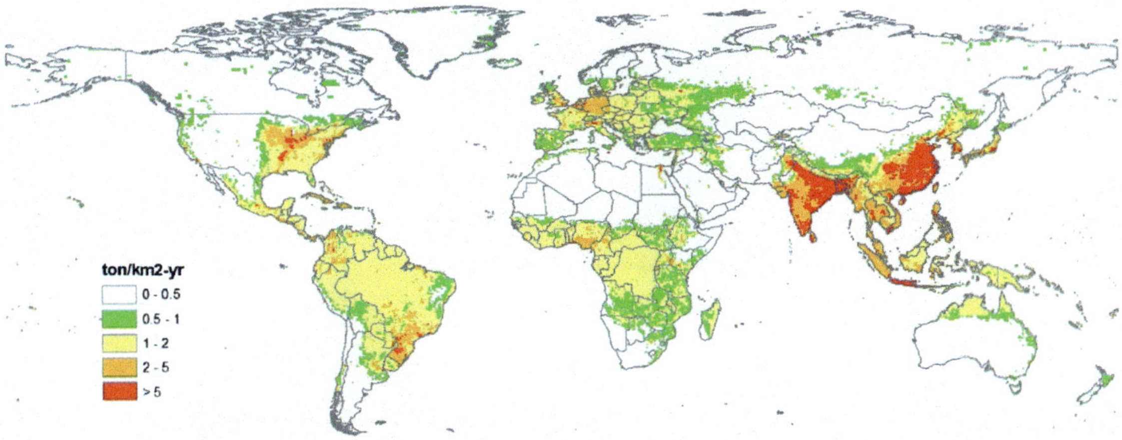

FIGURE 2.3 **Changes in large river basins.** Anticipated changes in large river basins in 2050 (compared to 2000–2005 data) on the basis of nitrogen compound content. *The Murky Future of Global Water Quality, 2015. Reproduced with permission from the International Food Policy Research Institute (IFPRI) www.ifpri.org. The original text is available online at https://hdl.handle.net/10568/150398.*

have been increasing due to climate change, which in turn increases the stochastic character of the hydrological processes. This, in turn, increases the cumulative impact, reducing the sustainability potential of the catchments (WBSR).

The concept of cumulative impact (Bedford and Preston, 1988) refers to the aggregate effect of multiple activities on the environment. These activities, though individually may be minor, can have a significant cumulative impact when their effects overlap, potentially resulting in substantial alterations to the environment. The complexity of predicting the cumulative effect stems from the numerous factors contributing to it, as well as the intricate interactions between these factors, which can be both synergistic and antagonistic. Therefore, mitigating and compensating for the growing cumulative impacts requires a novel approach and innovative solutions, based on a shift from structure-oriented to process-oriented thinking.

4. The hydrological mesocycle: Balancing water, ecosystems, and society

A comprehensive understanding of water circulation within the watershed is essential for enhancing sustainability. Water plays a fundamental role in regulating the movement of carbon, phosphorus, nitrogen, and other elements within the biosphere, particularly in terms of nutrient accumulation in ecosystems of the catchment. The temperature of the water determines its distribution between biomass and organic matter in the soil (Fig. 2.4).

Water is not only the driving force fostering the evolution of the biosphere, but also the main factor that regulates (or in situations of limitation restrains) the terrestrial ecosystem's ability to accumulate such biogenic substances as carbon, nitrogen, and phosphorus, all of which underpin the biodiversity of trophic networks, energy flow pathways, and species abundance. When the amount of water in the environment is sufficient, the temperature becomes another regulating (or restraining) factor as it dictates the rate of metabolism of microorganisms, plants, and ectothermic animals. Moreover, the assimilation of available nutrients is governed by stoichiometric relations in accordance with Liebig's Law of the Minimum. Also, the quantity of available water in terrestrial ecosystems within a given temperature range indirectly determines the quantity of carbon which the ecosystem can accumulate. The temperature determines the allocation of carbon between plant biomass and organic matter in the soil.

In high-temperature conditions, such as those found in tropical climates, nutrients are predominantly accumulated in biomass (Zalewski, 2002; United Nations Environment Programme, 2002). This relationship between water and nutrient accumulation in terrestrial ecosystems is a determining factor in bioproductivity and biodiversity (Zalewski, 2002). Also, resilience of all elements of the hydrological mesocycle should be widely recognized as a key element in management decisions (IWRM—Integrated Water Resources Management; WFD—Water Framework Directive of the European Commission). Consequently, management measures aimed at enhancing the sustainability potential of the catchments WBSR+CE+LPG and biosphere toward harmonization with the needs of catchment and global society has to adopt the hydrological mesocycle on the scale of catchments/river basins as their reference point. In this context, ecosystems should be recognized as essential, integrative components of water systems and water management, offering specific Water–Biodiversity–Ecosystem Services–Resilience (WBSR) potential for understanding the dynamics and feedback of key processes. Furthermore, the quantification of the mesocycle should serve to analyze and quantify processes, but also to

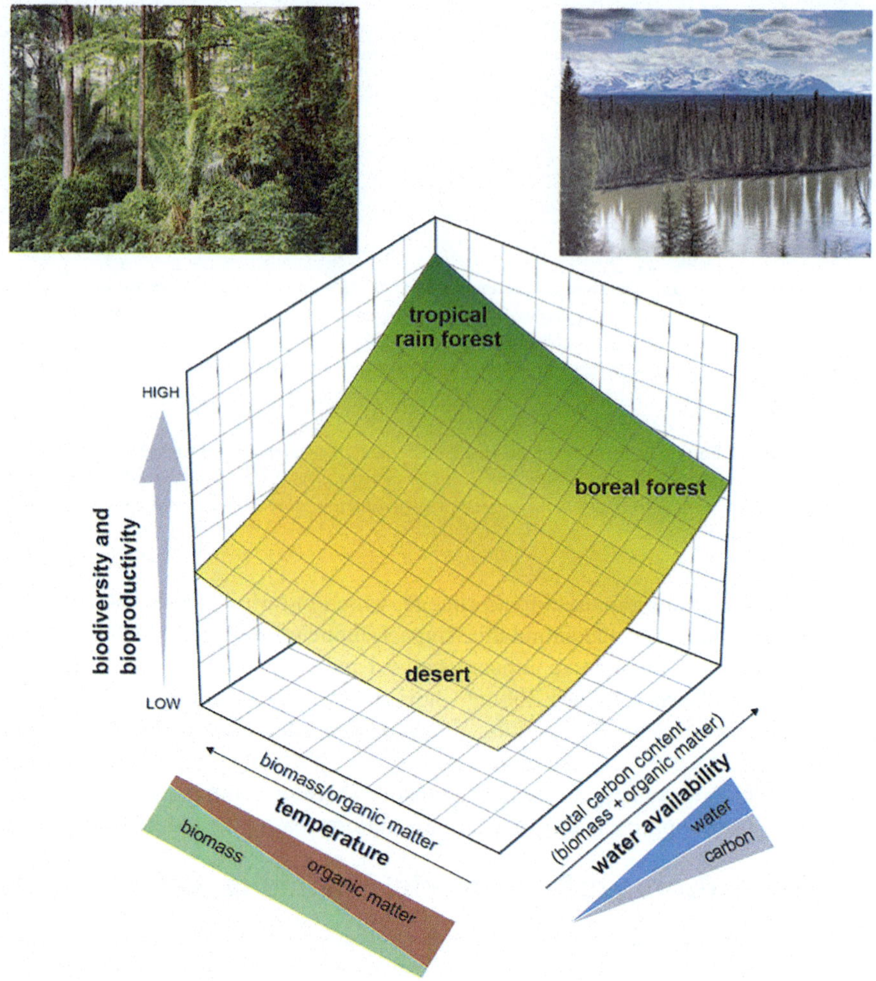

FIGURE 2.4 Deductive foundations of transdisciplinary ecohydrology This diagram illustrates the deductive foundations of ecohydrology, emphasizing the essential role of water and temperature in regulating terrestrial ecosystem productivity and carbon dynamics. The availability of water primarily determines the magnitude of primary production, and the total amount of carbon stored within an ecosystem. In contrast, temperature governs the seasonal duration of this productivity and influences the allocation of carbon between living biomass and soil organic matter. In colder environments, such as the taiga in Alaska, carbon tends to accumulate predominantly in dead organic matter due to slower metabolic rates and prolonged decomposition processes—which can take up to 40 years. Conversely, in tropical regions, elevated temperatures accelerate decomposition, resulting in more rapid cycling and a higher proportion of carbon being stored in biomass, with decomposition occurring in as little as 1 year. This ecohydrological framework demonstrates that maximum biodiversity and bioproductivity are typically achieved in ecosystems with both high-water availability and elevated temperatures. These principles form part of the deductive theoretical basis of ecohydrology, linking climate drivers with biogeochemical and ecological responses (modified from Zalewski (2002); photo credit: Maciej Zalewski (1) Steve Lyon (2) www.eso.org/public/images/armazonesparanal. *Modified from Zalewski, M., 2002. Ecohydrology—the use of ecological and hydrological processes for sustainable management of water resources. Hydrol. Sci. J. 47 (5), 823–832. https://doi.org/10.1080/02626660209492986.*

harmonize hydrotechnical solutions and ecohydrological nature-based solutions (NBS) into hybrid solutions. Given that this holistic approach encompasses the entire catchment, the effective utilization of ecohydrological NBS is contingent on the inclusion of society as a whole through the process of developing Sustainability Culture and Education (Wantzen et al., 2016; Zalewski and Wantzen, 2023).

It is evident that the reduction of nonpoint source (NPS) pollution must be grounded in the involvement of society as a whole. This is due to the fact that various forms of NPS (phosphorus, nitrogen, dioxins, xenobiotics, and heavy metals) are responsible for up to 50% of the adverse impacts on ecosystems such as the Baltic Sea (Skłodowski et al., 2014). These pollutants (N, P) pose a significant threat, particularly as drivers of eutrophication, anoxic zones, fisheries decline, and toxic algal blooms. This example underscores the imperative for societal education to mitigate a wide range of anthropogenic stressors. It is widely acknowledged that each individual contributes 1–2 g of phosphorus daily to the sewage system, with a town of 100,000 people generating 100 kg of phosphorus in untreated sewage daily. Assuming an efficiency of 90% for a sewage treatment plant in terms of phosphorus removal, the purified sewage provides 10 kg of phosphorus daily, equating to 3.5 tons annually. Conversely, if the efficiency of the sewage treatment plant in terms of energy and chemical conservation is reduced to 80%, the load of phosphorus will be doubled to 7 tons annually, entering recipient ecosystems, reservoirs, and coastal zones. The implications of this for the process of eutrophication are significant. According to limnological research, phosphorus is a limiting nutrient for freshwater ecosystems (Liebig Minimum Low), and it has been demonstrated that in eutrophic lake reservoirs, 1 kg of phosphorus generates 2 tons of algal bloom (Kajak, 2001).

It is an indisputable fact that water is the key asset and the most limiting of all the resources available in the environment. This is because it is indispensable for human civilization and the natural environment alike. Water is needed to satisfy basic societal needs such as food production (Mukherjee and Lal, 2013), industry, transport, recreation, and culture. The patchwork-like spatial distribution of tensions between societies and environmental capacity is driven by ecological benefits and societal demand, resulting from the varied distribution of geomorphological and climate conditions that underpin ecosystem bioproductivity. AI will be an important tool for solving these challenges.

Conventional water resources management largely relied on engineering solutions and did not consider the significance of degraded biological structures and ecosystem functions for the ecological benefits that determine the quality of people's lives and health. This approach often resulted in overly simplified environments, thereby reducing their resilience to anthropogenic stress, resulting in dying of lands (Fig. 2.5) and water shortages (Fig. 2.6). For example, the straightening of watercourses has been shown to accelerate water runoff to the sea, leading to the lowering of groundwater levels, the drying up of wetlands, and significant biodiversity loss. This process also impairs the rivers' capacity for self-purification, groundwater recharge, and their recreational potential (Box 2.1).

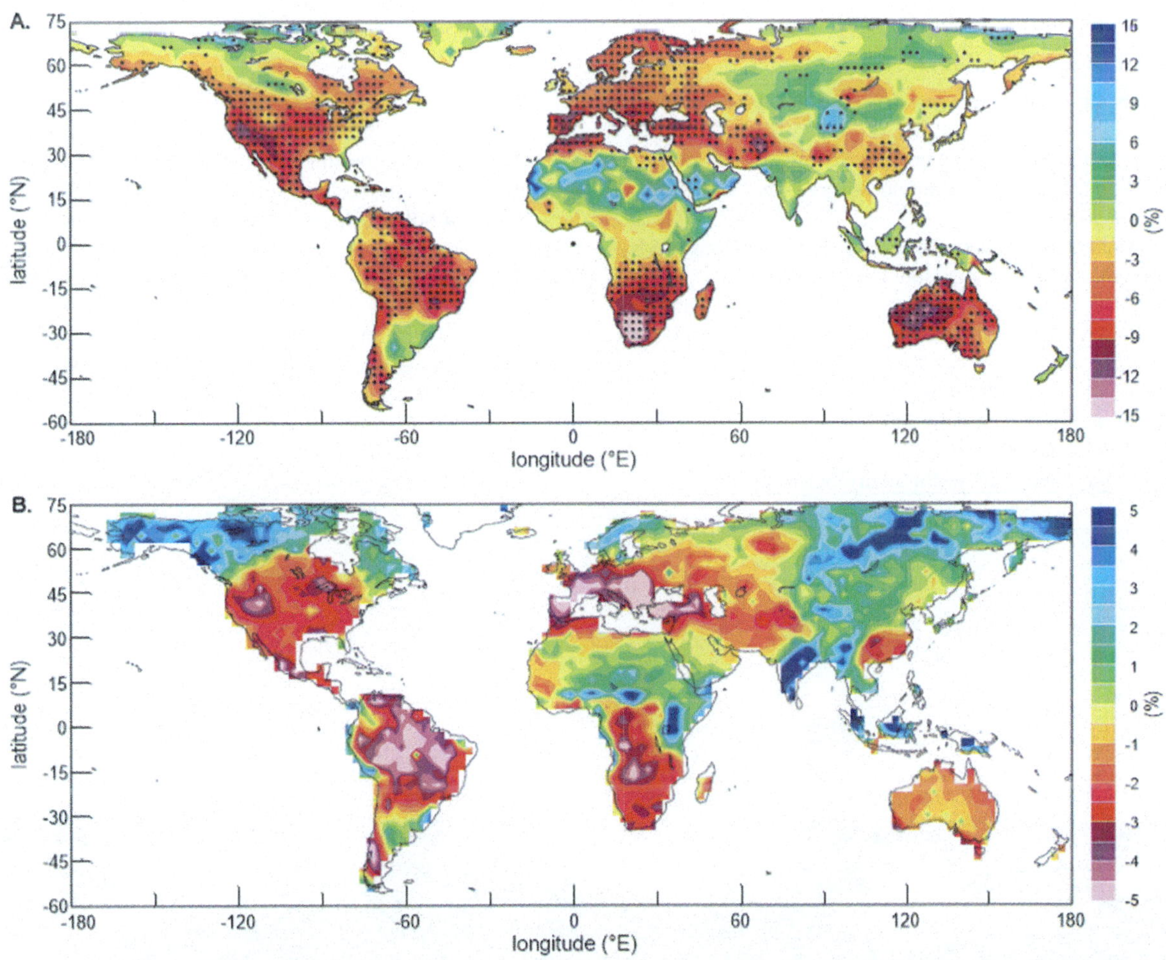

FIGURE 2.5 Projected changes in soil moisture. Anticipated changes in average soil moisture, in n topsoil, approx. 10 cm, expressed as a percentage: A—for the period 1980–1999, B—for 2080–2099. Forecast based on a set of projection models devised by the 11 Coupled Model Intercomparison Project Phase 5 (CMIP5). *Modified from: Dai, A. (2013). Increasing drought under global warming in observations and models. Nat. Clim, Change, 3(1), 52–58. https://doi.org/10.1038/nclimate1633.*

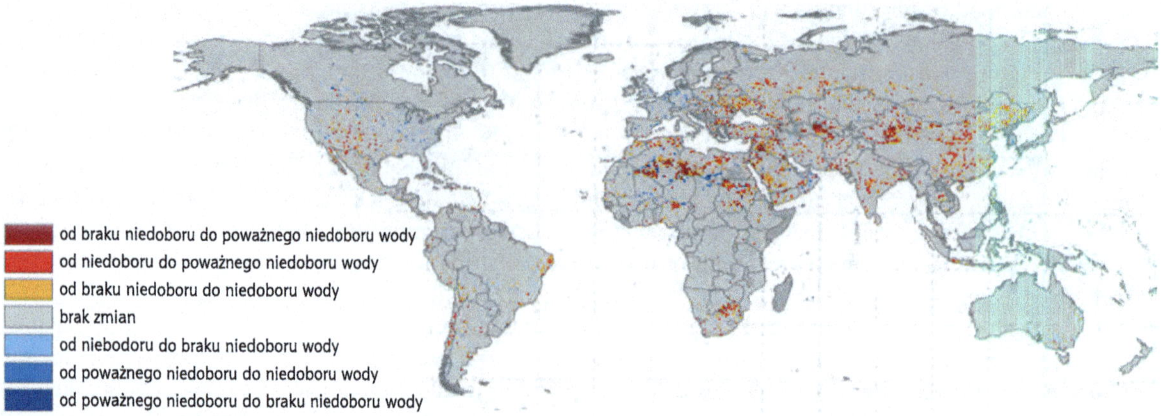

FIGURE 2.6 Anticipated changes in water shortages in 2050. Map showing anticipated changes in water shortages in 2050 (compared to the situation before 2010). Forecast made based on a middle-of-the-road scenario, which assumes that civilization will continue its development in line with the earlier trends and paradigms (the business-as-usual assumption). Here "water shortage" refers to a situation when the aggregate annual water intake represents 20%–40% of total available renewable surface water resources, while "serious water shortage" denotes intake of over 40%. *Burek, P., Satoh, Y., Fischer, G., Kahil, T., Scherzer, A., Tramberend, S., Nava, L. F., Wada, Y., Eisner, S., Flörke, M., Hanasaki, N., Magnuszewski, P., Cosgrove, W., Wiberg, D. (2016). Water Futures and Solutions: Fast Track Initiative – Final Report (Working Paper WP-16-006). International Institute for Applied Systems Analysis (IIASA), Laxenburg, Austria.*

BOX 2.1 Hydrological cycle

Why has the world become dry, and are biodiversity and ecosystem services declining?

1. Acceleration of hydrological cycles by
   1. Climate change amplified by green gas emission and deforestation
   2. Watershed fragmentation and overengineering (urbanization, industrialization and transport)
   3. River channelization and aquatic habitats degradation
2. Pollution and eutrophication freshwater and costal zones due to
   1. point source pollution
   2. nonpoint source pollution
   3. air deposition

What have we done to the hydrological cycle?

Anthropogenic evolution of the Danube River valley. Reduction of key drivers' sustainability WBSR—water retention capacity, bioproductivity, biodiversity, biocomplexity, services for society, resilience to climate change (Fig. 2.7).

Why is a holistic approach necessary?

1. Accelerating global changes
2. Increasing complexity of the interplay water/ecosystems/society
3. Fragmentation of knowledge, values, and efforts
4. "Financial gap" between needs and possibilities
5. Necessity of new paradigm, new concepts toward the development low-cost advanced nature-based solutions and systemic solutions (OECD, 2025).
6. Unequal potential for creation and adaptation of local and specific solutions

FIGURE 2.7 Anthropogenic evolution of the Danube River valley. *Photo credit: Curtesy of G. Janauer, 2010.*

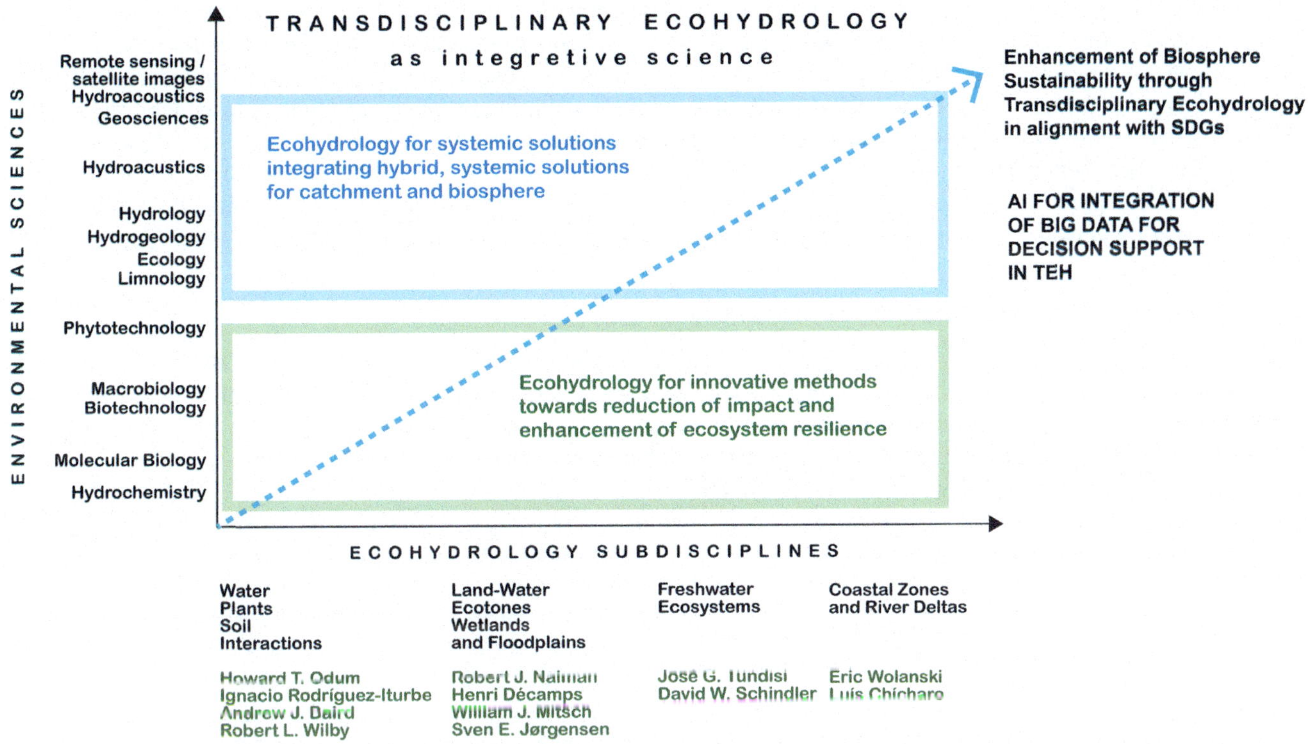

FIGURE 2.8 Transdisciplinary ecohydrology as an integrative science. It bridges environmental sciences (such as hydrology, ecology, and biotechnology) with ecohydrological subdisciplines (ranging from soil–water interactions to coastal ecosystems). It highlights two key levels of application: innovative methods for impact reduction and ecosystem resilience (green box) and systemic solutions integrating hybrid approaches for catchment and biosphere sustainability (blue box), ultimately contributing to the enhancement of biosphere sustainability in alignment with the UN Sustainable Development Goals (SDGs).

5. Breaking the mechanistic paradigm: Toward an evolutionary–ecosystemic approach

On the other hand, in cities, excessive coverage of the ground with impermeable surfaces hinders the replenishment of groundwater resources and leads to the emergence of depression cones, while highly polluted surface waters are directly discharged to river systems. Furthermore, the drastically shrinking of green areas and lack of water in the city landscape impair the quality of life and have an adverse impact on human health, increasing the incidence of allergies and asthma among residents (see part two of this book). The aforementioned measures, which stem from the widely prevalent mechanistic paradigm in environmental management, tend to degrade ecological processes and diminish the stabilizing role of ecosystems in the hydrological cycle. This, in turn, engenders a number of deleterious consequences, including flooding, deteriorating water quality (Fig. 2.6), increased secondary water pollution, and eutrophication. Consequently, these measures exacerbate the ecological condition and detract from society's quality of life, which is directly reflected in slower sustainable development.

Moreover, to provide comprehensive solutions to various natural environment and sustainable development issues, science must integrate various areas of knowledge and ensure rapport between academia and society as a prerequisite for developing systemic solutions. The kind of sustainable, stable socioeconomic development called for in nearly all environmental proactive management strategies and plans can only be achieved through a systemic approach founded on foresight (Rogut and Piasecki, 2011) The analysis of prognosis and hypothesis in a long-term perspective facilitates the formulation of projections about the future and the development of scenarios that enable the optimization of efforts to build the desired future.

Systems analysis has historically been employed to address large-scale sociotechnological problems in domains such as the power industry, human health, and enhancing living conditions and the environment. Under this model, potential action scenarios are formulated not only by researchers independently but also in consultation with decision-makers within the relevant field. The primary benefit of systems analysis is that it enables an objective perspective to be taken on subjective decision-making processes, thereby facilitating a balanced consideration of the conflicting interests of various societal groups.

The future of transdisciplinary ecohydrology lies in integrating enriched process-based regulation with interdisciplinary collaboration, ensuring that water management strategies consider both ecological function and societal well-being. As hydrology transitions from mechanistic to an evolutionary–ecosystemic paradigm, the focus must shift from the exploitation of water resources and flood prevention toward enhancement of the catchment sustainability potential WBSR by harnessing natural processes and enhancing biosphere sustainability.

This transformation requires breaking down disciplinary silos and fostering global cooperation. Only through a transdisciplinary approach, combining ecological insights, hydrological expertise, and socioeconomic considerations, can we effectively regulate the hydrological cycle and enhance the sustainability potential of catchments and supports IWRM.

6. Defining transdisciplinary ecohydrology

The concept of ecohydrology as a transdisciplinary science, first proposed by Zalewski (Zalewski et al., 1997) and further developed by Zalewski (2000), integrates hydrology and ecology to address sustainability challenges in both aquatic and terrestrial ecosystems. This approach emphasizes that understanding the interactions between water and ecological processes is essential for enhancing the sustainability potential of catchments and support IWRM.

To achieve this, ecohydrology operates at two key levels:

First, mitigating pollution at the source is crucial, as aquatic ecosystems—situated at the lowest points in the landscape—are the final recipients of point and diffuse pollutants from catchments. Reducing contamination before it reaches water bodies is essential for maintaining freshwater ecosystem health.

Second, enhancing ecosystem resilience through dual regulation recognizes that hydrological conditions and biological structures interact to determine ecosystem stability. This principle allows for optimizing natural processes, such as nutrient retention and water purification. For example, phosphorus—one of the primary drivers of eutrophication—can be blocked in unavailable pool by reacting with iron and calcium and assimilated by riparian vegetation. Additionally, when assimilated by algae and subsequently filtered through zooplankton, its availability in the water column is further reduced. These natural regulatory processes can be integrated with advanced nature-based solutions (ANbS), tailored to the specific characteristics of each ecosystem, and continuously monitored through environmental feedback to ensure their long-term effectiveness.

A variety of attempts to integrate ecology and hydrology have demonstrated the importance of understanding the interactions between these two disciplines and of synthesizing the knowledge they have accumulated. At the initial stage of the evolution of discipline, the term *ecohydrology* was also used to describe the dynamics of water cycling in the context of the relationships between climatic conditions, vegetation, and soil (Rodriguez-Iturbe, 2000; Baird and Wilby, 1999). On the other hand, the term *hydroecology* has been used to describe the hydrological perspective on the interactions that occur between the dynamics of flowing water and living organisms (Dunbar and Acreman, 2001).

Transdisciplinary ecohydrology, as a scientific discipline aimed at enhancing the sustainable development potential of catchments and the biosphere through the regulation of ecohydrological processes, is structured around a five-step approach:

1. Extending the traditional sociocentric–mechanistic paradigm, which emphasizes human-centered control and mechanical interventions, with **evolutionary–ecosystemic** one that is based on integration of hydrology and ecology by reduction to physics toward introducing ecosystem metabolism to water sciences and understanding the hierarchy of drivers of life-supporting processes in different ecosystems (ABRC model).
2. **Enhancement sustainability potential** and the biosphere WBRS+CE+LPG.
3. **Dual regulation as a framework** for understanding life-supporting processes and discovering the emerging properties of the system for innovative ANbS.
4. Advancement of Nature-based Solutions into cost-efficient ANbS, developed by integrating insights from multiple disciplines, including hydrology, ecology, ecohydraulics, molecular biology, biotechnology, and phytotechnology (Fig. 2.8).
5. Development of robust transdisciplinary science to develop **systemic and hybrid solutions** that combine ecohydrology principles with ANbS and hydroengineering techniques to be implemented at catchment and the biosphere scale.

To address today's sustainability challenges, transdisciplinary approaches must merge hydrology and ecology into a single integrated framework, ensuring sustainability at the catchment scale. Fig. 2.9 demonstrates the transformative impact of a new transdisciplinary and systemic approach to environmental thinking.

Transdisciplinary ecohydrology has evolved into a holistic, solution-oriented discipline capable of addressing the complex environmental challenges of the contemporary world. By integrating methodologies from diverse fields such as hydrology, ecology, engineering, and geomorphology, transdisciplinary ecohydrology provides a comprehensive framework for understanding water dynamics, ecosystem resilience, and sustainable resource management. This integrated approach enhances its capacity to tackle pressing ecological and hydrological challenges of the anthropocene, ensuring its applicability in developing ecohydrological advanced nature-based, hybrid, and systemic solutions for environmental sustainability (Fig. 2.10).

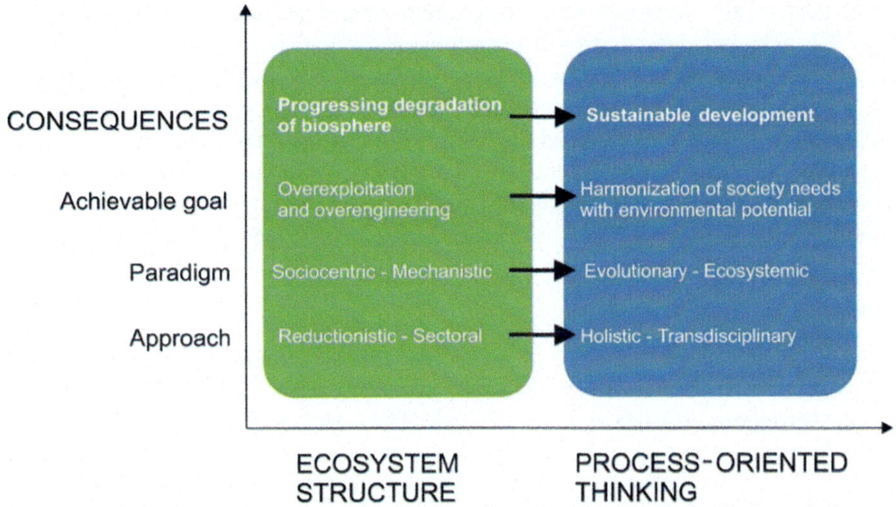

FIGURE 2.9 Transdisciplinary ecohydrology approach. Expected direction of a shift in environmental science paradigm from the currently prevailing structure-oriented paradigm (conservation and restoration actions, focused on the assessment of the effect of various human activities on biota) to process-oriented approach (regulation of ecosystem processes, i.e., water and nutrients cycling, and energy flow) to create a background for harmonization of societal needs with ecosystem potential, and ultimately the sustainable development; the arrows represent the expected shift in the paradigm at three levels, as follows: (1) scientific, (2) operational/practical, (3) political, and the expected results.

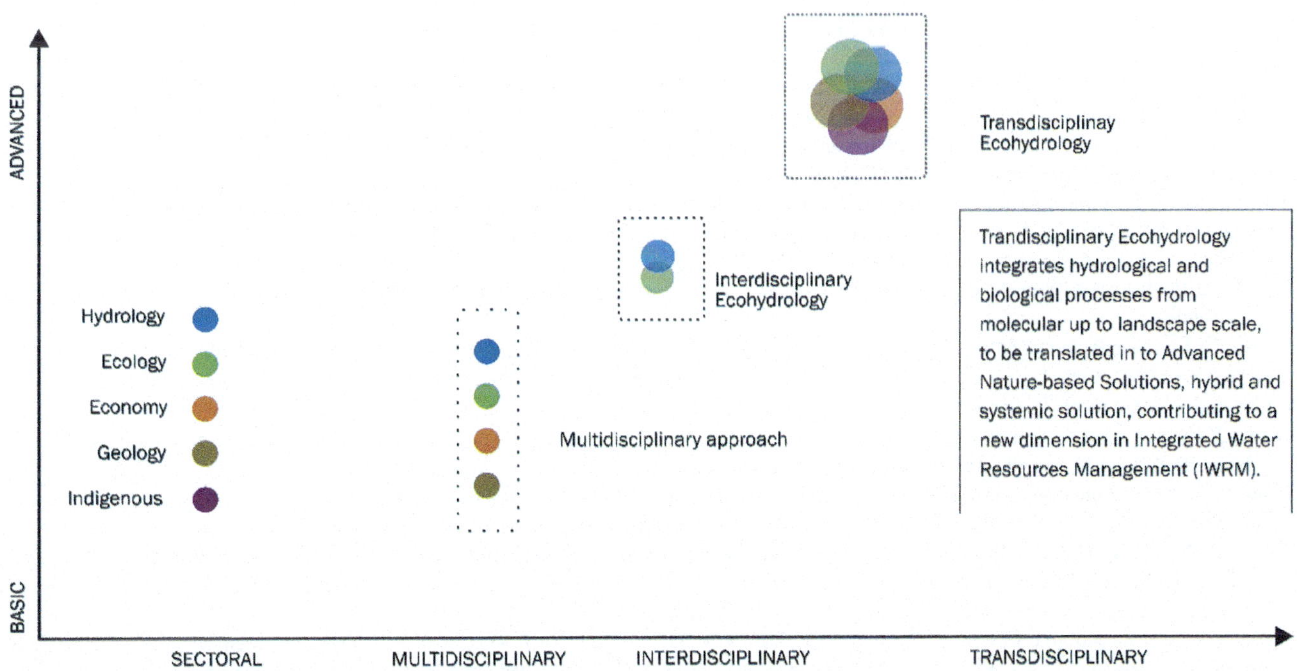

FIGURE 2.10 Methodology of science. The diagram illustrates the evolution of ecohydrology from a sectoral to a transdisciplinary approach, highlighting the increasing integration and complexity of knowledge across multiple fields. In the disciplinary stage, individual scientific domains, such as hydrology, ecology, and engineering, operate independently, each contributing to the understanding of environmental processes within its specialized framework. As ecohydrology progresses to the multidisciplinary phase, these fields begin to contribute knowledge in parallel, yet their interactions remain limited, lacking deep conceptual synthesis. The interdisciplinary stage, based on ABRC model, marks a significant shift, fostering stronger connections between disciplines and enabling the exchange of insights and methodologies. This phase facilitates the codevelopment of knowledge, where interactions between hydrological, biological, and socioenvironmental processes generate novel approaches to addressing water-related challenges (e.g., IWRM). At the most advanced level, transdisciplinary ecohydrology represents a fully integrated scientific framework, where hydrological and biological processes are comprehensively understood across multiple scales, ranging from the molecular level to the catchment and biosphere scale. This holistic integration enables the development of ecohydrological, hybrid, and systemic solutions, designed for management and enhancement of catchment and the biosphere sustainability potential WBSR. It should be noted that the increased size of the science discipline circles expresses the growing synergy and an increasing capacity for problem-solving.

From a broader perspective, this approach reflects an open-minded attitude grounded in the principle of "knowing what we do not know." It acknowledges that effective problem-solving, particularly within complex systems such as ecosystems or water management, demands a willingness to move beyond the confines of established disciplinary boundaries. Genuine progress requires engagement with unfamiliar fields, approached with the awareness that some dimensions and variables may remain unrecognized. By embracing this uncertainty—and acknowledging that "we do not know what we do not know"—researchers and practitioners are better positioned to develop truly transdisciplinary dialog that deepens understanding, transforms knowledge into wisdom, and, in turn, fosters innovation and promotes more resilient and holistic solutions.

AI disclosure

During the preparation of this work, the author(s) used OpenAI's ChatGPT. This document includes content generated and/ or edited with the assistance of OpenAI's ChatGPT, used to support language refinement and structural improvements. All final content was reviewed and verified by the author. After using this tool/service, the author(s) reviewed and edited the content as needed and take(s) full responsibility for the content of the publication.

References

Baird, Wilby, L., 1999. Eco-Hydrology: Plants and Water in Terrestrial and Aquatic Environments. Routledge.

Bedford, B.L., Preston, E.M., 1988. Developing the scientific basis for assessing cumulative effects of wetland loss and degradation on landscape functions: status, perspectives, and prospects. Environ. Manag. 12 (5), 751–771. https://doi.org/10.1007/BF01867550.

Beven, K., Germann, P., 1982. Macropores and water flow in soils. Water Resour. Res. 18 (5), 1311–1325. https://doi.org/10.1029/WR018i005p01311.

Blöschl, G., Nester, T., Komma, J., Parajka, J., Perdigão, R.A.P., 2013. The June 2013 flood in the Upper Danube Basin, and comparisons with the 2002, 1954 and 1899 floods. Hydrol. Earth Syst. Sci. 17 (12), 5197–5212. https://doi.org/10.5194/hess-17-5197-2013.

Blöschl, G., Hall, J., Viglione, A., Perdigão, R.A.P., Parajka, J., Merz, B., Lun, D., Arheimer, B., Aronica, G.T., Bilibashi, A., Boháč, M., Bonacci, O., Borga, M., Čanjevac, I., Castellarin, A., Chirico, G.B., Claps, P., Frolova, N., Ganora, D., Gorbachova, L., Gül, A., Hannaford, J., Harrigan, S., Kireeva, M., Kiss, A., Kjeldsen, T.R., Kohnová, S., Koskela, J.J., Ledvinka, O., Macdonald, N., Mavrova-Guirguinova, M., Mediero, L., Merz, R., Molnar, P., Montanari, A., Murphy, C., Osuch, M., Ovcharuk, V., Radevski, I., Salinas, J.L., Sauquet, E., Šraj, M., Szolgay, J., Volpi, E., Wilson, D., Zaimi, K., Živković, N., 2019. Changing climate both increases and decreases European river floods. Nature 573 (7772), 108–111. https://doi.org/10.1038/s41586-019-1495-6.

Burek, P., Satoh, Y., Fischer, G., Kahil, T., Scherzer, A., Tramberend, S., Nava, L. F., Wada, Y., Eisner, S., Flörke, M., Hanasaki, N., Magnuszewski, P., Cosgrove, W., & Wiberg, D. (2016). Water Futures and Solutions: Fast Track Initiative – Final Report (Working Paper WP-16-006). International Institute for Applied Systems Analysis (IIASA), Laxenburg, Austria.

Dai, A., 2013. Increasing drought under global warming in observations and models. Nat. Clim. Change 3 (1), 52–58. https://doi.org/10.1038/nclimate1633.

Dunbar, M.J., Acreman, M.C., 2001. Applied Hydro-Ecological Science for the Twenty-First Century, 266. IAHS-AISH Publication, pp. 1–18.

Eagleson, P.S., 1982. Hydrology and Climate Scientific Basis of Water-Resource Management. National Academy Press, pp. 31–40.

Eagleson, P.S., 2002. Ecohydrology: Darwinian Expression of Vegetation Form and Function. Cambridge University Press.

Fausch, K.D., Torgersen, C.E., Baxter, C.V., Li, H.W., 2002. Landscapes to riverscapes: Bridging the gap between research and conservation of stream fishes. BioScience 52 (6), 483–498. https://doi.org/10.1641/0006-3568(2002)052[0483:LTRBTG]2.0.CO;2.

Kajak, Z., 2001. Hydrobiologia–limnologia: Ekosystemy wód śródlądowych. Wydawnictwo Naukowe PWN.

Grodziński, W., Klekowski, R.Z., Duncan, A., 1975. Methods for Ecological Bioenergetics. Blackwell Scientific Publications.

Habersack, H., Piégay, H., 2007. River restoration in the Alps and their surroundings: past experience and future challenges. In Habersack, H., Piégay, H., Rinaldi, M. (Eds.), Gravel-BedRivers VI: From Process Understanding to River Restoration (Developments in Earth Surface Processes, Vol. 11, pp. 703–735). Elsevier. https://doi.org/10.1016/S0928-2025(07)11161-5.

Harms, T.K., Groffman, P.M., Aluwihare, L., Craft, C., Wieder, W.R., Hobbie, S.E., Baer, S.G., Blair, J.M., Frey, S.D., Remucal, C.K., Rudgers, J.A., Collins, S.L., Kominoski, J.S., Ball, B.A., LTER OM Working Group, 2021. Patterns and trends of organic matter processing and transport: Insights from the US long-term ecological research network. Clim. Change Ecol. 2, 100025. https://doi.org/10.1016/j.ecochg.2021.100025.

Kiedrzyńska, E., Wagner, I., Zalewski, M., 2008. Quantification of phosphorus retention efficiency by floodplain vegetation and a management strategy for a eutrophic reservoir restoration. Ecol. Eng. 33 (1), 15–25. https://doi.org/10.1016/j.ecoleng.2007.10.010.

Krebs, C.J., 1972. Ecology: The Experimental Analysis of Distribution and Abundance. New York: Harper & Row.

Kundzewicz, Z.W., Krysanova, V., Benestad, R.E., Hov, Ø., Piniewski, M., Otto, I.M., 2018. Uncertainty in climate change impacts on water resources. Environ. Sci. Pol. 79, 1–8. https://doi.org/10.1016/j.envsci.2017.10.008.

Mann, M.E., Bradley, R.S., Hughes, M.K., 1999. Northern hemisphere temperatures during the past millennium: Inferences, uncertainties, and limitations. Geophys. Res. Lett. 26 (6), 759–762. https://doi.org/10.1029/1999GL900070.

Meadows, D.H., Meadows, D.L., Randers, J., Behrens, W.W., 1972. The limits to growth. Universe Books.

Montanari, A., Young, G., Savenije, H.H.G., Hughes, D., Wagener, T., Ren, L.L., Koutsoyiannis, D., Cudennec, C., Toth, E., Grimaldi, S., Blöschl, G., Sivapalan, M., Beven, K., Gupta, H., Hipsey, M., Schaefli, B., Arheimer, B., Boegh, E., Schymanski, S.J., Di Baldassarre, G., Yu, B., Hubert, P., Huang, Y., Schumann, A., Post, D.A., Srinivasan, V., Harman, C., Thompson, S., Rogger, M., Viglione, A., McMillan, H., Characklis, G., Pang, Z., Belyaev, V., 2013. "Panta Rhei—Everything flows": change in hydrology and society—the IAHS scientific decade 2013–2022. Hydrol. Sci. J. 58 (6), 1256–1275. https://doi.org/10.1080/02626667.2013.809088.

Mukherjee, A., Lal, R., 2013. Biochar impacts on soil physical properties and greenhouse gas emissions. Agronomy 3 (2), 313–339. https://doi.org/10.3390/agronomy3020313.

Naiman, R.J., Décamps, H., Fournier, F. (Eds.), 1989. The role of land/inland water ecotones in landscape management and restoration: A proposal for collaborative research. UNESCO, Vendome, Paris.

OECD, 2025. Water Governance Outlook 2025. OECD Publishing.

PAGES 2k Consortium, 2013. Continental-scale temperature variability during the past two millennia. Nat. Geosci. 6, 339–346. https://doi.org/10.1038/ngeo1797.

Rodriguez-Iturbe, I., 2000. Ecohydrology: a hydrologic perspective of climate-soil-vegetation dynamies. Water Resour. Res. 36 (1), 3–9. https://doi.org/10.1029/1999wr900210.

Rogut, A., Piasecki, B., 2011. Foresight methodology as a tool for elaboration of plans for sustainable management of water, energy, the environment and society. J. Water Land Dev.

Skłodowski, M., Kiedrzyńska, E., Kiedrzyński, M., Urbaniak, M., Zielińska, K.M., Kurowski, J.K., Zalewski, M., 2014. The role of riparian willows in phosphorus accumulation and PCB control for lotic water quality improvement. Ecol. Eng. 70, 1–10. https://doi.org/10.1016/j.ecoleng.2014.03.088.

United Nations Environment Programme, 2002. Freshwater Management Series. UNEP, Nairobi.

Vannote, R.L., Minshall, G.W., Cummins, K.W., Sedell, J.R., Cushing, C.E., 1980. The river continuum concept. Can. J. Fish. Aquat. Sci. 37 (1), 130–137. https://doi.org/10.1139/f80-017.

Veolia, I., International Food Policy Research Institute, 2015. The murky future of global water quality: new global study projects rapid deterioration in water quality.International Food Policy Research Institute and Veolia Water North America, Washington, DC and Chicago, IL https://hdl.handle.net/10568/150398.

Wantzen, K.M., Aziz, B., Longuet, I., Bao, I., Bocoum, H., Cissé, L., Chauhan, M., Girard, P., Gopal, B., Kane, A., Marchese, M.R., Prakash, N., Teixeira, P., Zalewski, M., 2016. River Culture: an eco-social approach to mitigate the biological and cultural diversity crisis in riverscapes. Ecohydrol. Hydrobiol. 16 (1), 7–18. https://doi.org/10.1016/j.ecohyd.2015.12.003.

Zalewski, M., 2000. Ecohydrology—the scientific background to use ecosystem properties as management tools toward sustainability of water resources. Ecol. Eng. 16 (1), 1–8. https://doi.org/10.1016/s0925-8574(00)00071-9.

Zalewski, M., 2002. Ecohydrology—the use of ecological and hydrological processes for sustainable management of water resources. Hydrol. Sci. J. 47 (5), 823–832. https://doi.org/10.1080/02626660209492986.

Zalewski, M., 2006. Flood pulses and river ecosystem robustness. In I. Tchiguirinskaia, K. N. N. Thein, & P. Hubert (Eds.), Frontiers in Flood Research: Proceedings of the Kovacs Colloquium, UNESCO, Paris. Int. Assoc. Hydrol. Sci. 305, 143–154.

Zalewski, M., Wantzen, K.M., 2023. Ecohydrology and River Culture: Synergies for the Enhancement of River and Catchment Sustainability. UNESCO, pp. 807–819. https://doi.org/10.54677/dtuq3662.

Zalewski, M., Janauer, G.A., Jolankai, G., (Eds.), 1997. Ecohydrology: A new paradigm for the sustainable use of aquatic resources. IHP-V Technical Documents on Hydrology No 7. UNESCO, Paris, p. 58.

Zalewski, M., Naiman, R.J., 1985. The regulation of riverine fish communities by a continuum of abiotic–biotic factors. In: Alabaster, J.S. (Ed.), Habitat Modification and Freshwater Fisheries. Butterworths, London, pp. 3–9.

Zalewski, M., Schiemer, F., Thorpe, J.E. (Eds.), 2001. Catchment processes: land/water ecotones and fish communities. Int. J. Ecohydrol. Hydrobiol. 1 (1–2).

Chapter 3

Deductive framework for transdisciplinary ecohydrology

Maciej Zalewski
European Regional Centre for Ecohydrology of the Polish Academy of Sciences under the auspices of UNESCO, Łódź, Poland

PHOTO CREDIT: M. ZALEWSKI.

When we have defeated nature, we will find ourselves on the losing side

Hendrik Antoon Lorentz (Nobel Prize Winner).

Transdisciplinary Ecohydrology. https://doi.org/10.1016/B978-0-443-41596-8.00023-6

1. Understanding scientific theories: Paradigms and falsifiability

The term "theory" is derived from the Greek word theoría, meaning "to see" or "to contemplate." A theory is defined as a structured set of concepts, definitions, axioms, and propositions that systematically organize relationships between these elements, thereby creating a coherent framework for describing physical or abstract phenomena. Specifically, a scientific theory is composed of logical, cohesive generalizations grounded in established facts and aligned with the current state of knowledge. Its fundamental objective is to elucidate the underlying causes, conditions, and circumstances that precipitate and propel the occurrence and progression of a specific phenomenon. As the culmination of scientific inquiry, a theory is formulated through rigorous research and serves multiple purposes: to organize and rationalize existing knowledge, to elucidate causal relationships, to predict future outcomes, and to inspire the creation of new systems. When an unsolvable problem arises, the difficulty in resolving it is directly linked to the presence or absence of knowledge and value consensus (see Fig. 3.1).

The criterion for determining whether a theory qualifies as scientific is its falsifiability, defined as the possibility that it can be proven false through empirical evidence (Popper, 1934). This principle was introduced by Karl Popper (1934), who argued that a theory lacking empirical criteria for falsification cannot be considered scientific. In empirical research, a theory is constructed based on a hypothesis that explains observed facts. If subsequent experimental or practical observations contradict the hypothesis, the theory is either disproved or its applicability is restricted to specific conditions. To illustrate this point, consider the example of classical mechanics, which remains valid only for objects traveling at speeds below the speed of light.

The concept of a paradigm, defined by Thomas Kuhn (Kuhn and Schlegel, 1963) as the foundational set of concepts and theories forming the basis of a scientific field, has been further developed within the framework of this discussion. It is important to note that paradigms are typically not questioned during periods of cognitive creativity, when they support the development of new theories that align with the empirical data within the discipline. The interplay between falsifiability and paradigms is central to the evolution and refinement of scientific understanding Fig. 3.2.

It has been demonstrated that certain paradigms are quickly understood and easily assimilated because they fit seamlessly into existing scientific thinking. A good example is the River Continuum paradigm, which provided a logical and intuitive framework for understanding how river ecosystems work. Such paradigms are often perceived as the "missing piece" that resolves existing gaps in scientific thought, which makes their acceptance relatively easy.

In contrast, other paradigm shifts, such as the Copernican Revolution (Kuhn, 1957) are much harder for people to accept. When Copernicus proposed that the Earth revolves around the Sun rather than the other way around, the idea was met with skepticism and resistance. This was partly due to a lack of broadly accessible scientific evidence at the time, and partly because it conflicted with the dominant worldview—anthropocentrism—which placed humans (and the Earth) at the center of the universe.

A similar dynamic can be observed in the contemporary context of climate change, where the concept of human-induced global environmental change has faced significant challenges in terms of public and political acceptance. Initially, the complexity of scientific evidence and the limited understanding of the mechanisms of climate change have contributed to this difficulty. Moreover, entrenched economic and political interests, coupled with deeply ingrained

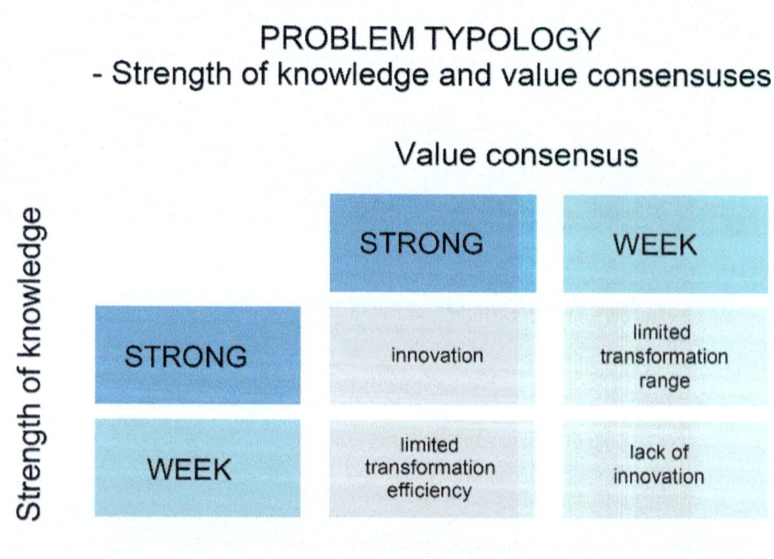

FIGURE 3.1 The problem typology framework categorizes challenges based on the strength of knowledge and value consensus. Problems are easier to solve when both are strong, as shared understanding and priorities facilitate action. However, lacking transformation knowledge can hinder solutions despite strong value alignment, while missing system knowledge limits effective change even with shared values. The most complex problems arise when both knowledge and value consensus are weak, requiring decision-makers, educators, and society at large to engage in transdisciplinary scientific dialog. This model is useful for addressing governance, sustainability, and systemic challenges. *After US Committee of Scientists (1999), p. 131, modified according to Transdisciplinary Ecohydrology tenets.*

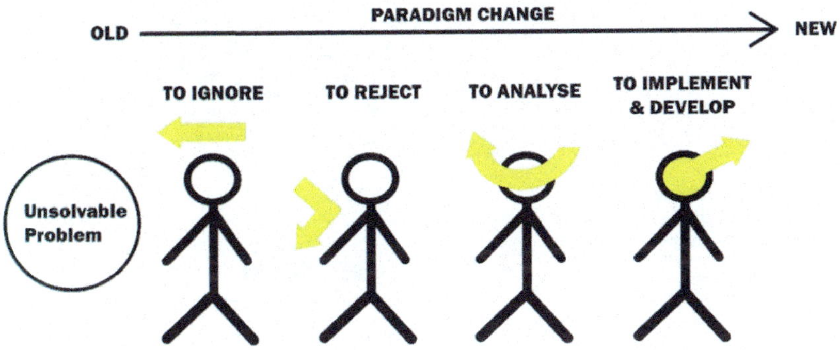

FIGURE 3.2 A new paradigm. The assimilation of a new paradigm typically unfolds in multiple phases before becoming common knowledge. It begins with ignorance and dismissal, followed by rejection and skepticism. As awareness grows, it moves into phases of exploration, criticism, and analysis, leading to experimentation and adaptation. Gradually, the paradigm undergoes slow assimilation and eventual implementation.

worldviews, have contributed to the emergence of resistance. As evidenced by previous paradigm shifts, widespread acceptance often lags behind scientific consensus, particularly when a new paradigm challenge established beliefs or requires major societal change.

Understanding the patterns of scientific development, particularly the evolution of paradigms, is fundamental for creative thinking in formulating new ones, which become frameworks for discovering emerging properties of the system and stimulating the development of innovative solutions.

The proposed approach is formulated as a framework for analyzing and anticipating patterns of scientific theory development. This approach can be considered a prototype of grounded theory (Strauss and Corbin, 2008), which is defined as "theoretical constructs derived from quality analysis" and "theoretical interpretation of phenomena generated from data using methodological guidance." This is a general methodology for developing theory grounded in data that are systematically gathered and analyzed. This methodology has been employed in the study of "perception of the subject under study," particularly in the context of ecohydrology, which examines the interplay between holistic concepts and empirical testing to anticipate future scientific trends and accelerate the advancement of discoveries.

Philosophy stimulates critical, analytical, and synthetic thinking, thereby broadening our capacity to perceive and interpret observed phenomena (Jones, 2022). This cognitive expansion enhances our capacity to understand diverse perspectives and engage in constructive discourse on controversial issues (Smith and Lee, 2023). Moreover, philosophy fosters a deeper awareness of the need to understand subjects more profoundly, leading to a more comprehensive grasp of both operational practices and strategic objectives. It also aids in formulating long-term visions for the future. These philosophical dimensions are instrumental in generating information, integrating knowledge, and transforming accumulated data into wisdom (Fig. 3.3).

2. From fragmentation to integration: Theoretical foundations of ecohydrology

Understanding of the integrated nature of ecological processes, such as water and nutrient cycling or the flow of energy through catchments, has historically been limited among narrowly specialized experts and decision-makers. These actors often failed to recognize the continuous influence of human activities and ecological succession on these interconnected systems. As a result, scientific research on aquatic environments, as well as practical water management strategies, frequently overlooked these dynamic linkages. Consequently, even in developed nations, decisions regarding the use and management of inland water resources were often made without sufficient scientific or empirical grounding. Such decisions tended to focus narrowly on individual problems and short-term outcomes, rather than embracing the systemic nature of catchment-scale processes.

A turning point emerged in the 1990s, when growing awareness of the limitations of reductionist approaches led to calls for more comprehensive strategies. The term "integrated catchment management" began to appear in academic and policy literature (Cullen, 1990), reflecting a shift toward more holistic thinking in land and water resource governance. Despite this conceptual progress, early applications remained largely theoretical. Although serious hydrological and ecological issues were increasingly being identified, the development of decision-making frameworks—such as cost–benefit analyses—was not yet embedded in broader programs that sought to integrate hydrology with ecology in a meaningful, systemic way.

In this context, long-term historical perspectives on hydrology have provided essential insights. A conceptual diagram originally developed by Starkel (1982) and later reintroduced by Zalewski (2002b). Fig. 3.4 illustrates climate-driven

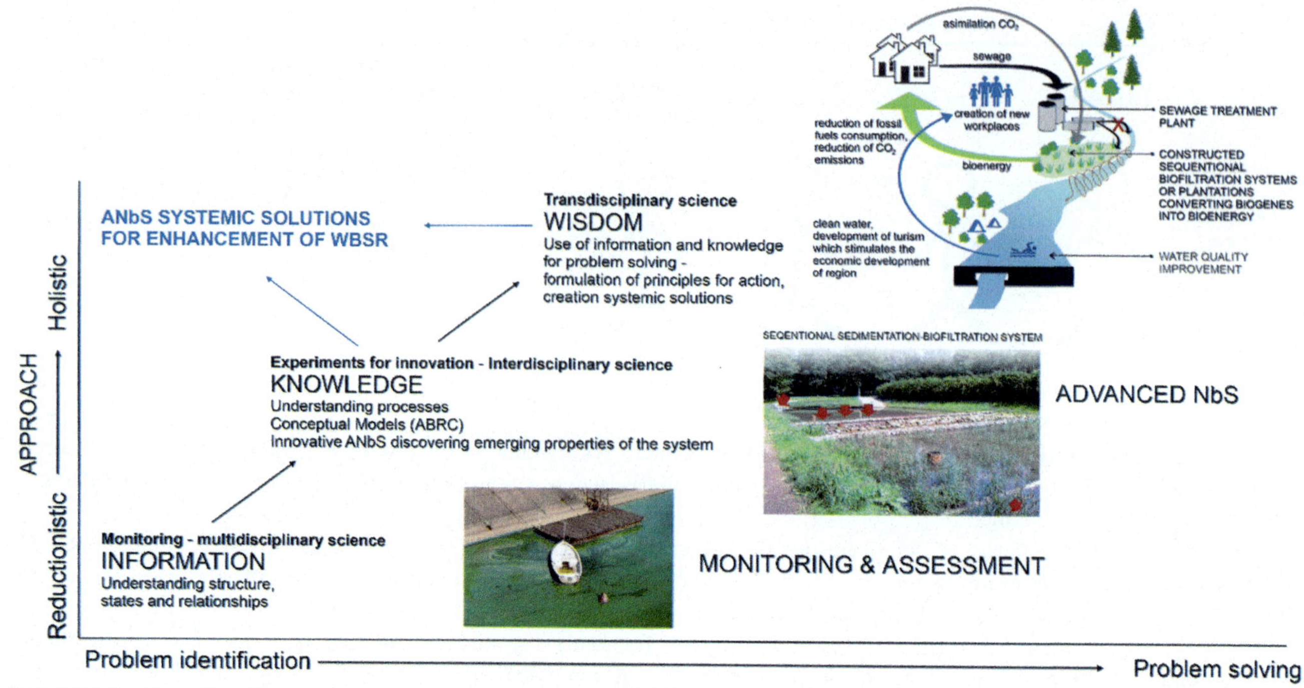

FIGURE 3.3 Transdisciplinary ecohydrology as a science. Transdisciplinary ecohydrology as a science focused on solving sustainability problems, from obtaining information (pure science), through expanding knowledge, to skills applying knowledge to problem-solving, strategy formulation, and planning activities. *Modified from: Zalewski, M., 2011a. Ecohydrology for implementation of the EU water framework directive. Water Manage. 164 (8), 375–385; Zalewski, M., 2014a. Ecohydrology and hydrologic engineering: regulation of hydrology-biota interactions for sustainability. J. Hydrol. Eng. 20 (1). https://doi.org/10.1061/(asce)he.1943-5584.0000999; bottom photo algal bloom: Mankiewicz-Boczek et al. (2006).*

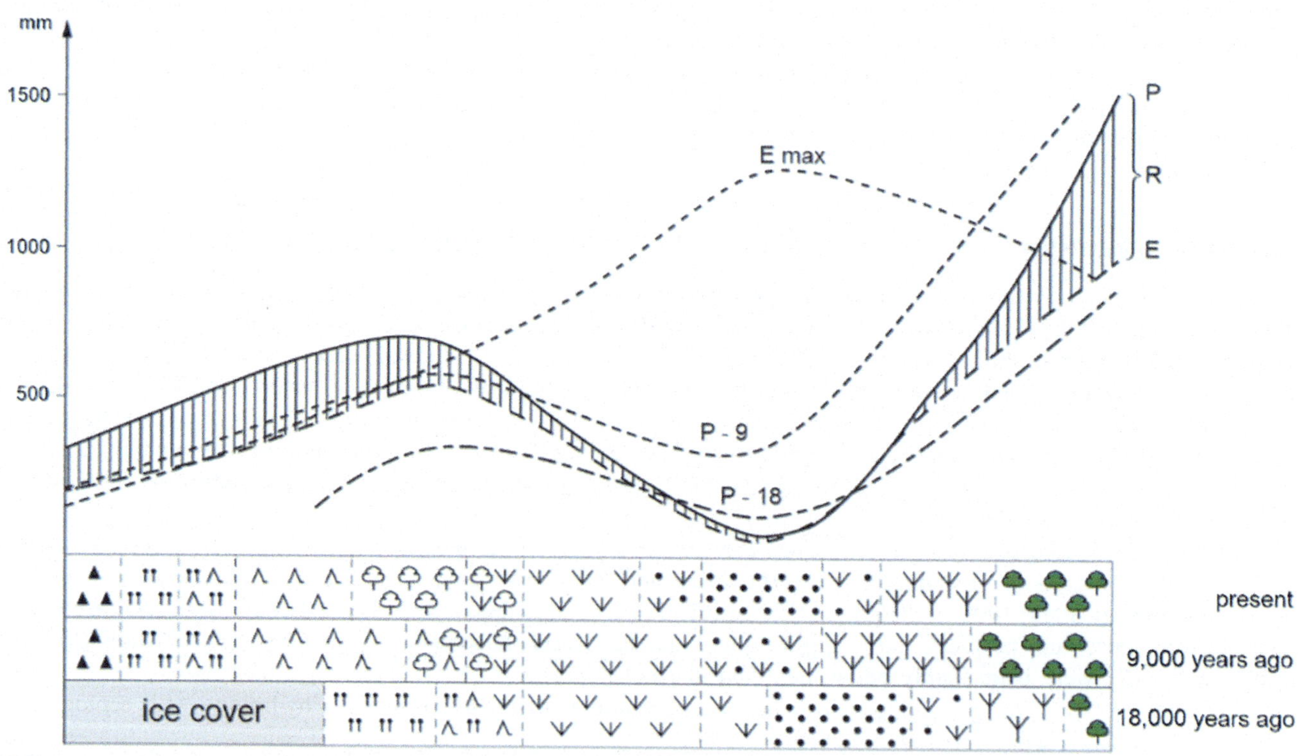

FIGURE 3.4 Interrelationships between changes in climate and vegetation cover: 18,000 years to present. Changes in rainfall, runoff and evapotranspiration along Euro-African transect from north to south 18,000 years ago, 9000 years ago, and now (P—precipitation, E—evapotranspiration, R—run-off). Emax = potential evapotranspiration now. According to Starkel climate change determines the distribution and structure of the vegetation cover, and the plant cover largely modifies the heat and water balance. *Modified from: Starkel, L., 1988. The evolution of the Vistula River valley during the last 15,000 years. In: Geographical Studies, 4. Polish Academy of Sciences, Institute of Geography and Spatial Organization, pp. 7–37.*

changes in precipitation, evaporation, and runoff in Central Europe over the past 18,000 years, alongside shifts in vegetation composition. During the Late Glacial period, low precipitation levels were counterbalanced by low evaporation, resulting in relatively high runoff. As the climate warmed in the Holocene, approximately 9000 years ago, evaporation increased significantly, peaking even as precipitation rose only moderately, leading to a sharp decline in runoff. Toward the present day, both precipitation and runoff have increased, while evaporation has slightly decreased. The shaded area between the precipitation and evaporation curves represents runoff, offering a visual representation of these changes. Beneath the graph, a vegetation panel illustrates ecological succession, depicting a gradual transition from sparse tundra vegetation to dense, biodiverse forest cover. These data offer compelling evidence of how long-term climate patterns shape not only water availability through runoff dynamics but also the structure and function of ecosystems over time.

These historical perspectives are further reinforced by contemporary understanding of how water resources within a drainage basin are influenced not solely by climate and geomorphological conditions, but also by biological evolution and ecological succession. Land cover, shaped by both natural processes and anthropogenic pressures, is now recognized as a central regulator of the hydrological cycle—one that ultimately determines the spatial and temporal patterns of human civilization (Wantzen, 2023). Over the past 300 years in particular, human activity has increasingly disrupted river basins, altering energy budgets, affecting both local and global climates, and significantly impacting the quantity and quality of freshwater resources. Starkel's observations thus provide a foundational perspective for understanding the interconnectedness of climate, vegetation, and hydrological processes, and how these relationships are essential for anticipating the long-term consequences of environmental change. The other side of the coin is the recent analysis of the role of terrestrial vegetation in stabilizing the hydrological cycle (Eagleson, 2002; Malcolm, 1999; Rodriguez-Iturbe, 2000) and important role of forest as the Biotic Pump (Makarieva and Gorshkov, 2010).

Fig. 3.5 underscores the critical role of land cover, especially vegetation, in maintaining local rainfall regimes. In high-recycling areas, deforestation or land degradation can significantly disrupt the water cycle, leading to reduced precipitation and long-term impacts on ecosystem sustainability. This visualization is characteristic of global moisture tracking studies, which emphasize the strong interdependence between terrestrial ecosystems and atmospheric processes in the regulation of regional hydrological cycles. This figure illustrates the essential function of forest ecosystems in regulating nutrient cycles, while simultaneously demonstrating the adverse effects of industrial agriculture, particularly the transfer of excess nutrients from terrestrial to freshwater systems and the consequent depletion of soil organic matter (Box 3.1).

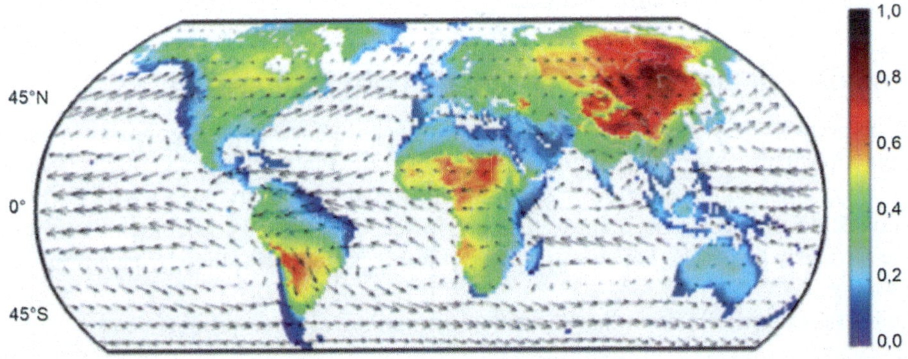

FIGURE 3.5 Precipitation recycling and ecosystem function in the global hydrological cycle. Ecosystems are water "recyclers," retainers and filters (WWAP, 2018). The aforementioned role of terrestrial ecosystems in stabilization of hydrological cycle will be increasing with climate change. The diagram presents a global map of precipitation recycling ratios, overlaid with wind vector fields to illustrate prevailing atmospheric circulation patterns. The precipitation recycling ratio represents the proportion of precipitation in a given region that originates from local evapotranspiration, rather than being transported from distant sources. The color gradient, ranging from 0.0 (blue) to 1.0 (red), indicates the degree of local moisture recycling. Areas shown in red, such as parts of Central Asia, the Amazon Basin, and Central Africa, exhibit high recycling ratios, meaning that a significant portion of their rainfall is generated locally. In contrast, blue regions rely more heavily on moisture transported from other areas. The wind vectors provide additional context by highlighting the dominant patterns of atmospheric moisture movement, such as the easterly trade winds near the equator and westerlies in mid-latitudes. *Taken from: Van der Ent, R. J., Savenije, H. H. G., Schaefli, B., & Steele-Dunne, S. C. (2010). Origin and fate of atmospheric moisture over continents. Water Resour. Res., 46(9), W09525. https://doi.org/10.1029/2010WR009127.*

BOX 3.1 Degradation of soils by agricultural landscapes and pastoral overgrazing

Homogenization of agricultural landscapes exemplified by representing N-NO_3- concentration (Fig. 3.6) in groundwater in the intensive agriculture area (F) (Fig. 3.6) and at the pine forest (P) (Fig. 3.6).

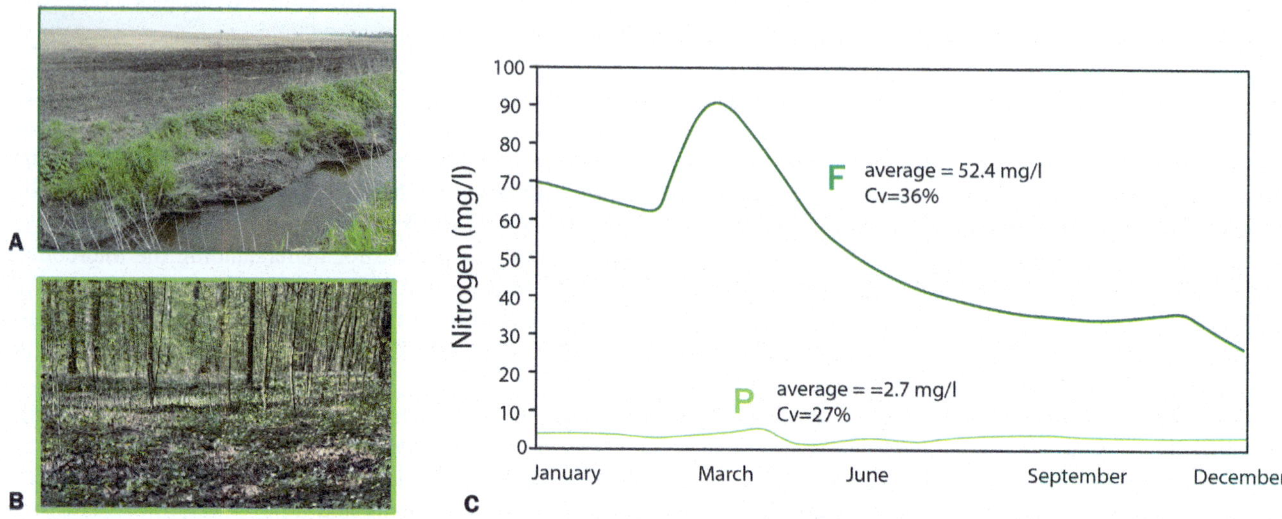

FIGURE 3.6 (A) Effect of the industrial agriculture: loss of organic matter in the soil due to lack of shelter belts and high wind speed moreover degraded bufering land/water ecotones, resulting in high nutrients loss to the chanlised stream of low selfpurification potential and eutrophication of reservoirs downstream. (B) Pine forest. The photograph resembles a natural pine forest ecosystem. (C) The role of forest in reduction N-NO_3- concentration in groundwaters of the catchment. Nitrogen-NO3- concentration in groundwater in the intensive agriculture area (F) and at the pine forest (P). *Source: (A and B) Photo credit: M. Zalewski. Modified from: Bartoszewicz, A. (1994). Skład chemiczny wód powierzchniowych zlewni intensywnie użytkowanych rolniczo w warunkach glebowo-klimatycznych Równiny Kościańskiej. Roczniki Akademii Rolniczej w Poznaniu, Rozprawy Naukowe, 250, 5–68.(C) Based on data from Bartoszewicz, 1994 dissertation/book*

3. The Gaia hypothesis and planetary-scale ecohydrology

The Gaia theory is a hypothesis that extends beyond the scope of ecosystem theory, encompassing the entire planet and its capacity for self-regulation. This concept was initially proposed by Lovelock (Lovelock, 1972, 1979, 1988; LOVELOCK and MARGULIS, 1974), although the notion of a superorganism had previously been advanced by Plato and, centuries later, by Leonardo da Vinci and geologist James Huston (Huston et al., 1988). Lovelock's seminal work posited that processes within the biosphere affect the homeostasis of the Earth's atmosphere, thereby ensuring suitable conditions for its continued functioning.

Initially, the theory faced criticism from the scientific community for three primary reasons:

1. It overlooked the seminal contributions of earlier works that had also demonstrated the impact of living organisms on their environment (Spencer, 1844; Huxley, 1877).
2. It failed to foresee the attendant homeostatic, teleonomic, and optimizing effects (Kirchner, 1989).
3. The metaphor of Gaia as a living being, albeit one which does not proliferate, did not align with the neo-Darwinism movement of the time, despite its ability to illustrate the second law of thermodynamics (Kopf and Lotka, 1925; Schrödinger, 1944).

Notwithstanding such critiques, the Gaia theory effectively demonstrated that living organisms can modulate the environment and ecosystem processes (in this case, the atmosphere) (Watson and Lovelock, 1983; Lenton, 1998; Lenton and Lovelock, 2000). Following that in Transdisciplinary Ecohydrology, the catchment—the framework for analysis and implementation of hybrid solutions (EH ANbS + Hydroengineering), due to dynamics of the water biota interplay, has to be considered as "Platonian superorganism" (Zalewski, 2000).

4. The emergence of integrated water management

The International Conference on Water and the Environment (International Conference on Water and the Environment (ICWE), 1992) and the United Nations Conference on Environment and Development (Report of the United Nations Conference on Environment and Development (UNCED), 1992), also known as the Rio Earth Summit, highlighted the inadequacy of contemporary water management solutions in achieving sustainability. This was further reinforced by Shiklomanov's UNESCO report (Shiklomanov, 1998), which warned of the ongoing depletion of water resources and the degradation of rivers and lakes, rendering many nonrenewable. A key conclusion from these global initiatives was the necessity for new, integrated approaches to water management that address hydrological, ecological, and socioeconomic dimensions.

The emergence of holistic ecological thinking was instrumental in shaping water management strategies. Odum's ecosystem perspective (1970) underscored the interconnectedness of biological and physical processes. Early studies in the 1970s and 1980s examined hydrological influences on biological structures, such as invertebrate communities (Stanford and Ward, 1979) and biogenic transfers from basins to river and lake systems (Hillbricht-Ilkowska, 1985, 2001, 2002; Lavorel and Garnier, 2002; Holmes and Likens, 2016). However, these studies lacked a predictive framework capable of integrating hydrological, ecological, and socioeconomic processes.

This gap led to a paradigm shift catalyzed by the UNESCO International Biological Program (IBP), which promoted an integrative approach to ecosystem studies. By quantifying ecological processes using thermodynamic principles (e.g., energy flux in calories or joules (Barnes and Minshall, 1983; Fisher and Likens, 1973; Grodziński et al., 1975) this research laid the foundation for systemic environmental management.

5. The development of river ecology paradigms

The evolution of river ecology followed a progression from structural to functional perspectives. The River Continuum Concept (RCC) (Vannote et al., 1980) expanded previous theories of river zonation, integrating physical and biological processes. Subsequent models refined this understanding:

- Abiotic–Biotic Regulatory Continuum (ABRC) Concept (Zalewski and Naiman, 1985).
- Nutrient spiraling (Webster, 1975).
- Flood Pulse Concept (Junk et al., 1989).
- The role of land–water ecotones in maintaining landscape functions and supporting ecological restoration (Naiman et al., 1989).
- Ecological Engineering and Ecosystem Restoration (Mitsch and Jørgensen, 1989).
- Regenerative ecology (Reed, 2007).
- Transdisciplinary ecohydrology (Zalewski et al., 1997; Zalewski, 2000, 2014; Elfithri et al., 2025).

These concepts, and in some cases theoretical frameworks, have contributed to the development of an integrated view of river systems, recognizing their interactions with catchments, land–water ecotones, and anthropogenic pressures. A significant shift occurred as researchers moved from a structural focus, centered on species composition and biodiversity, toward a functional approach, considering ecological processes within the broader context of catchment dynamics.

The conceptual framework of river ecology evolution (Fig. 3.7) presents the historical development of key ecological and hydrological theories. It maps their increasing predictive capacity against experimental testing over time. The trajectory begins with early structural approaches such as river zonation (Huet, 1949; Illies and Botosaneanu, 2017), which focused on fish community structure in relation to slope. This evolved into functional models such as nutrient spiraling (Webster and Patten, 1979; Newbold et al., 1981) and the ABRC model (Zalewski and Naiman, 1985); the last one highlights the shifts in regulatory drivers along Abiotic–Biotic Continuum by incorporating ecosystem metabolism into hydrological processes.

However, a major turning point came with the River Continuum Concept (Vannote et al., 1980), which introduced a dynamic perspective on shifts in the invertebrates' communities, forms of organic matter and production to respiration ratio (P/R) along the Strahler streams order. Subsequent theories, including the Flood Pulse Concept (Junk et al., 1989) and studies on land–inland water ecotones (Naiman and Décamps, 1989). All these advances have supported the development and application of ecological engineering (Mitsch and Jørgensen, 1989) and ultimately transdisciplinary

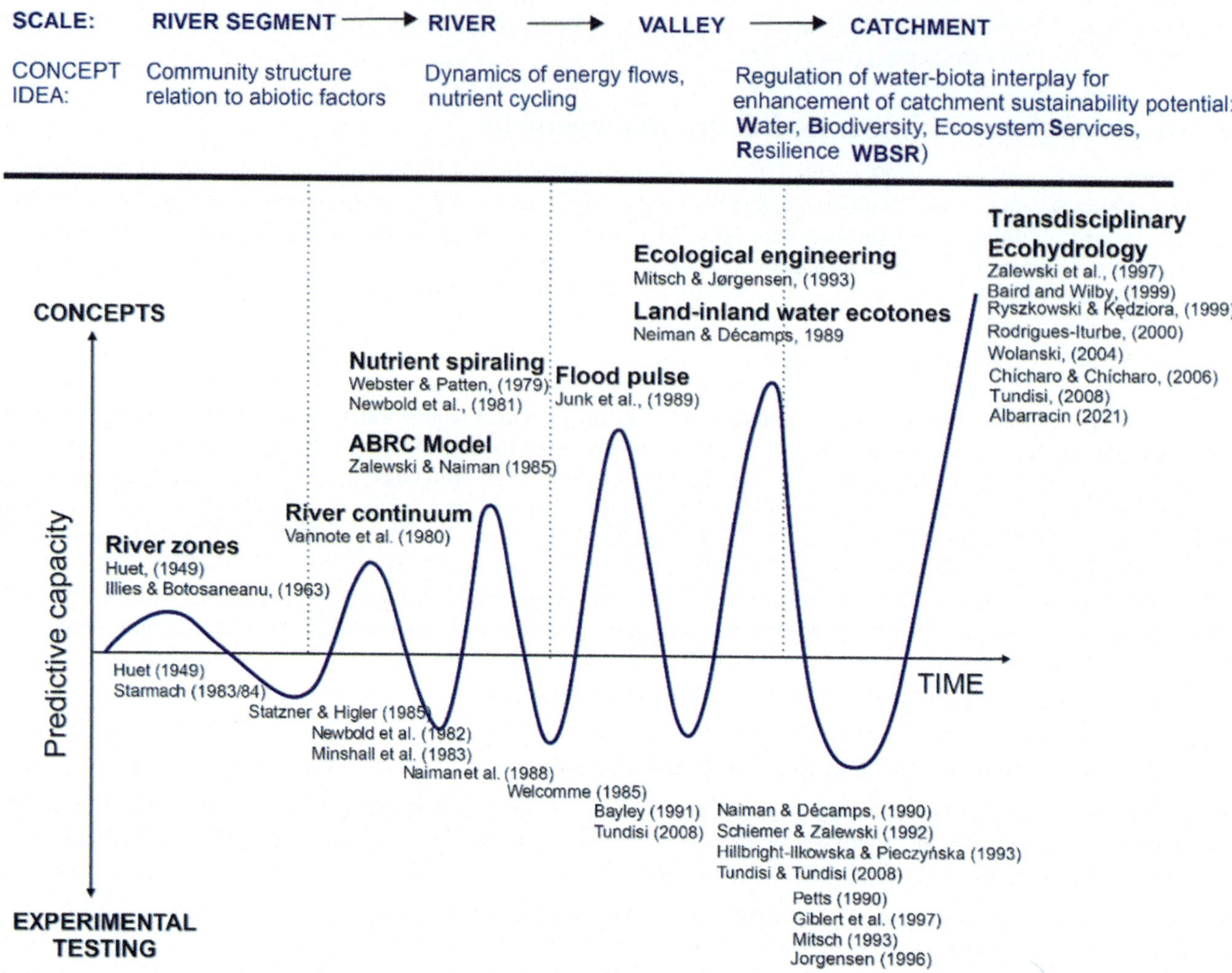

FIGURE 3.7 Evolution of river ecology paradigms. This diagram depicts the evolution of paradigms in river ecology, highlighting the progression of river science as a dynamic interplay between reductionist expertise and holistic approaches. Understanding this evolution is essential for anticipating changes, developing future scenarios, and formulating new concepts and innovative solutions. *Inspired by earlier concepts from: Zalewski, M., 2000. Ecohydrology—The scientific background to use ecosystem properties as management tools toward sustainability of water resources. Ecol. Eng. 16 (1), 1–8. https://doi.org/10.1016/S0925-8574(00)00071-9; and Zalewski, M., & Robarts, R. D. (2003). Ecohydrology – A new paradigm for integrated water resources management. SIL News, 40, 1–5.*

ecohydrology (Zalewski et al., 1997), which represents the most "hydrocentric" and comprehensive proactive scientific framework for regulating the water–biota interplay at the catchment scale.

6. The emergence of ecohydrology as a transdisciplinary science

By the 1990s, the need for a catchment-scale approach integrating hydrology and ecology became evident. The UNESCO Man and the Biosphere (MAB) Programme initiated a transdisciplinary perspective, which was further developed by highlighting the role of land–inland water ecotones in landscape management and restoration (Naiman and Décamps, 1989; UNEP Freshwater Management Series, 2002). The Dublin Conference on Water and the Environment (1992) postulated a new way of water management. Inspired by these challenges, transdisciplinary ecohydrology emerged providing a framework for "use of ecosystem properties as management tool" for systemic environmental management at the catchment scale (Zalewski, 2000). Two fundamental tenets were established:

1. Dual Regulation integrates water–biota interactions from the molecular to the catchment scale to enhance the sustainability potential of entire catchments, as defined by the Abiotic–Biotic Regulatory Continuum (deductive ABRC model), (Zalewski and Naiman, 1985) and inductive—hydrobiomanipulation (Zalewski et al., 1990a).

2. The use of ecosystem functions as biotechnological tools improves water and resource management (Zalewski and Wiśniewski, 1997; Zalewski et al., 1990b; Zalewski, 2014a).

These tenets were implemented through empirical research and the development of ecohydrological systemic solutions to regulate the cycling of water and biogens (NPC) within drainage basins (Zalewski, 2000b; Zalewski and Wagner, 2005; Zalewski et al., 2010). Hydrobiomanipulation has since been recognized as the foundation for nature-based solutions, which apply scientific understanding of ecosystems to water management challenges.

7. UNESCO's and European Union role in knowledge integration and policy development

UNESCO has played a pivotal role in fostering interdisciplinary collaboration by integrating ecological, hydrological, and socioeconomic knowledge. Programs such as the IHP–MAB initiative and subsequent projects have facilitated the exchange of insights across disciplines through dialog, supporting the development and implementation of integrated water resources management (IWRM) and Water Framework Directive of European Commission.

Notable examples of implementation of the aforementioned ideas were projects financed by European Commission such as the SWITCH Project, the Blue-Green Network in Łódź, and the ECOROB initiative (Ecohydrological Rehabilitation of Urbanized River Valleys), as well as EHREK—a holistic approach for improving water resources to support Water, Biodiversity, Ecosystem Services, and Resilience (WBSR) at the catchment scale.

8. Toward a holistic, planetary-scale approach

All aforementioned ideas were, to a great extent, inspired by the Gaia Hypothesis (Lovelock, 1972, 1979, 1988) who proposed that the biosphere functions as a self-regulating system, maintaining atmospheric and ecological homeostasis. Though initially controversial, this perspective has influenced contemporary ecohydrological thinking, demonstrating the profound interconnections between biotic and abiotic processes at a global scale (Watson and Lovelock, 1983; Lenton and Lovelock, 2000).

This recognition has driven the transition from analytical ecology to transdisciplinary ecohydrology, bridging fundamental science with practical environmental management. By developing applications of molecular biology, biotechnologies and development of conceptual background for systemic solutions, ecohydrology provides a strategic approach to sustaining ecosystem resilience, enhancing water resource management, and supporting socioeconomic development in an era of climate change.

While the integration of hydrology and ecology into ecohydrological model ABRC has advanced significantly, the development of truly innovative solutions requires more than empirical data analysis. The prevailing consensus in contemporary academic discourse is that artificial intelligence (AI) is poised to become an important support in shaping the future of scientific research. However, AI's ability to generate novel scientific insights is inherently limited—it relies on existing datasets rather than the creative conceptualization of new paradigms.

A holistic approach to scientific discovery necessitates creative problem-solving methodologies such as Thomas Kuhn's paradigm shifts, Karl Popper's falsification of hypotheses, Paul Feyerabend's "epistemological anarchism," and Edward de Bono's "lateral thinking". These frameworks disrupt prevailing thought patterns, inspire scientists to conceptualize new ecological functions, identify emergent ecosystem properties, and formulate transdisciplinary solutions.

This principle is particularly evident in transdisciplinary ecohydrological research, where the ABRC model (see Chapter 4) led to the discovery of calcium's role in intensifying nutrient spiraling—an insight that enhanced river self-purification and invertebrate biomass production. Such findings have directly contributed to innovative water management solutions, such as the sequential biofiltration system with a geochemical barrier for non-point source pollution reduction and urban stormwater purification (Zalewski et al., 2012). These systems possess great potential to increase water retention in urban landscapes, increases groundwater recharge, mitigates heat island effects, and fosters the development of blue-green city networks, promoting a resilient, sustainable urban future.

In the face of the decreasing availability of water globally, the emerging concept of ecohydrology was inspired by two main ideas. The first was the weakness of the traditional approach to water management (i.e., drawing solely on engineering solutions), based on a mechanistic vision of how nature operates. Solutions proposed based on that vision failed to ensure a lasting, self-sustainable ecosystem function, translating directly into both the quality and availability of resources. Water management issues stem from a plethora of cause-and-effect relationships, such as increasing urbanization, unsustainable agriculture, overexploitation of resources, and environmental pollution. Attempts at offsetting the consequences of human-made changes in the hydrological cycle solely with the aid of engineering methods have not proved effective, largely because they fail to address the causes of the deteriorating condition and availability of resources, but are instead geared toward reaping tangible benefits, and as efficiently as possible, from the environment or toward gaining effective control over it. The second idea, in turn, concerns financial and energy constraints: if integrating ecology with hydrology is meant to yield any realistic solutions for enhancing the WBSR+CE+LPG potential, it needs to be harmonized with society's cultural backgrounds and developmental needs. This is a prerequisite for ensuring the lasting and sustainable development of human civilization.

What poses the most daunting challenge to the integration of ecological and hydrological knowledge is the difference in the standard scales employed by the two sciences—concerning, for example, the minimum number of samples taken or the minimum size of the research site. The smallest spatial units employed in ecological research (e.g., microhabitats) are as a rule of thumb considerably smaller than the smallest spatial units analyzed in hydrology (e.g., experimental plots or research basins). By a similar token, ecological research can hardly be extrapolated to a larger area due to ecosystem heterogeneity, as a result of which conclusions from the observation of ecological processes made at a specific location cannot easily be transposed to other conditions. Since linking hydrological variables to the biological characteristics of an ecosystem on an appropriate scale is crucial for understanding the system (catchment) functioning, ecohydrological research begins by quantifying the catchment area's abiotic conditions, which constitute the basis for water and matter circulation and life-supporting processes (LSPs) and related ecological and socioeconomical processes.

Improvment of the ecological status of a catchment area, based on observed deviations from the "natural state," is one of the key objectives of water sciences and water resources management under Water Framework Directive, 2000 (European Parliament & Council of the European Union, 2000). The methods that have been devised by these disciplines provide information about individual indicators gauging the status of the environment. The transdisciplinary ecohydrological principles enable us to supply more such information, with a value greater than that of its individual component parts. By incorporating environmental impacts and interactions between living organisms and hydrological parameters, the deductive framework provides a more comprehensive understanding. Much as how ecosystem theory looks at the ecosystem as a living organism in terms of physiological traits, the task of diagnosing the condition of an ecosystem and catchment area can be compared to a medical diagnosis.

In both cases, "diagnosis" involves making use of a certain set of parameters, drawn from either environmental monitoring (e.g., water samples) or medical examinations (e.g., blood tests). However, before treatment can begin, the "diagnosing physician" needs to understand how the analyzed parameters are physiologically interrelated and what processes might underlie those interactions. In the case of a river basin, the individual components that are the focus of monitoring efforts need to be interpreted through a holistic perspective that integrates ecological and hydrological knowledge before restoration process can start and the environment's ecological potential can be successfully enhanced, in line with the WBSR strategy. These interactions shape how the biotic factors regulate hydrological processes and vice versa—how the hydrological factors affect the biocenosis of the analyzed system.

AI disclosure

During the preparation of this work the author(s) used OpenAI's ChatGPT. This document includes content generated and/ or edited with the assistance of OpenAI's ChatGPT, used to support language refinement and structural improvements. All final content was reviewed and verified by the author. After using this tool/service, the author(s) reviewed and edited the content as needed and take(s) full responsibility for the content of the publication.

References

Barnes, J.R., Minshall, G.W. (Eds.), 1983. Stream ecology: Application and testing of general ecological theory. Plenum Press.

Bartoszewicz, A., 1994. Skład chemiczny wód powierzchniowych zlewni intensywnie użytkowanych rolniczo w warunkach glebowo-klimatycznych Równiny Kościańskiej. Roczniki Akademii Rolniczej w Poznaniu. Rozprawy Naukowe 250, 5–68.

Cullen, 1990. The Public Inquiry into the Piper Alpha Disaster. HMSO.

Eagleson, P.S., 2002. Ecohydrology: A Darwinian Expression of Vegetation Form and Function. Cambridge University Press.

Elfithri, R., Zalewski, M., Arduino, G., Chícharo, L., 2025. Transdisciplinary ecohydrology for water management solutions and sustainability. Nat. Water 3 (4), 360–363. https://doi.org/10.1038/s44221-025-00395-x.

European Parliament & Council of the European Union, 2000. Directive 2000/60/EC of the European Parliament and of the Council of 23 October 2000 establishing a framework for Community action in the field of water policy. Off. J. Eur. Comm. L327, 1–72. https://eur-lex.europa.eu/legal-content/EN/TXT/?uri=CELEX:32000L0060.

Fisher, S.G., Likens, G.E., 1973. Energy flow in Bear Brook, New Hampshire: an integrative approach to stream ecosystem metabolism. Ecol. Monogr. 43 (4), 421–439. https://doi.org/10.2307/1942301.

Grodziński, W., Klekowski, R.Z., Duncan, A., (Eds.), 1975. Methods for ecological bioenergetics. Blackwell Scientific Publications.

Hillbricht-Ilkowska, A., 1985. The Role of Biotic Components in Lake Ecosystems: Functional Approach.

Hillbricht-Ilkowska, A., 2001. Biodiversity and functioning of freshwater ecosystems: a case study of Poland. Ecohydrol. Hydrobiol. 1, 113–122.

Hillbricht-Ilkowska, A., 2002. Structural and functional biodiversity of aquatic ecosystems in ecological modeling. Ecohydrol. Hydrobiol. 2, 77–89.

Holmes, R. T., & Likens, G. E. (2016). Hubbard Brook: The story of a forest ecosystem. Yale University Press.

Huet, M., 1949. Observations on the relationship between slope and fish populations in running waters [Aperçu des relations entre la pente et les populations piscicoles des eaux courantes]. Swiss J. Hydrol. 11, 322–351.

Huston, M.A., DeAngelis, D.L., Post, W.M., 1988. New computer models unite ecological theory. Bioscience 38 (10), 682–692.

Huxley, 1877. American Addresses and Physiography. Macmillan, Collected Essays. Vol. 3.

Illies, J., Botosaneanu, L., 2017. Problèmes et méthodes de la classification et de la zonation écologique des eaux courantes, considerées surtout du point de vue faunistique. SIL Commun. 1953-1996 12 (1), 1–57. https://doi.org/10.1080/05384680.1963.11903811.

International Conference on Water and the Environment (ICWE). ICWE (1992).

Jones, G., 2022. Proceedings of the National Academy of Sciences 9575 9580 Why Science Needs Philosophy 116.

Junk, W. J., Bayley, P. B., Sparks, R. E., 1989. Proceedings of the International Large River Symposium (LARS) 110 127 the Flood-Pulse Concept in River–Floodplain Systems 106.

Kirchner, J.W., 1989. The Gaia hypothesis: can it be tested? Rev. Geophys. 27 (2), 223–235. https://doi.org/10.1029/RG027i002p00223.

Kopf, E.W., Lotka, A.J., 1925. Elements of physical biology. J. Am. Stat. Assoc. 20 (151), 452. https://doi.org/10.2307/2965538.

Kuhn, T.S., 1957. The Copernican revolution: Planetary astronomy in the development of Western thought. Harvard University Press.

Kuhn, T.S., Schlegel, R., 1963. The structure of scientific revolutions. Phys. Today 16 (4). https://doi.org/10.1063/1.3050879, 69–69.

Lavorel, S., Garnier, E., 2002. Predicting changes in community composition and ecosystem functioning from plant traits: revisiting the Holy Grail. Funct. Ecol. 16 (5), 545–556. https://doi.org/10.1046/j.1365-2435.2002.00664.x.

Lenton, T.M., 1998. Gaia and natural selection. Nature 394 (6692), 439–447. https://doi.org/10.1038/28792.

Lenton, T.M., Lovelock, J.E., 2000. Daisyworld is Darwinian: constraints on adaptation are important for planetary self-regulation. J. Theor. Biol. 206 (1), 109–114. https://doi.org/10.1006/jtbi.2000.2105.

Lovelock, J.E., 1972. Preliminary specification of a physical–chemical biosphere model as applied to the Earth. J. Theor. Biol. 35 (3), 401–408.

Lovelock, J.E., 1979. Gaia: A New Look at Life on Earth. Oxford University Press.

Lovelock, J.E., 1988. The Ages of Gaia: A Biography of Our Living Earth. W. W. Norton.

Lovelock, J.e., Margulis, L., 1974. Atmospheric homeostasis by and for the biosphere: the gaia hypothesis. Tellus 26 (1–2), 2–10. https://doi.org/10.1111/j.2153-3490.1974.tb01946.x.

Makarieva, A.M., Gorshkov, V.G., 2010. The biotic pump: Condensation, atmospheric dynamics and climate. Int. J. Water 5 (4), 365–385. https://doi.org/10.1504/IJW.2010.038729.

Malcolm, C.M. (Ed.), 1999. Eco-hydrology: Plants and water in terrestrial and aquatic environments. Routledge.

Mitsch, W.J., Jørgensen, S.E., 1989. Ecological Engineering: An Introduction to Ecotechnology. John Wiley & Sons.

Naiman, R.J., Décamps, H., 1989. The ecology of interfaces: Riparian zones. Annu. Rev. Ecol. Systemat. 20, 591–620. https://doi.org/10.1146/annurev.es.20.110189.003111.

Naiman, R.J., Décamps, H., Fournier, F., 1989. Role of Land/Inland Water Ecotones in Landscape Management and Restoration: Proposals for Collaborative Research (MAB Digest 4). UNESCO.

Newbold, J.D., Elwood, J.W., O'Neill, R.V., Van Winkle, W., 1981. Measuring nutrient spiralling in streams. Can. J. Fish. Aquat. Sci. 38 (7), 860–863. https://doi.org/10.1139/f81-114.

Popper, K., 1934. The Logic of Scientific Discovery. Hutchinson.

Reed, B., 2007. Shifting from "sustainability" to regeneration. Build. Res. Inf. 35 (6), 674–680. https://doi.org/10.1080/09613210701475753.

Report of the United Nations, 1992. Conference on Environment and Development (UNCED). United Nations, pp. 3–14.

Rodriguez-Iturbe, I., 2000. Ecohydrology: a hydrologic perspective of climate-soil-vegetation dynamies. Water Resour. Res. 36 (1), 3–9. https://doi.org/10.1029/1999wr900210.

Schrödinger, E., 1944. What Is Life? the Physical Aspect of the Living Cell. Cambridge University Press.

Shiklomanov, I.A., 1998. World Water Resources: A New Appraisal and Assessment for the 21st Century. UNESCO.

Smith, J.A., Lee, R.B., 2023. Perspective-taking and its role in facilitating civil discourse. J. Educ. Psychol. 115 (4), 512–529.

Spencer, H., 1844. A new view of the functions of imitation and benevolence. The Zoist 1 (4), 369–385.

Stanford, J.A., Ward, J.V., 1979. Stream Regulation in North America. Springer Science and Business Media LLC, pp. 215–236. https://doi.org/10.1007/978-1-4684-8613-1_13.

Starkel, 1982. Evolution of the Vistula River Valley during the Last 15000 Years, 53. Folia Quaternaria.

Starkel, L., 1988. The evolution of the Vistula River valley during the last 15,000 years. In: Geographical Studies, 4. Polish Academy of Sciences, Institute of Geography and Spatial Organization, pp. 7–37.

Strauss, A., Corbin, J., 2008. Basics of Qualitative Research: Techniques and Procedures for Developing Grounded Theory. Sage Publications.

UNEP Freshwater Management Series, 2002. UNEP.

Van der Ent, Savenije, H.H.G., Schaefli, B., Steele-Dunne, S.C., 2010. Origin and fate of atmospheric moisture over continents. Water Resour. Res. 46 (9), W09525. https://doi.org/10.1029/2010WR009127.

Vannote, R.L., Minshall, G.W., Cummins, K.W., Sedell, J.R., Cushing, C.E., 1980. The river continuum concept. Can. J. Fish. Aquat. Sci. 37 (1), 130–137. https://doi.org/10.1139/f80-017.

Wantzen, K.M., 2023. River culture: living with the River, loving the River, taking care of the River. Blue Papers 2 (1). https://doi.org/10.58981/bluepapers.2023.1.06.

Watson, A.J., Lovelock, J.E., 1983. Biological homeostasis of the global environment: the parable of Daisyworld. Tellus 35 (4), 284–289.

Webster, J.R., 1975. Analysis of potassium and calcium dynamics in stream ecosystems of three southern Appalachian watersheds of contrasting vegetation (Doctoral dissertation). University of Georgia, Athens, GA.

Webster, J.R., Patten, B.C., 1979. Effects of watershed perturbation on stream potassium and calcium dynamics. Ecol. Monogr. 49 (1), 51–72. https://doi.org/10.2307/1942572.

WWAP, 2018. The United Nations World Water Development Report 2018: Nature-Based Solutions for Water. United Nations World Water Assessment Programme.

Zalewski, M., 1999. Minimizing the risk and amplifying the opportunities for restoration of shallow reservoirs. Hydrobiologia 395–396, 107–114.

Zalewski, M., 2000. Ecohydrology—The scientific background to use ecosystem properties as management tools toward sustainability of water resources. Ecol. Eng. 16 (1), 1–8. https://doi.org/10.1016/S0925-8574(00)00071-9.

Zalewski, M., 2002a. Ecohydrology – the use of ecological and hydrological processes for sustainable management of water resources. Hydrol. Sci. J. 47 (5), 825–834.

Zalewski, M. (Ed.), 2002b. Guidelines for the integrated management of the watershed – phytotechnology and ecohydrology, United Nations Environemant Programme. International Environmental Technology Centre. Freshwater Management Series. 5, 2002, 188.

Zalewski, M., 2011. Ecohydrology for implementation of the EU water framework directive. Water Management 164 (8), 375–385.

Zalewski, M., Naiman, R.J., 1985. The regulation of riverine fish communities by a continuum of abiotic–biotic factors. In: Alabaster, J.S. (Ed.), Habitat modification and freshwater fisheries. Butterworths, pp. 3–9.

Zalewski, M., Robarts, R.D., 2003. Ecohydrology – A new paradigm for integrated water resources management. SIL News 40, 1–5.

Zalewski, M., Wagner, I., 2005. Ecohydrology—The use of water and ecosystem processes for healthy urban environments. Ecohydrol. Hydrobiol. 5 (4), 263–268.

Zalewski, M., Wiśniewski, R.J. (Eds.), 1997. Application of ecosystem biotechnologies to improvement of the water body quality. Institute of Ecology Polish Academy of Science Press, Dziekanów Leśny, Polska.

Zalewski, M., 2014a. Ecohydrology and hydrologic engineering: regulation of hydrology-biota interactions for sustainability. J. Hydrol. Eng. 20 (1). https://doi.org/10.1061/(asce)he.1943-5584.0000999.

Zalewski, M., Brewinska-Zaraś, B., Frankiewicz, P., Kalinowski, S., 1990. The potential for biomanipulation using fry communities in a lowland reservoir: concordance between water quality and optimal recruitment. Hydrobiologia 200–201 (1), 549–556. https://doi.org/10.1007/BF02530372.

Zalewski, M., Janauer, G.A., Jolánkai, G., 1997. Ecohydrology: A New Paradigm for the Sustainable Use of Aquatic Resources. UNESCO IHP Technical Documents in Hydrology.

Zalewski, M., Harper, D.M., Wagner, I., 2010. Ecohydrology—why demonstration projects throughout the world? Ecohydrol. Hydrobiol. 10, 265–274.

Zalewski, M., Wagner, I., Frątczak, W., Mankiewicz-Boczek, J., Parniewski, P., 2012. Blue-Green City for Compensating Global Climate Change Conference Paper/Monograph. ASCE.

Chapter 4

The role of abiotic and biotic factors in regulating ecosystem processes

Maciej Zalewski

European Regional Centre for Ecohydrology of the Polish Academy of Sciences under the auspices of UNESCO, Łódź, Poland

PHOTO CREDIT: M. ZALEWSKI.

Transdisciplinary Ecohydrology. https://doi.org/10.1016/B978-0-443-41596-8.00008-X

1. Introduction and origin of the model

Ecohydrology is a transdisciplinary science. It not only integrates a wide range of scientific disciplines, many of which have traditionally focused on isolated aspects of environmental functioning, but also advances comprehensive hypotheses regarding the complexity and regulation of ecosystem processes related to reduction of impacts, restoration and enhancement by the society involvement.

As previously discussed in this book, the most pressing challenges of the Anthropocene include the depletion of natural resources and widespread environmental degradation, driven by a growing global population and climate change, both of which increase the demand for ecosystem services. These concerns are not new; they were anticipated as early as the 18th century. Thomas Malthus, reflecting on the Irish crisis, proposed a theory in 1798 stating that population growth is exponential, whereas food production increases linearly, inevitably leading to food limitations. This Malthusian theory was later expanded by the Club of Rome, whose projections introduced additional variables such as industrial output, consumption of nonrenewable resources, and environmental pollution. In their seminal report, The Limits to Growth (Meadows et al., 1972) the Club concluded that while human demands grow exponentially, the planet's resource base does not, setting the stage for inevitable resource depletion unless action is taken. However, the report also offered a more hopeful "sustainable" scenario, in which human development can proceed within ecological limits.

In light of the continuing rise in global population, sustained economic growth, and escalating consumption of energy and resources, future strategies must be based on two fundamental pillars: reducing risks and enhancing opportunities for sustainability. Addressing these issues is particularly complex because of demographic and economic constraints and cannot be achieved through traditional, conservative models of environmental protection alone. What is required is a transformative approach, such as the FACTOR FOUR concept proposed by (von Weizsäcker et al., 1997), which advocates for doubling human well-being while halving the use of natural resources and energy.

Like all species, the human population is subject to the basic laws of population ecology. In the absence of external regulatory pressures (e.g., predators), populations tend to grow until they reach the carrying capacity of their environment. At this point, resource scarcity and intensified intraspecific competition create stress, reducing population fitness. In human societies, such stress manifests in the form of social and political tensions. Therefore it is the responsibility of scholars and practitioners to develop innovative strategies that reconcile societal and economic development with ecological limitations—especially those aimed at enhancing environmental capacity and long-term resilience.

In ecohydrology, environmental capacity is conceptualized through four interrelated parameters—WBSRC: water (W), biodiversity (B), ecosystem services (S), resilience (R), which implementation is to a great extend depends on involvement of society through cultural and education (C) (Zalewski, 2014). Water (W) governs both the ecosystems ability to store carbon and the pathways of biogenic circulation/spiraling. Since both bioproduction and biodiversity (B) tend to increase with temperature and water availability, water and biodiversity form the foundational pillars upon which ecosystem services (S) depend. These services, in turn, sustain human well-being and development.

A central challenge in contemporary science is not only to develop mechanisms that support adaptation to climate change but also to enhance the environment's capacity to respond flexibly and effectively to various forms of anthropogenic pressure. This adaptive capacity—termed *resilience* (R)—is enabled through dual regulation, a principle intrinsic to ecohydrology that harmonizes ecological and hydrological processes. Importantly, such interventions must be developed with sensitivity to local societal needs and the protection of cultural heritage and education (C) (Wantzen et al., 2016). Cultural heritage, within this framework, extends beyond tangible assets such as architecture and art to encompass worldviews, values, and attitudes that shape human interaction, openness to dialogue, and educational approaches. These cultural dimensions influence the readiness of communities to collaborate across disciplines and cultures in a rapidly changing world.

Humans are unique in their capacity to shape their environment. Therefore the involvement of local communities is essential when designing strategies for protecting biodiversity and managing natural resources. Societal engagement must be viewed not merely as a component of implementation but as a determinant of environmental capacity itself.

The principle of Dual Regulation, central to ecohydrology, provides a methodological foundation for solving sustainability challenges through the recognition and management of interactions between water, resource dynamics, and biotic communities. These ecological processes operate at multiple scales, from bioenergetic and physiological levels to landscape-scale patterns and sociocultural dynamics. A comprehensive understanding of these processes allows for their quantification and application in the development of targeted solutions to environmental problems.

Once the operational mechanisms of ecological processes are understood and reduced to their fundamental components, drawing from chemistry, biology, physics, and mathematics, these solutions can often be analogously applied in other environmental contexts. This systems-based perspective underpins *process-oriented thinking*, which focuses on regulating ecological processes as a strategic means of resolving complex environmental challenges.

Deconstructing ecohydrological processes in this way reveals that nature operates through interconnected and nonrandom systems. Every process within an ecosystem contributes to the balance between biotic and abiotic components, shaping the ecosystem's stability and resilience. Only when biotic factors become stable and predictable they can be manifested, revealing the interactions between organisms and between their environment, which determines their species structure. Together, these factors define the productivity and trophic state of river ecosystems (Zalewski, 1986).

Biotic factors, on the other hand, are characterized by species composition, population structure, and interspecies interactions (Mahon, 1982). These dynamics are shaped by evolutionary history by the pressures exerted by abiotic conditions, e.g., postglacial colonization, which in turn has been shaping organisms' adaptations to prevailing conditions through a combination of physiological tolerances (e.g., resistance to oxygen depletion or elevated temperatures), compensatory mechanisms (e.g., behavioral strategies like migration), and reproductive strategies. Such adaptations influence colonization rates and determine the outcomes of intra- and interspecific interactions such as competition and predation (Zalewski, 1986; Thorpe et al., 1988).

This integrative understanding was already articulated in the foundational work of Eugene Odum, who empirically referred to natural systems and systematically quantified their processes. Odum's systems ecology perspective emphasized the regulation of ecosystem processes rather than their mere preservation. His insights underscored a key limitation of conventional conservation: while valuable, it alone cannot reconcile the demands of human societies with the constraints of natural systems.

Instead, exponentially increasing anthropogenic impact on environment requires active regulation of ecological processes to improve no only the adaptive and regenerative capacity of human-modified ecosystems but more advanced approach—enhancement sustainabilty potential. This is the essence of the WBSR framework in ecohydrology: an approach that integrates natural, social, and cultural dimensions to meet the urgent challenges of sustainable development.

2. Biodiversity in various river ecosystems as corroboration of the ABRC approach

On the holistic approach to ecology, the value of a given ecosystem is greater than the sum of its constituent parts. This is so because the relations between the constituent parts, i.e., biotic components and abiotic conditions, represent the ecosystem's added value and shape its characteristics in ways that researchers cannot anticipate if they only examine them in separation. The constituent parts of an ecosystem (and of practically any system analyzed in such a way) are hierarchically structured, with their new traits emerging at individual levels of the hierarchy (Solbrig, 1991).

The interactions between the ecosystem's abiotic conditions and biotic components underpin the concept of dual regulation via abiotic and biotic factors (the Abiotic-Biotic Regulatory Continuum Concept ABRC) (Zalewski and Naiman, 1985; Zalewski, 1986) which, by integrating hydrology and ecology and inducing process-oriented thinking, also becomes fundamental to ecohydrology. This particular idea was inspired by the already mentioned River Continuum Concept (RCC). The authors positing the ABRC model came to the conclusion that the RCC could not serve as a model to be extrapolated to other ecosystems e.g. (Statzner and Higler, 1986), since every river differs in abiotic and biotic conditions and in the role it plays for humans. For this reason, any quantification of ecological processes or deduction on their basis should begin at the most fundamental level of the laws of physics and hydraulics, because such basic physical-level transformations are preconditions of life in an aquatic environment. The ABRC approach also became an alternative so as to reconcile the dispute among ecologists regarding the density-independent (abiotic) and density-dependent (biotic) factors which influence the structure and functioning of river ecosystems (Krebs, 1972).

Figure 4.1 offers examples of various river ecosystems, structured according to the changing role of abiotic and biotic factors in regulating ecosystem operation. Fish biodiversity in rivers increases through a gradient, from rivers whose environment is shaped by abiotic factors (the environment is harsh and unstable), to rivers where biotic factors prevail. In those river sections where the hydrology—abiotic environment is stable and the fluctuations of its physical conditions can be foreseen (i.e., with a limited range of assumed values and/or duration), in the fish tolerance spectrum, e.g., in tropical rivers, biotic factors are the predominant regulators of both population and diversity (Zalewski and Naiman, 1985; Junk et al., 1989). Because of these interdependencies, the lowest biodiversity can be observed in environments regulated by abiotic factors first and foremost hydrology (e.g., in near-source segments of mountain, boreal, and desert rivers), where significant energy supply/expenditure ratio limitations can frequently be observed, mostly because of dynamic changes of stream hydraulics and low or very high temperature or short vegetation season. By contrast, the greatest biodiversity can be found in ecosystems regulated by biotic factors, where fluctuations in abiotic factors—hydrological patterns, occur within the physiological tolerance spectrum of living organisms. The ABRC is innovative in that it highlights the smoothly changing hierarchy between the regulating roles of abiotic and biotic factors, which shape the river ecosystem. In such a context, abiotic factors are superior to biotic ones since, as demonstrated above, they determine physiological processes that only partly undergo compensatory adaptations. Only when abiotic factors (e.g., water availability,

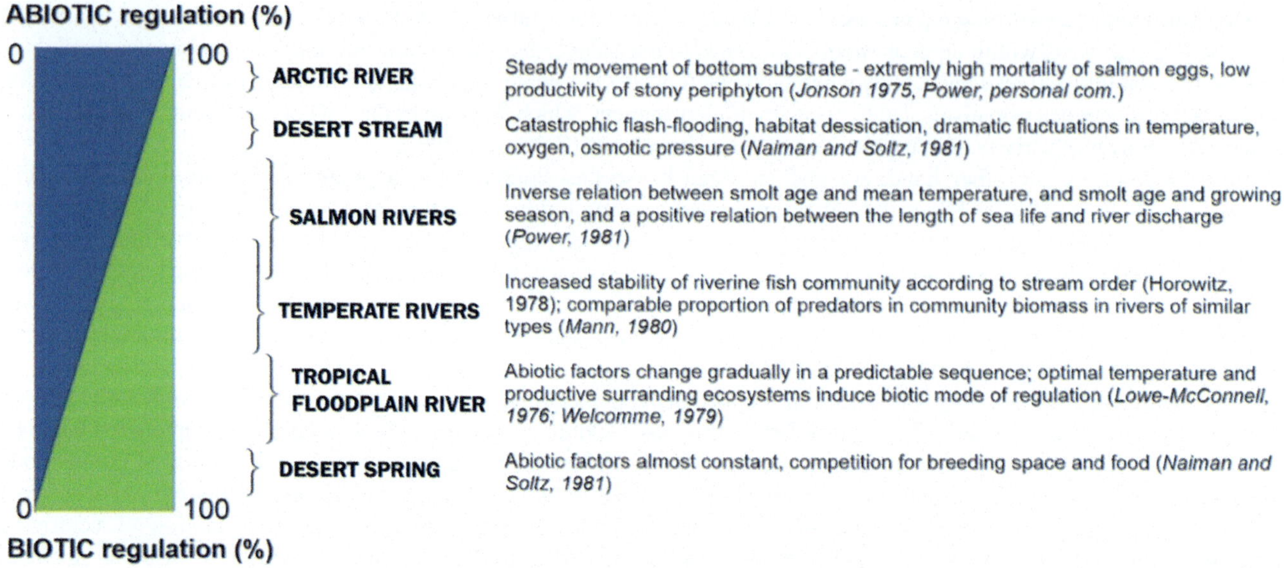

FIGURE 4.1 Hierarchy of abiotic and biotic factors. A gradient of changes in the hierarchy of abiotic and biotic factors regulating fish communities in rivers across different geographical zones illustrates how abiotic regulation, much like biodiversity, is shaped by geomorphological and climatic conditions—hydrology as well as the intensity of anthropogenic pressures. This challenges the broadly accepted view that biodiversity uniformly declines from south to north. The hierarchy of drivers shifts in response to the decreasing importance of abiotic pressures. *Modified from Zalewski, M., Naiman, R. J., 1985. The regulation of riverine fish communities by a continuum of abiotic–biotic factors. In: Alabaster, J. S. (Ed.),* Habitat modification and freshwater fisheries. *Butterworths for FAO, pp. 3–9.*

temperature, and flow dynamics) become stable and predictable are biotic factors manifested, revealing the interactions between organisms and between their environment, which determine their species structure.

The changing role of the factors that determine ecosystem functioning, oscillating between abiotic and biotic regulation, may also be visible in one river segment owing to seasonal variability. For instance, in rivers in temperate climate zones, processes of energy flow and circulation of matter are determined by abiotic factors in the winter season (e.g., low temperature inhibiting the rate of biophysical and biochemical reactions, considerable inflow of suspended matter coupled with reduced plant cover in the basin), while in the summer season they remain strongly influenced by biotic processes (e.g., decomposition and primary production) (Wagner and Zalewski, 2016). Explaining such environmental processes from molecular to catchment scale and biosphere is essential for efficient reduction of fragmentation of efforts, because it allows to explain the merit and hierarchy of drivers crucial for efficient problem solving by decision makers and also creating new dimensions for law, policy, and governance.

Deductive Abiotic-Biotic Regulatory Continuum model (Zalewski and Naiman, 1985) of changes in the hierarchy of ecological factors in rivers of different geographic zones as a background for regulation of abiotic and biotic ecological processes first became a framework for EIFAC FAO UN Program.

An ABRC analysis looks at the impact of the laws of chemistry and physics in specific physical conditions (on the functioning of the ecosystem's living constituents, partly discussed in Chapter 4). As mentioned earlier, according to Van't Hoff's rule, the temperature regulates the rate of chemical reactions, and thereby the rate of metabolism of ectothermic animals, while the laws of hydraulics, through the dynamics of the water flow, determine such physical parameters as the current velocity or amount of oxygen dissolved in the water (Grodziński et al., 1975). The slope and cross-section of the channel, in accordance with Bernoulli's equation Box 4.1 determines the river flow as turbulent if the elevation gradient is high and the channel cross-section is narrow, and as laminar when the slope is low, and the cross-section is wide. The morphology of the river channel plays a substantial role, since oxbow lakes form such environments where oxygen deficits are often encountered at advanced stages of ecological succession, when substantial amounts of organic matter have been accumulated. On the other hand, at early succession stages and when there is frequent water exchange with the river, they become preserves of biodiversity generating varied habitats, particularly as refugia for fish in the winter season. Therefore restitution of river biodiversity first and foremost ought to create oxbows with conditions for frequent water exchange (Obolewski et al., 2018).

Ambient temperature is another important factor as it regulates the activity of digestive enzymes. As temperature increases, so does the ecosystem productivity, the opportunities for securing food, and the ability to digest it. For this reason, ectothermic animals decrease their activity to minimize energy expenditure.

These considerations suggest that biocenotic processes are to a great extent shaped and dictated by hydrological and physical processes in the river. As mentioned above, the dominance of abiotic factors decreases with the river flow: the

BOX 4.1 Bernoulli's equation

Bernoulli's equation (formulated in 1738) is used to describe the flow of fluids in closed conduits and open channels. It assumes a steady flow whose parameters do not vary over time. The equation takes the following form:

$$z_1 + \frac{p_1}{\gamma} + \frac{\alpha_1 * V_1^2}{2 * g} = z_2 + \frac{p_2}{\gamma} + \frac{\alpha_2 * V_2^2}{2 * g} + \sum h_{str}$$

where z_1, z_2—height [m]; $\frac{p_1}{\gamma}$, $\frac{p_2}{\gamma}$—pressure height [m]; $\frac{\alpha_1 * V_1^2}{2*g}$, $\frac{\alpha_2 * V_2^2}{2*g}$—speed height [m]; α_1, α_2—de Saint-Venant coefficient (assumed often as 1.0), V_1, V_2—velocity [m/s]; g—earth acceleration; Σh_{str}—sum of linear losses and local losses [m]; p_1, p_2—pressure [N/m^2]; g—specific gravity of the liquid [N/m^3].

The first term of the equation represents the elevation of the center of the conduit above a certain reference level, the second represents the static pressure, and the third represents the velocity, which depends on the average speed in the conduit's cross-section. The equation implies that to maintain the continuity of flow and energy when a given channel's cross-section diminishes, the flow velocity must increase while the static pressure must fall.

Therefore, Bernoulli's equation describes the water dynamics in a river, which translates into the energy expenditure of the organisms living there, and thereby the structure and variability of organisms, and consequently, the variability of the energy flow channels (e.g., trophic relations between organisms). This has a direct bearing on the biodiversity of the aquatic environment and its ecosystem services. In particular, the parameter in question should be taken into account while assessing the impact of river regulation on the biocenosis and revitalization of rivers and streams (Chow, 1959).

How river currents shape habitat structure and biodiversity Ecohydraulics—fluid dynamics—currents distribution in river channel determines energy gain/energy expenditure ratio for fish, in consequence habitats availability—biocomplexity, bioproductivity, biodiversity.

Key reference points:

Bernoulli's equation is used to describe the flow of fluids in closed conduits and open channels. It assumes a steady flow whose parameters do not vary over time.

Shear stress is a measure of the force of friction from a fluid acting on a body in the path of that fluid. In the case of open channel flow, it is the force of moving water against the bed of the channel. This force of water is trying to drag the channel surface (bottom-dwelling fish and invertebrates) downstream.

Stream habitat complexity determines bioproductivity and biodiversity—through energy gain/loss, growth rate, and reproduction.

Nutrient states, thermodynamics, and biocomplexity determine biological productivity and self-purification of ecosystems.

slower the current, the more stable the abiotic factors, and the biotic factors are more active, thereby strongly affecting ecosystem operation. Water and temperature also affect river self-purification characteristics. The values of these two parameters vary during the seasonal cycle. Frame 10 describes a case study illustrating how the hierarchy of abiotic and biotic factors determining river self-purification changes with the seasons.

Similarly, biodiversity, defined as an indicator of good prospects for the future and human wellbeing (Millennium Ecosystem Assessment, 2005), is determined by the availability of water needed for plant growth (Visser, 1971; Rodríguez-Iturbe, 2000; Eamus et al., 2006), and by sun exposure, which affects the primary production of biomass (Kowalik and Eckersten, 1984). A comparison of biodiversity between the temperate and tropical zones shows a huge shift in the allocation of carbon from the soil to the plant biomass. This is explained by the Van't Hoff equation, stating that increased temperature stimulates the rate of nutrient circulation and energy flow. Therefore it can be assumed that sufficient water resources and high temperature will create suitable conditions (extended time of physiological optimum) for the differentiation of microorganism, plant, and animal communities, and thereby will consolidate mutations enhancing genotype adaptation through natural selection and environmental adaptation, producing in effect greater biological diversity and greater adaptation potential of living organisms. In the tropical zone, due to high productivity, surplus energy is stored in plants, enhancing their potential for creating new pathways for energy flow through the ecosystem. On the other hand, in the temperate zone, a harsher climate and shorter growing season combine to produce a much smaller number of pathways for energy flow and nutrient circulation. For this reason, ecosystems in the boreal temperate zone, as well as alpine and desert environments, have a lower potential for regeneration and are more prone to biodiversity decline due to natural calamities and human activity. One of the gravest consequences of such loss of species is the reduced biological ability of the ecosystem to change and adapt. A lower number of alternative pathways for the flow of energy and circulation of biogenic substances in the human devastated ecosystem may, in effect, reduce potential opportunities for using hydrological and biological processes to enhance the WBSR + CE potential.

The above-described increase in the species abundance of fish from environments regulated abiotically (A) to those regulated biotically (B) challenges the widely accepted hypothesis (Pianka, 1983; Magurran, 1988) that biodiversity increases in the gradient of latitude from north to south. A comparative analysis of the data showing shifts in biodiversity along the river continuum clearly corroborates such an increase in line with the ABRC model, i.e., the greatest

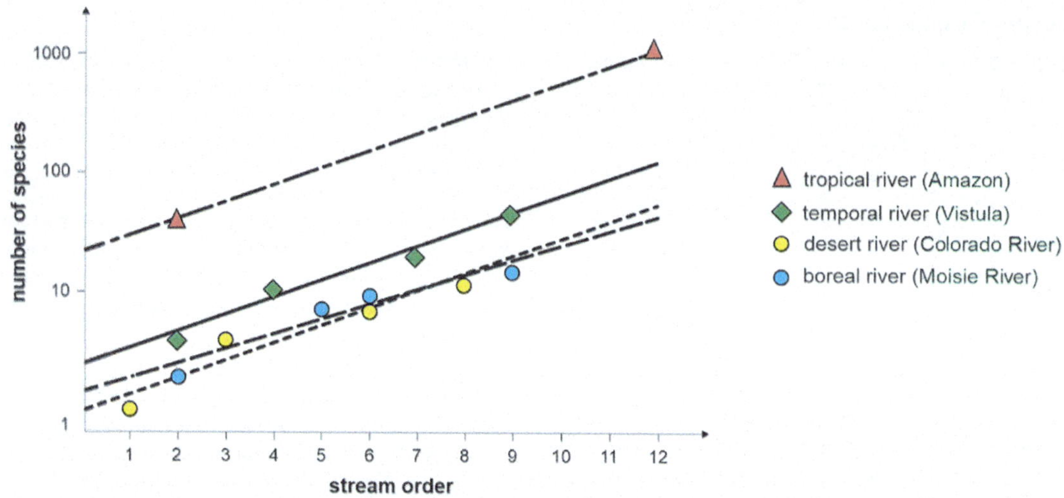

FIGURE 4.2 The influence of abiotic and biotic factors on fish biodiversity. Species abundance in a set of example rivers illustrates how biodiversity increases along the gradient of abiotic and biotic factors, and not from north to south. *Modified from Zalewski, M., Naiman, R. J., 1985. The regulation of riverine fish communities by a continuum of abiotic–biotic factors. In: Alabaster, J. S. (Ed.),* Habitat modification and freshwater fisheries. *Butterworths for FAO, pp. 3–9; Zalewski, M., 1986. Regulacja zespołów ryb w rzekach przez kontinuum czynników abiotycznych i biotycznych. Acta Universitatis Lodziensis 31, 3–68.*

biodiversity occurs in the Amazon river basin (A < B), while the lowest has been observed in abiotically regulated rivers (A > B), that is boreal or northern rivers, and a similar level in the hot Arizona deserts (Naiman and Soltz, 1981), whereas Poland's Vistula river basin (characterized by A = B equilibrium), situated further north than Arizona, manifests much greater biodiversity. This corroborates the hypothesis of increased biodiversity according to the A/B gradient, rather than from north to south. In boreal river ecosystems, A is the main determinant: the temperatures are low in winter, and the available energy (food) is scanty (Cunjak and Power, 1987), while catastrophic events can occur in spring such as meltwater floods Box 4.1 (Bernoulli's equation). These conditions represent the main mortality and natural selection factors for biodiversity in rivers. In contrast, the direst risk for biodiversity in desert areas is posed by other abiotic factors: long dry periods, decreased oxygen concentration in water, fluctuations of temperature, osmotic pressure, and catastrophic flood surges (Naiman and Soltz, 1981), (Fig. 4.2).

Despite the predominantly abiotic factors, the physiological compensation mechanism remains similar and involves the accumulation of the reserve material—fat—in body tissues and in eggs. The mechanism for storing fat compensates for low intake and limited food availability in low temperatures during long winters in the northern hemisphere, which in spring are followed by a period of extensive energy expenditure due to disturbed hydrological conditions during meltdown floods. On the other hand, in freshwater ecosystems in areas with prevalent high temperatures, e.g., the Arizona Desert, the high metabolism rate of living organisms, and in effect a considerable demand for oxygen, is confronted with low oxygen solubility in high temperatures. Oxygen availability is additionally reduced during nighttime hours (Naiman and Soltz, 1981) due to respiration processes being present in the entire biocenosis, including primary producers, thus affecting the efficiency of the Krebs cycle. Storing fat in tissues compensates for such night decrease in oxygen content, providing a source of energy for the process of anaerobic glycolysis (See the Krebs cycle and the respiratory chain in Chapter 5)—a metabolic process which provides energy in situations of oxygen deficit, but requires considerable energy reserves due to its low efficiency and generation of toxic lactic acid. However, it should be stressed that many fish species in tropical areas are morphologically adapted to breathing air, thus offsetting for low oxygen solubility in water (Weber and Kramer, 1983). These mechanisms evolved in a long evolutionary process in response to the operation of abiotic factors.

3. Reproductive strategies of fish as corroboration of the ABRC approach

Fish play an important role in regulating biological processes ("top-down") within aquatic ecosystems, particularly in functions such as nutrient spiraling, river self-purification, and the mitigation of eutrophication symptoms (bio-manipulation), including the occurrence of toxic algal blooms. Through mechanisms such as the top-down effect, fish can stabilize energy flows within ecosystems and thereby significantly influence ecological equilibrium (Northcote, 1988;

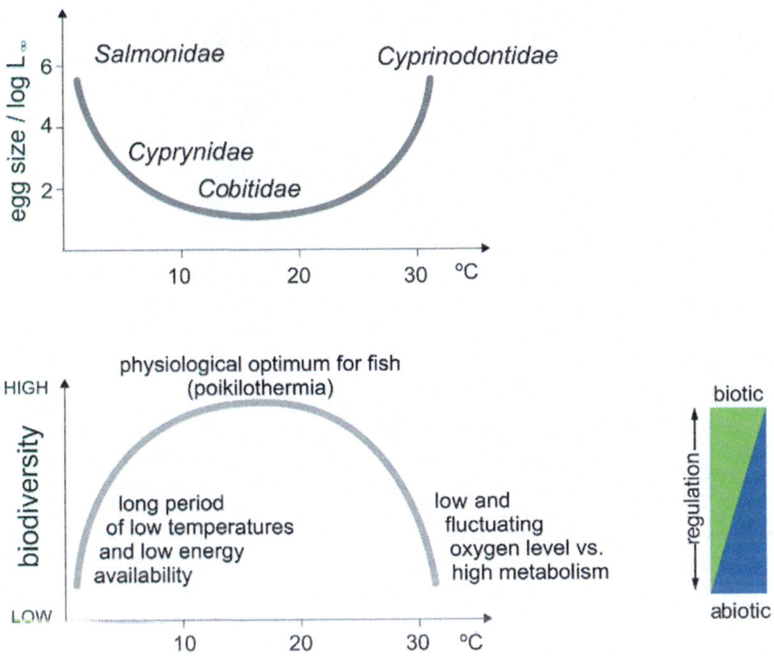

FIGURE 4.3 Adaptation of fish reproduction strategies. Adaptation of fish reproduction strategies to abiotic conditions by increasing the relative egg size. In boreal areas, due to long winters and limited nutrition, Salmonidae produce large eggs in relation to their size and reproduce late summer and autumn. In contrast, in Nevada deserts with temperatures reaching 42°C, *Cyprinodontidae* compensate nighttime oxygen deficits with large egg sizes. *Modified from Zalewski, M., 1997. Fish diversity and ecotonal habitats. In: Lachavanne, J.-B., Juge, R. (Eds.), Biodiversity in land–inland water ecotones. UNESCO, Paris / The Parthenon Publishing Group, London, pp. 183–203.*

Lyons et al., 1990); however, more extensive research on these processes is needed. The biodiversity of fish communities in rivers is shaped by a complex interplay of abiotic factors, such as climate, postglacial colonization patterns, and geological history, and biotic interactions including predation, competition, and mutualism. These interactions promote adaptive responses to fluctuating environmental conditions and have to be considered in the development of Advanced Nature-based Solutions (ANbS).

River ecosystems are characterized by spatial and temporal heterogeneity, which supports both high biodiversity and bioproductivity. However, these dynamics are increasingly subject to anthropogenic modulation, which can alter the balance between abiotic and biotic regulation. In contrast to rivers, lake ecosystems, especially under the top-down biomanipulation model proposed by Northcote, demonstrate strong functional linkages between fish and the lower trophic levels of Lindeman's ecological pyramid (Lindeman, 1942). In such cases, the trophic interactions are so tightly coupled that fish can be deliberately used to manage energy flow and nutrient cycling, improving water quality and reducing eutrophication (Hrbáček et al., 1961; With and Wright, 1984).

According to the abiotic–biotic regulation concept Abiotic Biotic Reguatory Continuum (ABRC), the number of fish species increases along with stream order and follows a gradient from abiotic-dominated systems, such as subarctic or desert rivers, to rivers where abiotic conditions remain within the physiological optimum for aquatic organisms. In these more stable systems, biotic factors play a dominant role in regulating ecosystem dynamics (Fig. 4.3). Tropical rivers exemplify ecosystems where biotic regulation: reproduction and early life startegy, adaptationas as far predation and competition, determines species composition and abundance (Chapman and Lowe-McConnell, 1976).

This observed increase in species richness with stream order is not solely a function of temperature but also reflects greater habitat heterogeneity and ecological stability (Horwitz, 1978; Schlosser, 1982). Where abiotic stress approaches or exceeds the physiological tolerance limits of organisms, only highly adapted species are able to persist. On the contrary, stable and predictable abiotic conditions, when combined with enhanced habitat productivity and diversity, promote biotic regulation and enable finely tuned species assemblages. Under such conditions, ecological processes such as competitive exclusion facilitate niche specialization and allow for the coexistence of multiple species in relatively confined spatial domains (Hutchinson, 1957).

It can therefore be stated that the physical conditions of a given river significantly influence the life strategies adopted by its resident fauna (Zalewski and Naiman, 1985; Kemp et al., 2000). In ecosystems predominantly regulated by abiotic factors, fish tend to produce relatively large eggs in proportion to their body size (Fig. 4.4). In subarctic climates, most fish species spawn in the autumn—likely an adaptive strategy to overcome the harsh winter conditions, characterized by low temperatures, limited food availability, and reduced metabolic activity. Under such constraints, fish would not be able to

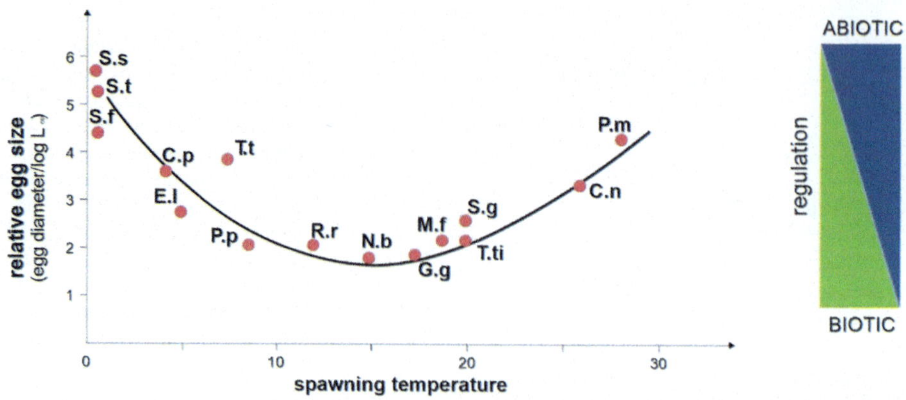

FIGURE 4.4 Spawning temperature as a key factor for fish reproductive strategy adaptation,to abiotic pressure. Reproduction strategies of fish living in different temperatures suggest that egg sizes increase in abiotically regulated ecosystems, both in the boreal zone and in hot deserts where the temperature tolerated by fish reaches 43°C (Naiman, 1980). *S.s.*, *Salmo salar*, salmon; *S.t.*, *Salmo trutta*, brown trout; *S.f.*, *Salmo fontinalis*, brook trout; *T.t.*, *Thymallus thymallus*, grayling; *C.p.*, *Cottus poecilopus*, alpine bullhead; *E.l.*, *Esox lucius*, pike; *P.p.*, *Phoxinus phoxinus*, common minnow; *R.r.*, *Rutilus rutilus*, roach; *N.b.*, *Nemacheillus barbatulus*, stone loach; *G.g.*, *Gobio gobio*, gudgeon; *M.f.*, *Misgurnus fossilis*, weatherfish; *T.ti.*, *Tinca tinca*, tench; *S.g.*, *Silurus glanis*, wels catfish; *C.n.*, *Cyprionodon nevadensis*, Amargosa pupfish (desert streams); *P.m.*, *Poeciliopsis monacha*, headwater livebearer—a small desert streams experiencing sudden changes in the quantity of water, food, and temperature. *On the basis of the data Naiman, R. J., Soltz, D. L. (Eds.), 1981. Fishes in North American deserts. Wiley-Interscience; Gerking, S. D., 1981. Feeding ecology of fish. Academic Press; Szczerbowski, J. A., 1985. Rybactwo jeziorowe i rzeczne. Państwowe Wydawnictwo Rolnicze i Leśne as the lower limit of the optimum temperature for a given species.*

produce eggs of sufficient quantity and size during spring to ensure reproductive success. Autumn spawning allows for hatching in the spring, and the resulting large larvae, equipped with substantial yolk reserves, can survive the critical transition from endogenous to exogenous feeding (Blaxter and Hempel, 1963). These energy reserves provide metabolic support over extended periods (Penczak et al., 1982), and the relatively large size of the fry enables them to access a broader range of food resources within the ecosystem—an advantage explained by the Optimal Foraging Strategy theory (Pyke, 1984). This strategy not only facilitates rapid growth but also supports energy storage needed for the following winter.

In contrast, in desert ecosystems, the production of large eggs is driven by high metabolic rates combined with low oxygen availability. In such environments, the yolk sac serves as a critical energy reserve for sustaining anaerobic glycolysis and the energetically demanding process of osmoregulation (Garrett et al., 1982).

The hierarchy of abiotic and biotic regulatory factors becomes particularly evident when comparing populations of the same species inhabiting distinctly different environmental conditions. For example, bullheads (Cottus gobio) in a warm, productive stream in southern England mature earlier, live shorter lives, and spawn multiple times per season (four to five spawning events), whereas conspecifics inhabiting a strongly acidified, abiotically regulated stream in Scottish moorlands exhibit delayed maturity, slower growth, and a single annual spawning event. This pattern reflects a strategic adjustment to harsh abiotic conditions, where reduced reproductive effort enhances long-term survival probabilities (Mann and Mills, 1979).

Similarly, the pupfish (*Cyprinodon nevadensis*) displays pronounced sexual dimorphism in the upper, groundwater-fed sections of Nevada streams—an environment characterized by stable, biotically regulated conditions. In this context, intraspecific competition for mates and nesting sites drives the evolution of distinct physical traits that enhance reproductive success. However, in the downstream, abiotically regulated sections—where environmental instability increases—these features are absent. Under such conditions, natural selection primarily occurs through environmental stressors (e.g., droughts or temperature extremes), and energy is not allocated to sexual dimorphism (Kodric-Brown, 1981).

ARBC model is a conceptual model describing the regulation of fish assemblages in rivers across various geographical regions, based on the relative dominance of abiotic or biotic factors. Following the hydrological methodology and RCC, the size of a river, expressed by its stream order (Horton, 1945; Kuehne, 1962), primarily determines the stability of hydrological and hydraulic conditions as well as habitat diversity. Water temperature, quantified in terms of cumulative degree-days per year, influences the rate of physiological processes and the energy budgets of individual organisms.

The third key parameter is geomorphology, specifically the river's longitudinal slope (visualized with green and blue dashed lines in the figure), which significantly influences hydrological characteristics—flow velocity, dissolved oxygen concentrations, and substrate composition. These physical parameters, in turn, determine both the energy expenditure

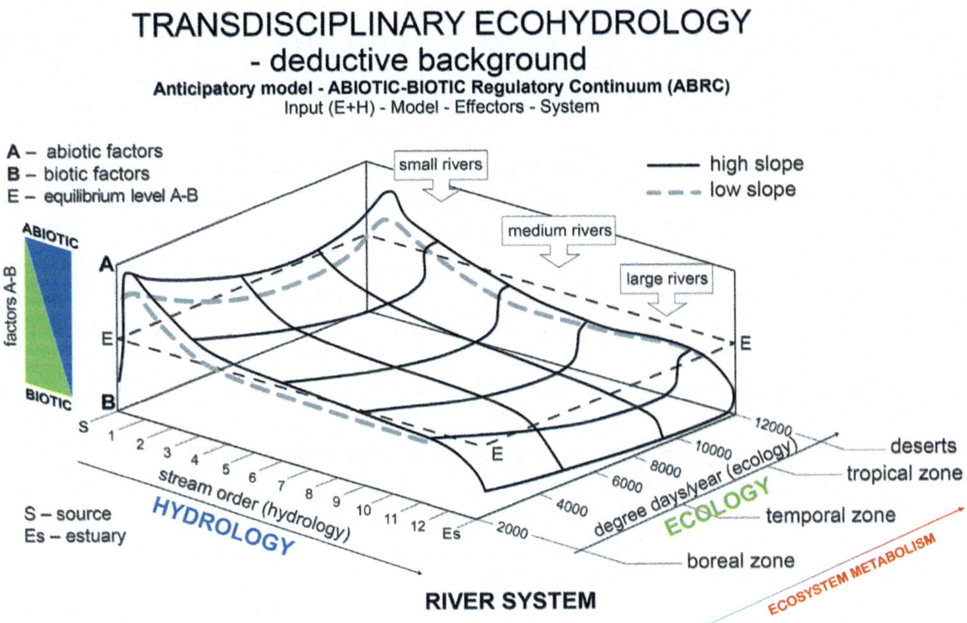

FIGURE 4.5 Deductive background of transdisciplinary ecohydrology. Model of changes in the hierarchy of abiotic and biotic factors along the river continuum and temperature gradient. The stream order determines the hydrodynamics and energy expenditure. The temperature dictates the rate of metabolism, rate of growth, and bioproductivity in the conditions of unrestricted trophy. Bioproductivity and biodiversity of fish communities depend on the relationship between energy intake from food, and energy expenditure described using the above parameters. The ABRC model (abiotic–biotic–regulation-continuum–concept model) carries important management implications. In temperate and terrestrial rivers, where the interplay between abiotic and biotic factors shifts seasonally, it is critical to ensure fish have access to different habitats at key times of the year. During the spawning season, access to floodplains is essential, while in winter, access to oxbow lakes and other low-current areas becomes crucial to reduce energy expenditure. Effective river management, restoration, and enhancement strategies must prioritize maintaining and restoring these habitat connections to support healthy aquatic ecosystems. Another key aspect of the ABRC model is river self purification, which depends on the balance between oxygen availability and the metabolic activity of the stream biocenosis. As water temperature rises, oxygen solubility declines while metabolic rates increase, creating a greater demand for oxygen. In temperate and hot rivers, enhancing channel and valley complexity with structures that increase flow turbulence is essential to boost oxygenation. In colder climates, improving sunlight access to streams raises water temperature, stimulates metabolic activity, enhances algal community productivity, and can triple fish biomass and self-purification capacity. This is fundamental for optimization and selection of appropriate measures, for example, in mountain (high-slope) and/or low-order boreal stream the habitat improvements (river pool sequence, diversified shoreline, large-wooded debris will be much more efficient than increasing number of stock fish). *Modified from Zalewski, M., Naiman, R.J., 1985. The Regulation of Riverine Fish Communities by a Continuum of Abiotic–Biotic Factors Habitat Modification and Freshwater Fisheries, 3–9.*

required for movement and respiration and the potential energy gain from food, i.e., the trophic efficiency (Brett and Groves, 1979; Kleiber, 1961; Grodziński et al., 1975).

Streams characterized by a pronounced longitudinal slope tend to exhibit regulatory dynamics dominated by abiotic factors. In contrast, streams of the same order but with a lower slope gradient often shift toward biotic regulation because of more favorable hydraulic conditions, lower energy expenditure, and greater habitat heterogeneity. Within the eco-hydrological model, the distribution of abiotic and biotic factors in the regulation of fish assemblages is influenced by three primary parameters: stream order, temperature, and longitudinal slope. This relationship can be visualized as a multidimensional surface (or "plane") that begins with low-order streams—such as spring-fed habitats or limnocrenes—where stable abiotic conditions permit dominant biotic control.

The model Fig. 4.5 then exhibits a sharp shift toward abiotic dominance in small streams with high gradient located in boreal or desert regions, where extreme climatic or hydrological stressors prevail. However, a U-shaped depression in the model occurs for similarly ordered and sloped streams located in temperate and tropical zones, where abiotic stress is relatively low, allowing greater influence of biotic interactions. This depression falls below the equilibrium line (E), which represents a balance between abiotic and biotic regulation.

As stream order or cumulative degree-days increase—typically moving downstream—abiotic pressure tends to diminish because of increasing environmental stability, higher biological diversity, and increasing productivity. However, beyond a certain thermal threshold (e.g., in arid or desert ecosystems), rising temperatures again amplify abiotic pressures. In such systems, temperature-driven declines in oxygen solubility coupled with increased mitochondrial oxygen demand result in metabolic stress on all aquatic organisms, algal communities' succession, invertebrates, and fish. This feedback is reflected in a renewed shift of the regulatory model toward abiotic dominance at the highest temperature regimes (Wetzel, 2001; Schlesinger and Bernhardt, 2013) (see Fig. 4.5).

The empirical study confirms that in the upper sections of the river continuum, abiotic factors (temperature, habitat structure, and food availability) primarily regulate fish populations, whereas in lower sections, biotic factors (predation and competition) become dominant (Zalewski et al., 1985). The findings support the ABRC model by demonstrating a dynamic shift from abiotic to biotic control along the river gradient, influencing the survival and growth patterns of brown trout.

However, numerous field observations conducted in ecosystems with similar climatic conditions, hydrological regimes, and regulatory hierarchies indicate that species abundance can vary widely. These variations are largely attributable to the specific characteristics of the aquatic environment. In river systems, biological diversity tends to increase with stream order, as abiotic factors become more stable and organisms are less exposed to environmental stress. Simultaneously, habitat heterogeneity also increases (Horwitz, 1978). Both of these phenomena, reduced abiotic stress and enhanced habitat diversity, are particularly evident in the lower sections of rivers, where extensive ecotone zones, such as those transitioning from riparian forests to meadows, floodplain, and wetlands, expand the available habitat mosaic. It is within these large and diverse inundation zones that the highest levels of biological diversity are often recorded (Lowe-McConnell, 2010; Welcomme, 1979).

Conversely, fish species richness is significantly reduced when river channels are straightened and disconnected from their natural floodplains (Portt et al., 1986; Schiemer and Spindler, 1989). This evidence underscores that the increase in biodiversity in downstream sections is not only a consequence of more favorable abiotic conditions within the river channel but also the result of greater habitat heterogeneity in the valley and floodplain systems.

Another critical determinant of fish species diversity is the geological age of the ecosystem. In general, ecosystems dominated by abiotic regulation tend to be geologically younger than those in which biotic regulation plays a stronger role. For example, subarctic rivers, colonized relatively recently following the retreat of the glaciers approximately 10,000 to 13,000 years ago (Griffiths, 2015), or desert rivers in arid regions exhibit markedly lower species diversity than their older tropical counterparts. Similarly, beta diversity (species turnover across environmental gradients) and gamma diversity (landscape-level diversity across habitats) are typically lower in these younger ecosystems (Cody and Diamond, 1975).

Understanding these interactions is of paramount importance for implementing effective measures to reverse negative outcomes caused by anthropogenic pressure, such as the loss of biodiversity and ecosystem productivity by river renaturalization. Further consequences of disturbances at the catchment scale include the depletion of organic matter in soils and a decline in self-purification capacity (United Nations Environment Programme, 2009) and the pollution and eutrophication of rivers, lakes, dammed reservoirs (Hillbricht-Ilkowska, 1993), and coastal zones (Wolanski and Richmond, 2008; Chícharo et al., 2001; Chícharo and Zalewski, 2012). Addressing these challenges requires transdisciplinary knowledge spanning physics, chemistry, plant and animal physiology, biochemistry, geography, ecology, evolutionary biology, and related fields. This integrative understanding will provide a scientific basis for designing and implementing ANbS, integrated with hydrological engineering and water management strategies, to reduce environmental degradation and promote systemic resilience towards efficient implementation IWRM.

AI disclosure

During the preparation of this work the author(s) used OpenAI's ChatGPT. This document includes content generated and/or edited with the assistance of OpenAI's ChatGPT, used to support language refinement and structural improvements. All final content was reviewed and verified by the author. After using this tool/service, the author(s) reviewed and edited the content as needed and take(s) full responsibility for the content of the publication.

References

Blaxter, J.H.S., Hempel, G., 1963. The influence of egg size on herring larvae (*Clupea harengus* L.). Journal du Conseil International pour l'Exploration de la Mer 28 (2), 211–240. https://doi.org/10.1093/icesjms/28.2.211.

Brett, J.R., Groves, T.D.D., 1979. Physiological energetics. Fish Physiol. 8, 279–352. https://doi.org/10.1016/S1546-5098(08)60029-1.

Chapman, D.W., Lowe-McConnell, R.H., 1976. Fish communities in tropical freshwaters. Copeia 1976 (3), 615. https://doi.org/10.2307/1443396.

Chícharo, L., Zalewski, M., 2012. Introduction to ecohydrology and restoration of estuaries and coastal ecosystems. Treatise on Estuarine and Coastal Science 10, 1–5. https://doi.org/10.1016/B978-0-12-374711-2.01001-9.

Chícharo, M.A., Chícharo, L.M., Galvão, H., Barbosa, A., Marques, M.H., Andrade, J.P., Esteves, E., Miguel, C., Gouveia, I., 2001. Status of the Guadiana estuary (South Portugal) during 1996-1998: an ecohydrological approach. Aquat. Ecosys. Health Manag. 4 (1), 73. https://doi.org/10.1080/146349801753569298.

Chow, V.T., 1959. Open-channel hydraulics. McGraw-Hill.

Cody, M.L., Diamond, J.M. (Eds.), 1975. Ecology and Evolution of Communities. Belknap Press of Harvard University Press.
Cunjak, R.A., Power, G., 1987. The feeding and energetics of stream-resident trout in winter. J. Fish. Biol. 31 (4), 493–511. https://doi.org/10.1111/j.1095-8649.1987.tb05254.x.
Eamus, D., Hatton, T., Cook, P., Colvin, C., 2006. Ecohydrology: Vegetation Function, Water and Resource Management. CSIRO Publishing. https://doi.org/10.1071/9780643094093.
Garrett, G.P., Naiman, R.J., Soltz, D.L., 1982. Fishes in North American deserts. Copeia 1982 (1), 237. https://doi.org/10.2307/1444310.
Gerking, S.D., 1981. Feeding Ecology of Fish. Academic Press.
Griffiths, A.M., 2015. Post-glacial colonization of northern rivers by fishes. Freshw. Biol. 60 (4), 785–803.
Grodziński, W., Klekowski, R.Z., Duncan, A., 1975. Methods for Ecological Bioenergetics. Blackwell Scientific Publications.
Hillbricht-Ilkowska, A., 1993. The dynamics and retention of phosphorus in lentic and lotic patches of two river-lake systems. Hydrobiologia 251 (1–3), 257–268. https://doi.org/10.1007/BF00007185.
Horton, R.E., 1945. Erosional development of streams and their drainage basins; Hydrophysical approach to quantitative morphology. Bull. Geol. Soc. Am. 56 (3), 275–370. https://doi.org/10.1130/0016-7606(1945)56[275:EDOSAT]2.0.CO;2.
Horwitz, R.J., 1978. Temporal variability patterns and the distributional patterns of stream fishes. Ecol. Monogr. 48 (3), 307–321. https://doi.org/10.2307/2937233.
Hrbáček, J., Dvořakova, M., Kořínek, V., Procházkóva, L., 1961. Demonstration of the effect of the fish stock on the species composition of zooplankton and the intensity of metabolism of the whole plankton association. SIL Proc. 14 (1), 192–195. https://doi.org/10.1080/03680770.1959.11899269.
Hutchinson, G.E., 1957. Concluding remarks. Cold Spring Harbor Symp. Quant. Biol. 22 (0), 415–427. https://doi.org/10.1101/SQB.1957.022.01.039.
Junk, W.J., Bayley, P.B., Sparks, R.E., 1989. The flood pulse concept in river–floodplain systems. In: Dodge, D.P. (Ed.), Proceedings of the International Large River Symposium, 106 Canadian Special Publication of Fisheries and Aquatic Sciences No, pp. 110–127.
Kemp, J.L., Harper, D.M., Crosa, G.A., 2000. The habitat-scale ecohydraulics of rivers. Ecol. Eng. 16 (1), 17–29. https://doi.org/10.1016/S0925-8574(00)00073-2.
Kleiber, M., 1961. The Fire of Life: An Introduction to Animal Energetics. Wiley.
Kodric-Brown, A., 1981. Variable Breeding Systems in Pupfishes (Cyprinodon spp.) Adapted to Desert Environments Fishes in North American Deserts. Wiley-Interscience, pp. 317–331.
Kowalik, P.J., Eckersten, H., 1984. Water transfer from soil through plants to the atmosphere in willow energy forest. Ecol. Model. 26 (3–4), 251–284. https://doi.org/10.1016/0304-3800(84)90072-3.
Krebs, C.J., 1972. Ecology: The Experimental Analysis of Distribution and Abundance. Harper and Row.
Kuehne, R.A., 1962. A classification of streams, illustrated by fish distribution in an Eastern Kentucky Creek. Ecology 43 (4), 608–614. https://doi.org/10.2307/1933450.
Lindeman, R.L., 1942. The trophic-dynamic aspect of ecology. Ecology 23 (4), 399–417. https://doi.org/10.2307/1930126.
Lowe-McConnell, R.H., 2010. Ecological Studies in Tropical Fish Communities. Cambridge University Press. https://doi.org/10.1017/CBO9780511721892.
Lyons, J., Karr, J.R., Angermeier, P.L., 1990. Fish communities as indicators of environmental degradation. In: Biological indicators of stress in fish, 8. American Fisheries Society Symposium, pp. 123–144.
Magurran, A.E., 1988. Ecological Diversity and Its Measurement. Princeton University Press.
Mahon, R., 1982. Review of Principles of fishery science. Can. Field Nat. 96 (2), 231–232.
Mann, R.H.K., Mills, C.A., 1979. Demographic aspects of fish fecundity. Proceedings of the Zoological Society of London Symposium Vol. 44, 161–180.
Meadows, D.H., Meadows, D.L., Randers, J., Behrens, W.W., 1972. The Limits to Growth. Universe Books.
Millennium Ecosystem Assessment, 2005. Ecosystems and Human Well-Being: Synthesis. Island Press.
Naiman, R.J., Soltz, D.L. (Eds.), 1981. Fishes in North American Deserts. Wiley-Interscience.
Northcote, T.G., 1988. Fish in the structure and function of freshwater ecosystems: a "Top-down" view. Can. J. Fish. Aquat. Sci. 45 (2), 361–379. https://doi.org/10.1139/f88-044.
Obolewski, K., Glińska-Lewczuk, K., Bąkowska, M., Szymańska, M., Mrozińska, N., 2018. Patterns of phytoplankton composition in coastal lakes differed by connectivity with the Baltic Sea. Sci. Total Environ. 631–632, 951–961. https://doi.org/10.1016/j.scitotenv.2018.03.112.
Penczak, T., Gallicka, W., Molinski, M., Kusto, E., Zalewski, M., 1982. The enrichment of a mesotrophic lake by carbon, phosphorus and nitrogen from the cage aquaculture of rainbow trout, Salmo gairdneri (Glebokie Lake). J. Appl. Ecol. 19 (2), 371–393. https://doi.org/10.2307/2403474.
Pianka, E.R., 1983. Evolutionary Ecology. Harper and Row.
Portt, C.B., Coker, G.A., Minns, C.K., 1986. Fish Habitat Associations in the Great Lakes Watershed: A Review of Existing Data. Fisheries and Oceans Canada.
Pyke, G.H., 1984. Optimal foraging theory: a critical review. Annu. Rev. Ecol. Systemat. 15 (1), 523–575. https://doi.org/10.1146/annurev.es.15.110184.002515.
Rodríguez-Iturbe, I., 2000. Ecohydrology: A hydrologic perspective of climate–soil–vegetation dynamics. Water Resour. Res. 36 (1), 3–9. https://doi.org/10.1029/1999WR900210.
Schiemer, F., Spindler, T., 1989. Endangered fish species of the Danube river in Austria. Regul. Rivers Res. Manag. 4 (4), 397–407. https://doi.org/10.1002/rrr.3450040407.
Schlesinger, W.H., Bernhardt, E.S., 2013. Biogeochemistry: An Analysis of Global Change, 3rd ed. Academic Press.

Schlosser, I.J., 1982. Fish community structure and function along two habitat gradients in a headwater stream. Ecol. Monogr. 52 (4), 395–414. https://doi.org/10.2307/2937352.

Shapiro, J., Wright, D.I., 1984. Lake restoration by biomanipulation: Round Lake, Minnesota, the first two years. Freshw. Biol. 14 (4), 371–383. https://doi.org/10.1111/j.1365-2427.1984.tb00161.x.

Solbrig, O.T., 1991. From Genes to Ecosystems: A Research Agenda for Biodiversity. International Union of Biological Sciences.

Statzner, B., Higler, B., 1986. Stream hydraulics as a major determinant of benthic invertebrate zonation patterns. Freshw. Biol. 16 (1), 127–139. https://doi.org/10.1111/j.1365-2427.1986.tb00954.x.

Szczerbowski, J.A., 1985. Rybactwo jeziorowe i rzeczne. Państwowe Wydawnictwo Rolnicze i Leśne.

Thorpe, J.E., Morgan, R.I.G., Pretswell, D., Higgins, P.J., 1988. Movement rhythms in juvenile Atlantic salmon, *Salmo salar* L. J. Fish. Biol. 33 (6), 931–940. https://doi.org/10.1111/j.1095-8649.1988.tb05542.x.

United Nations Environment Programme. (2009). UNEP Annual Report 2008. UNEP.

Visser, S., 1971. Ecology of the Fungal Populations of *Populus tremuloides* Litter. Doctoral dissertation, University of Alberta, Edmonton, Canada.

von Weizsäcker, Lovins, A.B., Lovins, L.H., 1997. Factor Four: Doubling Wealth, Halving Resource Use. Earthscan.

Wagner, I., Zalewski, M., 2016. Temporal changes in the abiotic/biotic drivers of selfpurification in a temperate river. Ecol. Eng. 94, 275–285. https://doi.org/10.1016/j.ecoleng.2016.05.073.

Wantzen, K.M., Aziz, B., Longuet, I., Bao, I., Bocoum, H., Cissé, L., Chauhan, M., Girard, P., Gopal, B., Kane, A., Marchese, M.R., Prakash, N., Teixeira, P., Zalewski, M., 2016. River Culture: an eco-social approach to mitigate the biological and cultural diversity crisis in riverscapes. Ecohydrol. Hydrobiol. 16 (1), 7–18. https://doi.org/10.1016/j.ecohyd.2015.12.003.

Weber, J.M., Kramer, D.L., 1983. Effects of hypoxia and surface access on growth, mortality, and behavior of juvenile guppies, *Poecilia reticulata*. Can. J. Fish. Aquat. Sci. 40 (10), 1583–1588. https://doi.org/10.1139/f83-183.

Welcomme, R.L., 1979. Fisheries Ecology of Floodplain Rivers. Longman.

Wetzel, R.G., 2001. Limnology: Lake and River Ecosystems, 3rd ed. Academic Press.

Wolanski, E., Richmond, R.H., 2008. Estuary restoration. Encyclopedia of Ecology, Five-Volume Set 1–5, 1422–1427. https://doi.org/10.1016/B978-008045405-4.00055-0.

Zalewski, M., 1986. Regulacja zespołów ryb w rzekach przez kontinuum czynników abiotycznych i biotycznych. Acta Universitatis Lodziensis 31, 3–68.

Zalewski, M., 2014. Ecohydrology and hydrologic engineering: Regulation of hydrology–biota interactions for sustainability. J. Hydrol. Eng. 19 (4), 01814001. https://doi.org/10.1061/(ASCE)HE.1943-5584.0000999.

Zalewski, M., Frankiewicz, P., Brewińska, B., 1985. The factors limiting growth and survival of Brown trout, *Salmo trutta m. fario* L. introduced to different types of streams. J. Fish Biol. 27 (A), 59–73.

Zalewski, M., Naiman, R.J., 1985. The regulation of riverine fish communities by a continuum of abiotic–biotic factors. In: Alabaster, J.S. (Ed.), *Habitat modification and freshwater fisheries*. Butterworths for FAO, pp. 3–9.

Zalewski, M., 1997. Fish diversity and ecotonal habitats. In: Lachavanne, J.-B., Juge, R. (Eds.), Biodiversity in land–inland water ecotones. UNESCO, Paris / The Parthenon Publishing Group, London, pp. 183–203.

Chapter 5

Empirical testing as a framework for optimizing river basin restoration

Maciej Zalewski
European Regional Centre for Ecohydrology of the Polish Academy of Sciences under the auspices of UNESCO, Łódź, Poland

PHOTO CREDIT: M. ZALEWSKI.

1. From big science to the metabolic theory of ecology

The International Biological Program (IBP) (1964–74) marked a transformative period in ecological research by introducing the paradigm of big science, a coordinated, interdisciplinary effort to understand ecosystem structure and function

Transdisciplinary Ecohydrology. https://doi.org/10.1016/B978-0-443-41596-8.00021-2

across spatial and temporal scales. The IBP's emphasis on quantifying energy flow, nutrient cycles, and ecosystem productivity laid the foundation for a systems-based approach to ecology. This era also catalyzed efforts to use as common denominator principles from physics, chemistry, physiology, and molecular biology, enabling researchers to conceptualize ecological processes through the lens of fundamental scientific laws.

Building upon this legacy, Kleiber's law, which describes how metabolic rate scales with body mass to the ¾ power, and the IBP publication extended the understanding of individual physiological processes to the functioning of entire ecosystems (Grodziński et al., 1975). These developments culminated in the Metabolic Theory of Ecology (MTE), a unifying framework linking body size, temperature, and metabolic rate to life-history traits (such as development rates, lifespan, and age at maturity), population dynamics (including carrying capacity, competition, and predation), and ecosystem processes (biomass production, respiration, patterns of energy flow, and nutrient cycling), thereby integrating ecological interactions across multiple scales—from individual organisms to the biosphere.

MTE draws upon Van't Hoff's law and the Van't Hoff–Arrhenius equation, which describe how metabolic rates increase exponentially with temperature. These principles reveal how biochemical activity governs not only individual performance but also broader ecological functions such as primary production, biogeochemical cycling, and trophic dynamics. By incorporating ecological biomechanics, as elaborated by Grodzinski et al., MTE also accounts for how organismal structure and energetics mediate responses to environmental conditions. This theoretical foundation is essential for optimizing advanced nature-based solutions (ANbS), particularly those in water management, where eco-hydrological interventions such as constructed wetlands and biofiltration systems must be tailored to local metabolic and climatic conditions for enhanced cost-effectiveness and functional resilience.

2. Dual regulation: Abiotic and biotic controls in ecosystems

Complementing this energy-centered perspective, the long-standing debate over density-dependent versus density-independent regulation provides further insight into ecosystem dynamics and stability. Density-dependent mechanisms, such as competition, predation, and resource limitation, intensify as population density increases and tends to stabilize populations in relatively stable environments. In contrast, density-independent factors, including temperature fluctuations, floods, and droughts, exert influence regardless of population size and often dominate in unstable or extreme ecosystems. While Krebs laid the foundation for understanding ecological systems (Krebs, 1972) and acknowledged the role of stochastic processes in population dynamics, Grossman et al. (1982) supported the idea of density independent regulation providing empirical evidence that in the North Fork of the Feather River in California, fish populations are regulated by stochastic factors.

The proposed Abiotic–Biotic Regulatory Continuum (ABRC) model, combined with the concept of Dual Regulation (Zalewski et al., 1990a; Zalewski, 2006) and supported by the three principles of Ecohydrology, integrates not only an understanding of hydrological and ecological processes, but also facilitates their translation into Integrated Water Resources Management (IWRM) at the catchment scale. This integration constitutes a key contribution of Transdisciplinary Ecohydrology. In this context, abiotic and biotic drivers are not mutually exclusive but act in concert to shape ecological outcomes. The growing frequency and intensity of climate-induced disturbances, such as extreme heatwaves, floods, and droughts, underscore the relevance of this dual perspective. Such events increasingly threaten to override established biotic controls, disrupt ecosystem services, and drive shifts in species distribution and productivity. Therefore, understanding the spatial and temporal variability of oxygen in aquatic ecosystems—particularly in the context of the inverse relationship between oxygen solubility in water and temperature—is essential for effective river management, especially when planning interventions aimed at enhancing ecological functioning and resilience under changing climatic and land-use conditions.

Nutrient cycling is significantly disrupted in hypoxic environments due to alterations in biogeochemical processes associated with low oxygen conditions. The mortality and migration of bioturbating organisms further impact the biogeochemistry of the water–sediment interface. Under hypoxic conditions, phosphorus release from sediments into the overlying water increases, which increase river trophic status and eutrophication of lakes and reservoirs, while the efficiency of nitrogen removal through denitrification and anaerobic ammonium oxidation (anammox) increase the removal of access of nitrogen from eutrophic ecosystem. These processes can elevate primary productivity in the water column; in case of high phosphorus release from sediments, one can elevate hypoxia through a positive feedback loop. This feedback mechanism, combined with the nonlinear relationship between dissolved oxygen concentrations and denitrification rates, contributes to hysteresis effects, whereby hypoxic conditions persist even after anthropogenic nutrient inputs have been reduced below the levels that initially triggered oxygen depletion (Fovet et al., 2015; Altieri and Gedan, 2019).

Together, the metabolic and regulatory perspectives underscore the importance of systems thinking in environmental science. By bridging fundamental metabolic laws with population dynamics and ecosystem regulation, transdisciplinary ecohydrology emerges as a critical scientific framework, one that supports the design of adaptive, scalable, and efficient Advanced Nature-based Solutions to meet the pressing challenges of the Anthropocene.

3. Energy budgets and bioenergetics in ecohydrology

Understanding how energy and matter flow through organisms, i.e., their energy budget (Duncan, 1989; Grodziński et al., 1975; Krebs and Davies, 1978) is the basis for regulating ecohydrological processes in the catchment area. The functioning of living organisms and their life strategies, as well as the manner in which they use resources and adapt to the changing environment, are dictated by the energy budget—reflecting the uptake and expenditure of energy within a given organism:

$$C = P + R + FU \tag{5.1}$$

where C—consumption, P—production, R—respiration, FU—egesta and excreta.

Consumption refers to the intake of food, but it encompasses more than the mere acquisition of nutrients from the external environment. The energy budget of an organism reflects not only the energy gained from food but also the energy costs associated with its ingestion and digestion. These costs include the mechanical work performed by muscles during feeding, as well as the metabolic processes that both supply energy for this muscular activity and require energy themselves.

In simplified terms, metabolic pathways can be categorized into four major processes: glycolysis, the link reaction, the Krebs cycle (also known as the citric acid or TCA cycle), and the respiratory chain (Fig. 5.1, Box 5.1). The primary substrate for these processes is glucose, a monosaccharide derived from dietary carbohydrates, which is stored in organisms in the form of starch (in plants) or glycogen (in animals). Glucose serves as the fundamental energy source for most organisms.

Adenosine-5′-triphosphate (ATP), an adenine nucleotide, is the principal energy carrier produced in these metabolic pathways. ATP stores energy within the high-energy phosphate bond linking adenosine to its phosphate groups. When one of these phosphate groups is removed, converting ATP to adenosine-5′-diphosphate (ADP), the energy released is harnessed by the cell to power various metabolic processes.

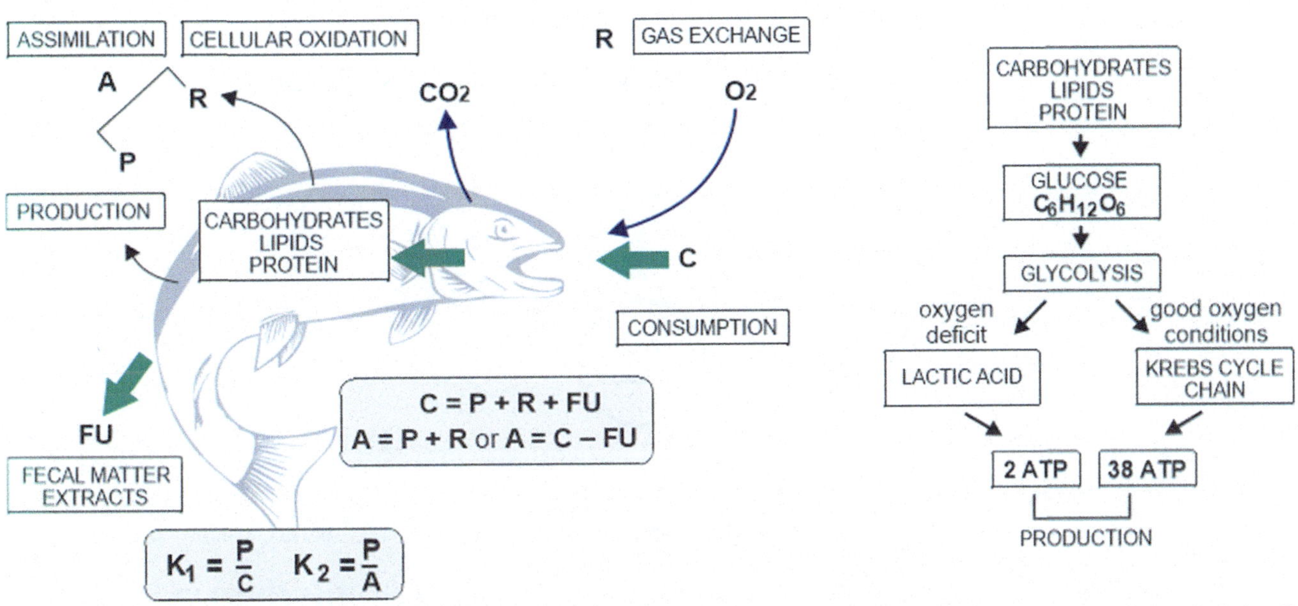

FIGURE 5.1 Energy budget in a fish. The energy budget of a heterotroph, illustrated using the example of a fish: *C*, consumption; *A*, assimilation; *P*, production; *R*, respiration, *FU*, egesta and excreta; *K1*, consumption to production ratio, which defines how much of the food intake is spent on the organism's growth; *K2*, production to assimilation ratio, which defines how much of the food intake is spent on the organism's energy costs. Basic physiological processes determine the flow of energy in an individual and in a population. *Modified from Zalewski, M., 1988. The role of riparian ecotones in river ecosystem functioning. Hydrobiologia 166, 3–11.*

BOX 5.1 The Krebs cycle and respiratory chain

Understanding the fundamental mechanism of the Krebs cycle, which generates energy for living organisms, is essential for interpreting key ecological processes such as ecosystem productivity, river self-purification, ecological succession, and eutrophication.

During glycolysis, a single glucose molecule is broken down into two molecules of pyruvic acid through a series of chemical reactions that initially consume two ATP molecules and generate four ATP and two NADH (nicotinamide adenine dinucleotide) molecules, resulting in a net gain of two ATP molecules. In the subsequent bridging reaction, the two pyruvic acid molecules are converted into two molecules of acetyl-CoA (acetyl coenzyme A), producing an additional two NADH molecules. Acetyl-CoA then enters the Krebs cycle. Two complete turns of the cycle (one for each acetyl-CoA molecule) generate two ATP (or GTP in plants), six NADH, and two $FADH_2$ (flavin adenine dinucleotide) molecules. NADH and $FADH_2$ are then oxidized in the respiratory chain: each NADH yields three ATP, and each $FADH_2$ yields two ATP.

Summarizing the energy yield from all four stages: one glucose molecule produces approximately 8 ATP molecules from glycolysis (2 directly and 6 via NADH), 6 ATP from the bridging reaction (via NADH), and 24 ATP from the Krebs cycle and oxidative phosphorylation (2 from GTP, 18 from NADH, and 4 from $FADH_2$), resulting in a total yield of 38 ATP molecules (see Fig. 5.1).

The initiation of the Krebs cycle and the electron transport chain requires the presence of oxygen. In oxygen-deficient aquatic environments—such as eutrophic lakes or heavily polluted rivers—only anaerobic glycolysis occurs. In this case, pyruvic acid is reduced to lactic acid, and the energy yield is significantly lower, providing the organism with only 8 ATP molecules per glucose molecule (Fig. 5.2).

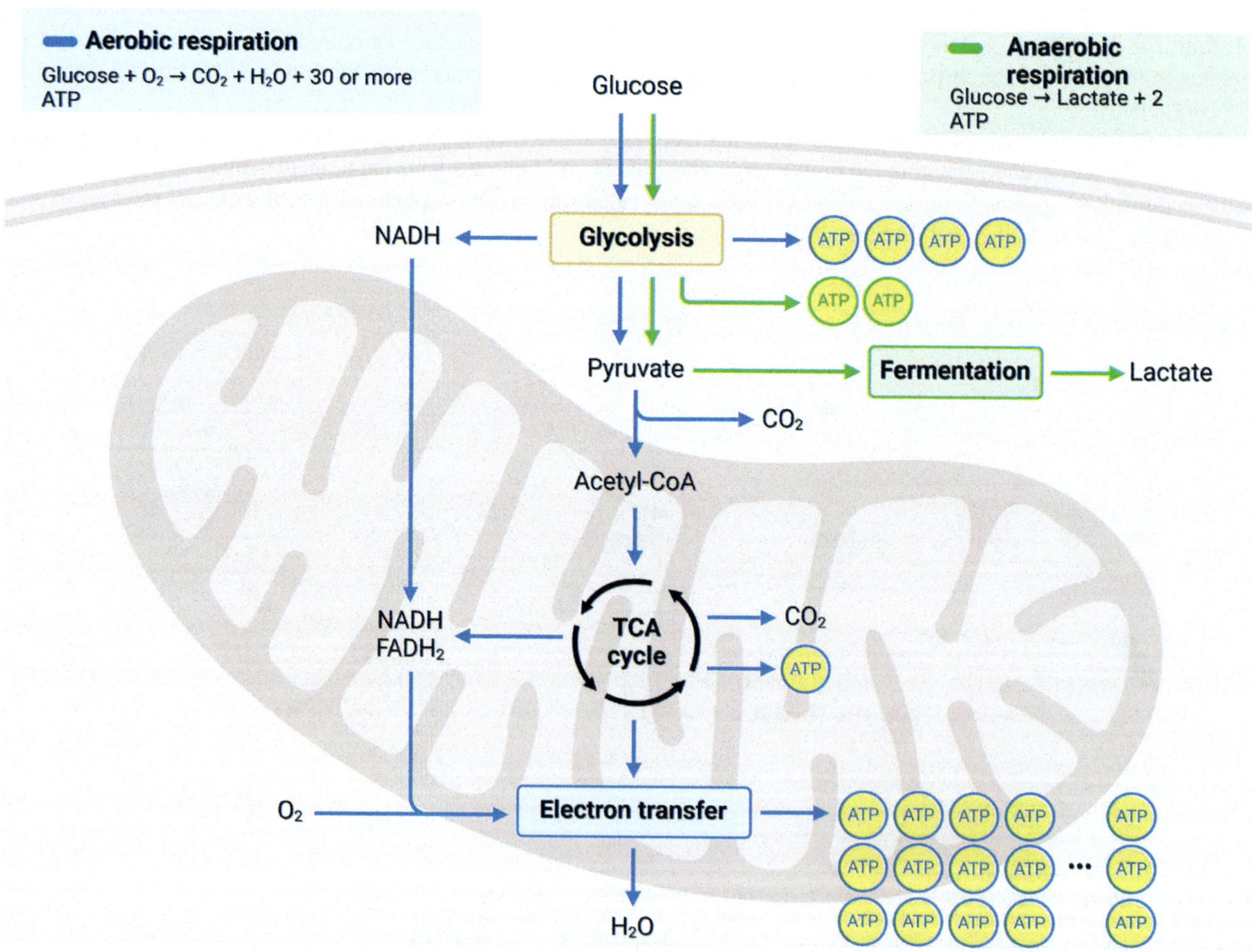

FIGURE 5.2 **Cellular respiration pathways.** The two pathways of cellular respiration are aerobic respiration (oxygen is required) and anaerobic respiration (oxygen is not required). *Source: Prof. Janusz Błasiak, created in https://BioRender.com.*

This simplified metabolic pathway illustrates how energy production (P), in the form of ATP molecules, must be balanced with respiration, i.e., the use of ATP, for metabolic processes to proceed efficiently. A positive energy balance indicates the quantity of assimilated energy (A), whereas a negative balance signifies that the organism is expending more energy than it obtains through consumption. In such cases, the organism typically draws on internal energy reserves, such as stored fat tissue, to meet its metabolic demands.

The final component of the energy budget consists of rejecta, which includes both egesta and excreta (FU). Egesta refers to substances ingested with food that are either undigested or only partially digested and subsequently expelled from the body. In contrast, excreta are substances secreted by cells and glands that serve specific physiological functions; unlike egesta, they are not purposeless metabolic by-products. Examples of excreta include sweat, which cools the body; tears, which lubricate the eyes; and gastric juices, which aid in digestion.

The energy budget of an organism varies according to environmental conditions, particularly temperature, which influences the rate of chemical and physiological processes, including those involved in food consumption and energy production. This variation is especially pronounced in ectothermic organisms such as fish. For example, at 10°C, the pumpkinseed (*Lepomis gibbosus*), a thermophilic freshwater species, can consume food equivalent to 1.2% of its body mass per day, allocating approximately one-third of this intake (0.4% of body mass) to somatic growth. At 30°C, with an increase in metabolic activity, the fish can ingest food equal to 9% of its body mass, with 3% allocated to tissue development. However, as the ambient temperature approaches the species' upper thermal limit (36°C), food intake decreases substantially, to only 4% of body mass, due to reduced oxygen availability. At this point, the energy obtained is insufficient to meet metabolic demands, and the fish begin to lose weight, at a rate of approximately 1% of body mass per day (Fig. 5.3). Understanding these bioenergetic dynamics is essential for optimizing aquaculture practices.

Such a state of affairs is caused by the pace of food digestion, which depends on the activity of the digestive enzymes dictating how fast the food passes through the digestive system; their activity in turn increases as the temperature rises.

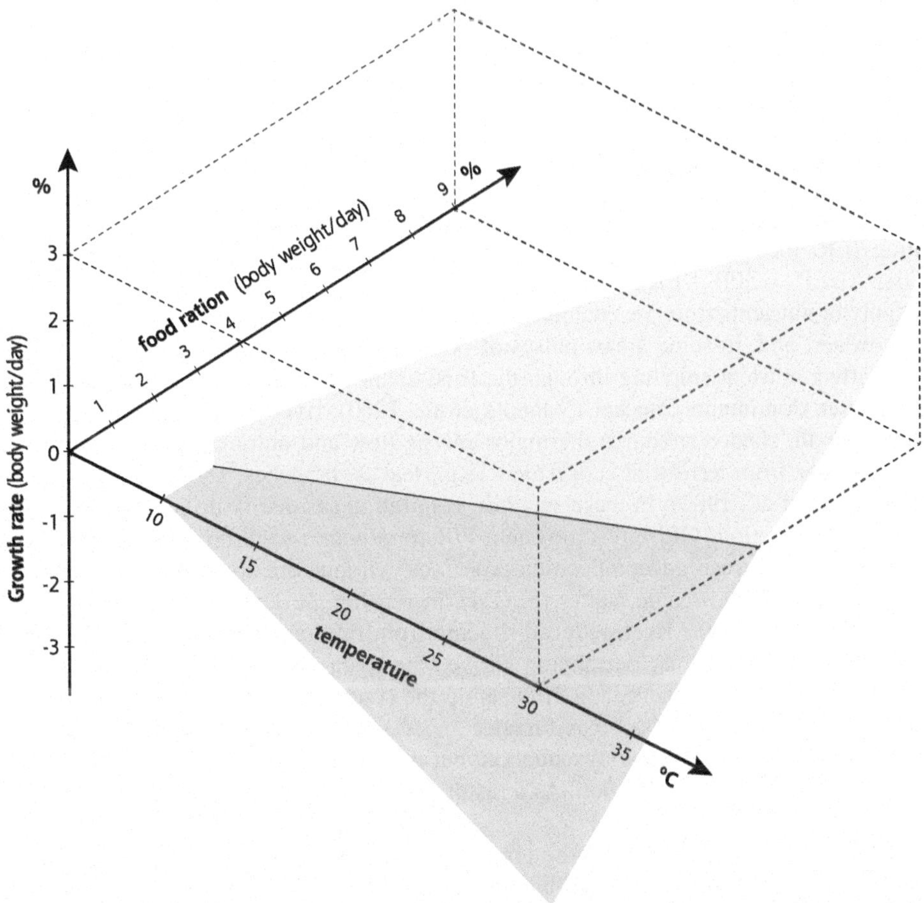

FIGURE 5.3 The energy budget. The energy budget, as illustrated based on the example of a thermophilic fish—as a manifestation of the dependency of the food assimilation on temperature (expressed as % of body growth per day). The shaded area shows the range of temperatures and food quantity, which is insufficient to ensure growth. The model can serve not only as a reference framework for aquaculture optimization but also as a tool to describe patterns of river self-purification and to help prevent excessive eutrophication in lakes and dammed reservoirs. *Adapted from Mitchell, R.E., Kirby A.M. Jr., 1976, Performance characteristics of pond aeration devices. Proc. Annu. Meet. World Maric. Soc. 7, 561–581.*

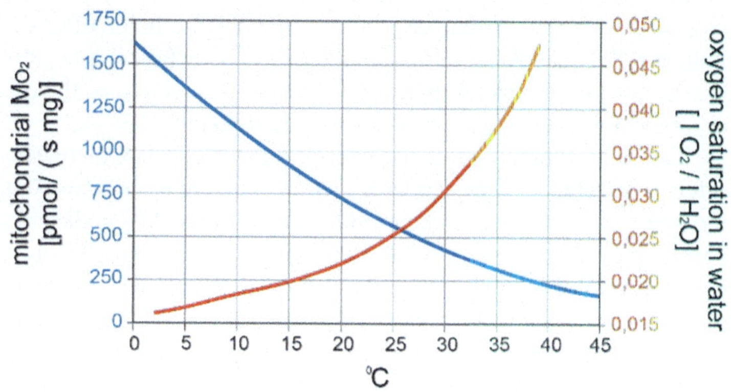

FIGURE 5.4 Oxygen solubility versus increased oxygen demand in mitochondrial processes. The significance of decreased oxygen solubility in water versus increased oxygen demand in mitochondrial processes along the gradient of temperatures for the functioning of aquatic organisms. As temperature rises, oxygen solubility decreases (blue line), while metabolic oxygen demand increases (red line). These insights are essential for interpreting long-term water quality monitoring data and for guiding the reduction of pollutants and river hydraulics optimization (renaturalization), especially in light of climate change, which alters temperature regimes and, consequently, oxygen dynamics in freshwater ecosystems. *Modified from Zalewski, M. (Ed.). (2020). Ekohydrologia. Warszawa: Wydawnictwo Naukowe PWN. ISBN 978-83-01-20804-2.*

The law formulated by the Dutch chemist Jacobus H. Van't Hoff describes the changes in the rate of chemical reactions depending on temperature. Under his equation, a temperature increase of 10°C increases the rate of chemical reactions from twofold to fourfold. At the same time, as the temperature increases the amount of oxygen dissolved in the water decreases, which augments the demand for oxygen (needed for glucose oxidation), whose content in the water is simultaneously decreasing, meaning that it is less accessible for aquatic organisms. The equilibrium between the amount of oxygen dissolved in water at a given temperature and the given organism's demand for oxygen at that temperature is the main factor, which determines the thermal tolerance of fish (Fig. 5.4).

To assess the bioenergetic efficiency of an organism, two indicators are used: K1 (representing the production to consumption ratio, P/C) and K2 (representing the production to assimilation ratio, P/A). These indicators can assume a wide spectrum of values. For instance, among omnivores, depending on the food, an Amur carp (*Cyprinus rubrofuscus*) can attain the bioenergetic efficiency of $K1 = 2.2\%$, and $K2 = 14\%$ (Fisher and Lavoy, 1972), in the case of plant food consumption, and, respectively, $K1 = 12.5\%$ and $K2 = 40\%$ in the case of animal food.

4. Understanding river metabolism

River metabolism is a function of energy generated during photosynthesis, gross primary productivity (GPP), and ecosystem respiration (ER, the use of a portion of GPP for plant life-supporting functions). It is expressed as net ecosystem production (NEP = GPP − ER) (Rodríguez-Castillo et al., 2019). However, river metabolism is greatly influenced by the supply of nutrients from the catchment via land–water ecotones, allochthonous organic matter (CPOM), surface fluxes, groundwater, and, in some areas, pulses of pollutants. These factors, along with the complexity of river channels and valleys, affect nutrient spiraling through the food chain and, in turn, river self-purification.

According to the River Continuum Concept (Vannote et al., 1980), river metabolism changes along the river continuum. In headwaters (small, shaded streams), the major energy flow and nutrient cycling pathways are dominated by allochthonous organic matter from terrestrial ecosystems (e.g., leaves, branches, and insects, coarse particulate organic matter, or CPOM) (Wallace et al., 1982). In these sections, respiration (primarily driven by microbial decomposition of CPOM) exceeds primary production (GPP), resulting in a P/R (production-to-respiration) ratio of less than 1.

As the river progresses downstream, decomposition processes continue due to nutrient spiraling. CPOM is converted (by food chain) into fine particulate organic matter (FPOM), increasing nutrient concentrations. With better light availability due to the increased width of the river, reduced shading from riparian ecotones, shallower depth, and good water transparency, primary producers such as periphyton thrive, increasing biomass and productivity on the substrate. Consequently, the P/R ratio becomes greater than 1, meaning the river section is autotrophic.

Further downstream, where the river becomes broader and deeper, it collects significant amounts of FPOM from upstream. Reduced transparency limits primary production, but the steady supply of FPOM and CPOM supports bottom sediments and filter feeders, increasing respiration. As a result, the river becomes heterotrophic again, with a P/R ratio of less than 1.

This sequence of processes is determined by the abiotic characteristics of the watershed, such as climate and geomorphology, and biotic watershed processes, including plant cover, land–water ecotones, biotic interactions (e.g., primary production and invertebrate communities), and bottom algal growth. Flow patterns and human impacts, such as dams, habitat degradation, and point and nonpoint source pollution, also play a significant role. These factors influence nutrient spiraling intensity along the river continuum, affecting water quality, bioproductivity, biodiversity, ecosystem services, and resilience to human impacts and climate change.

Understanding river metabolism where daily changes in oxygen's concentration are a fundamental indicator of ecosystem's functioning and a background for efficient implementation of ecohydrological nature-based solutions for enhancement river system sustainability potential WBSR. Also, it should be considered as the synthetic indicator of the progress achieved by the implementation of transdisciplinary ecohydrological nature-based solutions and their integration with the hydrotechnical infrastructure into systemic solutions in the catchment scale.

The integration of Advanced Nature-based Solutions (ANbS) with hydroengineering remains a key challenge, as hydrological research has traditionally focused on technical problem-solving (Blöschl et al., 2013) on various aspects of fluid dynamics modeling ecohydraulics which is fundamental for providing water for development and applying renaturalization. However, they did not take into consideration that the catchment and its ecosystems should be considered as a "Platonian superorganism" (Zalewski, 2000), which by water–biota feedback determines bioproductivity, biodiversity, ecosystem services, and resilience. Moreover, terrestrial and freshwater ecosystems play an important role in water retaining, groundwater recharging, and filtering in the catchment, also mitigating the effect of plethora of climate changes—by reducing stochastic character of climatic and hydrological processes.

5. The flow of energy and nutrients through the ecosystem

From the bioenergetic perspective, ecological processes in an ecosystem can also be described by an energy budget equation, comprising the amount of energy introduced into the ecosystem per unit of surface area and the efficiency of transforming that energy from one form into another. The solar energy introduced into an ecosystem is only marginally harnessed (95%–99% of solar energy reaching plants is lost as heat or vaporization or is deflected). Only the remaining 1%–5% is used in photosynthesis, which is the basic primary production process, defined as the quantity of chemical energy generated by primary producers (autotrophs) per unit of surface area and per unit of time. However, the energy generated by primary producers is not wholly transferred to higher trophic levels (consumers) because autotrophs need to allocate some of the generated energy to sustain their life processes (respiration, photosynthesis), which also require energy input. Therefore, the energy that is potentially available for higher trophic levels is expressed by the following formula:

$$\text{Pbrutto} - \text{R} = \text{Pnetto} \tag{5.2}$$

where Ptotal—total primary production, R—respiration (energy used for plant breathing and sustaining life processes), Pnet—energy available for higher trophic levels.

Energy losses also occur at higher trophic levels since, as explained before, heterotrophs assimilate 10%–50% of the energy from the food they eat. Such energy losses in the food chain were observed back in the 1940s by Raymond Lindeman, who formulated the so-called "10% Rule," stipulating that, if the costs of metabolic processes and of obtaining food from the environment are factored in, only about 10% of a living organism's energy can be conveyed up to the next higher trophic level (Lindeman, 1942). Considering the fact that trophic chains are estimated to consist of three to six levels, we can imagine how little energy from the total energy pool entering a particular ecosystem gets transferred upward through the trophic network. This also explains why the lowest levels of the trophic pyramids are the most numerous and have the greatest biomass. The volume of primary production determines the number of trophic levels, the biomass of apex predators, and to some extent the biodiversity level in the ecosystem. At the same time, the process highlights the major role that complexity and biodiversity play in the trophic chains for the flow of energy through the ecosystem, the circulation of chemical elements, and their impact on, for instance, river self-purification processes.

One indicator to describe ecological processes other than the energy budget is the circulation (cycle, rotation) and retention of biogenic elements in the ecosystem. Biogenic substances or nutrients refer to those chemical elements, which are needed for the functioning and growth of living organisms. The basic biogenic substances include carbon, phosphorus, nitrogen, and sulfur (C, P, N, S). According to Liebig's Law of the Minimum, biogenic availability is one of the factors that curtail the development of organisms in the natural environment. The law states that the resource (element) that is the least abundant in relation to what an organism needs is the factor that limits the development of that organism or of the entire population. In other words, Liebig's minimum means that a population will only develop to the level allowed by that resource which is the least abundant in the environment (carrying capacity).

The total quantity of a chemical element in a given ecosystem is termed a pool. For instance, the total phosphorus pool in a lake ecosystem is made up of the phosphorus content in the compounds dissolved in the water and present in the phytoplankton, zooplankton, fish, macrophytes, matter suspended in the pelagic zone, and in bottom sediments. The pool of a given element can be more readily or less readily accessible. The readily accessible pool refers to forms of a given element, which can easily be included into the circulation cycle of a given biogenic substance, e.g., sulfur in the form of hydrogen sulfide. On the other hand, the less accessible form for sulfur is iron sulfide (II) FeS, which is accumulated in bottom deposits and cannot be easily incorporated into the biological circulation cycle. Biogenic circulation or rotation refers to the cyclical movement of biogenic substances in an ecosystem. For instance, in a lake ecosystem, the biogenic pool contained in microphytes (such as algae) undergoes the most dynamic changes (fastest rotation), while it is the most stable pool in bottom sediments, where its recirculation is the slowest by accumulated release; however, it can be accelerated by bottom feeding fish such as carp (bioturbation). In turn, biogenic retention refers to the amount of time a given element is retained in the individual sections of its circulation cycle.

The nutrient cycle and energy flow through the ecosystem are prerequisites for understanding how an ecosystem works. In natural ecosystems with substantial biodiversity levels, there are many different pathways for the flow of energy and biogenic substances, which can be summarized as two main types. The first is a system in which the stream of energy and matter flows from primary producers through successive trophic levels of consumers—secondary producers. This is a state that occurs predominantly in water ecosystems. The second pathway type is utterly different since in this particular case both matter and energy flow directly from primary producers to detritivores. Such a state is typical of forest ecosystems or hypertrophic (overfertilized) lakes.

6. The river continuum concept—process-oriented thinking

The regulatory dynamics of fish communities in riverine ecosystems reflect the continuous interplay between abiotic and biotic factors, varying spatially and temporally along the river course. Early ecological studies in the 20th century, pioneered by Shelford (1929), described the zonal distribution of fish species from headwaters to river mouths, introducing the concept of fish "zonation" (Huet, 1954). This river zonation framework was later used to analyze allochthonous and autochthonous matter in the trophic dynamics of rivers and fish communities, and for quantification purposes (Penczak and Zalewski, 1974; Penczak et al., 1976), across different types of rivers (Penczak, 1992).

A paradigm shifts in river ecology occurred in the 1980s with the emergence of the River Continuum Concept (RCC), which redefined rivers as dynamic systems where ecosystem functioning with changes in the composition of invertebrate communities serves as the basis for ecological understanding (Vannote et al., 1980). The RCC conceptualizes rivers as longitudinal continua of gradual changes in abiotic conditions and biological communities. It emphasizes the flow and transformation of organic matter, primarily allochthonous ("from outside") inputs from riparian vegetation in the upper reaches and increasing autochthonous ("self-produced") production downstream where $P > R$ (production exceeds respiration), across successive stream orders. In large rivers, the system becomes dominated by fine particulate organic matter (FPOM), where sedimentation prevails and $P < R$ (respiration exceeds production). All of this transformation occurs through nutrient spiraling (Webster and Patten, 1979; Newbold et al., 1981).

The Nutrient Spiraling Concept (NSC) (Elwood et al., 1981) refines the understanding of biogeochemical cycling in lotic systems. Unlike circular nutrient cycles in closed systems, rivers exhibit spiral-shaped nutrient dynamics due to downstream transport. The length of the spiral reflects the efficiency of nutrient transfer to higher trophic levels (Newbold et al., 1981). Systems with long spirals and numerous intermediate consumers, such as protozoa and invertebrates, tend to be less efficient in transferring energy to fish, accompanied by intensified self-purification processes. In contrast, rivers with short food chains, where fish feed directly on primary producers such as periphyton, display high levels of secondary production and overall ecosystem productivity (Naiman, 1976).

In headwater streams (first- to second order [Box 5.2]), limited sunlight and dense riparian canopy result in highly heterotrophic conditions, where primary production is low and allochthonous coarse particulate organic matter (CPOM) dominates (Fig. 5.5). Invertebrate assemblages are rich in shredders, such as insect larvae with specialized chewing mouthparts, while fish communities include insectivores and benthic rheophilic species such as trout and bullhead. As the river widens in mid-order sections (third- to fourth-order), riparian shading decreases, light availability improves, and periphyton production becomes

BOX 5.2 The Strahler stream order system

The Strahler stream order system is a hierarchical classification used to define river networks. In this system, headwater streams (first order) are those with no tributaries, while a second-order stream forms when two first-order streams converge. Higher-order streams emerge as lower-order streams join, progressively increasing in complexity and discharge capacity. This classification is fundamental in understanding riverine ecology, as hydrological and ecological processes change with increasing stream order.

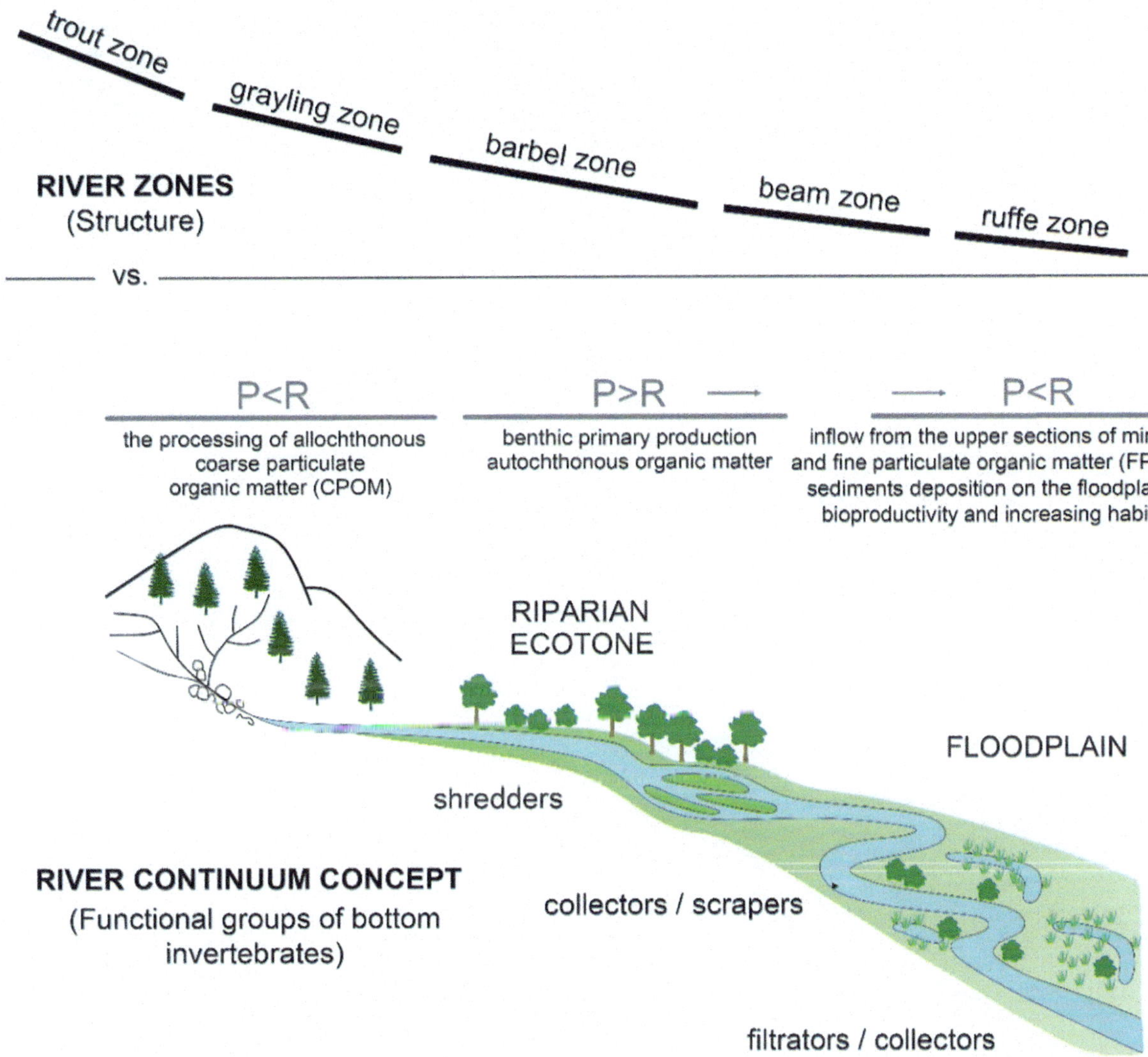

FIGURE 5.5 **Fish zones versus River Continuum Concepts.** Comparison of the structural concept of fish zones (Huet, 1949, 1954) versus the functional RCC, where the river's metabolic processes (P/R) are associated with invertebrate functional groups. *From Zalewski, M., Thorpe, J.E., Naiman, R.J., 2001. Fish and riparian ecotones—a hypothesis. Ecohydrol. Hydrobiol. 1 (1–2), 11–24.*

significant. Here, scrapers feeding on biofilms dominate the invertebrate community, and fish diversity increases, including predators. These reaches are increasingly autotrophic, with production-to-respiration (P/R) ratios exceeding one.

In high-order river sections ([Box 5.2] and above), depth and turbidity reduce light penetration, limiting primary production and shifting the system back to heterotrophy. Fine particulate organic matter (FPOM), transported from upstream and riparian inputs, sustains collector and filter-feeder invertebrates. Fish communities are dominated by benthivorous species such as bream and phytophilous species reliant on aquatic vegetation for reproduction.

The functional framework of the RCC has been used as a reference point by the Abiotic–Biotic Regulation Coninuum (ABRC) Concept, which posits that the relative influence of abiotic and biotic factors shifts along the river continuum. Long-term comparative studies in lowland and upland catchments demonstrate that the influence of abiotic factors, such as hydrology, temperature, and substrate, declines downstream, while biotic interactions, including competition and predation, become unceasingly important in structuring communities (Zalewski and Naiman, 1985; Zalewski, 1986).

However, neither the River Continuum Concept (RCC) nor the Abiotic–Biotic Regulation Concept (ABRC) explicitly accounted for habitat composition geological influences, despite the significant role of bedrock composition in regulating river trophic status. In particular, calcium ions (Ca^{2+}) are critical for enhancing bioproductivity and self-purification processes, as well as influencing sediment transport, river morphology, and overall ecosystem dynamics. The

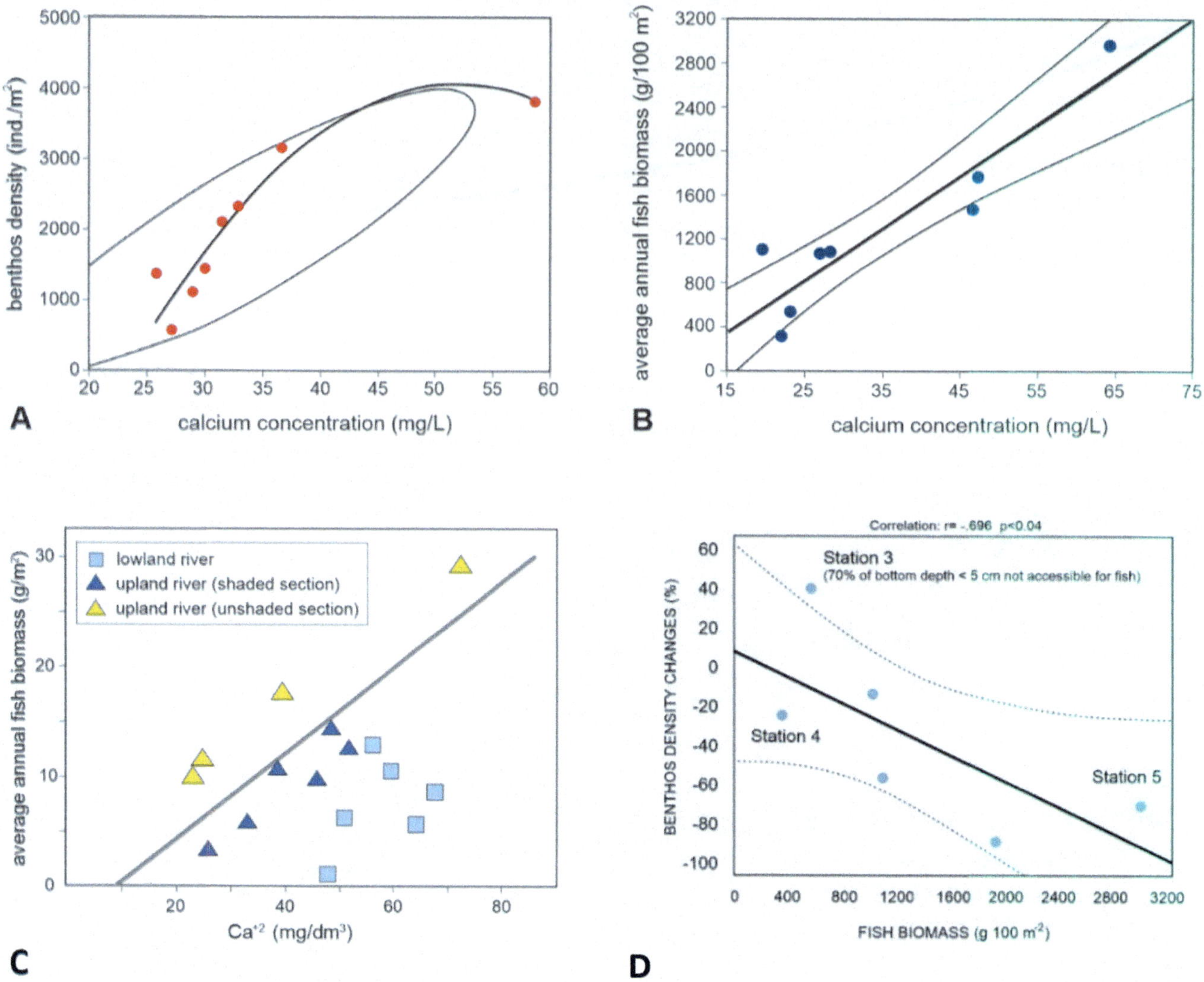

FIGURE 5.6 Discovering the ecosystem processes as management tools. (A and B) Importance of calcium concentration in upland river water as a factor increasing the density of benthos and the biomass of fish. The above values are proportional to the intensity of the river self-purification process, which underlines the role of calcium in development and implementation of geochemical barriers and the sequential sedimentation biofiltration system. (C) The influence of the shading of the watercourse on the average annual biomass of fish correlated with the concentration of calcium in the upland river, where the primary producer is periphyton, and in lowland, where there are large amounts of macrophytes. The benthos and fish are an indirect indicator of intensity of energy flow, and nutrient spiraling thus should be considered equivalent with the potential for self-purification processes. (D) The top-down effect in streams—the correlation between annual fish biomass (g 100 m^{-2}) and summer decline of benthos density (% change in numbers of individuals m^{-2}) (dashed lines show 95% confidence limits). This this fish/ invertebrate/periphyton top-down effect might seriously influence river selfpurification potential and need furter scientific investigation important for optimal ehancement stream ecosystem resilience and renaturisation. *Data from Zalewski, M., 1988. The role of riparian ecotones in river ecosystem functioning. Hydrobiologia 166, 3–11; Zalewski, M., Frankiewicz, P., Przybylski, M., Bańbura, J., Nowak, M., 1990b. Structure and dynamics of fish communities in termperate rivers in relation to the abiotic-biotic regulatory continuum concept. Polskie Archiwum Hydrobiologii 37 (1–2), 151–176.*

importance of Ca^{2+} became evident during field testing of the ABRC model as an emergent property of the system, see Fig. 5.6 (Boxes 5.3 and 5.4).

7. The role of flow in riverine systems in enhancement WBSR

A river in its natural state is an ecosystem that operates within a regime of intermediate disturbances that serve as the most effective drivers of biodiversity and bioproductivity, characterized by high regenerative potential. Why are bioproductivity and biocomplexity prioritized? Because human activities (e.g., deforestation, homogenization of agricultural landscapes, urbanization, channeling of rivers) and climate change amplify stochastic hydrological processes such as

BOX 5.3 Testing the abiotic–biotic regulation model: Influence of temperature on river metabolism

The processes of river enrichment with organic matter and biogenic elements from the catchment, along with the river's natural capacity for self-purification along the continuum, play a critical role in regulating phosphorus (P) concentrations and suspended matter (SM) transfer along the river continuum and eutrophication lakes, reservoirs and ecotone zones. These factors directly influence water quality in downstream reservoirs and estuarine environments. In a study on the ecohydrology of the Pilica River, Wagner and Zalewski (2016) demonstrated, through testing of the ABRC model, that the hierarchy of environmental drivers affecting various forms of phosphorus and suspended matter concentrations is subject to marked seasonal variation (Wagner and Zalewski, 2016).

During the summer, phosphorus concentrations tend to be negatively correlated with temperature, indicating strong dependence on intensive biological processes. Higher temperatures accelerate metabolic activity in aquatic organisms, enhance plant assimilation of nutrients, and intensify the spiraling of organic matter. In contrast, during the winter season, phosphorus and suspended matter concentrations are less influenced by temperature and are primarily driven by precipitation levels and river discharge.

This pattern can be attributed to drastically diminished biomass and low metabolic activity of stream channel vegetation and aquatic organisms in winter, resulting in reduced nutrient uptake and limited biogeochemical cycling. Additionally, in winter, higher nutrient export from land to water occurs, while the capacity for in-river self-purification is significantly weakened. These findings highlight the importance of hydrological factors, particularly during colder months, in governing nutrient and sediment transport.

The clear linkage between winter hydrodynamics and phosphorus/sediment flux underscores the necessity of catchment-scale planning, especially in agricultural landscapes. Land-use strategies should aim to minimize soil erosion and nutrient loss during winter, reducing phosphorus accumulation in reservoirs that could be reactivated in summer under low-oxygen conditions (Zalewski, 2013). Also, the necessity of urban stormwater purification using SSBS has also been recognized.

These processes are of critical importance for designing water management strategies in river valleys, particularly under climate change scenarios characterized by rising temperatures and more frequent extreme precipitation events. Understanding the seasonal variability in the dominance of biotic and abiotic controls is essential for implementing appropriate ecohydrological biotechnologies and systemic interventions to limit external nutrient loads and their downstream transfer. Such approaches are key to mitigating eutrophication and harmful algal blooms (see Part II of the book).

BOX 5.4 Hydrological seasonality, phosphorus removal, and organic matter in the Pilica river: Empirical evidence for the ABRC model

Discharge of the Pilica River over a 4-Year Observation Period and Statistical Analysis of Low (LQ) and High (HQ) Flow Conditions During Warm (Summer) and Cold (Winter) Months (Fig. 5.7)

Empirical validation of the Abiotic–Biotic Regulatory Continuum (ABRC) Concept model confirmed the dominant role of river metabolism, in nutrient spiraling and self-purification. In winter, low temperature reduced role of biocenosis in nutrient spiraling, and hydraulic processes prevail. In summer, natural channel and rich riparian vegetation and floodplain connectivity determines high ecosystem metabolism, which is resulting strong correlation between temperature and PO_4 concentration (Wagner and Zalewski, 2016).

floods and droughts, degrading habitats (biocomplexity) and reducing soil fertility. Together, these factors reduce the bioproductivity of drainage basins, leading to a reduction of resilience to climate change, decline in biodiversity and ecosystem services. Despite these challenges, the abiotic template, shaped by hydrology and governed by physical laws, creates a highly dynamic system, with each catchment uniquely defined by its climatic conditions (precipitation and temperature patterns), geomorphology and anthropogenic influences. This dynamic interplay between hydrology and biota is thus a critical element in the evolution of river systems influencing the dual regulation processess. The ecological structure and dynamics within the abiotic framework integrate with complex hydrological processes. A thorough understanding of river metabolism in the context of hydrology is essential to understand key processes such as nutrient spiraling, energy flow, river productivity, biodiversity enhancement, and self-purification. These processes are crucial tools for enhancement of water resources, bioproductivity, and biodiversity and strengthening catchment resilience.

Flood pulses, though often viewed as a natural hazard due to its impacts on human settlements, plays a vital ecological role in sustaining riverine and floodplain ecosystems (Junk et al., 1989). As a key element of the seasonal hydrological

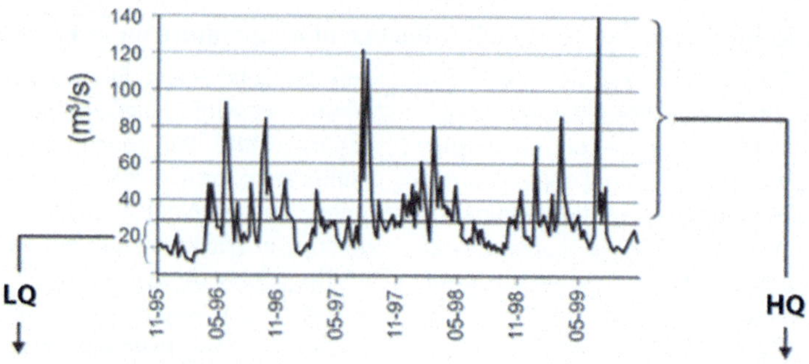

WINTER

LQ — COLD PERIOD (XI-IV)

	PPO4	DP	TP	TSM	MSM	OSM
ΔQ	0,436	0,540	0,425	0,628	0,581	0,478
Q_1	0,406	0,622		0,648	0,482	0,685
Q_3		0,576		0,580	0,412	0,644
Q_7		0,477		0,476		0,574
Q_30		0,496				0,496
T_1						0,401
T_3						0,448
T_7						0,429
P_1						
P_3						
P_7						
PS	-0,445	-0,498		-0,670	-0,447	-0,785

HQ — COLD PERIOD (XI-IV)

	PPO4	DP	TP	TSM	MSM	OSM	
ΔQ	0,355	0,442	0,459	0,488	0,500		hydrological factors
Q_1						-0,360	
Q_3						-0,454	
Q_7			-0,299	-0,617	-0,490	-0,609	
Q_30	-0,456	-0,573	-0,522	-0,770	-0,732	-0,425	
T_1							temperature
T_3							
T_7							
P_1						-0,371	precipitation
P_3	0,294	0,374	0,506	0,524	0,517		
P_7							
PS							

SUMMER

LQ — HOT PERIOD (V-X)

	PPO4	DP	TP	TSM	MSM	OSM
ΔQ						
Q_1				0,488	0,343	0,574
Q_3				0,453		0,575
Q_7	0,316			0,440		0,487
Q_30	0,406	0,399		0,369		0,404
T_1		0,538	0,359	0,377		
T_3		0,423		0,349		
T_7		0,439		0,366	0,347	
P_1						
P_3						
P_7						
PS						

HQ — HOT PERIOD (V-X)

	PPO4	DP	TP	TSM	MSM	OSM	
ΔQ				0,691	0,528	0,551	hydrological factors
Q_1							
Q_3							
Q_7				-0,550	-0,559		
Q_30							
T_1	0,482	0,548					temperature
T_3	0,594						
T_7	0,604	0,581	0,470				
P_1							precipitation
P_3			0,522				
P_7				0,685	0,449	0,610	
PS							

A. Winter

B Summer

FIGURE 5.7 Empirical evidence for the ABRC model. Discharge of the Pilica River over a 4-year observation period and statistical analysis of low (LQ) and high (HQ) flow conditions during warm (summer) and cold (winter) months (Wagner and Zalewski, 2016). *Modified from Wagner, I., Zalewski, M., 2016. Temporal changes in the abiotic/biotic drivers of self-purification in a temperate river. Ecol. Eng. 94, 275–285, https://doi.org/10.1016/j.ecoleng.2016.05.073.*

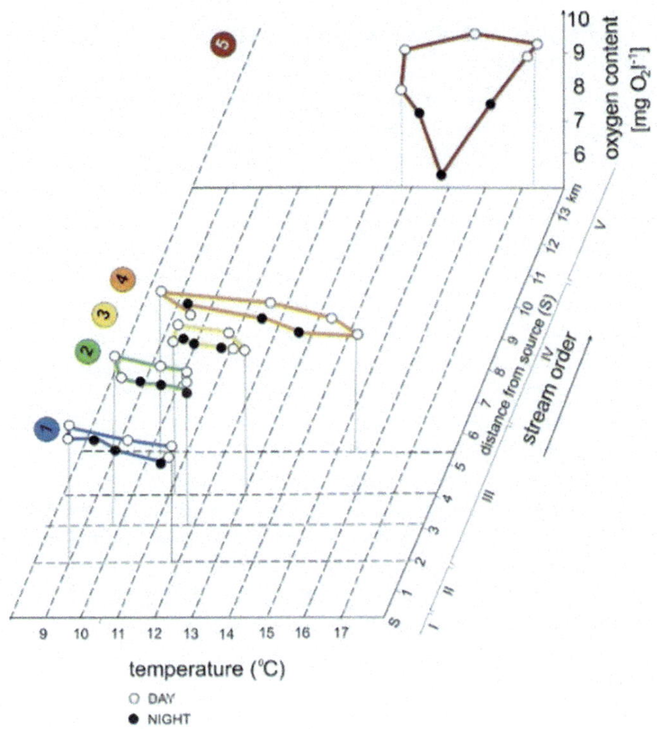

FIGURE 5.8 Oxygen fluctuations in rivers vary according to their stream order. This part of the figure shows how river metabolism (Lubrzanka River, Poland)—particularly oxygen levels—fluctuates over a 24-hour summer cycle along river continuum. During the day, sunlight promotes photosynthesis, increasing oxygen concentrations, while at night, respiration becomes dominant, leading to oxygen depletion. In forested catchments (sites 1–3), shaded streams maintain relatively stable temperature and oxygen levels because the canopy cover reduces sunlight and limits primary production. In contrast, open meadow areas (site 4) receive more sunlight, causing stronger daily swings in oxygen levels due to enhanced photosynthesis during the day and increased respiration at night. Further downstream, in a mosaic landscape with agricultural and urban areas (site 5), nutrient pollution stimulates excessive algal growth and daytime oxygen production but also leads to more severe oxygen drops at night due to the decomposition of organic matter, which is further increased by pollutants. *Modified from Zalewski M., Frankiewicz P., Brewińska B., 1985. The factors limiting growth and survival of Brown trout,* Salmo trutta *m. fario L. introduced to different types of streams. J. Fish Biol. 27 (A), 59–73.*

cycle, flood events redistribute organic matter, enhance nutrient cycling, and promote river self-purification by depositing suspended materials and nutrients onto floodplains (Kiedrzyńska et al., 2014). These processes not only support primary productivity but also expand the functional habitat of the land–water ecotone, enabling critical ecological services such as fish feeding, spawning, and refuge (Welcomme, 1979; Lowe-McConnell, 2010; Ross and Baker, 1983). In higher-order streams (e.g., orders IV–VI), where autotrophic production dominates and fish communities are more diverse than upstream, niche overlap is smaler. Flooded river valley in these systems provides essential microhabitats that support trophic specialization and reproductive success. The dynamic transformation of floodplains between terrestrial and aquatic phases throughout the seasons maintains plants communities in diverse successional stages, stimulates different types of primary producers, and sustains high biodiversity. Moreover, rivers with floodplains serve as important mitigation corridors for gene exchange and the maintenance of biodiversity, as well as for the adaptation of populations to changing climatic conditions.

Understanding stream metabolism along the river continuum requires simultaneous consideration of physical gradients, ecological structure, and temporal scales. By examining these dynamics across spatial (headwaters to large rivers), diurnal, and seasonal dimensions, scientists and water managers can better predict ecological responses to climate change, land use, and nutrient enrichment, particularly how oxygen availability, habitat quality, and biogeochemical cycling vary in time and space (see Fig. 5.8) (Box 5.5).

BOX 5.5 Shifts in the hierarchy of abiotic and biotic factors are reflected in the mortality and growth rates of brown trout along the upland river continuum

Zalewski et al. (1985) studied survival and growth rate of experimentally introduced trout fingerlings (approx. 8 cm)—the *Salmo trutta trout* m. *fario* along the upland reaches of the Lubrzanka River, testing the ABRC model. The obtained results conform the key assumption of the ABRC model. The growth rate of trout in the river with a natural catchment increased with the stream order, reflecting the stream trophy increase due to temperature and food conditions, Fig. 5.9.

At the same time, reduction in the survival rate of brown trout was observed in the lower sections (5) as results of predation by pike and eel (Fig. 5.10).

The analysis also considered water temperature and oxygen content, derived from field measurement and meteorological records supported by regression analysis between air and water temperature for this area. The most significant limitation on potential growth was observed in the near-source section, where strong abiotic pressures, including low temperatures and low trophic status, restricted trout growth. In this section, trout density seriously exceeded the environmental carrying capacity, further constraining growth. In the upper, shaded part of the stream (where $P/R < 1$), trout achieved only 50% of the growth rate theoretically possible at a given temperature (Elliott, 1975) due to food limitations (Figs. 5.9 and 5.11). The unavailability of optimally sized prey, especially in winter, forced trout to expend considerable energy capturing large numbers of smaller organisms. To obtain 1 gram of food, brown trout had to capture several hundred prey items (Fig. 5.12). Survival under such challenging conditions, characterized by low stream trophic status, was largely facilitated by the supplementation of prey originating from terrestrial ecosystems, which provided an additional food source.

With the increase in the stream order, the influence of abiotic factors as regulators of the abundance and growth rate of trout decreased. Increasing primary production thanks to better insolation, higher temperature and higher concentration of nutrients ($P/R > 1$) resulted in an increase in the biomass of invertebrates the optimal trout food, but on the other hand, the appearance of predatory fish (pike, eel). Under such conditions, the trout achieved a growth rate close to the maximum possible (Fig. 5.9), but their survival decreased dramatically as a result of predatory pressure (Fig. 5.10).

7.1 Integrating hydrodynamics to optimize habitat conditions by renaturalization

While foundational concepts significantly advanced riverine ecology, the emergence of ecohydraulics has further transformed the field by integrating hydrodynamic and ecological principles to optimize habitat conditions. Established as a distinct discipline in the late 20th century, marked notably by the First Symposium on Habitat Hydraulics in Norway in 1994,

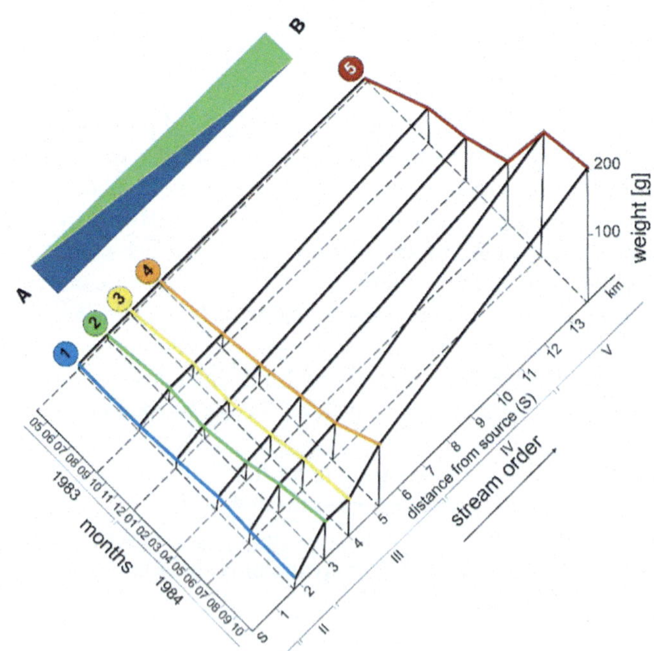

FIGURE 5.9 **ABRC on fish A.** Changes in the growth rate of brown trout introduced along the upland river continuum (Lubrzanka): Blue—site 1—source section, green—site 2—fir forest, yellow—site 3—forest-meadow ecotone strongly shaded with hazel, orange—site 4—meadow with intermediate complexity (riparian ecotone), red—site 5—mid-size stream $P/R > 1$. *Modified from Zalewski M., Frankiewicz P., Brewińska B., 1985. The factors limiting growth and survival of Brown trout,* Salmo trutta *m. fario L. introduced to different types of streams. J. Fish Biol. 27 (A), 59–73.*

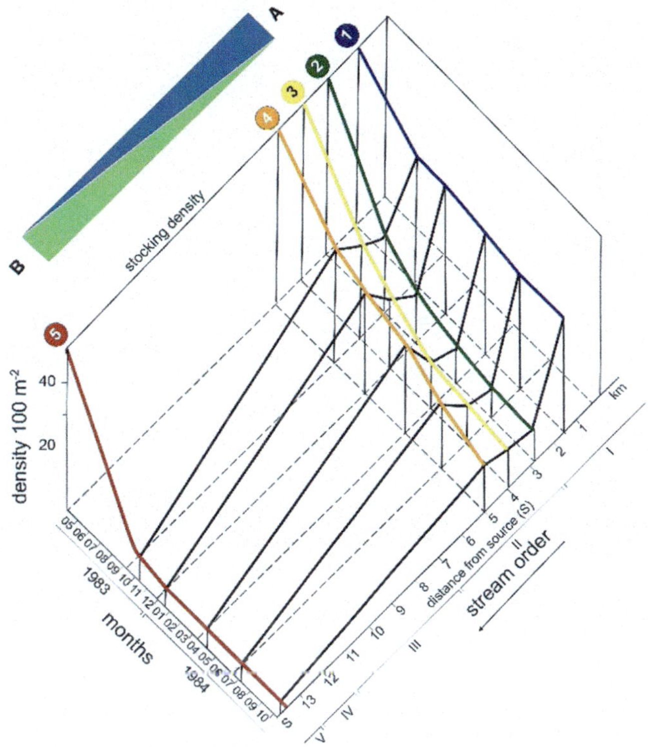

FIGURE 5.10 **ABRC on fish B.** The decline of survival of introduced brown trout as a function of increased predatory pressure (biotic regulation 2,3,4 occasionally eel, 5-pike + eel) along the river continuum (increase of biotic factor). *Modified from Zalewski M., Frankiewicz P., Brewińska B., 1985. The factors limiting growth and survival of Brown trout,* Salmo trutta *m. fario L. introduced to different types of streams. J. Fish Biol. 27 (A), 59–73.*

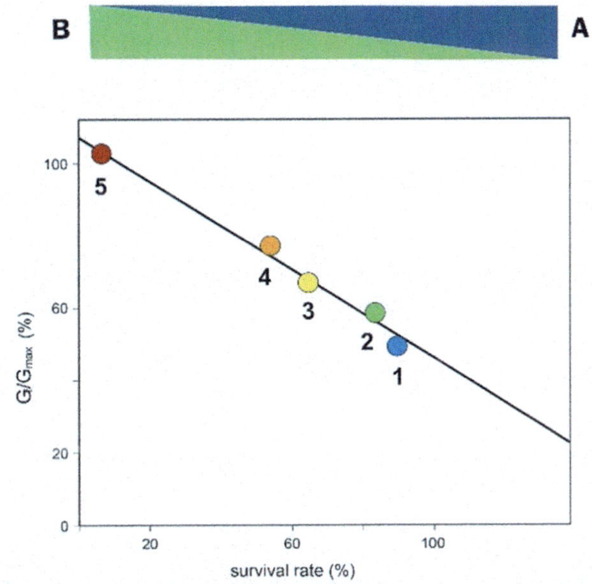

FIGURE 5.11 ABRC on fish C. Relationship between the survival of introduced brown trout along the upland river continuum and their growth rate as a percentage of growth possible to obtain at a given temperature: *Gf*, observed trout growth rate; *Gmax*, maximum for a given temperature growth rate according to Elliot data (Elliott, 1975). *Modified from Zalewski M., Frankiewicz P., Brewińska B., 1985. The factors limiting growth and survival of Brown trout,* Salmo trutta *m. fario L. introduced to different types of streams. J. Fish Biol. 27 (A), 59–73; according to Modified from Elliott, J.M., 1975. The growth rate of brown trout (Salmo trutta L.) fed on maximum rations. J. Animal Ecol. 44 (3), 805–821.*

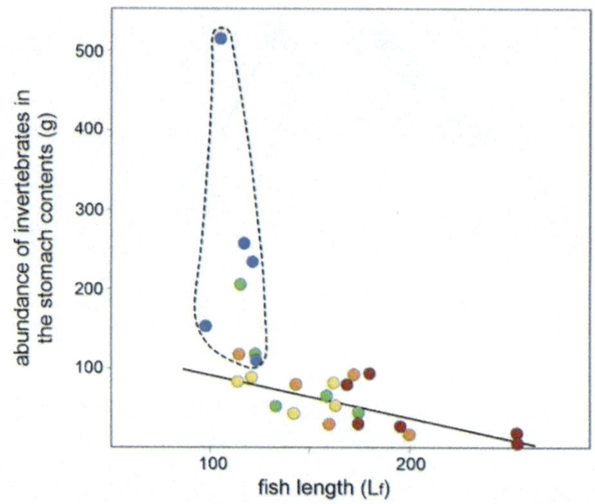

FIGURE 5.12 ABRC on fish D. Change in the feeding strategy of brown trout along the river continuum, expressed as the relationship between the length of the fish and the number of victims present in 1 gram of stomach contents: blue—site 1, green—site 2, yellow—site 3, orange—site 4, red—site 5, compare Fig. I5.10 A, B, and C. *Modified from Zalewski M., Frankiewicz P., Brewińska B., 1985. The factors limiting growth and survival of Brown trout,* Salmo trutta *m. fario L. introduced to different types of streams. J. Fish Biol. 27 (A), 59–73.*

ecohydraulics employs hydrodynamic modeling to predict how flow variability, turbulence, and substrate heterogeneity influence species distribution and energy expenditure (Baptist et al., 2010). However, the roots of ecohydraulics can be traced to earlier studies. Bachman (1982) conducted a seminal study on the foraging behavior of drift-feeding fish, particularly trout, in stream environments. They demonstrated that fish growth and maximum size are constrained by the balance between energy gained from drifting prey and energy expended in capturing it, which is influenced by water velocity and turbulence.

On the other hand, in the face of increasing climate change, the basic element of ecohydrological systemic solutions for increasing the catchment sustainability potential is reduction of homogenization of agricultural landscape (Levia et al., 2020) and increase of water retention in agricultural areas. According to Mioduszewski (2004), this includes measures to increase retention in the catchment landscape, including optimization of the spatial arrangement of arable fields, grasslands, and forests and shelter belts (Kedziora 2015). Forests stabilize the water cycle in the catchment area, reducing the outflow of water into rivers during rainfall and gradually releasing it during dry periods while simultaneously due to high transpiration, small forest areas may reduce the outflow to rivers (Fig. 5.13) (Box 5.6).

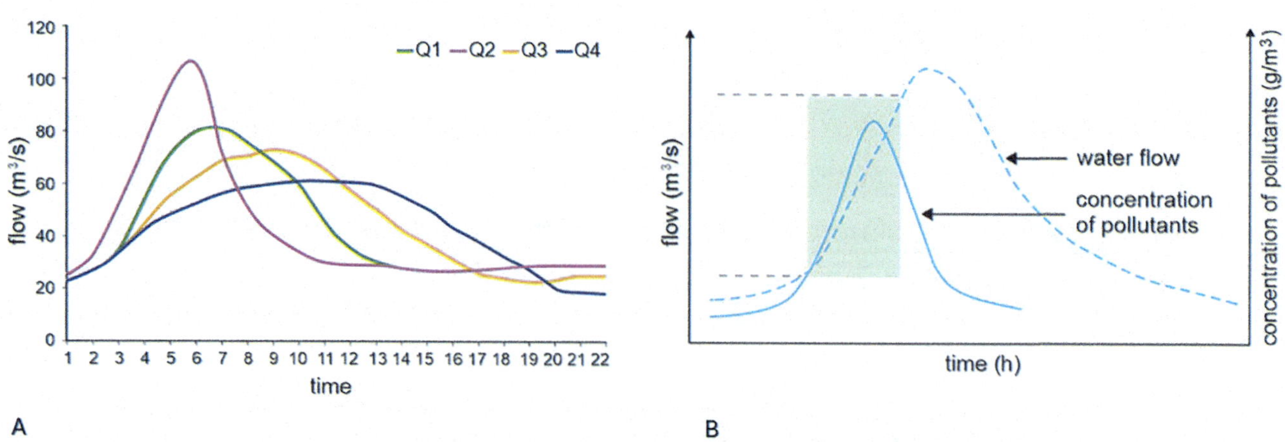

FIGURE 5.13 Examples of specific hydrographs and pollutant concentrations. (A) The relation of the shape of the flow curves to the characteristic of the catchment Q1 (green): an agricultural hydrograph Q2 (purple): an urban hydrograph Q3 (yellow): a hydrograph of floodplain with meadows. Q4 (blue): a forest hydrograph. (B) This exemplifies the different patterns of precipitation, discharge, and conduct of pollutants. The monitoring and the analysis of such pattern in the catchment is important input for the optimization of construction of SSBS. *(A) Modified from Mioduszewski, W., 2004. Modelling of Hydrological System of the Biebrza Valley This Work, Mioduszewski and Colleagues Applied the Groundwater Flow Model SIMGRO to Identify Hydrological Zones in the Lower Biebrza Basin; (B) Modified from Zalewski, M. (Ed.), 2002b. Guidelines for the integrated management of the watershed – phytotechnology and ecohydrology, United Nations Environemant Programme. International Environmental Technology Centre, Freshwater Management Series 5, 188.*

BOX 5.6 Ecohydraulics for enhancement of fish biodiversity

Biodiversity and river shoreline. The importance of diversified habitats and river hydraulics at large river shoreline zone as determinant for fish communities' recruitment, biodiversity and productivity (Fig. 5.16) (Schiemer and Spindler, 1989).

Microhabitats, hydraulics, and sediments granulometry determine

- organic matter accumulation
- energy expenditure by fish fry
- trophic conditions by:
- periphyton/phytoplankton status
- zooplankton on invertebrate quantity and distribution

The aforementioned factors determine fish fry's growth, survival, biodiversity.

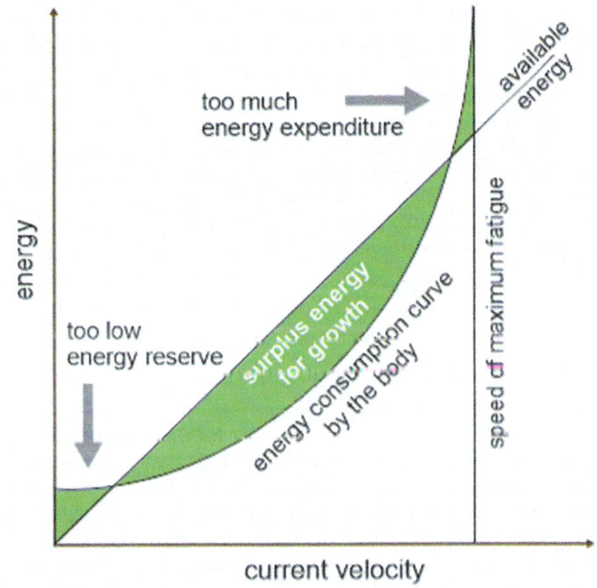

FIGURE 5.14 Relationship between energy expenditure and flow velocity. Dependence of energy expenditure on flow velocity for fish feeding on drift. The lower part of the graph represents regulated rivers with poor ecotone zones, while the upper fast-flowing streams with little habitat diversity, where the energy expenditure is greater than the energy obtained from the drift. This type of model is an example of the early stages of ecohydraulic analysis. Such relationships should be carefully considered when designing river restoration plans, for example, by introducing large wood debris to diversify the stream velocity in channel by enhancing stream chaanel and shoreline complexity (see Fig. I5.9). *Modified from Bachman R.A., 1982. Proceedings of the Salmon and Trout Migratory Behavior Symposium, A Growth Model for Drift-Feeding Salmonids: A Selective Pressure for Migration, 128, 135.*

A key conceptual advancement in ecohydraulics is the relationship between an organism's energy budget and flow velocity. As flow velocity increases, energy expenditure rises exponentially, creating an optimal range where energy intake exceeds metabolic costs. In regulated rivers with simplified habitats, flow velocities often surpass this optimal range, increasing energy costs and reducing habitat suitability. Conversely, in natural rivers with diversified shoreline ecotone structures, velocity gradients provide a range of microhabitats that enhance energy efficiency for aquatic organisms (Fig. 5.14).

The integration of ecohydraulics into river management underscores the need for flow regimes that maintain ecological functionality. By incorporating ecohydraulics methodology into restoration planning, ecohydrological strategies can optimize fish habitat, mitigate anthropogenic disturbances, and enhance river resilience.

The concept of natural flow regimes (see Fig. 5.15), which summarizes the relationship between aquatic biodiversity and natural flow regimes (Bunn and Arthington, 2002). Collectively, presented principles underscore how the integrity of natural flow regimes underpins biodiversity and ecosystem function across spatial and temporal scales (Schiemer and Waidbacher, 1992; Panchan et al., 2022).

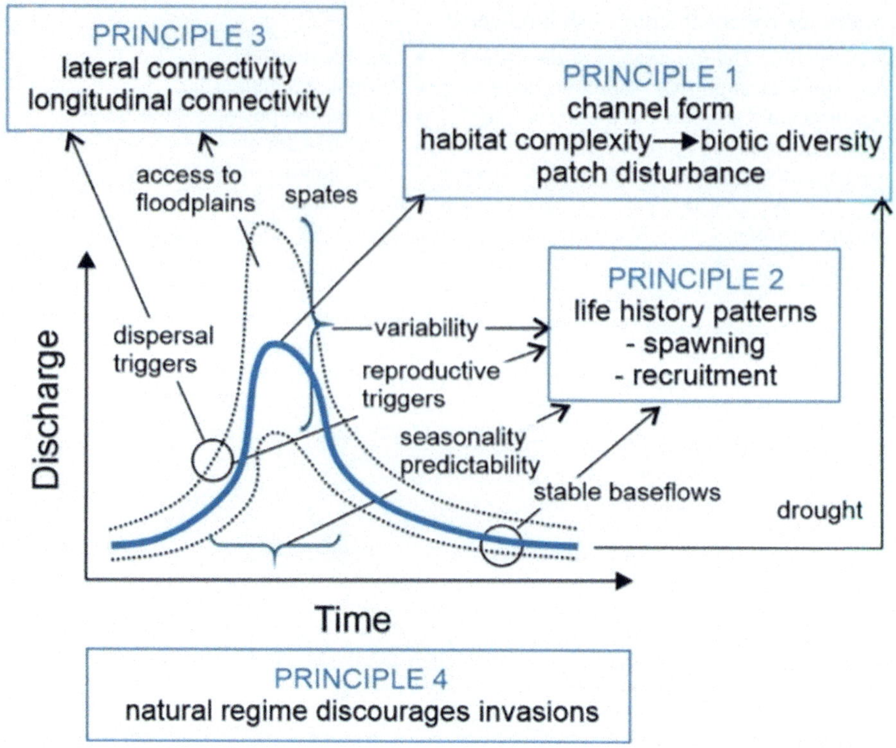

FIGURE 5.15 Aquatic biodiversity and natural flow regimes. This diagram presents the four interconnected principles: (1) high-discharge events shape channel morphology and habitat complexity, supporting biotic diversity; (2) predictable seasonal flow cues regulate species life histories, such as spawning and recruitment; (3) flood pulses enhance connectivity, enabling dispersal and access to off-channel habitats; and (4) natural flow variability discourages invasive species by maintaining ecological resilience. *Source: Bunn, S.E., Arthington, A.H., 2002. Basic principles and ecological consequences of altered flow regimes for aquatic biodiversity. Environ. Manag. 30 (4), 492–507, https://doi.org/10.1007/s00267-002-2737-0.*

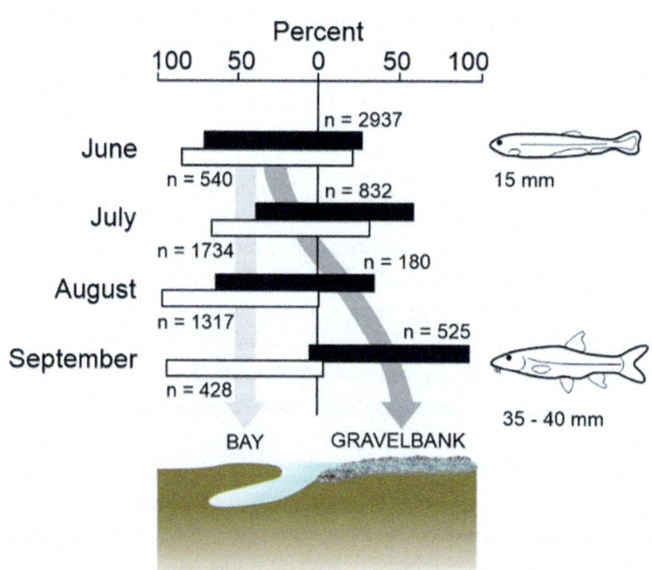

FIGURE 5.16 Biodiversity and productivity. Small juveniles rely heavily on structured, sheltered inshore habitats (bays). As they grow, they transition toward open, dynamic gravel zones where hydrological disturbance and spatial complexity support larger, more mobile fish. *Modified from Schiemer, F., Spindler, T., 1989. Endangered fish species of the Danube river in Austria. Regul. Rivers Res. Manag. 4 (4), 397–407, https://doi.org/10.1002/rrr.3450040407.*

FIGURE 5.17 The sequential sedimentation biofiltration system (SSBS). Results of the experimental testing of the ABRC model in construction context, inspired by emerging properties of the system such invertebrates and fish biomass (calcium concentration of major stimulant (nutrient spiraling and self-purification processes). (1) Enhanced sedimentation chamber. (2) Geochemical barriers (calcium, dolomite, biocare). (3) Constructed wetland. (4) Regeneration system at the gravel bed preventing gravel bed colmatation. *Modified from Zalewski, M., Wagner, I., Frątczak, W., Mankiewicz-Boczek, J., Parniewski, P., 2012. Blue-Green City for Compensating Global Climate Change.* The Parlament Magazine, 350, *2–3; Zalewski, M. (Ed.)., 2020.* Ekohydrologia. *Wydawnictwo Naukowe PWN.*

7.2 ABRC model—Discovering emerging properties of the system

A complex testing of ABRC model requires extensive experiments, which leads to a deeper understanding of the system's emerging properties and, therefore, to complex and innovative solutions that will be described in this chapter. During testing experiments, the strong correlation between invertebrate's biomass and calcium concentration appears as emerging properties of the system, which in turn inspire to incorporate to the constructed wetland the geochemical barrier (GB) or high-efficiency absorbent barrier (HEAB). The another phenomenon - the periodic acumulation the mineral and organic matter in pools and combined with observed lower efficency constructed wetland, when high load suspended matter appeare, resulted in adding before geochemical barrier the enhanced sedimentation chamber (ESC). By combining ESC+GB+HEB+ Constructed wetland all situated on gravel bed with regenerative system Fig. 5.6 the SSBS was constructed, recently used broadly to amplify urban stormwater purification (Zalewski et al., 2012; Szklarek et al., 2018) (see Chapter 21) and improvment small sewage treatment plants efficency.

These insights reinforce the necessity of integrating such ecohydrological processes into river management strategies. By accounting for ecotonal complexity, organic matter dynamics, and geochemical properties, sustainable water management practices can enhance biodiversity, optimize nutrient cycling, and improve the resilience of aquatic ecosystems in the face of environmental change (Kiedrzyńska et al., 2015).

Further testing of the ABRC model also revealed significant top-down regulatory effects of fish communities on lower trophic levels. A clear correlation between annual fish biomass and the summer decline of benthic invertebrate density was observed, highlighting how predation by fish can structure benthic communities and influence nutrient spiraling and ecosystem productivity (see Fig. 5.6). This relationship further supports the importance of maintaining balanced fish populations and habitat complexity as part of holistic river basin management. These processes require further in-depth investigation.

7.3 Hydrology → biota: Hydrogeology effect on biological structure

Methodological sequence from ABRC model to discover emerging properties of the system (Ca^{++}) into innovation (SSBS) and ANbS can be considered as an example for dual regulation hydrochemical properties (Ca++) → stimulate biota → intensification of biological processes is reflected in the improvement of water quality through enhancement of nutrient spiraling and biological productivity and thus self-purification. By accounting for ecotonal complexity, organic matter dynamics, and geochemical properties, sustainable water management practices can enhance biodiversity, optimize nutrient cycling, and improve the resilience of aquatic ecosystems in the face of environmental change (Fig. 5.17).

8. The role of woody debris in shaping riverine habitats

Fish species of different sizes occupy habitats with varying flow velocities and morphological features, reducing inter-specific competition. Optimal use of the river ecosystem requires diverse habitats with heterogenous velocities, provided by complex ecotone zones. In less complex environments, high flow velocity and elevated energy costs limit species richness (Thorpe, 1986). Accordingly, large woody debris and branches in the river channel play a crucial role in enhancing habitat heterogeneity, invertebrates, and fish productivity (Gregory et al., 1985; Gurnell et al., 1995; Piégay and Gurnell, 1997; Zalewski et al., 2001a; Zalewski and Robarts, 2003). This is somehow explained by experiments with wood placement in experimental stream (Fig. 5.18). The section of the channel with deposited wood debris was preferred by large fish species. After the removal of the wood, both the dominance structure and size distribution of fish changed completely. Following the reintroduction of wood debris, the channel structure returned to the state shown in Fig. 5.19. The amount of coarse particulate organic matter (CPOM), including large wood debris, not only shapes channel structure and habitat availability but also influences trophic dynamics. This is because CPOM becomes colonized by bacteria, algae, fungi, protozoa, and microinvertebrates, in proportion of CPOM deposited.

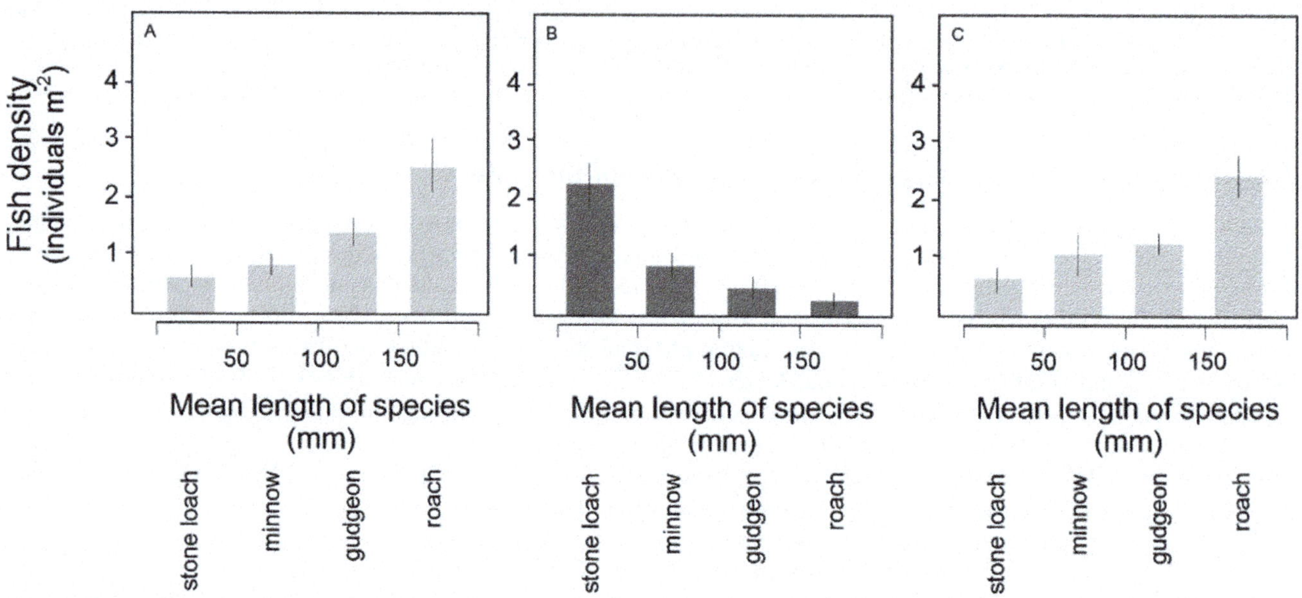

FIGURE 5.18 Wood placement and its effect on fish populations. The effect of the experimental wood placement on fish density and size of species (A), removal (B), and restoration (C) on density of four coexisting fish species in a river channel. *Modified from Lapińska, M., 1997. PhD Thesis Uniwersytet Łódzki Łódz Space as a Limiting Factor for Fish Communities in Lowland and Upland River Systems.*

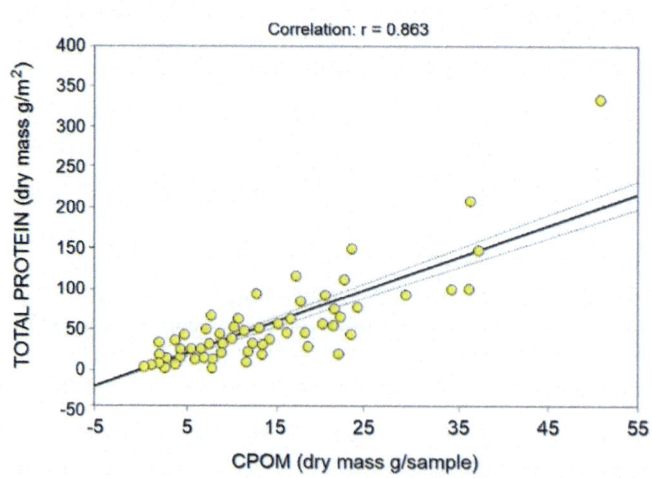

FIGURE 5.19 Relationship between CPOM and total protein in the river channel. Protein content increases in proportion to CPOM (coarse particulate organic matter) deposited in the stream channel, serving as an important component of stream trophic dynamics. *Data from: Lapińska, M., 1997. PhD Thesis Uniwersytet Łódzki Łódz Space as a Limiting Factor for Fish Communities in Lowland and Upland River Systems.*

Beyond organic matter origin, the geological and chemical properties of stream environments play a crucial role in regulating decomposition efficiency and energy flow. Streams with high calcium concentrations exhibit accelerated detritus breakdown, while those with lower calcium availability demonstrate reduced organic matter processing rates (Egglishaw, 1968). This variation in decomposition efficiency directly affects the nutrients available to primary consumers, such as benthic invertebrates, ultimately shaping the biomass and distribution of fish populations (Anderson et al., 1984).

Empirical studies have demonstrated a strong correlation between fish biomass and calcium ion concentration in riverbed sediments (Włodek, 1976; Cuinat, 1986; Zalewski et al., 1990a). A comparative analysis of upland and lowland rivers by Zalewski et al. (1990a) further indicated that this correlation is particularly pronounced in upland systems with high light availability, where primary production—particularly periphyton growth—is more intensive. Conversely, in lowland rivers dominated by macrophytes, nutrient transfer to higher trophic levels occurs at another pathway due to greater nutrient retention within plant biomass and sediments.

9. Riparian ecotones and nutrient spiraling

The interactions between the ecosystem's constituents and its resultant traits are determined by the physical structure of the environment and the way it is utilized by biocenosis (plants, animals, and fungi). Such interactions play a particularly significant role in the so-called ecotone zones, i.e., transition areas between two ecosystems. In terms of the quality of water resources, ecotone zones between terrestrial and aquatic ecosystems are especially important since they pose a barrier to diffuse pollutants permeating into water, e.g., plants in the riparian zone or in floodplains (Figs. 5.20 and 5.21) Plants in the ecotones convert nutrients in to organic matter. Such matter is the source of energy for becterials and fungi and nutrients by decomposition process for primary producers, both lay at the foundation of the stream's trophic pyramid. As its matter comes from two sources autochtonous and allochtonous, the ecosystem is more diversified, stable and more resilient, more habitats translate into more species populating various niches. This can serve as an example of a significant relationship between the environment's diversity and the abundance of its species (Winemiller, 1990). Regulatory mechanisms in such systems, founded on interactions between the physical environment and living organisms, can be applied to sustain or restore equilibrium and biodiversity in the ecosystem.

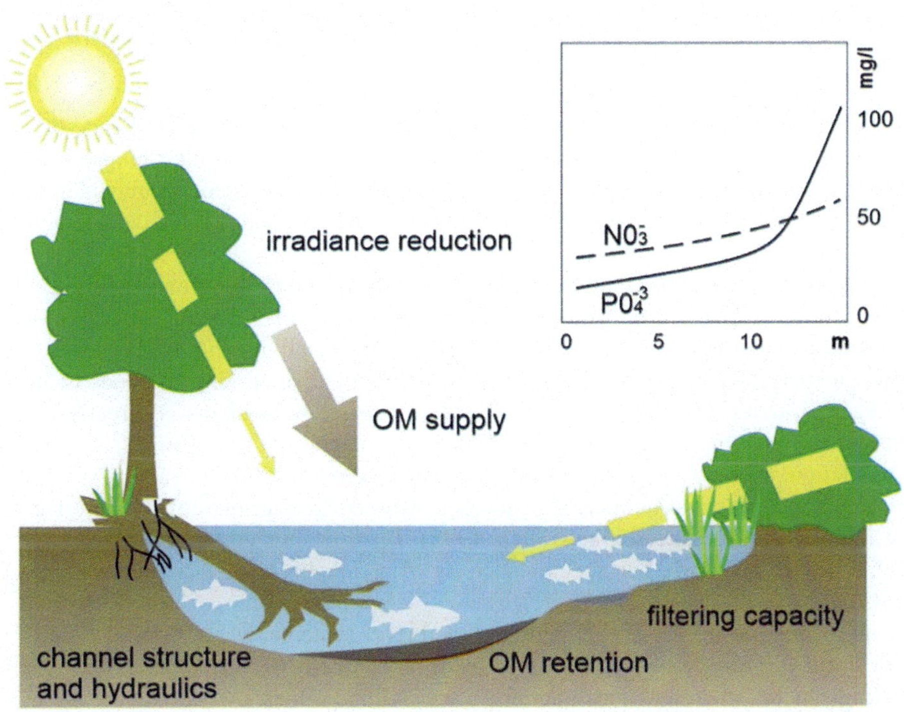

FIGURE 5.20 **(A) Ecotone functions.** This example illustrates how ecotone zones contribute to the purification of water bodies. *From Zalewski, M., 2002a. Ecohydrology – the use of ecological and hydrological processes for sustainable management of water resources. Hydrol. Sci. J. 47 (5), 825–834; Zalewski, M. (Ed.), 2002b. Guidelines for the integrated management of the watershed – phytotechnology and ecohydrology, United Nations Environemant Programme. International Environmental Technology Centre, Freshwater Management Series 5, 188.*

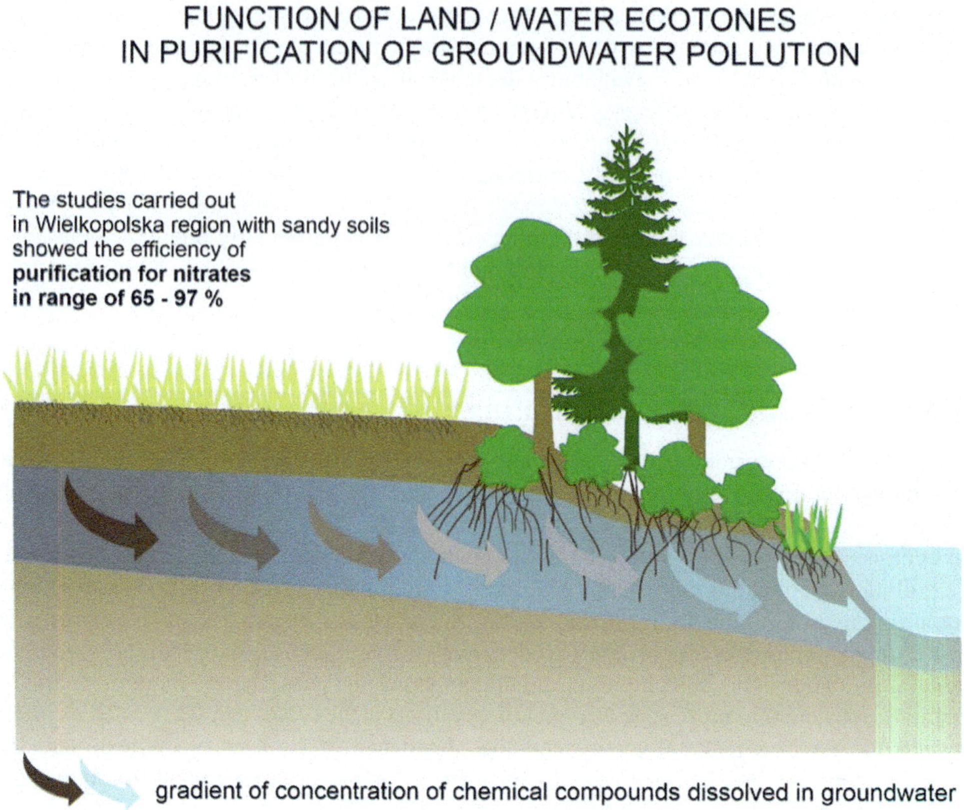

FIGURE 5.21 (B) Ecotone functions. This is an example of how ecotone zones contribute to groundwater purification. *From Zalewski, M., 2002a. Ecohydrology – the use of ecological and hydrological processes for sustainable management of water resources. Hydrol. Sci. J. 47 (5), 825–834; Zalewski, M. (Ed.), 2002b. Guidelines for the integrated management of the watershed – phytotechnology and ecohydrology, United Nations Environemant Programme. International Environmental Technology Centre, Freshwater Management Series 5, 188 based on Kedziora 2015.*

From an ecological perspective, ecotone zones are fundamental in enhancing the resilience of aquatic ecosystems to anthropogenic stressors, while simultaneously preserving biodiversity and maintaining bioproductivity (Zalewski et al., 1994; Zalewski, 2000). These transitional areas between terrestrial and aquatic environments act as buffers, regulating nutrient flows, mitigating pollution, and supporting complex biological interactions that contribute to ecosystem stability (Naiman et al., 1989).

Riparian ecotones play a fundamental role in buffering the impacts of nutrient inflows by intercepting and assimilating excess nutrients before they reach the river channel. They also regulate key instream processes such as organic matter retention, and sediment dynamics, all of which directly influence bioproductivity and fish biomass. Contributing to habitat complexity, they offer increased refugia for fish (see Fig. 5.22), while also serving as zones of nutrient exchange. The input of terrestrial organic matter (allochthonous material) into aquatic ecosystems is directly related to the structural complexity of the ecotone zone. In contrast, the contribution of autochthonous organic matter is indirectly influenced through ecotonal regulation of light penetration and nutrient retention. The availability of allochthonous material significantly enhances fish feeding opportunities, as fish can consume a greater quantity of invertebrates than is naturally produced within the aquatic system (Lancaster and Allan, 1995). Terrestrial invertebrates entering the water can constitute up to 50% of a fish's diet, particularly for species such as trout inhabiting low-order and low-productivity streams (Zalewski et al., 1985).

These issues are most pronounced in large, heavily polluted river systems, where reproduction of fish populations are confined to tributaries, making even small ecotone habitat losses critical for species survival. The situation is further worsened by escalating abiotic stressors, which intensify as ecotone zones shrink, increasing the risk of population extinction (Coelho and Zalewski, 1995; Poff et al., 1997). Empirical studies testing these ideas are presented in this diagram illustrating how shoreline complexity influences fish growth rates, showing that pikeperch (*Sander lucioperca*) thrives in moderately structured habitats, while higher complexity favors pike (*Esox lucius*), leading to the elimination of pikeperch due to efficient predation by pike. The findings support the concept of "intermediate riparian ecotone

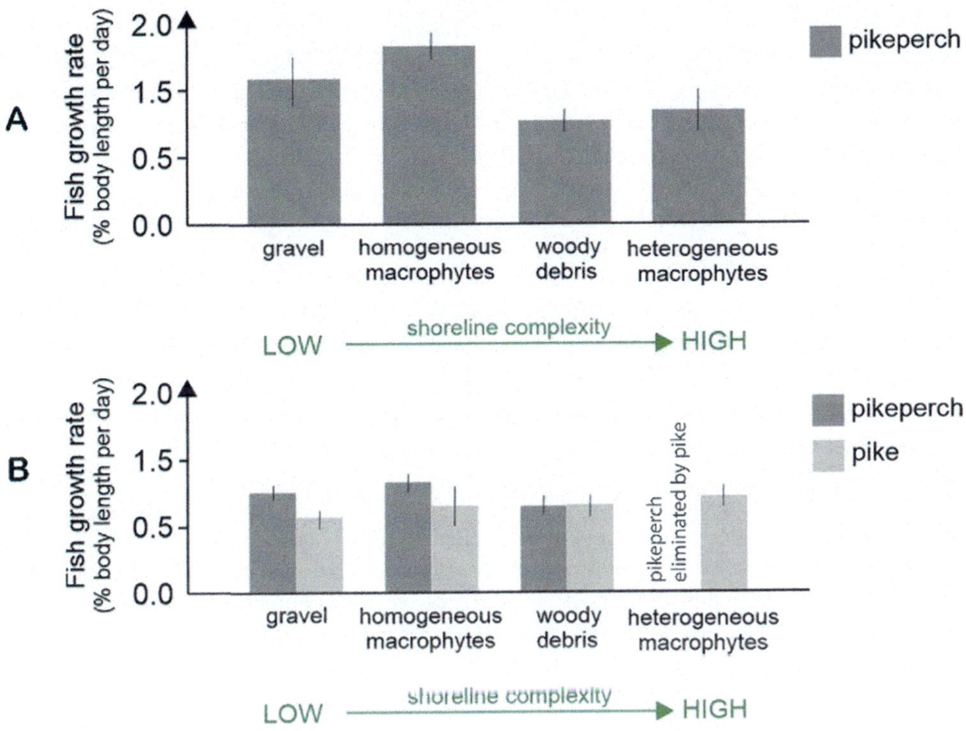

FIGURE 5.22 The importance of a diversified shoreline for the survival and growth rates of two predatory fish species. (A) Growth of pikeperch single species; (B) Growth and survival of two predatory species - effect of the competition. *Data from: Zalewski, M., Łapinska, M., Bayley, P.B., 2003. Fish relationships with wood in large rivers. Am. Fish. Soc. Symp. 2003 (37), 195–211.*

complexity," suggesting that habitat gradients with balanced structural diversity optimize fish biomass, regulate predator–prey interactions, and enhance overall ecosystem stability.

The decomposition dynamics of terrestrial organic matter entering aquatic ecosystems exhibit substantial variability, dictated by its biochemical composition and origin. For instance, organic inputs from meadow vegetation decompose more readily than those from coniferous forests, particularly pine-dominated systems (Minshall et al., 1985). The slow breakdown of pine needles is attributed to their thick cuticle and secondary metabolites that inhibit fungal colonization, particularly by aquatic hyphomycetes, in contrast to the more labile nature of deciduous leaf litter (Triska et al., 1975). Such variations in decomposition rates influence microbial activity, energy transfer through trophic levels, and overall ecosystem productivity.

Given the increasing pressures on aquatic ecosystems, the conservation and strategic management of ecotone zones are imperative for sustaining biodiversity and ecological functionality. Ecotones play an essential role in mitigating pollution, enhancing nutrient retention, and stabilizing aquatic habitats against both localized and global environmental changes. Future comparative research should focus on optimizing the design and management of land–water interface zones to enhance denitrification, nutrient absorption, and pollutant assimilation. Strengthening these self-purification processes can effectively reduce nutrient loads and downstream transfer, thereby mitigating the negative impacts of anthropogenic disturbances.

9.1 Riparian ecotones and fish communities

Traditionally, limnology and fish ecology have developed as distinct disciplines, with limnologists focusing on nutrient dynamics, primary production, and invertebrate communities, while ichthyologists have studied fish early ontogeny, reproduction, competition, foraging strategies, and behavioral responses to abiotic factors. However, understanding fish communities within the context of ecotone zones provides crucial complementary insights. Fish serve as effective indicators of ecosystem functioning due to their long-complicated life cycles dependent on diversity of habitat and migration opportunities.

Many fish species exhibit opportunistic behavior, adapting their habitat use at different life stages in response to ecological conditions (Schiemer and Spindler, 1989). The temperate rivers preferences are dictated primarily by bioenergetic requirements, with biotic interactions and behavioral factors playing secondary roles (Zalewski and Naiman, 1985). This adaptation is particularly evident in drift-feeding fish, where the relationship between water velocity and food availability determines their energy balance. The optimal flow velocity range for a given species varies with body size (Webb et al., 1984). In regulated rivers lacking ecotonal complexity, energy constraints become a limiting factor: in slow-flowing, channelized rivers, drifting food is insufficient to sustain rheophilic species, while in high-velocity, regulated rivers with degraded ecotones, fish must expend more energy foraging than they can obtain from available prey. Consequently, such environments exhibit high juvenile mortality, particularly during food-scarce periods (Cunjak and Power, 1987).

Fish are highly mobile organisms that migrate within river networks to exploit optimal habitat and foraging conditions. Large rivers provide environmental heterogeneity, with oxbow lakes functioning as stable refugia where fish expend less energy than in the main river channel. These areas serve as critical shelters during periods of abiotic stress, such as winter and flooding (Bouvet et al., 2017). Oxbows also act as genetic reservoirs for fish populations in heavily regulated rivers (Bouvet et al., 1991). Additionally, reproduction and early larval development occur preferentially in shallow water ecotone zones, where structurally complex habitats offer diverse food resources and protection from predation (Benke et al., 1984).

The structural complexity of ecotone zones enhances species coexistence by diversifying habitat conditions, including depth, current speed, substrate type, and refuge availability, thus reducing mortality due to competition or predation (Schiemer et al., 2017; Bardonnet et al., 1991).

The importance of shoreline ecotones and their impact on fish populations evolves with changes in environmental conditions, including variations in river morphology, trophic status, and hydrological connectivity. Riparian ecotones provide dynamic microhabitats that support biodiversity by influencing nutrient retention, food web dynamics, and fish community composition. By integrating hydrological, geological, and ecological factors, ecotone zones enhance resilience in aquatic ecosystems, contributing to habitat stability and species diversity. Their role in buffering anthropogenic disturbances highlights the necessity of preserving and restoring ecotonal habitats within broader river basin management frameworks.

Beyond chemical and biological recovery, riparian ecotone structure plays a crucial role in maintaining riverine biodiversity and stability. As shown in the diagram, the presence of woody debris, submerged vegetation, and complex habitat structures enhances water retention, nutrient cycling, and food web interactions. The degradation of riparian zones, however, is strongly correlated with declining fish density and biodiversity (Schiemer and Spindler, 1989; Schiemer and Waidbacher, 1992). Reduced habitat complexity leads to population declines in large aquatic species (Portt et al., 1986) alters trophic interactions (Schlosser, 1982), and diminishes available habitat for fish populations (Scholsser and Karr, 1981). This ecological imbalance reduces reproductive potential, limits genetic diversity, and increases species' vulnerability to environmental stressors (Dudgeon et al., 2006).

10. Ecotone complexity and riverine ecosystem productivity and resilience

The supply of energy to higher trophic levels depends on the rate of matter decomposition, but also in line with the nutrient spiraling concept, from the rate of primary production. Primary production is limited by the availability of light and is lower in places where ecotone zones create shading above watercourse (Fig. 5.23). Limited light access reduces the growth of algae (Zalewski et al., 1985) and significantly reduces the biomass and productivity of invertebrates and fish communities (Aho, 1977; Bilby and Bisson, 1987; Mann and Pęczak, 1986; Murphy and Hall, 1981). Decline of trophy results in reduced fry growth, survival and maturation and stimulates transition of smolt (last developmental stage) to adult form in salmonids (Thorpe, 1986).

In the shaded sections of the river, energy is transferred to the fish trophic level along two routes. The first depends on filtering invertebrates, such as *Trichoptera*, the biomass of which in turn depends on the content of proteins in seston (living and dead organic matter floating in water) (Valett and Stanford, 1987) and the efficiency of obtaining and processing polysaccharide molecules by periphyton (Lock and Williams, 1981). The second path includes many trophic levels of small heterotrophs: bacteria are eaten by *Flagellata* , which in turn are food for *Ciliata* (Fenchel, 1977; Taylor, 1981) filtered out by *Rotifera*, which are food for small *Chironomidae* (Moore and Gilbert, 1987). Due to the length of the food chain, loss of energy along the way to the fish level is considerable. Because most of the protozoa are too small to be a suitable food for fish (Dickie et al., 1987), fish biomass in shaded areas is small and largely depends on the influx of terrestrial food because low effectiveness of allohtonous matter processing in to primary productivity and invertebrate biomass. In turn, in the light exposed sections of the river the high biomass of algae, bacteria, and fungi provide source of energy for food webs (Zalewski et al., 1990a).

Further studies of effect of shading on algal, invertebrates, and fish communities showed that shaded streams were dominated by rhodophytes, which are inedible to invertebrates, resulting in low invertebrate biomass and, consequently, reduced fish populations (Fig. 5.24). This diminished energy transfer to higher trophic levels correlated with lower self-purification potential in shaded sections. Long-term catchment scale comparative studies (Fig. 5.25) confirmed this

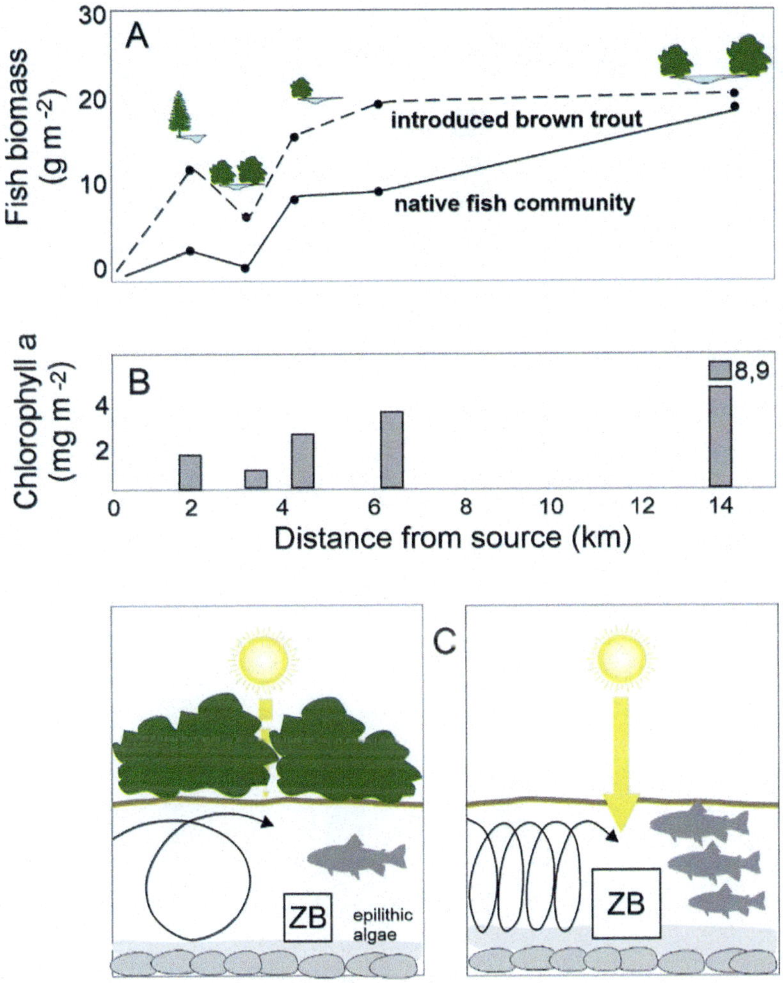

FIGURE 5.23 The effect of riparian ecotones on biomass (A, B, C). (A) Effect of riparian ecotones on biomass distribution of fish (native and introduced brown trout) in an upland river. (B) Algal biomass distribution (chlorophyll *a*, mg.m^{-2}). (C) Hypothetical differences in nutrient spiraling rate, primary production by edible algae, zoobenthos (ZB) and fish biomass that shaded and open-canopy stations. Empirical background for river renaturalization. Testing the ABRC model along the River Continuum revealed that shaded sections of the stream, despite their high habitat complexity, exhibited extremely low fish biomass. Further analysis of these "cold spots" indicated that shading significantly reduces primary productivity, which in turn diminishes nutrient spiraling and the stream's self-purification capacity. This finding inspired additional experimental studies aimed at explaining the balance between edible and inedible algae (A), as well as more advanced investigations into the broader ecological effects of shading. These studies challenge the prevailing assumption that autochthonous primary production plays a minor role in stream ecosystems. On the contrary, the findings demonstrate that primary productivity from autochthonous sources is essential for supporting in-stream biomass and bioproductivity, as well as maintaining self-purification processes. The most advanced experiments involved the removal of shading from selected stream sections, which confirmed and quantified these relationships. *(A) Data from Zalewski M., Frankiewicz P., Brewińska B., 1985. The factors limiting growth and survival of Brown trout, Salmo trutta m. fario L. introduced to different types of streams. J. Fish Biol. 27 (A), 59–73; scheme from Schimer V.E., Zalewski M., 1992, The importance of riparian ecotones for diversity and productivity of riverine fish communities, Netherland J. Zool. 42, 323–335.https://doi.org/10.1007/BF02530372.*

relationship and led to the concept of "intermediate land–water ecotone complexity" as suporting measure for enhancement of WBSR.

The above empirical data leads to conclusion that to reduce nutrient loads from agricultural catchments, denitrification and geochemical barriers should be implemented, with the willow strips and tree rows to enhance stream habitats complexity, trophic status and self-purification, while ensuring that at least 60% of the stream surface remains exposed to sunlight. This balance supports algal productivity, invertebrate populations, higher fish biomass, and self-purification.

Further testing of the aforementioned results is required across diverse ecosystem types and geographic regions to develop a more comprehensive understanding of these processes. Such research will contribute to optimizing land–water ecotone design, enhancing their efficiency in self-purification and nutrient load reduction, ensuring that nutrient spiraling will be enhanced to remain within the optimal absorbing capacity of the system. The ongoing reduction of land–water ecotone zones must be halted, particularly in small streams where intense pulses of nonpoint source pollution require mitigation. Enhancing nutrient spiraling through increased habitat diversity (renaturalization) and improved light availability by shaping intermediate complexity of land–water ecotones is essential for supporting self-purification processes (Fig. 5.25).

Streams with high levels of molecular and organic carbon losses are less productive than those in which carbon is used more efficiently by organisms in the respiration processes (R) (Fisher and Likens, 1973). In summer, during low flow levels, 83% of suspended organic matter in water can come from excreted products from invertebrates (Wallace et al., 1982). These observations were confirmed in both small and larger watercourses, with the quality of food matter suspended in water changes with the RC (Naiman, 1983b).

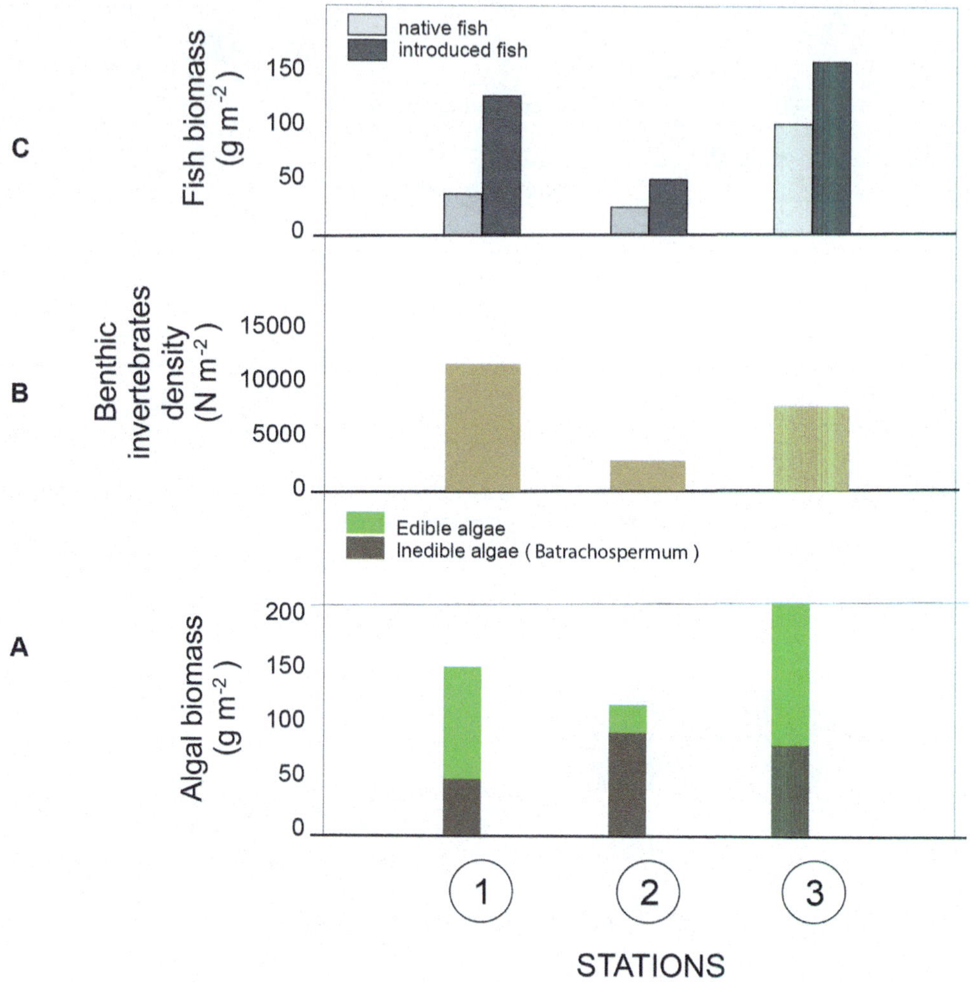

FIGURE 5.24 The effect of light reduction by riparian ecotone. The effect of light reduction by riparian ecotone on the structure and biomass of benthic algal communities (A), the biomass of benthic invertebrates (B), and the biomass of native fish and introduced trout (C) in upland streams (central Poland). Station 1: stream flowing through mixed forest; station 2: stream heavily shaded by hazelnut bushes; station 3: stream flowing thought meadows with sparse alder on its banks. Shading of the stream decreases algal biomass, moreover limited light access create favorable conditions for the inedible Rhodophytes. In stations 1 and 3, edible algae are two times higher than the inedible. This reflects on lower biomass of invertebrates and fish in heavily shaded station 2. *(C) Data from Zalewski M., Puchalski W., Frankiewicz Bis B., 1994. Rehabilitation of Freshwater Fisheries, Fishing News Books Riparian Ecotones and Fish Communities in Rivers—Intermediate Complexity Hypothesis, 152, 160.*

In the light of the cited observations considering the influence of shading on the food webs in the watercourse, one could conclude that a good method of increasing productivity and the biomass of the fish is to reduce the shaded areas in favor of the ones with high autotrophic production (Stockner and Shortreed, 1978). However, it is not always an appropriate course of action. Moderate removal of trees from the banks causes to some extent better insolation and raising of water temperature, which stimulates primary production (Murphy et al., 1986) and increases the growth rate of fish, allowing them to be more flexible in changing life strategies (Holtby, 1988; Thorpe, 1989; Thorpe et al., 1989). However, excessive removal of riparian vegetation makes the banks more exposed to erosion and also leads to the increased supply of the river with nutrients and matter from surface runoff (Scholsser and Karr, 1981) and accelerates macrophytes silting and shallowing of the watercourse (Pastor and Post, 1986). This changes the structure of fish and invertebrates' communities, eliminating the typical benthic species (Berkman and Rabieni, 1987). Hence, the highest productivity of fish communities falls on ecosystems in the middle of the scale, when it comes to the complexity of the ecotone zone (Fig. 5.25).

This conclusion was confirmed by the study of Zalewski (1988) on the Lubrzanka River, showing that the relationship of both fish density and biomass to the level of insolation of the watercourse are parabolic, diminishing with insufficient light availability in the watercourse (extensive, shading ecotone zones), as well as providing additional illumination exceeding the optimal spectrum (no ecotone zones or a zone with a very low density) (Figs. 5.24–5.26).

It is important to note that in unprotected river channels with high levels of solar exposure, the excessive growth of aquatic vegetation may occur. This can hinder water flow and increase the risk of flooding in adjacent areas of the river valley. But first and foremost, macrophytes removal reduced nutrients uptake, e.g., in Danish stream of ammonium was 34%–77% and phosphorous 50%–77% reduced by macrophytes. This is important for the development the strategy for

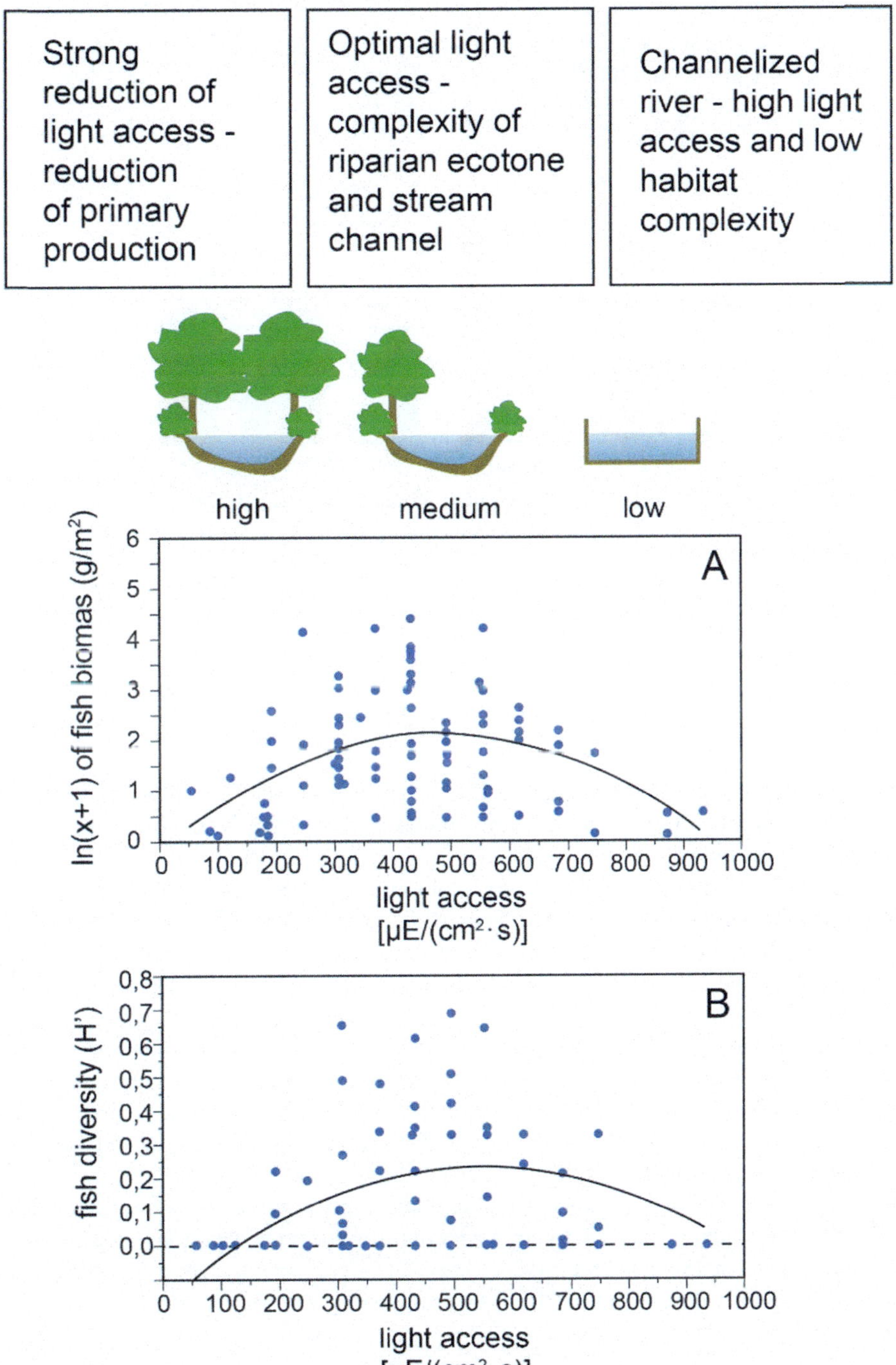

FIGURE 5.25 Light access to stream channel for fish and biodiversity. Empirical long-term data collected in the Lubrzanka catchment. The essential role of light for high biomass and species richness of fish communities and in consequence the self-purification process. *From Zalewski, M., Wagner, I., Frątczak, W., Mankiewicz-Boczek, J., & Parniewski, P., 2012. Blue-Green City for Compensating Global Climate Change.* The Parlament Magazine, 350, *2–3; Zalewski, M. (Ed.), 2020.* Ekohydrologia. *Wydawnictwo Naukowe PWN.*

reduction of rivers eutrophication. Furthermore, excessive sunlight and rising water temperatures may eliminate temperature-sensitive fish species, such as salmonids, from the ecosystem.

Achieving an optimal balance between shaded and sunlit conditions is therefore essential for maximizing the ecological benefits of land–water ecotone zones. However, the ideal state of this balance depends on local catchment conditions, including substrate characteristics and climate. Research indicates that in the northern banks of watercourses,

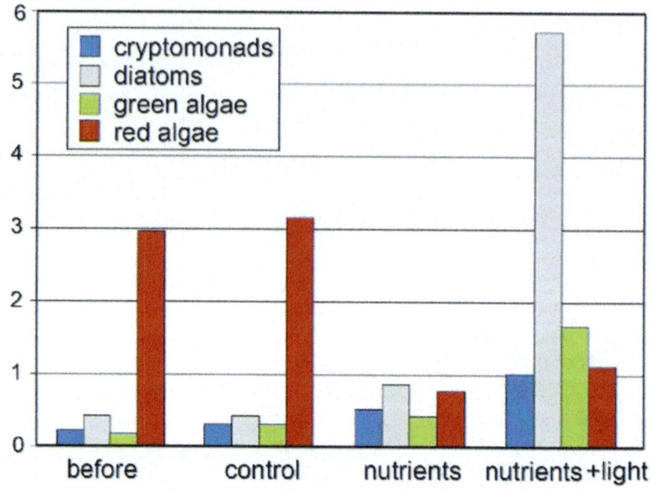

FIGURE 5.26 Experimental assessment of stream self-purification through the analysis of effect of the removal riparian shading (nutrients+light) and nutrient enrichment (nutrients) on algal communities. (A) Preexperiment conditions (before)—stream section prior to the removal of dense riparian hazelnut (Corylus) trees, serving as a reference for baseline ecological conditions. (B) Control section—an upstream segment of the stream unaffected by the experiment, monitored throughout to provide comparative data, during all periods of experiments. (C) Nutrient enrichment—examining the impact of nutrient addition in to shaded section (A) on algal community composition, biomass accumulation, and overall productivity in shaded section of stream (below control section B). (D) Nutrient and light—investigating the combined effects of riparian canopy removal and nutrient enrichment (D) on algal community dynamics and primary production (below section B). *Zalewski, M., Bis, B., Łapińska, M., Frankiewicz, P., Puchalski, W., 1998. The importance of theriparian ecotone and river hydraulics for sustainable basin-scale restoration scenarios. Aquat. Conserv. Mar. Freshw. Ecosyst. 8 (2), 287–307. https://doi.org/10.1002/(SICI)1099-0755(199803/04)8:2%3C287::AID-AQC274%3E3.0.CO;2-R.*

ecotone zones with tree cover are particularly beneficial, as they support habitat diversity, organic matter flow from terrestrial systems, and nutrient retention. In contrast, the southern stream banks should allow for 60% or more surface insolation with low riparian vegetation minimizing erosion and nutrient runoff.

The width of ecotone zones is another critical factor. Although ecotones are typically defined as narrow transition areas, some studies suggest that optimal widths range from 10 to 30 m (Petersen et al., 1992). In practice, however, it is often difficult to exclude such wide strips of fertile land from agricultural use. Generally, the required width should increase in proportion to river width, the complexity of the hydrological network, and the intensity of agricultural activity in the catchment. For example, Knauer and Mander (1989) found that nutrient filtration is most effective within 7 m of the shoreline, where phosphorus inflow can be reduced by up to 40%. Given that the prevention of eutrophication is a top priority in contemporary water management in agricultural catchment, this 7-meter minimum should be considered essential. However, when ecohydrological ANbS are applied—specifically through the use of denitrification barriers combined with geochemical barriers and biofilters, the required width can be further reduced to approximately 5 m, while still achieving even higher nutrient reduction of up to 70% (Izydorczyk et al., 2014).

At the landscape scale, ecotone zones between terrestrial and aquatic environments play a vital role in preserving biodiversity and ensuring water quality, especially in agriculturally dominated regions. These zones support not only aquatic species but also birds, mammals, and other terrestrial organisms by offering diverse habitats. From this perspective, maintaining ecotone widths of up to 30 m, where feasible, should remain a conservation target, at least in selected sections of the river corridor.

11. Role of diversity of stream channel and land ecotones for stream self-purification and renaturalization

The understanding of hyporheic zone role in the nutrient status of rivers is crucial for understanding the function of bioproductivity, self-purification, and resilience (Fig. 5.27), particularly their capacity to absorb nutrients and release sediments from small first-order streams (Triska et al., 1989) . As far as the role of aquatic microphytes around 30 μg/L—typical of natural, anthropogenically undisturbed streams in Poland, such as those in the premieval Białowieża Forest, UNESCO MAB Biosphere Reserve, the bottom sediment zone tends to release nutrients due to the sedimentation and accumulation of CPOM and FPOM. However, when phosphorus concentrations in water are higher (around 2000 μg/L, characteristic of streams in intensively agricultural catchments), the bottom sediments function similarly to activated

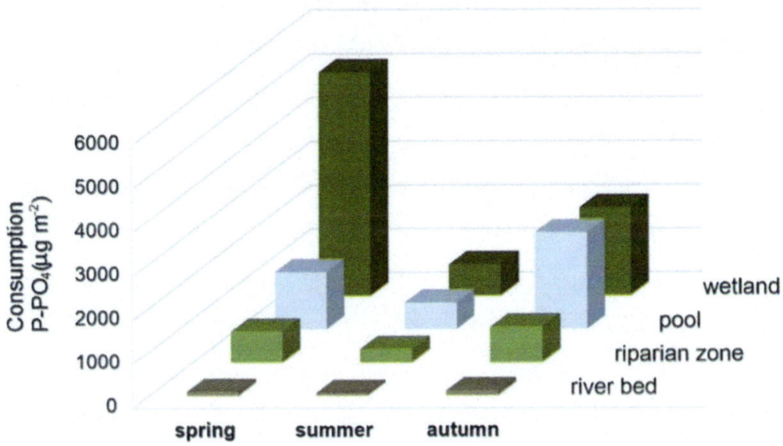

FIGURE 5.27 Phosphate uptake. Phosphate uptake by bottom sediments from autumn to spring, underlining the importance (wetland and riparian zones, pools, diversity of stream channel, and riparian zone in stream self-purification). *Zalewski, M., Bis, B., Łapińska, M., Frankiewicz, P., Puchalski, W., 1998. The importance of theriparian ecotone and river hydraulics for sustainable basin-scale restoration scenarios. Aquat. Conserv. Mar. Freshw. Ecosyst., 8 (2), 287–307. https://doi.org/10.1002/(SICI)1099-0755(199803/04)8:2%3C287::AID-AQC274%3E3.0.CO;2-R.*

sludge in a sewage treatment plant. In this case, all biotic components—including algae, fungi, protozoa, rotifers, and invertebrates—contribute to a high nutrient-absorbing capacity across the food chain, facilitating the conversion of absorbed nutrients into biomass and enhancing processes of nutrients spiraling and self-purification.

12. Self-purification process

The preceding comparison highlighted the role of nonpoint source pollution on stream habitas and importance of its complexity in river system ecological status. In contrast, point source pollution usually intensifies downstream, exerting greater influence on water quality along the river continuum. The diagram (Fig. 5.28) demonstrates changes in key chemical, biological, and ecological parameters along the river continuum, highlighting the river's ability to recover naturally through biogeochemical processes. Upon pollution entry, oxygen concentrations (O_2) decline sharply due to increased biological oxygen demand (BOD), while ammonium (NH_4^+) and phosphates (PO_4^{3-}) peak as organic matter decomposes. Over time, nitrification converts ammonium to nitrate (NO_3^-), signifying water quality recovery. Concurrently, microbial and macroinvertebrate communities shift, with pollution-tolerant species (e.g., Tubificidae worms) dominating the degradation zone, while more sensitive species (e.g., Asellus crustaceans) recolonize the system as conditions improve (Odum, 1969; Vannote et al., 1980; Allan and Castillo, 2007).

This natural self-purification mechanism, first described in the 20th century, occurs through a sequence of progressive ecological zones, each exhibiting improving water quality. As organic pollutants degrade, aquatic species with increasing oxygen requirements and decreasing tolerance to pollutants establish themselves. These self-purification zones are categorized using bioindicators (saprobes) and chemical markers, tracking organic matter decomposition and microbial activity (Kolkwitz and Marsson, 1909; Crisp and Hynes, 1971). This process has to be considered in the context of the River Continuum Concept (RCC), which describes changes in forms of organic matter (CPOM-FPOM), along the course of natural river, influencing the role of invertebrate's community structure and the balance between primary production and respiration (P/R) (Vannote et al., 1980).

However, the self-purification process primarily addresses organic pollutants. In contrast, heavy metals, pesticides, and other xenobiotics do not undergo natural biodegradation; instead, they acumulate in the bottom of the river and floodplain, bioaccumulate in aquatic organisms and transfer through trophic levels, posing long-term ecological risks (Mason, 2002; Luoma and Rainbow, 2008). These processes of river self-purification have driven the development of biological wastewater treatment plants, which partially mimic natural self-purification processes by employing microbial activity (biofilms) within activated sludge systems, enhanced by intensive aeration to accelerate the breakdown of organic matter (Tchobanoglous et al., 2003). Furthermore, future development in nutrient reduction through biological nutrient removal and chemical treatments in these systems help protect downstream ecosystems from excessive nutrient loading and contamination.

The interplay between self-purification mechanisms and riparian ecotone complexity is vital for sustaining riverine ecosystem health (Fig. 5.29). While rivers have a natural capacity to process organic pollutants, the conservation of riparian ecotones is equally essential for buffering abiotic stress, enhancing biodiversity, and maintaining ecological resilience. Future research should focus and optimize stream riparian buffer zones, improving wastewater treatment

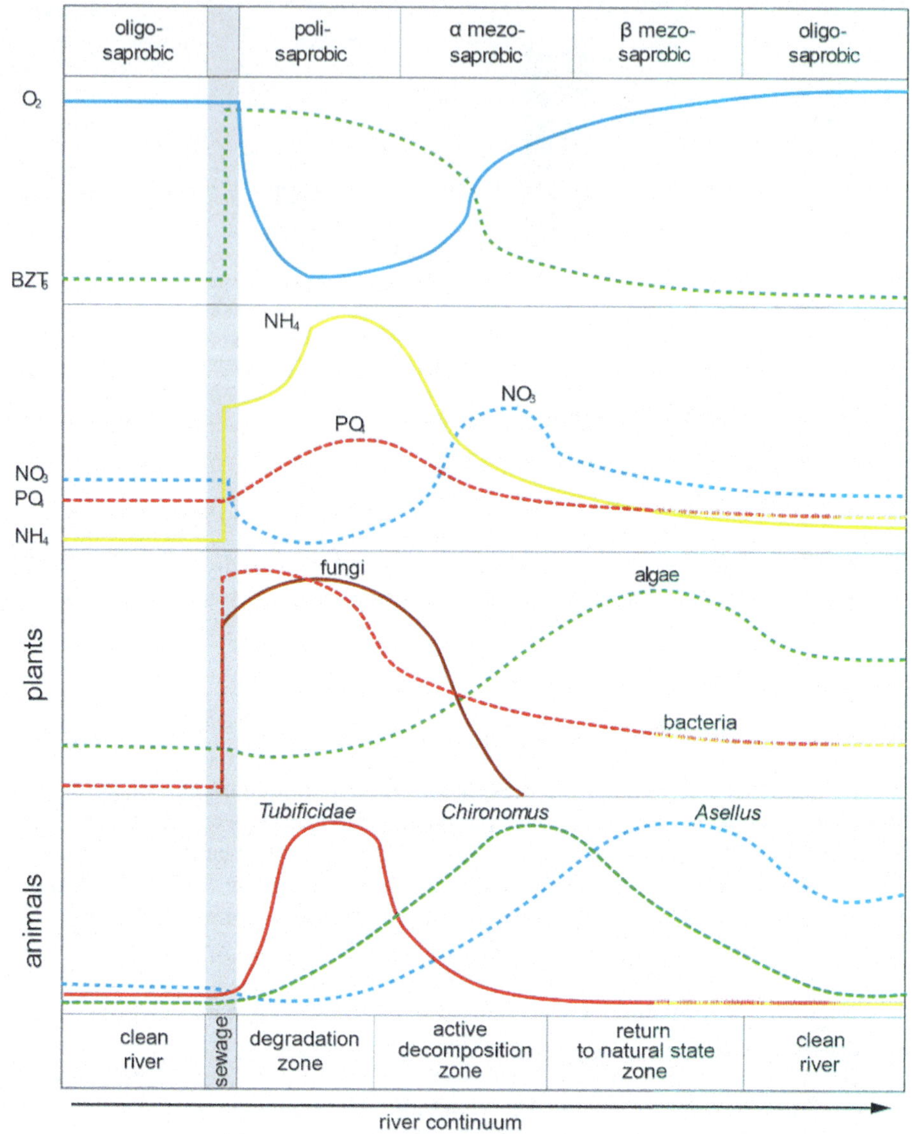

FIGURE 5.28 The river self-purification process. That diagram visualizes the saprobic zones in the context of the River Continuum Concept (RCC), showing how oxygene , BOD , nutrient concentration , primary producer groups, and key macroinvertebrate taxa shift from the point source of pollution along a stream's longitudinal gradient. *Adapted from Vannote, R.L., Minshall, G.W., Cummins, K.W., Sedell, J.R., Cushing, C.E., 1980. The river continuum concept. Can. J. Fish. Aquat. Sci. 37 (1), 130–137, https://doi.org/10.1139/f80-017; saprobic zones from Kolkwitz and Marsson (1908).*

efficiency, through adaptive management strategies to mitigate the cumulative impacts of pollution and habitat degradation (Palmer et al., 2009; Naiman et al., 2010).

13. Complexity of river channel and riparian ecotones—Functional factor for WBSR enhancement

Stream-valley ecotones complexity is essential for enchancing WBRS factors; Fig. 5.30 illustrates a model of a mosaic aquatic and riparian habitats ecotone, demonstrating how these zones regulate hydrological and biogeochemical processes that indirectly improve water quality. Tree roots, fallen trunks, and submerged woody debris enhance habitat diversity, retain organic matter, and support invertebrate populations that serve as prey for fish (Zalewski et al., 2003). Riparian vegetation complexity also ensures light access, thereby influencing primary production and trophic interactions within the watercourse (Zalewski et al., 1994).

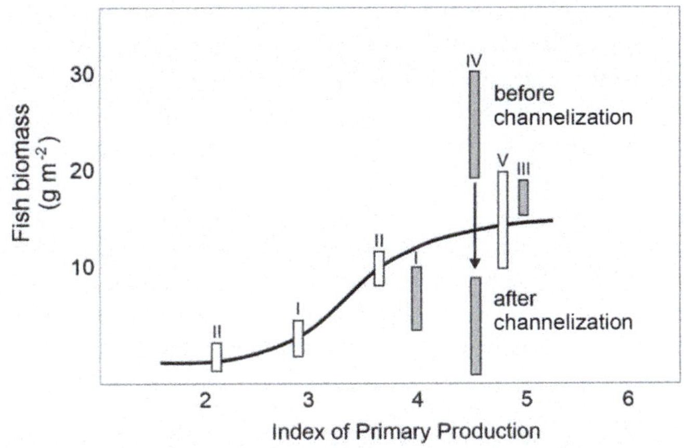

FIGURE 5.29 The long-term effect of river channelization on fish biomass. Relation between primary productivity (Zalewski et al., 1991)(and annual average fish biomass in an upland river. Open bars represent data from catchments of natural character, while shaded bars indicate agricultural catchments. Stream order is denoted by Roman numbers. *Data from Zalewski, M., Thorpe, J. E., & Gaudin, P., 1991. Fish and land/inland-water ecotones. In Man and the Biosphere Series (Vol. 4, pp. 81–96). UNESCO.*

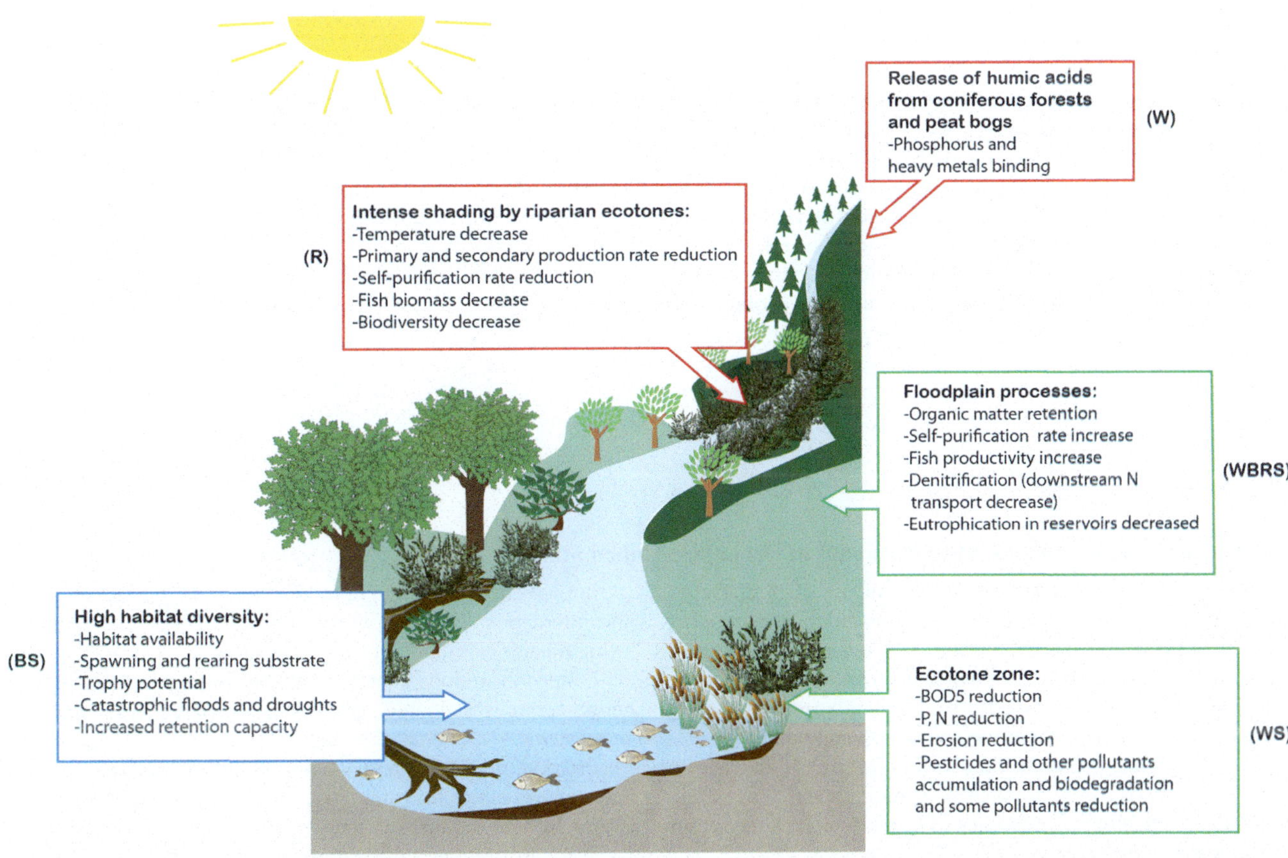

FIGURE 5.30 A model of a mosaic aquatic and riparian habitats ecotone. Optimal composition of river valley as a factor enhancing Water–Biodiversity–Ecosystem Services–Resilience (WBSR). *Modified from Zalewski, M., Puchalski, W., Frankiewicz Bis, B., 1994. Rehabilitation of Freshwater Fisheries 152 160 Fishing News Books Riparian Ecotones and Fish Communities in Rivers—Intermediate Complexity Hypothesis.; Zalewski, M., Frankiewicz, P., Nowak, M., 1995. Biomanipulation by ecotone management in a lowland reservoir. Hydrobiologia 303, 49–60.*

In addition to their structural role and self-purification, ecotones contribute to pollutant filtration and retention, acting as biochemical buffers that accelerate the decomposition of organic matter, reduce nutrient inflows, and mitigate persistent pesticide and heavy metals accumulation in the food web (Indika Herath and Vithanage, 2015). In areas dominated by coniferous forests and peat bogs, humic acid release during intense rainfall lowers phosphorus and heavy metal concentrations, reducing eutrophication risks (Górniak, 1996). Consequently, drainage ditches in intensively farmed landscapes should terminate in constructed wetlands, which effectively trap and reduce nutrient loads from agricultural runoff (Petersen et al., 1992).

Beaver dams and small retention reservoirs also play a significant role in water and nutrient retention and increasing riverbed complexity. However, while they enhance ecosystem diversity, they can also impede the upstream migration of anadromous fish species (Naiman et al., 1987). Despite such trade-offs, ecotone conservation and restoration remain fundamental to maintaining aquatic biodiversity and mitigating the adverse effects of anthropogenic and climatic pressures.

14. Strategic documents from international symposia

Strategic documents, which had been developed during UNESCO Program, formulated as conclusion of the World symposia (Boxes 5.7 and 5.8).

BOX 5.7 Floodplain declaration

The results of many scientific projects conducted within the framework of UNESCO's International Hydrological Programme (IHP), such as those described above, have raised awareness among scientists of the need to formulate strategic documents that support the transformation of knowledge into wisdom and, subsequently, into recommendations for decision-makers. Examples of such documents include the Floodplain Declaration and the EcoSummit Columbus Declaration. "**Sustainable floodplain management is vital for resilient water systems, healthy ecosystems, and thriving communities. We commit to integrating science, nature-based solutions, and inclusive governance to protect and restore floodplains for present and future generations.**" (Adapted from the Declaration on Floodplain Management—Change of Perspective. Based on the presentations and discussions at the International Conference on Ecohydrological Processes and Sustainable Floodplain Management: Opportunities and Concepts for Water Hazard Mitigation, and Ecological and Socioeconomic Sustainability in the Face of Global Changes.)

Organized at European Regional Centre for Ecohydrology under auspices of UNESCO—Institute Polish Academy of Science, 19–23 May 2008, Lodz, Poland.

BOX 5.8 Summary of the 2012 EcoSummit in Columbus, United States

Human activity has drastically altered nearly 80% of Earth's usable ecosystems, pushing the planet toward a global tipping point of declining natural resources, biodiversity, and environmental resilience. Past civilizations show that environmental degradation can cause societal collapse, but today's interconnected world faces the risk of a global collapse.

To avoid this, humanity must harmonize societal needs with natural systems through a shift in consciousness, policy, and action. The Scientific Committee of EcoSummit 2012 proposes a four-step strategy:

Adopt a Global Systems Paradigm—Recognize the interconnectedness of systems and move beyond narrow, reductionist thinking.

Implement Policies of Harmonization—Align policies with the interdependent relationship between society and nature, using meaningful quality of life indicators.

Foster Innovation—Invest in ecological engineering, restoration, and sustainable design to conserve and restore ecosystems.

Promote Education and Values—Build lifelong ecological literacy and ethics to foster a society that understands and respects its role within Earth's systems.

Through systemic thinking, responsible policy, innovation, and education, humanity can work toward a sustainable and habitable future.

AI disclosure

During the preparation of this work the author(s) used OpenAI's ChatGPT to support language refinement and structural improvements. After using this tool/service, the author(s) reviewed and edited the content as needed and take(s) full responsibility for the content of the publication.

References

Aho, J.M., 1977. A population study of parasites of fishes in freshwater ecosystems. Ecology 58 (3), 549–558.

Allan, J.D., Castillo, M.M., 2007. Stream Ecology: Structure and Function of Running Waters: Second Edition Stream Ecology: Structure and Function of Running Waters, second ed. Springer Netherlands, United States, pp. 1–436. https://doi.org/10.1007/978-1-4020-5583-6. Available from: http://www.springerlink.com/openurl.asp?genre=book&isbn=978-1-4020-5582-9.

Altieri, A.H., Gedan, K.B., 2019. Climate Change and Dead Zones Ocean Deoxygenation: Everyone's Problem. IUCN, pp. 221–232. https://doi.org/10.2305/IUCN.CH.2019.14.en.

Anderson, J.J., Copping, A., Jagielo, T., Postel, J., Peterson, W., Dumbauld, B., Haron, G., 1984. Section 4 in reknot sewage treatment plant project. Seahurst Baseline Study III.

Bachman, R.A., 1982 1982. Proceedings of the Salmon and Trout Migratory Behavior Symposium 128 135 A Growth Model for Drift-Feeding Salmonids: A Selective Pressure for Migration.

Baptist, M.J., Babovic, V., Rodríguez Uthurburu, J., Keijzer, M., Uittenbogaard, R.E., Mynett, A., Verwey, A., 2010. On inducing equations for vegetation resistance. J. Hydraul. Res. 45 (4), 435–450. https://doi.org/10.1080/00221686.2007.9521778.

Bardonnet, A., Gaudin, P., Persat, H., 1991. Microhabitats and diel downstream migration of young grayling (Thymallus thymallm L.). Freshw. Biol. 26 (3), 365–376. https://doi.org/10.1111/j.1365-2427.1991.tb01404.x.

Benke, A.C., Van Arsdall, T.C., Gillespie, D.M., Parrish, F.K., 1984. Invertebrate productivity in a subtropical blackwater river: the importance of habitat and life history. Ecol. Monogr. 54 (1), 25 63. https://doi.org/10.2307/1942455.

Berkman, H.E., Rabeni, C.F., 1987. Effect of siltation on stream fish communities. Environ. Biol. Fish. 18, 285–294.

Bilby, R.E., Bisson, P.E., 1987. Emigration and production of hatchery coho salmon (*Oncorhynchus kisutch*) stocked in streams draining an old growth and a clear-cut watershed. Canad. J. Fish. Aquat. Sci. 44, 1397–1407.

Blöschl, G., Nester, T., Komma, J., Parajka, J., Perdigão, R.A.P., 2013. The June 2013 flood in the upper Danube basin, and comparisons with the 2002, 1954 and 1899 floods. Hydrol. Earth Syst. Sci. 17 (12), 5197–5212. https://doi.org/10.5194/hess-17-5197-2013.

Bouvet, Y., Pattee, E., Meggouh, F., 2017. The contribution of backwaters to the ecology of fish populations in large rivers. Preliminary results on fish migrations within a side arm and from the side arm to the main channel of the Rhône. SIL Proc. 22 (4), 2576–2580. https://doi.org/10.1080/03680770.1983.11897728.

Bouvet, Y., Soewardi, K., Pattee, E., 1991. The discrimination of roach *Rutilus rutilus* (Linnaeus, 1758) populations in different parts of a river system. An investigation using biochemical markers. Hydrobiologia 09 (2), 161–167. https://doi.org/10.1007/bf00006928.

Bunn, S.E., Arthington, A.H., 2002. Basic principles and ecological consequences of altered flow regimes for aquatic biodiversity. Environ. Manag. 30 (4), 492–507. https://doi.org/10.1007/s00267-002-2737-0.

Coelho, M.M., Zalewski, M., 1995. Evolutionary adaptations by fish to ecotonal complexity in spatially variable landscapes—a perspective. Hydrobiologia 303 (1–3), 223–228. https://doi.org/10.1007/BF00034059.

Crisp, D.T., Hynes, H.B.N., 1971. The ecology of running waters. J. Anim. Ecol. 40 (3), 782. https://doi.org/10.2307/3451.

Cuinat, R., 1986. Principaux caractẽres demongraphiques observes sur 50 riviẽres ŕ truite françaises. Influence de la pente et du calcium. Ann. Hydrobiol. 2.

Cunjak, R.A., Power, G., 1987. The feeding and energetics of stream-resident trout in winter. J. Fish. Biol. 31 (4), 493–511. https://doi.org/10.1111/j.1095-8649.1987.tb05254.x.

Dickie, L.M., Kerr, S.R., Boudreau, P.R., 1987. Size-dependent processes underlying regularities in ecosystem structure. Ecol. Monog. 57 (3), 233–250.

Dudgeon, D., Arthington, A.H., Gessner, M.O., Kawabata, Z.I., Knowler, D.J., Lévêque, C., Naiman, R.J., Prieur-Richard, A.H., Soto, D., Stiassny, M.L.J., Sullivan, C.A., 2006. Freshwater biodiversity: Importance, threats, status and conservation challenges. Biol. Rev. 81 (2), 163–182. https://doi.org/10.1017/S1464793105006950.

Duncan, A., 1989. Food limitation and body size in the life cycles of planktonic rotifers and cladocerans. Hydrobiologia 186–187 (1), 11–28. https://doi.org/10.1007/BF00048891.

Egglishaw, H.J., 1968. The quantitative relationship between bottom Fauna and plant detritus in streams of different calcium concentrations. J. Appl. Ecol. 5 (3), 731. https://doi.org/10.2307/2401645.

Elliott, J.M., 1975. The growth rate of brown trout (*Salmo trutta* L.) fed on maximum rations. J. Animal Ecol. 44 (3), 805–821.

Elwood, N., Neill, O., Stark, S., 1981. The role of microbes associated with organic and inorganic substrates in phosphorus spiralling in a woodland stream. Verhand. Int. Vereinig. Limnol. 21, 818–824.

Fenchel, T., 1977. Aspects of the decomposition of seagrasses. In: McRoy, C.P., Helfferich, C. (Eds.), Seagrass Ecosystems: A Scientific Perspective. Marcel Dekker, pp. 123–145.

Fenchel, T., 1977. Proteoz. Filter Feed. Prog. Protistol. 1, 65–113.

Fisher, S.G., Lavoy, A., 1972. Differences in littoral Fauna due to fluctuating water levels below a hydroelectric dam. J. Fish. Res. Board Can. 29 (10), 1472–1476. https://doi.org/10.1139/f72-227.

Fisher, S.G., Likens, G.E., 1973. Energy flow in bear brook, new hampshire: An integrative approach to stream ecosystem metabolism. Ecol. Monog. 43 (4), 421–439.

Fovet, O., Ruiz, L., Hrachowitz, M., Faucheux, M., Gascuel-Odoux, C., 2015. Hydrological hysteresis and its value for assessing process consistency in catchment conceptual models. Hydrol. Earth Syst. Sci. 19 (1), 105–123. https://doi.org/10.5194/hess-19-105-2015.

Górniak, A., 1996. Humic Substances and Their Role in Functioning of Freshwater Ecosystems.

Gregory, K.J., Gurnell, A.M., Hill, C.T., 1985. The permanence of debris dams related to river channel processes. Hydrol. Sci. J. 30 (3).

Grodziński, W., Klekowski, R.Z., Duncan, A., (Eds.), 1975. In: Methods for Ecological Bioenergetics. Blackwell Scientific Publications.

Grossman, G.D., Moyle, P.B., Whitaker, J.O., 1982. Stochasticity in structural and functional characteristics of an Indiana stream fish assemblage: a test of community theory. Am. Nat. 120 (4), 423–454. https://doi.org/10.1086/284004.

Gurnell, A.M., Gregory, K.J., Petts, G.E., 1995. The role of coarse woody debris in forest aquatic habitats: implications for management. Aquat. Conserv. Mar. Freshw. Ecosyst. 5 (2), 143–166. https://doi.org/10.1002/aqc.3270050206.

Holmes, R.M., Peterson, B.J., Deegan, L.A., Hughes, J.E., Fry, B., 1994. Nitrogen biogeochemistry in rivers: An isotopic perspective. Biogeochemistry 27, 113–133.

Holmgren, M., Scheffer, M., Huston, M., 1997. The interplay of facilitation and competition in plant communities. Ecology 78 (7), 1966–1975.

Holtby, L.B., 1988. Effects of logging on stream temperatures in Carnation Creek, British Columbia, and associated impacts on the coho salmon (*Oncorhynchus kisutch*). Canad. J. Fish. Aquat. Sci. 45, 502–515.

Huet, M., 1949. Aperçu des relations entre la pente et les populations piscicoles des eaux courantes. Schweiz. Z. Hydrol. 11 (3–4), 332–351. https://doi.org/10.1007/bf02503356.

Huet, M., 1954. Biologie, profils en long et en travers des eaux courantes. Bull. Fran. Piscicult. 175, 41–53. https://doi.org/10.1051/kmae:1954001.

Indika Herath, Z., Vithanage, M., 2015. Phytoremediation in Constructed Wetlands Phytoremediation: Management of Environmental Contaminants, vol. 2. Springer International Publishing, Sri Lanka, pp. 243–263. https://doi.org/10.1007/978-3-319-10969-5_21.

Izydorczyk, K., Frątczak, W., Drobniewska, A., Cichowicz, E., Michalska-Hejduk, D., Gross, R., Zalewski, M., Wagner, I., 2014. A biogeochemical barrier to enhance a buffer zone for reducing diffuse phosphorus pollution—Preliminary results. Ecohydrol. Hydrobiol. 14 (2), 104–112.

Junk, W.J., Bayley, P.B., Sparks, R.E., 1989. The Flood Pulse Concept in River-Floodplain Systems, vol 106. Canadian Special Publication of Fisheries and Aquatic Sciences, pp. 110–127.

Kiedrzyńska, E., Kiedrzyński, M., Urbaniak, M., Magnuszewski, A., Skłodowski, M., Wyrwicka, A., Zalewski, M., 2014. Point sources of nutrient pollution in the lowland river catchment in the context of the Baltic Sea eutrophication. Ecol. Eng. 70, 337–348. https://doi.org/10.1016/j.ecoleng.2014.06.010.

Kiedrzyńska, E., Kiedrzyński, M., Zalewski, M., 2015. Sustainable floodplain management for flood prevention and water quality improvement. Nat. Hazards 76 (2), 955–977. https://doi.org/10.1007/s11069-014-1529-1.

Kolkwitz, R., Marsson, M., 1908. Ökologie der pflanzlichen Saprobien. Berichte der Deutschen Botanischen Gesellschaft 26a, 505–519.

Knauer, N., Mander, U., 1989. Untersuchungen ueber die Filterwirkung verschiedenen Saunbiotope am Gewassen in Schlezwig-Holstein. 1. Mittelung: Filterung von Stickstoff und Phosphor. Z. Kulturrechnik und Landentwicklung 30, 365–376.

Kolkwitz, R., Marsson, M., 1909. Ökologie der tierischen Saprobien. Beiträge zur Lehre von der biologischen Gewässerbeurteilung. Int. Rev. Hydrobiol. Hydrogr. 2 (1–2), 126–152. https://doi.org/10.1002/iroh.19090020108.

Krebs, C.J., 1972. Ecology: The Experimental Analysis of Distribution and Abundance. Harper & Row.

Krebs, J.R., Davies, N.B. (Eds.), 1978. Behavioural Ecology: An Evolutionary Approach. Blackwell Scientific Publications.

Lancaster, J., Allan, J.D., 1995. Stream ecology: structure and function of running waters. J. Ecol. 83 (4), 735. https://doi.org/10.2307/2261644.

Łapińska, M., 1997. Przestrzeń jako czynnik limitujący zespoły ryb w rzece nizinnej i wyżynnej (Space as a limiting factor for fish communities in lowland and upland river systems), (doctoral dissertation). Uniwersytet Łódzki, Łódź.

Levia, D.F., Creed, I.F., Hannah, D.M., Nanko, K., Boyer, E.W., Carlyle-Moses, D.E., van de Giesen, N., Grasso, D., Guswa, A.J., Hudson, J.E., Hudson, S.A., Iida, S., Jackson, R.B., Katul, G.G., Kumagai, T., Llorens, P., Ribeiro, F.L., Pataki, D.E., Peters, C.A., Carretero, D.S., Selker, J.S., Tetzlaff, D., Zalewski, M., Bruen, M., 2020. Homogenization of the terrestrial water cycle. Nat. Geosci. 13 (10), 656–658. https://doi.org/10.1038/s41561-020-0641-y.

Lindeman, R.L., 1942. The trophic-dynamic aspect of ecology. Ecology 23 (4), 399–417. https://doi.org/10.2307/1930126.

Lock, M.A., Williams, D.D., 1981. Perspectives in running water ecology: The importance of the hyporheic zone. J. North Am. Benthol. Soc. 1 (2), 86–97.

Lowe-McConnell, R.H., 2010. Ecological Studies in Tropical Fish Communities. Cambridge University Press. https://doi.org/10.1017/CBO9780511721892.

Luoma, S.N., Rainbow, P.S., 2008. Metal Contamination in Aquatic Environments: Science and Lateral Management. Cambridge University Press.

Mann, K.H., Pęczak, T., 1986. Fish production in rivers: a review. Polish Archiv. Hydrobiol. 33, 233–247.

Mason, J., 2002. Qualitative Researching. SAGE Publications.

Metcalfe, N.B., Thorpe, J.E., 1990. Determinants of geographical variation in the age of seaward-migrating Atlantic salmon (*Salmo salar*). J. Animal Ecol. 59 (1), 135–145.

Minshall, G.W., Cummins, K.W., Petersen, R.C., Cushing, C.E., Bruns, D.A., Sedell, J.R., Vannote, R.L., 1985. Developments in stream ecosystem theory. Can. J. Fish. Aquat. Sci. 42 (5), 1045–1055. https://doi.org/10.1139/f85-130.

Mioduszewski, W., 2004. Modelling of Hydrological System of the Biebrza Valley This Work, Mioduszewski and Colleagues Applied the Groundwater Flow Model SIMGRO to Identify Hydrological Zones in the Lower Biebrza Basin.

Mitchell, R.E., Kirby Jr., A.M., 1976. Performance characteristics of pond aeration devices. Proc. Annu. Meet. World Maric. Soc. 7, 561–581.

Moore, J.W., Gilbert, D.J., 1987. The trophic structure of freshwater ecosystems. Ecol. Monog. 57 (2), 101–124.

Murphy, M.L., Hall, J.D., 1981. Varied effects of clear-cut logging on predators and their habitat in small streams of Cascade Mountains, Oregon. Canad. J. Fish. Aquat. Sci. 38, 137–145.

Murphy, M.L., Hall, J.D., 1986. Habitat use by juvenile salmonids in streams. Transact. Am. Fish. Soc. 115 (4), 592–602.

Murphy, M.L., Heifetz, J., Johnson, S.W., Koski, K.V., Thedinga, J.F., 1986. Effects of clear-cut logging with and without buffer strips on juvienile salmonids in Alaska streams. Canad. J. Fish. Aquat. Sci. 43, 1521–1533.

Naiman, R.J., 1976. Productivity of a herbivore population in a freshwater ecosystem. Oecologia 23, 261–273.

Naiman, R.J., 1983. The influence of stream size on the food guality of seston. Canad. J. Zool 61, 1995–2010.

Naiman, R.J., Bechtold, J.S., Beechie, T.J., Latterell, J.J., van Pelt, R., 2010. A Process-based view of floodplain forest patterns in coastal river valleys of the Pacific Northwest. Ecosystems 13 (1), 1–31. https://doi.org/10.1007/s10021-009-9298-5.

Naiman R.J., Decamps H., Fourier F. (red.), 1989, Role of Land/Inland Water Ecotons in Landscape Management and Restoration, Proposals for Collaborative Research, UNESCO, Vendome.

Naiman, R.J., Melillo, J.M., Lock, M.A., Ford, T.E., Reice, S.R., 1987. Longitudinal patterns of ecosystem processes and community structure in a subarctic river continuum. Ecology 68 (5), 1139–1156. https://doi.org/10.2307/1939199.

Newbold, J.D., Elwood, J.W., O'Neill, R.V., Van Winkle, W., 1981. Measuring nutrient spiralling in streams. Can. J. Fish. Aquat. Sci. 38 (7), 860–863. https://doi.org/10.1139/f81-114.

Odum, E.P., 1969. The strategy of ecosystem development. Science 164 (3877), 262–270. https://doi.org/10.1126/science.164.3877.262.

Palmer, C.V., Modi, C.K., Mydlarz, L.D., 2009. Coral fluorescent proteins as antioxidants. PLoS One 4 (10). https://doi.org/10.1371/journal.pone.0007298UnitedKingdom. Available from: http://www.plosone.org/article/fetchObjectAttachment.action?uri=info%3Adoi%2F10.1371%2Fjournal.pone.0007298&representation=PDF.

Panchan, A., Pelz, D., Friedrich, T., Zeiringer, B., Hein, T., Unfer, G., 2022. Habitat use and spawning site fidelity of barbel and nase in large European rivers. Front. Environ. Sci. https://doi.org/10.3389/fenvs.2022.1006427.

Penczak, T., Molinski, M., Kusto, E., Galusiak, H., Panusz, H., Zalewski, M., 1976. The ecology of roach (*Rutilus rutilus* L.) in the barbel region of the polluted Pilica River. Part 2. Dry weight, ash and contents of some elements. Ekol. Pol. 24 (4), 623–638.

Penczak, T., Zalewski, M., 1974. Distribution of fish numbers and biomass in the barbel region of the river and the adjoining old river-beds. Ekol. Pol. 22, 107–119.

Pastor, J., Post, W.M., 1986. Nitrogen mineralization and nitrification in soils of four forest types in Minnesota. Biogeochemistry 2 (2), 137–162. https://doi.org/10.1007/BF02180100.

Penczak, T., 1992. Contribution to energy transformation by fish populations in a small tropical river, North Venezuela. Comp. Biochem. Physiol. Physiol. 101 (4), 791–798. https://doi.org/10.1016/0300-9629(92)90359-x.

Petersen, R.C., Petersen, L.B.M., Lacoursiere, J., 1992. A building-block model for stream restoration. River Conserv. Manag. 293–309.

Piégay, H., Gurnell, A.M., 1997. Large woody debris and river geomorphological pattern: examples from S.E. France and S. England. Geomorphology 19 (1–2), 99–116. https://doi.org/10.1016/s0169-555x(96)00045-1.

Poff, N.L.R., Allan, J.D., Bain, M.B., Karr, J.R., Prestegaard, K.L., Richter, B.D., Sparks, R.E., Stromberg, J.C., 1997. The natural flow regime: a paradigm for river conservation and restoration. Bioscience 47 (11), 769–784. https://doi.org/10.2307/1313099.

Portt, C.B., Balon, E.K., Naokes, D.L.G., 1986. Biomass and production of fishes in natural and channelized streams. Canad. J. Fish. Aquat. Sci. 45, 1436–1447.

Rodríguez-Castillo, T., Estévez, E., González-Ferreras, A.M., Barquín, J., 2019. Estimating ecosystem metabolism to entire river networks. Ecosystems 22 (4), 892–911. https://doi.org/10.1007/s10021-018-0311-8.

Ross, S.T., Baker, J.A., 1983. The response of fishes to periodic spring floods in a southeastern stream. Am. Midl. Nat. 109 (1), 1–14. https://doi.org/10.2307/2425509.

Schiemer, F., Spindler, T., Wintersberger, H., Schneider, A., Chovanec, A., 2017. Fish fry associations: important indicators for the ecological status of large rivers. SIL Proc. 24 (4), 2497–2500. https://doi.org/10.1080/03680770.1989.11899997.

Schiemer, F., Waidbacher, H., 1992. Strategies for Conservation of a Danubian Fish Fauna. River Conservation and Management, pp. 363–382.

Salonen, K., Jokinen, S., 1988. Flagellate grazing on bacteria in a small dystrophic lake. Hydrobiologia 161, 203–209.

Schiemer, F., Spindler, T., 1989. Endangered fish species of the Danube river in Austria. Regul. Rivers Res. Manag. 4 (4), 397–407. https://doi.org/10.1002/rrr.3450040407.

Schimer, V.E., Zalewski, M., 1992. The importance of riparian ecotones for diversity and productivity of riverine fish communities. Netherland J. Zool. 42, 323–335.

Schlosser, I.J., 1982. Trophic structure reproductive success and growth rate of fishes in natural and modified headwater stream. Canad. J. Fish. Aquat. Sci. 39, 968–978.

Scholsser, I.J., Karr, J.R., 1981. Riparian vegetation and channel morphology impact on special patterns of water quality in agricultural watersheds. Environ. Manage. 5, 232–243.

Shelford, V.E., 1929. Laboratory and Field Ecology: The Responses of Animals as Indicators of Correct Working Methods. Saunders.

Stanford, J.A., 1987. Ecology of regulated rivers. In: Craig, J.F., Kemper, J.B. (Eds.), Regulated Streams: Advances in Ecology. Plenum Press, pp. 1–18.

Stockner, J.G., Shortreed, K.S., 1978. Enhancement of autotrophic production in lakes. J. Fish. Res. Board of Canada 35 (7), 993–1004.

Szklarek, S., Wagner, I., Jurczak, T., Zalewski, M., 2018. Sequential Sedimentation-Biofiltration System for the purification of a small urban river (the Sokolowka, Lodz) supplied by stormwater. J. Environ. Manag. 205, 201–208. https://doi.org/10.1016/j.jenvman.2017.09.066.

Taylor, B.R., 1981. Detritus dynamics in stream ecosystems. Ann. Rev. Ecol. System. 12, 327–353.

Tchobanoglous, G., Burton, Stensel, 2003. Wastewater Engineering: Treatment and Reuse. McGraw-Hill.

Thorpe, J.E., 1986. Age at First Maturity in Atlantic Salmon, *Salmo salar*: Freshwater Period Influence and Conflict with Smolting, vol 89. Canadian Special Publication of Fisheries and Aquatic Sciences, pp. 7–14.

Thorpe, J.E., Mangel, M., Metcalfe, N.B., Huntingford, F.A., 1989. Modelling the proximate basis of salmonid life-history variation. J. Fish Biol. 35 (Suppl. A), 1–14.

Thorpe J.E., 1989, Development variation in salmonid populations, J. Fish Biol. 35, (Suppl. A), 295–303.

Triska, F.J., Kennedy, V.C., Avanzino, R.J., Zellweger, G.W., Bencala, K.E., 1989. Retention and transport of nutrients in a third-order stream. Ecology 70 (6), 1893–1905.

Triska, F.J., Sedell, J.R., Buckley, B.M., 1975. The processing of conifer and hardwood leaves in two coniferous forest streams: I. Weight loss and associated invertebrates. Verhand. Int. Ver. Limnol. 19, 1617–1627.

Valett, H.M., Stanford, J.A., 1987. Food quality and hydropsychid caddis fly density in a lake outlet stream in Glacier National Park, Monstana, USA. Canad. J. Fish. Aquat. Sci. 44, 77–82.

Vannote, R.L., Minshall, G.W., Cummins, K.W., Sedell, J.R., Cushing, C.E., 1980. The river continuum concept. Can. J. Fish. Aquat. Sci. 37 (1), 130–137. https://doi.org/10.1139/f80-017.

Wagner, I., Zalewski, M., 2016. Temporal changes in the abiotic/biotic drivers of self purification in a temperate river. Ecol. Eng. 94, 275–285. https://doi.org/10.1016/j.ecoleng.2016.05.073.

Wallace, J.B., Webster, J.R., Cuffney, T.F., 1982. Stream detritus dynamics: Regulation by invertebrate consumers. Oecologia 53 (2), 197–200.

Webb, P.W., Kostecki, P.T., Don Stevens, E., 1984. The effect of size and swimming speed on locomotor kinematics of Rainbow trout. J. Exp. Biol. 109 (1), 77–95. https://doi.org/10.1242/jeb.109.1.77.

Webster, J.R., Patten, B.C., 1979. Effects of watershed perturbation on stream potassium and calcium dynamics. Ecol. Monogr. 49 (1), 51–72. https://doi.org/10.2307/1942572.

Welcomme, 1979. Fisheries Ecology of Floodplain Rivers.

Winemiller, K.O., 1990. Spatial and temporal variation in tropical fish trophic networks. Ecol. Monogr. 60 (3), 331–367. https://doi.org/10.2307/1943061. Available from: http://esajournals.onlinelibrary.wiley.com/hub/journal/10.1002/%28ISSN%291557-7015/issues/.

Włodek, J.M., 1976. Wielkość biomasy ryb w rzekach w zależności od zawartości różnych składników chemicznych w wodzie. Arch. Ochr. Srodowiska 2.

Zalewski, M., 2002a. Ecohydrology – the use of ecological and hydrological processes for sustainable management of water resources. Hydrol. Sci. J 47 (5), 825–834.

Zalewski, M.. (Ed), 2002b. Guidelines for the integrated management of the watershed – phytotechnology and ecohydrology, United Nations Environemant Programme. International Environmental Technology Centre, Freshwater Management Series 5, 188.

Zalewski, M., 2006. Ecohydrology—engineering harmony for a sustainable world. Ecohydrol. Hydrobiol. 6, 61–66.

Zalewski, M., Puchalski, W., Frankiewicz Bis, B., 1994. Rehabilitation of Freshwater Fisheries 152 160 Fishing News Books Riparian Ecotones and Fish Communities in Rivers—Intermediate Complexity Hypothesis.

Zalewski, M., 1986. Regulacja Zespołów Ryb W Rzekach Przez Kontinuum Czynników Abiotycznych I Biotycznych [Regulation of Fish Communities in Rivers by a Continuum of Abiotic and Biotic Factors, vol 86, pp. 1–86.

Zalewski, M., 2012. Ecohydrology for implementation of the EU Water Framework Directive. In: Proceedings of the European Parliament and Water Policy Discussions on Sustainable River Basin Management. Brussels: European Parliament.

Zalewski, M., 2013. Ecohydrology: process-oriented thinking towards sustainable river basins. Ecohydrol. Hydrobiol. 13 (2), 97–103. https://doi.org/10.1016/j.ecohyd.2013.06.001.

Zalewski, M., 2000. Ecohydrology—the scientific background to use ecosystem properties as management tools toward sustainability of water resources. Ecol. Eng. 16 (1), 1–8. https://doi.org/10.1016/s0925-8574(00)00071-9.

Zalewski, M. (Ed.), 2020. *Ekohydrologia*. Wydawnictwo Naukowe PWN.

Zalewski, M., Brewinska-Zaraś, B., Frankiewicz, P., Kalinowski, S., 1990a. The potential for biomanipulation using fry communities in a lowland reservoir: concordance between water quality and optimal recruitment. Hydrobiologia 200–201 (1), 549–556. https://doi.org/10.1007/BF02530372.

Zalewski, M., Frankiewicz, P., Przybylski, M., Bańbura, J., Nowak, M., 1990b. Structure and dynamics of fish communities in termperate rivers in relation to the abiotic-biotic regulatory continuum concept. Polskie Archiwum Hydrobiologii 37 (1–2), 151–176.

Zalewski, M., Frankiewicz, P., Nowak, M., 1995. Biomanipulation by ecotone management in a lowland reservoir. Hydrobiologia 303, 49–60.

Zalewski, M., Łapinska, M., Bayley, P.B., 2003. Fish relationships with wood in large rivers. Am. Fish. Soc. Symp. 2003 (37), 195–211.

Zalewski, M., Naiman, 1985. The Regulation of Riverine Fish Communities by a Continuum of Abiotic–Biotic Factors Habitat Modification and Freshwater Fisheries, pp. 3–9.

Zalewski, M., 1988. The role of riparian ecotones in river ecosystem functioning. Hydrobiologia 166, 3–11.

Zalewski, M., Robarts, R.D., 2003. Ecohydrology—a new paradigm for integrated water resources management. SIL News 40, 1–5.

Zalewski M., Frankiewicz P., Brewińska B., 1985. The factors limiting growth and survival of Brown trout, *Salmo trutta* m. fario L. introduced to different types of streams. J. Fish Biol. 27 (A), 59–73.

Zalewski, M., Thorpe, J.E., Naiman, R.J., 2001. Fish and riparian ecotones—a hypothesis. Ecohydrol. Hydrobiol. 1 (1–2), 11–24.

Zalewski, M., Wagner, I., Frątczak, W., Mankiewicz-Boczek, J., Parniewski, P., 2012. Blue-Green city for compensating global climate change. Parlam. Magaz. 350, 2–3.

Zalewski, M., Thorpe, J.E., Gaudin, P., 1991. Fish and land/inland-water ecotones. In: Man and the Biosphere Series, 4, pp. 81–96. UNESCO.

Chapter 6

Dual regulation inductive framework for transdisciplinary ecohydrology

Maciej Zalewski
European Regional Centre for Ecohydrology of the Polish Academy of Sciences under the auspices of UNESCO, Łódź, Poland

PHOTO CREDIT: M. ZALEWSKI.

Transdisciplinary Ecohydrology. https://doi.org/10.1016/B978-0-443-41596-8.00025-X

1. Foundations of the transdisciplinary ecohydrology approach

Ecohydrology emerged as a scientific discipline grounded in both theoretical insight and empirical observation, particularly through the application of inductive reasoning, the process of drawing general conclusions from specific data. The development of ecohydrology under UNESCO's Intergovernmental Hydrological Program (IHP) was significantly influenced by several foundational frameworks, including the Abiotic–Biotic Regulattory Concept (ABRC) (see Chapter 4) and the UNESCO Man and the Biosphere (MAB) program "Role of land/inland water ecotones in landscape manegement and restoration" (Naiman et al., 1989). An additional pillar supporting the field's emergence is ecological engineering (Mitsch and Jørgensen, 1989), which stems from Odum's systems-based ecological perspective and is particularly focused on the functioning of both natural and constructed wetlands. These systems have proven effective in treating stormwater, agricultural runoff, and pollution from diffuse sources (Mitsch and Gosselink, 1993).

While ecological engineering typically uses the wetland ecosystem as the primary analytical unit, ecohydrology employs a broader catchment-scale framework. It is concerned with the hydrological mezzocycle as the regulatory backbone for ecosystem processes, including biogenic circulation, energy flow, competition, predation, and trophic cascades—conceptualized and structured by the ABRC model (see Chapter 4).

The first definition in ecohydrology was posited in the mid-1990s—as the science of integrating knowledge to better understand processes and harness them in sustainable management of water resources (Zalewski et al., 1997). The next step in the development of the discipline involved the use of ecosystem properties and processes as water management tools (Zalewski, 2000). Next, phytotechnology became an integral part of ecohydrology (Zalewski and Wagner, 2004), largely thanks to support from the United Nations Environment Program (UNEP).

Recently, in the course of collaboration between UNESCO IHP and IAHR to integrate Transdisciplinary Ecohydrology with hydroengineering and Ecosystem Ecohydraulics, the origins and development of ecohydrology can also be examined through the lens of two complementary theories operating at different spatial and temporal scales: the ecosystem theory and the Gaia theory. Both theories look at the interactions between live organisms and use the notion of "superorganism" to better understand the concept and illustrate the ecosystem's characteristics. Neither, however, should be associated with a superorganism in the sense of a group of organisms having the physiological traits, especially metabolism of an individual organism (Lincoln et al., 1982).

2. Ecosystem function and regulation in transdisciplinary ecohydrology

Ecosystem theory originates from the term coined by Arthur Tansley (1935), yet its conceptual foundations can be traced back to earlier reflections on the interactions between living organisms and their environment. One of the earliest contributions was made by Forbes (1887, 1925), who perceived a lake ecosystem as a microcosm where survival and natural selection create a state of equilibrium.

A major advancement in ecosystem theory came with the study of nutrient cycling, which introduced a geochemical perspective to ecological processes. Lotka (1925) and Kopf and Lotka (1925) viewed the Earth as an organized system powered by solar energy, with interconnected components linked through chemical processes. This idea was further developed by Lindeman (1942), who analyzed the flow of energy through trophic levels in lakes and wetlands, demonstrating the correlation between biotic and abiotic factors. A decade later, Eugene Odum (1953) and Burt and Odum (1954) expanded the ecosystem model by incorporating biogeochemical cycles and describing how ecosystems evolve through different stages of maturity. He examined structural and energy balances within ecological communities, linking concepts such as life history, natural selection, and homeostasis. His later works (Odum, 1969, 1985) provided early frameworks for predicting ecosystem responses to human-induced changes.

Odum's theories on the anthropogenic impact on ecosystems were later validated through empirical research. Schindler (1987, 1990) in whole-lake experiments demonstrated that excessive nutrient loading, especially phosphorus, is the primary driver of lake eutrophication, and that long water residence times exacerbate nutrient retention. These findings underscore the need for stringent phosphorus control and the incorporation of hydrologic factors (such as residence time) into ecosystem management and ecohydrological strategies for effective lake restoration. Meanwhile, Gates (1963) investigated the role of biotic factors in energy flow, and Leopold and Langbein (1962) applied thermodynamic principles to ecological processes. Odum (1971, 1983) and Dickinson and Odum (1972) further developed these ideas, integrating geochemical models into ecosystem studies.

As ecosystem theory matured, researchers began considering longer time scales and linking ecological succession to system stability. Margalef (1960, 1963) proposed that ecosystems naturally progress toward more stable states. Southwood (1977, 1988) took this further, introducing the habitat templet concept, which suggests that species traits are shaped by environmental gradients. This framework was refined by Townsend and Hildrew (1994), who categorized ecological niches into biotic (competition and predation), chemical (nutrient levels, pollution), and physical (soil, relief, and climate) factors (Townsend and Hildrew, 1994).

Despite these advancements, ecosystem theory largely remained mechanistic, treating ecological processes as individual components rather than an interconnected system. This limitation made it difficult to explain self-sustainability and self-regulation within ecosystems, necessitating a paradigm shift (O'Neill, 2001).

The transition from mechanistic to cybernetic ecosystem models began with the understanding that ecosystems regulate themselves through feedback loops. Norbert Wiener (Johnson and Wiener, 1949) introduced the concept of cybernetics, which refers to circular causality, regulation, and feedback within a system. Odum (1956) later applied this to ecology, arguing that ecosystems function as cybernetic systems where energy flow, nutrient cycles, and communication networks interact dynamically.

Unlike human-made devices, where control functions are externally imposed, ecosystems rely on internal self-regulation. Feedback mechanisms allow ecosystems to adapt to external pressures, mitigating disturbances through dynamic equilibrium. For example, in aquatic ecosystems, shifts in water temperature, nutrient availability, and biotic interactions can regulate trophic dynamics and productivity levels.

Every catchment area has a unique hierarchy of drivers, shaped by geomorphology, climate, and vegetation that influence its hydrological and ecological dynamics. However, human activities such as deforestation, urbanization, industrialization, and transportation have altered these natural drivers, accelerating hydrological processes and disrupting ecological balance. These changes have led to an increased transfer of minerals, organic matter, nutrients, and pollutants from terrestrial to aquatic systems, causing severe environmental degradation (Chicharo and Zalewski, 2012; Kiedrzyńska et al., 2014; Meybeck, 2003; Wolanski, 2007). To mitigate these impacts, tailored measurement tools are required to analyze site-specific changes and to develop appropriate ecohydrological strategies. These strategies should be based on the identification and understanding of all relevant "Dual Regulation" processes, from the molecular to the catchment scale, translated into the development of a transdisciplinary ecohydrology. Although these solutions must be adapted to local conditions, their cumulative application at a global scale has the potential to drive transformative improvements, ultimately enhancing biosphere sustainability.

In transdisciplinary ecohydrology, ecosystems are viewed not as mechanical systems but as superorganisms, where interconnected components function similarly to the organs of a living being. Energy and nutrient flows, once studied as isolated processes, are now seen as part of a broader metabolic system. Disruptions to these natural flows, such as excessive nutrient runoff leading to eutrophication, can be likened to toxicity afflicting an organism.

3. Inductive background of transdisciplinary ecohydrology: Water level regulation for reduction of toxic algal blooms in reservoir

The importance of a thorough understanding of water quality–biota interactions for water management was first defined by a body of empirical evidence from lakes (e.g., biomanipulation (Brooks and Dodson, 1965; Hrbáček et al., 1961)), which show that eutrophication symptoms (algal blooms) can be reduced or eliminated by manipulating predator–prey relationships—predatory fish reduce zooplankton-feeding fish or fish fry, thus reducing the pressure on filtering zooplankton, increasing biofiltration many times over, and eliminating algal blooms despite high nutrient loads. Zalewski et al. (1990) were the first to demonstrate that by understanding the hydrological processes, the interplay between biota structure and hydrology dynamics can replace the role of predatory fish stocking. By lowering the water level during the spawning season (spring), the density of zooplanktonivorous fish fry in the Sulejów Reservoir can be reduced by as much as an order of magnitude. As a result, the filtering pressure of zooplankton on algae is increased, and the intensity of algal blooms in relation to phosphorus concentration can be reduced by half (see Chapter 15). The important effect is that such hydro-biomanipulation increases fish yield, because the overdense population of juvenile fish in the spring was starved due to intense predation, and elimination of the large zooplankton filter feeders, and did not grow sufficiently to survive the winter.

Research on river basins, including analyses of temperature and water regulation, was derived from extensive field studies (Zalewski et al., 1989, 1990). The Sulejów Reservoir serves as an example of early findings that contributed to the development of ecohydrology's principles and methodology Box 6.1, Fig. 6.2. This summarizes the empirical findings on

BOX 6.1 Regulation of water levels in lowland reservoir for change of biological structure—reduction of toxic algal blooms

The series of diagrams presents the empirical data for hydrobiomanipulaiton that became the basis for the dual regulation concept Fig. 6.1.

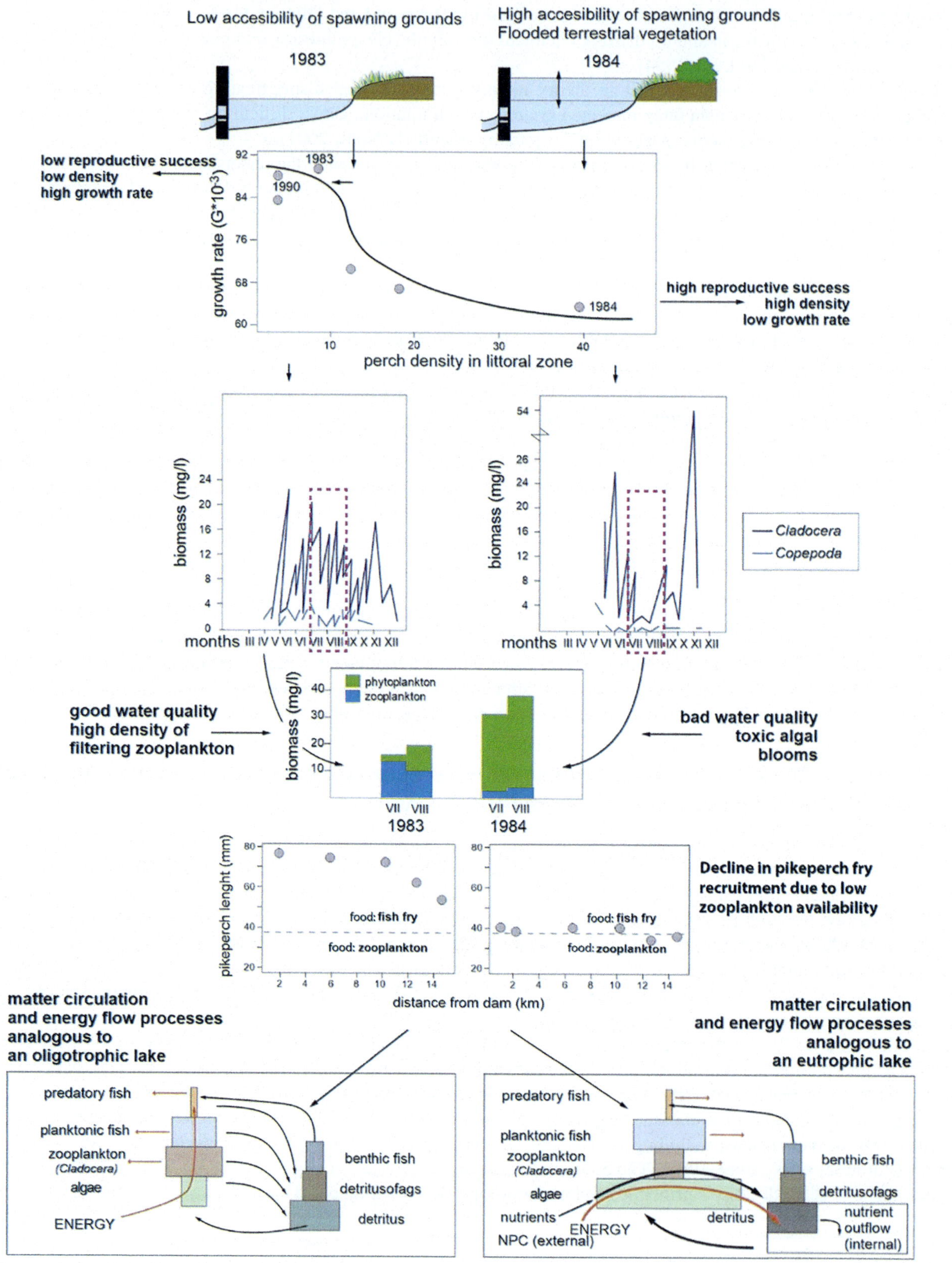

FIGURE 6.1 Empirical background for dual regulation. These empirical findings exemplify an inductive ecohydrological model linking hydrological conditions to biological dynamics. *Modified from Zalewski M., Brewińska-Zaraś, B., Frankiewicz,P., Kalinowski, S. 1990 , Hydrobiologia 200/ 2001,549–556.*

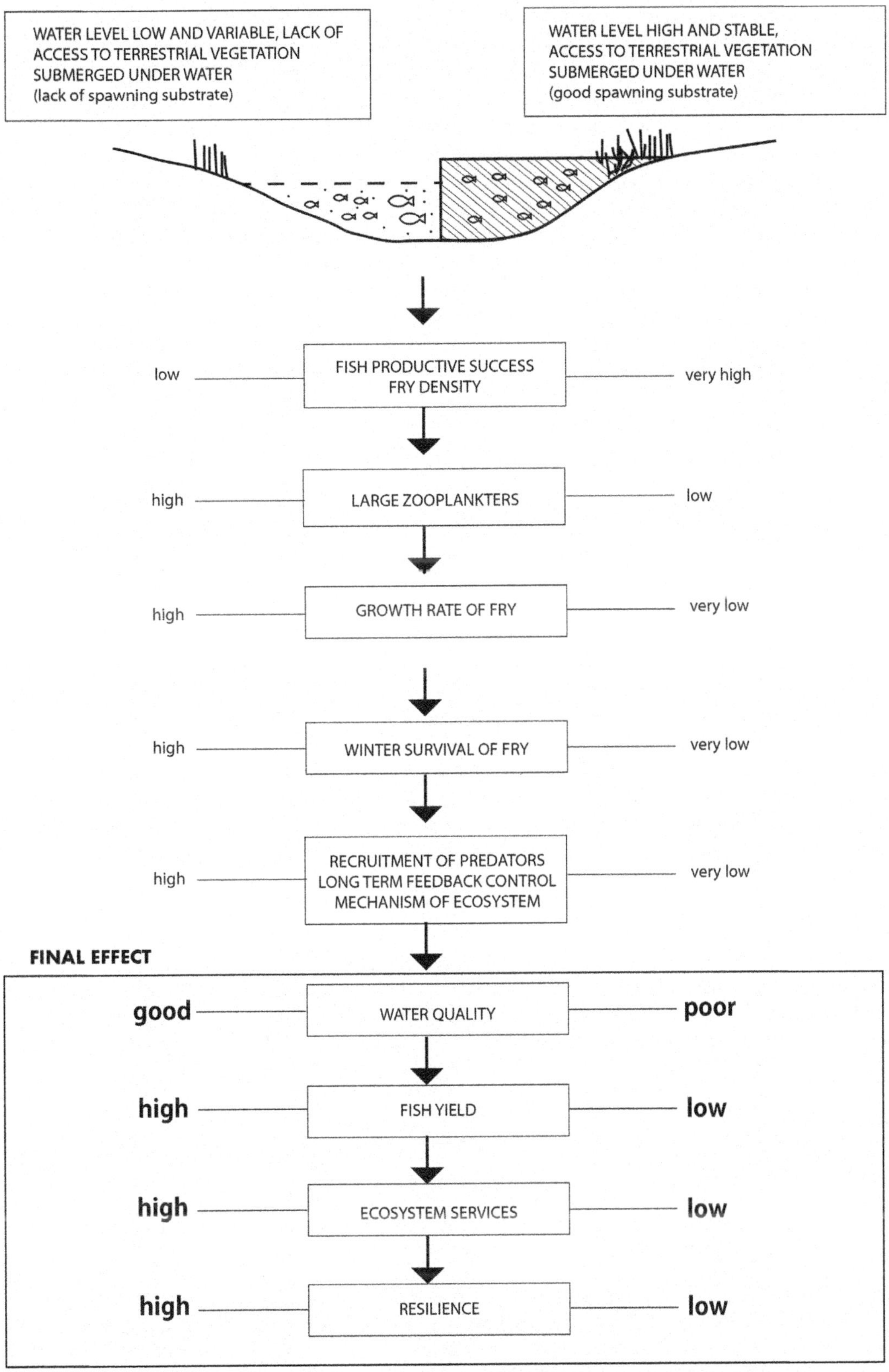

FIGURE 6.2 Synthesis of the role of water level regulation on biological structure in a lowland reservoir. This case demonstrates how targeted water level management can serve as an effective ecohydrological tool to regulate aquatic food webs and enhance ecosystem resilience. *Modified from Zalewski M., Brewinska-Zaras, B., Frankiewicz,P., Kalinowski, S. 1990, Hydrobiologia 200/2001, 549–556.*

the relationship between water levels and ecological processes. The regulation of the water level in spring within the Sulejów Reservoir serves as a regulatory factor in the population dynamics of planktivorous fish fry. This reduction in fish fry population, in turn, increases filtering zooplankton density, which significantly reduces algal blooms. However, when zooplankton populations are suppressed by zooplanktivorous fish fry, the growth rate and winter survival of pike-perch fry are drastically, and in consequence, reduced. This is an example of an inductive model of ecohydrological processes, which involves a comparison of the density of the filtering zooplankton in years of low versus high water levels during spawning. This results in fluctuations between low and high fry populations and determines the intensity of blooms in the reservoir.In the year with a low damming level (1983), the fry population density was $4/m^2$, the biomass of the filtering zooplankton (e.g., Daphnia) was high, 15 mg/L, and this intense biofiltration considerably reduced algal blooms in the reservoir. In contrast, in the year with a high damming level and high spawning habitat availability (1984), considerable reproductive success of planktivorous fish fry (40 mc^2) resulted in a drastic decrease in zooplankton density (<2 mg/L)—low biofiltration, and the occurrence of toxic blooms. Also, presented are changes in the structure of the trophic pyramid and energy pathways (red lines), and circulation of matter (black lines). The elimination of zooplankton by abundant fry (1984) drastically reduced the growth and survival rate of both zooplankton fish and pike-perch fry, and the survival rate in winter (Zalewski, 1997).

3.1 Hydrobiomanipulation—An advanced, low-cost nature-based solution

From the perspective of integrated water resources management (IWRM), the examples above illustrate not only the concept of dual regulation, referring to the use of water level regulation as a key factor for shaping the structure of biotic communities toward a positive effect on water quality, but also demonstrates how the regulation of hydrological processes can become a highly effective tool for improving water quality, fish community bioproductivity, and ecosystem services for society. This is achieved through enhanced resilience of biocenosis, even in the presence of high nutrient inflow from the catchment to the river–reservoir system.

It is worth emphasizing that the recreational potential of the areas surrounding the reservoir was once estimated at 50,000 visitors during the summer season. However, there were past periods when intense algal blooms appeared agrotourism activity dropped significantly, by nearly an order of magnitude. Given the economic losses associated with algal blooms, the region's sustainable development was severely limited. Although recent nutrient loads to the reservoir have decreased somewhat because of the development of sewage treatment plants in the catchment, and sometimes toxic algal blooms may appear, primarily because of increasing agricultural intensity, resulting in greater nonpoint source pollution, the expansion of recreational infrastructure in the surrounding areas, and a steady increase in periods of high temperature during summertime.

4. AI disclosure

During the preparation of this work, the author(s) used OpenAI's ChatGPT. This document includes content generated and/or edited with the assistance of OpenAI's ChatGPT, used to support language refinement and structural improvements. All final content was reviewed and verified by the author. After using this tool/service, the author(s) reviewed and edited the content as needed and take(s) full responsibility for the content of the publication.

References

Brooks, J.L., Dodson, S.I., 1965. Predation, body size, and composition of plankton. Science 150 (3692), 28–35. https://doi.org/10.1126/science.150.3692.28.

Burt, W.H., Odum, E.P., 1954. Fundamentals of ecology. J. Mammal. 35 (3), 462. https://doi.org/10.2307/1376006.

Chicharo, L., Zalewski, M., 2012. The importance of ecohydrology for the achievement of the United Nations Millennium Development Goals (MDGs). Ecohydrol. Hydrobiol. 12 (1), 8–24.

Dickinson, J.C., Odum, H.T., 1972. Environment, power, and society. Geogr. Rev. 62 (2), 294. https://doi.org/10.2307/213225.

Forbes, S.A., 1887. In: The lake as a microcosm. Bulletin of the Scientific Association of Peoria, pp. 77–87.

Forbes, S., 1925. The lake as a microcosm. Ill Nat. Hist. Surv. Bull. 15 (1–9), 537–550. https://doi.org/10.21900/j.inhs.v15.303.

Gates, D.M., 1963. Energy exchange in the biosphere. Soil Sci. 96 (1), 76. https://doi.org/10.1097/00010694-196307000-00019.

Hrbáček, J., Dvořakova, M., Kořínek, V., Procházkóva, L., 1961. Demonstration of the effect of the fish stock on the species composition of zooplankton and the intensity of metabolism of the whole plankton association. SIL Proc. 14 (1), 192–195. https://doi.org/10.1080/03680770.1959.11899269.

Johnson, H.G., Wiener, N., 1949. Cybernetics: control and communication in the animal and the machine. Econ. J. 59 (236), 573. https://doi.org/10.2307/2226579.

Kiedrzyńska, E., Kiedrzyński, M., Urbaniak, M., Magnuszewski, A., Skłodowski, M., Wyrwicka, A., Zalewski, M., 2014. Point sources of nutrient pollution in the lowland river catchment in the context of the Baltic sea eutrophication. Ecol. Eng. 70, 337–348. https://doi.org/10.1016/j.ecoleng.2014.06.010.

Kopf, E.W., Lotka, A.J., 1925. Elements of physical biology. J. Am. Stat. Assoc. 20 (151), 452. https://doi.org/10.2307/2965538.

Leopold, L.B., Langbein, W.B., 1962. The Concept of Entropy in Landscape Evolution. U.S. Geological Survey Professional Paper 500-A . U.S. Government Printing Office.

Lincoln, R.J., Boxshall, G.A., Clark, P.F., 1982. In: A Dictionary of Ecology, Evolution and Systematics.

Lindeman, R.L., 1942. The trophic-dynamic aspect of ecology. Ecology 23 (4), 399–417. https://doi.org/10.2307/1930126.

Lotka, A.J., 1925. Elements of Physical Biology. Williams & Wilkins Company, Baltimore.

Margalef, R., 1960. Ideas for a synthetic approach to the ecological niche. Publ. Inst. Biol. Apl. 28, 7–14.

Margalef, R., 1963. On certain unifying principles in ecology. Am. Nat. 97 (897), 357–374. https://doi.org/10.1086/282286.

Meybeck, M., 2003. Global analysis of river systems: from Earth system controls to Anthropocene syndromes. Phil. Trans. Biol. Sci. 358 (1440), 1935–1955. https://doi.org/10.1098/rstb.2003.1379.

Mitsch, W.J., Gosselink, J.G., 1993. Wetlands, 2nd ed. Van Nostrand Reinhold, New York.

Mitsch, W.J., Jørgensen, S.E., 1989. Ecological Engineering: An Introduction to Ecotechnology. John Wiley & Sons.

Odum, E.P., 1969. The strategy of ecosystem development. Science 164 (3877), 262–270. https://doi.org/10.1126/science.164.3877.262.

Odum, E.P., 1985. Trends expected in stressed ecosystems. Bioscience 35 (7), 419–422. https://doi.org/10.2307/1310021.

Naiman, R.J., Decamps, H., Fournier, F. (Eds.), 1989. The role of land/inland-water ecotones in landscape management and restoration: A proposal for collaborative research. MAB Digest, 4. UNESCO, Paris, pp. 94–pp.

Odum, E.P., 1953. Fundamentals of Ecology. W. B. Saunders Company, Philadelphia.

Odum, H.T., 1956. Power and circulation of energy through the ecosystem. Science 114, 343–349.

Odum, E.P., 1971. Fundamentals of Ecology, 3rd ed. W. B. Saunders Company, Philadelphia.

Odum, H.T., 1983. Systems Ecology: An Introduction. John Wiley & Sons.

O'Neill, R.V., 2001. Is it time to bury the ecosystem concept? (with full military honors, of course!). Ecology 82 (12), 3275–3284. https://doi.org/10.2307/2680151.

Schindler, D.W., 1987. Detecting ecosystem responses to anthropogenic stress. Can. J. Fish. Aquat. Sci. 44 (S1), s6. https://doi.org/10.1139/f87-276.

Schindler, D.W., 1990. Experimental perturbations of whole lakes as tests of hypotheses concerning ecosystem structure and function. Oikos 57 (1), 25–41. https://doi.org/10.2307/3565733.

Southwood, T.R.E., 1977. Habitat, the templet for ecological strategies? J. Anim. Ecol. 46 (2), 336. https://doi.org/10.2307/3817.

Southwood, T.R.E., 1988. Tactics, strategies and templets. Oikos 52 (1), 3–18. https://doi.org/10.2307/3565974.

Tansley, A.G., 1935. The use and abuse of vegetational concepts and terms. Ecology 16 (3), 284–307. https://doi.org/10.2307/1930070.

Townsend, C.R., Hildrew, A.G., 1994. Species traits in relation to a habitat templet for river systems. Freshw. Biol. 31 (3), 265–275. https://doi.org/10.1111/j.1365-2427.1994.tb01740.x.

Wolanski, E., 2007. Estuarine Ecohydrology Estuarine Ecohydrology. Elsevier, Australia. https://doi.org/10.1016/B978-0-444-53066-0.X5001-6.

Zalewski, M., 1997. Ecohydrology — the scientific background to use ecosystem properties as management tools toward sustainability of water resources. In: 1997 Proceedings of the International Symposium on Ecohydrology, pp. 9–18.

Zalewski, M., 2000. Ecohydrology — the scientific background to use ecosystem properties as management tools toward sustainability of water resources. Ecol. Eng. 16 (1), 1–8. https://doi.org/10.1016/s0925-8574(00)00071-9.

Zalewski, M., Bretschko, G., Morawska, K., Piekarska-Strońska, B., Hughes, R., Bravard, J.P., 1990. Evolution of ecological approaches in hydraulic engineering and river basin management. Proceedings IAHS Publication 193, 247–256.

Zalewski, M., Czerniawski, R., Frankiewicz, 1989. The role of hydrological and biotic factors in shaping the fish community structure in the lowland reservoir. In: 1989 Proceedings of the International Workshop on Reservoir Limnology and Water Quality, pp. 119–126.

Zalewski, M., Janauer, G.A., Jolankai, G., 1997. Ecohydrology: a new paradigm for the sustainable use of aquatic resources. In: Technical Documents in Hydrology.

Zalewski, M., Wagner, I., 2004. Ecohydrology – the use of ecological and hydrological processes for sustainable management of water resources. Hydrol. Sci. J. 49 (3), 467–478.

Chapter 7

Principles of transdisciplinary ecohydrology: System-based solutions for water management

Maciej Zalewski
European Regional Centre for Ecohydrology of the Polish Academy of Sciences under the auspices of UNESCO, Łódź, Poland

PHOTO CREDIT: M. ZALEWSKI.

In the Anthropocene era, numerous environments required protection due to their unique natural and cultural heritage. However, in the case of novel ecosystems (Hobbs et al., 2006), landscapes that have been entirely transformed by human activity, such as agricultural areas, traditional conservation strategies are often inadequate. These ecosystems have been so extensively altered that strict conservation measures become both impractical and ineffective. Moreover, demographic and economic factors further complicate conservation efforts. For instance, agriculture occupies approximately 48% of Europe's land area (Wood et al., 2019), highlighting the need for more pragmatic and sustainable land management strategies.

Transdisciplinary Ecohydrology. https://doi.org/10.1016/B978-0-443-41596-8.00011-X

To enhance the catchment sustainability potential (WBSR) of a given environment, the fundamental steps should involve the characteristics and quantification of: geomorphology, catchment hydrological processes, and the analysis of varies ecosystem types, natural, degraded, and modified, followed by an evaluation of the specific actions required for each. However, even the protection of highly valuable ecosystems cannot rely solely on legal restrictions for interventions and impacts. Instead, conservation efforts must integrate continuous monitoring of adjacent developed areas to mitigate their negative impacts. For example, controlling drainage from agricultural land adjacent to primeval forests or reducing the inflow of pollutants and nutrient loads into protected waters via groundwater and surface runoff are essential measures. Thus, even considering traditional conservation approaches, holistic perspective and comprehensive and hybrid, systemic solutions are necessary.

A distinct category of ecosystems, "novel ecosystems", comprises highly modified landscapes undergoing secondary succession (Hobbs et al., 2006). The challenge for ecohydrology is to enhance the sustainability potential—water resources and ecological status (biodiversity) of these ecosystems through Dual Regulation strategy that emphasizes both resilience-building (R) to withstand environmental stress and the optimization of ecosystem services (S) for societal benefit. Strengthening these adaptive capacities is essential to ensuring that novel ecosystems contribute meaningfully to broader ecological and socioeconomic systems.

In the assessment of ecological processes, the identification of potential threats, and the selection of appropriate management strategies, recognizing the spatial scale of environmental processes is crucial. Given the vast range and diverse forms of anthropogenic impacts on ecosystems, threats at local, regional, and global levels interact cumulatively. Local threats include point- and non-point source pollution affecting specific sections of rivers. Regional threats involve broader environmental disturbances, such as the emission of pollutants into the atmosphere, which can lead to acid rain and subsequently impact entire catchment areas. Global threats, such as the overall emission of CO_2 and other greenhouse gases, contribute to the intensification of the greenhouse effect, which alters the dynamics of the water cycle, leading to increased occurrences of floods, droughts, landslides, and hurricanes. Understanding these interconnected scales of impact is fundamental to developing effective, multi-scalar environmental management approaches.

Mutual interaction between hydrological conditions and living organisms can be moderated by creating ecotone zones, that are plantings near the banks of water courses, wetlands, as well as shaping vegetation and forest cover, and biomanipulation at particular trophic levels in aquatic ecosystems. Such regulation improves the condition of the environment and its self-purification potential. The creation of ecotone zones has a positive effect on increasing biodiversity in a given area, because it is such transition zones that constitute migratory pathways and refuge for many species. In turn, the creation of wetlands and ponds significantly increase water retentiveness and reduces nitrogen runoff from agricultural land and urbanized areas, improving land complexity and biodiversity.

Creating ponds and small reservoirs in natural depressions within the catchment landscape enhances water retention in the surrounding areas by replenishing groundwater, sustaining wetlands, and mitigating drought effects through evaporation from the reservoir and lateral precipitation Fig. 7.1.

The interwoven processes described above can be managed and designed using ecohydrological methodology. The innovation of ecohydrology lies in its dual approach: it not only integrates transdisciplinary sciences to deepen the understanding of interactions between aquatic environments and living organisms but also establishes a methodological framework for leveraging these interactions, ecosystem properties, and natural processes as tools for environmental management. This approach aims to enhance environmental capacity and resilience. This methodology integrates biocenotic solutions, which rely on the biological structure of ecosystems, such as land-water ecotones to reduce impact of nonsource pollution, with hydrological interventions, including adjustments to reservoir water levels to increase resilience. The goal is to create synergistic effects that enhance water quality, increase freshwater availability, and support both bioproductivity and biodiversity. Furthermore, Transdisciplinary Ecohydrology emphasizes the integration of innovative ecohydrological strategies with existing management frameworks and hydrotechnical infrastructure (Zalewski, 2006; Zalewski et al., 2003, 2004).

In addition to its overarching concept and objective of integrating existing knowledge on biological and hydrological processes, ecohydrology is founded on three key principles that guide both analysis and action. These principles foster holistic and comprehensive approaches to address ecological challenges within a catchment area. They also represent distinct stages of strategic planning and dimensions of data analysis, facilitating a deeper understanding of ecohydrological processes and the implementation of appropriate solutions.

By adhering to these principles, a catchment area can be conceptualized as a superorganism, as discussed in Chapter 3 Deductive framework for transdisciplinary ecohydrology. The principles establish a structured framework, define objectives, and provide a methodological basis for interventions within the catchment, enabling the synergy and integration

FIGURE 7.1 Hydrological benefits of small Ponds. A small pond has formed at the bottom of the dry stream valley increases air and soil moisture through lateral precipitation, which helps mitigate drought in adjacent fields while also supporting biodiversity. This process serves as one of the natural adaptations of the catchment landscape to climate change. *Photo credit: M. Zalewski.*

of hydrological, biological, and environmental engineering solutions. The primary goal of these actions is to enhance the quality, resilience, and capacity of the water resources and environment while ensuring its alignment with societal needs.

Ecohydrology-based interventions are ecologically aligned with the natural functions of the environment. Moreover, they are more cost-effective than conventional management approaches and can complement existing technical solutions. Furthermore, these principles serve as the foundation for integrating ecohydrology with Integrated Water Resource Management (IWRM), promoting a comprehensive and sustainable approach to water and ecosystem management.

The catchment-scale environment serves as the primary reference for formulating the principles of ecohydrology, where the integration of ecological processes is particularly evident, and the hydrological cycle provides a foundation for quantifying these processes. From a catchment-scale perspective, which considers water circulation, matter and energy flow, the hydrological cycle can be divided into three distinct phases:

Atmospheric phase- Climatic conditions, both present and future, regulate rainfall frequency, precipitation intensity, and overall water availability (Kaczmarek et al., 1996; Starkel, 1982), see Chapter 3.

Terrestrial phase- Vegetation cover plays a fundamental role in modulating the heat budget, evapotranspiration, infiltration, and groundwater recharge. It also influences surface runoff, and the quality of water transported through the landscape (Kabat, 2002; Rodriguez-Iturbe, 2000; Zalewski et al., 2003). The Food and Agriculture Organization of the United Nations (FAO) has published a wide range of reports and studies on water and soil (e.g., The State of the World's Land and Water Resources for Food and Agriculture–SOLAW, 2021).

Aquatic phase–In closed ecosystems (e.g., lakes), water circulation governs nutrient cycling, while in open ecosystems (e.g., rivers), flow conditions interact with habitat characteristics (regulated vs. natural land-water ecotones, floodplains, oxbow lakes) to influence biological productivity, self-purification processes, and biodiversity. Reflecting these dynamics, the *United Nations World Water Development Report 2024*, prepared by UNESCO WWAP on behalf of UN-Water, underscores the vital importance of sustainable and cooperative water management.

Although the nature of ecosystems, including plant biomass specific to different climatic zones, plays a crucial role in regulating nutrient cycling and the water cycle within catchments, much of the Earth's land surface, excluding desert, mountainous, and temperate regions, is now used for agriculture. The widespread loss of biomass and soil organic matter has significantly altered the hydrological cycle, disrupting the biological mechanisms that help stabilize hydrological dynamics and heat balance at levels suitable for plant growth.

In catchments with high plant biomass, water conditions and temperature fluctuations tend to be moderate, reducing the likelihood of catastrophic droughts and floods. Furthermore, diverse vegetation cover plays a key role in minimizing nutrient drainage from terrestrial to aquatic ecosystems, thereby enhancing carbon retention and maintaining low nutrient outflow, which supports overall ecosystem stability.

In the terrestrial phase, surface runoff plays a critical role, with the ecotone zone between terrestrial and aquatic ecosystems serving two essential functions. First, it acts as a natural filter, capturing nutrients and contaminants from surface and groundwater runoff converting a portion of these substances in to biomass and organic matter. Second, it retains organic matter, nutrients, and pollutants on the floodplain, thereby contributing to the self-purification processes of river ecosystems (Hillbricht-Ilkowska and Pieczyńska, 1993; Kiedrzyńska et al., 2014; Naiman and Decamps, 1990; Wagner and Zalewski, 2016; Zalewski et al., 1998; Zalewski and Wagner, 1998).

Stagnant water ecosystems, such as lakes and reservoirs, play a crucial role in the final stage of the terrestrial hydrological cycle by regulating river runoff. These ecosystems perform essential functions, including mineral and sediment accumulation, nutrient retention, and the transformation of transported nutrients and mineralized organic matter (Tundisi and Staskraba, 1999). Their water quality can be actively managed through dual regulation and hydrological processes, further enhancing their ecological stability and function (Zalewski et al., 1990a,b).

The terrestrial phase, especially agricultural landscape but also urbanized areas of the hydrological cycle play a crucial role in implementing ecohydrological solutions aimed at improving water quality and enhancing catchment sustainability. When abiotic factors remain stable, living organisms can thrive, and over time, these organisms begin to influence and shape the abiotic environment through their interactions.

In the terrestrial phase, plant cover serves as the primary water-retentive and nutrient-transforming system, exerting a significant influence on surface runoff, water quality, and evaporation rates. In contrast, in the aquatic phase, bank, coastal, and aquatic vegetation regulates nutrient flow, sediment accumulation, and eutrophication levels, ensuring ecological balance.

In 2004, the Pilica River Demonstration Project was carried out in Poland under the auspices of UNESCO and UNEP. The initiative explored the integration of ecohydrology and phytotechnology as a means to improve water resource management. By combining ecological and hydrological processes, the project aimed to enhance water quality and strengthen ecosystem health. The findings and methodologies were published in "Environment and Economy—Dual Benefit of Ecohydrology and Phytotechnology in Water Resources Management: Pilica River Demonstration Project" by Wagner-Łotkowska et al. (2004). The strategies proposed in this work focus on stabilizing hydrological dynamics by reducing runoff fluctuations, enhancing water retention across catchment areas, and supporting the restoration of natural hydrological cycles. By mitigating these fluctuations, ecohydrological approaches help to prevent issues such as erosion and nutrient leaching, ultimately contributing to more resilient and sustainable landscapes.

1. Hydrological principle of TEH

The first principle of ecohydrology, known as the hydrological principle, focuses primarily on processes related to the hydrological cycle within a catchment Fig. 7.2. It was formulated based on the works of Hynes (1970), Crisp and Hynes (1971), Naiman (1983), Hillbricht-Ilkowska (1985), Bajkiewicz-Grabowska and Mikulski (1993), Kaczmarek et al. (1996), and Starkel (1982).

This principle establishes the scale of analysis, emphasizing that the evaluation of catchment processes should begin with a comprehensive quantitative and qualitative assessment of hydrological dynamics. Activities should be planned in an integrated manner, considering the entire catchment as a cohesive system. The quantification of these processes is based on the circulation of water between the atmosphere, terrestrial, and aquatic ecosystems. This includes analyzing water flow along the river continuum and identifying sources of anthropogenic pressures within the catchment, such as point and non-point pollution sources.

A key issue in Transdisciplinary Ecohydrology (TEH) is understanding the interactions between biotic and abiotic factors, and more importantly, the hierarchical relationships that determine how these factors influence one another. These interactions shape ecosystem structure and function, from molecular processes to landscape-scale dynamics.

The Hydrological Principle is based on the premise that water and temperature are the primary factors regulating nutrients cycling and energy flow in terrestrial and freshwater ecosystems. Water availability influences bioproductivity and biodiversity, while temperature determines the length of the growing season and the metabolic rates of living

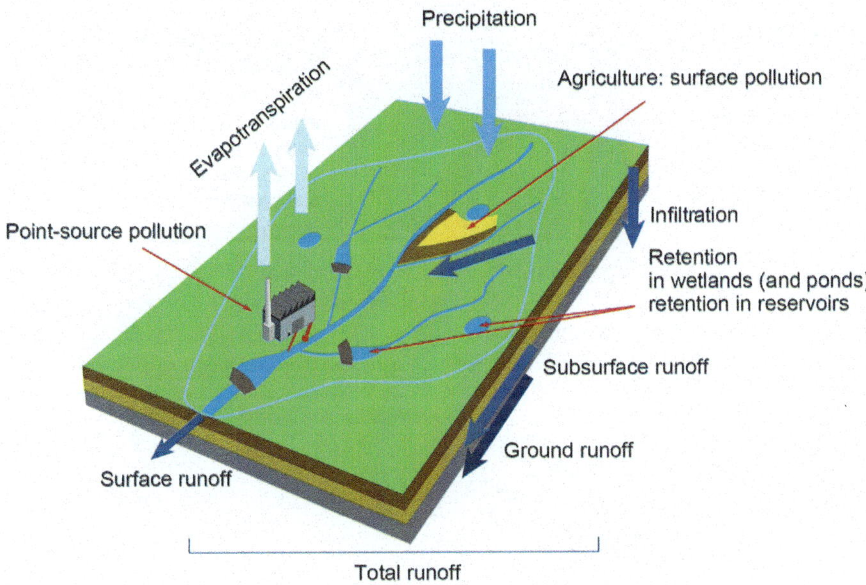

FIGURE 7.2 1st hydrological principle of TEH. The first principle of transdisciplinary ecohydrology—hydrological: quantification of the hydrological cycle and spatial distribution of various forms of anthropogenic pressure within the catchment. *Based on earlier concepts from Zalewski, M. (Ed.), 2020. Ekohydrologia. Wydawnictwo Naukowe PWN, Warszawa.*

organisms. Consequently, these factors affect an ecosystem's ability to accumulate retain and release essential nutrients such as carbon, nitrogen, and phosphorus.

Abiotic factors, such as hydrology, play a fundamental role in shaping ecosystems. Only when these factors remain stable and within the physiological tolerance range of aquatic organisms do biotic interactions begin to influence and modify ecosystem dynamics (Zalewski and Naiman, 1985). Due to this hierarchical structure, hydrological processes at the mesoscale level of a catchment provide a framework for describing and quantifying ecological processes, as well as for developing spatial management and environmental resource management strategies.

Quantifying these processes involves systematically describing abiotic conditions within the catchment, including precipitation levels, evaporation rates, groundwater and surface water quality, and sources of contamination. These sources can be categorized as point sources, such as sewage discharge (Kiedrzyńska et al., 2014), or non-point sources, such as nutrient runoff from agricultural areas (Izydorczyk et al., 2019).

Precipitation and infiltration at the given geological structure determine the rate of groundwater recharge, bioproductivity, and the ecosystem's resilience to pollution. These conditions are further influenced by landscape structure, including vegetation cover and its associated effects, such as wind speed and evaporation.

By considering abiotic conditions and incorporating landscape complexity at the planning level, land use can be optimized to maximize ecosystem services while minimising potential negative impacts on the natural environment Figs. 7.3 and 7.4.

As an example of the application of the first principle of ecohydrology, the modelling and quantification of nitrogen and phosphorus emission processes in the Pilica River basin provides aluable insights, see Chapter 3 for more details Figs. 7.5–7.7.

The results of SWAT modelling revealed a pronounced spatial variability in nitrogen and phosphorus emissions across the Pilica catchment. The analysis identified specific areas where land-water ecotones should be created, covering a total of 31,962 ha, that contribute 36% of the total nitrate load and 51% of the total phosphorus load. Notably, these areas constitute only 6.6% of the entire Pilica catchment and 16.3% of the arable land within it (Izydorczyk et al., 2019).

The analyses conducted demonstrated that SWAT modelling can be utilized not only for estimating emissions but also for identifying priority areas. Precisely pinpointing the focal sources of pollutant emissions within the catchment enables the selection of effective mitigation measures and their targeted, cost-efficient application (Program działań dla ograniczenia zanieczyszczeń obszarowych w zlewni Pilicy powyżej Zbiornika Sulejowskiego 2015). The insights gained from this research were used for the development of strategic documents for water resource management, such as the Catchment Management Plans required under the Water Framework Directive.

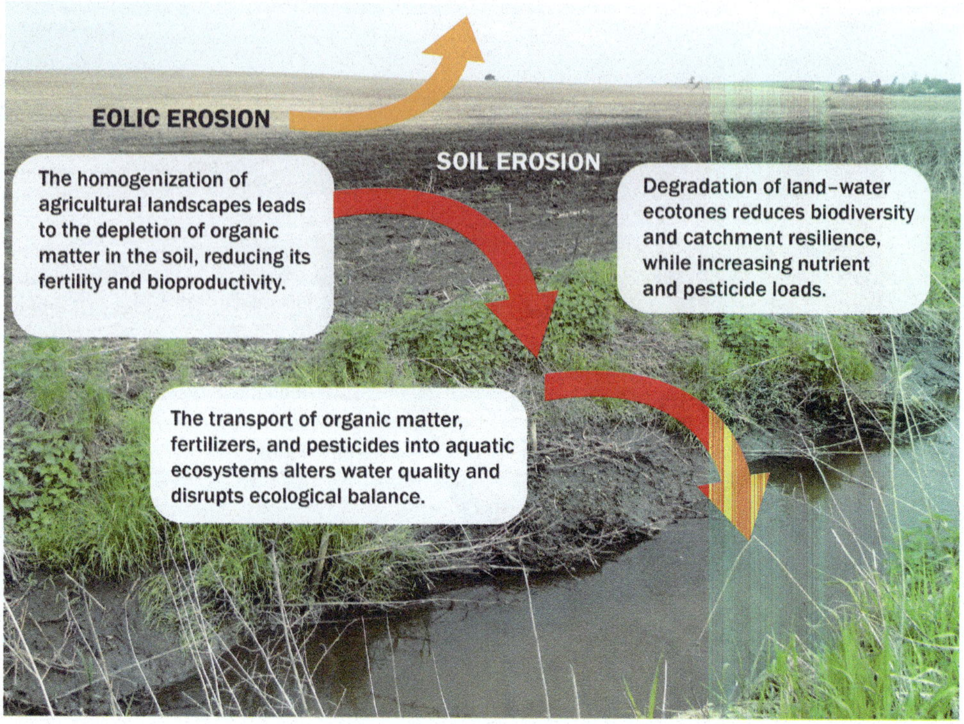

FIGURE 7.3 A drastic reduction in the catchment landscape complexity. The loss of organic matter in intensely cultivated and agricultural landscape—a drastic reduction in the complexity (reduction of tree rows and land/water ecotone buffer zones between)—the results are drying of soils and loss of organic matter as an effect of aeolian erosion and surface runoff. Additionally, fertilizers flowing down from the field to a stream and to a large extent contribute to an increase in the frequency and intensity of algal blooms in reservoirs located below. *Based on concepts from Zalewski, M., 2014; Ecohydrology and hydrologic engineering: Regulation of hydrology–biota interactions for sustainability. Journal of Hydrologic Engineering 19 (5), 1080–1088. https://doi.org/10.1061/(ASCE)HE.1943-5584.0000999.*

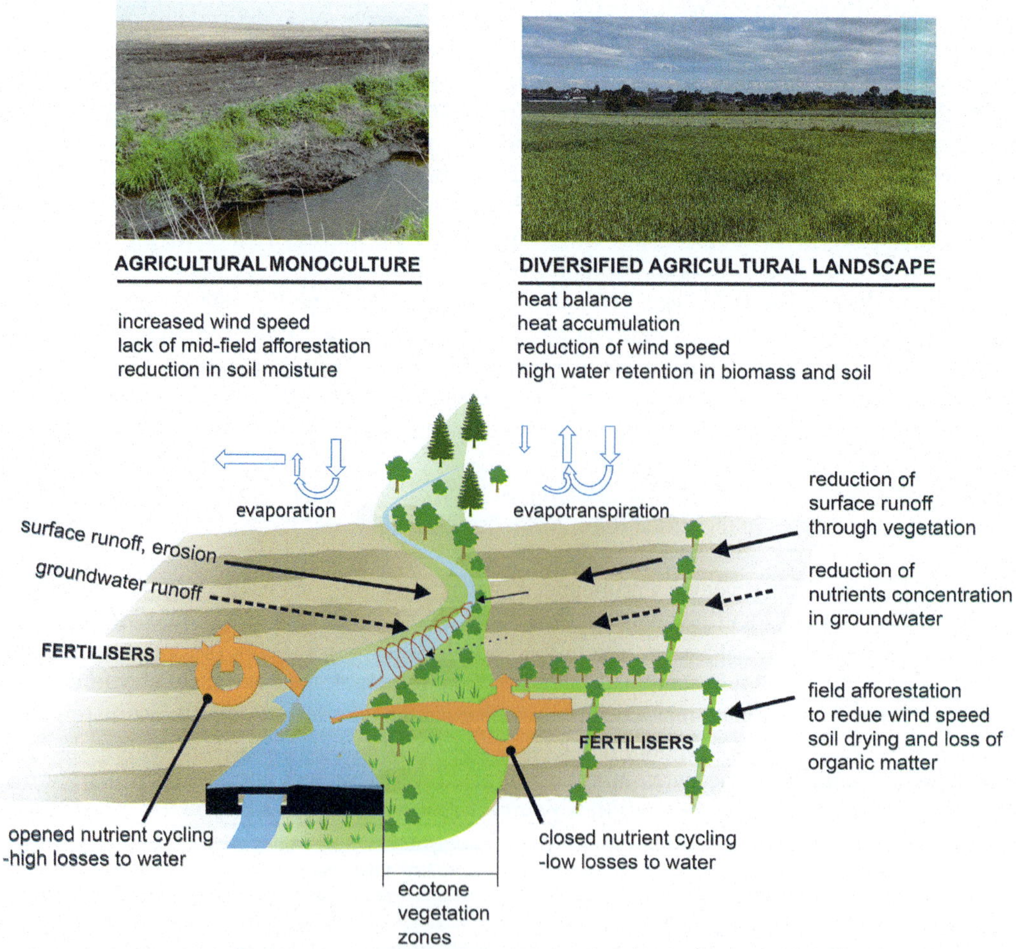

FIGURE 7.4 The importance of agricultural landscape complexity. The importance of agricultural landscape complexity (forests, trees and small mid-field reservoirs, highly effective land/water ecotones) for water retention, sensitivity to drought, productivity of agricultural areas and water quality (reduction of eutrophication and elimination of toxic blooms). *Based on earlier concepts from Zalewski, M. (Ed.), 2020. Ekohydrologia. Wydawnictwo Naukowe PWN, Warszawa.*

FIGURE 7.5 The importance of geology in assessing hydrological processes in Tatra Mountains streams. (A) The stream 'Dry Water,' which, due to its geological structure and a low water table during summer months with high evapotranspiration, remains dry for part of the year. (B) The photographs illustrates the importance of geology in assessing hydrological processes in Tatra Mountains streams. Chocholowska stream, which is characterized by limestone bedrock with high retention capacity maintains year-round flow. *Photo credit: M. Zalewski.*

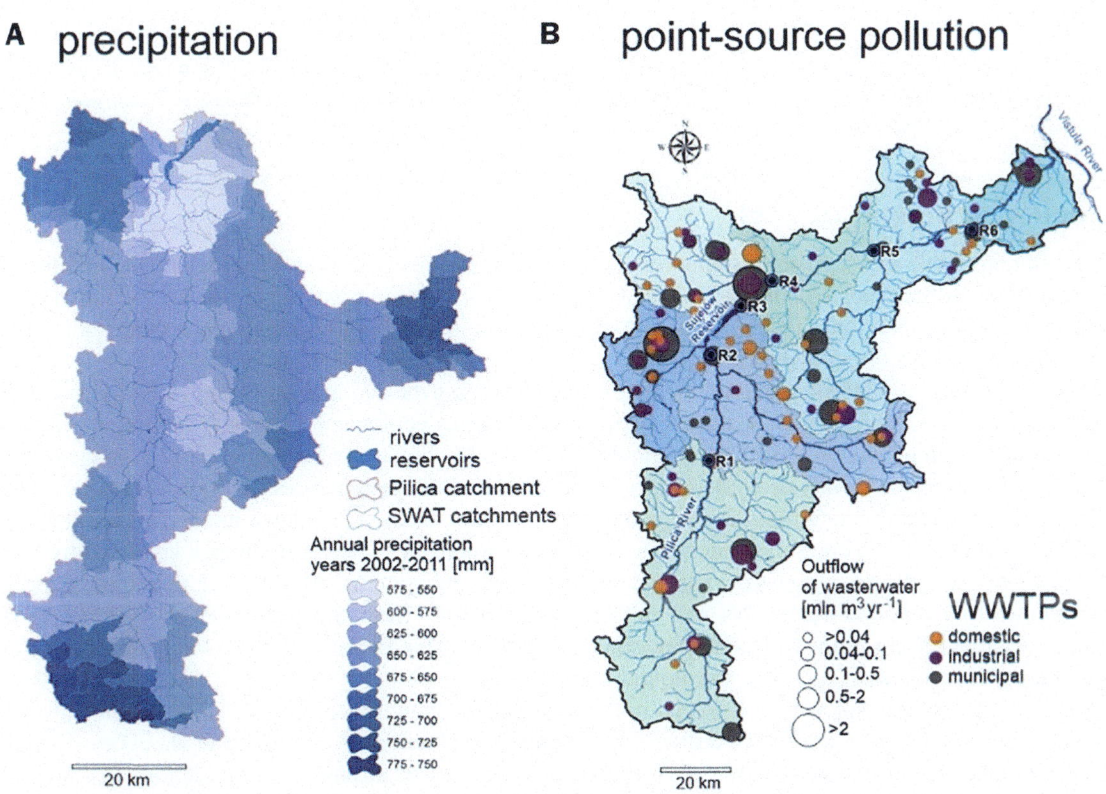

FIGURE 7.6 I hydrological principle of TEH: Precipitation and point-source Pollution. (A) Distribution of precipitation (B) Point-source pollution (Kiedrzyńska et al., 2014). *Reproduced from Frątczak, W., Izydorczyk, K. (Eds.), 2015. Action Programme for Reducing Diffuse Pollution in the Pilica River Catchment Upstream of the Sulejów Reservoir. ERCE PAS, ISBN 978-83-928245-2-7, (in Polish).*

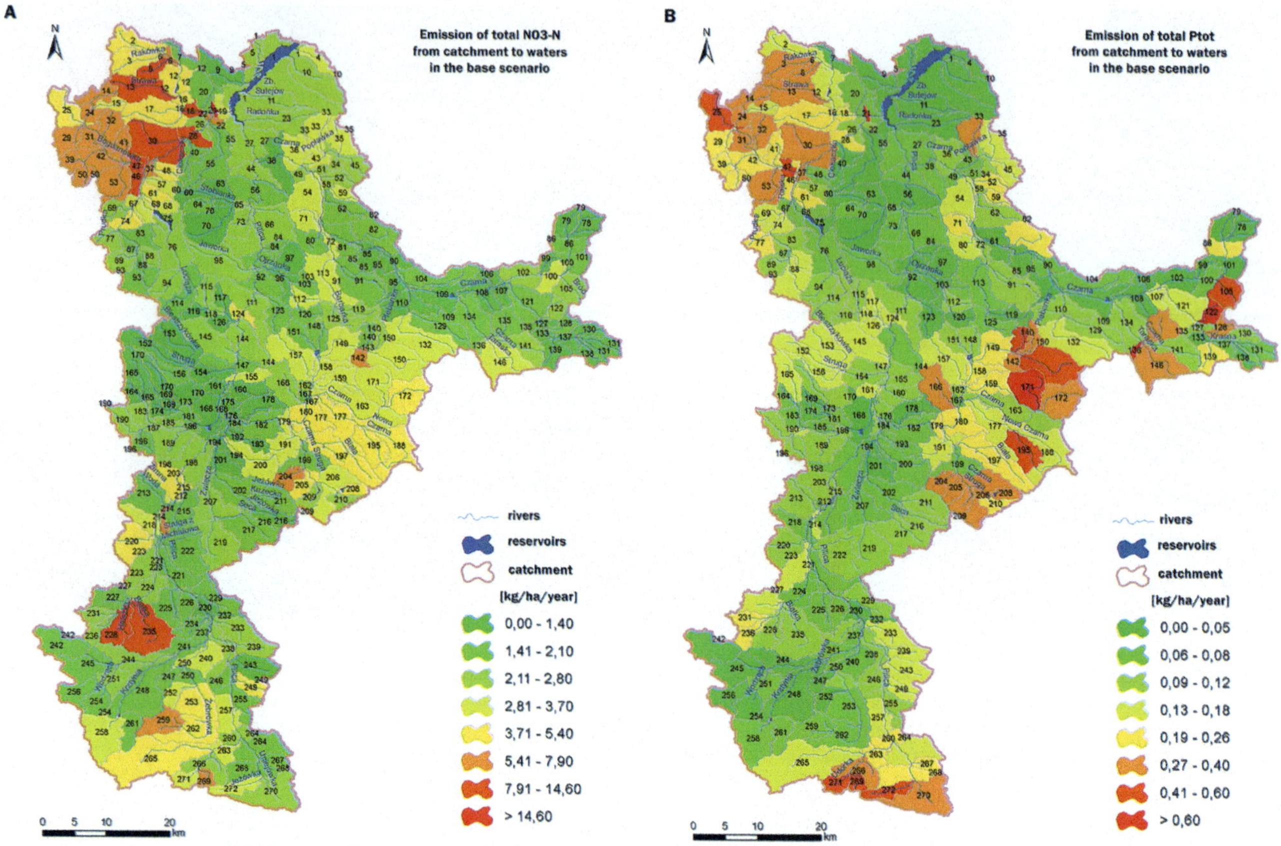

FIGURE 7.7 Pilica river basin nonpoint and point-source pollution distribution. Distribution of areas contributing the highest loads of total phosphorus (A) and nitrate nitrogen (B) within the Pilica River basin, shown in relation to administrative divisions. *Piniewski, M., Marcinkowski, P., Kardel, I., Giełczewski, M., Izydorczyk, K., Frątczak, W., 2015. Spatial quantification of nonpoint source pollution in a meso-scale catchment for an assessment of buffer zones efficiency. Water 7 (5), 1889–1916, https://doi.org/10.3390/w7051889.*

2. Ecological principle of TEH

According to this principle Fig. 7.8, ecosystems function as superorganisms with inherent resilience and the capacity to adapt to environmental pressures. At the catchment scale, this resilience emerges from the interactions between biotic communities and abiotic conditions. This principle serves as the foundation for enhancing environmental potential, which is the overarching objective of ecohydrology.

Moreover, this principle emphasizes the significance of understanding the evolutionary relationships between biocenoses and the hydrological cycle. It highlights the necessity of quantifying key ecological processes, such as nutrient cycling and energy flow and integrating them with the hydrological cycle at the catchment scale, in alignment with the first principle.

The quality of water in freshwater ecosystems largely depends on biological processes. Biological processes that drive the nutrient cycle in aquatic ecosystems are regulated by five key factors: habitat complexity, nutrient availability, temperature, light, and water cycling dynamics. The course and intensity of water dynamics significantly influence the biotic structure and activity of organisms. This is because these ecosystems are situated in terrain depressions and therefore continue to receive nutrients (N) and organic matter (OM) from the terrestrial ecosystems located above (allochthonous sources), from point and non-point source pollution. This organic matter is decomposed by aquatic organisms, including invertebrates, bacteria, and fungi.

Additionally, certain biotic factors play a crucial role in regulating abiotic processes. Wetlands and floodplain vegetation Fig. 7.9, for example, are key determinants of a river's self-purification potential (Kiedrzyńska et al., 2008). The nutrient load entering a river or lake originates from two primary sources: nutrients released through the degradation of biological structures (such as the decomposition of dead organisms) and those derived from point and diffuse source pollutants transported via surface runoff, which is exacerbated by erosion. Erosion and accelerated surface runoff result from the simplification of biological structures, the shortening of biological cycles (matter, water, and energy), and a decline in biodiversity.

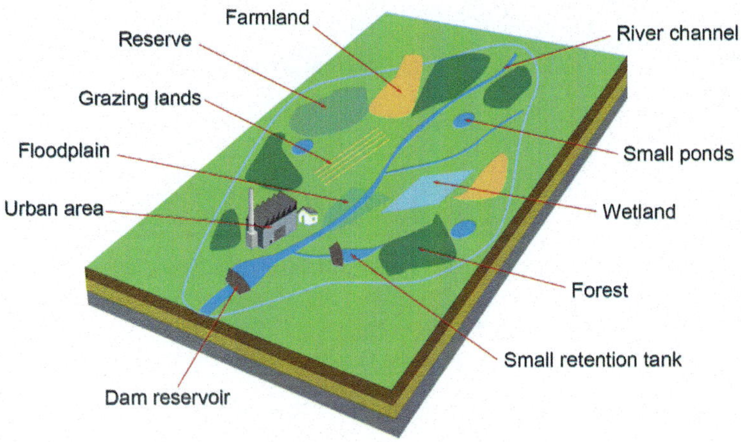

FIGURE 7.8 II ecological principle of TEH. The spatial distribution of ecosystems serves as the basis for identifying priority areas for the conservation of natural ecosystems and implementation of advanced nature-based solutions (ANbS), integrated synergistically with hydroengineering. This approach aims to enhance water- biodiversity-ecosystem services- resilience (WBSR), supported by the application of the dual regulation concept. *Based on earlier concepts from Zalewski, M. (Ed.), 2020. Ekohydrologia. Wydawnictwo Naukowe PWN, Warszawa.*

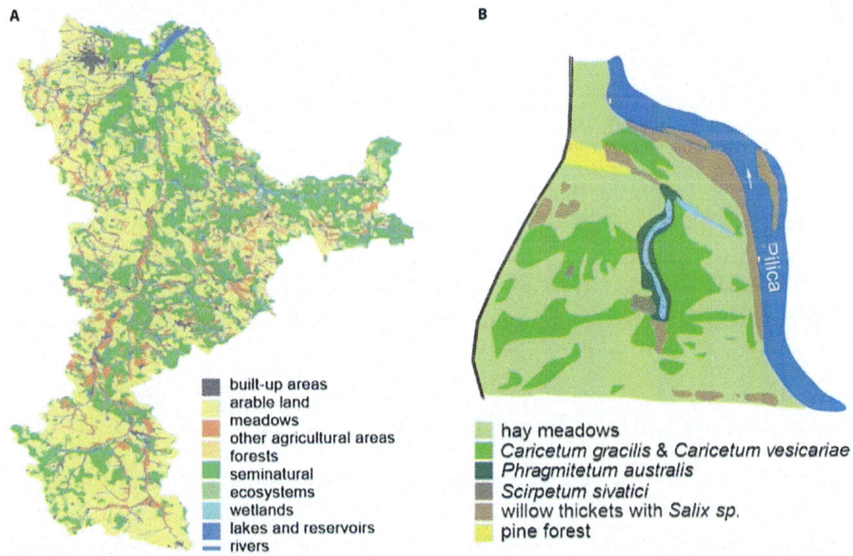

FIGURE 7.9 Diverse ecosystems and urbanization. Fig. I.7.7. (A) Diverse ecosystems and urbanization distribution in the Pilica river cachment, Poland. (B) Plant communities types distribution at the Pilica river floodplain, Poland. *From Frątczak, W., Izydorczyk, K. (Eds.), 2015. Action Programme for Reducing Diffuse Pollution in the Pilica River Catchment Upstream of the Sulejow Reservoir. ERCE PAS, ISBN 978-83-928245-2-7, (in Polish).*

This phenomenon can be mitigated through Dual Regulation. By employing ecological strategies to restore biological structures, it is possible to significantly influence the availability of nitrogen in the soil Fig. 7.10. As biocomplexity of the catchment landscape and biodiversity increases, interspecific competition for nitrogen resources intensifies, while the mechanisms by which different species source and utilise nitrogen also diversify. Efficient nitrogen uptake and strong retention in the soil through biological structures, such as plant roots, help prevent nitrogen loss caused by leaching via surface and groundwater runoff (Tilman et al., 1996). Conversely, reduced plant biodiversity leads to a simplified nitrogen cycle, decreased competition between plant species, and an increase in readily available nitrogen, thereby promoting its leaching into water bodies.

It is important to note that under natural conditions, nitrogen is a limiting factor for plants and algal productivity. A rapid increase in nitrogen availability within aquatic environments can have severe consequences, leading to the proliferation of toxic algal blooms, particularly in coastal zones.

The concentration of nitrates in the soil decreases with increasing species diversity. Tilman et al. (1996) demonstrated that environments with a greater number of species tend to have lower nitrate concentrations in the root zone.

This relationship highlights that biodiversity is not only of cultural and aesthetic significance but also serves as a crucial tool for enhancing the ecological status of river basins. For instance, it should be considered a key role in meeting the requirements of the Nitrates Directive Fig. 7.10.

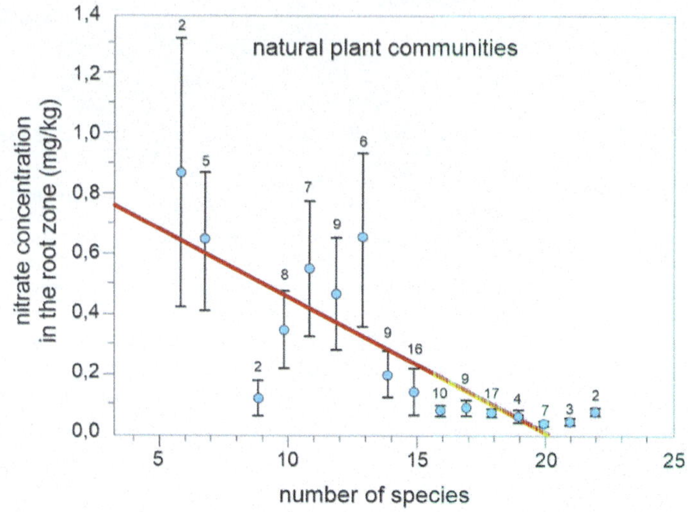

FIGURE 7.10 Dependence of nitrate concentration in the root zone on the number of plant species in the community. Dependence of nitrate concentration in the root zone on the number of plant species in the community as an example of the importance of biodiversity of plant communities, a potential element dual regulation and NBS for the reduction of diffuse source pollution (Tilman et al., 1996). *From Tilman, D., Wedin, D., Knops, J., 1996. Productivity and sustainability influenced by biodiversity in grassland ecosystems.* Nature *379 (6567), 718–720. https://doi.org/10.1038/379718a0.*

Preventing the inflow of external nutrients and sequestering those already present within plant biomass, making them less bioavailable, is essential for water reclamation and quality improvement, as it inhibits the growth of phytoplankton responsible for toxic blooms.

The findings presented above indicate that addressing water quality issues at the catchment scale requires a comprehensive understanding of the interdependencies between hydrological and biological processes, particularly in agricultural and developed areas (see Chapter 11 for more details). Another key conclusion is that biodiversity conservation in a given area is not solely intended to protect species and enhance landscape value and aesthetics. Rather, it is a fundamental component in increasing ecosystem productivity, biodiversity, and stability, thereby enhancing its resilience to pollution and strengthening its overall environmental potential.

3. Ecological engineering principle of TEH

The third principle of ecohydrology, also known as the ecological engineering principle, focuses on practical applications, expressing the merit of Transdisciplinary Ecohydrology. It is based on the application of the two preceding principles, demonstrating that the processes and activities described above function as tools for regulating catchment hydrology and ecosystem dynamics.

Design thinking, a creative, iterative process focused on human-centered innovation Box 7.1, also guides the development of Advanced Nature-Based Solutions (ANbS) in Transdisciplinary Ecohydrology. ANbS are designed to work in harmony with existing water infrastructure, such as dams, irrigation systems, and sewage treatment plants, creating integrated, systemic responses to environmental challenges. Developed according to the three core principles of ecohydrology (hydrological, ecological, and ecotechnological), ANbS apply the problem-solving spirit of design thinking to engineer solutions that are adaptive, holistic, and resilient.

Traditionally, biotechnologies involve the transformation of one substance into another through the use of living organisms (e.g., the conversion of sugar into alcohol by yeast). In the context of ecohydrology, the foundational knowledge required for the application of biotechnology begins with an understanding of hydrological cycle dynamics and the biotic potential of plants and bacteria to convert nutrients from mineral to organic forms. Environmental management strategies that harness such natural processes are referred to as ecosystemic biotechnologies (Zalewski et al., 1997).

BOX 7.1 Design thinking

Design Thinking is a human-centered, iterative methodology that fosters innovation by deeply understanding user needs, generating creative ideas, rapidly prototyping, and continuously refining solutions based on feedback (Brown, 2009) and Stanford d.school (2020).

The primary challenges in developed catchments include the excessive acceleration of runoff and increased water evaporation, accompanied by a reduced proportion of evapotranspiration. The loss of water leads to landscape desiccation and, consequently, the depletion of organic matter from the soil, an essential factor in succession processes and moisture retention within the landscape. Under these conditions, characterized by water and organic matter loss, the xerosere process occurs, resulting in the formation of a dry, nutrient-poor ecosystem with low biodiversity and minimal biomass production.

Ecohydrological biotechnologies seek to counteract negative processes by increasing functional plant biomass (phytotechnology), reducing runoff, and shifting evaporation dynamics in favor of evapotranspiration. This, in turn, enhances water retention within the landscape (Kędziora and Ryszkowski, 1999). Retaining water facilitates the restoration of organic matter, which serves as the foundation for reversing the xerosere process, restoring ecosystem resilience, and enhancing bioproductivity.

Traditional hydroengineering solutions have often demonstrated a lack of consideration of fundamental ecological processes and signal process thinking. However, the growing complexity of contemporary environmental issues necessitates the integration of ecological knowledge within ANbS and hydrological engineering, leading to the development of hybrid solutions (Zalewski, 2015; UNESCO International Hydrological Program doc. 7). This approach enables the regulation of ecological processes through the use of existing infrastructure (Zalewski, 2013) while ensuring that these processes, when integrated with ecohydrological methods and based on understanding the natural systems functioning, serve as effective tools for enhancement and the sustainable use of environmental resources at the catchment Fig. 7.11.

The integration of various engineering interventions at the catchment scale, combined with the application of ecohydrological processes as regulatory tools, offers the potential for a significant reduction in nutrient and pollutant loads entering river systems (Zalewski, 2000). The restoration of eutrophic and polluted water bodies by limiting pollutant inflows and reallocating excess nutrients into inaccessible pools, such as biomass, timber, and deep sediments, through various biotechnological approaches exemplifies the ecohydrological approach to river restoration at the catchment scale.

This process begins with enhancing nutrient retention within the catchment through afforestation, the establishment of highly efficient ecotone zones (see Chapter 11 for more details), and the optimization of agricultural practices. Ecotone zones play a crucial role in reducing both surface and subsurface runoff by increasing retention and evapotranspiration along the river corridor (Izydorczyk et al., 2013; Kędziora and Ryszkowski, 1999). These zones can decrease nutrient outflow from agricultural areas by up to 90% (Kędziora and Ryszkowski, 1999) primarily by facilitating the transformation of nutrients into plant biomass and intercepting nutrients transported via subsurface flow.

Wetlands within river valleys also function as natural buffer zones, mitigating the transport of mineral sediments, organic matter, and biogenic substances carried by rivers during flood events (Kiedrzyńska et al., 2008; Lowland floodplain rivers: Geomorphological perspectives, 1992; Mitsch and Gosselink, 1993). In both natural and constructed wetlands, nitrogen loads are significantly reduced through the elevation of the water table, which stimulates anaerobic denitrification processes (Bednarek et al., 2010).

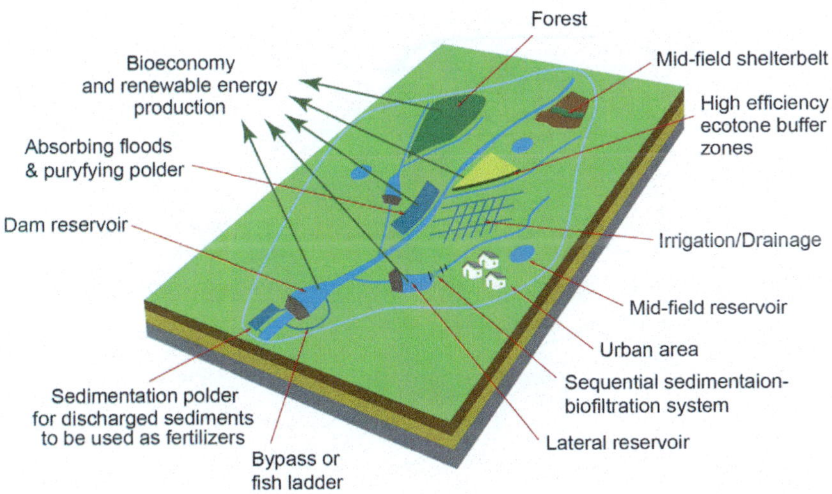

FIGURE 7.11 The third principle of ecohydrology—ecological engineering. Identification of catchment areas for the application of various forms of dual regulation and ANbS, i.e., sequential sedimentation-biofiltration system (see Chapter II.7.3), reservoir (see chapter I.7.2), polder (see chapter I.8., II.6) and integrated with hydrotechnical infrastructure. *Zalewski, M., 2014. Ecohydrology and hydrologic engineering: Regulation of hydrology–biota interactions for sustainability. Journal of Hydrologic Engineering 19 (5), 1–8. https://doi.org/10.1061/(ASCE)HE.1943-5584.0000999*

According to those three fundamental principles of ecohydrology, a comprehensive understanding of the interactions between aquatic environments and living organisms, alongside the development of effective ecological biotechnologies, must encompass a broad spectrum of ecological processes. These range from the molecular biology level, such as microbial loop processes (Chróst, 2009) and genetic analyses—through to the ecosystem level, including biomanipulation, and ultimately to the catchment scale, where biocomplexity can be enhanced through Nature-Based Solutions (NBS), such as afforestation, the creation of ecotone zones, and the establishment of shelterbelts within the landscape. Furthermore, this approach must integrate knowledge from other environmental sciences, including molecular biology.

The principles of ecohydrology not only provide a framework for the holistic implementation of NBS but also facilitate the alignment of ecohydrological strategies with existing hydrotechnical infrastructure. Additionally, they enable the integration of a diverse range of decision-making processes and educational initiatives, ensuring a more cohesive and sustainable approach to environmental management.

An example of the harmonization of ecohydrological solutions with societal needs is illustrated in Fig. 7.12. Sewage treatment plants should be complemented by ecohydrological biotechnologies, such as sequential sedimentation–biofiltration systems (see invited chapters) or the establishment of bioenergy plantations irrigated with pretreated sewage water. These plantations should be developed using local willow species to enhance biodiversity functions and optimize ecosystem services.

By integrating traditional technologies with ecohydrological biotechnologies, "hybrid systems" are created. In these systems, the biological conversion of phosphorus into plant biomass prevents eutrophication, while simultaneously stimulating bioenergy production through willow biomass cultivation, which reduces costs of energy for society. Additionally, CO_2 emissions are absorbed into biomass through photosynthesis within bioenergy plantations, contributing to climate change mitigation. Furthermore, these sites generate employment opportunities related to their maintenance and operation.

This approach to economic development is known as the bioeconomy, an economy that relies on renewable resources, such as biomass. The bioeconomy encompasses the application of molecular biology mechanisms and processes (e.g., genetic research) to drive advancements in agriculture, healthcare, and the chemical and energy industries (Smyth et al., 2011; Wesseler et al., 2010).

It is important to acknowledge that the methods described have certain limitations. Their effectiveness can be broadly illustrated through analogy: the efficiency of process regulation increases as long as it is reinforced by ecohydrological biotechnologies and the ecosystem's capacity to absorb pollutants and nutrients remains greater than the external load introduced. However, once the pollutant and nutrient load surpasses the ecosystem's absorptive capacity, the efficacy of ecohydrological methods declines significantly. This phenomenon is particularly evident in hypertrophic lakes, where extreme alterations in abiotic conditions, including physical, chemical, and hydrological factors, as well as severe oxygen depletion, lead to the degradation of biotic structures and a sharp decline in biodiversity. In such cases, biomanipulation proves ineffective (Dawidowicz, 1988; Gliwicz, 2017; Zalewski et al., 2004).

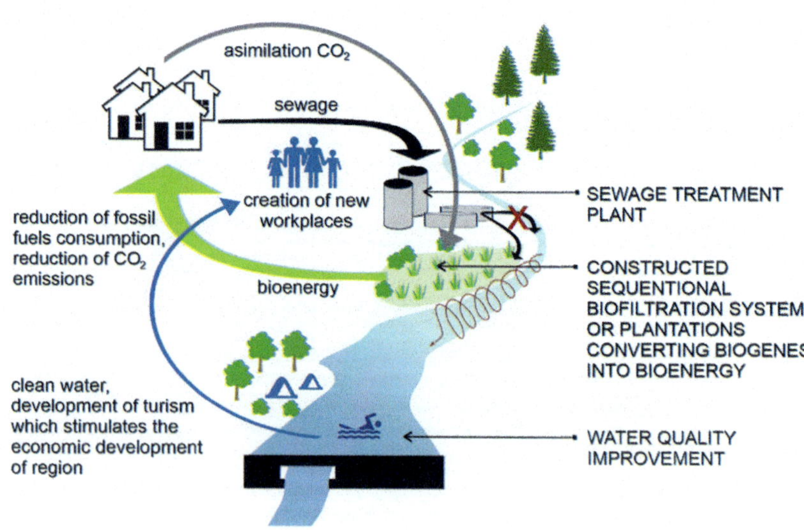

FIGURE 7.12 An example of ecohydrological system solutions. A functional example of enhancing water-biodiversity-ecosystem services-resilience (WBSR) through transdisciplinary ecohydrological systemic solutions—use pretreated sewage water as irrigation and fertilizer for bioenergy plantations (circular economy). Biomass is used to heat schools and other buildings, thus reducing fossil fuels consumption and CO_2 emissions. Improving the ecological status of the river and the reservoir has a positive effect on the development of tourism. Both bioenergy plantation and tourism development create jobs and stimulate the economic development of the region. *Basen on Zalewski, M. (Ed.), 2002b. Guidelines for the integrated management of the watershed – phytotechnology and ecohydrology, United Nations Environemant Programme. International Environmental Technology Centre. Freshwater Management Series 5, 188.*

One of the negative consequences of the degradation of biological structures (forests, shelterbelts, gardens, parks) within a catchment is the increased likelihood of extreme hydrological events, such as droughts and floods. These events, in turn, intensify physical processes that alter geochemical structures and trigger adverse responses in biological systems and ecosystem dynamics.

According to the Intermediate Disturbance Hypothesis (IDH) (Connell, 1978), ecosystems achieve their highest levels of biodiversity under moderate disturbances. However, increased runoff from catchments, resulting from deforestation and river regulation, creates instability in physical conditions, reducing the catchment's capacity to retain water and nutrients within soil and vegetation biomass. This leads to a decline in both biodiversity and terrestrial ecosystem productivity, while simultaneously exacerbating nutrient enrichment in aquatic ecosystems. Consequently, water quality deteriorates, and reservoirs and lakes become eutrophic, as the likelihood of toxic algal blooms increases proportionally with the intensification of nutrient accumulation.

It is important to emphasize that many aspects of aquatic ecosystem functioning and organismal interactions within a changing biosphere still require deeper investigation. Empirical testing of transdisciplinary hypotheses at the mesoscale of the hydrological cycle within catchments contributes to scientific advancement and the development of systemic solutions by:

1. Enhancing the understanding of ecohydrological processes, including the identification and quantification of potential synergies and antagonistic relationships between terrestrial and aquatic processes within the catchment. This involves analyzing nutrient and pollutant cycles under the influence of various anthropogenic factors across different catchment types.
2. Aligning societal expectations with the goal of increasing catchment sustainability potential (WBSR + CE) while simultaneously limiting polutants emission, resource consumption and enhancing environmental awareness through education and public participation.

The implementation of biotechnological solutions for regulating water flow and nutrient dynamics within catchments must be guided by the principles of adaptive assessment and management (Holling, 1978), ensuring that strategies remain flexible and responsive to evolving ecological and socioeconomic conditions.

In summary, ecohydrology is underpinned by three integrative principles that collectively define its methodological and operational framework.

4. Summary

4.1 The hydrological principle

This principle designates catchment as the fundamental spatial unit for analysis and management. It emphasizes the quantification of abiotic processes, especially geomorphology and hydrology, and impacts in determining ecosystem functioning. Biotic processes regulate effectively only once dynamics of these foundational physical conditions are identified and quantified.

4.2 The ecological principle

The distribution of pristine, novel, and highly modified ecosystems within the catchment, along with the identification of their functionality, provides a basis for their use as management tools and reveals their potential for the implementation of ANbS and systemic solutions.

4.3 The ecological engineering principle

This principle encourages the integration of Dual Regulation based on I and II principles of ecohydrological approaches with existing hydrotechnical infrastructure. It supports the combined application of Advanced Nature-based Solutions and engineering technologies to achieve synergistic, systemic enhancement of catchment sustainability potential WBSR.

Together, these principles enable the development of transdisciplinary, systemic approaches to environmental challenges, positioning ecohydrology as both a scientific discipline and a practical tool for promoting river-basin sustainability and climate resilience Fig. 7.13.

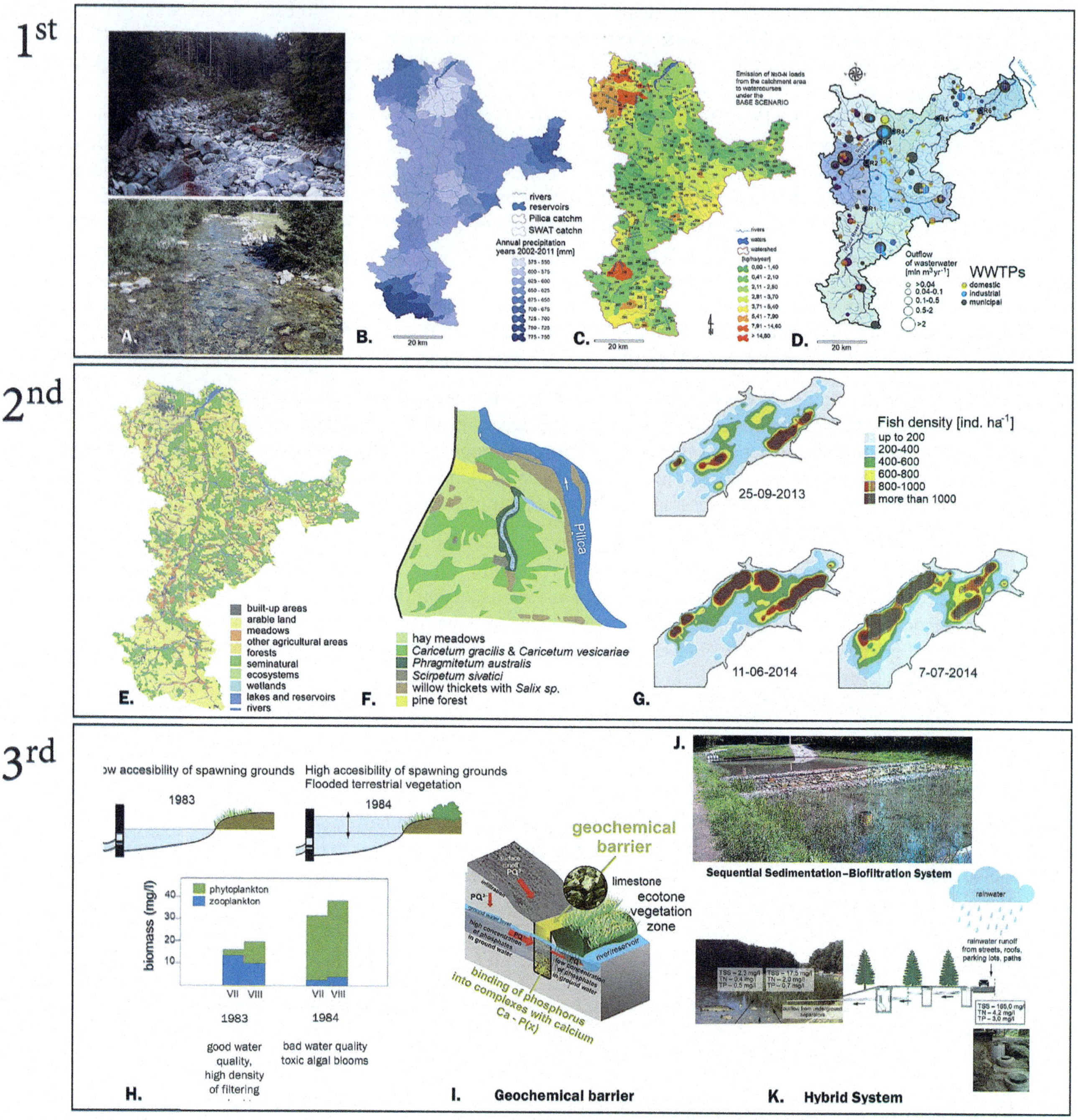

FIGURE 7.13 Principles of ecohydrology as framework for systemic catchment scale solutions for increasing the potential of sustainable development in the catchment area economy. Hydrological principle I: (A) The importance of geology for the stability of flows—comparison of the Sucha Woda (Dry Water) valley (sedimentary rocks and granite substrate with low retention capacity, water is flowing only after heavy rains, top photo) and the Potok river Chochołowski (limestone substrate with high retention, river flow stable whole year, bottom photo), (photo by Jerzy Opioła); (B) Spatial differentiation of annual rainfall in the Pilica valley; (C) Emission of phosphorus loads from agriculture, non-point source pollution (see Chapter 14) and (D) the distribution of point source pollution in the Pilica basin (Kiedrzyńska et al., 2014). **Ecological principle II:** (E) Map of the distribution of various types of ecosystems in the Pilica river basin (Izydorczyk et al., 2019); (F) Map of ecosystems with different potential for absorbing pollutants on the terrace the floodplain in the Pilica river basin (Kiedrzyńska et al., 2008); (G) Hydroacoustic analysis of bloom dynamics cyanobacteria in the reservoir (see Chapter 16). **Ecological Engineering principle III:** (H) Hydrobiomanipulation—regulation trophic cascade by changing the water level in the reservoir to reduce the bloom intensity (see Chapter 6); (I) High-efficiency ecotone zone for the reduction of diffuse pollution (see Chapter 14); (J) Sequential (see Chapter 22) and K Hybrid system of stormwater purification from areas urbanized areas (see Chapter 26). The context for the implementation of these rules should provide design thinking process (Brown, 2009), by identifying the real needs and the given community empathy—diagnosis of needs, generating ideas, building prototypes. *Inspired by earlier concepts (Zalewski, 2014).*

AI disclosure

During the preparation of this work the author(s) used OpenAI's ChatGPT in order to generate content. This document includes content generated and/or edited with the assistance of OpenAI's ChatGPT, used to support language refinement and structural improvements. All final content was reviewed and verified by the author. After using this tool/service, the author(s) reviewed and edited the content as needed and take (s) full responsibility for the content of the publication.

References

Adaptive environmental assessment and management. In: Holling (Ed.), 1978. International Institute for Applied system Analysis. John Wiley & Sons, p. 420.

Bajkiewicz-Grabowska, E., Mikulski, Z., 1993. Hydrometria [Hydrometry]. Wydawnictwo Naukowe PWN.

Bednarek, A., Stolarska, M., Ubraniak, M., Zalewski, M., 2010. Application of permeable reactive barrier for reduction of nitrogen load in the agricultural areas — preliminary results. Ecohydrol. Hydrobiol. 10 (2–4), 355–361. https://doi.org/10.2478/v10104-011-0007-6.

Brown, T., 2009. Change by Design: How Design Thinking Creates New Alternatives for Business and Society. Harvard Business Press.

Chróst, R.J., 2009. Microbial food webs in aquatic ecosystems: structure and functioning. In: Encyclopedia of Inland Waters. Elsevier, pp. 395–404.

Connell, J.H., 1978. Diversity in tropical rain forests and coral reefs. Science 199 (4335), 1302–1310. https://doi.org/10.1126/science.199.4335.1302.

Crisp, D.T., Hynes, H.B.N., 1971. The ecology of running waters. J. Anim. Ecol. 40 (3), 782. https://doi.org/10.2307/3451.

Dawidowicz, P., 1988. Can Daphnia prevent a blue-green algal bloom in hypertrophic lakes? A laboratory test. Limnologica 19, 21–26.

Frątczak, W., Izydorczyk, K. (Eds.), 2015. Action Programme for Reducing Diffuse Pollution in the Pilica River Catchment Upstream of the Sulejow Reservoir. ERCE PAS. ISBN 978-83-928245-2-7, (in Polish).

Gliwicz, Z.M., 2017. Food web interactions: why are they reluctant to be manipulated? SIL Proc. 1922-2010 29 (1), 73–88. https://doi.org/10.1080/03680770.2005.11901976.

Hillbricht-Ilkowska, A., 1985. Assessment of trophic impact on the lake environment in Poland: a proposal and case study. In: Ecological Risks: Perspectives from Poland and the United States. National Academy Press, pp. 283–295.

Hillbricht-Ilkowska, A., Pieczyńska, E., 1993. Ecosystem Functioning and Eutrophication in the Lakes of the Polish Lowland. Institute of Ecology, Polish Academy of Sciences.

Hobbs, R.J., Arico, S., Aronson, J., Baron, J.S., Bridgewater, P., Cramer, V.A., Epstein, P.R., Ewel, J.J., Klink, C.A., Lugo, A.E., Norton, D., Ojima, D., Richardson, D.M., Sanderson, E.W., Valladares, F., Vilà, M., Zamora, R., Zobel, M., 2006. Novel ecosystems: theoretical and management aspects of the new ecological world order. Global Ecol. Biogeogr. 15 (1), 1–7. https://doi.org/10.1111/j.1466-822X.2006.00212.x.

Hynes, H.B.N., 1970. The Ecology of Running Waters. Liverpool University Press, Liverpool.

Izydorczyk, K., Frątczak, W., Drobniewska, A., Cichowicz, E., Michalska-Hejduk, D., Gross, R., Zalewski, M., 2013. A biogeochemical barrier to enhance a buffer zone for reducing diffuse phosphorus pollution – preliminary results. Ecohydrol. Hydrobiol. 13 (2), 104–112. https://doi.org/10.1016/j.ecohyd.2013.06.003.

Kabat, P., 2002. Biospheric aspects of the hydrological cycle (BAHC) [Editorial]. BAHC News 9.

Kaczmarek, Z., Strzepek, Somlyódy, L., 1996. Water Resources Management in the Face of Climatic/hydrologic Uncertainties. Kluwer Academic Publishers.

Kędziora, A., Ryszkowski, L., 1999. Does plant cover structure in rural areas modify climate change effects? Geogr. Pol. 72 (2), 85–87.

Kiedrzyńska, E., Kiedrzyński, M., Urbaniak, M., Magnuszewski, A., Skłodowski, M., Wyrwicka, A., Zalewski, M., 2014. Point sources of nutrient pollution in the lowland river catchment in the context of the Baltic Sea eutrophication. Ecol. Eng. 70, 337–348. https://doi.org/10.1016/j.ecoleng.2014.06.010.

Kiedrzyńska, E., Kiedrzyński, M., Zalewski, M., 2008. Flood sediment deposition and phosphorus retention in a lowland river floodplain: impact on water quality of a reservoir, Sulejów, Poland. Ecohydrol. Hydrobiol. 8 (2–4), 281–289. https://doi.org/10.2478/v10104-009-0022-z.

Lowland floodplain rivers: Geomorphological Perspectives, 1992. John Wiley & Sons.

Mitsch, W.J., Gosselink, J.G., 1993. Wetlands, 2nd ed. Van Nostrand Reinhold, New York.

Naiman, R.J., 1983. The annual pattern and spatial distribution of aquatic oxygen metabolism in boreal forest watersheds. Ecol. Monogr. 53 (1), 73–94. https://doi.org/10.2307/1942588.

Naiman, R.J., Decamps, H., 1990. The ecology and management of aquatic-terrestrial ecotones. Man Biosph. Ser. 4, 316.

Plattner, H., 2020. Institute of Design at Stanford University (d.school). In: Design thinking bootleg: An introduction to design thinking and creativity tools. Stanford University.

Program działań dla ograniczenia zanieczyszczeń obszarowych w zlewni Pilicy powyżej Zbiornika Sulejowskiego [Action program to reduce diffuse pollution in the Pilica River catchment above the Sulejów Reservoir]. In: European Regional Centre for Ecohydrology, 2015. Polish Academy of Sciences.

Rodriguez-Iturbe, I., 2000. Ecohydrology: a hydrologic perspective of climate-soil-vegetation dynamies. Water Resour. Res. 36 (1), 3–9. https://doi.org/10.1029/1999wr900210.

Smyth, S.J., Aerni, P., Castle, D., Demont, M., Falck-Zepeda, J.B., Paarlberg, R., Phillips, P.W.B., Pray, C.E., Savastano, S., Wesseler, J.H.H., Zilberman, D., 2011. Sustainability and the bioeconomy: synthesis of Key themes from the 15th ICABR conference. AgBioforum 14 (3), 180–186.

Starkel, L., 1982. Ewolucja doliny Wisły podczas ostatnich 15 000 lat, cz. 1 (Evolution of the Vistula river valley during the last 15 000 years, pt. 1). Zakład Narodowy im. Ossolińskich, Wydawnictwo PAN.

Tilman, D., Wedin, D., Knops, J., 1996. Productivity and sustainability influenced by biodiversity in grassland ecosystems. Nature 379 (6567), 718–720. https://doi.org/10.1038/379718a0.

Tundisi, J.G., Staskraba, M. (Eds.), 1999. Theorethical reservoir ecology and its application. Backhuys Publisher / International Institute of Ecology, pp. 585–pp.

Wagner, I., Zalewski, M., 2016. Temporal changes in the abiotic/biotic drivers of selfpurification in a temperate river. Ecol. Eng. 94, 275–285. https://doi.org/10.1016/j.ecoleng.2016.05.073.

Wagner-Łotkowska, I., Bocian, J., Pypaert, P., Santiago-Fandino, V., Zalewski, M., 2004. Environment and economy - dual benefit of ecohydrology and phytotechnology in water resources management: Pilica River Demonstration Project under the auspices of UNESCO and UNEP. Ecohydrol. Hydrobiol. 4 (3), 345–352.

Wesseler, J., Spielman, D.J., Demont, M., 2010. The future of governance in the global bioeconomy: policy, regulation, and investment challenges for the biotechnology and bioenergy sectors. AgBioforum 13 (4), 288–290.

Wood, C.M., Popescu, V.D., Klinck, H., Keane, J.J., Gutiérrez, R.J., Sawyer, S.C., Peery, M.Z., 2019. Detecting small changes in populations at landscape scales: a bioacoustic site-occupancy framework. Ecol. Indic. 98, 492–507. https://doi.org/10.1016/j.ecolind.2018.11.018.

Zalewski, M., 2000. Ecohydrology - the scientific background to use ecosystem properties as management tools toward sustainability of water resources. Ecol. Eng. 16 (1), 1–8. https://doi.org/10.1016/S0925-8574(00)00071-9.

Zalewski, M., 2002a. Ecohydrology – the use of ecological and hydrological processes for sustainable management of water resources. Hydrol. Sci. J. 47 (5), 825–834.

Zalewski, M. (Ed.), 2002b. Guidelines for the integrated management of the watershed – phytotechnology and ecohydrology, United Nations Environemant Programme. International Environmental Technology Centre. Freshwater Management Series 5, 188.

Zalewski, M., 2006. Flood pulses and river ecosystem robustness. IAHS-AISH Pub. 305, 143–154.

Zalewski, M., 2013. Ecohydrology: process-oriented thinking towards sustainable river basins. Ecohydrol. Hydrobiol. 13 (2), 97–103. https://doi.org/10.1016/j.ecohyd.2013.06.001.

Zalewski, M., 2014. Ecohydrology and hydrologic engineering: Regulation of hydrology–biota interactions for sustainability. Journal of Hydrologic Engineering 19 (5), 1–8.

Zalewski, M., 2015. Ecohydrology and hydrologic engineering: regulation of hydrology-biota interactions for sustainability. J. Hydrol. Eng. 20 (1). https://doi.org/10.1061/(ASCE)HE.1943-5584.0000999.

Zalewski, M., 2017. Ecohydrology measures for sustainable catchment management: An outline. Ecohydrol. Hydrobiol. 17 (3), 205–220.

Zalewski, M. (Ed.), 2020. Ekohydrologia. Wydawnictwo Naukowe PWN, Warszawa.

Zalewski, M., Bis, B., Łapińska, M., Frankiewicz, P., Puchalski, W., 1998. The importance of the riparian ecotone and river hydraulics for sustainable basin-scale restoration scenarios. Aquat. Conserv. Mar. Freshw. Ecosyst. 8 (2), 287–307. https://doi.org/10.1002/(sici)1099-0755(199803/04)8:2<287::aid-aqc274>3.0.co;2-r.

Zalewski, M., Brewinska-Zaraś, B., Frankiewicz, P., Kalinowski, S., 1990a. The potential for biomanipulation using fry communities in a lowland reservoir: concordance between water quality and optimal recruitment. Hydrobiologia 200–201 (1), 549–556. https://doi.org/10.1007/BF02530372.

Zalewski, M., Brewińska-Zaras, B., Frankiewicz, P., 1990b. Fry communities as a biomanipulating tool in a temperate lowland reservoir. Beihefte Ergebnisse der Limnologie. 33, 763–774.

Zalewski, M., Janauer, G.A., Jolankai, G., 1997. Ecohydrology: A New Paradigm for the Sustainable Use of Aquatic Resources. UNESCO International Hydrological Programme.

Zalewski, M., Naiman, 1985. The regulation of riverine fish communities by a continuum of abiotic–biotic factors. In: 1985 Habitat Modification and Freshwater Fisheries. Butterworths Scientific Ltd, pp. 3–9.

Zalewski, M., Wagner, I., 1998. Temperature and nutrient dynamics in eutrophic freshwater ecosystems. Geogr. Pol. 71, 90–92.

Zalewski, M., Santiago-Fandino, V., Neate, J., 2003. Energy, water, plant interactions: 'green feedback' as a mechanism for environmental management and control through the application of phytotechnology and ecohydrology. Hydrol. Process. 17 (14), 2753–2767. https://doi.org/10.1002/hyp.1432.

Zalewski, M., Wagner-Lutowska, I., Robarts, R.D., 2004. In: Ecohydrology and Phytotechnology Integrated Watershed Management. UNESCO.

Chapter 8

Methodology of transdisciplinary ecohydrology

Maciej Zalewski
European Regional Centre for Ecohydrology of the Polish Academy of Sciences under the auspices of UNESCO, Łódź, Poland

PHOTO CREDIT: M. ZALEWSKI.

Transdisciplinary Ecohydrology. https://doi.org/10.1016/B978-0-443-41596-8.00016-9

1. From global challenges to transdisciplinary scientific frameworks

This chapter introduces ecohydrology as a transdisciplinary, integrative approach that offers a scientifically grounded and cost-effective methodology for addressing contemporary environmental challenges. By optimizing resource use and enhancing ecosystem resilience, ecohydrology facilitates a systemic transition toward sustainable development. It does so through a stepwise method that links theory with practical application and emphasizes regulation over exploitation.

Rooted in the scientific method, defined as "a mathematical and experimental technique employed in the sciences … used in the construction and testing of a scientific hypothesis" (*Encyclopedia Britannica*), ecohydrology applies rigorous inquiry to real-world environmental problems. Defined as a problem-solving, transdisciplinary science, ecohydrology seeks to mitigate the detrimental effects of anthropogenic pressure through the regulation of ecological processes. It explains the underlying mechanisms of life-supporting processes (LSPs), driving large-scale phenomena such as increasing biospheric floods and droughts, degradation of habitat complexity, freshwater pollution, ocean acidification, nitrogen overfertilization, and soil fertility loss. Furthermore, it offers a repertoire of strategies aimed at reversing or, at the very least, alleviating these adverse processes.

Currently, transdisciplinary ecohydrology is challenged by several interrelated objectives:

1. Reducing water loss via surface runoff, accelerated river outflow at channelized rivers, particularly in the context of flood and drought mitigation.
2. Enhancing the landscape's capacity to retain water.
3. Improving both water quality and biodiversity.
4. Closing biogeochemical cycles by retaining nutrients in the landscape and minimizing their discharge into aquatic systems, by increasing landscape complexity, biodiversity and bioproductivity.

To fulfill these aims, transdisciplinary ecohydrology must address the following research and management priorities:

1. Comprehensive quantification of hydrological processes and biological structure and processes at the catchment scale.
2. Analysis of nutrient cycles and opportunities to use phytotechnology and trophic cascade regulation.
3. Regulation of hydrology-biota interplay to enhance water–biodiversity–ecosystem services–resilience (WBSR) of the catchment to harmonize enhanced ecosystem sustainability potential with society's needs.
4. Creation of culture and education of water and sustainability (CE) to involve society in catchment-scale active support of transdisciplinary ecohydrology (TEH) for Integrated Water Resources Management (IWRM).
5. Transdisciplinary science based on proactive law, policy, governance (LPG).

Sustainable human–nature coexistence depends on the convergence of ecohydrology with sociotechnological progress, particularly through the implementation of circular economy and bioeconomy strategies. This dual imperative involves reducing per capita resource consumption while simultaneously enhancing the environmental potential of WBRS to anthropogenic stressors. These principles are echoed in the Factor Four concept (Weizsäcker et al., 1997) and have been adapted within ecohydrology as articulated by Zalewski (2011). Central to this paradigm is the enhancement of environmental capacity—not merely through risk mitigation, but by fostering the potential of ecosystems enhancing catchment sustainability philosophy through the water, biodiversity, ecosystem services, resilience + culture and education + law, policy and governance (WBSR + CE + LPG). Fig. 8.1 illustrates the evolution of ecosystem management approaches from structure-oriented strategies, such as conservation and restoration, to process-oriented frameworks such as ecological engineering and TEH. It highlights a shift from preserving or restoring ecosystem structure to actively regulating ecological and hydrological processes at the catchment scale (Dual Regulation) to enhance the sustainability potential of the catchment and biosphere and align with societal needs.

2. Transdisciplinary ecohydrology: From theory to solutions

This diagram Fig. 8.2 illustrates the adaptive, cyclical process of transdisciplinary ecohydrology for the development of advanced nature-based solutions (ANbS) as a transition from pure science to applied transdisciplinary practice. It is based on a framework proposed by Zalewski (2020), with examples from catchment-based implementations such as those in Ethiopia (Zalewski and Zerihun, 2010). Key components of this diagram include:

Structure-oriented Thinking → Process-oriented Thinking

	Conservation	Restoration	Ecological + Regenerative Engineering Ecology	Transdisciplinary Ecohydrology
Goal	Maintaining biodiversity and the natural character of the ecosystem	Reversing the degradation of ecosystem structures and functions	"Design of ecosystem for mutual benefit of man and nature" (Mitsch, 1992)	Enhancing the catchment and biosphere sustainability potential (WBSR) thought trandlsciplinary ecohydrology
Methods	No ecosystem management	Managing natural succession	Ecological Engineering + Regenerative Ecology	Principles of transdisciplinary ecohydrology
Unit	Ecosystem or population	Ecosystem or part of it	Ecosystem	Drainage basin
Status	Preserving "status quo"	Enhancement of secondary succession process in terrestrial ecosystem	Shaping of ecosystem function; biomanipulation, river self purification	Regulation of interactions between the hydrological mesocycle and biocenosis, enhancing ecosystem potential (WBSR), and harmonizing it with societal needs
Examples	 Bialowieza National Park	 Reclaimed spoil heap of the Bełchatów Mine	 Constructed wetlands for water purification and biodiversity enhancement - Ohio State University (Photo: William J. Mitsch)	 Catchment-based, hybrid, systemic ecohydrological approach

FIGURE 8.1 From structure-oriented thinking to process-oriented thinking. This diagram presents a conceptual progression in ecosystem management, transitioning from structure-oriented to process-oriented approaches. It compares four stages of humankind and the biosphere relations—conservation, restoration, ecological engineering, regenerative ecology, and transdisciplinary ecohydrology—along multiple dimensions: goals, methods, scale of application, ecological status, and representative examples. Conservation emphasizes the preservation of biodiversity and ecosystem integrity through conventional management practices, aiming to maintain the ecological "status quo" at the population or ecosystem level, as exemplified by UNESCO MAB, Białowieża National Park. Restoration shifts the focus toward reversing ecosystem degradation by supporting natural succession processes, often applied to partially degraded systems such as the reclaimed spoil heap of the Bełchatów mine. Ecological engineering introduces a more dynamic perspective by actively designing ecosystems to serve both human and ecological functions, including biomanipulation and water purification, as demonstrated by constructed wetlands such as those at Ohio State University. At the most integrative end of the spectrum, transdisciplinary ecohydrology operates at the catchment scale, creating advanced nature-based and systemic solutions, employing principles of TEH to regulate biogeochemical cycles, enhance water and biodiversity-related ecosystem services, and align environmental processes with societal needs. This shift from conserving structure to regulating processes underscores the growing recognition of ecosystems as active regulators of sustainability, rather than passive entities to be preserved or restored in isolation. *Based on Zalewski, M., 2020. Ekohydrologia. Wydawnictwo Naukowe PWN, Warszawa. ISBN:978-83-01-20804-2.*

2.1 Pure science and big data

The process begins with pure scientific research, grounded in big data. This foundational knowledge includes empirical findings, environmental monitoring, and theoretical advancements across disciplines (ecology and hydrology).

2.2 Development of holistic models (E + H)

This scientific understanding is then used to develop holistic models of ecological and hydrological processes, such as the abiotic–biotic regulation concept (ABRC).

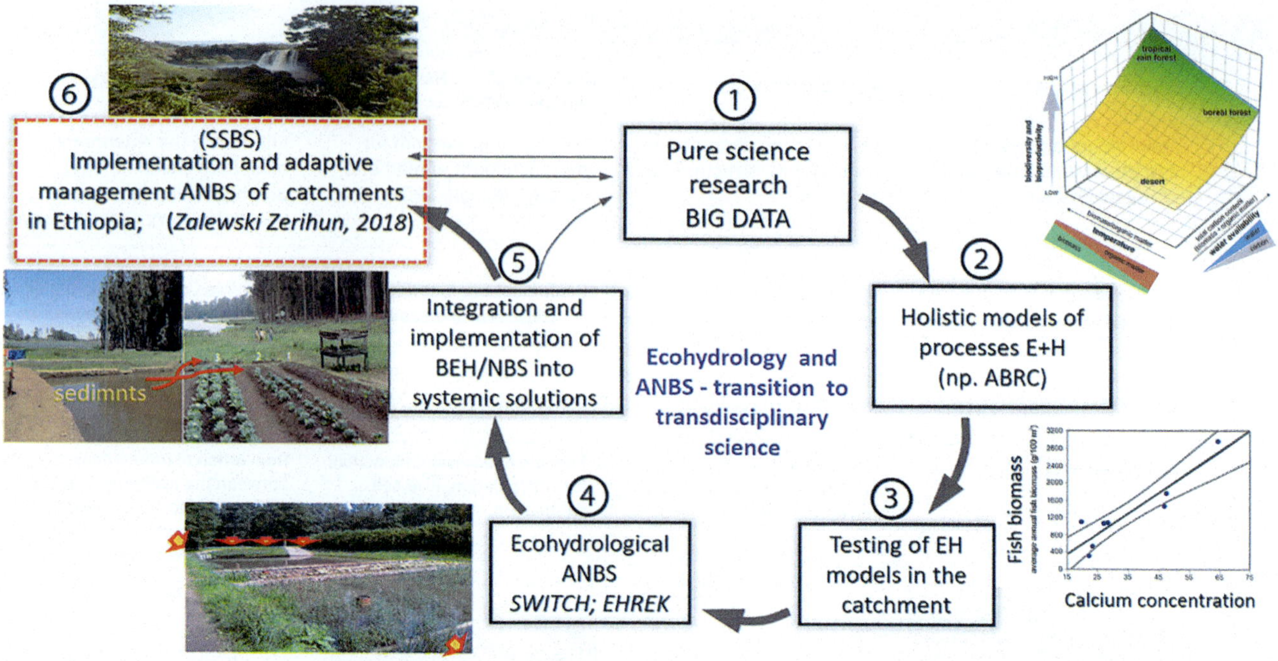

FIGURE 8.2 Methodology of transdisciplinary ecohydrology. The scientific findings of transdisciplinary ecohydrology are translated into the development of advanced nature-based solutions (ANbS), illustrating the use of pure scientific research formulated in (1) big data and (2) holistic model of processes for (3) testing of models in catchment will allow to identify emerging properties of the system, which are the key for development of ANbS (4) SSBS, which are the element of systemic solutions integrated with hydroengineering and indigenous knowledge (5) adopted to broad scope of ecosystems at different continents (6), which inspire further pure science research. This iterative process, exemplified by catchment projects in Ethiopia, generates feedback that refines scientific understanding and fosters the development of new transdisciplinary approaches, advancing ecohydrology as both a theoretical and applied science. The scientific background of systemic solutions is translated towards implementation by principles of transdisciplinary ecohydrology (Chapter 7) supporting IWRM. *Based on Zalewski, M., 2020. Ekohydrologia. Wydawnictwo Naukowe PWN, Warszawa. ISBN:978-83-01-20804-2*

2.3 Testing models in the catchment

The next step is the field testing of these TEH models at the catchment scale. This phase involves experimental research to validate the model's predictions, identify emerging properties, and assess outcomes such as water quality, biodiversity, or productivity. These models explain the interactions between ecological structures (e.g., fish biomass) and abiotic conditions (e.g., calcium concentration in water)—discovery of new emerging properties of systems (NEPS).

2.4 Application of ecohydrological advanced nature-based solutions

The development of the complex ABRC model necessitates the design of equally complex experiments, which in turn increases the likelihood of discovering new emergent system properties. These insights can be translated into ANbS, as demonstrated by EU projects such as SWITCH, EHREK, and EKOROB. Such interventions are designed to regulate key ecological functions, like nutrient cycling and water retention, by harnessing natural processes within systemic frameworks. Notable applications include geochemical barriers and sequential sedimentation-biofiltration systems (SSBS).

2.5 Integration into systemic solutions (ANbS + foresight)

Successful ANbS prototypes are then integrated into systemic catchment-scale solutions, harmonizing ecological processes with societal and infrastructural needs. This involves the implementation of ecohydrological solutions in real-world landscapes, as shown in agricultural and water management settings (see EHREK). Effective implementation requires a comprehensive understanding of the processes underpinning ecosystem health, as well as the regulatory mechanisms that sustain them, guided by foresight-based approaches.

2.6 Adaptive management and implementation through design thinking

Finally, solutions are implemented and adaptively managed, as demonstrated by real-world examples such as catchment management initiatives in Ethiopia. Feedback from these applications informs ongoing data collection and research, establishing a dynamic feedback loop that continuously refines the science–practice interface. Central to this process is the application of design thinking—a human-centered, iterative approach that prioritizes deep understanding of users' needs and the development of innovative, practical solutions. Through this lens, the evaluation and optimization of ANbS are embedded within the framework of Integrated Water Resources Management (IWRM), ensuring that interventions remain responsive and effective across diverse continental contexts.

2.7 Feedback—Inspirations from implemented innovations to pure science

Feedback from implemented innovations provides critical insights that inspire and refine pure scientific research, fostering the development of new hypotheses, models, and transdisciplinary approaches within transdisciplinary ecohydrology.

3. Transdisciplinary ecohydrology: A new dimension of Integrated Water Resources Management

Integrated Water Resources Management, an internationally recognized framework, takes a holistic approach to water governance, ensuring that resources are managed sustainably to support ecological balance while advancing the sustainable development goals (SDGs). As climate change accelerates and ecosystems continue to decline, reimagining water management is not just necessary, it is essential to securing the future of both human and planetary ecosystems. The concept of IWRM is defined as "a process that promotes the coordinated development and management of water, land, and related resources to maximize economic and social welfare equitably, without compromising the sustainability of vital ecosystems" (IWRM Action Hub). By addressing global challenges holistically, IWRM emphasizes transdisciplinary solutions that acknowledge the complex interdependencies between human health and planetary well-being Fig. 8.3.

This approach aligns with key recommendations from global environmental frameworks, including the United Nations Sustainable Development Goals (UN SDGs) and European Union policies on water management and biodiversity conservation. It emphasizes not only the protection of natural ecosystems through reduced consumption but also the mitigation of anthropogenic pressures on the environment. Strategies such as the circular economy, bioeconomy, ecosystem-based management play, EU Nature Restoration Law and Water Resilience Strategy a crucial role in this paradigm shift.

In response to these growing concerns, numerous voices, writers, social activists, and scientists are calling for urgent action. Global organizations have formed alliances and established benchmarks to guide nations in adopting sustainable practices. The United Nations Environment Program (UNEP), the leading global authority on environmental issues, was

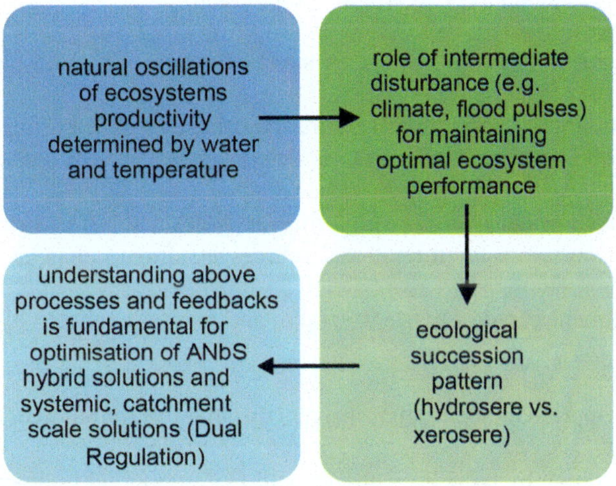

FIGURE 8.3 Conceptual model of implementing TEH. This diagram presents in a conceptual model of implementation of dual regulation, illustrating how understanding the natural dynamics of ecosystems can support the optimization of advanced nature-based solutions with hydrotechnical and hybrid solutions for IWRM.

established to inspire, inform, and support nations in their pursuit of sustainable development. Today, 118 countries have committed to its initiatives.

Recognizing the growing urgency of climate change, the Intergovernmental Panel on Climate Change (IPCC) was created in 1988 by UNEP and the World Meteorological Organization (WMO). Its mission is to provide scientific assessments on climate change, its impacts, and potential adaptation and mitigation strategies, ensuring that policymakers have access to reliable, evidence-based information.

In 2010, the Intergovernmental Science-Policy Platform on Biodiversity and Ecosystem Services (IPBES) was established as an independent body to assess biodiversity and ecosystem services. Through scientific evaluations, policy guidance, and capacity-building initiatives, IPBES helps bridge the gap between science and decision-making, ensuring that global leaders have the knowledge needed to protect and sustain natural ecosystems. Complementing this effort, the economics of ecosystems and biodiversity (TEEB) is a global initiative that highlights the economic value of ecosystems and biodiversity while underscoring the financial and social costs of their degradation. By linking ecological health with economic decision-making, TEEB provides a framework for recognizing, valuing, and integrating nature's contributions into policy and business strategies.

Fundamental ecohydrological processes form the scientific basis for designing and implementing systemic solutions in line with the Water Framework Directive (WFD), which aims to achieve good ecological and chemical status of all water bodies in the EU. These processes focus on the functional interactions between water dynamics and biological systems at the catchment scale, and are crucial for enhancing the sustainability and resilience of aquatic ecosystems under anthropogenic pressure and climate change.

Here are the key ecohydrological processes that must be considered:

1. **Catchment-scale systemic thinking**
   - Viewing the entire catchment as a functional unit, where upstream interventions influence downstream outcomes.
   - Ensuring spatial integration of nature-based solutions, hydrotechnical infrastructure, and socioeconomic/cultural considerations for holistic water governance.
   - Foresight-analysis water and environment status, potential and global regional and local trends/technologies for creation the desirable vision of the future, unique for every catchment.
2. **Hydrological regulation**
   - Understanding the water cycle within the catchment, including precipitation, infiltration, evapotranspiration, runoff, retention, and groundwater recharge.
   - Managing hydrological pulses (e.g., flood and drought events) to support ecological processes, which increase catchment retention, recharge of groundwater, filtration, considering functions of the forests, floodplains and wetlands, and regulation by improved catchment complexity by ANbS
3. **Biotic regulation**
   - Biological processes such as primary production for phytoremediation, and resilience enhancement.
   - Trophic interactions, particularly nutrient cycling and food web dynamics, play a key role in maintaining water quality (e.g., through biomanipulation or hydrobiomanipulation).
4. **Nutrient cycling and retention**
   - Controlling the flow and transformation of biogenic substances (nitrogen, phosphorus, carbon) in soil-water-plant systems.
   - Enhancing nutrient retention, especially in agricultural landscapes by increasing the complexity of the landscape with forest patches, riparian zones, wetlands, and floodplains to prevent eutrophication and improve water quality.
5. **Ecotone functionality**
   - Preserving and restoring land–water interfaces e.g., riparian zones and buffer strips, which act as natural biofilters, biodiversity hotspots and corridors for migrations, refuges at climate changes phenomena and gene exchange pathways.
   - These zones regulate sediment capture, nutrient uptake, and habitat connectivity, serving as transition areas that stabilize the landscape.
6. **Self-purification capacity**
   - Enhancing the natural self-purification capacity of rivers through the ANbS, hybrid, and shaping of habitats for the enhancement of WBSR.
   - Enhanced regulation of ecology by reducing nutrient outflows, supported by interventions such as high efficiency land/water ecotones, sequential biofiltration systems (e.g., SSBS) and geochemical barriers.
7. **Dual regulation (Abiotic–biotic integration)—"water for ecosystem, ecosystem for water"**

- Implementing dual regulation, the central concept in ecohydrology, which simultaneously manages hydrological (abiotic) and ecological (biotic) processes to optimize system performance.
- For example, modifying water levels to influence fish spawning or zooplankton dynamics, thereby improving water quality, regulation of storm water fluxes to constructed wetlands or SSBS to optimize their efficiency.

3.1 Frame 26. Integrated Water Resources Management

IWRM is a process that promotes the coordinated development and management of water, land, and related resources to maximize economic and social welfare in an equitable manner while ensuring the sustainability of vital ecosystems.

IWRM is founded on three core principles:

1. Social equity
2. Economic efficiency
3. Environmental sustainability.

Applying these principles involves addressing the following question:

How will my decision or action impact other users' access to water or their ability to benefit from its use?

There are four central components of IWRM: stormwater **management, wastewater treatment, water supply, and conservation of existing water sources.**

IWRM implementation is done by assessing the four key dimensions of IWRM: **enabling environment, institutions and participation, management instruments, and financing**. IWRM aims to support countries in their efforts to tackle specific water challenges, e.g., water scarcity, waterborne diseases, floods, droughts, and access to water and sanitation, and thus sustain their development to achieve the goals such as poverty alleviation, food security, economic growth, and ecological sustainability **to maximize economic welfare without compromising the sustainability vital ecosystems.**

Ecohydrology adds to the IWRM the aspect of enhancing the multidimensional sustainability potential WBSR + CE + LPG, through a transdisciplinary ecohydrology methodology Fig. 8.4.

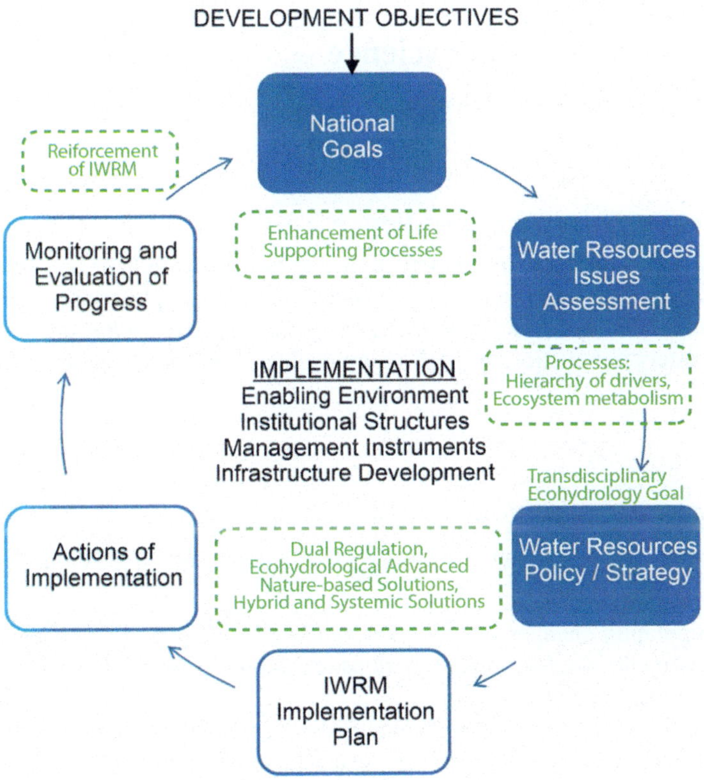

FIGURE 8.4 Expended cycle of Integrated Water Resources Management (IWRM). This diagram illustrates how transdisciplinary ecohydrology expands the cycle of IWRM, linking national development goals to water policy, implementation, and monitoring. It emphasizes how ecohydrological solutions, such as advanced nature-based and hybrid interventions, enhance catchment sustainability (WBSR + CE + LPG) and improve the harmony between human needs, water systems, and the environment.

While IWRM is globally recognized as a holistic framework for the sustainable management of water, land, and related resources, its conventional application often limits the role of ecosystems to the principle of "not compromising their integrity." In the context of accelerating global ecosystem degradation, this passive stance is no longer adequate. Ecosystems are not only passive recipients of protection but active providers of essential **life-supporting processes**, such as water retention, nutrient cycling, energy flow, and climate regulation. Therefore the long-term sustainability of humanity is fundamentally dependent on the health and functionality of both terrestrial and freshwater ecosystems.

To accelerate progress toward the SDGs, the WFD, IWRM must be expanded through the integration of **transcapillary** ecohydrology that focuses on the functional interplay between hydrological and ecological processes. Ecohydrology enriches IWRM by offering a **scientific and practical framework** for:

1. **Understanding how the quantity and quality of water resources** depend on ecosystem functions such as retention, purification, cooling, and biogeochemical cycling at the catchment scale, and how LSPs can be redistributed and enhanced.
2. **Developing and implementing advanced nature-based solutions (ANbS) that use ecosystem properties as management tools**, and integrating them with hydrotechnical infrastructure to generate synergies at the catchment level.
3. **Applying a systemic perspective to the water–biota interface—from molecular interactions to biosphere-scale processes**—to enhance catchment sustainability across environmental, cultural, and governance dimensions (WBSR + CE + LPG).

IWRM is founded on three guiding principles: social equity, economic efficiency, and environmental sustainability. Implementing these principles requires an inclusive approach that asks: How will decisions impact access to water and its benefits for all users—present and future?

The operationalization of IWRM involves four core dimensions: an enabling policy environment, institutional arrangements and stakeholder participation, appropriate management instruments, and sustainable financing mechanisms. When reinforced by ecohydrology, IWRM is better equipped to address today's most pressing water challenges, including water scarcity, waterborne diseases, flooding, drought, and access to clean water and sanitation. It becomes a powerful instrument for achieving broader development goals, such as poverty alleviation, food security, economic growth, and ecological integrity, not merely by preventing harm to ecosystems, but by actively enhancing their regenerative and regulatory potential in service of a sustainable future.

4. Why transdisciplinary ecohydrology becomes rocket science

Transdisciplinary ecohydrology framework: AI inputs for the development of IWRM.

4.1 Goal

4.1.1 Enhancement of water–biodiversity–ecosystem services–resilience

The overarching objective of the TEH framework is to enhance the resilience of water systems, biodiversity, and ecosystem services under increasing environmental pressures. This ambitious goal is founded on transdisciplinary science, combining hydrology, ecology, biotechnology, molecular biology, system modeling, engineering, and socioeconomic analysis. By breaking down disciplinary silos, TEH fosters a holistic approach to managing freshwater resources in dynamic social-ecological contexts.

4.2 Dual regulation

4.2.1 Defining the specific interplay between water and biota

A core concept of ecohydrology, dual regulation refers to the reciprocal feedback mechanisms between hydrological processes and biotic components. This involves both:

- Hydrological regulation of biota (e.g., hydrobiomanipulation changes the access of nutrients allocated in the trophic pyramid for elimination of algal blooms, flow regimes to enhance floodplain functions, groundwater levels to optimize phytotechnology, sediment transport to feed floodplain and ecotone zones)
- Biotic regulation of hydrology (e.g., structure of vegetation to shape river channel and ecotones in catchment, influencing water retention capacity, biofiltration, evapotranspiration, infiltration, and heat budget of catchment based on principles of TEH, quantification and testing of different scenarios)

AI tools can be employed to model and visualize these bidirectional feedbacks across spatial and temporal scales, allowing for adaptive management strategies under uncertain future conditions.

4.3 Transdisciplinary ecohydrological principles

Assessment framework for hydrological and ecological processes and implementation of ANbS.
This operational layer translates theory into practice through:

- Systematic and continuous monitoring for quantification of dynamic hydrological and ecological processes across river systems and their catchments (remote sensing, satellite imaging).
- Scenario-based modeling is supported by AI for simulating the performance of various ecohydrological interventions.
- Integration of ANbS with hydrological engineering to codesign multifunctional landscapes that mitigate floods, enhance water retention, support biodiversity, and improve ecosystem services for society.
- Design thinking is a holistic, iterative approach that generates innovative solutions.

The algorithm acts as a decision-support tool in aligning technical, ecological, and community-driven knowledge.

4.4 Catchment-scale foresight as a reference for transdisciplinary ecohydrology development

4.4.1 Artificial intelligence for optimization of Integrated Water Resources Management via transdisciplinary ecohydrology

The catchment serves as the natural unit for understanding complex water-ecosystem dynamics. TEH leverages AI to provide foresight and strategic planning by:

- Focus on creating a desirable future for the catchment.
- Forecasting socioeconomic processes at the catchment scale using foresight approaches that support the implementation of the SDGs and the WFD.
- Forecasting hydrological and ecological responses to climate change, land use shifts, and socioeconomic drivers.
- Detecting ecological tipping points and risk factors.
- Optimizing resource allocation across competing needs (e.g., agriculture, conservation, and urban development) while enhancing resilience.
- Facilitating participatory governance and scenario cocreation with leaders and stakeholders.

4.4.2 Summary artificial intelligence-supported transdisciplinary ecohydrology

Building upon the TEH framework, the key methodological pillars where artificial intelligence can enhance ecohydrological approaches are summarized below Fig. 8.5:

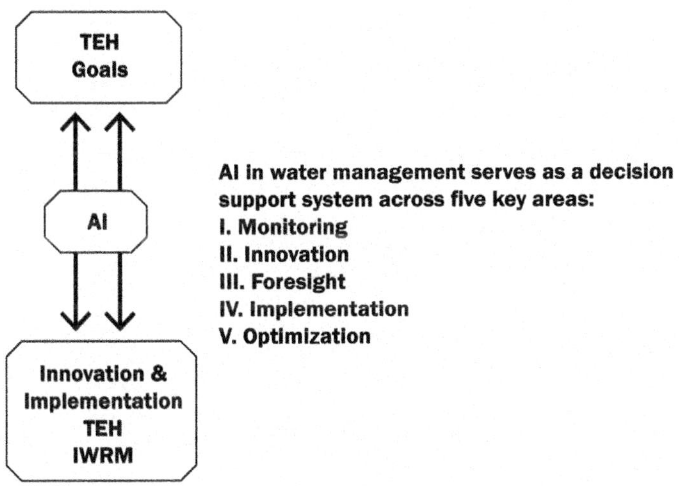

FIGURE 8.5 AI-supported transdciplinary ecohydrology. The diagram presents a flexible and intelligent system for water management, where AI-powered decision-making aligns technical analysis with ecological, social, and governance principles, supporting adaptive and forward-thinking water resource planning.

1. Ecohydrological biotechnologies—leveraging physical processes, with temperature as a key driver of biocenotic dynamics, and optimizing hydrological regulation in the context of climate change.
2. Stoichiometry and nutrient regulation, guided by Liebig's Law of the Minimum, including management of nutrient ratios (e.g., N:P) to prevent ecological disturbances such as cyanobacterial blooms.
3. Integration of hybrid systems, combining engineering and ANbS, where bioprocess stoichiometry informs AI-supported innovation and system design.
4. Landscape optimization, enhancing complexity, bioproductivity, and biodiversity to strengthen Water-Biodiversity-Ecosystem Services -Resilience (WBSR) under changing societal and climatic conditions.
5. Cultural and educational dimensions, fostering societal understanding of landscape system processes to stimulate public engagement, adaptive governance, and long-term stewardship.

AI disclosure

During the preparation of this work the author(s) used OpenAI's ChatGPT. This document includes content generated and/or edited with the assistance of OpenAI's ChatGPT, used to support language refinement and structural improvements. All final content was reviewed and verified by the author. After using this tool/service, the author(s) reviewed and edited the content as needed and take(s) full responsibility for the content of the publication.

References

Weizsäcker, Lovins, A.B., Lovins, L.H., 1997. Factor Four: Doubling Wealth, Halving Resource Use—The New Report to the Club of Rome. Earthscan.
Zalewski, M., 2011. Towards engineering harmony between water, ecosystem and society: Editorial. Ecohydrol. Hydrobiol. 11 (3–4), 137–140.
Zalewski, M., 2020. Ekohydrologia. Wydawnictwo Naukowe PWN, Warszawa, ISBN 978-83-01-20804-2.
Zalewski, M., Zerihun, Y., 2010. Ecohydrology for Ethiopia: regulation of water–biota interactions for sustainable water resources and ecosystem services for societies. In: Proceedings of the International Symposium Ecohydrology for sustainable water ecosystems and society in Ethiopia UNESCO–European Regional Centre for Ecohydrology, Polish Academy of Sciences.

Chapter 9

Foresight: A necessity for a sustainable future

Maciej Zalewski
European Regional Centre for Ecohydrology of the Polish Academy of Sciences under the auspices of UNESCO, Łódź, Poland

PHOTO CREDIT: M. ZALEWSKI.

1. Toward a holistic water management paradigm

Water management must be based on long-term actions and investments. That is why, to achieve synergy and efficient money use, there is a necessity to develop catchment foresight as a fundamental step in planing the sustainable water resource management.

Transdisciplinary Ecohydrology. https://doi.org/10.1016/B978-0-443-41596-8.00012-1

Foresight is the systemic approach to the evaluation of new trends based on knowledge of climate, environmental, sociological processes, and technologies from the point of view of the economy, quality of life, and sustainable development.

Foresight methodology has three main goals:

1. FORECAST OF FUTURE—enabling the adaptive attempts, preparation for unpredictable events, and reduction of negative consequences of events that cannot be changed
2 MANAGEMENT OF FUTURE—means the proactive (management of probable crisis) and positive (management by goals)
3. CREATION OF DESIRABLE FUTURE—means mainly the proactive creation of a needed vision of future

Foresight on the catchment scale has to be based on the three principles of ecohydrology (Zalewski, 2011). The vision and strategy for water management, formulated on the basis of catchment foresight, should align with and take as reference points the UN Sustainable Development Goals (SDG), and the European Commission's Water resilience strategy. Such documents provide a framework for improving environmental quality and human well-being while simultaneously creating conditions that foster innovation, which in turn generates the potential for dynamic development and competitive advantage. It is important to remember that we are currently observing changes in the strategies, goals, and priorities of international systems in which the advantages established so far are changing.

A good example of the ambitious holistic approach is the European Green Deal, addressing key sectors and areas such as transportation, agriculture and food, climate, energy, closed-loop economy, ecosystems and biodiversity, water, and soil and air pollution. Such an acceleration of goals needs integrative, systemic, low-cost, and low-energy solutions based on natural processes, aimed at reducing emissions and energy consumption, while improving the environment and quality of life of society, which is an attractive alternative to the traditional economic model. As part of the Green Deal strategy, biological sciences and biotechnologies so far used mainly in the food and pharmaceutical industries are increasingly being adapted and developed in agriculture to reduce pollutant emissions and increase the potential of ecosystems. Sustainable and integrative objectives should be rooted in solutions that take catchment foresight as a primary reference point. Transdisciplinary ecohydrology supports such an approach by providing a hydrocentric concept for enhancing the sustainability potential of catchments through the WBSR framework, which is particularly important in the face of progressing climate change. This framework is further expanded by CE—Culture and Education for Sustainability—and proactive LPG—Law, Policy, and Governance.

It is fundamental to base water management on catchment foresight. Each drainage basin must be treated as a unique superorganism with specific features such as climate conditions, geomorphology, hydrogeology, resistance to anthropogenic stress, and the way it is used by people. Therefore it is necessary to analyze how a given catchment will be sensitive to climate change because of the above-mentioned features. Actions taken in each catchment area should be based on an action plan consisting of three elements:

1) Monitoring—analysis and quantification of the processes that determine the catchment potential
2) Innovation—creating integrative, hybrid, site-specific solutions
3) Foresight—economic and sociological, technological, and climatic
4) Implementation of the systemic solutions by a transdisciplinary ecohydrology approach because the catchment area is a complex system with complex interactions (natural and social), and understanding its operation requires knowledge from various disciplines of science, and solving system problems with solutions that will provide:
   - Stabilize and increase water resources through various forms of retention, beginning with enhancing the complexity of the catchment landscape.
   - Increasing water resources and their resilience is closely linked to catchment landscape complexity, biomass, biological productivity, and biodiversity. These factors stabilise the hydrological cycle and create positive feedbacks that enhance ecosystem resilience and the provision of ecosystem services. According to the Intermediate Disturbance Hypothesis, increasing stochasticity of abiotic processes may lead to reductions in biodiversity.

As water demand will continue to increase and climate change will continue to cause increasing stochasticity of processes, i.e., resource variability, the most important actions will be to:

1. Increase various forms of water retention in the catchment area to restore natural processes (slowing the outflow of water to the sea and reducing evaporation), e.g., constructed wetlands, rewetting, reconnection of floodplains with rivers, stream and river renaturalisation, and the construction of lateral reservoirs.
2. Increased water retention should contribute to the stabilisation and restoration of groundwater resources, while enabling society to adapt its needs to the available potential of groundwater and surface water systems.

Therefore, action must begin with a hydrogeological assessment, first and foremost identifying opportunities to increase groundwater recharge in catchments. In Poland, in the June July, intensive rainfall occurs, while the evaporation and evapotransiration and demand for water is highest. That is why also water from autumn - low temperature months should be stored in groundwater resources. In connection with the above, the following research directions for fine-tuning the broad scope of ANbS, particularly in agriculture, may be proposed:

1. Analysis of the effects of soil and geological formations on water retention in soils and groundwater resources, and the possibilities for their enhancement.
2. Comparison of the impacts of industrial agriculture—characterised by large open fields exposed to intensive winds that accelerate evaporation, soil compaction caused by heavy agricultural machinery vs. spatially diversified regenerative agriculture, including no-till farming, cover cropping, and livestock integration, on the restoration of soil organic matter, water retention, and groundwater resources. The study should also examine how irrigation prectices, different types of soils, crops and agricultural practices influence groundwater recharge at the given hydrogeological and climatic conditions. As the bioeconomy is expected to generate increasing demand for agricultural production, it is necessary to optimise agricultural systems by considering climatic and geological processes as well as soil types. Based on these factors, a strategy for diversified agriculture should be proposed to limit the negative impacts of industrial agriculture and enhance landscape resilience.
3. Comparison of forests and forest plantations of the different sizes in terms of water retention, heat balance, convection rains, and groundwater renewal.
4. Retention in river valleys by promoting semi-intensive aquaculture with a low carbon footprint.
5. Water retention in lateral reservoirs that do not disrupt the river continuum procesess but increase water retentivness in the river valley.

Sustainable development requires forecasting to be based on the desired vision of the future and to identify phenomena such as demographic processes, priorities in technology, economy, and factors that may prevent its implementation, e.g., climate change, global economy, and mega trends (Naisbitt, 1982).

Foresight, when considered in the context of managing natural resources, represents a culture of thinking about the future (Rogut and Piasecki, 2007). When conducted at a national or regional scale, it systematizes and integrates knowledge for the benefit of science and the economy stimulating exchange of information, experiences, and views between scientists, industry representatives, the business sector, and decision-makers, which allows for maximization of opportunities.

Foresight is also a tool and a starting point for developing a management strategy on a catchment scale. One of the most important factors in the Anthropocene, generating a competitive advantage of the region is environmentally diversified and undegraded space, as well as water resources and freshwater ecosystems. It not only ensures the production of food that ensures healthy development of society, but is also increasingly associated with high quality of life deeply dependent on opportunities for regeneration in green areas of high biodiversity. Accessible and attractive blue-green living space has become an extremely important resource in the 21st century, comparable to innovative technologies because support life suporting process in Biopshere and in paralell is rising opportunities and awareness of the active life, good health, well-being.

2. Implementation: Methodology and strategy of foresight implementation

Due to the transdisciplinary ecohydrology approach and the inclusion of various parties interested in the process, system analysis introduces an important component of objectivity to the often-subjective decision-making process. It also enables reconciliation of the distant interests of various social groups interested in the solution to the problem. It is precisely because of the inclusion of not only specialists from different fields of science, but also stakeholders concerned by the problem that systems analysis is referred to as a transdisciplinary approach.

Due to the high complexity of the relationship between the individual tasks, as in the changing situation pictured above, the proposed network-based method for presentation of foresight results should be a network of interrelated elements. Networks, thanks to their graphic form, better than a linear system illustrate both the hierarchy and the relations between individual elements (Zalewski, 2008). They are also flexible illustrations and allow for easy application of corrections. Even though one item is eliminated or changed, the entire network continues to function as visual ilustration the interconnectedness. The networks support the development of flexible strategies and analysis of various scenarios in terms of shaping the future instead of predicting it by pointing out activities that currently best stimulate sustainable development of the given cachment.

3. Principles of implementing a foresight vision

Formulating the vision with the support of foresight should begin with: (1) Analysis of the effects of long-term anthropogenic and climatic changes on biota and hydrological patterns in the river and its catchment. (2) Assessment of the existing research potential in a given region. To do this, it is necessary to identify strengths of research units with extensive knowledge of new research areas, knowledge in the field of research application, and cooperation with the economic sector and enterprises. Institutions with the highest scientific achievements, but without implementation experiences, should be supported through joint projects with teams with implementation achievements.

This important supporting element of creating a vision of the sustainable future is the analysis of development dynamics in a given field of science and an attempt to formulate a developmental algorithm. Such algorithms are helpful in formulating anticipatory research hypotheses. An example of algorithm of the scientific evolution of paradigm as effect of the positive feedbacks between the theoretical concepts and empirical tests for river and its catchment is modelled by (Zalewski, 2000; Zalewski and Robarts, 2003), which highlight the change of paradigms in river ecology from zonation , River Continuum up to Transdisciplinary ecohydrology - enhancement of the river system and its catchment sustainability potential WBSR. Proposed algorithm increasing the scale extending the research, activities and forecasts from the river section to the entire catchment area and taking into account the process dynamics, (describing the relationship between organisms/ecosystems and hydrological dynamics to develop the innovative ANbS and systemic solutions.

(3) Dialogue and cooperation transdisciplinary ecohydrology-based water management, cooperation between the so-called implementers (specializing in implementations), pure science resercher and inventors specializing in developing new, innovative technologies and systemic solutions is vital. Existing knowledge and traditional nature conservation measures represent the "hardware," while understanding ecosystem processes as a tool is employed as "software" to develop innovative and systemic solutions.

4. Adaptive management

Foresight is a tool for setting the direction of the region's development and its strategic goals in the face of increasing competition, both at the national and global levels. Just like in nature, the social and economic situation is changing fast; therefore in these areas, success is related to skillful adapting to changing conditions, as well as changing methods, plans, and goals at the level of implementing the recommended solutions (Rogut and Piasecki, 2007).

Adaptive assessment management (Holling, 1978; Olsson et al., 2006) defined as: planing, acting, monitoring, evaluating and adjusting, is helpful to reduce uncertainty in long term large scale projects such as cachment mangemnt ,where considering the monitoring of management resultas as reference point allowing to correct the further actions to achieve startegic goals. It refers to the theory of decision-making and is used in the absence of full information on economic, social, and environmental processes, future changes, and factors that may change the form of activities affecting the final effect.

Adaptative management - "learning by doing" , used to manage programs under uncertainty, assumes continuous monitoring of the plan implementation process, decisions, and their consequences, thus reducing the risk of undesirable effects. This monitoring also allows you to register changes in the socioeconomic and environmental situation (climate change impacts) and adequately modify the goals and methods for planned activities. It also allows for continuous updating of information and databases concerning individual implementations (Zalewski, 2008).

Depending on the available information about the environment, the methods used for adaptive management can be passive or active. Passive management is when there is not enough information about the natural state and socioeconomic systems and directions of change. In this situation leaders often avoid proactive goal setting this leads to fragmentation of efforts and low effciency of measures. In the case of areas and processes that are difficult to forecast such as cachment the implementation of innovative solutions, in the holistic framwork the active management should be used. Proactive management is about developing and testing new strategies and relying on new and unconventional assumptions. Such management allows to learn new strategies through experimenting, developing new approaches, and choosing the best of several available options (Zalewski, 2008).

In proactive and adaptive management requires an interactive decision-making process, i.e., assessing the effects of activities and adjusting them to the knowledge gained in the implementation process. For this, in turn, there is a need for the appropriate development of a feedback loop between monitoring the situation and the decision-making process. Another important element is risk definition, as well as defining the unknown variables by developing a multilevel model and identifying interactions between its elements. This allows the probability of success to be assessed for each of the actions taken at each implementation level. This approach also implies the recognition of risk and a certain level of uncertainty as permanent parts of the decision-making and management process; however, they allow learning through experience and improvement thanks to this (Williams, 2011; Zalewski, 2008).

All the above-described elements of the implementation and adaptive management process allow for optimization of solutions for the large scale in the long term horizon to achieve the best possible results. It is also good to present the goals, methods, and directions in the form of a network, which—as mentioned above—is an easily modifiable, and supports adaptive management.

5. Changing the paradigm and the method of dialogue

The second half of the 20th century turned environmental protection into a kind of battlefield. Human actions in the industrial age caused a severe degradation of ecosystems, and at the end of this era, we saw the emergence of environmental organizations that put pressure on decision-makers and often block their activities as part of further economic and industrial progress. This state of affairs has made the paradigm of environmental protection restrictive, sometimes prohibiting restoration activities, causing deterioration of the quality of the environment.

The problems of the 21st century pose challenges to civilization that it is unable to cope with in the restrictive environmental paradigm. The process of effectively implementing WBSR approach and the ANbS as far as hybrid technologies identified by foresight, requires awareness of the goal and its acceptance by the society and parties interested in the solution of a given problem, and the basis for achieving this awareness and acceptance is a modern form of dialog based on culture and education (CE) of water and sustainability. Therefore, in the domain of Law, Policy, and Governance (LPG), a paradigm shift is also needed, from a restrictive approach to a creative one, based on dialogue among different stakeholder groups: scientists, policymakers, engineers, and local communities. Only such a state of affairs creates conditions for research and implementation of solutions driven by the demand of individual groups (demand-led). This creates an opportunity to enhance the sustainability potential of WBSR reverse degradation of ecosystems and to protect the pristine natural environment. To enable dialog, stakeholder platforms are created and adopted under various names by international programs and projects, including multistakeholder platforms (United Nations Development Programme, 2006) or Learning Alliance (Morrison and Mezentseff, 1997).

6. Building a vision for the future, scenarios, and strategies

The first step in the planning and management process, with the help of stakeholder platforms, is to build a common vision of the future by analyzing the current situation in regard to challenges, opportunities, threats, and uncertainties, and considering possible scenarios of changes in the situation in different time horizons, with the determination of the probable events.

This allows for the emergence of extreme factors that may affect the preparation of vision and create scenarios of situations corresponding to the combinations of the impact of these factors.

Building a vision begins with identifying the most important problems to be included in the vision (this identification can take place with the use of many tools, such as problem tree analysis, problem list, and brainstorming). To answer these problems and needs, a vision is formulated in a narrative convention, with its goals, and then it is assessed in terms of the feasibility of achievement. The SMART criteria are used here to determine if the vision for the future:

S—specific;
M—measurable;
A—acceptable;
R—realistic;
T—time-bound.

In addition, it should be checked whether it is consistent with, e.g., other visions resulting from administrative or legal processes. If the vision of the future meets all requirements, it can proceed to the next phase—building a strategy to achieve the worked-out vision in previously developed scenarios. It is important to give scenarios and strategies a chance to lead to the realization of the vision. It is vital to remember that some scenarios may be so complicated that it is difficult to devise a realistic strategy for achieving the vision, and this may result in a need to reverify it.

Building a strategy is the stage of basic decision-making, strategic direction, and the identification of a broad spectrum of possible actions and solutions. Creating a strategy starts with identifying the practical options and opportunities for the realization of the vision, and social, technical, and political evaluation, the economic and environmental acceptability, and the feasibility of these options and opportunities. The constraints and risk factors that will affect the process also need to be identified, as well as the chances of success of the strategy. Such limitations include, e.g., slow turnaround cost of actions taken or lack of administrative capacity and skills for their implementation, but also many others, such as corruption and assigning a small importance to environmental matters. As with the scenarios developed, there are several possible strategies that allow the vision to be realized with consideration of information gathered earlier in the process; this is being done through evaluation of the strategy in terms of potential for realizing the vision. On this basis, one strategy with the broadest spectrum is selected in consultation with the stakeholders' actions. The final stage is to assess whether the implementation of the strategy is in agreement with budgetary possibilities and has political backing.

At this stage, facilitators play an important role. They coordinate communication, ensuring the internalization of the dialog approach, stakeholder engagement, and a high probability of achieving goals identified by stakeholders, as well as quick response and resolution of possible conflicts. Coordinators also play a supportive and advisory role in managing integrated actions and resolving conflicts by observing and monitoring the entire build process.

Based on the above analysis of the vision, scenario, and strategy building process, one can see how serious and complicated the process of dialog between the parties and the ability to cooperate is. The ability to engage in dialog allows the achievement of goals on the way to targeted development, and it is also a good tool for Transdisciplinary social education.

7. Summary of foresight

Because any strategy for success should include both the maximization of opportunities for success and the elimination of possible threats, foresight includes these elements, as comprehensive and strategic thinking about the future. Elimination of threats requires adaptation to climate change based on a new evolutionary ecosystemic paradigm, supporting the development of a broad range of Advanced Nature-based Solutions (ANbS) based on Dual Regulation and hybrid technologies across regions and economic sectors, thereby contributing to the enhancement of Water–Biodiversity–Ecosystem Services–Resilience (WBSR), and improving health and quality of life. The problem to be addressed is how to maximize opportunities for overcoming the challenges arising from the process of globalization, in which the key factors for sustainable development are resources, energy, water, and intellectual potential. Hence, the emphasis on developing and acquiring technologies and optimal use of renewable sources, as well as informatics, telecommunications mechatronics, and nanotechnologies, contributes to the reduction of energy consumption. Technologies with great potential to stimulate development are biotechnologies used for bioeconomy , health, now more and more used for ecohydrological processes regulation according to the philosophy of green economy for SDGs.

Why Foresight?

Coastal Zones and Future Sea Level Rise

Most of humanity lives in the coastal zones Fig. 9.1. The rise of sea levels will create new challenges for which we need to be ready.

FIGURE 9.1 Calpe, Spain, on the Mediterranean coast. This photograph clearly illustrates how city infrastructure has approached the coastline without adequate foresight to anticipate the consequences of sea-level rise. *Photo credit: M. Zalewski.*

AI disclosure

During the preparation of this work the author(s) used OpenAI's ChatGPT. This document includes content generated and/or edited with the assistance of OpenAI's ChatGPT, used to support language refinement and structural improvements. All final content was reviewed and verified by the author. After using this tool/service, the author(s) reviewed and edited the content as needed and take(s) full responsibility for the content of the publication.

References

Holling, C.S., 1978. In: Adaptive Environmental Assessment and Management. Wiley.

Morrison, M., Mezentseff, L., 1997. Learning alliances — a new dimension of strategic alliances. Manage. Decis. 35 (5), 351–357. https://doi.org/10.1108/00251749710173715.

Naisbitt, J., 1982. Megatrends: Ten New Directions Transforming Our Lives. Warner Books.

Olsson, P., Gunderson, L.H., Carpenter, S.R., Ryan, P., Lebel, L., Folke, C., Holling, C.S., 2006. Shooting the rapids: navigating transitions to adaptive governance of social-ecological systems. Ecol. Soc. 11(1) (18). https://doi.org/10.5751/ES-01595-110118.

Rogut, A., Piasecki, B., 2007. In: Foresight as a Tool for Building the Knowledge-Based Economy. Wydawnictwo Biblioteka, pp. 7–28.

United Nations Development Programme, (UNDP), 2006. Human development report 2006: Beyond scarcity: Power, poverty and the global water crisis. Palgrave Macmillan.

Williams, B.K., 2011. Adaptive management of natural resources-framework and issues. J. Environ. Manage. 92 (5), 1346–1353. https://doi.org/10.1016/j.jenvman.2010.10.041.

Zalewski, M., 2000. Ecohydrology — the scientific background to use ecosystem properties as management tools toward sustainability of water resources. Ecol. Eng. 16 (1), 1–8. https://doi.org/10.1016/s0925-8574(00)00071-9.

Zalewski, M., 2008. Rationale for the "Floodplain Declaration" from environmental conservation toward sustainability science. Ecohydrol. Hydrobiol. 8 (2 4), 107 111.

Zalewski, M., 2011. Ecohydrology for implementation of the EU Water Framework Directive. Proc. Inst. Civ. Eng. Water Manag. 164 (8), 375–385. https://doi.org/10.1680/wama.1000030.

Zalewski, M., Robarts, R., 2003. Ecohydrology – a new paradigm for integrated water resources management. SIL News 40, 1–5.

Chapter 10

Linking catchment and floodplain dynamics for climate-resilient river landscapes: An ecohydrological approach

Edyta Kiedrzyńska[1, 2] and Marcin Kiedrzyński[3]

[1]*European Regional Centre for Ecohydrology of the Polish Academy of Sciences under the auspices of UNESCO, Łódź, Poland;* [2]*University of Lodz, Faculty of Biology and Environmental Protection, UNESCO Chair on Ecohydrology and Applied Ecology, Łódź, Poland;* [3]*University of Lodz, Department of Biogeography, Paleoecology and Nature Conservation, Łódź, Poland*

Photo credit: M. Zalewski.

Transdisciplinary Ecohydrology. https://doi.org/10.1016/B978-0-443-41596-8.00015-7

1. Introduction

Every part of the Earth's surface is part of a catchment. River catchments with river valleys, floodplains, and riverbeds are key hydrosystems that constitute global reservoirs of fresh water. Floodplains are an integral part of river valley systems and often function as natural wetlands, forming a transitional "dynamic spatial mosaic" (Amoros and Bornette, 2002; Thoms, 2003) between the river and its valley. Floodplains play an essential role in the exchange of water and matter between the river and the ecosystems in the valley, affecting water quality (Kiedrzyńska et al., 2008, 2015, 2021; Mitsch et al., 2008; Mitsch, 2014). In the floodplain, surface and groundwater act as links between the various abiotic (soil) and biotic (vegetation) elements, and at the same time as a connection between the riverbed and the valley along the river continuum from source to estuary (Kiedrzyńska et al., 2008, 2014). These connections, operating on various scales and levels, are of key importance for maintaining the function and integrity of river valley systems (Kiedrzyńska and Zalewski, 2012; Thoms, 2003). These systems function as interfaces for biodiversity, both terrestrial and aquatic, within the space-time dynamics of the catchment landscape (Zalewski, 2008, 2009; Zalewski and Kiedrzynska, 2010).

Catchments—particularly river floodplains—are exposed to pressure in the form of changes in land management and intensification of agricultural, industrial, or urban land use, which lead to disturbances in the water and nutrient cycles within the ecosystem. Rivers and floodplains are located in landscape depressions; thus, various substances of anthropogenic origin naturally flow into them (Fig. 10.1), e.g., mineral and organic matter, biogenic compounds (Altinakar et al., 2006; Kiedrzyńska et al., 2008b; Magnuszewski et al., 2005, 2007), dioxins (Urbaniak et al., 2012, 2014, 2019), and microbiological contamination (Harnisz et al., 2020; Koniuszewska et al., 2020; Serwecińska et al., 2021). The nature of catchment and river valley use largely determines the functioning of floodplains, as well as the frequency and magnitude of flood pulses. Consequences of changes in catchment use—often negatively amplified by changes in the hydrological

FIGURE 10.1 The floodplain of the Pilica River, central Poland. Floodplain of the Pilica River upstream of the Sulejów Reservoir: A – high-flow conditions ($Q = 83.2\ \mathrm{m}_3\ \mathrm{s}_{-1}$), spring 2006 (photo by P. Wysocki); B – low-flow conditions ($Q = 6.7\ \mathrm{m}_3\ \mathrm{s}_{-1}$), summer 2006 (photo by M. Koch). *From: Kiedrzyńska, E., Zalewski, M., 2012. Water Quality Improvement through an Integrated Approach to Point and Non-point Sources Pollution and Management of River Floodplain Wetlands Ecological Water Quality – Water Treatment and Reuse. InTech, pp. 325–342. https://doi.org/10.5772/31374. Photo credit: M. Koch., P. Wysocki.*

cycle and the loss of integrity between river ecosystems and floodplains—lead to a drastic intensification of eutrophication symptoms, a reduction in biodiversity, and the loss of ecosystem services for society.

From the perspective of ecohydrology, floodplains are extremely important ecosystems that periodically capture pollution loads transported with floodwaters, but also attenuate flood peaks and reduce the risk of flood occurrence (Fig. 10.1). An increase in the natural water retention capacity of floodplains and across the catchment under progressing climate change conditions is possible through three steps:

An increase in the natural water retention capacity of floodplains and across the catchment under progressing climate change conditions is possible through three key actions:

1. The sustainable management of water resources within the catchment area,
2. The optimal use of existing hydrotechnical infrastructure, and
3. The implementation of ecohydrological biotechnologies and nature-based solutions (NBS).

From a holistic perspective, river valleys with their associated floodplains enhance the overall resilience of the region to climate change and anthropogenic pressures. At the same time, they contribute to increased flood safety, improved water quality, and a higher provision of ecosystem services for society.

2. Impacts of climate change on river systems and catchment dynamics

Global warming, recognized as a scientifically proven phenomenon (Climate Change, 2023; Climate Change Evident across Europe, 2012; C3S & WMO, 2024; Steffen et al., 2018), together with the aforementioned human-induced environmental changes, leads to significant disruptions of the carbon, nitrogen, phosphorus, and silicon cycles, as well as alterations in the biological structure and productivity of ecosystems. It also modifies the energy balance of the Earth's surface (Hassan, 2005) and the water cycle (Kundzewicz et al., 2007; Li et al., 2019). These processes raise serious concerns about the future and the capacity to maintain ecosystem services essential for sustaining human civilization (Steffen et al., 2015, 2020; Zalewski, 2015).

Under conditions of climate warming, evapotranspiration becomes more intensive, and water within the hydrological cycle undergoes accelerated circulation, potentially triggering more frequent extreme hydrological events (Chang and Jung, 2010; Huntington, 2006; Rockström et al., 2014; Steffen et al., 2015; Vörösmarty et al., 2013). Climate change, coupled with an increasing frequency of extreme weather phenomena—such as floods, tornadoes, and prolonged droughts that cause the periodic drying of streams and rivers (Climate Change, 2023)—and anthropogenic degradation of natural environments in many river catchments and floodplains, has led to a reduction in ecosystem carrying capacity and their ability to buffer adverse changes (Freudenberger et al., 2012; Hobson, 2010; Kiedrzyńska et al., 2015; Zhang et al., 2014), consequently contributing to biodiversity loss and, ultimately, to environmental dysfunction (Hansen et al., 2005; Hirsch, 2010).

Current climate change adaptation strategies are primarily focused on mitigating the effects of potential catastrophic weather events and minimizing associated risks for society (Climate Change, 2023; Climate Change Evident across Europe, 2012). However, it is important to note that these risks are determined not only by climatic and meteorological factors but also by human activities (Climate Change, 2023). Therefore the effectiveness of adaptation and risk prevention strategies depends on a thorough understanding of exposure and vulnerability dimensions (e.g., droughts and floods) as well as on accurate assessment of changes within these parameters (Climate Change, 2023; C3S & WMO, 2024).

In the context of accelerating global change, passive ecosystem protection is insufficient. A key requirement is the active regulation of landscape structure and ecological processes to enhance resilience and improve the capacity to buffer anthropogenic pressures. This capability can be defined as the ability to maintain ecological balance and ecosystem functioning under increasing and unavoidable environmental stress. It is therefore necessary to improve water quality, restore biodiversity, enhance ecosystem services for society, and implement mechanisms and solutions that reduce pollutant loads entering river systems and, consequently, coastal zones of seas and oceans (Kiedrzyński et al., 2014; Zalewski, 2011).

Floods are recognized as one of the most severe natural disasters, both in Europe and worldwide, because of their large-scale destruction and economic consequences (Glaser et al., 2010). Historical records document significant flood-s—for instance, those on the Nile in ancient Egypt, or in ancient Greece and the Roman Empire—occurring at a time when human activities and settlements were still relatively limited (Ganoulis, 2008). Today, flood risk is a serious concern in many regions of Europe and beyond. When assessing and managing flood risk, it must be remembered that floods are an inherent component of the natural hydrological cycle. Catastrophic flooding events over the past 2 decades in Europe—such as:

- Central Europe (Storm Boris, 14–21 September 2024): Austria, Czechia, Poland, Germany, Hungary, Romania, Slovakia, and Moldova were affected. At least 27 fatalities were reported, with total damages estimated at over €2 billion.
- Spain (Valencia region, 29 October–1 November 2024): One of the deadliest recent events, with at least 232 deaths and economic losses amounting to approximately €10.7 billion.
- Western and Central Europe (July 2021): Severe flooding in Germany (Rhineland-Palatinate, North Rhine-Westphalia) and Belgium caused more than 220 fatalities and extensive destruction, especially in the Ahr and Erft river valleys — the deadliest flood in Germany since World War II.
- Slovenia and Austria (August 2023): The worst floods in Slovenia's history, prompting large-scale international assistance, including support from the EU and NATO.
- Northern Italy (floods in Emilia-Romagna in May 2023): Following record rainfall over a short period, widespread floods and landslides occurred, causing significant evacuations and damage.
- Germany (June 2013): Extensive flooding along the Danube and Elbe rivers.
- Poland, Czech Republic, Hungary, Slovakia (May–June 2010): Severe floods in the Vistula, Oder, and Danube basins.
- Romania (2006), Danube and Elbe (2002), Vistula and Oder (1997), Rhine (1995): Previous decades also witnessed large-scale flood disasters with major human and economic losses.

Comparable events outside Europe—such as the Mississippi River floods (USA, 1993 and 2011), the Yangtze, Yellow, and Songhua Rivers (China, 2010), and the Queensland floods (Australia, 2011)—have demonstrated that human activity and conventional, poorly planned river engineering can exacerbate flood frequency and amplify negative social and economic impacts (C3S & WMO, 2024; Fendler, 2008; Ganoulis, 2008).

In August 2002, an extreme flood event in Central Europe caused severe damage and loss of life in Austria, the Czech Republic, and southeastern Germany. One hundred people lost their lives, and approximately 100,000 were evacuated. The total economic loss was estimated at €15–16 billion (Ganoulis, 2008). The flood of the Elbe River in Dresden was classified as a 500-year event. In Austria, the flood wave along the Danube was estimated to have a recurrence interval of 70–100 years, whereas in some tributary catchments of the Danube, 1000-year flood events were recorded (Ganoulis, 2008). According to the European Environment Agency (2010), between 1998 and 2009, Europe experienced 213 major floods that caused 1126 fatalities, the evacuation of 500,000 people, and insured property losses exceeding €52 billion.

Prolonged droughts, which lead to severe water shortages, are another major category of extreme events affecting societies worldwide (C3S & WMO, 2024). Currently, 40% of the global population does not have access to sufficient water resources, and according to a UNESCO report, global water availability will decline by approximately 30% over the next 20 years (UNESCO WWDR1 2003). It is estimated that only around 2.5% of the world's water is freshwater, and less than 1% is suitable for drinking (UNESCO WWDR4 2012). The UN WWDR (2018) projects that by 2050, more than five billion people could experience water scarcity, a problem further exacerbated by climate change, increasing demand, and water pollution. Globally, humanity currently consumes approximately 4600 km^3 of water annually, 70% of which is used for agriculture, 20% for industry, and 10% for households (UN WWDR 2018). Global water demand has increased sixfold over the past century and continues to grow at a rate of approximately 1% per year. The world's population is projected to continue increasing, potentially reaching between 9.4 and 10.2 billion by 2050 (up from 7.7 billion today), with two-thirds of people expected to live in urban areas (UN WWDR 2018). The scarcity of drinking water is aptly illustrated by the quote from the English poet Samuel Taylor Coleridge: "Water, water everywhere, not a drop to drink."

Extreme events such as catastrophic floods and prolonged droughts remain among the greatest challenges for long-term management of the catchment–valley–river continuum.

3. Management of the catchment area

Success in the sustainable management of a catchment depends on a thorough understanding of the diversity of ecological processes taking place across both time and space, ranging from molecular processes, through valley-scale processes, up to the basin-wide scale. Such an approach to river system analysis provides an opportunity to effectively cope with anthropogenic pressure and enhance the buffering capacity of the floodplain, which plays a key role in floodwater retention and flood protection. It also contributes to the reduction of nutrient and pollutant loads transported by the river, through processes such as sedimentation and retention on the floodplain, thereby helping to improve water quality in the river.

Flood risk in river catchments is primarily generated by large-scale changes in land use, deforestation, and the development of impermeable surfaces in the catchment and river valleys, as well as by the transformation of natural, meandering riverbanks into regulated, engineered channels Fig. 10.2. Therefore the crucial assumption in all river management systems should be that natural floodplain areas remain as intact and undeveloped as possible along the entire course of the river, from its source to its mouth. These natural floodplains function as temporary reservoirs during floods, where water can spread freely, be retained, and naturally purified, without posing a threat or causing significant damage to the economy or society.

Modified river basin surfaces with a high degree of development and extensive impermeable areas are characterized by an almost instantaneous rainfall-runoff response, due to the limited capacity for water infiltration and retention in the catchment area (Barnett et al., 2008; Kundzewicz et al., 2007; Vörösmarty et al., 2000; Yang and Tian, 2009). This significantly increases the probability of catastrophic flood or drought events (Fig. 10.3). Of course, the characteristics of the hydrological regime of a river system depend on the local climate and the specific features of the catchment. Geology, geomorphology, and land use are among the most influential factors modulating the catchment's response to precipitation, including flood concentration time and the magnitude of runoff. In addition, catchments where a large portion of runoff originates from groundwater are often less sensitive to climate change (Chang and Jung, 2010; Tague et al., 2008). Understanding the role of human activity in shaping land use and development within river catchments—and thereby

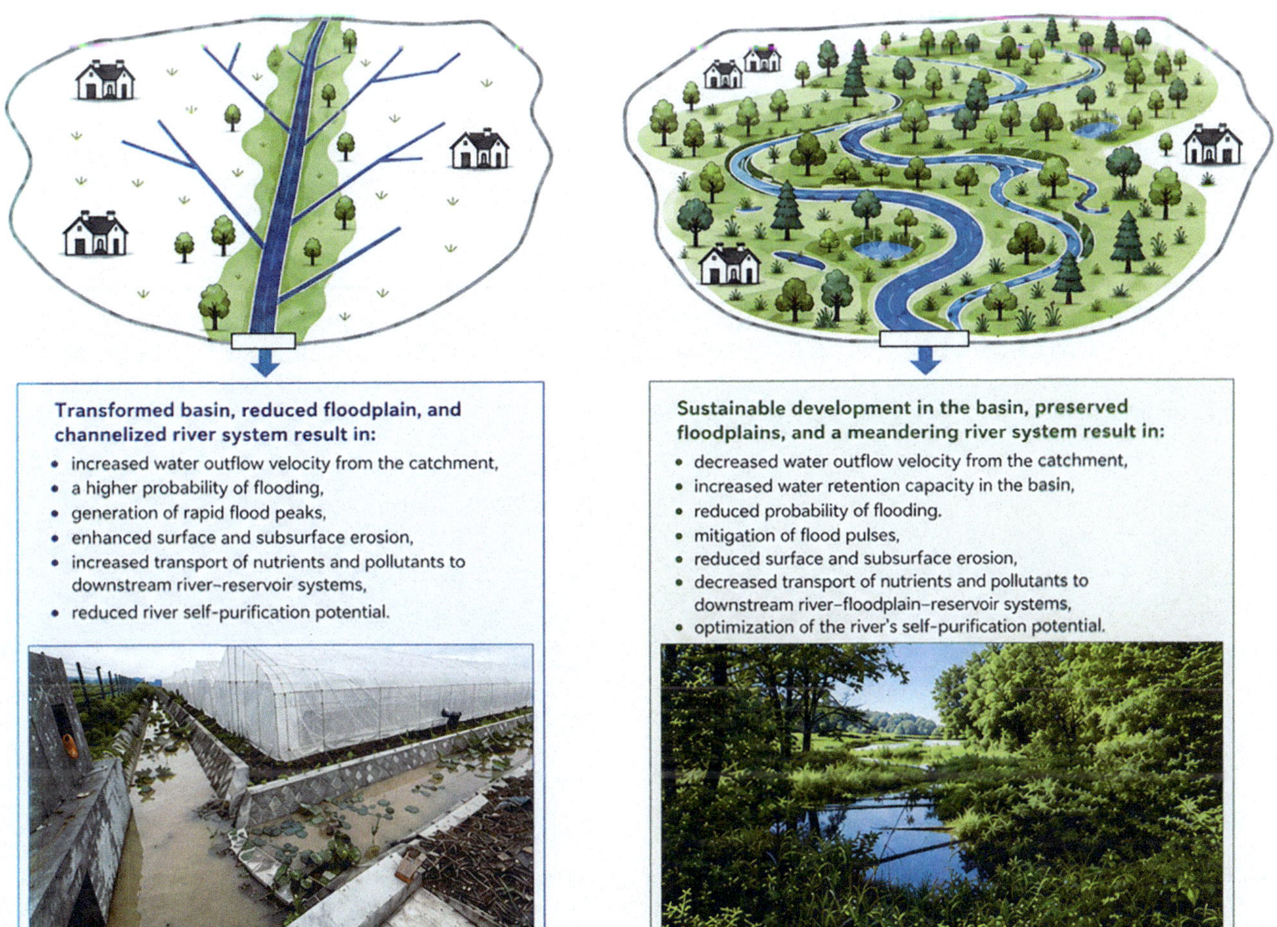

FIGURE 10.2 Impact of catchment and floodplain condition on hydrological dynamics and water quality. The role of catchment characteristics in determining flood probability and the river's ability to self-purify and improve water quality. *Modified from Kiedrzyńska, E., Kiedrzyński, M., Zalewski, M., 2015. Sustainable floodplain management for flood prevention and water quality improvement. Nat. Hazards. 76 (2), 955–977. https://doi.org/10.1007/s11069-014-1529-1. Photo credit: E. Kiedrzyńska.*

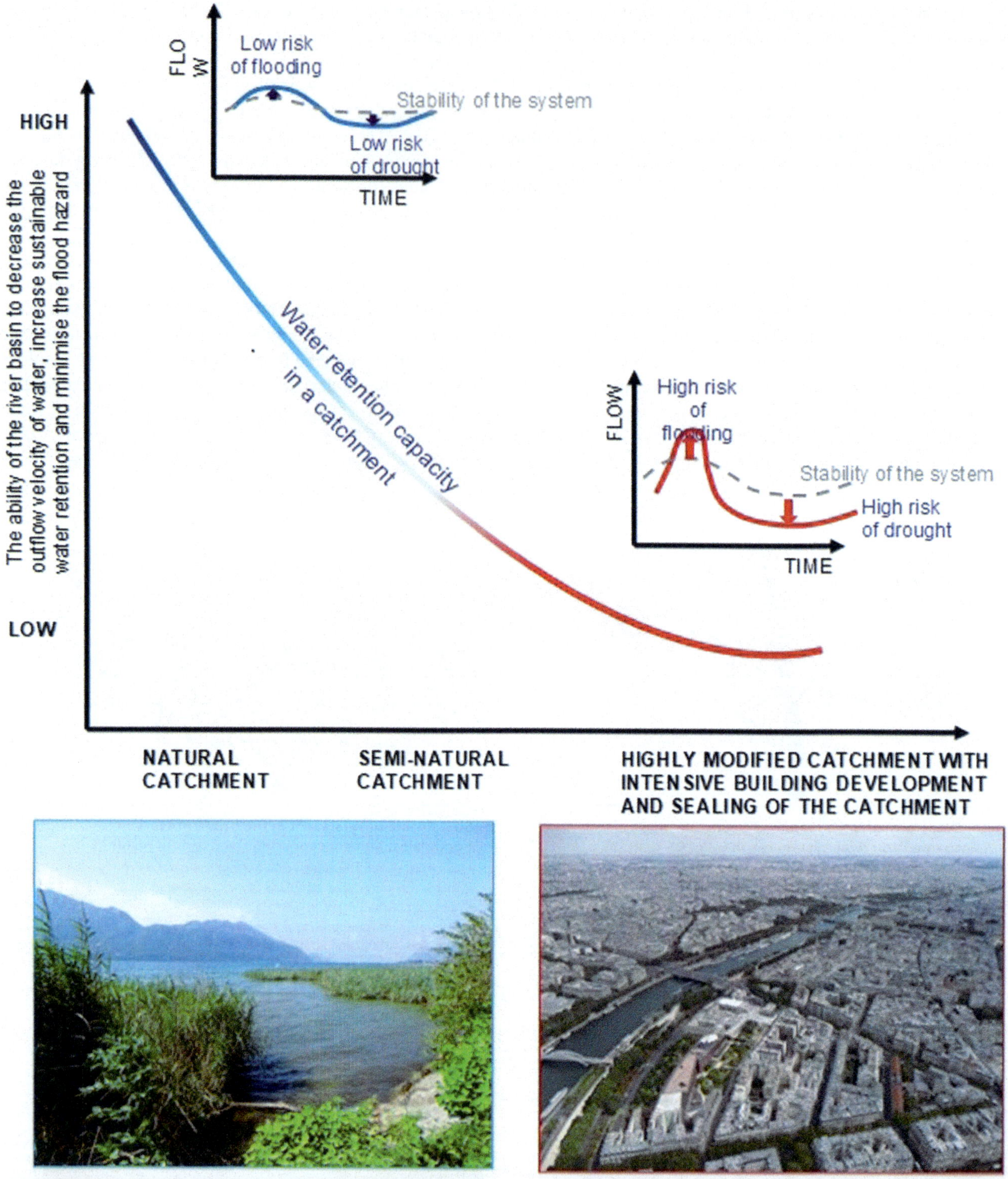

FIGURE 10.3 **The relationship between water retention capacity and catchment characteristics.** The influence of catchment characteristics on runoff generation, water retention capacity, and the risk of catastrophic floods and droughts. *Modified from Kiedrzyńska, E., Kiedrzyński, M., Zalewski, M., 2015. Sustainable floodplain management for flood prevention and water quality improvement. Nat. Hazards. 76 (2), 955–977. https://doi.org/10.1007/s11069-014-1529-1. Photo credit: E. Kiedrzyńska.*

modifying the hydrological cycle (Fig. 10.3) —is crucial for sustainable and safe water resource management and effective flood risk assessment.

Most research on the effects of extreme flood events has focused primarily on damage to buildings and industrial infrastructure, often overlooking the risks of chemical and microbiological contamination. However, many hazardous substances contained in flood sediments pose a serious threat to human health, as well as to livestock, wildlife, vegetation, and soil.

Although a broad range of flood risk management strategies exists in theory, floods continue to have catastrophic consequences because of technical, political, economic, social, and environmental limitations (UNISDR 2013, ESOTC 2024). Knowledge of past catastrophic events can enhance the effectiveness of policy measures by supporting the development of appropriate institutions, improving education, and increasing public awareness of associated risks. Nonetheless, this is not sufficient; in practice, a comprehensive and systematic set of coordinated actions is required—supported by well-designed institutional frameworks—to provide effective assistance to flood-prone areas.

Ecohydrology, based on its three foundational principles—hydrological, ecological, and ecotechnological (Zalewski, 2006; Zalewski and Janauer, 1997; Zalewski et al., 2021)—offers a framework for integrated management that aims to minimize flood risk in catchments and improve water quality in river systems. Implementing sustainable ecohydrological management of floodplains requires the restoration of their natural structure and function within river valleys.

4. Ecohydrology for flood prevention and management

The development of a flood prevention and risk management strategy that incorporates the Principles of Ecohydrology requires:

1. A shift in thinking—from technical to ecohydrological—emphasizing the value and self-regulating capacity of natural ecosystems.
2. A shift in operational practice—from narrow, sectoral approaches to integrated, multidimensional ones—drawing on cutting-edge biological science and advanced digital technologies.

Sustainable land-use planning and management should be implemented across the entire catchment area, including:

1. Promoting the mosaic character of land cover throughout the upper, middle, and lower sections of the catchment.
2. Expanding buffer zones that reduce surface and subsurface runoff from the catchment into aquatic ecosystems, while simultaneously limiting pollutant transfer through phytoremediation processes—such as phytoaccumulation, phytodegradation, phytostabilization, rhizofiltration, and rhizodegradation (Bhandari et al., 2007; Kiedrzyńska et al., 2008b; Piwowarska et al., 2024; Skłodowski et al., 2014; Sumorok and Kiedrzyńska, 2007)—which occur with particular intensity in ecotone zones.
3. Developing technologies and practices that enhance water infiltration and retention, for example, by using permeable materials in the construction of parking lots and paved areas, or through rainwater harvesting in retention reservoirs—thereby reducing runoff intensity from the catchment.
4. Increasing the catchment's retention capacity through afforestation, agrotechnical practices, river and wetland restoration, and the construction of cascade flood protection lateral reservoir systems in urban areas, preceded by sequential sedimentation-biofiltration systems (Kiedrzyńska et al., 2021).
5. Restricting the expansion of residential and industrial development in river floodplains.

The ecohydrological approach to developing an action plan for flood risk reduction and water quality improvement includes five sequential phases Fig. 10.4 (Kiedrzyńska et al., 2015):

PHASE I. Catchment monitoring—understanding land-use structure, catchment condition, cause-effect relationships, and quantification of hydrological and ecological processes and threats—supported by GIS-based spatial data analysis.
PHASE II. Ecohydrological assessment of floodplains—assessment of the ecohydrological state of the river valleys' floodplains, determination of the flood water retention potential, and the ability to improve water quality.
PHASE III. Process regulation and capacity enhancement—ecohydrological regulation aimed at increasing the capacity for floodwater and nutrient retention—to minimize flood risk and improve river water quality.
PHASE IV. Assessment of the system solutions' functioning—assessment of ecohydrological measures applied to floodplains, with possible adjustments to optimize effectiveness—a process of adaptive assessment and management. Adaptive management is viewed as a policy experiment, where interventions are tested and adjusted to achieve intended outcomes.

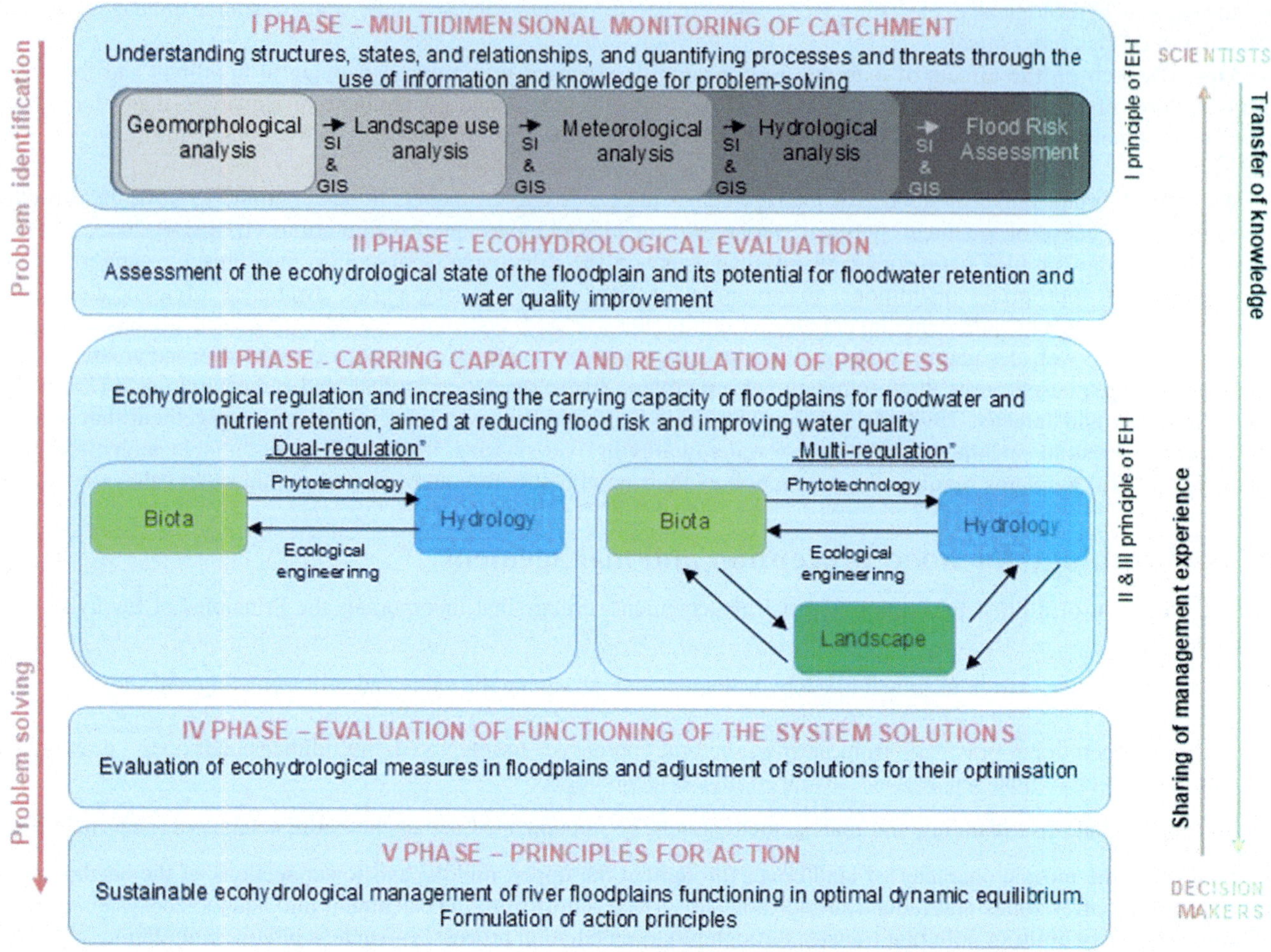

FIGURE 10.4 Phases of ecohydrological adaptive management of river valley floodplains. Phases of ecohydrological adaptive management of river valley floodplains to prevent flooding and improve water quality in catchments. *Modified from Kiedrzyńska, E., Kiedrzyński, M., Zalewski, M., 2015. Sustainable floodplain management for flood prevention and water quality improvement. Nat. Hazards. 76 (2), 955–977. https://doi.org/10.1007/s11069-014-1529-1.*

PHASE V. Sustainable ecohydrological floodplain management—Development of a long-term ecohydrological management plan for river valley floodplains that operate in an optimal "dynamic equilibrium" accounting for continuous adaptation to climate change and socioeconomic development. This includes formulating future-oriented principles of action (Fig. 10.4).

Flood risk analysis and assessment require a multidisciplinary approach and the integration of multiple sources of information, combining geomorphological and landscape analysis with meteorological and hydrological data, along with the use of tools such as satellite imagery (SI) and geographic information systems (GIS) (Fig. 10.4). In the development of flood protection strategies and conceptual models for water quality improvement, the following GIS thematic layers are typically employed: subcatchment and catchment boundaries, digital terrain model (DTM), hydrographic networks, systems of causeways and canals, hydrometric and meteorological station networks, rain gauge networks, transport infrastructure (roads, railways, and airports), urban areas, and land cover/land use maps derived from updated satellite data and historical flood event records (Stancalie et al., 2006). This integrated information, combined with flood probability data, supports the identification of areas at future flood risk. Scaled SI provides insight into flood zones, dynamics, and extents, highlighting the areas affected.

Methods for identifying, delineating, and mapping flooded areas are based on various classification techniques that utilize both optical and radar satellite data. A major advantage of high-resolution SI is its ability to deliver precise spatial information by combining multiple image layers and accurately detecting and outlining flooded or flood-prone areas (Stancalie et al., 2006).

Specialized satellite applications for flood risk mapping—such as land cover and land use maps, flood extent maps, flood risk zone maps, and thematic flood hazard maps—have been developed over the past 2 decades as part of NATO-led projects (e.g., monitoring extreme floods in Romania, Hungary, and other countries). These tools contribute significantly to the prevention of extreme flood events by enabling more rational spatial planning, supporting flood mitigation planning (including infrastructure development in flood-prone areas), and improving spatial data distribution for decision-makers and stakeholders (Balint, 2004; Stancalie et al., 2006).

In addition, forecasting efforts based on foresight methodologies, conducted by interdisciplinary expert teams, play a key role in advancing sustainable floodplain development, as well as flood prevention and management (Rogut and Piasecki, 2011; Piasecki, 2011; Zalewski, 2008). Foresight is defined as "the process of discovering a shared space for open-minded thinking about the future and the development of strategic approaches" (Butter and Popper, 2008). It represents a systemic approach to identifying and evaluating emerging trends, grounded in environmental, technological, and sociological knowledge (Zalewski, 2012; Zalewski et al., 2021). The exchange of knowledge, experience, and perspectives between scientists, industry, business stakeholders, and policymakers helps systematize and integrate awareness across key scientific and economic sectors (Fig. 10.4).

5. Ecohydrology for water quality improvement

The concept of ecosystem regulation, which is the key assumption of ecohydrology (Zalewski and Janauer, 1997; Zalewski and Wagner, 2004; Zalewski, 2011; Zalewski et al., 2021), is a crucial component of water resource management aimed at reducing nutrient and pollutant runoff into river systems, thereby limiting eutrophication and pollution of rivers and ultimately the Baltic Sea. Many European countries, particularly those within the Baltic Sea catchment, face the challenge of limiting pollutant loads from anthropogenic sources. As a member state of the European Union, Poland is obligated to implement the provisions of the Water Framework Directive (2000/60/EC), Urban Wastewater Treatment Directive (2024/3019), the Nitrates Directive (91/676/EEC), and the Marine Strategy Framework Directive (2008/56/EC), all of which aim to protect aquatic ecosystems, reduce nutrient and pollutant loads, and achieve good ecological and chemical status of water bodies.

The role of natural river floodplains in self-purification and water quality enhancement has been explored by numerous researchers (Kiedrzyńska et al., 2008b; Mitsch et al., 2009; Mitsch and Gosselink, 2007; Skłodowski et al., 2014; Wang et al., 2019). Floodplains can retain and filter flood sediments, nutrients, and micropollutants, such as dioxins, under both stable hydrological conditions and flood events. Thus along the river continuum, floodplains serve as a valuable tool for pollution reduction (Kiedrzyńska et al., 2008a; Skłodowski et al., 2014; Urbaniak et al., 2012). Since the highest pollutant loads typically occur during the rising limb of a flood, diverting floodwaters onto floodplains early in the event allows for retention and sedimentation of pollutants, contributing to the self-purification of river systems.

Ecohydrological research conducted in the floodplain of the middle section of the Pilica River has demonstrated the potential for enhancing these processes through both sedimentation (Altinakar et al., 2006; Kiedrzyńska et al., 2008b; Magnuszewski et al., 2007) and nutrient uptake by vegetation (Kiedrzyńska et al., 2008a). Studies using a digital terrain model, a hydrological model, and physicochemical analyses have shown that sedimentation of flood-borne material on a 26.6-hectare floodplain substantially limits downstream transport to the Sulejów Reservoir. On a 40 km stretch of the Pilica River with a near-natural floodplain, the flood area during high water stages may reach up to 1007 ha (Kiedrzyńska and Zalewski, 2012). According to Kiedrzyńska and Zalewski (2012), a single flood event in this area can result in the deposition of approximately 500 tons (0.5 t/ha) of fine-grained sediment and 1.5 tons (1.5 kg/ha) of phosphorus (P).

Moreover, ecohydrological studies have shown that rush vegetation communities are capable of absorbing and storing significant amounts of phosphorus, particularly in spring and summer. In the 26.6 ha experimental floodplain, summer biomass accumulated 255 kg P, equivalent to 9.6 kg P/ha/year (Kiedrzyńska et al., 2008a). Transforming 24% or 48% of the area into willow (*Salicetum*) communities could further increase phosphorus retention to 332 kg P/year or even 399 kg P/year, corresponding to 12.5–15 kg P/ha/year, respectively (Kiedrzyńska et al., 2008a).

Additional studies in the same section of the Pilica valley examined the role of *Salicetum triandro-viminalis* communities in phosphorus removal from river water and phosphorus accumulation in biomass. The findings indicated a 41% reduction in suspended solids, a 54% decrease in phosphate concentrations, and a 36% reduction in total phosphorus. Furthermore, total PCB concentrations declined by 39%, and toxic equivalent (TEQ) toxicity by 49% (Skłodowski et al., 2014). In this stretch of the river, natural willow communities covered around 30 ha, accumulating approximately 431 kg P in total biomass, or an average of 14.44 kg P/ha. Proper management of such riparian willow communities, including periodic biomass harvesting, may help reduce nutrient loads, mitigate eutrophication, and control toxic algal blooms in the Sulejów Reservoir (Skłodowski et al., 2014). Notably, 1 kg of phosphorus can potentially stimulate the development of 1–2 tons of toxic phytoplankton in the reservoir (Zalewski, 2005).

Collectively, these findings confirm that floodplains function as sedimentation, biofiltration, and water purification systems, effectively lowering concentrations of suspended matter, nutrients, and various micropollutants derived from the entire catchment. Adopting a balanced, ecohydrological approach to the planning and management of river valleys can therefore generate numerous benefits—ecological, social, and economic—at local, regional, and even transboundary scales. The solutions presented in this chapter, based on sustainable and adaptive floodplain management, may serve as a foundation for improving environmental quality and supporting the dynamic development of regions.

AI Disclosure

During the preparation of this work the author(s) used OpenAI's ChatGPT. This document includes content generated and/or edited with the assistance of OpenAI's ChatGPT, used to support language refinement and structural improvements. All final content was reviewed and verified by the author. After using this tool/service, the author(s) reviewed and edited the content as needed and take(s) full responsibility for the content of the publication.

References

Altinakar, M., Kiedrzyńska, E., Magnuszewski, A., 2006. Modelling of inundation patterns on the Pilica River flood plain, Poland. IAHS-AISH Publ. 308, 579–585.

Amoros, C., Bornette, G., 2002. Connectivity and biocomplexity in waterbodies of riverine floodplains. Freshw. Biol. 47 (4), 761–776. https://doi.org/10.1046/j.1365-2427.2002.00905.x.

Balint, Z., 2004. Flood protection in the Tisza basin. Proceedings of the NATO Advanced Research Workshop – Flood Risk Management: Hazards, Vulnerability, Mitigation Measures 171–183.

Barnett, T.P., Pierce, D.W., Hidalgo, H.G., Bonfils, C., Santer, B.D., Das, T., Bala, G., Wood, A.W., Nozawa, T., Mirin, A.A., Cayan, D.R., Dettinger, M.D., 2008. Human-induced changes in the hydrology of the Western United States. Science 319 (5866), 1080–1083. https://doi.org/10.1126/science.1152538.

Bhandari, A., Surampalli, R.Y., Champagne, P., Ong, S.K., Tyagi, R.D., Lo, I.M.C., 2007. Remediation Technologies for Soils and Groundwater Remediation Technologies for Soils and Groundwater. American Society of Civil Engineers (ASCE), United States, pp. 1–449. https://doi.org/10.1061/9780784408940.

Butter, M., Popper, R., 2008. How are foresight methods selected? Foresight 10 (6), 62–89. https://doi.org/10.1108/14636680810918586.

Chang, H., Jung, I.W., 2010. Spatial and temporal changes in runoff caused by climate change in a complex large river basin in Oregon. J. Hydrol. 388 (3–4), 186–207. https://doi.org/10.1016/j.jhydrol.2010.04.040.

Climate Change Evident across Europe, Confirming Urgent Need for Adaptation. EEA. (2012),

Climate Change, 2023. Synthesis Report Contribution of Working Groups I, II and III to the Sixth Assessment Report of the Intergovernmental Panel on Climate Change. https://doi.org/10.59327/IPCC/AR6-9789291691647.

Copernicus Climate Change Service (C3S), & World Meteorological Organization (WMO) European State of the Climate Report. (2024). https://doi.org/10.24381/14j9-s541,

Fendler, R., 2008. Floods and safety of establishments and installations containing hazardous substances. Nat. Hazards 46 (2), 257–263. https://doi.org/10.1007/s11069-007-9209-z.

Freudenberger, L., Hobson, P.R., Schluck, M., Ibisch, P.L., 2012. A global map of the functionality of terrestrial ecosystems. Ecol. Complex. 12, 13–22. https://doi.org/10.1016/j.ecocom.2012.08.002.

Ganoulis, J., 2008. Flood risk analysis and management. In: Linkov, I., Ferguson, E.A. (Eds.), Real-Time and Deliberative Decision Making. An Integrated Approach for Flood Risk Management. Springer, pp. 109–132. https://doi.org/10.1007/978-1- 4020-9026-4_6.

Glaser, R., Riemann, D., Schönbein, J., Barriendos, M., Brázdil, R., Bertolin, C., Camuffo, D., Deutsch, M., Dobrovolný, P., van Engelen, A., Enzi, S., Halíčková, M., Koenig, S.J., Kotyza, O., Limanówka, D., Macková, J., Sghedoni, M., Martin, B., Himmelsbach, I., 2010. The variability of European floods since AD 1500. Clim. Change 101 (1), 235–256. https://doi.org/10.1007/s10584-010-9816-7.

Hansen, J., Nazarenko, L., Ruedy, R., Sato, M., Willis, J., Del Genio, A., Koch, D., Lacis, A., Lo, K., Menon, S., Novakov, T., Perlwitz, J., Russell, G., Schmidt, G.A., Tausnev, N., 2005. Earth's energy imbalance: confirmation and implications. Science 308 (5727), 1431–1435. https://doi.org/10.1126/science.1110252.

Harnisz, M., Kiedrzyńska, E., Kiedrzyński, M., Korzeniewska, E., Czatzkowska, M., Koniuszewska, I., Jóźwik, A., Szklarek, S., Niestępski, S., Zalewski, M., 2020. The impact of WWTP size and sampling season on the prevalence of antibiotic resistance genes in wastewater and the river system. Sci. Total Environ. 741, 140466. https://doi.org/10.1016/j.scitotenv.2020.140466.

Hassan, R., 2005. Ecosystems and Human Well-Being: Current State and Trends, vol. 1. Island Press.

Hirsch, T., 2010. Secretariat of the Convention on Biological Diversity. Montreal, Quebec, Canada. Canada Global Biodiversity Outlook 3.

Hobson, 2010. An alternative conceptual framework for sustainability: Systemics and thermodynamics. Interdependence of Biodiversity and Development Under Global Change 54, 127–148.

Huntington, T.G., 2006. Evidence for intensification of the global water cycle: review and synthesis. J. Hydrol. 319 (1–4), 83–95. https://doi.org/10.1016/j.jhydrol.2005.07.003.

Kiedrzyńska, E., Zalewski, M., 2012. Water Quality Improvement through an Integrated Approach to Point and Non-point Sources Pollution and Management of River Floodplain Wetlands Ecological Water Quality – Water Treatment and Reuse. InTech, pp. 325–342. https://doi.org/10.5772/31374.

Kiedrzyńska, E., Belka, K., Jarosiewicz, P., Kiedrzyński, M., Zalewski, M., 2021. The enhancement of valley water retentiveness in climate change conditions. Sci. Total Environ. 799, 149427. https://doi.org/10.1016/j.scitotenv.2021.149427.

Kiedrzyńska, E., Kiedrzyński, M., Zalewski, M., 2008a. Flood sediment deposition and phosphorus retention in a lowland river floodplain: impact on water quality of a reservoir, Sulejów, Poland. Ecohydrol. Hydrobiol. 8 (2–4), 281–289. https://doi.org/10.2478/v10104-009-0022-z.

Kiedrzyńska, E., Wagner, I., Zalewski, M., 2008b. Quantification of phosphorus retention efficiency by floodplain vegetation and a management strategy for a eutrophic reservoir restoration. Ecol. Eng. 33 (1), 15–25. https://doi.org/10.1016/j.ecoleng.2007.10.010.

Kiedrzyńska, E., Kiedrzyński, M., Zalewski, M., 2015. Sustainable floodplain management for flood prevention and water quality improvement. Nat. Hazards 76 (2), 955–977. https://doi.org/10.1007/s11069-014-1529-1.

Kiedrzyński, M., Kiedrzyńska, E., Witosławski, P., Urbaniak, M., Kurowski, J.K., 2014. Historical land use, actual vegetation, and the hemeroby levels in ecological evaluation of an urban river valley in perspective of its rehabilitation plan. Pol. J. Environ. Stud. 23 (1), 109–117.

Koniuszewska, I., Korzeniewska, E., Harnisz, M., Kiedrzyńska, E., Kiedrzyński, M., Czatzkowska, M., Jarosiewicz, P., Zalewski, M., 2020. The occurrence of antibiotic-resistance genes in the Pilica River, Poland. Ecohydrol. Hydrobiol. 20 (1), 1–11. https://doi.org/10.1016/j.ecohyd.2019.09.002.

Kundzewicz, Z.W., Mata, L.J., Arnell, N.W., Döll, K., Jiménez, B., Miller, K.A., Oki, T., Şen, Z., Shiklomanov, I.A., 2007. Freshwater Resources and Their Management Climate Change 2007: Impacts, Adaptation and Vulnerability. Cambridge University Press,, pp. 173–210.

Li, Y., Tao, H., Su, B., Kundzewicz, Z.W., Jiang, T., 2019. Impacts of 1.5 °C and 2 °C global warming on winter snow depth in Central Asia. Sci. Total Environ. 651, 2866–2873. https://doi.org/10.1016/j.scitotenv.2018.10.126.

Magnuszewski, A., Kiedrzyńska, E., Wagner, I., Zalewski, M., 2007. Numerical modeling of material fluxes on the floodplain wetland of the Pilica River, Poland. Proceedings of the International Conference W3M "Wetlands: Modelling, Monitoring, Management" 205–210.

Magnuszewski, A., Kiedrzyńska, E., Wagner-Łotkowska, I., Zalewski, 2005. Immobilising of sediments in a lowland river floodplain. Computational Modeling for the Development of Sustainable Water-Resources Systems in Poland, vol. 5, pp. 239–260.

Mitsch, W.J., Gosselink, J.G., 2007. Wetlands. John Wiley & Sons.

Mitsch, W.J., Gosselink, J.G., Anderson, C.J., Zhang, L., 2009. Wetland Ecosystems. John Wiley & Sons.

Mitsch, W.J., Zhang, L., Fink, D.F., Hernandez, M.E., Altor, A.E., Tuttle, C.L., Nahlik, A.M., 2008. Ecological engineering of floodplains. Ecohydrol. Hydrobiol. 8 (2–4), 139–147. https://doi.org/10.2478/v10104-009-0010-3.

Mitsch, W.J., 2014. Unifying a city with its natural riverine environment for the benefit of both: Extending Ohio's only wetland of international importance to a much larger river ecosystem corridor. Ecol. Eng. 72, 138–142. https://doi.org/10.1016/j.ecoleng.2014.09.104.

Piwowarska, D., Kiedrzyńska, E., Jaszczyszyn, K., 2024. A global perspective on the nature and fate of heavy metals polluting water ecosystems, and their impact and remediation. Crit. Rev. Environ. Sci. Technol. 54 (19), 1436–1458. https://doi.org/10.1080/10643389.2024.2317112.

Rockström, J., Falkenmark, M., Folke, C., Lannerstad, M., Barron, J., Enfors, E., Gordon, L., Heinke, J., Hoff, H., Pahl-Wostl, C., 2014. Water Resilience for Human Prosperity. Cambridge University Press. https://doi.org/10.1017/CBO9781139162463.

Rogut, A., Piasecki, B., 2011. Foresight methodology as a tool for elaboration of plans for sustainable management of water, energy, the environment and society. Ecohydrol. Hydrobiol. 11 (3–4), 261–272. https://doi.org/10.2478/v10104-011-0049-9.

Serwecińska, L., Kiedrzyńska, E., Kiedrzyński, M., 2021. A catchment-scale assessment of the sanitary condition of treated wastewater and river water based on fecal indicators and carbapenem-resistant Acinetobacter spp. Sci. Total Environ. 750, 142266. https://doi.org/10.1016/j.scitotenv.2020.142266.

Skłodowski, M., Kiedrzyńska, E., Kiedrzyński, M., Urbaniak, M., Zielińska, K.M., Kurowski, J.K., Zalewski, M., 2014. The role of riparian willows in phosphorus accumulation and PCB control for lotic water quality improvement. Ecol. Eng. 70, 1–10. https://doi.org/10.1016/j.ecoleng.2014.03.088.

Stancalie, G., Catana, S., Irimescu, A., Savin, E., Diamandi, A., Hofnar, A., Oancea, S., 2006. Contribution of earth observation data supplied by the new satellite sensors to flood management. In: Marsalek, J., Stancalie, G., Balint, G. (Eds.), Transboundary Floods: Reducing Risks through Flood Management, vol 72. Springer Science and Business Media LLC, pp. 287–304. https://doi.org/10.1007/1-4020-4902-1_27.

Steffen, W., Richardson, K., Rockström, J., Schellnhuber, H.J., Dube, O.P., Dutreuil, S., Lenton, T.M., Lubchenco, J., 2020. The emergence and evolution of Earth system science. Nat. Rev. Earth Environ. 1 (1), 54–63. https://doi.org/10.1038/s43017-019-0005-6.

Steffen, W., Broadgate, W., Deutsch, L., Gaffney, O., Ludwig, C., 2015. The trajectory of the anthropocene: the great acceleration. Anthr. Rev. 2 (1), 81–98. https://doi.org/10.1177/2053019614564785.

Steffen, W., Rockström, J., Richardson, K., Lenton, T.M., Folke, C., Liverman, D., Summerhayes, C.P., Barnosky, A.D., Cornell, S.E., Crucifix, M., Donges, J.F., Fetzer, I., Lade, S.J., Scheffer, M., Winkelmann, R., Schellnhuber, H.J., 2018. Trajectories of the Earth system in the anthropocene. Proc. Natl. Acad. Sci. 115 (33), 8252–8259. https://doi.org/10.1073/pnas.1810141115.

Sumorok, B., Kiedrzyńska, E., 2007/12. Mycorrhizal status of native willow species in the Pilica River floodplain along the moisture gradient. Proceedings of the International Conference W3M "Wetlands: Modelling, Monitoring, Management" 281–286.

Tague, C., Grant, G., Farrell, M., Choate, J., Jefferson, A., 2008. Deep groundwater mediates streamflow response to climate warming in the Oregon Cascades. Clim. Change 86 (1–2), 189–210. https://doi.org/10.1007/s10584-007-9294-8.

Thoms, M.C., 2003. Floodplain-river ecosystems: lateral connections and the implications of human interference. Geomorphology 56 (3–4), 335–349. https://doi.org/10.1016/S0169-555X(03)00160-0.

Urbaniak, M., Kiedrzyńska, E., Wyrwicka, A., Zieliński, M., Mierzejewska, E., Kiedrzyński, M., Kannan, K., Zalewski, M., 2019. An ecohydrological approach to the river contamination by PCDDs, PCDFs and dl-PCBs – concentrations, distribution and removal using phytoremediation techniques. Sci. Rep. 9 (1). https://doi.org/10.1038/s41598-019-55973-3.

Urbaniak, M., Kiedrzyńska, E., Zalewski, M., 2012. The role of a lowland reservoir in the transport of micropollutants, nutrients and the suspended particulate matter along the river continuum. Hydrol. Res. 43 (4), 400–411. https://doi.org/10.2166/nh.2012.108Poland.

Urbaniak, M., Kiedrzyńska, E., Kiedrzyński, M., Mendra, M., Grochowalski, A., 2014. The impact of point sources of pollution on the transport of micropollutants along the river continuum. Hydrol. Res. 45 (3), 391–410. https://doi.org/10.2166/nh.2013.242.

Vörösmarty, C.J., Green, P., Salisbury, J., Lammers, R.B., 2000. Global water resources: vulnerability from climate change and population growth. Science 289 (5477), 284–288. https://doi.org/10.1126/science.289.5477.284.

Vörösmarty, C.J., Pahl-Wostl, C., Bhaduri, A., 2013. Water in the anthropocene: new perspectives for global sustainability. Curr. Opin. Environ. Sustain. 5 (6), 535–538. https://doi.org/10.1016/j.cosust.2013.11.011.

Wang, Q., Mei, D., Chen, J., Lin, Y., Liu, J., Lu, H., Yan, C., 2019. Sequestration of heavy metal by glomalin-related soil protein: Implication for water quality improvement in mangrove wetlands. Water Res. 148, 142–152. https://doi.org/10.1016/j.watres.2018.10.043.

Yang, Y., Tian, F., 2009. Abrupt change of runoff and its major driving factors in Haihe River Catchment, China. J. Hydrol. 374 (3–4), 373–383. https://doi.org/10.1016/j.jhydrol.2009.06.040.

Zalewski, M., 2009. Ecohydrology for engineering harmony between environment and society. Danube News 11 (19).

Zalewski, M., 2005. Engineering harmony. Academia 1 (5), 4–7.

Zalewski, M., 2006. Flood pulses and river ecosystem robustness. IAHS-AISH Publ. 305, 143–154.

Zalewski, M., Janauer, G.A., 1997. Ecohydrology: a new paradigm for the sustainable use of aquatic resources. In: Conceptual Background, Working Hypothesis, Rationale and Scientific Guidelines for the Implementation of the IHP-V Projects.

Zalewski, M., Kiedrzynska, E., 2010. System approach to sustainable management of inland floodplains – declaration on sustainable floodplain management. CAB Rev. 5, 1–8. https://doi.org/10.1079/PAVSNNR20105056.

Zalewski, M., Wagner, I., 2004. Integrated Watershed Management – Ecohydrology and Phytotechnology: Manual. UNESCO/UNEP.

Zalewski, M., 2012. Ecohydrology – process oriented thinking for sustainability of river basins. Ecohydrol. Hydrobiol. 12 (2), 89–92. https://doi.org/10.2478/v10104-012-0012-4.

Zalewski, M., 2015. Ecohydrology and hydrologic engineering: regulation of hydrology-biota interactions for sustainability. J. Hydrol. Eng. 20 (1). https://doi.org/10.1061/(asce)he.1943-5584.0000999.

Zalewski, M., 2011. Ecohydrology for implementation of the EU water framework directive. Proc. Inst. Civil Eng. Water Manag. 164 (8), 375–385. https://doi.org/10.1680/wama.1000030.

Zalewski, M., 2008. Rationale for the "Floodplain Declaration" from environmental conservation toward sustainability science. Ecohydrol. Hydrobiol. 8 (2–4), 107–113. https://doi.org/10.2478/v10104-009-0008-x.

Zalewski, M., Kiedrzyńska, E., Wagner, I., Izydorczyk, K., Mankiewicz Boczek, J., Jurczak, T., Krauze, K., Frankiewicz, P., Godlewska, M., Wojtal-Frankiewicz, A., Łapińska, M., Urbaniak, M., Bednarek, A., Kaczkowski, Z., Gągała, I., Serwecińska, L., Szklarek, S., Włodarczyk-Marciniak, R., Font-Nájera, A., Mierzejewska, E., Połatyńska-Rudnicka, M., Belka, K., Jarosiewicz, P., 2021. Ecohydrology and adaptation to global change. Ecohydrol. Hydrobiol. 21 (3), 393–410. https://doi.org/10.1016/j.ecohyd.2021.08.001.

Zhang, Z., Lu, W.X., Zhao, Y., Song, W.B., 2014. Development tendency analysis and evaluation of the water ecological carrying capacity in the Siping area of Jilin Province in China based on system dynamics and analytic hierarchy process. Ecol. Model. 275, 9–21. https://doi.org/10.1016/j.ecolmodel.2013.11.031.

Further reading

United Nations Office for Disaster Risk Reduction, 2019. United Nations Office for Disaster Risk Reduction Annual Report 2018. UNISDR.

Chapter 11

How to manage water cycle at the watershed scale

Andrzej Kędziora
Institute for Agricultural and Forest Environment of the Polish Academy of Sciences, Poznań, Poland

IMAGE CREDIT: M. ZALEWSKI.

1. Landscape changes

Rapidly growing human population, technical knowledge and technologies which did not keep pace with the development of knowledge about the functioning of nature. A lack of understanding of the processes of water cycling, energy flow and

Transdisciplinary Ecohydrology. https://doi.org/10.1016/B978-0-443-41596-8.00007-8

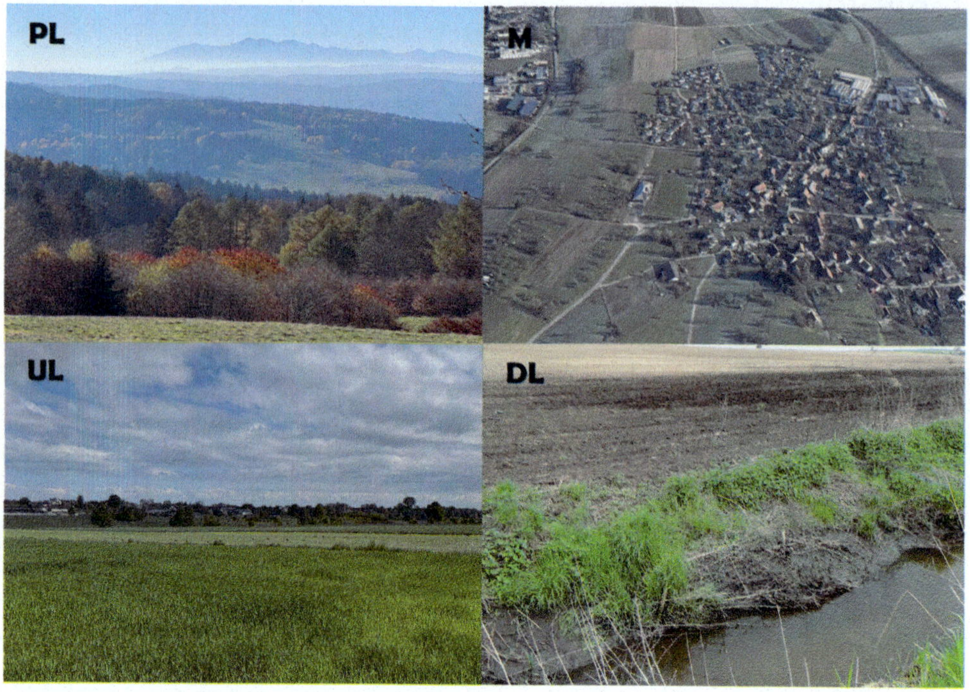

FIGURE 11.1 Types of landscapes. Stable: PL—primary landscapes in which structure dominated forests, meadows and waters, ML—mosaic landscapes, cultural landscapes with a very developed structure, many small elements, UL-uniform landscapes, industrial, agricultural landscapes like monoculture, DL—Devoid plants landscapes, like urban areas, bare soil, poor pastures, savannas. *Photo credit: M. Zalewski.*

matter cycling in the agricultural landscape has been the cause decline in sustainability potential in rural areas (Kędziora, 2011). Primary landscapes, whose structures were predominantly stable from the point of view of thermodynamics, ecosystems (streams of energy and matter optimal in terms of the intensity and direction of flow, which guaranteed the stability of the system) have been transformed into more and more unstable. In landscapes with a very simplified spatial structure and a poorly closed circulation of matter Fig. 11.1.

The rapid development of civilization and population growth in recent centuries have led to such a significant, adverse change in the nature of the surface of the earth that there has been a strong distortion of these processes which are fundamental for the functioning and stability of the Earth system.

Changes of landscape surface

The changes of landscape surface unfavorable for ecosystem services are:

1. Deforestation and fragmentation of forest ecosystems and changes them into agroecosystems.
2. Changing wet meadows into dry pasture.
3. Decreasing number of water bodies are mainly small postglacial ponds, streams and ditches.
4. Simplification of landscape structure.
5. Straightening of river course and its shortening (elimination of meanders and floodplains, embankments of riverbank).
6. Worsening soil structure and diminishing organic matter content.
7. Increasing of share of impermeable surface in total surface area of landscape.

One of the key elements contributing to the stability of the Earth system is forest ecosystems, which play a critical role in atmospheric and hydrological regulation through a mechanism known as the biotic pump (Makarieva and Gorshkov, 2007). The biotic pump is a theoretical concept that explains how forests generate and control wind patterns by releasing water vapor through evapotranspiration, a process in which trees absorb water from the soil and emit excess moisture into the atmosphere via microscopic pores in their leaves. This release of vapor induces pressure gradients that drive the movement of moist air from oceanic regions inland. As a result, the biotic pump facilitates the transport of atmospheric moisture over long distances, delivering precipitation to continental interiors and sustaining rainfall in remote forested regions. This mechanism underscores the fundamental role of forests in maintaining the water cycle and supporting climate stability at regional and global scales.

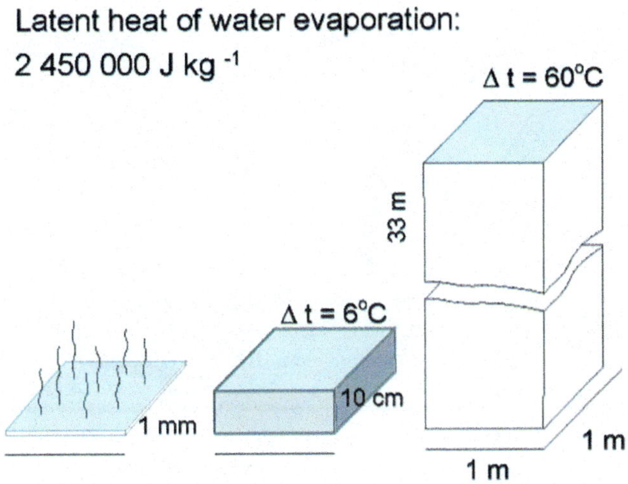

FIGURE 11.2 **Role of water in temperature regulation.** This diagram provides a visual comparison between the energy required for water evaporation—represented by the latent heat of vapourization, and the energy required to raise the temperature of air by specific increments. © *United Nations Environment Programme (UNEP), Integrated Watershed Management – Ecohydrology & Phytotechnology Manual, 2004. Reproduced with permission.*

Conversion of more stable ecosystems like forest, meadow and wetlands into less stable like arable land causes the increase of threats for such fundamental processes like energy flow and matter cycling in environment. Such profound changes of land use and plant cover structure lead straight to acceleration of water cycle, which, in turn brings to many threats for ecosystem services and finally for human existence (Levia et al., 2020). For example, it causes the increase of the rate of surface runoff and decrease in the infiltration rate and time of water storing in the landscape.

Water is a fundamental component of all living organisms. It not only forms their internal medium enabling the chemical reactions regulating life but also secures the maintenance of definite cell shapes and the whole body—conditions for their proficient functioning.

Being a good solvent for many compounds, water leaches them when percolating through soils and nonsoil materials (e.g., rocks), and then transfers them together with subsurface runoff to rivers or lakes. Insoluble materials can be transferred in the form of a suspension. No wonder that water being displaced in the landscape significantly influences the spread of various substances in the environment.

The ability of water to absorb large amounts of heat determines its significant role in temperature regulation of not only man's body, but also the environment surrounding him. Thus, for instance, evaporation of 1 liter of water, i.e., a 1 mm thick film of one square meter, absorbs as much energy as is necessary to heat a 33 m high air column by 60°C Fig. 11.2.

The structure of catchment water balance depends mainly on: an amount of energy available for:

- evapotranspiration;
- variability and time distribution of precipitation;
- a parameter that is discrete in time and space;
- physiographical characteristics of a catchment (slope, denivelation, soil cover);
- density and type of plant cover and its development stage; and land use.

Generalizing, it can be proved that plant cover within a catchment causes Fig. 11.3

- increased evapotranspiration.
- reduced surface runoffs, both due to increased infiltration to soil and evaporation.
- a slowing and increasing time extension of subsurface runoff from soils characterized by higher contents of humus (in underflows situated in ground covered by forest water flows all year round while ditches, situated among fields under cultivation, are dry in summer, even in a year of average precipitation);
- modification of microclimatic conditions, as in the case of fields protected against wind by forests or shelterbelts where evapotranspiration is lower than in open spaces Fig. 11.4.

Water shortage is observed in many regions of the world where low precipitation and high evapotranspiration occur. Water deficits may often happen during the summer season, mainly because of the prevalence of light soils, low

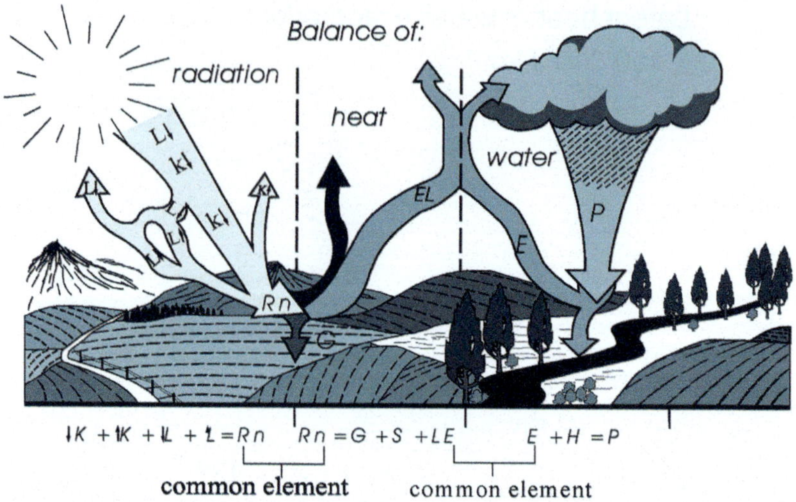

FIGURE 11.3 **Heat and water balance.** Balance of radiation (left part), heat balance (central part) and water balance (right part). © *United Nations Environment Programme (UNEP), Integrated Watershed Management – Ecohydrology & Phytotechnology Manual, 2004. Reproduced with permission.*

precipitation and very high atmospheric water demands. Fig. 11.5 shows an example of water shortage in the Wielkopolska region of Poland.

Proper water management in a landscape can improve these unfavorable conditions. It can be attained mainly through:

- increasing small water retention aided by artificial reservoirs storing excess thaw waters;
- increasing soil retention;
- forming plant cover structure.

An increase of small water retention can be obtained mainly through:

- the exploitation of existing small field water reservoirs;
- reconstruction of destroyed postglacial ponds;
- interceptions of draining waters at the time of their greatest runoff in local depressions;
- introducing swelling equipment (gates) in the network of drainage ditches.

Small field reservoirs not only store water in their basins, but also increase retention in the soil surrounding the reservoir Fig. 11.6. Increases of soil retention near small field reservoirs can be even higher than retention increases in the reservoir itself. Small water reservoirs contribute to the rise of ground waters in neighboring areas, increase the humidity of soils and, subsequently, decrease soil drifting.

The exploitation of small field reservoirs in the spring season can increase water availability of rural catchments by an amount equivalent to 20 mm of precipitation.

The best way for improving soil water retention is by increasing the content of organic matter in the soil. Soil organic matter plays an essential part in improving water conditions in agricultural landscapes. Organic matter increases soil retention because it retains more water than nonorganic matter. Specifically this means an improvement of soil structure by increasing the average size pores, which determine the amount of water accessible for plants.

In some cases, a 1% increase of organic matter increases water supply in a 30 cm ploughed layer by 10 mm or by 100 $m^3\ ha^{-1}$.

An increase of water retention by 100 $mm^3\ ha^{-1}$ has significant economic meaning because the increase is not a single event but refers to each rain event, which can be accumulated in the soil. Thus, during a year, the amount of retention should increase several times.

If retention in a ploughed layer is repeated only three times during a year, the increase of soil water supply during a summer season will reach about 30 mm. This makes an essential saving, even without taking into account the improvement of soil moisture—thermal conditions favorable for vegetative growth and activity of microorganisms and soil fauna.

The proper structure of plant cover within agricultural landscapes exerts a strong positive effect on water cycling. The structure of plant cover, especially shelterbelts, plays a particular part in improving water conditions. They exert a

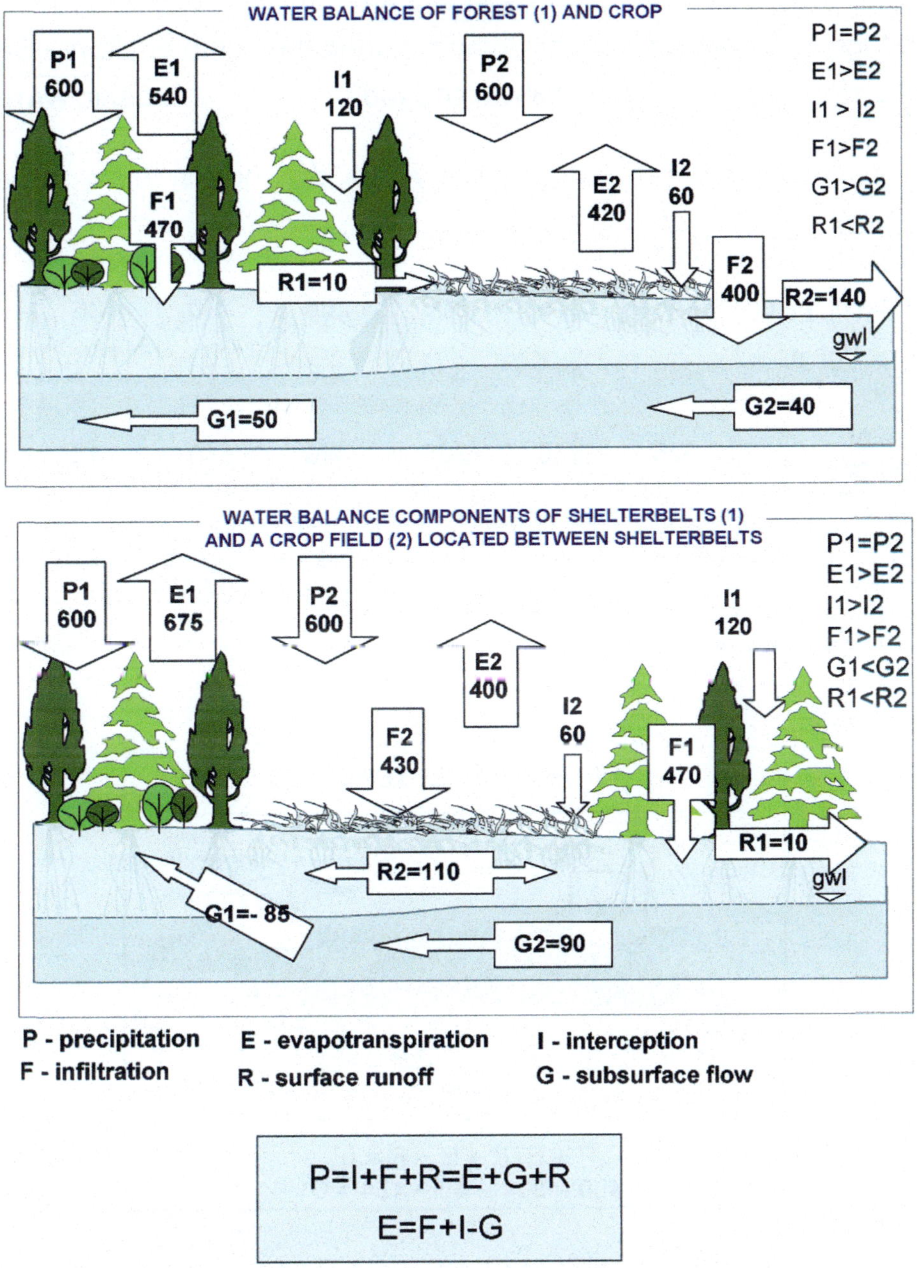

FIGURE 11.4 The importance of shelterbelts. Water balance structure of shelterbelts (left part of the upper picture), crop field (right part of the upper picture), and crop field between shelterbelts (bottom picture). © *United Nations Environment Programme (UNEP), Integrated Watershed Management – Ecohydrology & Phytotechnology Manual, 2004. Reproduced with permission.*

favorable influence on the microclimate by reducing wind speed by 35%–40%, increasing relative air humidity, decreasing potential evaporation, increasing snow depth, and reducing the melting rate of snow in spring. When taken altogether, these increase the percolation rate by 300 m^3 ha^1 in areas covered with shelterbelts compared to open areas Fig. 11.6.

Improvement of water cycling in the landscape requires:

- developing landscape complexity by introduction of shelterbelts, meadow strips restoration of midfield ponds.
- increasing organic matter content in the soil;
- keeping as much water as possible in the landscape for as long as possible, taking care that it is properly allocated;

TABLE 9.2

Water shortage in Turew landscape (Poland) during the summer (April - September)

CLASS NUMBER	SOIL TEXTURE	UNIFORM WATERSHED				MOSAIC WATERSHED			
		normal	dry	v. dry	ex. dry	normal	dry	v. dry	ex. dry
1	Pl, pl.ps, pl:ps, ps.pl,	425	359	286	246	10	-56	-129	-169
2	ps, ps:pl, ps:gl	445	379	306	266	30	-36	-109	-149
3	ps.gl, pgl.gl, pgl:gl	485	419	346	306	70	4	-69	-109
4	pgm.gl, pgm:gl	525	459	386	346	110	44	-29	-69

One dot between symbols (pl.ps) means thickness of upper layer equals 30 cm, and bottom layer 70 cm
Two dots means inversely - upper layer thickness equals 70 cm but bottom layer 30 cm
pl, ps, - sand, pgl - loamy sand, pgm - sandy loam, gl - loam

FIGURE 11.5 Water shortage. Water shortage in Turew landscape in Poland during the summer (April–September). *Data from Kędziora A.*

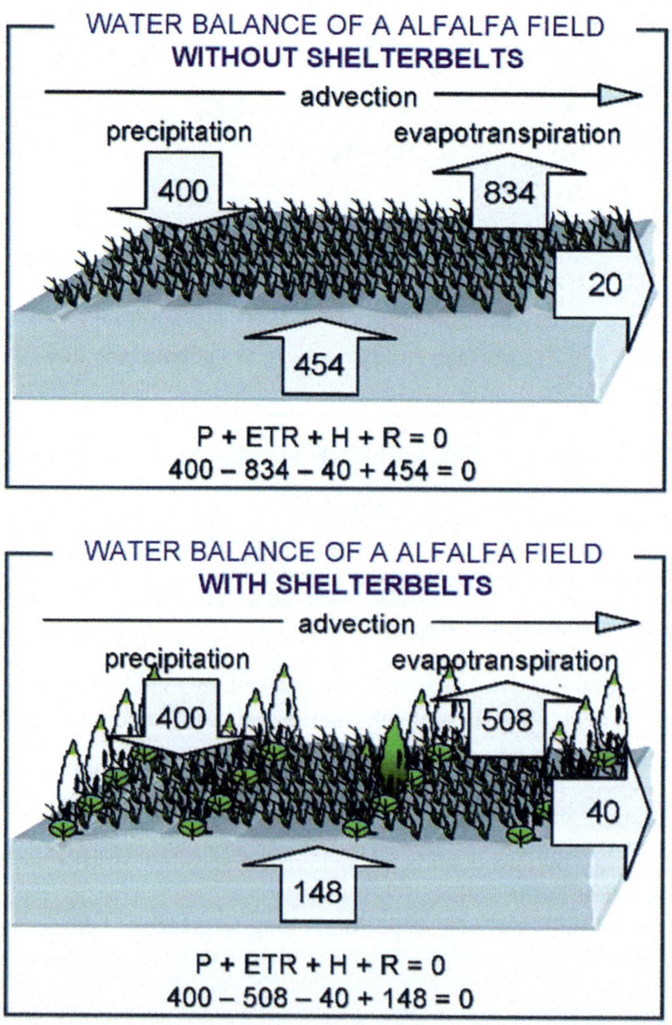

FIGURE 11.6 The importance of shelterbelts. Impact of shelterbelts on alfalfa field water balance under advection. © *United Nations Environment Programme (UNEP), Integrated Watershed Management – Ecohydrology & Phytotechnology Manual, 2004. Reproduced with permission.*

- ensuring that as much water as possible moves from the soil into the atmosphere via plant transpiration, but not as evaporation from the soil to the atmosphere.

For this purpose:

- unsystematic and partial draining should be used more widely and every opportunity for retaining draining runoffs in a catchment area should be utilized;
- supplementary to drainage retention, agromelioration measures for improving the physical water properties of soils and increasing their retention capacities and, consequently, decreasing water deficits for plants during the summer, should be widely applied;
- the scope of necessary agromelioration must take into account negative interactions of farm work mechanization for soil structure by condensing surface soil layers;
- proper landscape management of catchments by optimizing arable land structure and adjustment of agricultural output to the natural resources of the environment, as well as introduction of shelterbelt networks, are fundamental conditions for increasing the effectiveness of water resource exploitation Fig. 11.7.

2. Control of diffuse pollutant inputs to water bodies

To increase production farmers, simplify plant cover structure, both within cultivated fields (selection of genetically uniform cultivars and weed elimination) and within agricultural landscapes (elimination of hedges, stretches of meadows and wetlands, small midfield ponds). Animal communities in cultivated fields are also impoverished (Karg, 1996; Ryszkowski and Życzyńska-Bałoniak, 1985). Farmers interfere with matter cycling in agroecosystems directly by inputs of fertilizers, pesticides, etc., or indirectly by changing water cycling and decreasing holding capacities of soils for chemical compounds. In addition, agricultural activity often leads to decreased humus contents. Increased use of power equipment enables not only deeper soil plowing, but also land surface leveling, modification of water drainage systems, etc., which leads to changes in the geomorphological characteristics of the terrain. These effects of farming activity result in the development of a less complex network of interactions among agroecosystem's components. Relationships between agroecosystem's components are altered so that there are fewer tie-ups of local matter cycles. Thus, increased leaching, wind erosion, volatilization and escape of various chemical components and materials from agroecosystems have been observed (Ryszkowski, 1992; Ryszkowski and Balazy, 1994)

Many environmentally significant effects of agriculture intensification are connected with the impoverishment or simplification of agroecosystem structure. However, a farmer in order to obtain high yields must eliminate weeds, control herbivores and pathogens, ensure that nutrients are easily accessible only for cultivated plants during their growth, increase mechanization efficiency, amongst other things. Therefore, agricultural activity aimed at higher and higher yields leads inevitably to the simplification of agroecosystem structure, which in turn causes further environmental hazards.

Such an ecological analysis leads to a conclusion of major significance for the sustainable development of rural areas. Applying intensive means of production, farmers cannot prevent the threats to arable fields, as noted above, and these increase the risks of diffuse pollution to ground and surface waters, evolution of greenhouse gases (N_2O, CO_2) and water or wind erosion. It must be clearly said that although farmers can moderate the intensity of these processes through proper selection of crops and tillage technologies, they are not able to eliminate them entirely. A higher control efficiency of environmental threats evoked by agriculture could be achieved by structuring agricultural landscapes with various nonproductive components like, e.g.:

- hedges;
- shelterbelts;
- stretches of meadows;
- riparian vegetation strips;
- small ponds.

Therefore, any activity to maintain or increase landscape diversity is important not only for aesthetics and recreational reasons, but even more so for environment protection and for the protection of living resources in the countryside Fig. 11.8.

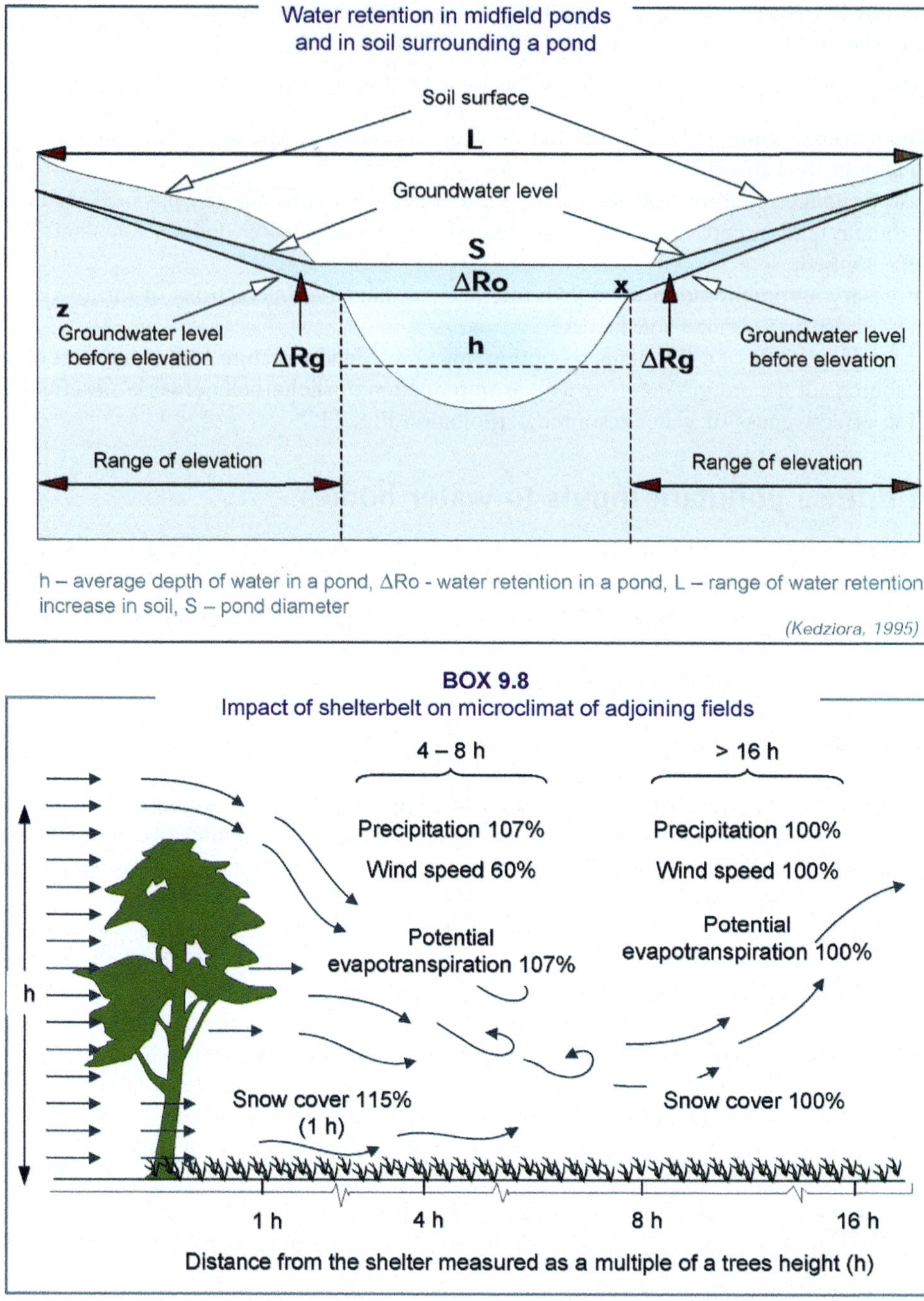

FIGURE 11.7 Reduction wind speed for reduction evaporation and improvement water retention in soil. © *United Nations Environment Programme (UNEP), Integrated Watershed Management – Ecohydrology & Phytotechnology Manual, 2004. Reproduced with permission.*

Recent developments in agroecology and, especially in studies on agroecosystems and rural landscape functions like solar energy flows, matter cycling, and maintenance of biodiversity, help to tackle the problems of environmental threats. Studies on impacts of plant cover patterns on agricultural landscape functions are especially relevant in this respect.

There is an increasing amount of evidence that permanent vegetation strips can control the ground water under some cultivated fields very high concentrations of nitrates, reaching 50 mg NO_3–N per liter, were detected, while in the stream draining this watershed the average concentration of NO_{-3}-N over many years did not exceed 1.5 mg NO_3–N per liter Fig. 11.9. The stream is separated from fields by dispersion of chemical compounds leached out of cultivated fields

FIGURE 11.8 **Diversified agricultural landscape.** The aerial photograph of the diversified landscape. *Photo credit: A.Kędziora.*

TABLE 9.3

Mean concentrations of NO_3-N (mg dm^{-3}) in ground water under cultivated fields, shelterbelts, small forests and meadows in the Turew agricultural landscape, Poland

Period of sampling	Culitivated field (a)	Shelterbelt or forest	Meadow (b)	Reduction (a-b):a (percent)	Reference
1982-1986	22,2	1,0	-	95	*Bartoszewicz & Ryszkowski, 1996*
1982-1986	37,6	1,1	-	97	*Bartoszewicz & Ryszkowski, 1996*
1972-1973	12,6	0,3	-	98	*Margowski & Bartoszewicz, 1976*
1984-1986	33,1	8,1	-	75	*Ryszkowski et al., 1997*
1994	52,4	2,7	-	94	*Ryszkowski et al., 1997*
1995	13,1	4,9	-	63	*Ryszkowski et al., 1996*
1986-1989	48,3	-	6,5	87	*Bartoszewicz, 1990*
1987-1989	15,9	-	0,7	95	*Bartoszewicz, 1990*
1987-1991	13,1	-	2,8	79	*Szpakowska & Zyczynska-Baloniak, 1994*
1993	18,7	-	1,4	92	*Ryszkowski et al., 1996*
1993	22,1	-	2,0	91	*Ryszkowski et al., 1996*
1994	19,1	-	1,2	94	*Ryszkowski et al., 1996*
1994	13,4	-	2,4	82	*Ryszkowski et al., 1996*
1995	18,3	-	0,6	97	*Ryszkowski et al., 1996*

FIGURE 11.9 **Mean concentrations of nitrates NO_{-3}-N in mg dm^3.** Mean concentrations of nitrates NO_{-3}-N under varying conditions: cultivated fields, shelterbelt or small forest and meadows. *From Bartoszewicz, A., Ryszkowski, L., 1996. Influence of shelterbelts and meadows on the chemistry of ground water. In: Ryszkowski, L., French, N. R., Kędziora, A. (Eds.), Dynamics of an Agricultural Landscape. Państwowe Wydawnictwo Rolnicze i Leśne, Poznań, pp. 98–109.*

(Haycock et al., 1996). To describe the effectiveness of buffer zones for controlling diffuse pollution, the results of studies carried out by the Research Center for Agricultural and Forest Environment in Poznan, Poland will be presented.

Nitrate concentrations in ground water from beneath meadows and shelterbelts studied in the Wielkopolska region (Poland) were significantly lower than those in ground water under adjoining fields. In some areas the reduction in the mean nitrate concentrations in ground water under the biogeochemical barrier (shelterbelt) was 34-fold (from 37.6 to 1.1 mg L^{-1}) (Bartoszewicz, 1996). But usually, the decrease in nitrate concentrations under the biogeochemical barrier was lower, in the range of 10–20-fold. In stretches of meadows, hedges, and shelterbelts. Thus, the strong controlling effect of these biogeochemical barriers can be observed (Bartoszewicz, 1994; Ryszkowski et al., 1997). The decrease of phosphate concentrations in ground water under the fields and biogeochemical barriers was less striking; usually the concentration decrease was 10–50%.

Concentrations of many chemical compounds migrating from the ground in outflows from neighboring cultivated fields are seriously reduced when the water passes under such biogeochemical barriers as shelterbelts, midfield forests and riparian vegetation strips.

The great influence of plant cover structure on the output of elements from watersheds was shown by Bartoszewicz (1994) and Bartoszewicz and Ryszkowski (1996). The studies were carried out in two small watersheds. The first one, called a uniform watershed, was 99% composed of cultivated fields and 1% small forests. In the second one, called a mosaic watershed, cultivated fields made up 70% of the area, meadows 14% and riparian forest 16%. The mean annual water output from the mosaic watershed during a 3 year period was 70.2 mm^2 and in the uniform one, 102.0 mm^2. The mean annual precipitation for both watersheds was the same; 514 mm^2 Fig. 11.10. From a uniform arable watershed, 20.4 kg of inorganic nitrogen leached from 1 ha annually, 20% of which was in the form of ammonium ions. When the migration of mineral components from a mosaic watershed was analyzed, a low leaching rate of nitrogen constituents and different ratio of nitrate to ammonia ions were observed. The annual leaching rates of N from 1 ha of this watershed amounted to about 2 kg (ten times less than in the uniform watershed), and both ionic forms of N were represented in almost identical proportions. Even more striking were the differences between the uniform arable watershed and the mosaic one with respect to seasonal variations in the migration of nitrogen. The majority of both nitrogen ion forms (86%) had leached from the mosaic watershed during winter, while during the summer period, the leaching of both nitrogen forms (particularly nitrate) was negligible.

Naturally compatible structures that assist in controlling matter cycles in agricultural landscapes are of great importance for enhancing a countryside's resistance to degradation.

Various plant cover structures like hedges, shelterbelts, stretches of meadows and riparian vegetation strips are of special interest. Application of these structures has several benefits, of which the most important are:

- they can be easily planted; they are not expensive and could provide economic benefits (e.g., timber, herbs, honey etc.);
- they can fulfill some societal needs (hunting, photography, mushroom and berry picking, etc.);
- they are very important from the point of view of enhancement of catchment sustainability potential WBSR + CE.

TABLE 9.4
Annual mean water output (mm) and nutrient loss (g m^{-2} $year^{-1}$) from two small watersheds in Poland, from November 1988 – October 1991

SEASON		UNIFORM WATERSHED			MOSAIC WATERSHED		
	Precipitation (mm)	Water output (mm)	NO_3-N	NH_4-N	Water output (mm)	NO_3-N	NH_4-N
Winter season Nov.-April	220,7	60,8	12,3	3,0	56,8	0,90	0,95
Summer season May-Oct.	292,9	41,2	4,0	1,1	13,4	0,05	0,25
Whole year	513,6	102,0	16,3	4,1	70,2	0,95	1,20

(Bartoszewicz, 1994)

FIGURE 11.10 Annual mean water output. Annual mean water output from two small watersheds in Poland from November 1988–October 1991. *From Bartoszewicz, A., 1994. The chemical compounds in surface waters of agricultural catchments. Roczniki Akademii Rolniczej 250, 5–68.*

Midfield water reservoirs also intercept chemical substances, immobilizing them in bottom deposits where they are subjected to transformation by biogeochemical processes. The role of small midfield ponds, lately neglected and often treated as wastelands, is particularly significant for more efficient use of fertilizers. They may serve as a tool for modification of matter cycling because the chemical compounds leached from fields could be returned to arable fields with sediment. Such forms of field fertilization were applied in the past on a fairly wide scale, as was described by General Dezydery Chlapowski in his book on agriculture published in 1843.

3. Wind and water erosion

Among many unfavorable environmental phenomena, from human civilization's stability point of view, are processes leading to deterioration of soil properties. Especially wind and water erosion are dangerous as they quickly lead to degradation of productive top soil layers. Intensive wind erosion usually occurs in the spring time when arable lands are bare or covered by very poor plant cover and solar radiation and wind speed are very intensive. Water erosion is very intensive in the plant less landscape having rich relief and sloping surface. Two such examples are presented below. Of key importance in shaping the conditions conducive to wind erosion are water vapor pressure deficit, air temperature and wind speed. The most important factors in shaping the conditions of water erosion are air and soil temperature, as well as the frequency and intensity of rainfall. These factors (in relation to wind as well as to water erosion) do not work individually, but comprehensively. Their effect is largely controlled by the abundance and structure of the plant cover.

In spring, with low air temperatures and strong solar operation, there is high deficit of air humidity in the air layer near the ground. The simultaneous high instantaneous heat balance on the surface of soil aggregates causing their intense evaporation, which in the case of clay and dusty soils leads to strong surface stresses and crumbling of soil lumps. If there are strong winds and no vegetation cover or terrain obstacles in the form of trees or shrubs that reduce the wind speed, then spring dust storms appear in these regions.

It is well known that wind erosion is mainly served by the conditions leading to drying of the soil surface. It is also known that such conditions are lack of rainfall and high temperature. However, these conditions occur normally during the growing season when soil is usually covered with vegetation. In an annual course of the weather in climatic conditions of our country (Poland), however, there is a short period in the early spring season when specific conditions favoring strong wind erosion developing. In March or at the beginning of April, and sometimes at the end of February, in sunny weather, the intensity of radiation is so strong that it heats the surface of small, several millimeter soil aggregates to the temperature of up to above 30°C. At the same time the air is still cold and has about 0°C. This results in the formation of very large water vapor pressure deficit around these aggregates, which causes very fast drying of the outer layer of these aggregates which are shrinking and cracking. The soil grains formed in this process are highly susceptible to erosion. The intensity of wind erosion can be so high that can fulfill roadside ditch by sand during several hours Table 11.1.

TABLE 11.1 Wind erosion.

Frame 27. Example illustrates this situation
Time and place of observation: Beginning of April, field before sowing, bare soil.
Weather conditions: Clean and cold Arctic air prevails above the ground. The sun rays fall on the soil at the angle of 30O.
Data: • Air temperature (ta) is 0°C. • Water vapor pressure (e) is 4 hPa • Wind speed at the level of 2 m above the ground (u) is 0.2 m * s^{-1} • Diameter of the analyzed soil lump (d) is 1 cm • Initial moisture of this soil lump (w) is 0.5 m^3 * m^{-3} • Soil density (g) is 1.6 g * cm^{-3} • Specific heat of soil (cp) is 1000 J * kg^{-1} * K^{-1} • Specific heat of water (cw) 4200 J * kg^{-1} * K^{-1} • Soil albedo (a) is 0.2 • Coefficient of the atmospheric transparency (p) is 0.8 • Coefficient of the optical mass of the atmosphere at the sun altitude of 30 o (o) is 2. • Solar constant (S) is 1400 W * m^{-2}.

Heat balance of the soil lump can be written down as Rn=S + LE, and calculations of the components of that heat balance showed that the radiation balance (Rn) will reach the value of 535 W * m^{-2}, the sensible heat flux (S) will be 235 W * m^{-2} and the heat flux for water evaporation from the sample (LE) will be 300 W * m^{-2}. Water vapor pressure deficit in the air surrounding the lump will reach as much as 34 hPa and the intensity of water evaporation from the surface of the sample will be about 11 mm/day. The outer layer of the lump, 1 mm thick (containing half the water), would dry out after less than 4 hours of solar operation. The result of this process will be creation of a layer of sand and silt particles covering surface of soil which can be eroded.

However, the possibility of wind erosion occurrence mostly depends on wind conditions. A characteristic feature of the vertical distribution of the wind profile is that speed increases with the height increase. There is a thin layer of air standing still or showing laminar flow directly above the ground surface. The thickness of this layer depends on wind speed and roughness of ground surface. In the case of bare soil and wind speed equal to 4 m * s^{-1} (at 1 m above the ground) this thickness is equals to 0.4 mm. All soil particles with a diameter smaller than the thickness of this layer cannot be lifted by the wind. For this reason, wind speeds lower than 4 m * s^{-1} measured at meteorological stations do not cause wind erosion. Sand grains with diameters from 0.2 to 0.5 mm exposed to the wind, lifted by the wind, when fall activate smaller particles embedded in the laminar layer, and only then are these small particles transported by the wind over long distances (Ziemnicki, 1968). For grains of soil with a diameter of 0.2, 0.3; 0.4 and 0.5 mm the calculated values of the critical wind speed are respectively: 2.16; 2.65; 3.05 and 3.41 m * s^{-1}, measured at the height of 1 meter above the ground or: 3.4; 4.2; 4.8; and 5.4 m * s^{-1} measured at 12 m above the ground. Borówka (1980) states that the transport of material begins basically at an average wind speed of about 4 m * s^{-1}.

Although in the blowing process fine and medium sand particles are first activated, generally dust is the main material blown out from the field. It constitutes from 47% to 88% of all blown material during the weakest wind, from 52% to 73% during the medium wind and from 32% to 47% during the strongest winds (Uggla and Nożyński, 1962).

Regarding water erosion, the most important problem is decreasing the infiltration rate and increasing the surface runoff rate. In winter, the most dangerous phenomenon is the variability of the temperature of the soil surface not covered with vegetation and the occurrence of heavy rainfall in the periods of warming following frosty periods.

Water erosion is becoming an increasing threat to human existence, a phenomenon whose frequency and intensity is increasing at terrible pace. Surface runoff, which is driving force of erosion, is strongly impacted by simplification of landscape structure. From the point of view of ecohydrology, the possibility of controlling the proportion of surface to ground runoff is a cardinal problem for management of quality and quantity of water resources necessary for existence of human civilization (Zalewski et al., 2021). The smaller the ratio of surface to underground runoff the less intensity of erosion and more water remains in the soil, longer is time of supplying watercourses in water and lower their contamination. The amount of water outgoing from the catchment depends on many factors, the most important of which are intensity and spatial distribution of rain, in any scale, local, regional and global. In forests or in meadows the topsoil is usually loose (low density and high porosity) and overgrown by plants. Big infiltration capacity of such soils and resistance put up by plants to flowing water reduces significantly runoff. Especially in lowlands the runoff from such ecosystems is reduced to minimum excluding unusual rainstorms. In these areas the subsurface outflow is considerably high in comparison with surface runoff and is very stable in time. In contrast to this, in row crop field or even in grain crop field, and especially in the case of bare soil the surface runoff is very high. These areas, because of compaction of topsoil layer by machinery, very often have low infiltration rate, even in light soil. Also, the furrows in row crop fields, and small grooves in grain crop fields, especially when they are oriented from the top to the foot of a hill, increases water outflow. Thus, surface runoff from the cultivated fields can be observed even in the case of the mild rain intensity, and it usually exceeds the subsurface runoff. In general, the total runoff as well as its partition into surface and subsurface outflow depends on many factors among which the plant cover appears as the most important. Consequently, in an agricultural landscape, the intensity, timing and structure of runoff change as landscape diversification is changed. In mosaic landscape having a rich shelterbelt network, many meadow strips and bushes, total runoff as well as the ratio of surface to subsurface outflow is low. As a result of it, the water outflow from such landscape is slow, stable and lasts long time creating favorable conditions for rebuilding soil water supply on the one hand, preventing the small streams against drying up on the other, and by the same, allows to survive many animals and plants living in or near these streams. Within uniform agricultural landscape, the runoff is born quickly, lasts short time, is very intensive and mainly in form of surface outflow. Such situation is unfavorable for improving water conditions in the environment but may bring about serious soil erosion.

An example of a change in the run-off pattern is Ethiopia. Primary forests covered 64% of the country's area and 90% in the mountainous Gumara catchment area. In 1950 these figures were 16% and 20% and in 2000 they were 3% and 5% respectively.

Average daily discharge in period 1959–2006 was 30 m^3/s (Fig) and within 50 years increased from 23 to 44 m^3/s. Fig. 11.11. So, annual volume of discharge is 946 million m^3. The area of catchment is 1464 km^2. Thus, the depth of water outflows from the catchment is 646 mm. Average precipitation is 1400 mm. So, coefficient of outflow (ratio of outflow to precipitation) is 0.46. Finally, the evapotranspiration is 754 mm, which is 54% of precipitation. In Poland this parameter is 0.28.

The result of these changes was a significant increase of water erosion, mainly gully erosion Fig. 11.12. Such an increase in runoff caused a rapid increase in the inflow of sediment to Lake Tana, from 0.16 to 2.5 million tons per day, which is more than 15 times over a 50-year period.

One of the most important factors impact process of runoff creation is the relation between infiltration rate and intensity of rainfall Fig. 11.13. Thus, in the case of low infiltration capacity, rainfall with intensity exceeding basic infiltration rate cannot infiltrate into the soil, and it becomes wholly or partly surface runoff. In analyzed case, the intensity of rainfall during rainstorm lasting 15 hours was oscillating between 3 and 6 mm per hour. Sometimes it was higher than infiltration rate, which was changed from 4 mm per hour (at the beginning of the rain) to 2.4 mm per hour at the end (basic infiltration rate). As a result of such relation, from 42 mm of rainfall only 31 mm of rain infiltrated into the soil, but 11 mm formed the surface runoff (24% of precipitation).

A problem such as growing threat of erosion, both wind and water, cannot be solved in isolation from other environmental problems. The smallest spatial unit for solving some problems, such as improving the hydropedological

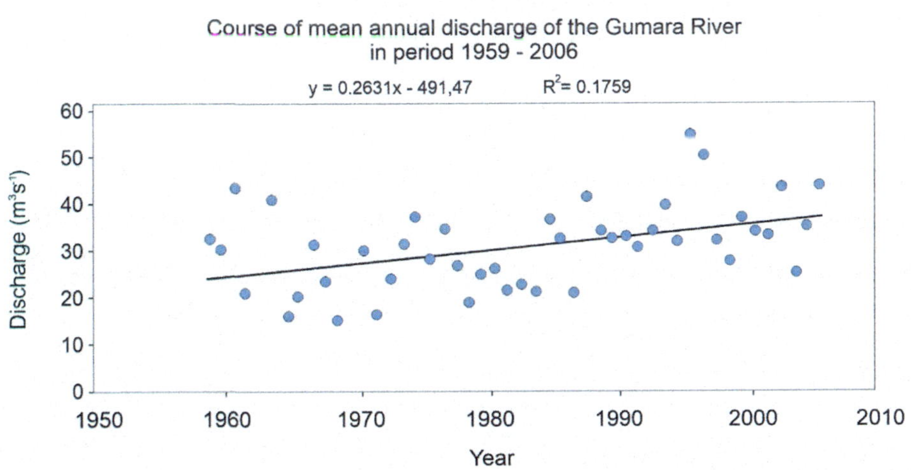

FIGURE 11.11 Mean annual discharge of Gumara River. Course of mean annual discharge of Gumara River as the result of deforestation of the catchment. (Ethiopia).

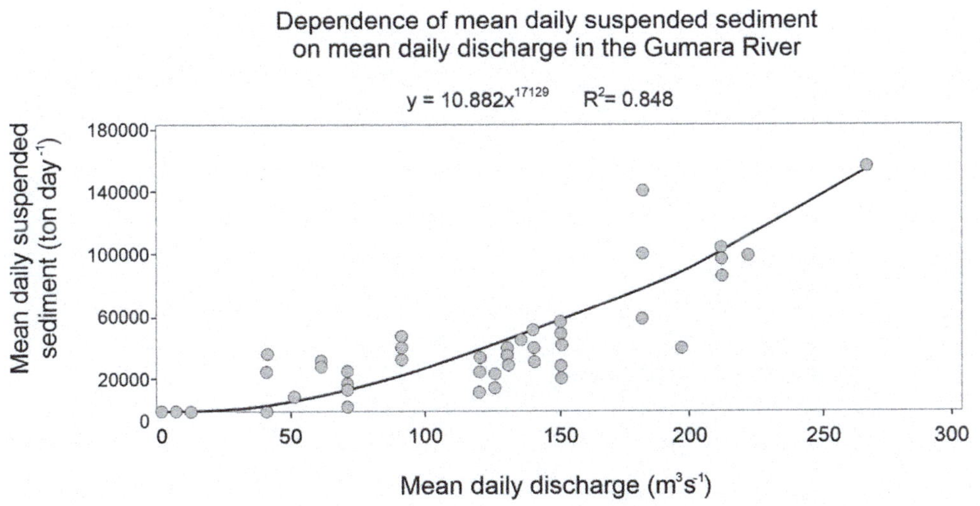

FIGURE 11.12 Suspended sediment. Dependance of suspended sediment on discharge of Gumara River. *Modified from Mequanint, T. & Awulachew, S. B. (2008). Soil and Water Assessment Tool (SWAT)-based runoff and sediment yield modelling: A case of the Gumera watershed in Lake Tana sub-basin. MSc thesis, Arba Minch University, Ethiopia/International Water Management Institute, Addis Ababa, Ethiopia.*

properties of the soil, can be a field or a landscape. However, some problems, such as erosion and water pollution, need to be dealt with on a river basin scale (Zalewski, 2000).

Long term investigation carried out by Institute for Agricultural and Forest Environment PAS i Poznań as well as by European Regional Center on Ecohydrology under auspices of UNESCO in Łódź showed that increasing complexity of agricultural landscape mainly by introduction of nonproductive elements like shelterbelts, strips of meadows, bushes and small midfield ponds is one of the best tool for controlling water cycling and chemical pollution of surface and ground water in agricultural Lanscape (Zalewski, 2000; Kędziora, 2015).

Such findings open new frontiers for conservancy of living resources outside protected areas and open new prospect for reconciliation of agriculture with nature protection. Paradigms that are presently in force should be changed as follow:

Paradigms comparison: Present versus future

Present paradigms	Future paradigms
Conservation	Management of protected ecosystems
Protection by isolation	Reconciliation with economic activities
Species and communities	Ecosystems
Inventory of species and recognition of their adaptation to environment	Understanding life supporting processes (energy flows, matter cycling and their control mechanisms, transmission of information).
Rare, endangered or protected species	Guilds of functional groups, key stone species.
Homogeneity	Heterogeneity (diversity).

The content of organic matter is one of the most important hydropedological characteristic of the soil. Organic matter increases water storage in the soil because it can retain more water than nonorganic material does. Specifically, this means an improvement of the soil structure by increasing the medium size pores, determining the amount of water readily available for plants and macrofauna living in the soil. In turn, macrofauna improve soil porosity, structure and water holding capacity. The increase of organic matter content by 1% in the upper, plowed layer of sandy loam soil having normally about 1% of organic matter causes the increase of the available water by 30 mm which in the scale of the country gives the increase of water supply equal to the volume of all artificial reservoirs in Poland (Kędziora and Olejnik, 2002). It should be also kept in mind that this soil reservoir can be emptied and recharged several times during a year. The average content of organic matter in arable land in Poland is equals to 2.26% changing from 1.65% in Wielkopolska to 3.13 in Lubuskie. A better structure of the upper layer also results in increase of infiltration rates which allows catching more water from precipitation and decreasing the surface runoff. The increase of organic natural fertilization is only one way to increase the organic matter in the soil (Derpsch, 2005).

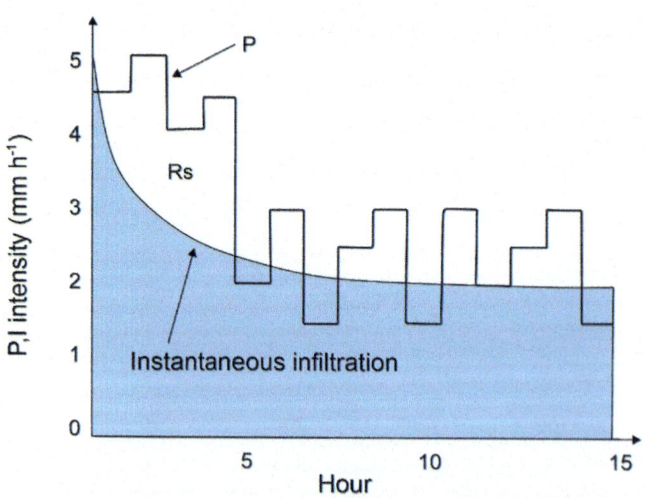

FIGURE 11.13 Precipitation. The course of precipitation (stair line) and infiltration (continuous line) during one case of rainfall lasting 15 hours.

The ability to collect water (water holding capacity—WHC) needed for agricultural production is one of the most important functions of the soil. Maintaining the relevant structure and an appropriate level of soil organic matter is necessary to ensure this function. Long-term tillage of soil leads to a systematic deterioration of its structure and to reduction of the organic matter content. The use of heavy farm equipment causes excessive compaction, especially the top layers of the soil. On the other hand, since the past century began to be used more and more chemical fertilizers and less organic fertilizers. All these errors led in some cases even to double decreasing retention capacity of the soil. It's time for searching other than the traditional methods of cultivation of the soil. One such technique is the use of simplified or no-tillage ways of soil cultivation. No tillage system stimulates biodiversity in soil through increased organic matter content Fig. 11.14.

There is intensive development of soil fauna, both beneficial and harmful to the crop (Bailey and Lazarovits, 2003; Holland et al., 2004). Microorganisms on the one hand can improve the circulation of matter in the soil and help to combat pests and diseases, on the other hand are the cause of diseases of agricultural plants. An important measure of the biological activity of the soil is the presence and concentration of dehedrogenazy and urease enzyme, which take an active part in the processes of synthesis and analysis. Doran (1987) showed in their study increased activity of both substances in the surface layers of the soil on objects with direct sowing than in conventional tillage. Growing conservative, especially direct seeding, favorable effect on the growth of the population of earthworms in the soil compared to plow cultivation (Kladivko, 2001; Tebrügge and Düring, 1999). Increased abundance of soil fauna positively affects the productivity of plants by mineralization of organic matter and the formation of humus layer and the loosening of the soil. Kladivko (2001) presents in his paper the synthesis of years of research with the use of direct sowing and its effects on the soil environment. The author shows that after 14 years of use of direct sowing earthworms' size 0–30 cm layer and on the surface of 1 m^2 was 138, whereas in the conventional tillage 64, and thus was more than two-fold higher. Increased availability of plant debris and weed seeds remaining on the surface of the field under no tillage system is also a perfect food for insects, birds and small mammals (Holland et al., 2004).

For checking effectiveness of new agrotechnics in acting against threats such as erosion and impoverishment of mezofauna the following experiment was carried out: Two fields on the same farm were chosen. Matrix material of two soils was the same.

On the left field traditional agrotechnics was used (Tillage). On the right field, for 15 years new agrotechnics had been used (No tillage or simplified). Topsoil on no-tillage field had thickness of 35 cm, while on the tillage field only 22 cm. The thickness of topsoil of no-tillage was growing nearly 1 cm per year. Bulk density individual layers in soil profile on no-tillage field was near the same, while in profile on tillage field bulk density of soil is clearly higher in under plow layer. This is the effect of heavy plow using. But very important is it, that water and organic matter content in no-tillage soil is much higher than in tillage field Table 11.2. Organic matter content in topsoil of no-tillage field was 1.73, that is 2.2 times higher than in topsoil of tillage field—0.78. As a result of it field capacity of water was 0.275 and 0.200 m^3m^{-3} respectively in no-tillage and tillage fields. It means that in 0.5 m thick soil layer on no-tillage field can be stored 38 liters on 1 m^2 more water than on tillage field. It is equal of 38 mm precipitation. However, the most important from the point of

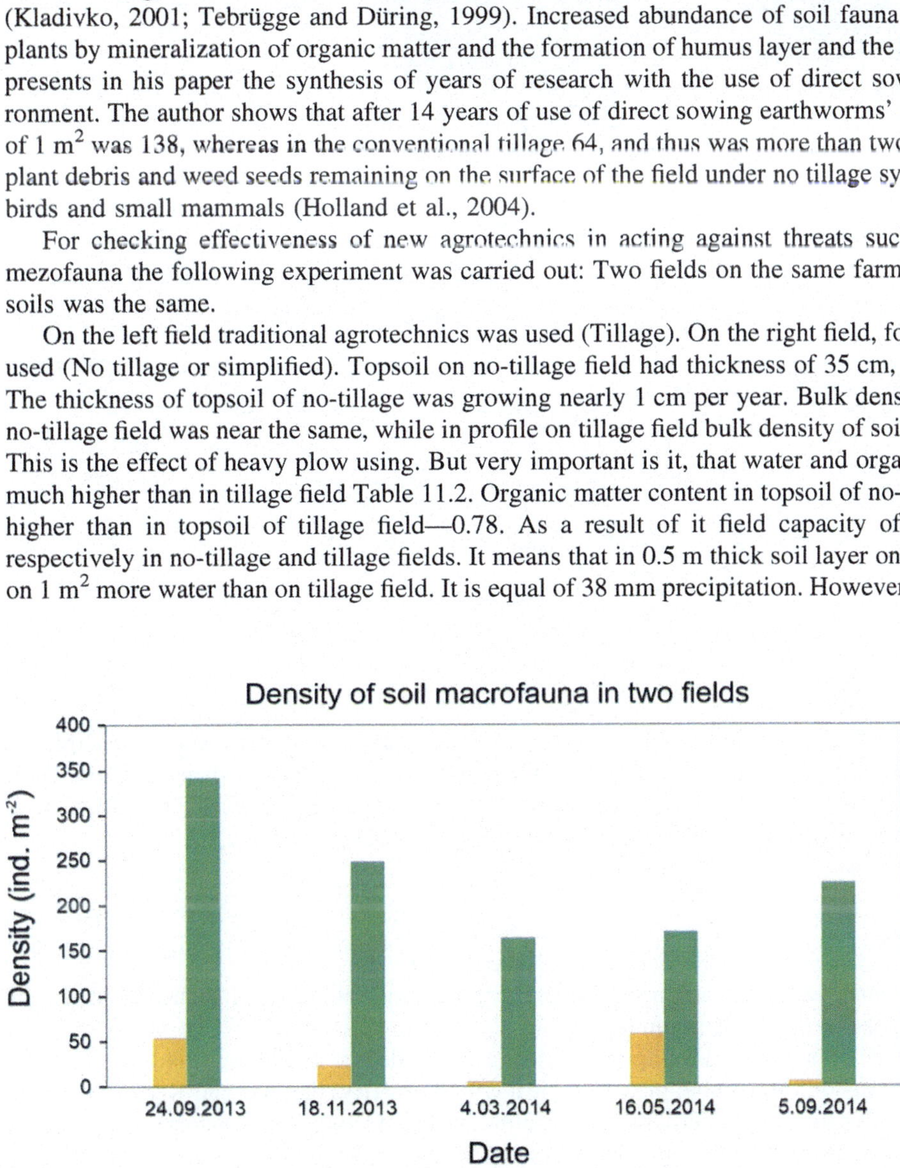

FIGURE 11.14 Density of soil macrofauna in tillage and no-tillage areas. Density of soil macrofauna in two fields being under different agrotechnics: tillage—yellow, no-tillage—green.

TABLE 11.2 Soil profiles and their characteristic of two fields being under different agrotechnics

Tillage versus no-tillage. Photo credit: A. Kedziora.

Soil cultivation	Layer cm	Bulk density g/cm^3	Moisture cm^3/cm^3	Organic matter %	Infiltration rate m/24 h	
					Value	Classe
Tillage	0–25	1.75	0.20	0.78		
	25–50	1.84	0.20		1.9	Very slow
	50–75	1.83	0.22			
	75–100	1.78	0.24			
No-tillage	0–25	1.74	0.28	1.73	9.6	Slow
	25–50	1.76	0.27			
	50–75	1.77	0.25			
	75–100	1.78	0.24			

view of water erosion is much higher (five times) infiltration rate on no-tillage field. This is the result of macrofauna activity, mainly earthworm. The abundance of soil macrofauna on no-tillage field was much higher than on tillage Table 11.2. The ratio (no-tillage/tillage) of abundance of soil macrofauna on that field were changing between 2.6 and 14.3 ind./m^2 (on average it was 7.1) for 1 year observation. In relation to earthworm this ratio was 10.2.

4. Conclusion

During the last decades a lot of negative changes in environment have been observed.

The most important of them are connected with water and kinetic energy of atmosphere like increase of water and wind erosion impoverishment of biodiversity due to agriculture.

Appropriate shaping of the landscape structure, consisting in increasing of the degree of diversity and mosaic character, can significantly mitigate the effects of unfavorable meteorological conditions (Ryszkowski and Kędziora, 2007).

Enriching the landscape with such elements as shelterbelts and meadow bushes, restoration of natural water bodies, improvement of soil retention capacity by increasing organic matter content and improving the structure of the soil, is a prerequisite for improving the quality and quantity of water at the local and regional levels (Kędziora et al., 2011)

For adaptation and counteraction to these effects human civilization must:

1. Improve the structure of the agricultural landscape by increasing its complexity by introduction of elements which control fluxes of energy and matter, as well as wind erosion (Kędziora, 2010)
2. Increase content of organic matter in the soil and improve soil structure by the use of new agrotechnics like, for example, no-tillage systems.
3. Increase surface water retention of the earth by restoration of small water bodies and better managing of the melioration means (ditches, weirs, water valves and so on).
4. Reduce and slow down surface runoff by improving infiltration of soil surface and enrich the landscape in elements which control the speed of surface runoff and water erosion (Kędziora, 2015).

To reach these goals the previous sociocentric/mechanistic paradigm should be changed into new evolutionary/ecosystemic one. Urgent need is "generate innovative systematic solutions for enhancement of sustainability potential WBSRCE. That is mean that many aspects of our space of live as well as aur behavior should be taken under consideration as one holy system: Water, Biodiversity, Services of ecosystems, Resilience to climate and various anthropogenic impact and Culture and Education" (Zalewski et al., 2021).

References

Bailey, K.L., Lazarovits, G., 2003. Suppressing soil-borne diseases with residue management and organic amendments. Soil Till. Res. 72 (2), 169–180.

Bartoszewicz, 1996. Influence of Shelterbelts and Meadows on the Chemistry of Groundwater Dynamics of an Agricultural Landscape, pp. 98–109.

Bartoszewicz, A., 1994. The chemical compounds in surface waters of agricultural catchments. Roczniki Akademii Rolniczej. 250, 5–68.

Bartoszewicz, A., Ryszkowski, L., 1996. Influence of shelterbelts and meadows on the chemistry of ground water. In: Ryszkowski, L., French, N.R., Kędziora, A. (Eds.), Dynamics of an Agricultural Landscape. Państwowe Wydawnictwo Rolnicze i Leśne, Poznań, pp. 98–109.

Borówka, R. K., 1980. Present day dune processes and dune morphology on the Łeba Barrier, Polish coast of the Baltic. Geogr. Ann. Ser. A., 62A (1-2), 75–82.

Derpsch, R., 2005. Third World Congress on Conservation Agriculture the Extent of Conservation Agriculture Adoption Worldwide: Implications and Impact.

Doran, J.W., 1987. Microbial biomass and mineralizable nitrogen distributions in no-tillage and plowed soils. Biol. Fert. Soils 5, 68–75.

Haycock, N.E., Burt, T.P., Goulding, 1996. Proceedings of the international conference on buffer zones. Buffer Zones: Their Processes and Potential in Water Protection.

Holland, J.D., Bert, D.G., Fahrig, L., 2004. Determining the spatial scale of species' response to habitat. BioScience 54 (3), 227–233.

Karg, J., 1996. Państwowe Wydawnictwo Rolnicze i Leśne animals in arable land. Dynamics of an Agricultural Landscape 138–172.

Kędziora, A., 2011. The network of shelterbelts as an agroforestry system controlling the water resources and biodiversity in the agricultural landscape. In: Institute for the Agricultural and Forest Environment. Polish Academy of Sciences.

Kędziora, A., 2010. Landscape management practices for maintenance and enhancement of ecosystem services in a countryside. Ecohydrol. Hydrobiol. 10 (2–4), 133–152. https://doi.org/10.2478/v10104-011-0006-7.

Kędziora, A., 2015. The network of shelterbelts as an agroforestry system controlling the water resources and biodiversity in the agricultural landscape. In: Proceedings of the 22nd International Grassland Congress.

Kędziora, A., Negussie, Y.Z., Asres, M.T., Zalewski, M., 2011. Shaping of an agricultural landscape to increase water and nutrient retention. Ecohydrol. Hydrobiol. 11 (3–4), 205–222. https://doi.org/10.2478/v10104-011-0048-x.

Kędziora, A., Olejnik, J., 2002. Water balance in agricultural landscape and options for its management by change in plant cover structure of landscape. In: Ryszkowski, L. (Ed.), Landscape Ecology in Agroecosystems Management. CRC Press, pp. 57–110.

Kladivko, E.J., 2001. Tillage systems and soil ecology. Soil Till. Res. 61 (1–2), 61–76.

Levia, D.F., Creed, I.F., Hannah, D.M., Nanko, K., Boyer, E.W., Carlyle-Moses, D.E., van de Giesen, N., Grasso, D., Guswa, A.J., Hudson, J.E., Hudson, S.A., Iida, S., Jackson, R.B., Katul, G.G., Kumagai, T., Llorens, P., Ribeiro, F.L., Pataki, D.E., Peters, C.A., Carretero, D.S., Selker, J.S., Tetzlaff, D., Zalewski, M., Bruen, M., 2020. Homogenization of the terrestrial water cycle. Nat. Geosci. 13 (10), 656–658. https://doi.org/10.1038/s41561-020-0641-y.

Makarieva, A.M., Gorshkov, V.G., 2007. Biotic pump of atmospheric moisture as driver of the hydrological cycle on land. Hydrol. Earth Syst. Sci. 11 (2), 1013–1033. https://doi.org/10.5194/hess-11-1013-2007.

Mequanint, T., Awulachew, S.B., 2008. Soil and Water Assessment Tool (SWAT)-Based Runoff and Sediment Yield Modelling: A Case of the Gumera Watershed in Lake Tana Sub-Basin. MSc Thesis. Arba Minch University, Ethiopia/International Water Management Institute, Addis Ababa, Ethiopia.

Ryszkowski, L., 1992. Energy and Material Flows across Boundaries in Agricultural Landscapes. Springer Nature, pp. 270–284. https://doi.org/10.1007/978-1-4612-2804-2_13.

Ryszkowski, L., Bartoszewicz, A., Kędziora, A., 1997. The Potential Role of Mid-field Forests as Buffer Zones Buffer Zones: Their Processes and Potential in Water Protection, pp. 171–191.

Ryszkowski, L., Życzyńska-Bałoniak, I., 1985. Homeostaza ekosystemów [Homeostasis of ecosystems]. Zesz. Probl. Postepow Nauk. Rol. (305), 105–125.

Ryszkowski, L., Kędziora, A., 2007. Modification of water flows and nitrogen fluxes by shelterbelts. Ecol. Eng. 29 (4), 388–400. https://doi.org/10.1016/j.ecoleng.2006.09.023.

Ryszkowski, L., Balazy, S., 1994. Research Centre for agricultural and forest environment Strategy for increasing countryside resistance to environmental threats. Functional Appraisal of Agricultural Landscape in Europe 9–18.

Tebrügge, F., Düring, R.A., 1999. Reducing tillage intensity — A review of results from a long-term study in Germany. Soil Till. Res. 53 (1), 15–28.

Uggla, H., Nożyński, A., 1962. Deflametr – przyrząd do badań nad erozją wietrzną [Deflameter: an apparatus for studying wind erosion]. Zeszyty Naukowe WSR w Olsztynie. 13 (24), 567–570.

Zalewski, M., 2000. Ecohydrology — The scientific background to use ecosystem properties as management tools toward sustainability of water resources. Ecol. Eng. 16 (1), 1–8.

Zalewski, M., Kiedrzyńska, E., Wagner, I., Izydorczyk, K., Boczek, J.M., Jurczak, T., Krauze, K., Frankiewicz, P., Godlewska, M., Wojtal-Frankiewicz, A., Łapińska, M., Urbaniak, M., Bednarek, A., Kaczkowski, Z., Gągała, I., Serwecińska, L., Szklarek, S., Włodarczyk-Marciniak, R., Font-Nájera, A., Mierzejewska, E., Połatyńska-Rudnicka, M., Belka, K., Jarosiewicz, P., 2021. Ecohydrology and adaptation to global change. Ecohydrol. Hydrobiol. 21 (3), 393–410. https://doi.org/10.1016/j.ecohyd.2021.08.001.

Ziemnicki, S., 1968. Melioracje przeciwerozyjne. Państwowe Wydawnictwo Rolnicze i Leśne.

Part II

Implementation

Chapter 12

Ecohydrological quantification of factors affecting nutrient loads to the Baltic Sea and sustainable management of its drainage basin

Edyta Kiedrzyńska[1,2], Marcin Kiedrzyński[3] and Maciej Zalewski[1]

[1]European Regional Centre for Ecohydrology of the Polish Academy of Sciences under the auspices of UNESCO, Łódź, Poland; [2]University of Lodz, UNESCO Chair on Ecohydrology and Applied Ecology, Łódź, Poland; [3]University of Lodz, Department of Biogeography, Paleoecology and Nature Conservation, Łódź, Poland

PHOTO CREDIT: M. ZALEWSKI.

Transdisciplinary Ecohydrology. https://doi.org/10.1016/B978-0-443-41596-8.00005-4

1. Introduction

The Baltic Sea, as the world's largest semienclosed brackish sea, is exceptionally sensitive to anthropogenic pressures and climate change. It receives freshwater inflows from nearly 200 rivers draining territories of nine coastal and six upstream countries. Over the latter half of the 20th century, continuous nutrient enrichment from these rivers has led to significant eutrophication and the widespread occurrence of toxic cyanobacterial blooms (Kiedrzyńska et al., 2014a).

These phenomena not only diminish the recreational and tourism value of the Baltic coastline but also pose serious threats to human and animal health (Aleksandrov, 2010; Diaz and Rosenberg, 2008; HELCOM, 2020; Lips et al., 2014). One of the most severe consequences of nutrient overenrichment is the degradation of benthic habitats and the decline of marine biodiversity, including fish stocks. Biodiversity preservation is crucial not only for ecological stability but also for ensuring long-term economic resilience in the region (Arheimer et al., 2012), aligning with the European Green Deal Strategy.

The intensity and seasonal dynamics of nutrient and pollutant inputs vary between years, influenced by both natural conditions and human land-use practices in the drainage basin (Kiedrzyńska et al., 2014b; Wulff et al., 2007). Nutrient inputs originate from both point and diffuse sources. Point sources encompass municipal wastewater treatment plants, industrial discharges, intensive livestock farming, and aquaculture facilities. Diffuse sources include nutrient runoff from agricultural lands, atmospheric deposition, stormwater overflows, scattered dwellings, and natural background losses (HELCOM, 2011, 2018, 2020).

These cumulative inputs—especially when combined with rising water temperatures—significantly affect the ecological balance of the sea (Conley et al., 2011; Eilola et al., 2012; Gustafsson et al., 2012). Warmer waters reduce oxygen solubility and accelerate organic matter decomposition, intensifying oxygen demand at the seabed and limiting oxygen renewal (Meier et al., 2012). Additionally, acidification lowers pH and reduces calcium carbonate saturation, further expanding hypoxic zones (Omstedt et al., 2012). Excessive nutrient inputs can disrupt ecological functioning by promoting filamentous algal blooms, altering species composition, and reducing nutrient retention (Conley et al., 2011; Rönnberg and Bonsdorff, 2004). According to HELCOM (2013), nearly 4% of marine species and around 25% of biotopes in the Baltic are at risk of extinction.

Future climate scenarios predict further warming of Baltic waters. When combined with sustained high nutrient loads, this will likely worsen eutrophication and reduce ecosystem service delivery. Since climate change cannot be fully reversed, reducing nutrient inputs across all contributing countries must become a top priority. In this context, effective nutrient reduction strategies and tools for quantifying nutrient transport are urgently needed—particularly in relation to catchment characteristics (Kiedrzyńska et al., 2014b).

Developing targeted and innovative management actions for rivers of various scales across the Baltic basin requires identifying the main socioecological pressures and the activities most responsible for nutrient emissions (Kiedrzyńska et al., 2014a). It is therefore essential to apply ecohydrological principles to assess the dominant catchment-level and socio-economic factors influencing nutrient fluxes to the sea and to address key questions, including:

1. Which anthropogenic pressures dominate nutrient exports in different countries?
2. How can nutrient contributions from different sources be measured and compared?
3. What opportunities does the EU Green Deal Strategy present for addressing eutrophication through ecohydrological Nature-based Solutions (EH NbS)?

How can the EH NbS framework be aligned with EU water policy instruments, such as those reviewed under the Water Framework Directive Fitness Check?

2. Characteristics of the Baltic Sea basin

The Baltic Sea, the world's largest semienclosed brackish water body linked to an ocean, is characterized by its low and relatively uniform salinity ranging from 7 to 9‰. It spans an area of 415,000 km^2 and holds approximately 22,000 km^3 of water (Köhn, 1998). Due to the absence of tides, water exchange is slow, resulting in a residence time of 25–40 years. The drainage basin that feeds the Baltic Sea is around four times larger than the sea itself (Voss et al., 2006) and is hydrologically structured into 117 distinct watersheds (Vogt et al., 2007). With inputs from nearly 200 rivers and annual precipitation, the basin receives 400–500 km^3 of freshwater each year (Köhn, 1998), which contributes significantly to the input of nutrients, trace metals, and micropollutants (Kiedrzyńska et al., 2014b).

The entire drainage area (Fig. 12.1A and B) extends over 1,720,270 km^2, with about 93% located within the HELCOM Contracting Parties—namely Poland, Russia, Sweden, Finland, Denmark, Lithuania, Germany, Latvia, and Estonia—and

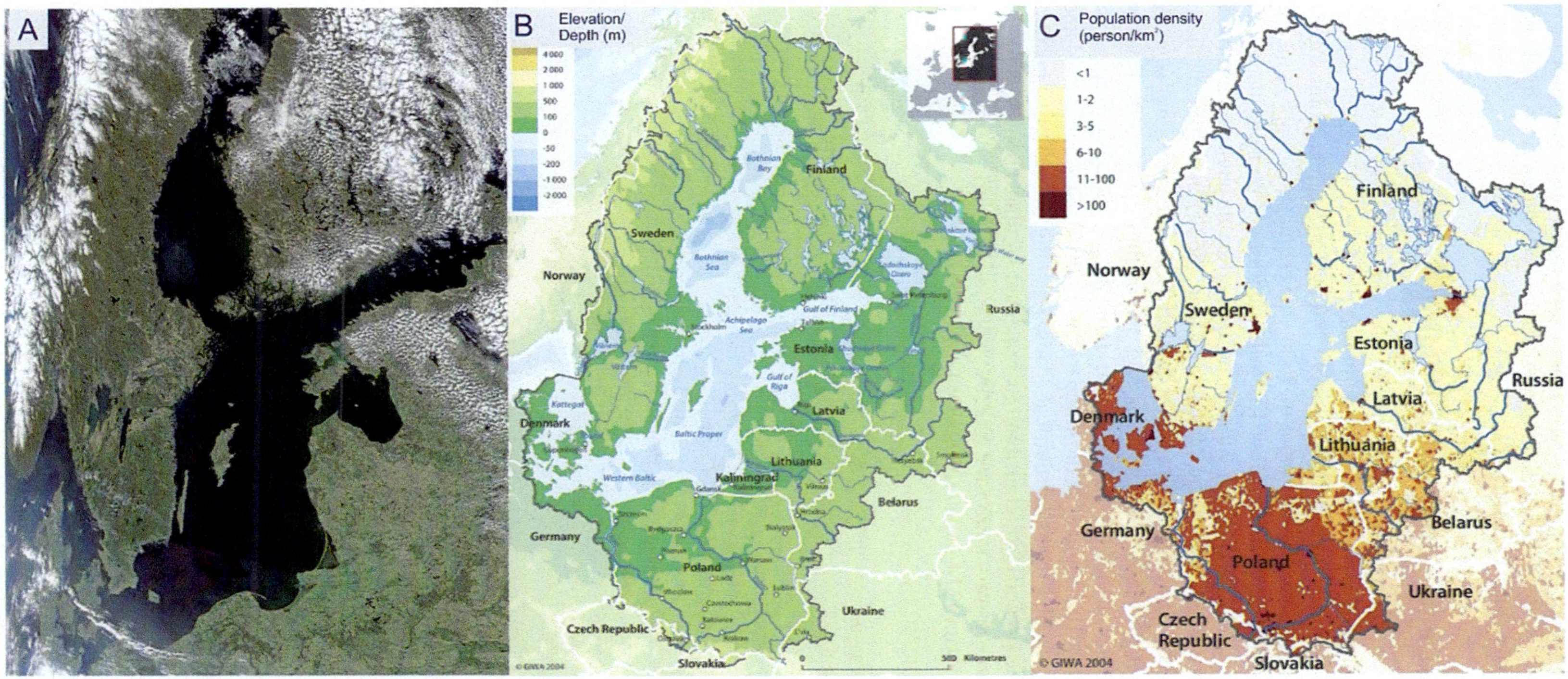

FIGURE 12.1 Baltic Sea drainage basin A, B, C. Baltic Sea drainage basin (A) NASA Satellite image (B) elevation/depth (m) and (C) density of human population (person/km). *(A) Source: SeaWiFS Project, NASA/Goddard Space Flight Center; (B and C) Modified from HELCOM, 2006. Development of tools for assessment of eutrophication in the Baltic Sea. In: Baltic Sea Environment Proceedings No. 104. Helsinki Commission.*

the remaining 7% lying in noncontracting countries such as Belarus, Ukraine, the Czech Republic, Slovakia, and Norway (HELCOM, 2011, 2013). The southern and southwestern parts of the basin are notable for their high population density and intensive agricultural land use (Fig. 12.1C), with agricultural land accounting for roughly 60%–70% of the basin area in Germany, Denmark, and Poland (Fig. 12.1B).

3. Main sources of nutrients supply of the Baltic Sea

The nutrient input model for phosphorus (P) and nitrogen (N) developed by Kiedrzyńska et al. (2014a) offers a comprehensive insight into the long-term dynamics and dominant pathways of nutrient delivery to the Baltic Sea from both atmospheric and waterborne sources, including riverine inflows and direct point-source discharges (Fig. 12.2). According to HELCOM (2011), the total nutrient input in 2006 was estimated at approximately 29,250 t of phosphorus and 807,000 t of nitrogen. As shown in the model (Fig. 12.2), the majority of phosphorus (about 97%, or 28,400 t) and a significant portion of nitrogen (74%, or 638,000 t) entered the Baltic Sea through waterborne transport. In contrast, atmospheric deposition contributed about 3% of the phosphorus load (852 t) and 26% of the nitrogen load (169,000 t) (Kiedrzyńska et al., 2014a). Within the waterborne pathway, around 92% of the phosphorus and 95% of the nitrogen loads were transported via rivers, with only 8% of phosphorus and 5% of nitrogen coming from point sources discharging directly into the sea (Kiedrzyńska et al., 2014b). In 2006, considered a hydrologically average year, the total riverine flow to the Baltic Sea was estimated at 16,230 $m^3\ s^{-1}$. By comparison, in 2010—a wetter year—the river flow increased by about 9%–17,698 $m^3\ s^{-1}$. This resulted in higher nutrient loads, reaching 38,300 t of phosphorus and 977,000 t of nitrogen. Despite this increase, the proportional contributions remained similar: about 95% of phosphorus (36,385 t) and 78% of nitrogen (762,060 t) came from waterborne sources, while atmospheric deposition accounted for 5% of phosphorus (1915 t) and 22% of nitrogen (214,940 t) (HELCOM, 2013).

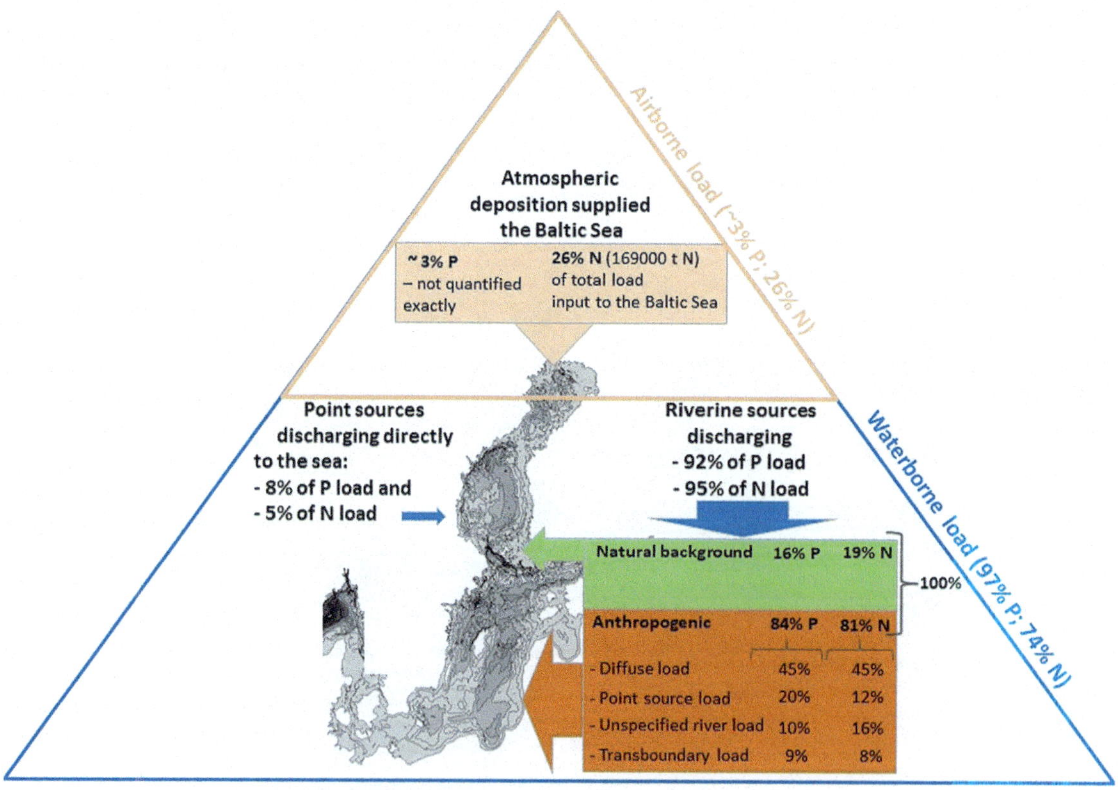

FIGURE 12.2 The phosphorus and nitrogen budget model. The budget model of phosphorus and nitrogen inputs from different sources to the Baltic Sea. *The model is based on data from HELCOM, Helsinki Commission, 2011. The Fifth Baltic Sea pollution load Compilation (PLC-5). Baltic Sea Environment Proceedings. 128; Modified from Kiedrzyńska, E., Jóźwik, A., Kiedrzyński, M., Zalewski, M., 2014a. Hierarchy of factors exerting an impact on nutrient load of the Baltic Sea and sustainable management of its drainage basin. Mar. Pollut. Bull. 88 (1–2), 162–173. https://doi.org/10.1016/j.marpolbul.2014.09.010.*

4. Hierarchy of factors causing nutrient load input

Multiple anthropogenic pressures—including expanding human populations, deforestation, intensified agriculture, and widespread changes in land use—collectively drive the degradation of marine environments (European Environmental Agency, 2006, 2008. "Intergovernmental Panel on Climate Change, 2007; SEPA, 2009). However, the diverse environmental and socio-economic contexts among the Baltic States make it challenging to establish a clear hierarchy of the drivers behind this degradation. To better address this issue, it is crucial to identify and rank the key anthropogenic nutrient sources and evaluate their relative contributions to phosphorus (P) and nitrogen (N) loads exported via rivers to the Baltic Sea.

A clear understanding of the interactions between different nutrient sources is fundamental for applying ecohydrological processes as effective management tools aimed at improving the sustainability of catchments in the Baltic Sea Region. This requires consideration of the functional relationships between hydrology, biocenoses, and nutrient dynamics across terrestrial, fluvial, and marine systems. Kiedrzyńska et al. (2014a) carried out an in-depth analysis exploring the correlation between total phosphorus (TP) loads and 23 socioecological variables. Among these, five parameters showed statistically significant correlations (Fig. 12.3). Notably, total nitrogen (TN) load exhibited the strongest correlation ($r = 0.883$). Positive associations were found with livestock density (particularly pigs), the share of the population connected to wastewater treatment plants (WWTPs), and areas classified as heavily impacted by human activity. In contrast, forest area displayed a negative correlation with TP load ($r = -0.544$) (Kiedrzyńska et al., 2014a; Fig. 12.3).

Total nitrogen (TN) loads entering the Baltic Sea are influenced by a combination of socioecological factors, among which six statistically significant contributors were identified by Kiedrzyńska et al. (2014b), as illustrated in Fig. 12.3. The strongest correlations were found with total phosphorus (TP) load, livestock density (pigs and cattle), the share of population connected to wastewater treatment plants (WWTPs), and the extent of agricultural land. In contrast, forest area showed a negative relationship with TN load. To illustrate the environmental burden posed by livestock production, it is useful to consider pollutant outputs per animal. A single cow produces, on average, 74.2 kg of nitrogen and 14.4 kg of phosphorus annually, while a pig produces 14.6 kg of nitrogen and 9.1 kg of phosphorus (Skorupski et al., 2012). When comparing these nutrient ratios, pigs contribute proportionally more phosphorus, and because they outnumber cattle in most countries, their overall impact is considerable. Large-scale animal farms are estimated to contribute approximately 60% of nitrogen and 40% of phosphorus loads to the Baltic Sea (Skorupski et al., 2012). Excessive manure production in such systems often exceeds the absorptive capacity of soils and ceases to function as fertilizer—becoming a form of wastewater and posing a serious environmental hazard (Ten Hoeve et al., 2014). According to Skorupski et al. (2012), one cow can produce as much nitrogen and phosphorus as 16.4 people, while one pig is equivalent to 1.9 people. Kiedrzyńska et al. (2014a,b) identified two factors—pig population and population served by WWTPs—that were positively associated with both TP and TN loads. Forest area, by contrast, was negatively associated with both nutrient loads. Additionally, TP loads were closely linked with areas under strong anthropogenic pressure, while TN loads showed higher correlation with the number of cattle and the extent of agricultural land. Forested landscapes, according to the analysis, are associated with lower nutrient loads (Kiedrzyńska et al., 2014b). This is in line with earlier findings by Knight et al. (1991), who noted the key role of trees in nutrient immobilization. Tree species also influence nutrient cycling through differences in litter and throughfall chemistry (Carnol and Bazgir, 2013; Skłodowski et al., 2014; Yin et al., 2014). For example, a study in

Correlation between TP load and parameters

Parameter	N	R	*p*
Total waterborne load of TN	**9**	**0.883**	***0.0016***
Pigs	**9**	**0.583**	***0.0992***
Forest area	**9**	**-0.544**	***0.1300***
Population served by WWTPs	**9**	**0.533**	***0.1392***
Heavy environmental impact area	**9**	**0.522**	***0.1497***
Rural population	9	0.517	*0.1544*
Population in catchment	9	0.500	*0.1705*
Municipal load	9	0.467	*0.2054*
Cattle	9	0.450	*0.2242*
Agriculture area	9	0.443	*0.2328*
Number of WWTPs	9	0.400	*0.2861*

Correlation between TN load and parameters

Parameter	N	R	*p*
Total waterborne load of TP	**9**	**0.883**	***0.0016***
Forest area	**9**	**-0.611**	***0.0805***
Pigs	**9**	**0.600**	***0.0876***
Cattle	**9**	**0.567**	***0.1116***
Population served by WWTPs	**9**	**0.567**	***0.1116***
Agriculture area	**9**	**0.536**	***0.1366***
Fertilizer consumption – Organic P	9	0.524	*0.1827*
Rural population	9	0.483	*0.1875*
Population in catchment	9	0.450	*0.2242*
Number of WWTPs	9	0.433	*0.2439*
Fertilizer consumption – Organic N	8	0.452	*0.2604*

FIGURE 12.3 Correlation coefficients between total phosphorus and nitrogen loads and selected socioecological factors. Correlation coefficients between the total phosphorus (TP) load, and also total nitrogen (TN) load, and the mentioned parameters according to countries of the Baltic Sea region (Rank Spearman Test), respectively. Bold values with $P < .15$. *From: Kiedrzyńska E., Jóźwik A., Kiedrzyński M., Zalewski M., Hierarchy of factors exerting an impact on nutrient load of the Baltic Sea and sustainable management of its drainage basin Mar. Pollut. Bull. 2014a 162-173.*

Norway reported nitrogen returns of around 50 kg ha^{-1} year^{-1} under alder, birch, willow, and rowan, 33 kg ha^{-1} year^{-1} under spruce, and about 20 kg ha^{-1} year^{-1} under beech and oak.

Losses of nutrients from actively growing forests remain close to natural background levels, while losses from intensively managed areas are significantly higher (Knight et al., 1991). In recent decades, forest ecosystems in Europe have been affected by nutrient enrichment due to the abandonment of traditional practices and increased atmospheric deposition. Consequently, artificial fertilization is now rarely practiced in forestry—mainly in Finland, Norway, and Sweden (Røsberg et al., 2006). Additionally, better forest management practices, such as buffer strips and modified clear-cutting, have helped reduce nutrient losses (HELCOM, 2011). Agricultural land per unit area is more strongly correlated with nitrogen load than with phosphorus load, consistent with findings from other studies (e.g., Arheimer et al., 2012; Hong et al., 2012). Agriculture accounts for 70%–90% of diffuse nitrogen emissions and 60%–80% of phosphorus, which explains its slightly lower impact on phosphorus loads. Nitrogen emissions depend on multiple factors, including fertilizer rate and type, crop species, soil organic matter, texture, drainage, and pH (Bouwman et al., 2002). In contrast, phosphorus emissions are more complex to estimate due to various interacting factors—such as soil moisture, temperature, pH, and redox conditions (Shen et al., 2011). High phosphorus levels in soils increase the risk of losses, as noted by McDowell et al. (2003).

Reducing nutrient losses from agriculture is more challenging than addressing point sources, due to nutrient storage and retention in soils and water bodies. Thus, local reductions may not yield immediate effects at the Baltic Sea scale. The interdependencies between geographical and socioenvironmental variables are crucial. Livestock density and the extent of population covered by WWTPs appear to exert the greatest influence on nutrient loads. Strong correlations with agricultural areas and inverse correlations with forest cover highlight the role of land use change in modifying the natural nutrient balance. Reactivation of the phosphorus cycle—through intensified land use—has resulted in higher particulate phosphorus runoff into aquatic systems. Nitrogen, however, cycles primarily via the atmosphere, which explains its stronger association with industrial and heavily impacted areas. As shown by Hong et al. (2012), there is a distinct north–south gradient in land use and fertilizer consumption across the Baltic region. Despite similar spatial patterns, nitrogen fluxes are generally much greater than phosphorus fluxes. Moreover, changes in one nutrient affect the other due to their interconnected roles in ecosystem dynamics. For instance, reducing fertilizer use lowers both crop production and livestock numbers, which in turn limits manure output, atmospheric deposition, and ultimately, nutrient flow into the Baltic Sea (Hong et al., 2012).

5. Clustering countries by factors correlated to nutrient input to the Baltic Sea

Kiedrzyńska et al. (2014a) conducted a k-means clustering analysis using 14 socioecological indicators from national subcatchments, which allowed the identification of three distinct country groups (clusters) sharing similar characteristics (Fig. 12.4). The results revealed that Cluster 1 comprised countries with high forest cover ($\geq$50 ha/km^2) and moderate agricultural land use (7–32 ha/km^2), whereas Cluster 3 included states with low forest cover (<20 ha/km^2) and intensive agriculture (>60 ha/km^2) (Kiedrzyńska et al., 2014b). Significant variation in demographic and farming pressures was also observed across the clusters. For example, the population connected to wastewater treatment systems was below 25 inhabitants/km^2 in Cluster 1 but exceeded 100 inhabitants/km^2 in Cluster 3. Livestock density followed a similar trend: Cluster 3 exhibited the highest numbers of pigs and cattle per km^2, while Cluster 1 had the lowest (Fig. 12.4). Poland was classified into Cluster 2, characterized by intermediate values: 75 inhabitants/km^2 served by WWTPs, 60 pigs/km^2, and 17 cattle/km^2. In contrast, Denmark and Germany (Cluster 3) showed extremely high pig densities—429 and 73 pigs/km^2, respectively—and high cattle densities—61 and 49 heads/km^2 (Kiedrzyńska et al., 2014a).

The average annual loads of total phosphorus (TP) and total nitrogen (TN) per km^2 also increased across the clusters, from Cluster 1 to Cluster 3 (Fig. 12.5), reflecting cumulative anthropogenic pressure in these catchments. Nutrient load assessments often rely on national-level data due to the availability of consistent monitoring and reporting. This makes country-based evaluations more feasible for practical implementation compared to those based solely on hydrological subcatchments (Kiedrzyńska et al., 2014a). The spatial clustering approach using data normalized per unit area proved effective for recognizing transboundary patterns in nutrient loads and catchment characteristics. These findings align with those of Hong et al. (2012), who observed a clear latitudinal gradient across the Baltic Sea basin—northern regions tend to be less densely populated and more forested, while southern parts are more intensively farmed and exposed to over-fertilization. Clustering thus offers a useful framework for regional assessments of pressures on aquatic ecosystems.

Although Poland is often identified as a significant source of nutrient pollution to the Baltic Sea (Eriksson et al., 2007; Hong et al., 2012), the cluster-based approach shows that its contribution, when compared to neighboring countries within the same group, is less dominant (Kiedrzyńska et al., 2014a). The clustering results reflect biogeographic land use

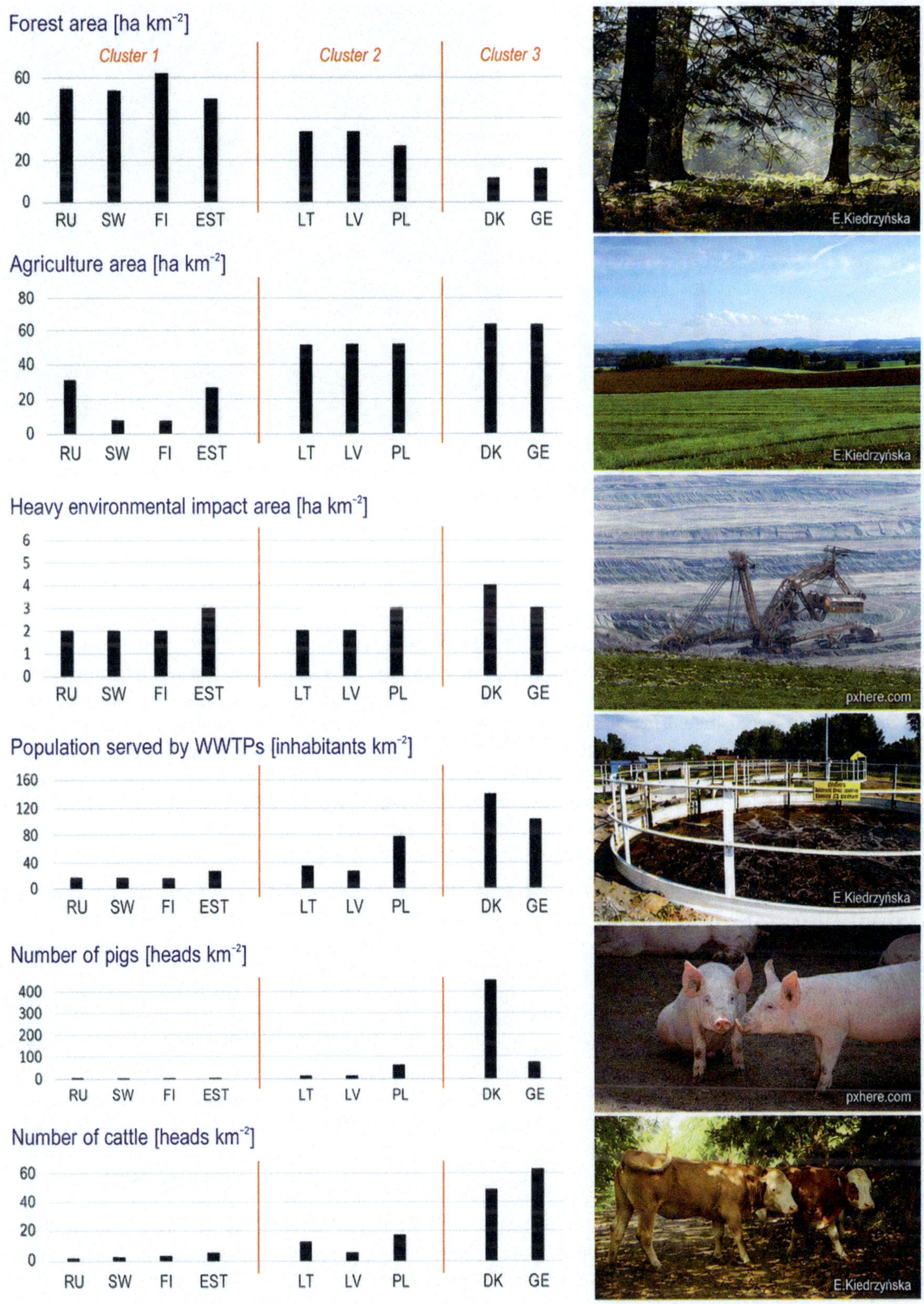

FIGURE 12.4 General characteristics of groups of countries from the Baltic sea catchment. General characteristics of groups of countries (clusters) from the Baltic sea catchment. Clusters were identified according to the k-means clustering method using 14 socioecological parameters. The figure presents the six catchment parameters most closely correlated to TP and TN loads. *Modified from Kiedrzyńska, E., Jóźwik, A., Kiedrzyński, M., Zalewski, M., 2014a. Hierarchy of factors exerting an impact on nutrient load of the Baltic Sea and sustainable management of its drainage basin. Mar. Pollut. Bull. 88 (1–2), 162–173. https://doi.org/10.1016/j.marpolbul.2014.09.010.*

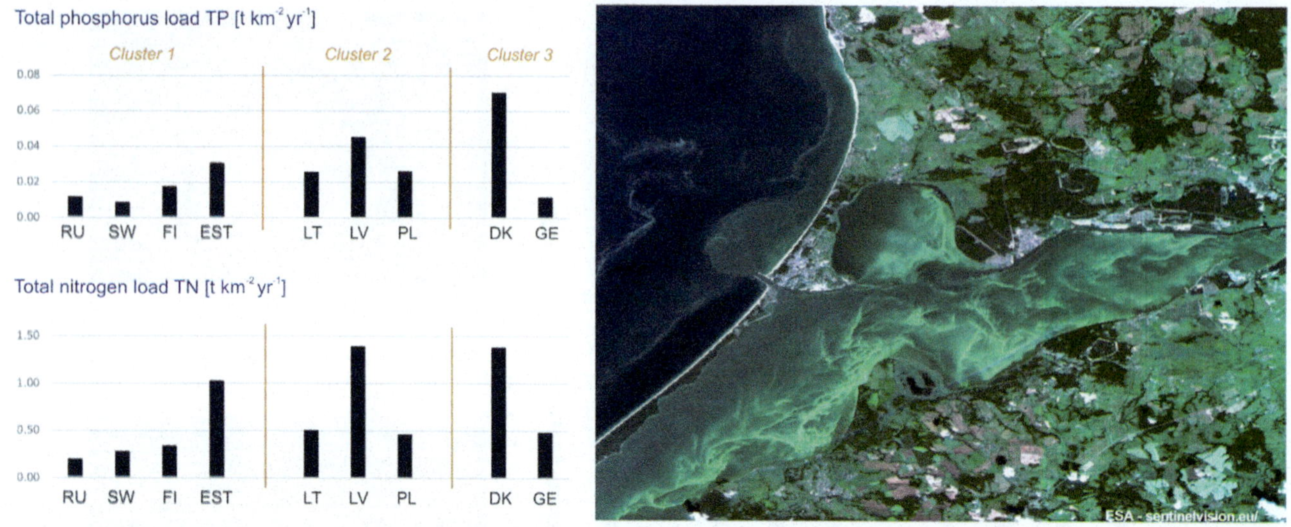

FIGURE 12.5 Annual TP and TN loads by country clusters in the Baltic Sea basin. Total phosphorus (TP) and total nitrogen (TN) loads per km^2 per year in groups of countries (clusters) from the Baltic Sea catchment. Clusters were identified according to the k-means clustering method using 14 socioecological parameters (Kiedrzyńska et al., 2014a). Sentinel image: Massive algal blooms in the Vistula lagoon & the Gdansk Bay on July 30, 2018. Source: Kiedrzyńska, 2014a. Sentinel Vision Portal (VisioTerra), based on Copernicus Sentinel data. Used with permission.

patterns—Central Europe is predominantly agricultural, whereas Southern and Central Fennoscandia are dominated by forests (Kiedrzyńska et al., 2014b). Overall, intensified land cultivation (Eriksson et al., 2007; Kremser and Schnug, 2002) and high-density livestock farming (Hoeve et al., 2014) were identified as key drivers of nutrient inputs to the Baltic, especially from southern agricultural landscapes, while the sparsely populated, forest-rich northern basins—such as those in Finland and Sweden—contributed less to the nutrient load (Gustafsson et al., 2012), Fig. 12.5. In several Baltic Sea countries, a considerable share of nutrient inputs comes from rivers crossing national borders. A prominent example is the Neman River, which originates in Belarus and discharges into the Baltic Proper through Lithuania. Another key transboundary river is the Daugava, which traverses Russia and Belarus before reaching the Gulf of Riga in Latvia (HELCOM, 2013, 2018; *The United Nations Economic Commission for Europe*, 2014).

6. How to reverse the degradation of the Baltic Sea? – Enhancement of catchment sustainability potential WBSR+CE

The Baltic Sea, recognized as one of the largest brackish water ecosystems in the world (Lundberg, 2013), is facing serious environmental degradation. One of the central questions today is how to effectively reverse this process. A range of EU environmental directives plays a crucial role in shaping both inland and coastal water quality. The Nitrates Directive (ND 91/676/EEC) aims to reduce nitrogen runoff from agriculture, while the Water Framework Directive (WFD 2000/60/EC) mandates the achievement of good ecological status in all waters via integrated catchment-based management. The Urban Wastewater Treatment Directive (UWWTD 91/271/EEC) addresses the collection and treatment of wastewater in urban areas. Meanwhile, the Marine Strategy Framework Directive (MSFD 2008/56/EC), together with its 2017 annex (2017/845), complements the WFD by extending environmental objectives into marine waters. Adopted in 2009, the EU Strategy for the Baltic Sea Region (EUSBSR) was the first of the EU's macroregional strategies. It fosters coordinated actions among eight EU countries (Poland, Sweden, Denmark, Germany, Finland, Estonia, Latvia, and Lithuania), and also encourages cooperation with nonEU countries such as Belarus, Russia, Norway, and Iceland. Its accompanying Action Plan, regularly updated, addresses key regional challenges including climate change, pandemics, demographic shifts, and migration, in alignment with the 2021–27 EU Multiannual Financial Framework (EUSBSR, 2020).

HELCOM's Baltic Sea Action Plan (BSAP), launched in 2007, is another vital strategic framework for the region. It sets out goals and measures to achieve good ecological status for the Baltic Sea, with eutrophication being one of its four main focus areas. The original target year for reaching BSAP goals was 2021. However, while many reductions in nutrient inputs were achieved (14% for nitrogen and 24% for phosphorus), the targets of an 18% nitrogen and 42% phosphorus reduction were not fully met (BSAP, 2007, 2020). Consequently, a revised BSAP was adopted in 2021, reaffirming the vision of a Baltic Sea with a balanced, healthy ecosystem that supports sustainable human activity. Under the revised BSAP, the maximum allowable inputs (MAI) for nutrients—thresholds for nitrogen and phosphorus loads that still permit the achievement of good environmental status—were set at 792,209 tons of nitrogen and 21,716 tons of phosphorus (BSAP, 2020). Although many indicators show gradual improvement in eutrophication status, about 96% of the Baltic Sea area (including 86% of coastal waters) remains below good status. The Baltic Proper subbasin continues to face the greatest reduction demands (BSAP, 2020). Progress so far has largely come from reducing point source discharges, such as those from municipal and industrial wastewater treatment. By contrast, diffuse sources—particularly from agriculture—remain largely unaddressed despite their major contribution (around 35% of riverine inputs) (Izydorczyk et al., 2019; Kiedrzyńska et al., 2014b). Future nutrient reduction efforts should therefore target both agriculture and point sources in the upper catchment areas, including those outside the HELCOM region (Kiedrzyńska et al., 2014b).

Efforts to reduce nutrient pollution from point sources rely heavily on regulatory enforcement and monitoring, while limiting diffuse pollution requires the wide-scale adoption of sustainable farming practices (Kiedrzyńska et al., 2014b; SEPA, 2008, 2009; Zalewski, 2009). Changes in EU agricultural policy, such as the decoupling of subsidies from production under the Common Agricultural Policy, are expected to further reduce nutrient emissions. Nevertheless, poor manure management remains an issue across the Baltic region, with excessive animal waste turning into pollution rather than fertilizer. In terms of airborne nitrogen, significant reductions have been made in energy and transport sectors, but ammonia emissions from agriculture persist and may even be rising (BSAP, 2020).

Another challenge lies in nutrient-rich wastewater discharged by inefficient or overloaded WWTPs, particularly during heavy rainfalls. These plants, as point sources, significantly impact freshwater systems and, ultimately, the Baltic Sea (Harnisz et al., 2020; Kiedrzyński et al., 2014b, 2017; Serwecińska et al., 2021). To mitigate this, sustainable wastewater management is needed not only in large cities but also in small towns across the basin.

According to HELCOM (2020), in 2017, the total nitrogen (TN) load from municipal WWTPs reported in the PLC database was 69,800 tons (Table 12.1). Poland (28%), Russia (23%), and Sweden (23%) were the largest contributors. Per capita nitrogen discharge was highest in Russia, Finland, and Sweden, while Germany and Denmark demonstrated high removal efficiency, reflected in low nitrogen concentrations per person.

For phosphorus (TP), the 2017 reported total was 4220 tons. Over half came from Poland, the most populous country in the region. Finland, Germany, Sweden, and Estonia had the lowest per capita TP discharges, while Russia showed the highest, pointing to substantial potential for improvement (HELCOM, 2020). If all municipal WWTPs complied with HELCOM Recommendation 28E/5, nitrogen discharges to inland and marine waters could be cut by 13,600 tons, with the

TABLE 12.1 Total nitrogen (TN) and total phosphorus (TP) load of municipal WWTPs in 2017.

Country	Population	Population connectivity	TN load	%	Concentration of TN	TP load	%	Concentration of TP
	Million	Million	t		mg/inhabitant	t		mg/inhabitant
Germany	2.7	2.484	1419	2	571	64	2	24
Denmark	5.1	4.335	3010	4	694	305	7	60
Estonia	1.3	1.066	746	1	700	36	1	28
Finland	5.5	4.51	9566	14	2121	124	3	23
Lithuania	2.9	2.32	1986	3	856	151	4	52
Latvia	2.2	1.672	1320	2	789	157	4	71
Poland	38.5	27.72	19,903	28	718	2180	52	57
Russia	8.4	6.972	16,031	23	2299	963	23	115
Sweden	9.5	8.265	15,855	23	1918	243	6	26
Sum	76.1	59.344	**69,836**	100		**4223**	100	

Source: HELCOM, 2020. Input of Nutrients: Potential to Reduce Input from Point Sources. HELCOM ACTION project. Eutrophication. Baltic Marine Environment Protection Commission.

greatest potential in Russia, Finland, and Sweden. For phosphorus, emissions could be reduced by 2050 tons, particularly in Poland and Russia (HELCOM, 2020). In Poland, for example, phosphorus discharges are more heavily weighted toward point sources (42%) than agriculture (34%), while nitrogen is more influenced by agriculture (45%) than point sources (31%) (HELCOM, 2018). Natural background contributions account for 16% of nitrogen and 18% of phosphorus inputs, while atmospheric deposition plays a comparatively minor role (3% and 1%, respectively) (HELCOM, 2018).

Reducing the inflow of pollutants, particularly nutrients, from both diffuse and point sources to the Baltic Sea requires a deep understanding of the intricate interactions among ecological, hydrological, and geochemical processes. Effective mitigation strategies must adopt a holistic perspective that also considers how pollutants are transferred and transformed across different environmental compartments. Integrating this knowledge into practice creates opportunities for implementing Ecohydrology-based Nature-Based Solutions (EH NBS), which are cost-effective, environmentally sustainable, and socially accepted (Zalewski, 2020). These approaches should be applied in synergy with hydrotechnical infrastructure (Kiedrzyńska et al., 2017). The concept of harnessing natural processes for ecosystem restoration has been formalized as *ecological engineering* (Mitsch et al., 2008).

Adopting such an integrated and synergistic approach, complemented by basin-scale regulatory measures, can significantly reduce nutrient and pollutant loads to aquatic systems. To support this, a five-step strategy grounded in Ecohydrology, ecological engineering, and biotechnological principles was proposed by Kiedrzyńska et al. (2014a):

- **Step I: Enhancing landscape heterogeneity in small catchments**, especially along headwater streams where diffuse pollution is most intense. Bowes et al. (2003) observed that in lowland river sections, phosphorus is predominantly transported in particulate form, and phosphorus export tends to increase downstream. They estimated that 85% of the total phosphorus load originated in lowland zones, with the lowest exports occurring in summer. Promoting mosaic vegetation structures, such as native forest and rush assemblages, significantly enhances nutrient retention capacity (Kiedrzyńska et al., 2008b, 2014b; Skłodowski et al., 2014).
- **Step II: Increasing nutrient retention potential in small watercourses** by boosting ecological productivity and biodiversity. Creating complex land–water ecotones improve natural self-purification processes (Izydorczyk et al., 2019; Zalewski, 2013, 2015).
- **Step III: Enhancing the retention capacity of river floodplains**. According to Skłodowski et al. (2014), young willow stands in floodplains can retain an average of 14.44 kg P ha^{-1}, while meadows and rushes in lowland floodplains may accumulate over 10 kg P ha^{-1} yr^{-1} (Kiedrzyńska et al., 2008b). This could potentially reduce phosphorus inputs to the Baltic by more than 10%, with further improvements possible via sequential biomass harvesting for silage or bioenergy. Floodplains also trap sediment (Altinakar et al., 2006; Magnuszewski et al., 2007) and associated nutrients—fine sediment retention was estimated at 0.63 t ha^{-1} yr^{-1}, and phosphorus retention at 1.81 kg ha^{-1} yr^{-1} (Kiedrzyńska et al., 2008a).
- **Step IV: Constructing reservoirs that preserve river continuum functions.** These reservoirs act as sinks for nutrients and micropollutants due to reduced flow velocity and enhanced flocculation and sedimentation (Urbaniak et al., 2012, 2014a,b). For example, the Sulejów Reservoir achieved reductions of 45% in suspended particulate matter, 28% in total phosphorus, 34% in total nitrogen, and 29% in dioxin-like PCBs (Urbaniak et al., 2012). Additional declines in toxic compounds such as PCDDs/PCDFs and dl-PCBs were observed along the reservoir gradient, attributed to sedimentation and subsequent biological or photochemical degradation. Although large reservoirs can disrupt river systems (Miller, 2005; Ward and Stanford, 1983), and may foster eutrophication and harmful algal blooms (Howarth, 2008; Kiedrzyńska et al., 2014b), their role in flood protection and water storage is increasingly vital under climate change.
- **Step V: Enhancing WWTP efficiency using Hybrid Sequential Biofiltration Systems (HSBS).** This ecohydrological biotechnology integrates hydrotechnology with nature-based filtration systems (Kiedrzyńska et al., 2017). Annual removal rates observed in HSBS systems include: 25.3 kg TP (18%), 37.1 kg SRP (26%), 191.3 kg TN (17%), 178 kg NO_3^-–N (19%), 4.5 kg NO_2^-–N (33%), 8.8 kg NH_4^+–N (7%), and 13.6 g PCB Eq. (49%). The most effective component was the macrophyte bed, where 83% of TP and 72% of NO_3^-–N were removed. Due to their low cost and high efficiency, these systems are particularly well suited to small WWTPs to reduce point-source pollution and improve river and coastal water quality.

These examples demonstrate that major improvements in the ecological condition of the Baltic Sea are achievable without imposing high economic burdens. While notable progress has been made through the implementation of EU directives like the WFD, further efforts are required. Broad-scale adoption of low-cost, efficient EH NBS for nutrient reduction should become a cornerstone of regional environmental policy.

Achieving this requires a strong foundation of scientific knowledge, strategic communication, and the creation of multi-stakeholder learning platforms. Policymakers across the Baltic region must be made aware of how climate change, land use, and river basin management are interconnected (Eisenack et al., 2007). Moreover, clear, accessible, and consistent information linking environmental problems with actionable solutions is urgently needed (Lorenzoni and Pidgeon, 2006; Moser and Tribbia, 2006). These efforts should be integrated into multilevel governance structures—from local to international—and contribute to building a collective understanding of the need for nutrient reduction in the context of climate change adaptation.

Acknowledgments

The analyses were partially conducted within the Polish Ministry of Science and Higher Education project—NN305 365738 'Analysis of point sources pollution of nutrients, dioxins and dioxin-like compounds in the Pilica River catchment and draw up of reclamation methods.

AI disclosure

During the preparation of this work the author(s) used OpenAI's ChatGPT in order to generate contenmt. This document includes content generated and/or edited with the assistance of OpenAI's ChatGPT, used to support language refinement and structural improvements. All final content was reviewed and verified by the author. After using this tool/service, the author(s) reviewed and edited the content as needed and take (s) full responsibility for the content of the publication.

References

Aleksandrov, S.V., 2010. Biological production and eutrophication of Baltic Sea estuarine ecosystems: the Curonian and Vistula lagoons. Mar. Pollut. Bull. 61 (4–6), 205–210. https://doi.org/10.1016/j.marpolbul.2010.02.015.

Altinakar, M., Kiedrzyńska, E., Magnuszewski, A., 2006. Modelling of inundation patterns on the Pilica River flood plain, Poland. IAHS-AISH Publ. 308, 579–585.

Arheimer, B., Dahné, J., Donnelly, C., 2012. Climate change impact on riverine nutrient load and land-based remedial measures of the Baltic Sea Action plan. Ambio 41 (6), 600–612. https://doi.org/10.1007/s13280-012-0323-0.

Bouwman, A.F., Boumans, L.J.M., Batjes, N.H., 2002. Emissions of N_2O and NO from fertilized fields: Summary of available measurement data. Global Biogeochem. Cy. 16 (4), 1–13. https://doi.org/10.1029/2001gb001811.

Bowes, M.J., House, W.A., Hodgkinson, R.A., 2003. Phosphorus dynamics along river systems—interaction of source, biota, and dissolved inorganic phosphorus concentration. J. Hydrol. 281 (1–4), 70–88.

BSAP, 2020. First Draft of the Updated BSAP – Baltic Sea Action Plan. HOD 59-2020. Baltic Marine Environment Protection Commission.

BSAP, HELCOM Baltic Sea Action Plan Adopted by Poland and Other Baltic States at the Ministerial Conference Held in Krakow on 15.11.2007.

Carnol, M., Bazgir, M., 2013. Nutrient return to the forest floor through litter and throughfall under 7 forest species after conversion from Norway spruce. For. Ecol. Manag. 309, 66–75. https://doi.org/10.1016/j.foreco.2013.04.008.

Conley, D.J., Carstensen, J., Aigars, J., Axe, P., Bonsdorff, E., Eremina, T., Haahti, B.M., Humborg, C., Jonsson, P., Kotta, J., Lännegren, C., Larsson, U., Maximov, A., Medina, M.R., Lysiak-Pastuszak, E., Remeikaitė-Nikienė, N., Walve, J., Wilhelms, S., Zillén, L., 2011. Hypoxia is increasing in the coastal zone of the Baltic Sea. Environ. Sci. Technol. 45 (16), 6777–6783. https://doi.org/10.1021/es201212r.

Diaz, R.J., Rosenberg, R., 2008. Spreading dead zones and consequences for marine ecosystems. Science 321 (5891), 926–929. https://doi.org/10.1126/science.1156401.

Eilola, K., Rosell, E.A., Dieterich, C., Fransner, F., Höglund, A., Meier, H.E.M., 2012. Modeling nutrient transports and exchanges of nutrients between shallow regions and the open Baltic Sea in present and future climate. Ambio 41 (6), 586–599. https://doi.org/10.1007/s13280-012-0322-1.

Eisenack, K., Tekken, V., Kropp, J.P., 2007. Stakeholders perception of climate change in the Baltic Sea region. Coastline Reports 8, 245–255.

Eriksson, H., Pastuszak, M., Löfgren, S., Mörth, C.M., Humborg, C., 2007. Nitrogen budgets of the Polish agriculture 1960–2000: implications for riverine nitrogen loads to the Baltic Sea from transitional countries. Biogeochemistry 85 (2), 153–168. https://doi.org/10.1007/s10533-007-9126-y.

European Environmental Agency, 2008. Impacts of Europe's Changing Climate: 2008 Indicator Based Assessment, vol 246.

European Environmental Agency, 2006. The Changing Faces of Europe's Coastal Areas, vol 107.

EUSBSR, 2020. EU Strategy for the Baltic Sea Region Action Plan Agreed by the NC Group.

Gustafsson, B.G., Schenk, F., Blenckner, T., Eilola, K., Meier, H.E.M., Müller-Karulis, B., Neumann, T., Ruoho-Airola, T., Savchuk, O.P., Zorita, E., 2012. Reconstructing the development of Baltic Sea eutrophication 1850–2006. Ambio 41 (6), 534–548. https://doi.org/10.1007/s13280-012-0318-x.

Harnisz, M., Kiedrzyńska, E., Kiedrzyński, M., Korzeniewska, E., Czatzkowska, M., Koniuszewska, I., Jóźwik, A., Szklarek, S., Niestępski, S., Zalewski, M., 2020. The impact of WWTP size and sampling season on the prevalence of antibiotic resistance genes in wastewater and the river system. Sci. Total Environ. 741, 140466. https://doi.org/10.1016/j.scitotenv.2020.140466.

HELCOM, Helsinki Commission, 2013. Review of the Fifth Baltic Sea pollution load Compilation for the 2013 HELCOM Ministerial Meeting. Baltic Sea Environment Proceedings.

HELCOM, Helsinki Commission, 2011. The Fifth Baltic Sea pollution load Compilation (PLC-5). Baltic Sea Environment Proceedings 128.

HELCOM, 2006. Development of tools for assessment of eutrophication in the Baltic Sea. In: Baltic Sea Environment Proceedings No. 104. Helsinki Commission.

HELCOM, 2018. Input of Nutrients by the Seven Biggest Rivers in the Baltic Sea Region. Baltic Sea Environment Proceedings.

HELCOM, 2020. Input of Nutrients: Potential to Reduce Input from Point Sources. HELCOM ACTION project. Eutrophication. Baltic Marine Environment Protection Commission.

Hong, B., Swaney, D.P., Mörth, C.M., Smedberg, E., Eriksson Hägg, H., Humborg, C., Howarth, R.W., Bouraoui, F., 2012. Evaluating regional variation of net anthropogenic nitrogen and phosphorus inputs (NANI/NAPI), major drivers, nutrient retention pattern and management implications in the multinational areas of Baltic Sea basin. Ecol. Model. 227, 117–135. https://doi.org/10.1016/j.ecolmodel.2011.12.002.

Howarth, R.W., 2008. Coastal nitrogen pollution: a review of sources and trends globally and regionally. Harmful Algae 8 (1), 14–20. https://doi.org/10.1016/j.hal.2008.08.015.

Intergovernmental Panel on Climate Change. Mitigation of Climate Change. In: Contribution of Working Group III to the Fourth Assessment Report of the IPCC 3 23, 2007. Cambridge University Press.

Izydorczyk, K., Piniewski, M., Krauze, K., Courseau, L., Czyż, P., Giełczewski, M., Kardel, I., Marcinkowski, P., Szuwart, M., Zalewski, M., Frątczak, W., 2019. The ecohydrological approach, SWAT modelling, and multi-stakeholder engagement – a system solution to diffuse pollution in the Pilica basin, Poland. J. Environ. Manag. 248, 109329. https://doi.org/10.1016/j.jenvman.2019.109329.

Kiedrzyńska, E., Kiedrzyński, M., Zalewski, M., 2008a. Flood sediment deposition and phosphorus retention in a lowland river floodplain: impact on water quality of a reservoir, Sulejów, Poland. Ecohydrol. Hydrobiol. 8 (2–4), 281–289. https://doi.org/10.2478/v10104-009-0022-z.

Kiedrzyńska, E., Wagner, I., Zalewski, M., 2008b. Quantification of phosphorus retention efficiency by floodplain vegetation and a management strategy for a eutrophic reservoir restoration. Ecol. Eng. 33 (1), 15–25. https://doi.org/10.1016/j.ecoleng.2007.10.010.

Kiedrzyńska, E., Jóźwik, A., Kiedrzyński, M., Zalewski, M., 2014a. Hierarchy of factors exerting an impact on nutrient load of the Baltic Sea and sustainable management of its drainage basin. Mar. Pollut. Bull. 88 (1–2), 162–173. https://doi.org/10.1016/j.marpolbul.2014.09.010.

Kiedrzyńska, E., Kiedrzyński, M., Urbaniak, M., Magnuszewski, A., Skłodowski, M., Wyrwicka, A., Zalewski, M., 2014b. Point sources of nutrient pollution in the lowland river catchment in the context of the Baltic Sea eutrophication. Ecol. Eng. 70, 337–348. https://doi.org/10.1016/j.ecoleng.2014.06.010.

Kiedrzyńska, E., Urbaniak, M., Kiedrzyński, M., Jóźwik, A., Bednarek, A., Gągała, I., Zalewski, M., 2017. The use of a hybrid Sequential Biofiltration System for the improvement of nutrient removal and PCB control in municipal wastewater. Sci. Rep. 7 (1). https://doi.org/10.1038/s41598-017-05555-y.

Knight, D.H., Yavitt, J.B., Joyce, G.D., 1991. Water and nitrogen outflow from lodgepole pine forest after two levels of tree mortality. For. Ecol. Manag. 46 (3–4), 215–225. https://doi.org/10.1016/0378-1127(91)90233-L.

Köhn, J., 1998. An approach to Baltic Sea sustainability. Ecol. Econ. 27 (1), 13–28. https://doi.org/10.1016/s0921-8009(97)00099-2.

Kremser, U., Schnug, E., 2002. Impact of fertilizers on aquatic ecosystems and protection of water bodies from mineral nutrients. Landbauforschung Volkenrode 52 (2), 81–90.

Lips, I., Rünk, N., Kikas, V., Meerits, A., Lips, U., 2014. High-resolution dynamics of the spring bloom in the Gulf of Finland of the Baltic Sea. J. Mar. Syst. 129, 135–149. https://doi.org/10.1016/j.jmarsys.2013.06.002.

Lorenzoni, I., Pidgeon, N.F., 2006. Public views on climate change: European and USA perspectives. Clim. Change 77, 73–95. https://doi.org/10.1007/s10584-006-9072-z.

Lundberg, C., 2013. Eutrophication, risk management and sustainability. The perceptions of different stakeholders in the northern Baltic Sea. Mar. Pollut. Bull. 66 (1–2), 143–150. https://doi.org/10.1016/j.marpolbul.2012.09.031.

Magnuszewski, A., Kiedrzyńska, E., Wagner, I., Zalewski, M., 2007. Proceedings of the international Conference W3M "Wetlands: modelling, monitoring, management". In: Numerical Modeling of Material Fluxes on the Floodplain Wetland of the Pilica River, Poland, pp. 205–210.

McDowell, W.H., Likens, G.E., Aber, J.D., 2003. Biogeochemistry: An Analysis of Global Change, second ed. Academic Press.

Meier, H.E.M., Müller-Karulis, B., Andersson, H.C., Dieterich, C., Eilola, K., Gustafsson, B.G., Höglund, A., Hordoir, R., Kuznetsov, I., Neumann, T., Ranjbar, Z., Savchuk, O.P., Schimanke, S., 2012. Impact of climate change on ecological quality indicators and biogeochemical fluxes in the Baltic Sea: a multi-model ensemble study. Ambio 41 (6), 558–573. https://doi.org/10.1007/s13280-012-0320-3.

Miller, J.G.T., 2005. Environmental Science, fifth ed.

Mitsch, W.J., Zhang, L., Fink, D.F., Hernandez, M.E., Altor, A.E., Tuttle, C.L., Nahlik, A.M., 2008. Ecological engineering of floodplains. Ecohydrol. Hydrobiol. 8 (2–4), 139–147. https://doi.org/10.2478/v10104-009-0010-3.

Moser, S.C., Tribbia, J., 2006. Vulnerability to inundation and climate change impacts in California: coastal managers' attitudes and perceptions. Mar. Technol. Soc. J. 40 (4), 35–44. https://doi.org/10.4031/002533206787353169.

Omstedt, A., Edman, M., Claremar, B., Frodin, P., Gustafsson, E., Humborg, C.H., Hägg, H., Mörth, M., Rutgersson, A., Schurgers, G., Smith, B., Wällstedt, T., Yurova, A., 2012. Future changes in the Baltic Sea acid base (pH) and oxygen balances. Tellus B 64.

Rönnberg, C., Bonsdorff, E., 2004. Finland Baltic Sea eutrophication: area-specific ecological consequences. Hydrobiologia 514, 227–241. https://doi.org/10.1023/B:HYDR.0000019238.84989.7f.

Røsberg, I., Frank, J., Stuanes, A.O., 2006. Effects of liming and fertilization on tree growth and nutrient cycling in a Scots pine ecosystem in Norway. For. Ecol. Manag. 237 (1–3), 191–207. https://doi.org/10.1016/j.foreco.2006.09.045.

SEPA, Swedish Environmental Protection Agency, 2008. The costs of environmental improvements in the Baltic Sea and Skagerrak. Economic Marine Information. Report. 5876.

SEPA, 2009 Swedish Environmental Protection Agency What's in the Sea for Me? Ecosystem Services provided by the Baltic Sea and Skagerrak vol 5872.

Serwecińska, L., Kiedrzyńska, E., Kiedrzyński, M., 2021. A catchment-scale assessment of the sanitary condition of treated wastewater and river water based on fecal indicators and carbapenem-resistant Acinetobacter spp. Sci. Total Environ. 750, 142266. https://doi.org/10.1016/j.scitotenv.2020.142266.

Shen, J., Yuan, L., Zhang, J., Li, H., Bai, Z., Chen, X., Zhang, W., Zhang, F., 2011. Phosphorus dynamics: from soil to plant. Plant Physiol. 156 (3), 997–1005. https://doi.org/10.1104/pp.111.175232.

Skłodowski, M., Kiedrzyńska, E., Kiedrzyński, M., Urbaniak, M., Zielińska, K.M., Kurowski, J.K., Zalewski, M., 2014. The role of riparian willows in phosphorus accumulation and PCB control for lotic water quality improvement. Ecol. Eng. 70, 1–10. https://doi.org/10.1016/j.ecoleng.2014.03.088.

Skorupski, J., Kowalewska-Łuczak, I., Kulig, H., Roggenbuck, A., 2012. Large-Scale Animal Production in Poland in the Context of Environmental Protection of the Baltic Sea.

Ten Hoeve, M., Hutchings, N.J., Peters, G.M., Svanström, M., Jensen, L.S., Bruun, S., 2014. Life cycle assessment of pig slurry treatment technologies for nutrient redistribution in Denmark. J. Environ. Manag. 132, 60–70. https://doi.org/10.1016/j.jenvman.2013.10.023.

The United Nations Economic Commission for Europe, 2014. Drainage Basin of the Baltic Sea. UNECE.

Urbaniak, M., Kiedrzyńska, E., Zieliński, M., Tołoczko, W., Zalewski, M., 2014a. Spatial distribution and reduction of PCDD/PCDF toxic equivalents along three shallow lowland reservoirs. Environ. Sci. Pollut. Control Ser. 21 (6), 4441–4452. https://doi.org/10.1007/s11356-013-2401-7.

Urbaniak, M., Kiedrzyńska, E., Kiedrzyński, M., Mendra, M., Grochowalski, A., 2014b. The impact of point sources of pollution on the transport of micropollutants along the river continuum. Hydrol. Res. 45 (3), 391–410. https://doi.org/10.2166/nh.2013.242.

Urbaniak, M., Kiedrzyńska, E., Zalewski, M., 2012. The role of a lowland reservoir in the transport of micropollutants, nutrients and the suspended particulate matter along the river continuum. Hydrol. Res. 43 (4), 400–411. https://doi.org/10.2166/nh.2012.108.

Vogt, J., Soille, P., Jager, Rimaviciute, E., Mehl, W., Foisneau, S., et al., 2007. A Pan-European River and Catchment Database. EC-JRC (Report EUR 22920 EN). Office for Official Publications of the European Communities.

Voss, M., Deutsch, B., Elmgren, R., Humborg, C., Kuuppo, P., Pastuszak, M., Rolff, C., Schulte, U., 2006. Source identification of nitrate by means of isotopic tracers in the Baltic Sea catchments. Biogeosciences 3 (4), 663–676. https://doi.org/10.5194/bg-3-663-2006.

Ward, J.V., Stanford, J.A., 1983. The Serial Discontinuity Concept of Lotic Ecosystems River Stream Ecosystem Theory. Flow Regulation, pp. 29–42.

Wulff, F., Savchuk, O.P., Sokolov, A., Humborg, C., Mörth, C.M., 2007. Management options and effects on a marine ecosystem: Assessing the future of the baltic. Ambio 36 (2–3), 243–249. https://doi.org/10.1579/0044-7447(2007)36[243:MOAEOA]2.0.CO;2.

Yin, H., Wheeler, E., Phillips, R.P., 2014. Root-induced changes in nutrient cycling in forests depend on exudation rates. Soil Biol. Biochem. 78, 213–221. https://doi.org/10.1016/j.soilbio.2014.07.022.

Zalewski, M., 2009. Ecohydrology: a framework for reversing the degradation of the Baltic Sea. Baltex Newsletter.

Zalewski, M., 2013. Ecohydrology: process-oriented thinking towards sustainable river basins. Ecohydrol. Hydrobiol. 13 (2), 97–103. https://doi.org/10.1016/j.ecohyd.2013.06.001.

Zalewski, M., 2020. Ekohydrologia. PWN, Warszawa, pp. 149–151.

Zalewski, M., 2015. Ecohydrology and hydrologic engineering: Regulation of hydrology-Biota interactions for sustainability. J. Hydrol. Eng. 20 (1). https://doi.org/10.1061/(asce)he.1943-5584.0000999.

Chapter 13

Municipal wastewater as a vector of environmental contaminants

Edyta Kiedrzyńska[1, 2], Marcin Kiedrzyński[3] and Maciej Zalewski[1]
[1]European Regional Centre for Ecohydrology of the Polish Academy of Sciences under the auspices of UNESCO, Łódź, Poland; [2]University of Lodz, UNESCO Chair on Ecohydrology and Applied Ecology, Łódź, Poland; [3]University of Lodz, Department of Biogeography, Paleoecology and Nature Conservation, Łódź, Poland

IMAGE CREDIT: E.KIEDRZYNSKA

Transdisciplinary Ecohydrology. https://doi.org/10.1016/B978-0-443-41596-8.00019-4

1. Introduction

Globally, over 80% of all wastewater is discharged into the environment without any treatment (The United Nations World Water Development Report 2017; Wastewater Management: A UN-Water Analytical Brief, 2015). Current global trends indicate that high-income countries treat about 70% of the municipal and industrial wastewater they generate. This proportion drops to 38% in upper middle-income countries, to 28% in lower middle-income countries, and in low-income countries, only 8% of wastewater undergoes any form of treatment (Sato et al., 2013; The United Nations World Water Development Report, 2017). Therefore, the water and sanitation situation represents an urgent and interconnected challenge that demands a collective and coordinated response. Water use and sanitation-related issues are escalating at an alarming rate, threatening people, the planet, peace, and prosperity (UN-Water 2030 Strategy, 2020). All human activities that use water produce wastewater. As overall water demand increases, the volume of wastewater generated—and its associated pollution load—is steadily rising worldwide Fig. 13.1A. According to the AQUASTAT database, global freshwater withdrawals are estimated at 3928 km^3 per year, and approximately 44% (1716 km^3 per year) of this water is consumed, mainly by agriculture (Fig. 13.1 B). The remaining 56% (2212 km^3 per year) is released into the environment as wastewater in the form of municipal (8%), industrial (16%), and agricultural (32%) effluents (Fig. 13.1B) (Mateo-Sagasta et al., 2015; The United Nations World Water Development Report, 2017). The consequences of releasing untreated or inadequately treated wastewater can be grouped into three categories: i) detrimental effects on human health; ii) negative environmental impacts; and iii) adverse consequences for economic activities (The United Nations World Water Development Report, 2017). According to United Nations data, by 2050, at least one in four people is likely to live in a country experiencing chronic or recurring freshwater shortages. In addition, it is estimated that at least 1.8 billion people globally use drinking water sources that are contaminated with fecal matter (The United Nations World Water Development Report, 2017). Contaminants are transported via two primary pathways: point sources, which refer to pollutants entering a waterway from a single, identifiable origin such as a pipe or ditch (e.g., discharges from a wastewater treatment plant (WWTP), factory, municipal storm drain, domestic wastewater, or industrial effluent), and nonpoint (diffuse) sources, where pollution originates from the broader agricultural landscape.

Wastewater is a crucial component of the water cycle and must be managed throughout the entire water management continuum: from freshwater abstraction, treatment, distribution, and use, to collection, posttreatment, reuse, and eventual return to the environment (Fig. 13.2) (The United Nations World Water Development Report, 2017). However, the management of water after it has been used remains an underfunded and often overlooked aspect of the water cycle. Compared to water supply issues, wastewater management typically receives limited social and political attention, particularly in the context of growing water scarcity and climate change. Since wastewater is an integral part of the water cycle (Fig. 13.2), neglecting its sustainable management can have serious negative consequences for water quality and availability, public health, economic development, and environmental integrity.

The primary objective of a WWTP is to transform raw sewage into effluent clean enough to be discharged into receiving water bodies without posing risks to human health or causing environmental and ecological harm. Specifically, WWTPs are designed to remove organic matter—measured as chemical oxygen demand (COD) and biological oxygen demand (BOD)—as well as nutrients such as nitrogen and phosphorus, which contribute to eutrophication and the degradation of natural aquatic ecosystems (Flores-Alsina et al., 2010; Harnisz et al., 2020; Kiedrzyńska et al., 2014a,b; Kiedrzyńska et al., 2017). Treated effluents must comply with the regulatory standards set forth in Council Directive 91/271/EEC (as amended) concerning urban wastewater treatment. However, even when treated wastewater meets these regulatory thresholds, it is not necessarily safe for the environment. In many cases, effluent discharges exceed the permissible limits for nutrient pollution and contain additional contaminants that are transported to aquatic ecosystems. These include heavy metals (Carletti et al., 2008; Xu et al., 2017), microplastics (Kay et al., 2018; Sun et al., 2019) dioxins and dioxin-like compounds (Urbaniak and Kiedrzyńska, 2015; Urbaniak et al., 2017), pesticides (Bunzel et al., 2013), pharmaceuticals (Paíga et al., 2019), fecal bacteria (Serwecińska et al., 2021), and antibiotics (Felis et al., 2020; Harnisz et al., 2020; Kümmerer, 2009). Effluents from WWTPs also contain antibiotic-resistant bacteria (ARB) and antibiotic resistance genes (ARGs), which pose a serious threat to human and animal health (Czatzkowska et al., 2020; Harnisz et al., 2020; Koniuszewska et al., 2020). The long-term anthropogenic discharge of such pollutants from WWTPs is one of the major threats to riverine and coastal marine ecosystems, a phenomenon observed globally (Han et al., 2011; Harnisz et al., 2020; Kiedrzyńska et al., 2014a,b, 2017; Russell et al., 2008). Preventing water contamination should begin with the identification and quantification of pollution sources within each catchment, followed by a thorough understanding of the migration, transport, and transformation of contaminants in the environment. Subsequent steps must include sustainable water and wastewater management at local, regional, and transboundary scales, alongside the implementation of advanced wastewater treatment technologies in WWTPs. Therefore, it is essential to analyze the size and spatial distribution of point

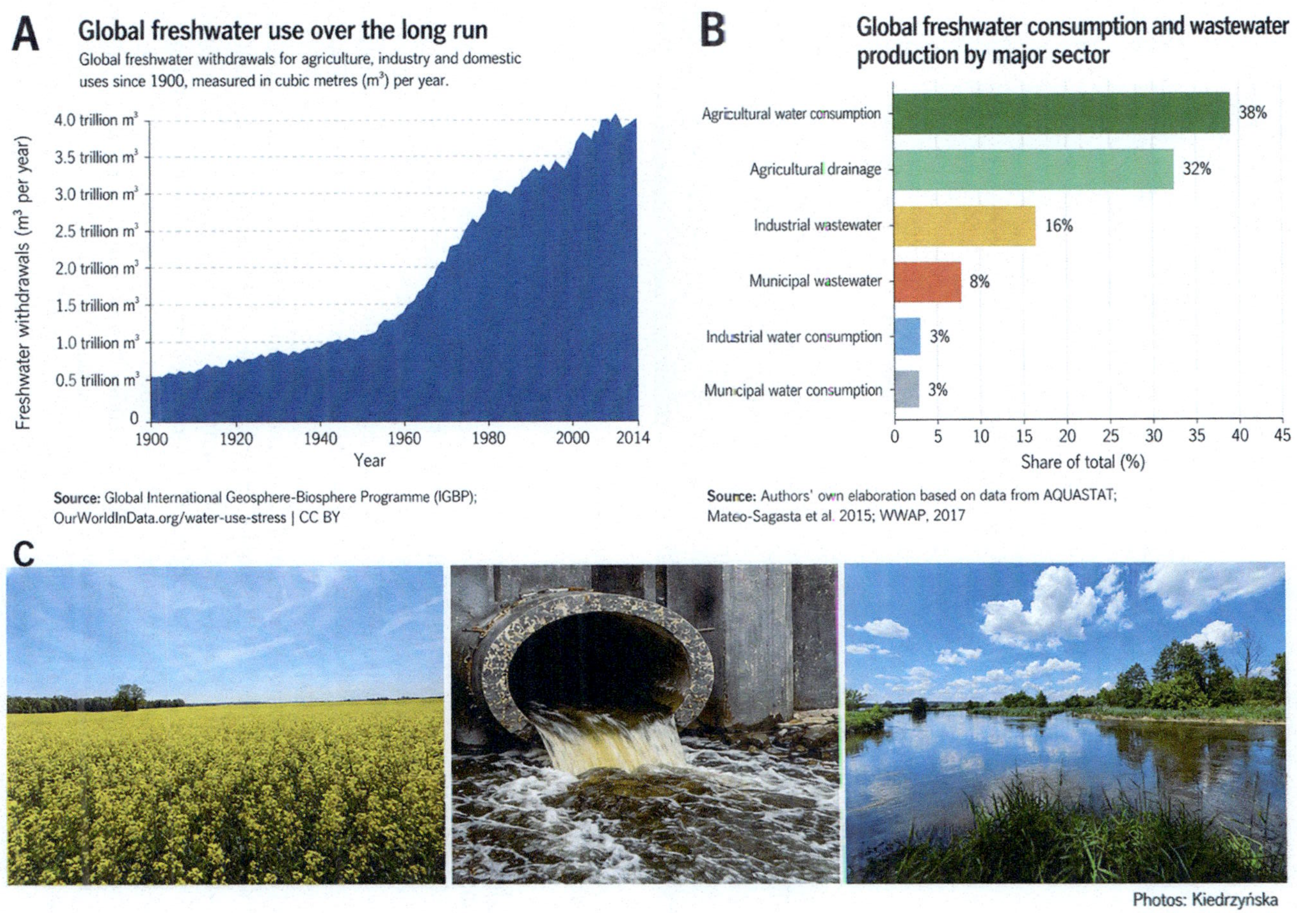

FIGURE 13.1 Global trends in water use, consumption, and wastewater generation Global situation regarding water and wastewater: (A) Global freshwater use over the long run; (B) Global freshwater consumption and wastewater production by major sector. *Source: (A) Global International Geosphere-Biosphere Programme (IGB), OurWorldinData.org/water-use-stress | CC BY. (B) Authors' own elaboration based on data from AQUASTAT; Mateo-Sagasta, J., et al. 2015. Global Wastewater and Sludge Production, Treatment and Use Wastewater: Economic Asset in an Urbanizing World. Springer Netherlands, Sri Lanka, pp. 15–38. https://doi.org/10.1007/978-94-017-9545-6_2; WWAP, 2017; Photo credit: E.Kiedrzyńska.*

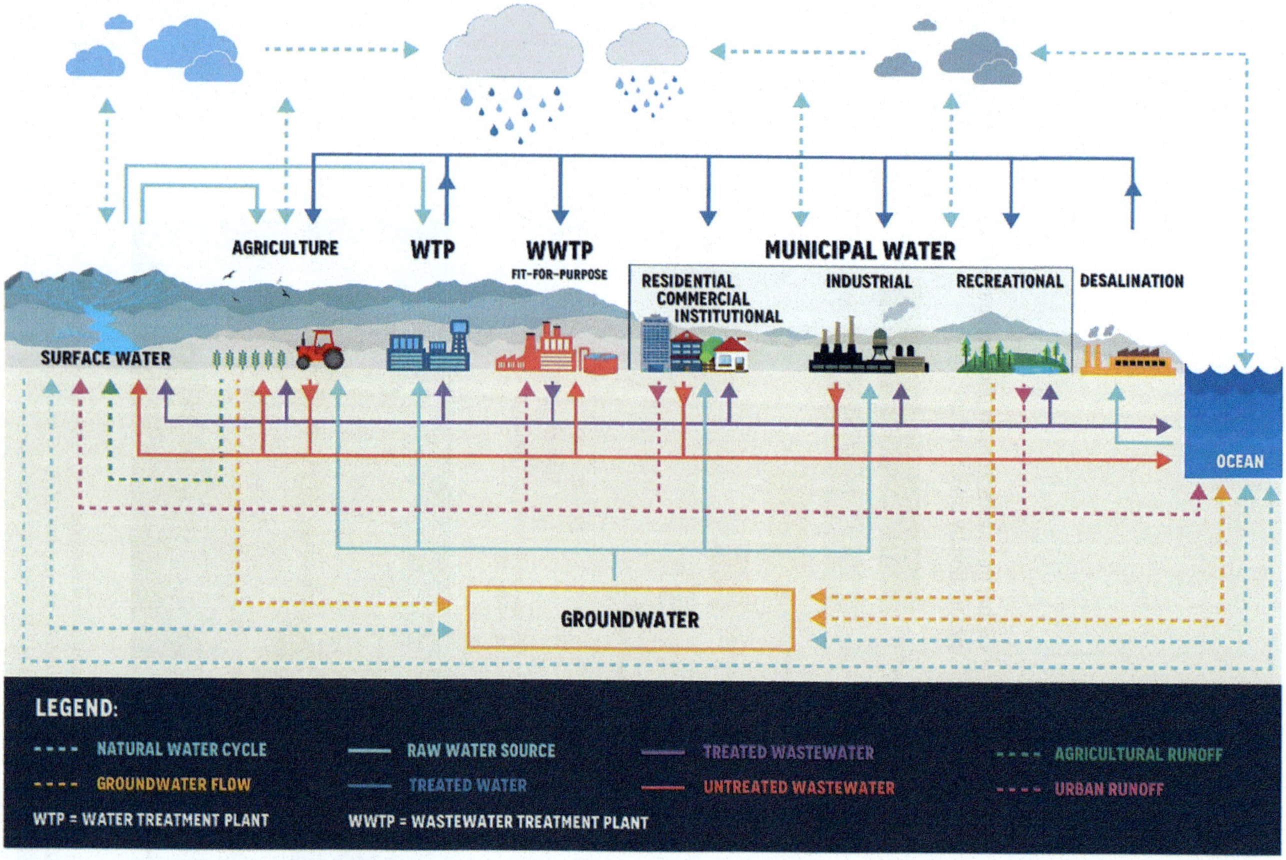

FIGURE 13.2 Wastewater in the water cycle. Fate of wastewater in the anthropogenically modified water cycle. *Source: UNESCO World Water Assessment Programme (WWAP), 2017. The United Nations World Water Development Report 2017: Wastewater – The Untapped Resource. UNESCO, Paris, ISBN 978-92-3-100201-4.*

sources such as WWTPs within each catchment to assess and quantify their contribution to the total contaminant load entering aquatic systems. In this chapter, a catchment-scale assessment of the impact of WWTPs (Fig. 13.3B and C) was conducted, focusing on their role as point sources of nutrients, dioxins, ARB, ARGs, and fecal bacteria, including Acinetobacter. The main objectives of this ecohydrological research in Poland were to evaluate the quality of treated effluents discharged from municipal WWTPs and to assess the role of wastewater in contaminating the Pilica River along its continuum from the source to the estuary with: (1) nutrients; (2) ARB and ARGs; and (3) fecal bacteria, including antibiotic-resistant Acinetobacter spp.

The study was conducted across the entire Pilica River catchment (Fig. 13.3A–C) in accordance with the first principle of ecohydrology (Zalewski, 2015, 2020), which emphasizes the need to quantify ecological, biological, and hydrological processes within river basins. This is crucial for improving our understanding of ecosystem health and the extent of pollution. The Pilica River (342 km in length; 9258 km^2 catchment area) is one of the largest tributaries of the Vistula River (1092 km), which drains into the Baltic Sea (Fig. 13.3A) (Buszewski et al., 2005). The Vistula and Pilica rivers flow through some of the most industrialized and contaminated regions of Europe, particularly in Silesia. The Pilica catchment is predominantly agricultural, with over 60% of its area occupied by farmland, while forests cover approximately 31%. On average, 59% of the human population within the Pilica River catchment is connected to WWTPs, with regional variation ranging from 51.3% to 70.5%. Advanced nutrient removal technologies are implemented in 43.6% of existing WWTPs (Kiedrzyńska et al., 2014a,b; PCSO, 2010).

2. Contamination by nutrients

Spatial analysis of nutrient loads in river catchments is often challenging due to the combined influence of point and diffuse sources. The complex dynamics of water and nutrient flow can be simplified by dividing the entire catchment into a cascade of differentiated subcatchments, each controlled by monitoring stations. This approach allows for the interpretation of changes in water quality along the river continuum and facilitates the quantification of self-purification processes. According to Kiedrzyńska et al. (2014a,b), wastewater in WWTPs is often insufficiently treated, as

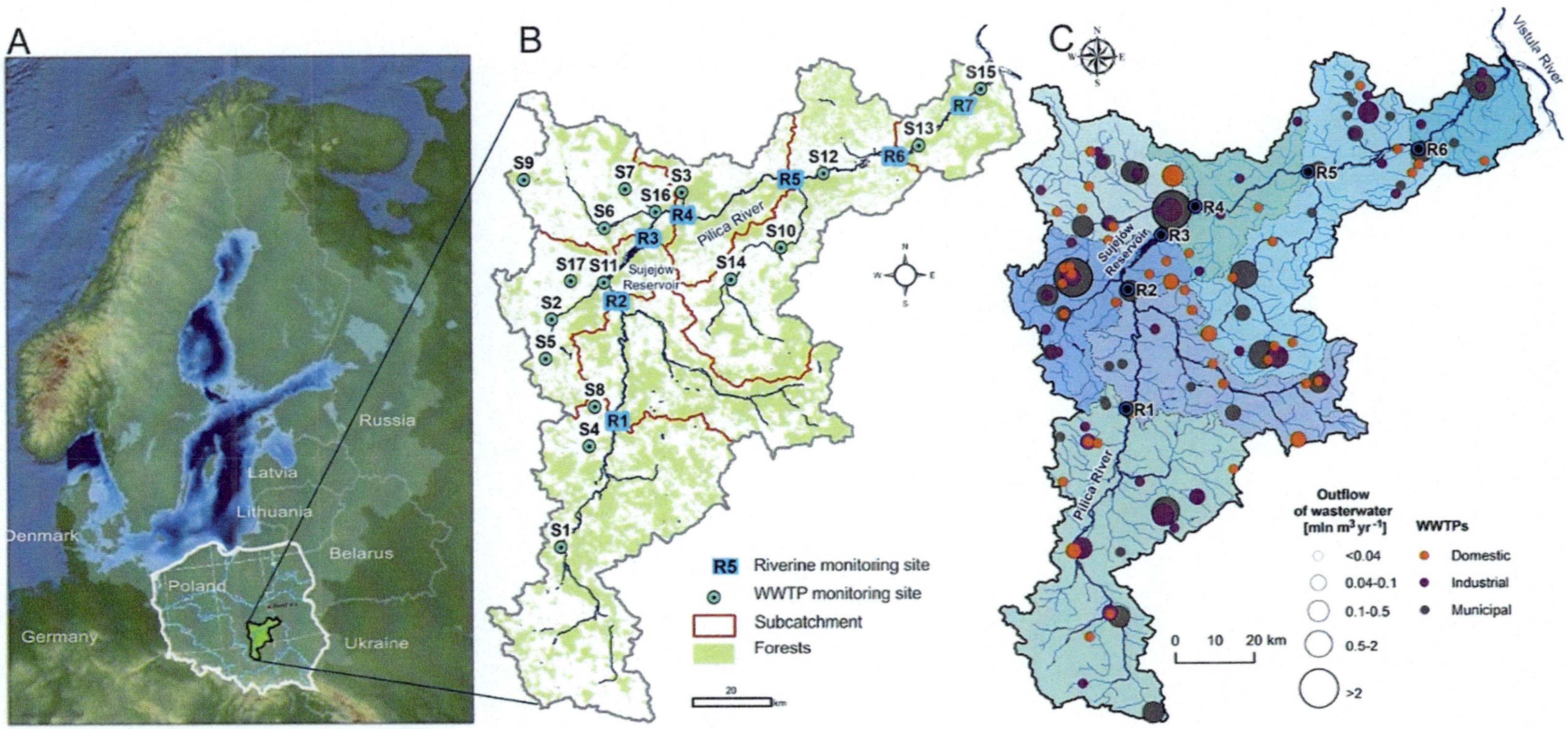

FIGURE 13.3 Geographical context and wastewater sources in the Pilica River basin. (A) Baltic Sea basin with location of the Pilica River catchment study case (Poland); (B) riverine (R1–R6) and WWTPs (S1–S17) monitoring stations along the Pilica River continuum; (C) distribution and volume flow of wastewater from domestic, industrial, and municipal WWTPs. *Modified from Kiedrzyńska, E., Kiedrzyński, M., Urbaniak, M., Magnuszewski, A., Skłodowski, M., Wyrwicka, A., Zalewski, M.. 2014. Point sources of nutrient pollution in the lowland river catchment in the context of the Baltic Sea eutrophication. Ecol. Eng. 70, 337–348, https://doi.org/10.1016/j.ecoleng.2014.06.010.*

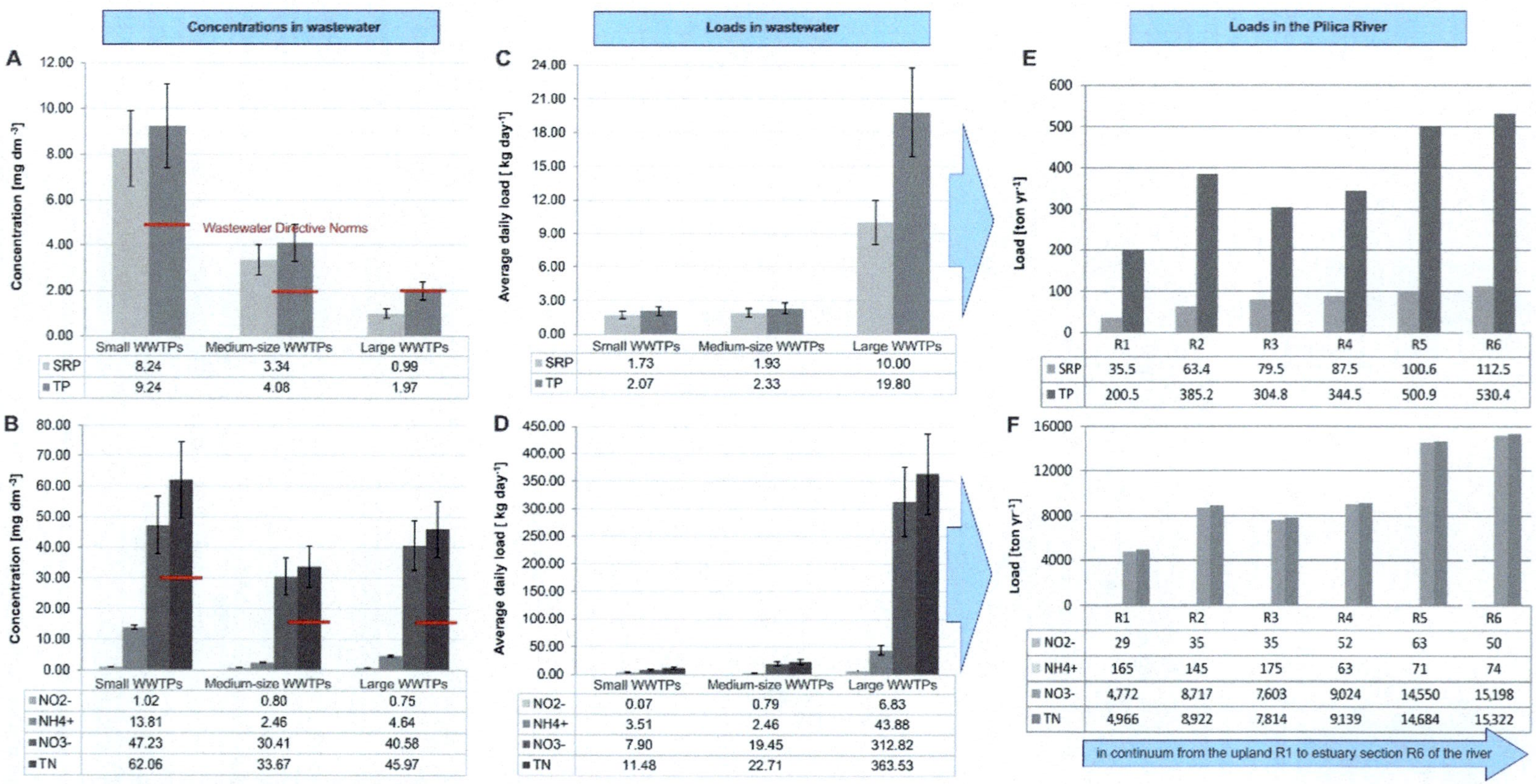

FIGURE 13.4 Nutrient loads and transport from WWTPs and in the Pilica River system. Average concentrations of (A) total phosphorus (TP), soluble reactive phosphorus (SRP), and (B) total nitrogen (TN), nitrate (NO_3^-), nitrite (NO_2^-), and ammonium (NH_4^+) compounds in wastewater discharged from WWTPs of various size classes. Daily loads of phosphorus (C) and nitrogen (D) discharged from WWTPs. Transport of phosphorus (E) and nitrogen (F) loads along the Pilica River continuum from the upland section (R1) to the downstream section (R6). *Modified from Kiedrzyńska, E., Kiedrzyński, M., Urbaniak, M., Magnuszewski, A., Skłodowski, M., Wyrwicka, A., Zalewski, M., 2014. Point sources of nutrient pollution in the lowland river catchment in the context of the Baltic Sea eutrophication. Ecol. Eng. 70, 337–348, https://doi.org/10.1016/j.ecoleng.2014.06.010.*

concentrations of total phosphorus (TP) and total nitrogen (TN) frequently exceed permissible levels (Fig. 13.4). Maintaining proper treatment parameters is more difficult in small WWTPs compared to larger facilities, which are typically equipped with more advanced nutrient removal technologies.

Nevertheless, it is primarily large WWTPs that are responsible for significant nutrient load contributions to the river (Fig. 13.4C and D), due to the high volume of wastewater discharged. The study also revealed a gradual increase in TP and TN loads transported along the Pilica River continuum (Fig. 13.4E and F), driven by nutrient inflows from the catchment—including, among others, municipal WWTPs. The cumulative discharge of large nutrient loads from the Pilica River catchment into the Vistula River (Fig. 13.4E and F) contributes, at least in part, to the eutrophication of the Baltic Sea (Kiedrzyńska et al., 2014a,b). The presence of numerous point sources of pollution in the Pilica River catchment (Fig. 13.4C) is typical of many river systems in Poland and across Europe, making these findings broadly applicable for characterizing the condition of similar catchments. According to Kiedrzyńska et al. (2014a,b), the average TP and TN loads discharged from the Pilica catchment amounted to 0.057 t TP km^{-2} and 1.655 t TN km^{-2}, respectively. These values were approximately two and three times higher, respectively, than the annual average loads recorded for the Polish territory and the Baltic States. Sustainable management of point sources of pollution should be based on frequent and comprehensive assessments of WWTP performance across the entire catchment, rather than being limited by administrative boundaries. Reducing nutrient concentrations in wastewater and minimizing WWTP failures should be considered among the highest priorities in modern wastewater management. WWTPs typically employ physical, chemical, and biological methods to improve wastewater quality; however, additional technologies such as wastewater hygienization, enhanced nutrient capture, and nutrient recovery should be implemented across all treatment plants. Integrating novel wastewater treatment technologies with ecological engineering, biotechnological innovations, and ecohydrological approaches—such as the Hybrid Sequential Biofiltration System (HSBS), a nature-based solution for the advanced treatment of effluents (Kiedrzyńska et al., 2017)—offers promising prospects for improving water quality and mitigating the eutrophication of the Baltic Sea.

3. Contamination by antibiotic-resistant bacteria and antibiotic resistance genes

Total antibiotic use in particular countries is high and especially drugs from beta-lactams (blaTEM) group (Fig. 13.5) (Versporten et al., 2014; WHO, 2017). According to the World Health Organization (2017), antimicrobial resistance represents an increasingly serious global health threat.

Wastewater discharged from WWTPs contains ARB and ARGs, which pose health risks to both humans and animals (Huang et al., 2012; Niestępski et al., 2019; Rizzo et al., 2013). Rivers located downstream of WWTPs can also become enriched with mobile genetic elements (MGEs), including conjugative plasmids, integrative and conjugative elements, transposons, and integrons (Cacace et al., 2019; Pallares-Vega et al., 2019). MGEs play a key role in the acquisition and dissemination of ARGs. However, the transport and fate of ARGs in aquatic systems—particularly at the catchment scale—remain poorly understood. Harnisz et al. (2020) hypothesized that WWTP effluents may serve as a source of ARGs and MGEs that facilitate the spread of resistance in receiving water bodies. The goal of this innovative study was to explore potential differences in the abundance of ARGs in treated wastewater from small, medium-sized, and large WWTPs within the catchment, as well as to assess seasonal variations in ARGs in both WWTPs and river water—an analysis not previously conducted at the basin scale. The study also examined the impact of treated effluent on ARG concentrations along the river continuum—from upland to lowland segments—up to the confluence with the Vistula River. Genes encoding resistance to five groups of drugs, including beta-lactams (blaTEM), tetracyclines (tet(A)), macrolides, lincosamides and streptogramins (ermF), sulphonamides (sul1), and fluoroquinolones (aac(6')-Ib-cr), as well as the integrase 1 (intI1) gene and the 16S rRNA gene, were analyzed by quantitative PCR. The results showed that the highest absolute concentrations of the studied genes occurred in effluents from small WWTPs, especially for blaTEM (Fig. 13.6). ARG concentrations (expressed as gene copies/mL) peaked in winter and spring samples, likely reflecting increased antibiotic use during these seasons (Harnisz et al., 2020). Statistical analysis further indicated that, in small WWTPs, ARG concentrations could be predicted based on biochemical oxygen demand (BOD), a parameter routinely measured in wastewater monitoring. However, none of the studied parameters allowed for reliable prediction of ARG abundance in medium-sized or large WWTPs, or in river water (Harnisz et al., 2020). These findings underscore the need for continuous monitoring of microbiological risks in river ecosystems, with special attention to small WWTPs. In addition, effective sewage hygienization strategies should be implemented to prevent the environmental spread of resistance determinants from WWTP effluents.

4. Contamination by fecal bacteria and carbapenem-resistant *Acinetobacter* spp

Human population growth and urbanization have created significant sanitation challenges in cities and communities across Europe and globally. Consequently, it is essential to identify the most environmentally hazardous microbiological sources of pollution within river catchments. A study conducted by Serwecińska et al. (2021) involved a catchment-scale

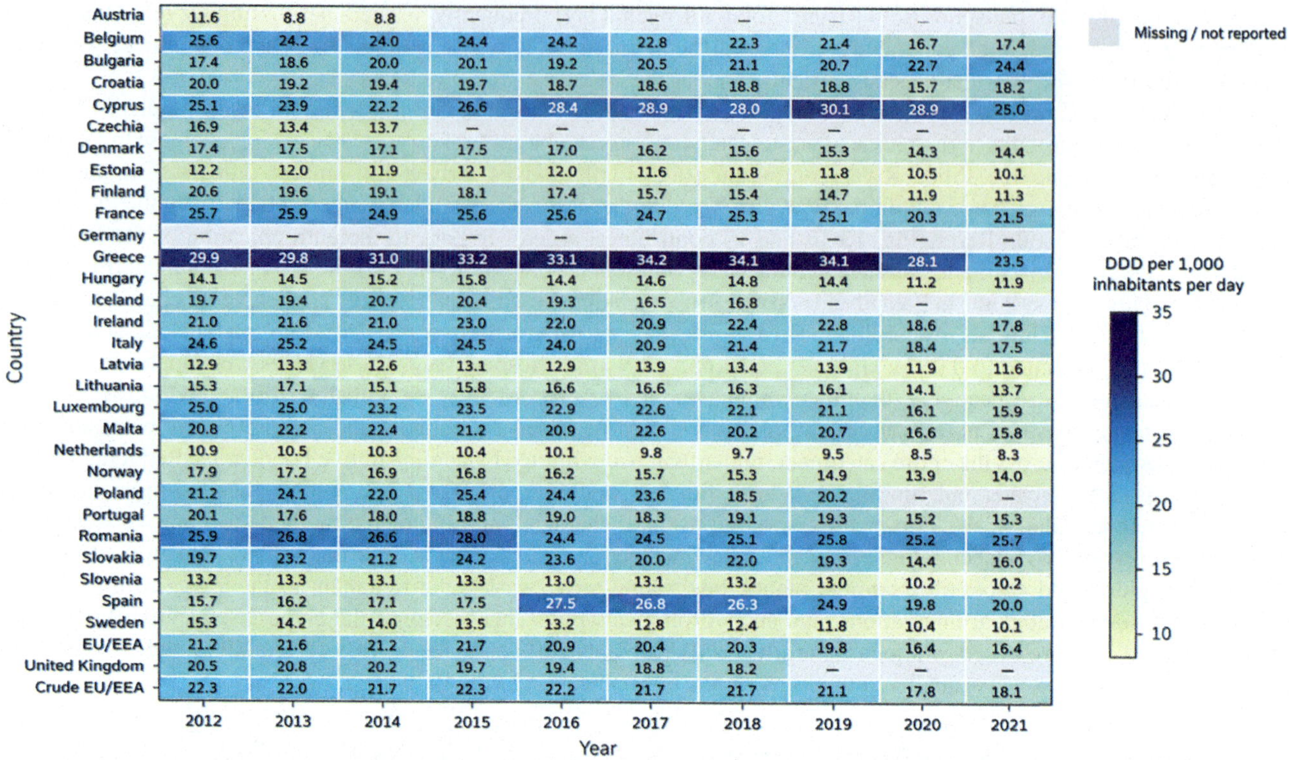

Grey cells indicate missing or not reported data. Spain changed data source in 2016; Luxembourg changed data process in 2020.

FIGURE 13.5 Antibiotic consumption in Europe. Total consumption (community and hospital sector) of antibacterials for systemic use (ATC group J01), EU/EEA and UK, 2012–2021 (expressed as DDD per 1000 inhabitants per day). *Source: Authors' own elaboration based on data from Antimicrobial consumption in the EU/EEA (ESAC-Net) - Annual Epidemiological Report 2021. European Centre for Disease Prevention and Control. Stockholm: ECDC; 2022.*

assessment of the sanitary condition of treated wastewater from 17 WWTPs of various sizes, located throughout the Pilica River catchment. The analysis was based on fecal indicator bacteria and carbapenem-resistant Acinetobacter spp. The primary objective was to identify "hot spots" representing the most critical sources of sanitary risk. Bacteriological risks to the river ecosystem were assessed using classical microbiological methods, focusing on fecal indicator bacteria (FIB) such as coliforms, *Escherichia coli*, enterococci, *Clostridium perfringens*, and carbapenem-resistant Acinetobacter spp. (CRA). In addition, physicochemical parameters were also measured. The results demonstrated that WWTPs—particularly small facilities (<2000 population equivalent, PE)—exhibited considerable variability in physicochemical characteristics. Alarmingly, carbapenem-resistant Acinetobacter spp. capable of growing at 42°C were detected in effluents from all tested municipal WWTPs, as well as in most Pilica River water samples, posing a serious threat to public health. Moreover, a positive correlation was found between the abundance of *E. coli* and CRA in treated wastewater, although no such relationship was observed in river water. Seven small WWTPs were identified as discharging effluents with highly variable microbiological characteristics. Strikingly, three of these plants—serving only 0.56% of the population in the study area—were responsible for continuously releasing extremely high levels of microbiological contamination, contributing between 54% and 82% of the total FIB load in the catchment (Serwecińska et al., 2021). These findings underscore the value of comprehensive, catchment-scale sanitary assessments for identifying critical threats to surface water quality and informing targeted improvements in the performance of poorly functioning WWTPs.

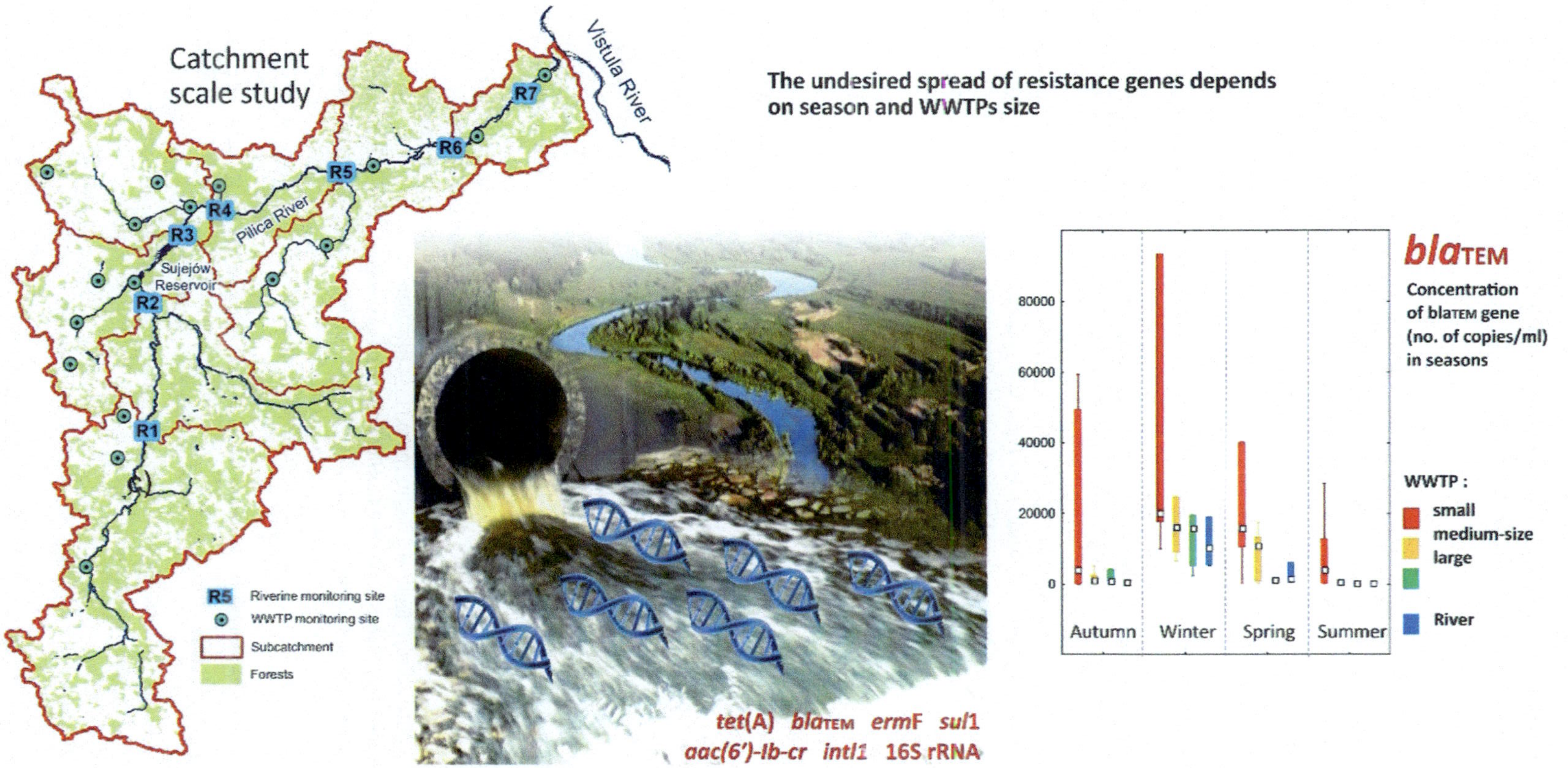

FIGURE 13.6 Distribution of ARGs in WWTPs and the Pilica River. The occurrence of antibiotic resistance genes (ARGs) including bla_{TEM} in treated wastewater from small, medium-sized, and large WWTPs and in the Pilica River. *Modified from Harnisz, M., Kiedrzyńska, E., Kiedrzyński, M., Korzeniewska, E., Czatzkowska, M., Koniuszewska, I., Jóźwik, A., Szklarek, S., Niestępski, S., Zalewski, M., The impact of WWTP size and sampling season on the prevalence of antibiotic resistance genes in wastewater and the river system. Sci. Total Environ. 741 (2020), 140466, https://doi.org/10.1016/j.scitotenv.2020.140466.*

Acknowledgments

This research was funded by the Polish Ministry of Science and Higher Education (Project No. NN305 365738) and by the Polish National Science Centre (Project Nos. 2015/19/B/ST10/02167 and 2021/43/B/ST10/01076).

AI disclosure

During the preparation of this work, the author(s) used OpenAI's ChatGPT. This document includes content generated and/or edited with the assistance of OpenAI's ChatGPT, used to support language refinement and structural improvements. All final content was reviewed and verified by the author. After using this tool/service, the author(s) reviewed and edited the content as needed and take(s) full responsibility for the content of the publication.

References

Bunzel, K., Kattwinkel, M., Liess, M., 2013. Effects of organic pollutants from wastewater treatment plants on aquatic invertebrate communities. Water Res. 47 (2), 597–606. https://doi.org/10.1016/j.watres.2012.10.031.

Buszewski, B., Buszewska, T., Chmarzyński, A., Kowalkowski, T., Kowalska, J., Kosobucki, P., Zbytniewski, R., Namieśnik, J., Kot-Wasik, A., Pacyna, J., Panasiuk, D., 2005. The present condition of the Vistula river catchment area and its impact on the Baltic Sea coastal zone. Reg. Environ. Change 5 (2–3), 97–110. https://doi.org/10.1007/s10113-004-0077-8.

Cacace, D., Fatta-Kassinos, D., Manaia, C.M., Cytryn, E., Kreuzinger, N., Rizzo, L., Karaolia, P., Schwartz, T., Alexander, J., Merlin, C., Garelick, H., Schmitt, H., de Vries, D., Schwermer, C.U., Meric, S., Ozkal, C.B., Pons, M.N., Kneis, D., Berendonk, T.U., 2019. Antibiotic resistance genes in treated wastewater and in the receiving water bodies: a pan-European survey of urban settings. Water Res. 162, 320–330. https://doi.org/10.1016/j.watres.2019.06.039.

Carletti, G., Fatone, F., Bolzonella, D., Cecchi, F., 2008. Occurrence and fate of heavy metals in large wastewater treatment plants treating municipal and industrial wastewaters. Water Sci. Technol. 57 (9), 1329–1336. https://doi.org/10.2166/wst.2008.230.

Czatzkowska, M., Harnisz, M., Kiedrzyńska, E., Kiedrzyński, M., Koniuszewska, I., Korzeniewska, E., Szklarek, S., Zalewski, M., 2020. Catchment scale analysis of occurrence of antibiotic resistance genes in treated wastewater. Ecohydrol. Hydrobiol. 20 (1), 12–20. https://doi.org/10.1016/j.ecohyd.2019.06.002.

Felis, E., Kalka, J., Sochacki, A., Kowalska, K., Bajkacz, S., Harnisz, M., Korzeniewska, E., 2020. Antimicrobial pharmaceuticals in the aquatic environment—occurrence and environmental implications. Eur. J. Pharmacol. 866, 172813. https://doi.org/10.1016/j.ejphar.2019.172813.

Flores-Alsina, X., Gallego, A., Feijoo, G., Rodriguez-Roda, I., 2010. Multiple-objective evaluation of wastewater treatment plant control alternatives. J. Environ. Manag. 91 (5), 1193–1201. https://doi.org/10.1016/j.jenvman.2010.01.009.

Food and Agriculture Organization of the United Nations (FAO), 2015. The State of Food and Agriculture 2015: Social Protection and Agriculture: Breaking the Cycle of Rural Poverty. FAO, Rome, ISBN 978-92-5-108861-6.

Han, H., Bosch, N., Allan, J.D., 2011. Spatial and temporal variation in phosphorus budgets for 24 watersheds in the Lake Erie and Lake Michigan basins. Biogeochemistry 102 (1), 45–58. https://doi.org/10.1007/s10533-010-9420-y.

Harnisz, M., Kiedrzyńska, E., Kiedrzyński, M., Korzeniewska, E., Czatzkowska, M., Koniuszewska, I., Jóźwik, A., Szklarek, S., Niestępski, S., Zalewski, M., 2020. The impact of WWTP size and sampling season on the prevalence of antibiotic resistance genes in wastewater and the river system. Sci. Total Environ. 741, 140466. https://doi.org/10.1016/j.scitotenv.2020.140466.

Huang, J.J., Hu, H.Y., Lu, S.Q., Li, Y., Tang, F., Lu, Y., Wei, B., 2012. Monitoring and evaluation of antibiotic-resistant bacteria at a municipal wastewater treatment plant in China. Environ. Int. 42 (1), 31–36. https://doi.org/10.1016/j.envint.2011.03.001.

Kay, P., Hiscoe, R., Moberley, I., Bajic, L., McKenna, N., 2018. Wastewater treatment plants as a source of microplastics in river catchments. Environ. Sci. Pollut. Control Ser. 25 (20), 20264–20267. https://doi.org/10.1007/s11356-018-2070-7.

Kiedrzyńska, E., Jóźwik, A., Kiedrzyński, M., Zalewski, M., 2014a. Hierarchy of factors exerting an impact on nutrient load of the Baltic Sea and sustainable management of its drainage basin. Mar. Pollut. Bull. 88 (1–2), 162–173. https://doi.org/10.1016/j.marpolbul.2014.09.010.

Kiedrzyńska, E., Kiedrzyński, M., Urbaniak, M., Magnuszewski, A., Skłodowski, M., Wyrwicka, A., Zalewski, M., 2014b. Point sources of nutrient pollution in the lowland river catchment in the context of the Baltic Sea eutrophication. Ecol. Eng. 70, 337–348. https://doi.org/10.1016/j.ecoleng.2014.06.010.

Kiedrzyńska, E., Urbaniak, M., Kiedrzyński, M., Jóźwik, A., Bednarek, A., Gągała, I., Zalewski, M., 2017. The use of a hybrid Sequential Biofiltration System for the improvement of nutrient removal and PCB control in municipal wastewater. Sci. Rep. 7 (1). https://doi.org/10.1038/s41598-017-05555-y.

Koniuszewska, I., Korzeniewska, E., Harnisz, M., Kiedrzyńska, E., Kiedrzyński, M., Czatzkowska, M., Jarosiewicz, P., Zalewski, M., 2020. The occurrence of antibiotic-resistance genes in the Pilica River, Poland. Ecohydrol. Hydrobiol. 20 (1), 1–11. https://doi.org/10.1016/j.ecohyd.2019.09.002.

Kümmerer, K., 2009. Antibiotics in the aquatic environment – a review – part I. Chemosphere 75 (4), 417–434. https://doi.org/10.1016/j.chemosphere.2008.11.086.

Mateo-Sagasta, J., Raschid-Sally, L., Thebo, A., 2015. Global Wastewater and Sludge Production, Treatment and Use Wastewater: Economic Asset in an Urbanizing World. Springer Netherlands, Sri Lanka, pp. 15–38. https://doi.org/10.1007/978-94-017-9545-6_2.

Niestępski, S., Harnisz, M., Korzeniewska, E., Aguilera-Arreola, M.G., Contreras-Rodríguez, A., Filipkowska, Z., Osińska, A., 2019. The emergence of antimicrobial resistance in environmental strains of the Bacteroides fragilis group. Environ. Int. 124, 408–419. https://doi.org/10.1016/j.envint.2018.12.056.

Paíga, P., Correia, M., Fernandes, M.J., Silva, A., Carvalho, M., Vieira, J., Jorge, S., Silva, J.G., Freire, C., Delerue-Matos, C., 2019. Assessment of 83 pharmaceuticals in WWTP influent and effluent samples by UHPLC-MS/MS: hourly variation. Sci. Total Environ. 648, 582–600. https://doi.org/10.1016/j.scitotenv.2018.08.129.

Pallares-Vega, R., Blaak, H., van der Plaats, R., de Roda Husman, A.M., Hernandez Leal, L., van Loosdrecht, M.C.M., Weissbrodt, D.G., Schmitt, H., 2019. Determinants of presence and removal of antibiotic resistance genes during WWTP treatment: a cross-sectional study. Water Res. 161, 319–328. https://doi.org/10.1016/j.watres.2019.05.100.

PCSO, 2010. Environment Protection. Polish Central Statistical Office.

Rizzo, L., Manaia, C., Merlin, C., Schwartz, T., Dagot, C., Ploy, M.C., Michael, I., Fatta-Kassinos, D., 2013. Urban wastewater treatment plants as hotspots for antibiotic resistant bacteria and genes spread into the environment: a review. Sci. Total Environ. 447, 345–360. https://doi.org/10.1016/j.scitotenv.2013.01.032.

Russell, M.J., Weller, D.E., Jordan, T.E., Sigwart, K.J., Sullivan, K.J., 2008. Net anthropogenic phosphorus inputs: spatial and temporal variability in the Chesapeake Bay region. Biogeochemistry 88 (3), 285–304. https://doi.org/10.1007/s10533-008-9212-9.

Sato, T., Qadir, M., Yamamoto, S., Endo, T., Zahoor, A., 2013. Global, regional, and country level need for data on wastewater generation, treatment, and use. Agric. Water Manag. 130, 1–13. https://doi.org/10.1016/j.agwat.2013.08.007.

Serwecińska, L., Kiedrzyńska, E., Kiedrzyński, M., 2021. A catchment-scale assessment of the sanitary condition of treated wastewater and river water based on fecal indicators and carbapenem-resistant Acinetobacter spp. Sci. Total Environ. 750, 142266. https://doi.org/10.1016/j.scitotenv.2020.142266.

Sun, J., Dai, X., Wang, Q., van Loosdrecht, M.C.M., Ni, B.J., 2019. Microplastics in wastewater treatment plants: Detection, occurrence and removal. Water Res. 152, 21–37. https://doi.org/10.1016/j.watres.2018.12.050.

The United Nations World Water Development Report 2017. Wastewater: The Untapped Resource. (2017).

UN-water 2030 Strategy Geneva, Switzerland, 2020

UNESCO World Water Assessment Programme (WWAP), 2017. The United Nations World Water Development Report 2017: Wastewater – The Untapped Resource. UNESCO, Paris, ISBN 978-92-3-100201-4.

Urbaniak, M., Kiedrzyńska, E., 2015. Concentrations and toxic equivalency of polychlorinated biphenyls in polish wastewater treatment plant effluents. Bull. Environ. Contam. Toxicol. 95 (4), 530–535. https://doi.org/10.1007/s00128-015-1631-4.

Urbaniak, M., Kiedrzyńska, E., Grochowalski, A., 2017. The variability of PCDD/F concentrations in the effluent of wastewater treatment plants with regard to their hydrological environment. Environ. Monit. Assess. 189 (2). https://doi.org/10.1007/s10661-017-5794-9.

Versporten, A., Bolokhovets, G., Ghazaryan, L., Abilova, V., Pyshnik, G., Spasojevic, T., Korinteli, I., Raka, L., Kambaralieva, B., Cizmovic, L., Carp, A., Radonjic, V., Maqsudova, N., Celik, H.D., Payerl-Pal, M., Pedersen, H.B., Sautenkova, N., Goossens, H., 2014. Antibiotic use in eastern Europe: a cross-national database study in coordination with the WHO Regional Office for Europe. Lancet Infect. Dis. 14 (5), 381–387. https://doi.org/10.1016/s1473-3099(14)70071-4.

Wastewater Management: A UN-water Analytical Brief. (2015). Available from: www.unwater.org/fileadmin/user_upload/unwater_new/docs/UNWater_Analytical_Brief_Wastewater_Management.pdf.

WHO, 2017. Antimicrobial Resistance: Fact Sheets on Sustainable Development Goals: Health Targets.

Xu, L., Wang, T., Wang, J., Lu, A., 2017. Occurrence, speciation and transportation of heavy metals in 9 coastal rivers from watershed of Laizhou Bay, China. Chemosphere 173, 61–68. https://doi.org/10.1016/j.chemosphere.2017.01.046.

Zalewski, M., 2015. Ecohydrology and hydrologic engineering: regulation of hydrology-biota interactions for sustainability. J. Hydrol. Eng. 20 (1). https://doi.org/10.1061/(ASCE)HE.1943-5584.0000999.

Zalewski, M., 2020. Ecohydrologia. PWN, Warsaw.

Chapter 14

Enhancement of lowland reservoirs through ecohydrological catchment management—Case study: Sulejów reservoir

Katarzyna Izydorczyk, Wojciech Frątczak and Maciej Zalewski
European Regional Centre for Ecohydrology of the Polish Academy of Sciences under the auspices of UNESCO, Łódź, Poland

Transdisciplinary Ecohydrology. https://doi.org/10.1016/B978-0-443-41596-8.00002-9

PHOTO CREDIT: K. IZYDORCZYK.

1. Introduction

Reservoirs are artificial lakes formed by damming rivers to store water for cities, agriculture and other recipients and protect against flooding. The construction of reservoirs disrupts the continuity of the river environment, interrupting the natural continuity of physical conditions in the river and biological processes occurring along its course (Ward and Stanford, 1983). Reservoirs also cause drastic changes in biotic communities, changing the living environment from unstable (flowing water) to stagnant (standing water) and posing an obvious obstacle to migrating fish (Backiel, 1985; Belletti et al., 2020). The change from a flowing to a standing environment also leads to an increased risk of sedimentation of mineral organic matter and in consequence accelerated eutrophication. However, despite all these threats, a well-designed and managed reservoir can be an important and positive element in stabilizing the hydrological cycle within an anthropogenically transformed catchment area, Fig. 14.1, (Zalewski, 2015).

In the case of reservoirs, the most important processes that have been modified as a result of anthropogenic pressure are the nutrient cycles of carbon, nitrogen and phosphorus, as they determine the fertility of the reservoir and, in the case of degraded reservoirs, excessive fertilization (hypertrophy). Accelerated eutrophication can lead to secondary

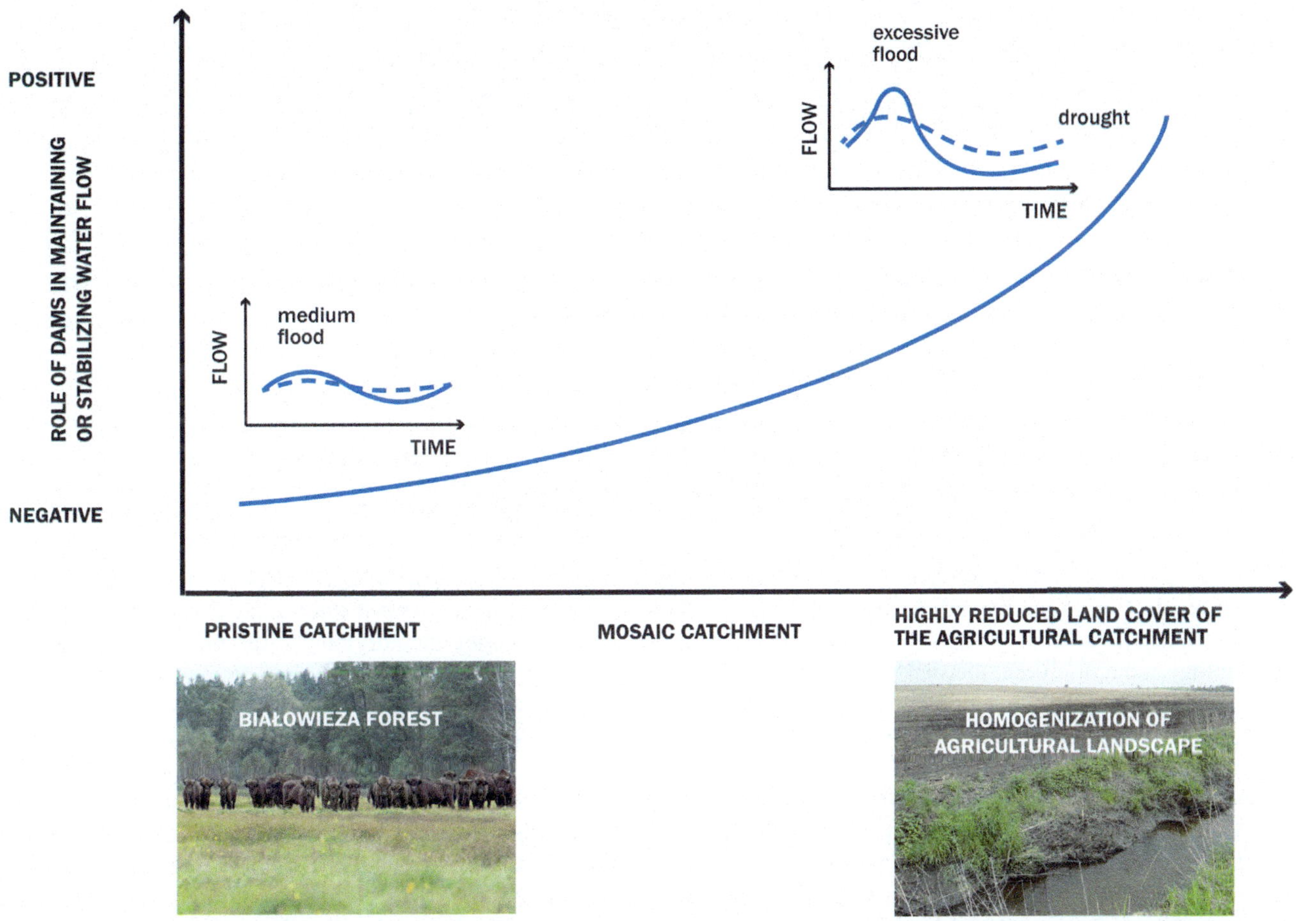

FIGURE 14.1 Reservoirs and dams. How can threats be transformed into opportunities for sustainable river basin management through ecohydrology (EH) process-oriented thinking? When river outflow is accelerated and landscape evaporation is intensified by human activities, reservoirs become necessary to compensate for disruptions to the natural water cycle. *Inspired by concepts Zalewski, M., 2011. Ecohydrology for implementation of the EU water framework directive. Water Manag. 164 (8), 375–385.*

contamination of the reservoir with toxic cyanobacterial blooms, which are dangerous to human health and the proper functioning of the reservoir, and can also cause secondary release of nutrients into the water, mainly phosphorus accumulated in bottom sediments.

Reservoirs are more complex than natural lakes. They are characterized by dynamic water inflow, large amounts of sediments and a significant load of nutrients and pollutants. In most cases, lakes are a closed ecosystem in which the accumulated pool of nutrients is subject to circulation. In the case of natural lakes, cutting off sources of diffuse pollution from the immediate catchment area and blocking internal sources, e.g., by covering bottom sediments with phosphorus-binding compounds and removing eutrophic bottom waters, e.g., using the Olszewski Jack device, is usually sufficient to achieve good ecological results. However, this is often not sufficient in the case of reservoirs.

Reservoirs, unlike natural lakes, are open ecosystems that are supplied with water volumes several orders of magnitude greater than those of lakes. Moreover, due to frequent water-level fluctuations, they usually have poorly developed littoral zones. This results in lower resilience, as a limited littoral area reduces the system's capacity to absorb nutrients and provides fewer refuges for zooplankton filter feeders. In addition, the lack of macrophytes diminishes allelopathic inhibition of algal growth.

Therefore, to protect and reverse the process of eutrophication, the first step within the Transdisciplinary Ecohydrology (TEH) framework is to reduce nutrient pollution across the entire catchment. The second step involves enhancing ecosystem resilience by increasing nutrient absorption capacity—for example, through the shaping of littoral zones, hydrobiomanipulation, and reduction of nonpoint source pollution. Both steps require a comprehensive diagnosis and quantification of underlying processes.

A strategy for the restoration of water bodies based on the concept of transdisciplinary ecohydrology should include four elements: (1) monitoring of threats and their dynamics, (2) analysis of cause-and-effect relationships, (3) development of fine-tuned ecohydrological methods and tools, and (4) planning of systemic solutions leading to sustainable development and good ecological status (Zalewski, 2002). Due to the complexity of hydrological, biological, social, and economic processes influencing the dynamics and use of water resources within a catchment, reducing pressures at the catchment scale enhances the resilience of the reservoir. Therefore, adopting a systemic restoration approach that considers the entire catchment area is essential.

2. Determining the hierarchy of factors influencing the intensity of cyanobacterial blooms in the reservoir

2.1 Hydrodynamics—Factors driving phytoplankton succession

In the case of flow-through reservoirs, hydrological processes are the key factor determining phytoplankton succession and interactions between phytoplankton and higher trophic levels (Wojtal et al., 2008; Zalewski et al., 1990). The increasing scale and frequency of extreme hydrological events (floods and droughts), which destabilize surface runoff in the catchment area, mean that hydrological conditions will increasingly play a dominant role in regulating the dynamics of biotic processes in reservoirs.

In the case of reservoirs, the decisive factor for the spatial diversity of the phytoplankton community is the variable hydrodynamic conditions prevailing in different parts of the reservoir. We distinguish three main zones in a reservoir: *riverine*, *transitional* and lac*ustrine* Fig. 14.2. The size of the zones varies in individual reservoirs and depends on morphology, retention time, thermal stratification, season and geographical location. In deep reservoirs of the temperate zone, if the retention time in summer is less than 10 days, the riverine zone may cover the entire reservoir, while if the retention time is longer than 200 days, the riverine zone is short and most of the reservoir is lacustrine in nature (Kimmel and Groeger, 1984).

Individual parts of the dammed reservoir differ in their morphological, physical and chemical parameters. As one moves toward the dam, the reservoir widens and deepens. The morphological variability of the reservoir determines the variation in flow intensity in its individual zones. As one moves away from the river mouth, the flow rate decreases. The amount of suspended matter carried by the tributary is also reduced, which improves the light conditions in the water. In the river section, the nutrient load is high, mainly due to external inputs. The lower content of biogenic substances in the

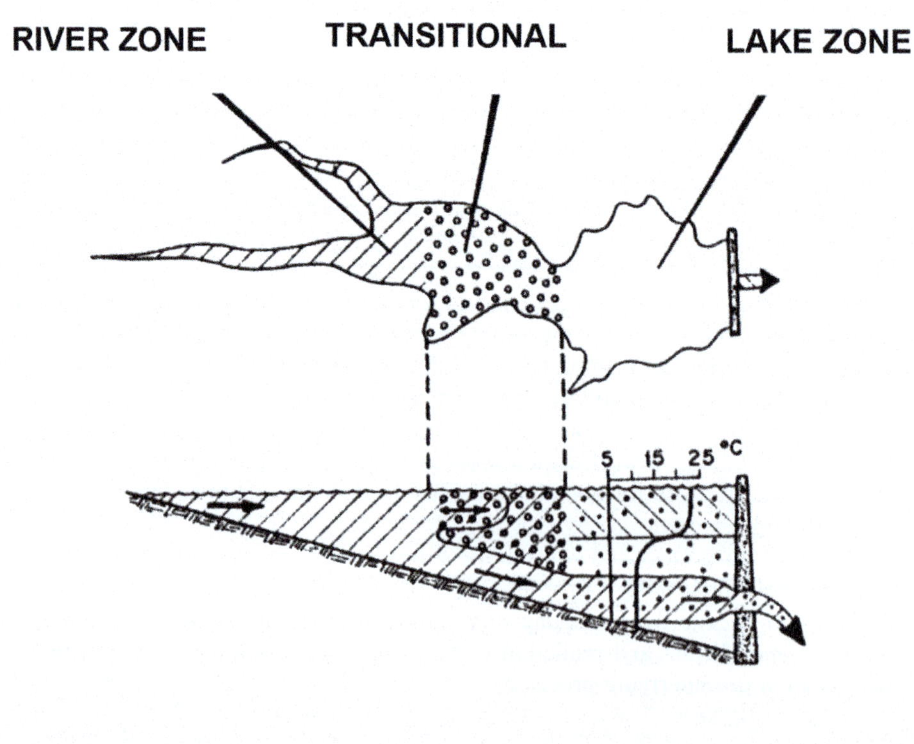

FIGURE 14.2 Zones in a reservoir. Distribution of individual zones in dammed reservoir. *Tundisi, J.G., Straškraba, M., 1999. Theoretical Reservoir Ecology and Its Applications. Backhuys Publishers.*

lake section of the reservoir is supplemented by internal inputs. Different physicochemical conditions shape the phytoplankton community. Primary production in the river section is mainly limited by light and cell sedimentation. In the lake section, however, limitation by nutrients and grazing by herbivores intensifies (Tundisi and Straškraba, 1999).

In the case of seasonal succession of phytoplankton communities in a dammed reservoir, another important factor influencing its course is **the water retention time** in the reservoir, which determines the flow velocity. A period of several days to a week is necessary for the development of phytoplankton communities. Too short a water retention time prevents the development and growth of the phytoplankton population (Kajak, 1995). According to Kawara et al. (1998), a retention time of less than 14 days is insufficient for phytoplankton growth. A short water retention time in the reservoir also contributes to increased water turbulence, which has a beneficial effect on the development of species that prefer unstable conditions, i.e., diatoms. Extending the retention time during periods of warm, windless weather contributes to increased stratification of the reservoir and the creation of stable conditions in the water column, which stimulates the development of cyanobacteria and promotes the formation of blooms (Ressom et al., 1994). A sudden reduction in retention time, e.g., as a result of a flood wave, causes the physical flushing of phytoplankton from the reservoir (Tundisi and Straškraba, 1999). The indirect impact of retention time on phytoplankton results from its influence on environmental parameters. As the retention time of a reservoir increases, so does the retention time of nutrients. This means that eutrophication occurs more frequently in reservoirs with long retention times than in flow-through reservoirs with retention times of less than 20 days (Tundisi and Straškraba, 1999). A multi-year analysis showed that in the Sulejów reservoir, the growth of cyanobacterial blooms was favored by the extension of water retention time. Extending the water retention time from 10 to 100 days resulted in a six-fold increase in cyanobacterial biomass (Tarczyńska et al., 2001), see Chapter 18 and 19, while shortening the water retention time to less than 30 days seems to be a factor limiting the development of cyanobacteria (Izydorczyk, 2003).

The hydrodynamic conditions in the reservoir depend on the water retention time as well as the morphometry of the reservoir. The varied shape of the Sulejów reservoir determines the horizontal variation **in flow velocity**, which may affect the spatial variation of the phytoplankton community Fig. 14.3a and c. In spring, the main factor influencing phytoplankton composition is low water temperature. Diatoms, especially colonial feather-like forms, are strong potential candidates for dominance in phytoplankton composition in spring due to their photosynthetic behavior, cell size, surface-to-volume ratio, and ability to maintain overwintering vegetative populations in suspension in ice-free lakes (Reynolds, 1980). In spring, the phytoplankton community throughout the reservoir was dominated by diatoms. Heterogeneous hydrodynamic conditions did not cause horizontal variability in the phytoplankton community. Nevertheless, during the summer, the composition of the phytoplankton community differed at the two stations analyzed. In the lake part (station I), characterized by lower flow velocity, cyanobacteria (mainly *Microcystis aeruginosa)* dominated. Despite the high water temperature – the most important factor contributing to the dominance of cyanobacteria – in the upper, riverine part (station II), despite the same temperature, diatoms dominated the phytoplankton. The higher flow velocity caused increased water turbulence and prolonged suspension of immobile diatoms, which reach their maximum biomass in the Sulejów reservoir.

Intensive mixing of the water column prevents the occurrence of cyanobacterial blooms, therefore increasing turbulence in a reservoir of type a may reduce both the formation and accumulation of cyanobacterial blooms. Water column mixing can be increased by:

(1) shortening the water retention time (e.g., adjusting the reservoir volume to climatic conditions in newly constructed reservoirs),
(2) increasing the flow velocity (e.g., by constructing barriers that extend the water flow path within the same water retention time in an existing reservoir); and
(3) maximizing wind-induced turbulence (e.g., by positioning a newly constructed reservoir to maximize wind mixing) (Wagner et al., 2004).

2.2 Dynamics of phosphorus availability in water reservoirs and consequences for cyanobacteria

Long-term analysis has shown that phosphorus availability is a key factor determining the intensity of cyanobacterial blooms in the Sulejów reservoir (fig. AW-F or PF graph from hydro manipulation, TP vs. cyanobacterial biomass). Many authors confirm that high nutrient levels lead to increased intensity of cyanobacterial blooms (see Chapters 23 and 24).

Flow-through reservoirs (such as the Sulejów reservoir) are characterised by **a continuous inflow of nutrients**, which is not constant from year to year or throughout the year Fig. 14.3b. The concentration of nutrients in the reservoir differed significantly from the water outflow pattern, which in turn affected the intensity of cyanobacterial blooms. Studies by

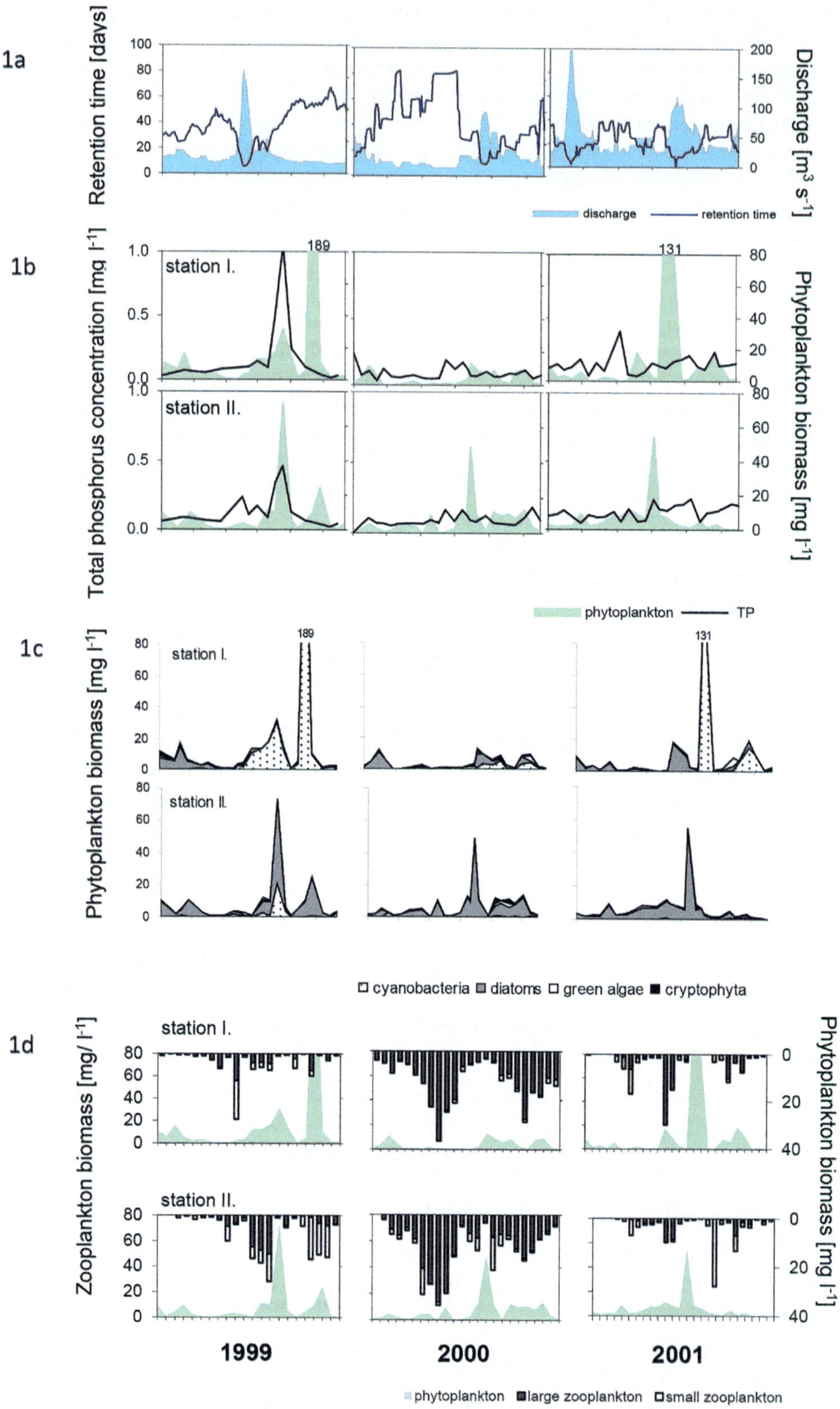

FIGURE 14.3 Seasonal dynamics of water retention time and outflow. Seasonal dynamics of water retention time and discharge (1a), total phosphorus concentration (1b), phytoplankton (1c) and zooplankton (1d) composition for Sulejow Reservoir between 1999 and 2001. *Modified from Izydorczyk, K. (2003). Wpływ czynników środowiskowych na czasoprzestrzenne zróżnicowanie fitoplanktonu w Zbiorniku Sulejowskim w latach 1999–2001: Rola temperatury wody, natężenia przepływu oraz turbulencji wywołanych przez wiatr w kształtowaniu dynamiki zespołu fitoplanktonu (Doctoral dissertation). University of Łódź.*

Wagner and Zalewski (2000) showed that the highest concentrations and loads of phosphorus reach the reservoir through the main tributaries during short periods of peak river flow. The hydrochemical composition of the inflowing waters, although dependent on several factors, differs significantly from the water outflow pattern. The highest concentrations usually occur during the rising phase of medium floods and in the very early stages of rapid floods (Wagner and Zalewski, 2000). The hierarchy of factors influencing the concentration of nutrients in water also varies seasonally (Wagner and Zalewski, 2016). Furthermore, in the case of agricultural catchment areas, such as the Pilica River catchment area, which feeds the Sulejowski Reservoir, the amount of precipitation will be a determining factor in the load discharged from the catchment area to the reservoir.

As a result of the increase in the load from the catchment area, high concentrations of nutrients are observed in the reservoir. According to *the bottom-up* concept, this may stimulate phytoplankton growth and increase the intensity of cyanobacterial blooms (Fig. 14.3a and b). In 1999, as a result of a flood wave, high phosphorus concentrations were recorded, which, combined with long water retention times, high temperatures and reduced zooplankton pressure, stimulated the formation of a particularly large cyanobacterial bloom. In contrast, in the spring of 2001, despite increased nutrient concentrations after the spring flood, phytoplankton biomass remained low due to strong zooplankton feeding pressure Fig. 14.3d. The floods in July 2000 and August 2001, which were smaller in amplitude and lasted longer, did not lead to an increase in phosphorus concentration. The short water retention time (15–30 days) combined with high water outflow diluted phosphorus loads and prevented phytoplankton growth (Izydorczyk, 2003).

In the case of flow-through reservoirs, suspended organic matter supplied to the reservoirs settles at the bottom of the reservoir due to the reduced water flow velocity (Tundisi and Straškraba, 1999) and constitutes a potential **internal source** of nutrients. Phosphorus recovered from sediments often contributes significantly to the phosphate resources in water, which promotes phytoplankton blooms (Jensen and Andersen, 1992; Rydin, 2000). Physical and chemical processes, depending on the chemical properties of sediments and their changes (pH, redox potential, organic matter, Al and Fe, inorganic P equilibrium), are responsible for the retention of phosphorus in sediments and its release into the water column (Golterman, 1995; Psenner et al., 1988). However, the activity of microorganisms, especially hydrolysis catalyzed by one of the microbial enzymes, alkaline phosphatase, seems to be crucial for phosphorus transformation in sediments. Phosphatase is mainly released by bacteria in response to a lack of soluble phosphates in water, which is why it is considered very important in the phosphorus cycle in aquatic and soil environments (Chróst and Siuda, 2006; Jansson et al., 1988). Trojanowska and Izydorczyk (2010) showed that high alkaline phosphatase activity in sediments in spring may be an important factor influencing the formation of cyanobacterial blooms in summer, which was also hypothesized by Zhou et al. (2001, 2002). Analysis of sediments from the Sulejów reservoir showed higher organic matter content and total phosphorus concentration in spring than in summer, which was due to the deposition of phosphorus supplied by floods Fig. 14.4. In spring, an increase in the total number of bacteria and alkaline phosphatase activity in the sediments was observed, which resulted in the release of nutrients into the water column and promoted the development of cyanobacterial blooms in summer (Trojanowska and Izydorczyk, 2010).

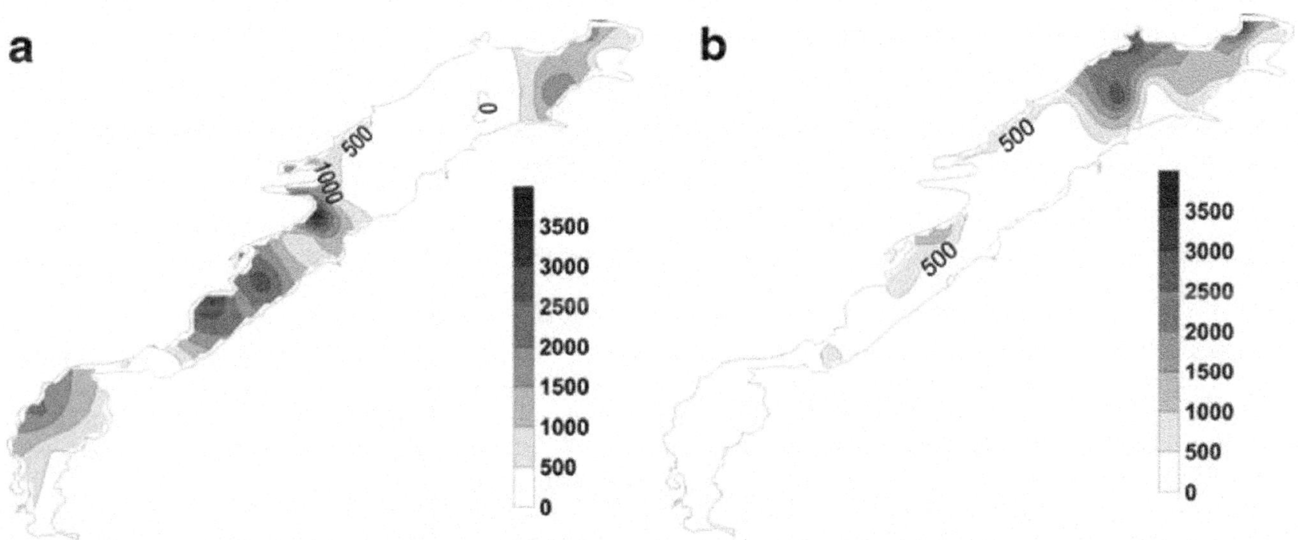

FIGURE 14.4 Seasonal variations in phosphorus content in the Sulejów reservoir Seasonal variation in phosphorus content [μg TP/g d.w.] in the surface layer of bottom sediments in the Sulejowski reservoir in May (a) and August (b) 2005. *Trojanowska, A.A., Izydorczyk, K., 2010. Phosphorus fractions transformation in sediments before and after Cyanobacterial bloom: implications for reduction of eutrophication symptoms in dam reservoir. Water Air Soil Pollut. 211 (1–4), 287–298. https://doi.org/10.1007/s11270-009-0299-y.*

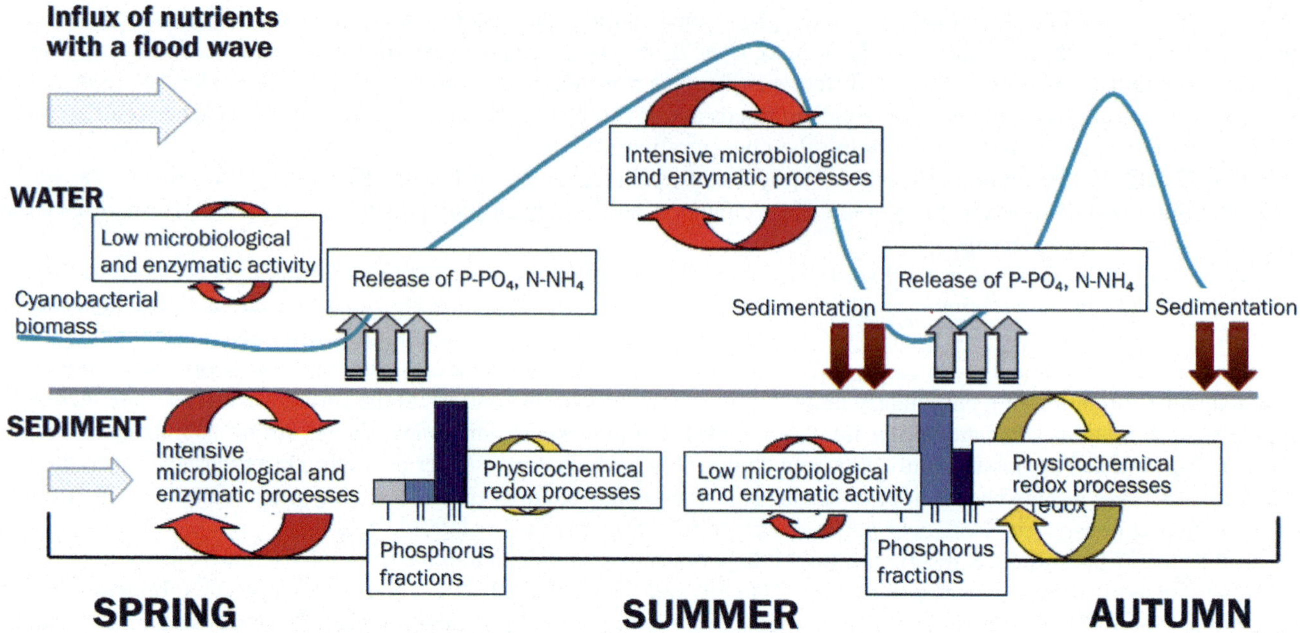

FIGURE 14.5 Seasonal dynamics of microbiological and physicochemical processes in sediment–water interactions. Seasonal variability of microbiological and physicochemical processes at the sediment–water interface, showing shifts in the dominant mechanisms of nutrient cycling and self-purification within reservoir ecosystems. *Trojanowska (2010) project report; ERCE PAN*

In spring, an increase in the number of bacteria and phosphatase activity in sediments was observed, which may indicate intensive decomposition processes and the release of labile biogenic components into the water column, which could have contributed to the formation of cyanobacterial blooms. In summer, however, sedimentation and physicochemical release of phosphorus into the water column associated with a local decrease in redox potential played a greater role. This process was observed in areas characterized by high deposition of organic matter from dying cyanobacterial blooms, the decomposition of which caused oxygen depletion Fig. 14.5.

2.3 Quantification of pollution sources on a catchment scale

According to the first principle of ecohydrology, the processes occurring in a reservoir, especially the water balance and nutrient flows, should be considered in the context of the entire catchment area. The Sulejów reservoir catchment area covers approximately 4900 km^2. Reducing the inflow of nutrients from the catchment area is a key element in the process of reversing the eutrophication of the reservoir. Estimating the inflow of nutrients from various point and diffuse sources is the first step toward regulating processes for sustainable water use and ecosystem protection. GIS-based hydrological models are tools used to quantify nutrient flow paths and identify priority areas for nutrient leaching.

The results of the MONERIS model showed that only about 7% of nitrogen and phosphorus loads came from point sources in the Pilica river basin, and over 90% came from diffuse sources, Fig. 14.6 (Kowalkowski, 2012). Assuming that approximately 8% of nitrogen loads and 20% of phosphorus loads came from urban sources, agriculture could have been one of the most important factors responsible for water pollution, resulting from the high (61.3%) share of agricultural land in the Pilica catchment area, including 40.7% of arable land and 11.2% of meadows and pastures. Many studies have shown that agricultural activity is the main source of elevated nitrate concentrations in soils (Howarth, 2004; Pastuszak et al., 2012). Nitrogen and phosphorus, as elements essential for plants, are used as natural and artificial fertilizers to increase yields. However, nutrient loss from agricultural land occurs when intensive fertilizer use is not balanced by plant demand and consumption. Nitrogen is a highly mobile element due to its high solubility and presence in gaseous and ionic forms. Nitrogen losses from agricultural soils include: (1) volatilization of ammonia from soil and manure stores into the atmosphere; (2) leaching of mobile nitrates from the soil and their transfer through groundwater, surface drainage water and drainage ditch water to surface waters; and (3) surface runoff to surface waters. In the Pilica River basin, 65% of nitrogen emissions to river basins occurred *through* groundwater and pipe drainage Fig. 14.7a. Unlike nitrogen, the movement of phosphorus in soil or water is one of the slowest biochemical cycles. Phosphorus occurs in soil in organic and inorganic forms and can be dissolved in water or deposited in minerals and organic materials. Phosphorus losses to

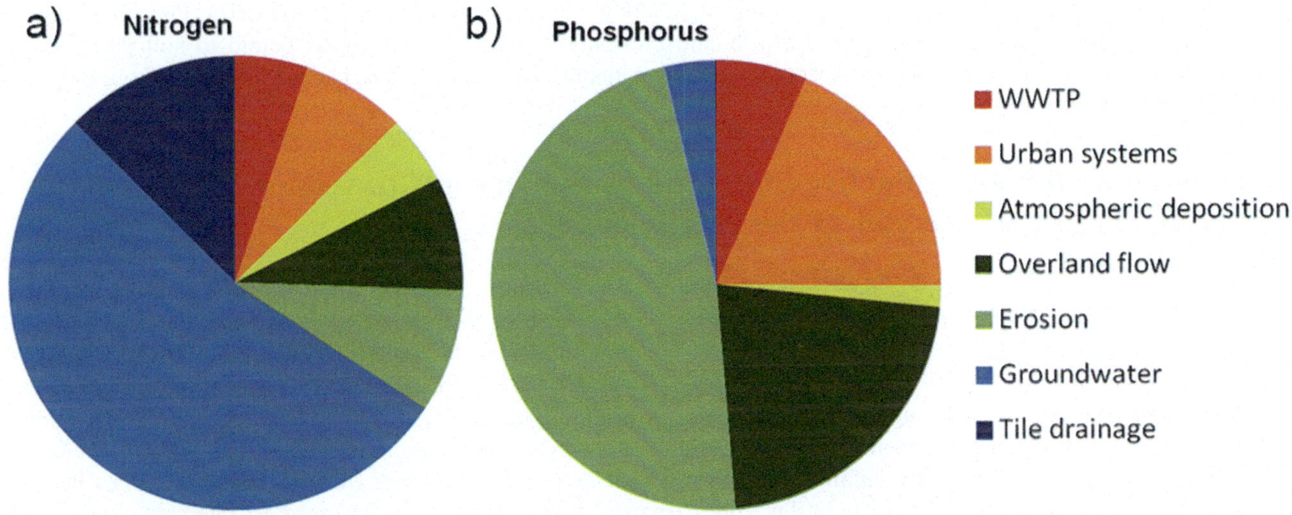

FIGURE 14.6 Different pathways of nitrogen and phosphorus emissions. Percentage share of different pathways of nitrogen and phosphorus emissions to the Sulejów reservoir from the Pilica River basin. *Data from LIFE + EKOROB project report, ERCE PAN.*

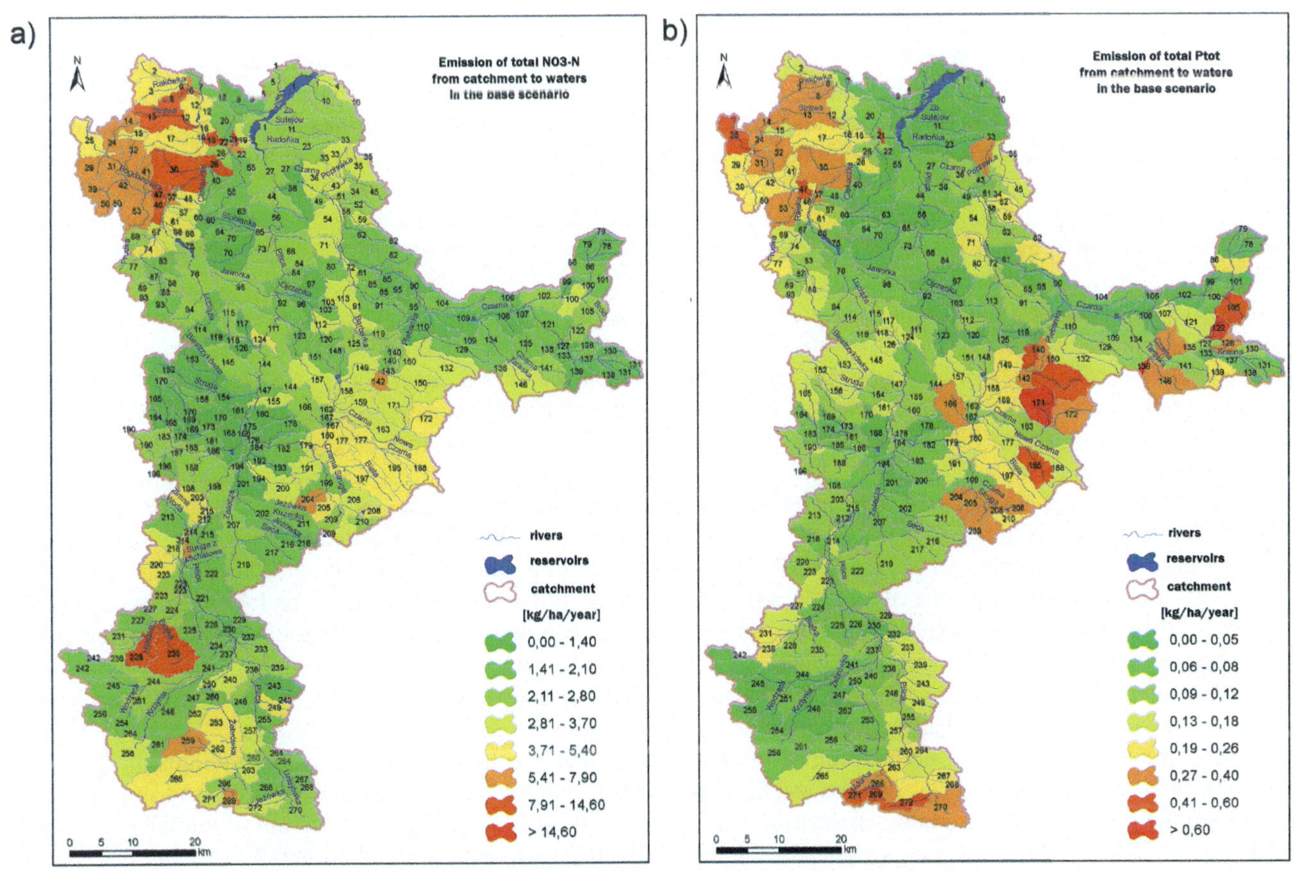

FIGURE 14.7 Pilica river basin nonpoint and point-source pollution distribution. Distribution of areas contributing the highest loads of total phosphorus (a) and nitrate nitrogen (b) within the Pilica River basin, shown in relation to administrative divisions. *Piniewski, M., Marcinkowski, P., Kardel, I., Giełczewski, M., Izydorczyk, K., Frątczak, W., 2015. Spatial quantification of nonpoint source pollution in a meso-scale catchment for an assessment of buffer zones efficiency. Water 7 (5), 1889–1916. https://doi.org/10.3390/w7051889.*

groundwater and surface water can occur as a result of leaching or surface runoff with soil particles. Phosphorus losses due to leaching are insignificant, while sediment-bound phosphorus includes phosphorus bound to soil particles and organic material eroded during flows and accounts for approximately 51% of the phosphorus transported from the Pilica river basin Fig. 14.7b.

3. Ecotones and highly effective buffer zones as measures to reduce the transfer of pollutants to the reservoir at the catchment scale

3.1 Riparian ecotones

The response to the need to reduce eutrophication of reservoirs, and consequently toxic cyanobacterial blooms, as well as to increase biodiversity and adaptation to climate change in agricultural areas, is ecohydrological biotechnology, which is part of the current mainstream research on the development of nature-based solutions (NBS). These solutions are defined as solutions that are inspired by or supported by nature, are cost-effective and provide environmental, social and economic benefits, and help to increase the resilience of ecosystems. Nature-Based Solutions (NBS) in agricultural landscapes include structurally and functionally significant elements such as riparian ecotones (natural transition zones between terrestrial and aquatic systems), riparian buffer zones (managed vegetated strips designed to intercept nutrient and sediment fluxes), field margins, scattered trees and shrubs, field ponds, oxbow lakes, and wetlands. This group can be expanded to include ecohydrological biotechnologies—installations that harness natural processes and enhances their performance such as sequential sedimentation–biofiltration systems (Chapter 17), or highly effective buffer zones (Chapter 23) representing an Advanced Nature-based Solutions (ANbS).

It is necessary to reverse the trend of degradation of the resilience of the biogeochemical cycles of river basins through industrial agricultural development, "concrete" urbanization. An integral part of such overengineering development is the unification of the agricultural areas by eliminating land-water buffer zones, as well as forest areas and wetlands, where intensive denitrification of the overload of nitrogen by the conversion of nitrogen compounds into gaseous form occurs, thereby reducing the nitrogen load in the landscape. These mosaic elements of the landscape are also places where phosphorus compounds accumulate. Disruption and reduction of these natural components of the agricultural landscape and processes cause also the release of organic and mineral compounds into the water, which in turn leads to the overfertilization of aquatic ecosystems. The homogenization of the agricultural landscape also leads to soil drying and further loss of organic matter and nutrients (phosphorus and nitrogen) as a result of increased wind and water erosion (Levia et al., 2020).

Measures to reduce the transfer of nutrients (phosphorus and nitrogen) from terrestrial ecosystems to water should be the first and fundamental element of any water body restoration. One of the tools for this purpose is the maintenance of riparian buffer zones in river valleys, as well as the creation of new ecotones as an effective tool for nutrient retention and sedimentation (Ryszkowski and Kędziora, 1993; Hillbricht-Ilkowska and Pieczyńska, 1993; Hillbricht-Ilkowska et al., 1995; Doskkey et al., 2010; Pärn et al., 2012), which significantly reduces the load of trophic substances entering reservoirs along with rivers.

Studies conducted in the Pilica river basin have shown that buffer zones consisting of extensive meadows contribute to a reduction in the concentration of nutrients in shallow groundwater, even with a high load entering the zone. The high concentration of nitrates, twice exceeding the level of 50 mg $NO_{((3))/l}$ introduced by the Nitrates Directive as the threshold value for polluted waters, was reduced by 68% and 99% as a result of passing through a 25- and 45-meter-wide meadow Fig. 14.8. In the case of heavy pollution of waters with phosphorus compounds (above 1.5 mg $PO((4)_{)/l})$, the effectiveness was 81% and 76% for zones 45 and 47 m wide, respectively (Izydorczyk et al., 2018), see Fig. 14.8.

Buffer zones effectively reduce nitrogen, and phosphorus loads from diffuse pollution sources through biogeochemical processes occurring at the soil-water-organism interface (Dosskey et al., 2010; Pärn et al., 2012).

One of these processes is the assimilation of inorganic compounds, including nitrogen and phosphorus, by plants and their conversion into biomass. Strips of grass and herbaceous plants, with or without deep-rooted trees and shrubs, along a stream can help improve water quality (Balestrini et al., 2011; Hefting et al., 2006). The amount of nutrients that can be taken up and stored in plant tissues varies depending on the plant species and can range from 0.2 to 50 kg P ha^{-1} $year^{-1}$ and from 10 to 350 kg N ha^{-1} $year^{-1}$ (Hefting et al., 2006; Kiedrzyńska et al., 2008; Mander et al., 1997). However, it should be noted that plants retain nutrients only for a limited time, mainly during the growing season, which requires biomass harvesting (e.g., mowing or cutting) to remove the accumulated material (Kuusemets and Lohmus, 2005; Uusi-Kämppä, 2005). This practice can reduce the risk of nutrient release during dormancy and their transport by surface runoff or accumulation in the topsoil (Räty et al., 2010).

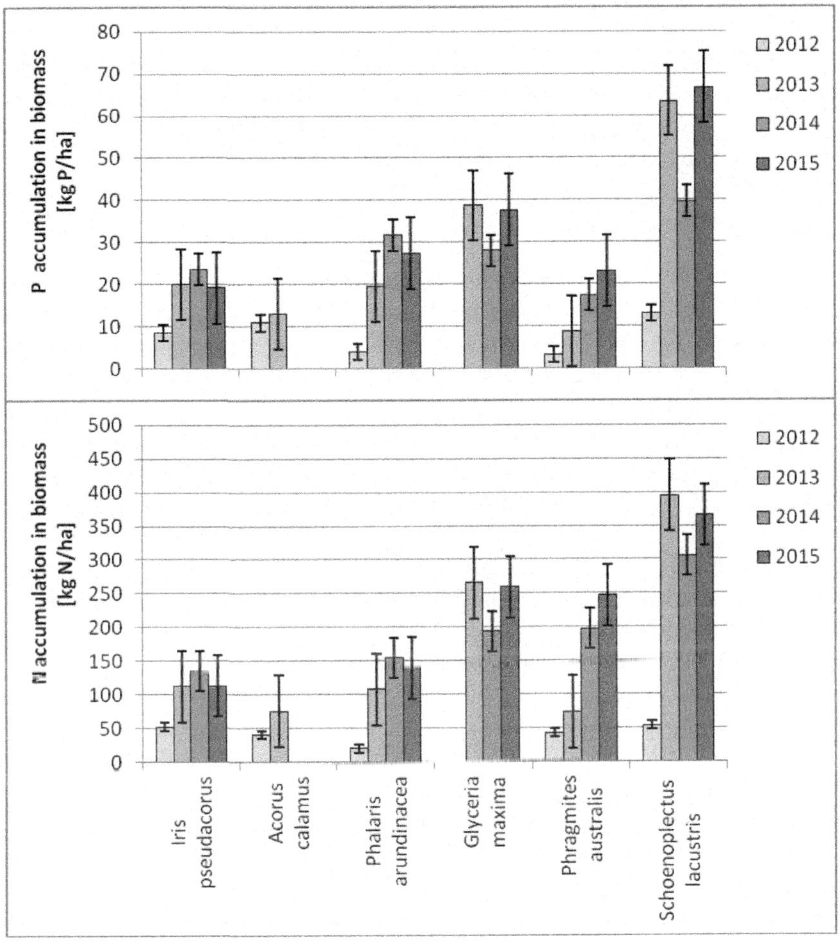

FIGURE 14.8 The effectiveness of plant species in phosphorus and nitrogen removal over a 4-year experimental period. Plant phosphorus and nitrogen uptake obtained in the area where selected plant species were dominant. *Frątczak, W., Michalska-Hejduk, D., Zalewski, M., Izydorczyk, K., 2019. Effective phosphorous reduction by a riparian plant buffer zone enhanced with a limestone-based barrier. Ecolog. Eng., 130, 94–100. https://doi.org/10.1016/j.ecoleng.2019.01.015.*

The role of buffer zones in phosphorus capture is related not only to the process of nutrient accumulation in plant tissues, but also to the binding of soil by the plant root system, which reduces erosion and blocks the flow of matter into rivers and water bodies. Insoluble forms of phosphorus dominate in surface runoff from agricultural areas, and their share increases with increasing flow (Johannesson et al., 2011). Therefore, sedimentation within the buffer zone is important for reducing the phosphorus load in the runoff from the river basin. Dense vegetation (especially tall, stiff grasses with dense stems) increases hydraulic roughness and reduces surface flow velocity and sediment transport capacity. A 0.6 m wide and 20–30 cm high strip of Bermuda grass reduced sediment runoff from a cultivated area by an average of 63% compared to an area cultivated without a barrier (Dillaha et al., 1989). Soil particles with phosphorus absorbed on the surface of molecular sediments are retained in buffer zones. Sedimentation can account for phosphorus concentrations of up to 128 kg P ha^{-1} $year^{-1}$ (Hoffmann et al., 2009).

Buffer zones are also a place where nitrogen compounds are converted by denitrifying bacteria. Denitrification is a key remediation process that results in the release of nitrogen. Denitrification can be intensified by regulating the water level in the reservoir itself, which contributes to increased sedimentation of organic matter and reduced redox potential (reduction and oxidation) in the riparian and littoral zones (Bednarek et al., 2010).

The vegetation of buffer zones has a beneficial effect on the microclimate by regulating water temperature and light availability in the riverbed. In addition, these zones also significantly accelerate the self-cleaning process of rivers. Developed riparian vegetation accelerates the decomposition of organic matter, both native and brought in with the pollutant load, provided that this load does not exceed the ecological resistance of the ecosystem appropriate for the habitat type and stage of succession. These processes depend on the availability of light, productivity and biodiversity in a given habitat (Zalewski et al., 1997; Zalewski and Naiman, 1985).

Furthermore, it should be emphasized that in a monocultural agricultural landscape, ecotones are biodiversity refuges, forming a network of ecological corridors along river valleys. Furthermore, the diversity of plants in buffer zones stimulates the richness of soil microorganism communities and controls the rate of element circulation, while species abundance initiates many processes in the ecosystem, which in turn further enhances its multifunctionality. Thus, riparian ecotones provide environmental benefits, help increase resilience, and at the same time support the provision of ecosystem services, bringing social and economic benefits to agricultural production areas.

3.2 Enhanced buffer zones for reducing diffuse pollution

One of the challenges associated with the construction of ecotones is the nature of the coastline, which is often too narrow for the zone to function effectively. This problem was one of the challenges for the project "EKOROB – Ecotones for the reduction of diffuse pollution" (LIFE08 ENV/PL/00518), funded by the European Commission and implemented by the European Center for Ecohydrology of the Polish Academy of Sciences and the Regional Water Management Authority in Warsaw (Frątczak and Izydorczyk, 2015). The aim of the project was to develop ecohydrology-based solutions for the reduction of diffuse pollution, mainly from agriculture, using highly efficient buffer zones created in small areas. Prototypes of advanced, demonstrative and typologically diverse buffer zones were developed in areas constantly exposed to the inflow of nutrients into inland waters directly from the Sulejów reservoir catchment area) (Izydorczyk et al., 2015).

The basis for the construction of highly efficient buffer zones was planting along the shoreline. Natural plant species found in the Pilica valley were used in the created zones. Biodegradable geotextiles were used for planting to improve the rooting process. When creating new, constructed buffer zones, the habitat preferences of plants, their resistance to abiotic conditions and their effectiveness in converting nutrients into biomass were taken into account. As demonstrated in studies on the Sulejowski reservoir, of the five willow species considered, only the native species *Salix fragilis* was able to withstand the hydrological stress present in the area, and the most effective community was *Phragmitetum australis* (Kiedrzyńska et al., 2008). In addition, it is recommended to use native species in plantings, as well as multi-species plantings, which increase landscape value and terrestrial biodiversity. Furthermore, a multi-species zone allows for an extension of the period during which the buffer zone functions most effectively. For example, in central Poland, *Acorus calamus* and *Iris pseudacorus* grow intensively immediately after the snow melts (April/May), while common reed *Phragmites australis* is characterized by high biomass, which appears relatively late (June/July) and peaks in autumn (September). Furthermore, the use of a multi-species approach in the newly created buffer zone takes into account the time needed for plants to fully develop in accordance with natural ecological succession. In the case of the highly effective zone above the Sulejów reservoir, most of the selected species grew rapidly in the second year after planting. The exception was *Phragmites australis,* which took 3 years to take root. *Acorus calamus,* on the other hand, disappeared in the third growing season, Fig. 14.8 (Frątczak et al., 2019).

The essence of the reinforced buffer zones was the use of additional installations within the plant zones, in accordance with the principle of adapting solutions to previously identified threats.

In the case of nitrate contamination in shallow groundwater (approximately 200 mg NO_3/l), the plant buffer zone was reinforced with a denitrification wall Fig. 14.9. This is a structure used to intensify the denitrification process, which significantly reduces the amount of nitrogen in the environment by releasing it into the atmosphere. A denitrification wall is constructed by creating a trench perpendicular to the direction of nitrogen-contaminated water flow underground and filling it with slowly decomposing organic material, such as sawdust. This provides a source of carbon for heterotrophic denitrifying bacteria. The effectiveness of denitrification barriers ranges from 50% to 95%, and the amount of nitrogen removed ranges from 2 to 20 g N*m (−3) *day (−1) (Bednarek et al., 2010; Schipper et al., 2010). The results of 4 years of operation of a high-efficiency buffer zone above the Sulejów reservoir showed the effectiveness of the denitrification wall at 67%, Fig. 14.10, (Frątczak et al., 2013).

In an area with high phosphorus concentrations in shallow groundwater (average 5.5 mg PO_4/l), mainly resulting from leaks from septic tanks, the use of a limestone barrier as part of the buffer zone was proposed Fig. 14.11. The limestone barrier was constructed to increase phosphorus removal from water by increasing its adsorption capacity in the soil through an increase in calcium ion content. It should be emphasized that the chemical binding of phosphorus by calcium is more durable than in the case of reactions with aluminum or iron. Furthermore, it poses less risk to the environment. The prototype barrier, as part of the buffer zone, was constructed in the form of a ditch filled with limestone (5–8 cm aggregate), protected against clogging with geotextile. In the first year of the barrier's operation, the concentration of phosphates in groundwater decreased by 58% (Izydorczyk et al., 2012).

An important element of the implementation of buffer zones at the Sulejów reservoir was raising the environmental awareness of local communities, (Izydorczyk et al., 2015). Despite the undeniable benefits, the construction of buffer

FIGURE 14.9 Enhanced buffer zone for reducing diffuse nitrogen pollution. Denitrification wall was constructed by digging a trench perpendicular to groundwater flow and by mixing pine sawdust with soil. *Izydorczyk, K., Michalska-Hejduk, D., Frątczak, W., Bednarek, A., Łapińska, M., Jarosiewicz, P., Kosińska, A., Zalewski, M., 2015. Strefy buforowe i biotechnologie ekohydrologiczne w ograniczaniu zanieczyszczeń obszarowych.*

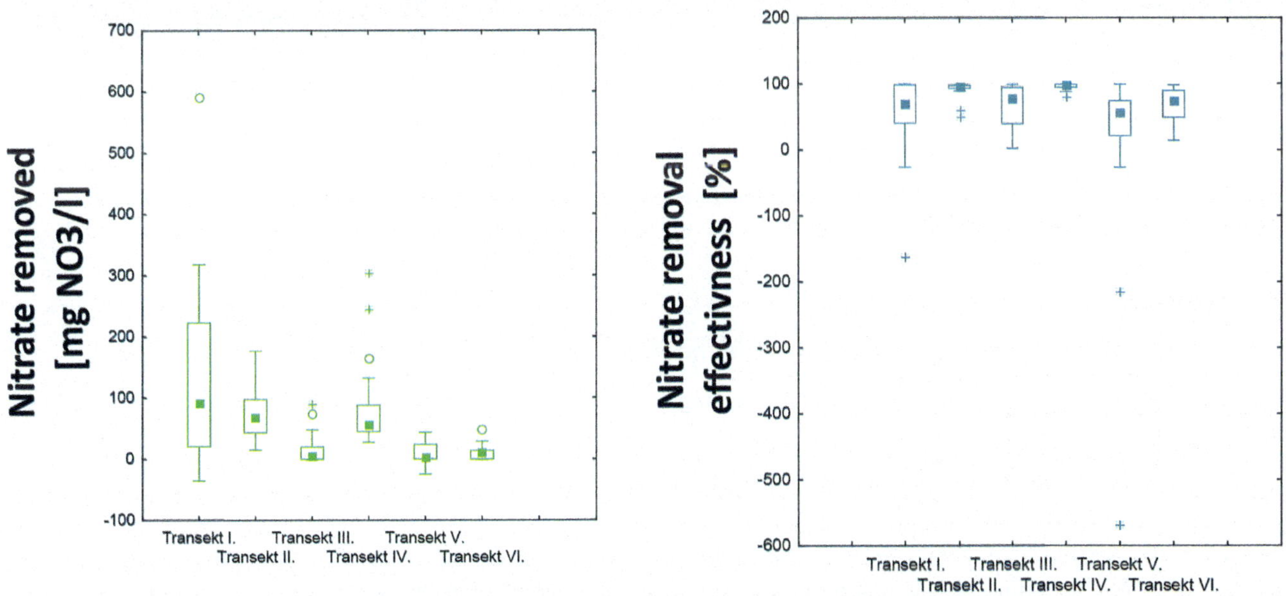

FIGURE 14.10 Empirical evidence for the effectiveness of a denitrification wall. Box plot notation: squares represent median, boxplots illustrate 25th and 75th percentiles, whiskers indicate points which are not outliers, points represent outlying observations and a cross represents extremes. These diagrams show that the effectiveness of a denitrification wall in a buffer zone was approximately 67%. The difference between nitrate concentrations in the inflow and those recorded within the denitrification wall was analysed over a four-year period. *Data from LIFE + EKOROB project report, ERCE PAN.*

zones in river valleys and water areas may meet with resistance from the inhabitants of the area. Therefore, it is important to include buffer zones in recreational infrastructure projects, such as footbridges, piers, designated picnic areas, beaches and nature trails. The creation of a buffer zone also creates new jobs for residents, as these areas require maintenance. The periods of nutrient release and accumulation are linked to the stages of succession. At the beginning, when plants are in the growth phase, the ability to retain nutrients is high, but over time it stabilizes and reaches a balance between retention and release. Therefore, it is necessary to maintain the buffer zone in the early stages of succession through careful treatment. Recreational infrastructure, footbridges and piers also require maintenance. The connected platforms along the shore not only serve as recreational areas for water sports enthusiasts on the Sulejów reservoir but also act as a breakwater for the buffer zone, protecting it from erosion and damage to vegetation.

FIGURE 14.11 Reduction of diffuse phosphorus pollutions. Enhanced buffer zone for reduction of diffuse phosphorus pollution (riparian plant buffer zone with biogeochemical limestone-based barriers). (a) The first barrier was constructed in the stream valley mouth to limit pollution of the groundwater and the stream's surface water, which directly flows into the reservoir. Apart from its purifying function, the biogeochemical barrier also performs the role of damming, which extends the time of flow of polluted water into the reservoir. (b) The second barrier forms a scarp along the reservoir shoreline. Limestone and dolomite were used to purify seepage water. This barrier serves the dual purposes of adsorption of phosphorus and protection of the scarp from washout. *Izydorczyk, K., Michalska-Hejduk, D., Frątczak, W., Bednarek, A., Łapińska, M., Jarosiewicz, P., Kosińska, A., Zalewski, M. 2015. Strefy buforowe i biotechnologie ekohydrologiczne w ograniczaniu zanieczyszczeń obszarowych.*

4. Summary conclusions

4.1 Catchment use of ecohydrological biotechnology in water management

In Europe, 38% of total area is farmland and reducing the load of nitrogen flowing with surface waters (leading to eutrophic and toxic algal blooms) from the catchment is one of the main implementation tasks of the European Union's Nitrogen Directive and Water Framework Directive (WFD). In this case only wide-scale, low-cost and highly effective solutions integrating engineering and biological methods (Statzner and Sperling, 1993), such as ecohydrological biotechnologies, are able to cope with the problem.

As a variety of forms of diffuse pollution occur in every catchment area, it is necessary to optimize and harmonize various measures, including best agricultural practices, ecohydrological biotechnologies (Zalewski, 2000) and nature-based solutions (WWAP, 2018), in order to achieve synergy between them and develop them into Advanced Nature-based Solutions to enhance catchment and the biosphere sustainability potential WBSR.

It should be emphasized that the on a catchment scale implementation of Advanced Nature-based Solutions is particularly important in conditions of high nonpoint source pollution areas. At the same time, it requires a thorough analysis of natural conditions, sources of pollution emissions and the pathways of their transfer to the aquatic ecosystem. The selection of the most effective tools, their adaptation to the prevailing specific conditions in the catchment area and the optimization of their effectiveness requires quantification basic nutrients cycling processes prior to the implementation phase.

Mathematical modeling should be applied to identify priority areas for reduction measures. In the case of the Pilica catchment area, the use of the SWAT model allowed for the identification of very strong spatial variation in the amount of biogenic compound emissions Fig. 14.7. The three priority areas identified differed significantly from each other. In the first area, physiographic characteristics played a dominant role, as loess soils and moderate slopes contributed to high total phosphorus emissions as a result of surface water runoff. In contrast, high nitrogen emissions in the second area were mainly due to excessive fertilization, including excessive use of mineral fertilisers and intensive pig production. The main source of pollution in PA3 was meadows on dry peat soils. The selected priority areas, which together account for 36% of the nitrogen load and 51% of the total phosphorus load, cover only about 6.6% of the Pilica catchment area and 16.3% of the arable land located in the catchment area. (320 km^2 of arable land located in the 4900 km^2 Pilica catchment area (Piniewski et al., 2015).

FIGURE 14.12 Cooperation with stakeholders for building capacity for the RBMP implementation. Second Catchment Area Meeting of Stakeholders of the LIFE Pilica Integrated Project 9–10.10.2023 (130 in person). *Image credit: M. Kuzior.*

The designated priority area differed in terms of threats, sources of pollution and pathways of transfer to the aquatic ecosystem, which resulted from the diversity of both natural conditions and agricultural pressure. Therefore, preventive and mitigating measures should also be well targeted in terms of area and type of action (tailored to the threats and needs specific to a given location) in order to achieve maximum environmental improvement at minimum cost and to ensure that they are accepted by farmers (Frątczak and Izydorczyk, 2015; Izydorczyk et al., 2019).

For implementation of catchment scale Advanced Nature-based Solutions (ANbS) biogenic substance management is therefore the involvement of stakeholders in the planning and implementation of measures to reduce area-wide pollution. This collaborative model was demonstrated in the LIFE Pilica Integrated Project, where in October 2023 more than 130 stakeholders participated in an in-person multi-sectoral meeting Fig. 14.12, marking a significant step toward coordinated action and collective ownership of catchment restoration outcomes. Successful implementation in agricultural landscapes depends largely on the acceptability and practical applicability of measures by farmers; therefore, educational programs and strong cooperation with agricultural advisers are essential components of the management strategy. For this reason, educational campaigns and cooperation with agricultural advisers are important elements of management strategy. Complementary economic tools, such as subsidies and payment schemes, can additionally incentivize farmers to adopt practices that reduce nutrient losses. Therefore, catchment-wide nutrient management programs should consist of a combination of legislative, regulatory and economic initiatives, combined with targeted information campaigns. For this reason, cooperation between policy makers, scientific experts and stakeholders is essential to reverse the process of eutrophication in freshwater bodies.

References

Backiel, T., 1985. Fall of migratory fish populations and changes in commercial fisheries in impounded rivers in Poland. In: Alabaster, J.S. (Ed.), Habitat Modification and Freshwater Fisheries. Butterworths, pp. 28–41.

Balestrini, R., Arese, C., Delconte, C.A., Lotti, A., Salerno, F., 2011. Nitrogen removal in subsurface water by narrow buffer strips in the intensive farming landscape of the Po River watershed, Italy. Ecol. Eng. 37 (2), 148–157. https://doi.org/10.1016/j.ecoleng.2010.08.003.

Bednarek, A., Stolarska, M., Urbaniak, M., Zalewski, M., 2010. Applying a denitrification wall for groundwater protection. Ecohydrol. Hydrobiol. 10 (4), 355–362.

Belletti, B., Garcia de Leaniz, C., Jones, J., Bizzi, S., Börger, L., Segura, G., Castelletti, A., van de Bund, W., Aarestrup, K., Barry, J., Belka, K., Berkhuysen, A., Birnie-Gauvin, K., Bussettini, M., Carolli, M., Consuegra, S., Dopico, E., Feierfeil, T., Fernández, S., Fernandez Garrido, P., Garcia-Vazquez, E., Garrido, S., Giannico, G., Gough, P., Jepsen, N., Jones, P.E., Kemp, P., Kerr, J., King, J., Łapińska, M., Lázaro, G., Lucas, M. C., Marcello, L., Martin, P., McGinnity, P., O'Hanley, J., Olivo del Amo, R., Parasiewicz, P., Pusch, M., Rincon, G., Rodriguez, C., Royte, J., Schneider, C., Tummers, J.S., Vallesi, S., Vowles, A., Verspoor, E., Wanningen, H., Wantzen, K.M., Wildman, L., Zalewski, M., 2020. More than one million barriers fragment Europe's rivers. Nature 588 (7838), 436–441. https://doi.org/10.1038/s41586-020-3005-2.

Chróst, R.J., Siuda, W., 2006. Microbial production, utilization, and enzymatic degradation of organic matter in the upper trophogenic layer in the pelagial zone of lakes along a eutrophication gradient. Limnol. Oceanogr. 51 (1), 749–762. https://doi.org/10.4319/lo.2006.51.1_part_2.0749.

Dillaha, T.A., Sherrard, J.H., Lee, D., Mostaghimi, S., Shanholtz, V.O., 1989. Evaluation of vegetative filter strips as a best management practice for feedlots. J. Water Pollut. Control Fed. 61 (7), 1231–1238.

Dosskey, M.G., Vidon, P., Gurwick, N.P., Allan, C.J., Duval, T.P., Lowrance, R., 2010. The role of riparian vegetation in protecting and improving chemical water quality in streams. J. Am. Water Resour. Assoc. 46 (2), 261–277. https://doi.org/10.1111/j.1752-1688.2010.00419.x.

Frątczak, W., Izydorczyk, K. (Eds.), 2015. Program działań dla ograniczenia zanieczyszczeń obszarowych w zlewni Pilicy powyżej Zbiornika Sulejowskiego. European Regional Centre for Ecohydrology PAS.

Frątczak, W., Izydorczyk, K., Zalewski, M., 2013. Wysokoefektywne strefy buforowe dla zwiększenia potencjału ekologicznego i turystycznego zbiornika Sulejowskiego. Gospodarka Wodna. 12, 479–483.

Frątczak, W., Michalska-Hejduk, D., Zalewski, M., Izydorczyk, K., 2019. Effective phosphorous reduction by a riparian plant buffer zone enhanced with a limestone-based barrier. Ecol. Eng. 130, 94–100. https://doi.org/10.1016/j.ecoleng.2019.01.015.

Golterman, H.L., 1995. The role of the ironhydroxide-phosphate-sulphide system in the phosphate exchange between sediments and overlying water. Hydrobiologia 297 (1), 43–54. https://doi.org/10.1007/bf00033500.

Hefting, M.M., Bobbink, R., Janssens, M.P., 2006. Spatial variation in denitrification and N_2O emission in relation to nitrate removal efficiency in a N-stressed riparian buffer zone. Ecosystems 9 (4), 550–563. https://doi.org/10.1007/s10021-006-0160-8.

Hillbricht-Ilkowska, A., Pieczyńska, E. (Eds.), 1993. Nutrient Dynamics and Retention in Land/Water Ecotones of Lowland, Temperate Lakes and Rivers. Hydrobiologia, 251. Kluwer Academic Publishers.

Hillbricht-Ilkowska, A., Wiśniewski, R.J., Pieczyńska, E., 1995. Nutrient dynamics and retention in land–water ecotones of lowland, temperate lakes and rivers. Hydrobiologia 310, 1–15.

Hoffmann, C.C., Kjaergaard, C., Uusi-Kämppä, J., Hansen, H.C.B., Kronvang, B., 2009. Phosphorus retention in riparian buffers: review of their efficiency. J. Environ. Qual. 38 (5), 1942–1955. https://doi.org/10.2134/jeq2008.0087.

Howarth, R.W., 2004. Human acceleration of the nitrogen cycle: drivers, consequences, and steps toward solutions. Water Sci. Technol. 49 (5–6), 7–13. https://doi.org/10.2166/wst.2004.0731.

Izydorczyk, K. (2003). Wpływ czynników środowiskowych na czasoprzestrzenne zróżnicowanie fitoplanktonu w Zbiorniku Sulejowskim w latach 1999–2001: Rola temperatury wody, natężenia przepływu oraz turbulencji wywołanych przez wiatr w kształtowaniu dynamiki zespołu fitoplanktonu (Doctoral dissertation). University of Łódź.

Izydorczyk K., Michalska-Hejduk D., Frątczak W., Bednarek A., Łapińska M., Jarosiewicz P., Kosińska A., Zalewski M.. 2015. Strefy buforowe i biotechnologie ekohydrologiczne w ograniczaniu zanieczyszczeń obszarowych. ERCE PAN, ISBN 978-83-928245-1-0

Izydorczyk, K., Krauze, K., Sodol, A., 2012. Regulation of matter circulation in watersheds by the application of buffer zones. In: Zalewski, M., Urbaniak, M. (Eds.), Adaptation of ecohydrological system solutions and biotechnologies for Africa. II PAS ERCE, pp. 39–48.

Izydorczyk, K., Michalska-Hejduk, D., Jarosiewicz, P., Bydałek, F., Frątczak, W., 2018. Extensive grasslands as an effective measure for nitrate and phosphate reduction from highly polluted subsurface flow – case studies from Central Poland. Agric. Water Manag. 203, 240–250. https://doi.org/10.1016/j.agwat.2018.03.021.

Izydorczyk, K., Piniewski, M., Krauze, K., Courseau, L., Czyż, P., Giełczewski, M., Kardel, I., Marcinkowski, P., Szuwart, M., Zalewski, M., Frątczak, W., 2019. The ecohydrological approach, SWAT modelling, and multi-stakeholder engagement – a system solution to diffuse pollution in the Pilica basin, Poland. J. Environ. Manag. 248, 109329. https://doi.org/10.1016/j.jenvman.2019.109329.

Jansson, M., Olsson, H., Pettersson, K., 1988. Phosphatases; origin, characteristics and function in lakes. Hydrobiologia 170 (1), 157–175. https://doi.org/10.1007/BF00024903.

Jensen, H.S., Andersen, F.O., 1992. Importance of temperature, nitrate, and pH for phosphate release from aerobic sediments of four shallow, eutrophic lakes. Limnol. Oceanogr. 37 (3), 577–589. https://doi.org/10.4319/lo.1992.37.3.0577.

Johannesson, K., Smolarz, K., Grahn, M., André, C., 2011. The future of Baltic Sea populations: local extinction or evolutionary rescue? Ambio 40 (2), 179–190. https://doi.org/10.1007/s13280-010-0129-x.

Kajak, A., 1995. The role of soil predators in decomposition processes. Eur. J. Entomol. 92 (3), 573–580.

Kawara, O., Iwami, N., Fujii, S., Matsui, T., Matsuda, M., 1998. The relationship between water quality and phytoplankton in a shallow river. Water Res. 32 (2), 543–550.

Kiedrzyńska, E., Kiedrzyński, M., Zalewski, M., 2008a. Flood sediment deposition and phosphorus retention in a lowland river floodplain: impact on water quality of a reservoir, Sulejów, Poland. Ecohydrol. Hydrobiol. 8 (2–4), 281–289. https://doi.org/10.2478/v10104-009-0022-z.

Kiedrzyńska, E., Wagner, I., Zalewski, M., 2008b. Quantification of phosphorus retention efficiency by floodplain vegetation and a management strategy for a eutrophic reservoir restoration. Ecol. Eng. 33 (1), 15–25. https://doi.org/10.1016/j.ecoleng.2007.10.010.

Kimmel, B.L., Groeger, A.W., 1984. Factors controlling primary production in lakes and reservoirs: a perspective. Lake Reserv. Manag. 1 (1), 277–281. https://doi.org/10.1080/07438148409354524.

Kowalkowski, T., 2012. Nitrogen and phosphorus emission into the Vistula and Oder basins—Modeling studies (MONERIS). In: Pastuszak, M., Igras, J. (Eds.), Temporal and spatial differences in emission of nitrogen and phosphorus from Polish territory to the Baltic Sea. National Marine Fisheries Research Institute; Institute of Soil Science and Plant Cultivation – State Research Institute; Fertilizer Research Institute, pp. 265–308.

Kuusemets, V., Lohmus, K., 2005. Nitrogen and phosphorus accumulation and biomass production by Scirpus sylvaticus and Phragmites australis in a horizontal subsurface flow constructed wetland. J. Environ. Sci. Health Part A 40 (6–7), 1167–1175. https://doi.org/10.1081/ese-200055629.

Levia, D.F., Creed, I.F., Hannah, D.M., Nanko, K., Boyer, E.W., Carlyle-Moses, D.E., van de Giesen, N., Grasso, D., Guswa, A.J., Hudson, J.E., Hudson, S.A., Iida, S., Jackson, R.B., Katul, G.G., Kumagai, T., Llorens, P., Ribeiro, F.L., Pataki, D.E., Peters, C.A., Carretero, D.S., Selker, J.S., Tetzlaff, D., Zalewski, M., Bruen, M., 2020. Homogenization of the terrestrial water cycle. Nat. Geosci. 13 (10), 656–658. https://doi.org/10.1038/s41561-020-0641-y.

Mander, Ü., Kuusemets, V., Lõhmus, K., Mauring, T., 1997. Efficiency and dimensioning of riparian buffer zones in agricultural catchments. Ecol. Eng. 8 (4), 299–324. https://doi.org/10.1016/s0925-8574(97)00025-6.

Pärn, J., Pinay, G., Mander, Ü., 2012. Indicators of nutrients transport from agricultural catchments under temperate climate: a review. Ecol. Indic. 22, 4–15. https://doi.org/10.1016/j.ecolind.2011.10.002.

Pastuszak, M., Kowalkowski, T., Igras, J., 2012. Nitrogen and phosphorus emission into the Vistula and Oder basins—Modeling studies (MONERIS). In: Pastuszak, M., Igras, J. (Eds.), Temporal and spatial differences in emission of nitrogen and phosphorus from Polish territory to the Baltic Sea. National Marine Fisheries Research Institute; Institute of Soil Science and Plant Cultivation – State Research Institute; Fertilizer Research Institute, pp. 265–308.

Piniewski, M., Marcinkowski, P., Kardel, I., Giełczewski, M., Izydorczyk, K., Frątczak, W., 2015. Spatial quantification of nonpoint source pollution in a meso-scale catchment for an assessment of buffer zones efficiency. Water 7 (5), 1889–1916. https://doi.org/10.3390/w7051889.

Psenner, R., Boström, B., Dinka, M., Pettersson, K., Pucsko, R., Sager, M., 1988. Fractionation of phosphorus in suspended matter and sediment. Archiv für Hydrobiologie–Beiheft Ergebnisse der Limnologie 30, 83–112.

Räty, M., Uusi-Kämppä, J., Yli-Halla, M., Rasa, K., 2010. Phosphorus and nitrogen cycles in the vegetation of differently managed buffer zones. Nutr. Cycl. Agroecosystems 86 (1), 121–132. https://doi.org/10.1007/s10705-009-9277-4.

Ressom, R., Soong, F.S., Fitzgerald, J., Turczynowicz, L., El Saadi, O., Roder, D., Maynard, T., Falconer, I.R., 1994. Health Effects of Cyanobacteria (Blue-green Algae), p. 108.

Reynolds, C.S., 1980. Phytoplankton assemblages and their periodicity in stratifying lake systems. Ecography 3 (3), 141–159. https://doi.org/10.1111/j.1600-0587.1980.tb00721.x.

Rydin, E., 2000. Potentially mobile phosphorus in Lake Erken sediment. Water Res. 34 (7), 2037–2042. https://doi.org/10.1016/s0043-1354(99)00375-9.

Ryszkowski, L., Kędziora, A., 1993. The role of shelterbelts in agroecosystems. Lands. Urban Plan. 25 (1–2), 121–130.

Schipper, L.A., Robertson, W.D., Gold, A.J., Jaynes, D.B., Cameron, S.C., 2010. Denitrifying bioreactors-An approach for reducing nitrate loads to receiving waters. Ecol. Eng. 36 (11), 1532–1543. https://doi.org/10.1016/j.ecoleng.2010.04.008.

Statzner, B., Sperling, F., 1993. Potential contribution of system-specific knowledge (SSK) to stream-management decisions: ecological and economic aspects. Freshw. Biol. 29 (2), 313–342. https://doi.org/10.1111/j.1365-2427.1993.tb00767.x.

Tarczyńska, M., Romanowska-Duda, Z., Jurczak, T., Zalewski, M., 2001. Toxic cyanobacterial blooms in a drinking water reservoir-causes, consequences and management strategy. Water Sci. Technol. Water Supply 1 (2), 237–246. https://doi.org/10.2166/ws.2001.0043.

Trojanowska, A.A., Izydorczyk, K., 2010. Phosphorus fractions transformation in sediments before and after Cyanobacterial bloom: Implications for reduction of eutrophication symptoms in dam reservoir. Water Air Soil Pollut. 211 (1–4), 287–298. https://doi.org/10.1007/s11270-009-0299-y.

Tundisi, J.G., Straškraba, M., 1999. Theoretical Reservoir Ecology and Its Applications. Backhuys Publishers.

Uusi-Kämppä, J., 2005. Phosphorus purification in buffer zones in cold climates. Ecol. Eng. 24 (5), 491–502. https://doi.org/10.1016/j.ecoleng.2005.01.013.

Wagner, A., Hülsmann, S., Dörner, H., Janssen, M., Kahl, U., Mehner, T., Benndorf, J., 2004. Initiation of the midsummer decline of Daphnia as related to predation, nonconsumptive mortality and recruitment: a balance. Archiv fur Hydrobiologie 160 (1), 1–23. https://doi.org/10.1127/0003-9136/2004/0160-0001.

Wagner, I., Zalewski, M., 2000. Effect of hydrological patterns of tributaries on biotic processes in a lowland reservoir—consequences for restoration. Ecol. Eng. 16 (1), 79–90. https://doi.org/10.1016/s0925-8574(00)00092-6.

Wagner, I., Zalewski, M., 2016. Temporal changes in the abiotic/biotic drivers of selfpurification in a temperate river. Ecol. Eng. 94, 275–285. https://doi.org/10.1016/j.ecoleng.2016.05.073.

Ward, J.V., Stanford, J.A., 1983. The Serial Discontinuity Concept of Lotic Ecosystems Dynamics of Lotic Ecosystems, pp. 29–42.

Wojtal, A., Bogusz, D., Menshutkin, V., Izydorczyk, K., Frankiewicz, P., Wagner-Łotkowska, I., Zalewski, M., 2008. A study of Daphnia-Leptodora-juvenile Percids interactions using a mathematical model in the biomanipulated Sulejow Reservoir. Annales de Limnologie—Int. J. Limnol. 44 (1), 25. https://doi.org/10.1051/limn:2008020.

WWAP (United Nations World Water Assessment Programme), 2018. The United Nations World Water Development Report 2018: Nature-Based Solutions for Water. UNESCO, Paris.

Zalewski, M., 2011. Ecohydrology for implementation of the EU water framework directive. Water Manag. 164 (8), 375–385.

Zalewski, M., 2015. Ecohydrology and hydrologic engineering: regulation of hydrology-biota interactions for sustainability. J. Hydrol. Eng. 20 (1). https://doi.org/10.1061/(ASCE)HE.1943-5584.0000999.

Zalewski, M., Brewinska-Zaras, B., Frankiewicz, 1990. Fry communities as a biomanipulating tool in a temperate lowland reservoir. Reservoir limnology and water quality: Proceedings of the International Conference on Reservoir Limnology and Water Quality (Arch. Hydrobiol. Beih., Ergebn. Limnol.) 33, 763–774. https://doi.org/10.1007/978-94-017-0924-8_49.

Zalewski, M., Janauer, G.A., Jolankai, G., 1997. Conceptual Background. Ecohydrology: A New Paradigm for the Sustainable Use of Aquatic Resources.

Zalewski, M., Naiman, R.J., 1985. The regulation of riverine nutrient dynamics: controlling factors, exchange processes, and ecological significance. Ecol. Stud. 60, 143–162. https://doi.org/10.1007/978-1-4612-5046-6_8.

Zalewski, M., 2000. Ecohydrology—the scientific background to use ecosystem properties as management tools toward sustainability of water resources. Ecol. Eng. 16 (1), 1–8. https://doi.org/10.1016/s0925-8574(00)00071-9.

Zalewski, M., 2002. Ecohydrology—the use of ecological and hydrological processes for sustainable management of water resources. Hydrol. Sci. J. 47 (5).

Zhou, Q., Gibson, C.E., Zhu, Y., 2001. Evaluation of phosphorus bioavailability in sediments of three contrasting lakes in China and the UK. Chemosphere 42 (2), 221–225. https://doi.org/10.1016/S0045-6535(00)00129-6.

Zhou, Q., Gibson, C.E., Zhu, Y., 2002. Evaluation of phosphorus bioavailability in sediments of three contrasting lakes in China and the UK. Water Res. 36 (2), 371–380.

Chapter 15

Biomanipulation and hydrobiomanipulation

Adrianna Wojtal-Frankiewicz[1], Piotr Frankiewicz[1] and Maciej Zalewski[2]
[1]University of Lodz, Faculty of Biology and Environmental Protection, UNESCO Chair on Ecohydrology and Applied Ecology, Łódź, Poland;
[2]European Regional Centre for Ecohydrology of the Polish Academy of Sciences under the auspices of UNESCO, Łódź, Poland

PHOTO CREDIT: M. ZALEWSKI.

The enhancement of biochemical processes in the riparian buffer zone, such as the incorporation of groundwater-transported phosphates into plant biomass or the conversion of nitrates to gaseous form by denitrifying bacteria, is called landscape-scale biotechnology. The basis for their implementation is an understanding of the dynamics of surface runoff in relation to water level oscillations in a reservoir, lake or river and the potential of plants and bacteria to convert mineral nutrients into biomass or gaseous forms (Zalewski, 2014). However, in order to maintain the dynamic equilibrium of a lake or dam reservoir, while keeping these ecosystems in a nonadvanced state of succession and sustaining fish

Transdisciplinary Ecohydrology. https://doi.org/10.1016/B978-0-443-41596-8.00006-6

biodiversity, it is necessary to reduce the level of internal nutrient loads (mainly phosphorus) by manipulating the food web (Opuszyński, 1987; Zalewski et al., 1990a, 1990b). One of the most important methods of food web regulation is biomanipulation, which has been used worldwide for over 30 years and is based on so-called fishery methods of controlling the species composition and biomass of ichthyofauna in the ecosystem (top-down effect; Carpenter et al., 1985). The specific role of predatory fish in the grazing chain was exploited—predatory fish at the top of the trophic pyramid naturally control the biomass of planktivorous fish (Benndorf, 1995) meaning that the dynamics of zooplankton abundance are indirectly dependent on the presence and density of predatory fish (Hrbácek et al., 1961). Consequently, predatory fish act as key species in aquatic ecosystems, not only directly affecting the neighboring trophic level, but also the levels below it (Paine, 1969). This effect has been used to carry out activities called "biomanipulation" by Shapiro and coauthors (1975); these were aimed at reducing algal biomass (and thus improving water quality) through controlled changes in fish community structure. An important effect of biomanipulation is supposed to be to reduce the number of zooplanktivorous fish and to maintain a strong population of filter feeding Daphnia.

The importance of Daphniidae for the functioning of the aquatic ecosystem is not only linked to their influence on the abundance, species composition and structure of phytoplankton communities. They also influence the dynamics of bacterial and protozoan populations, and thus the cycles of matter and nutrients that occur in a reservoir—which is why these animals are particularly useful in lake and reservoir restoration activities. However, maintaining strong populations of zooplankton in hydrologically unstable ecosystems such as dammed reservoirs can be difficult. This is due to a hierarchy of factors: abiotic processes (including the hydrological regime) have a dominant influence on biota; when they are stable and predictable, biotic interactions can manifest themselves (Zalewski, 2000). Consequently, Daphnia density in reservoirs fluctuates strongly and depends on both abiotic factors and biotic interactions. Daphniids' populations may remain low in early spring when, despite the abundance of suitable food (phytoplankton), low water temperature and dynamic changes in the water regime are limiting factors. Another important factor affecting the abundance and structure of plankton in reservoirs is the so-called flushing effect, observed with large changes in water level and significant reduction of retention time (Wojtal et al., 2008). In late spring, when temperature and insolation are optimal for small forms of phytoplankton, the population of filter-feeders increases rapidly. This results in intensive grazing of herbivorous zooplankton and the "clear water phase". This phenomenon is characterized by a low density of phytoplankton subjected to strong consumer pressure by zooplankton, in particular by Daphnia (Sommer et al., 1986). The length of the clear water phase is variable over time and usually ends in summer (Izydorczyk et al., 2008), when Daphnia population declines again and when cyanobacterial blooms are most frequent (see Fig. II.3.1.). Due to their morphological structure (filamentous, colonial forms), low nutritional value and high toxicity, cyanobacteria are not a preferred food for Daphnia and may significantly limit the development of filter feeding zooplankton populations (DeMott, 1999). In summer, abiotic factors such as temperature and hydrology become stable, thus that the zooplankton population depends mainly on dynamically changing biotic factors, including the presence of cyanobacteria, which, preferring high temperatures (>20°C), gain a competitive advantage over other forms of phytoplankton (Elliott, 2010), and predation pressure from vertebrate and invertebrate predators (Wissel et al., 2000; Wojtal et al., 2004). To determine the potential of planktonic filter feeders in water quality improvement efforts in a reservoir, the long-term dynamics of the density of these animals in relation to environmental conditions should be analyzed. An example of this can be found in studies carried out during 10 years monitoring of the Sulejow Reservoir. Kohonen's self-organizing map (SOM), was applied in this research, because this method is effective in recognizing patterns in complex data sets, even when the data distribution is not normal and/or the relationships between variables are nonlinear. Furthermore, SOMs, unlike classical statistical methods, are applicable to data expressed at different measurement scales, which is often the case in ecological studies. In relation to data from the Sulejow Reservoir, SOM grouped sampling dates with similar biotic states, which allowed the identification of key abiotic factors responsible for the seasonal sequence of events in the reservoir (Wojtal-Frankiewicz et al., 2015):

1. The CL–ExSp (extremes/spring) cluster grouped early spring and spring dates corresponding to hydrologically unstable cold periods in which the biota was controlled by abiotic factors. They were characterized by short retention times and low temperatures as well as low zooplankton density and absence of cyanobacteria;
2. The CL–StSm (stable/summer) cluster grouped late spring and summer dates, which were characterized by stable abiotic conditions, making biotic interactions dominant during this time. Long retention times and high temperatures allowed an intense increase in Daphnia density in late spring and the occurrence of a clear water phase as well as cyanobacterial blooms in summer.
3. In addition, the CL–ExSm (extremes/summer) cluster was distinguished, corresponding to dates during spring and summer when thermal or hydrological extremes occurred, thus weakening the role of biotic factors Fig. 15.1

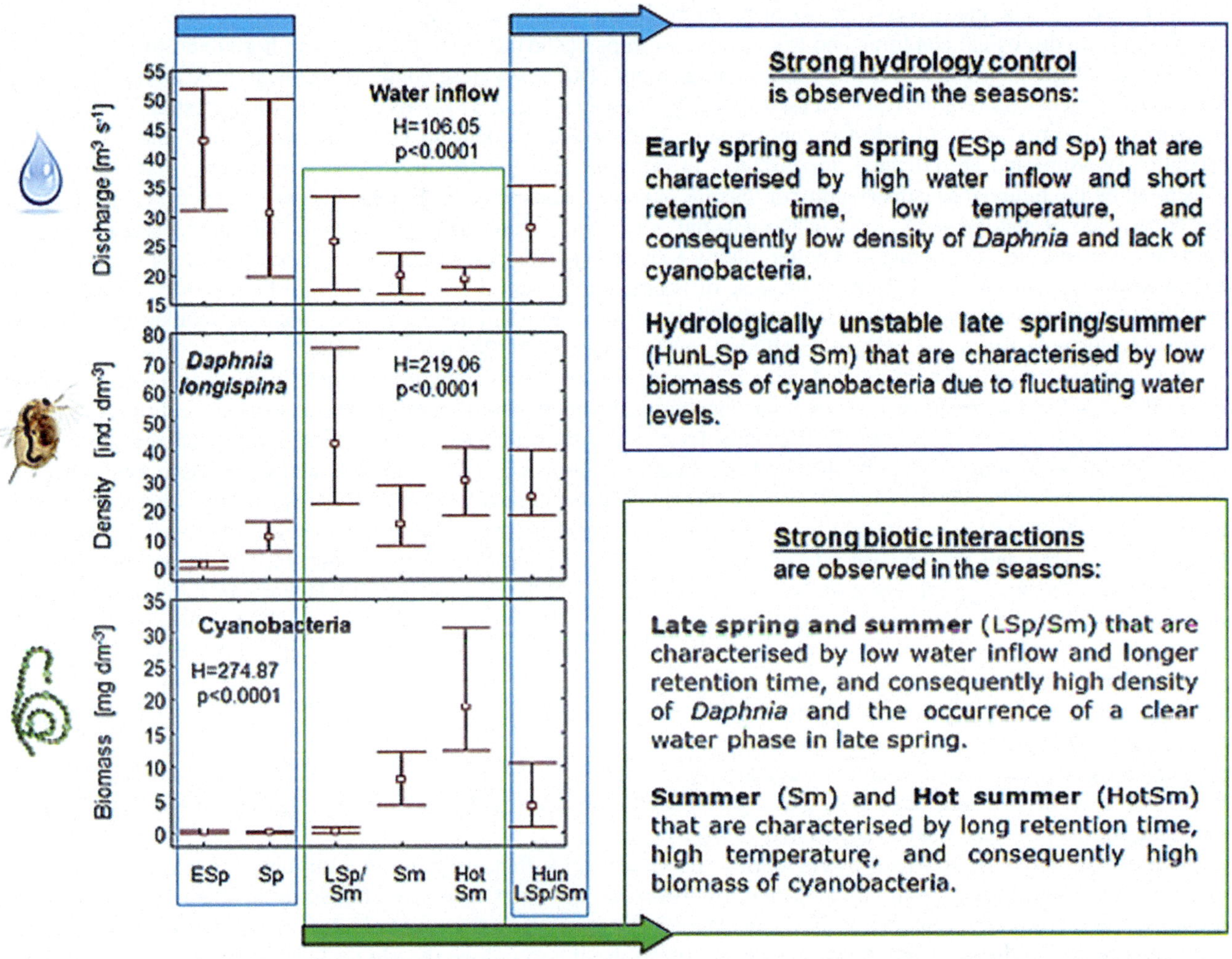

FIGURE 15.1 Seasonal sequence of events. Seasonal sequence of events with similar abiotic and biotic states grouped by SOM, based on 10 years sampling dates from the Sulejow Reservoir. The parameters used: water inflow, densities of Daphnia longispina and biomass of cyanobacteria. Point—median; whiskers—interquartile range; H—statistics of the Kruskal–Wallis test (df = 4, N = 308). Dominant seasons of the year: ESp = early spring (April), Sp = spring (May), LSp/Sm = late spring and summer (June), Sm = summer (July), HunLSp/Sm = hydrologically unstable late spring and summer (June–September). *According to Wojtal-Frankiewicz, A., Kruk, A., Frankiewicz, P., Oleksińska, Z., Izydorczyk, K., Miki, T., 2015. Long-term patterns in the population dynamics of Daphnia longispina, Leptodora kindtii and cyanobacteria in a Shallow reservoir: a self-organising map (SOM) approach. PLoS One. 10 (12), e0144109. https://doi.org/10.1371/journal.pone.0144109.*

The SOM showed that the density of Daphnia was dependent on abiotic factors (temperature, inflow, retention time) during spring. Biotic conditions (mainly phytoplankton species composition and abundance) were more important in shaping Daphnia population dynamics during summer and autumn. SOM also confirmed observations that cyanobacteria biomass is closely related to water inflow and retention time in all seasons. Fluvial part of the reservoir was characterized by a lower incidence of blooms than lacustrine parts, which was in contrast to the spatial distribution of Daphnia (Wojtal-Frankiewicz et al., 2015). The results clearly indicate that the importance of hydrology as a factor regulating plankton may be crucial for formulating management decisions in dam reservoirs. Biological methods to protect and improve water quality in dam reservoirs should be intensified when biotic impacts are strong:

1. in spring/late spring to enhance the Daphnia population (e.g., by water level regulation of the density of zooplanktivorous fry = hydrobiomanipulation);
2. in summer to exploit the potential of Daphnia for natural control of phytoplankton biomass (e.g., by enhancing populations of predatory fish = classical biomanipulation).

In the case of dammed reservoirs, when planning biomanipulation measures, in addition to effects related to the variability of hydrological condition and trophic relationships occurring along the reservoir, horizontal links between the littoral and pelagic zones should also be taken into account. There is a clear coupling between these zones that strongly influences nutrient cycles in the reservoir (see review by Stewart et al., 2018). This coupling is manifested, for example, in horizontal daily migrations of fish and zooplankton. These are dependent on both food availability and differences in predation pressure and are reflected, for instance, in different growth and survival rates of juvenile fish. This may be illustrated by the bimodal size distribution of juvenile pikeperch described by Frankiewicz et al. (1996), characterized by the occurrence of smaller individuals in the pelagial zone, where food quality was poor but compensated for by low predation pressure. In the littoral, with better quality food, pikeperch grew faster, at the price of greater pressure, usually cannibalistic, from older fish. In turn, in the case of zooplankton, clear diurnal migration of Daphnids observed between the littoral and pelagic zone, resulted from a trade-off between avoiding predation pressure from cladoceran Leptodora kindtii and juvenile fish in the pelagic and littoral zones, respectively (Wojtal et al., 2003).

The coupling between the littoral and pelagic zones is well visible in many aspects fish life history. The water level in reservoirs is subject to frequent and irregular changes that prevent the formation of a permanent eulittoral zone with emersed and submerged vegetation. Fish usually feed in the limnetic zone (open water) but reproduce mainly in the littoral. Under eutrophic conditions, the food of planktivorous fish is not a limiting factor for them, but rather their spawning site, which largely depends on access to flooded coastal areas (Ploskey, 1985). If water levels remain high in the reservoir and the shoreline vegetation remains flooded, then fish have access to suitable spawning substrate. They spawn in the flooded area and the survival rate of the hatching fry is high. Due to predation pressure of fry, the density of zooplankton in the reservoir decreases drastically and consequently the density of phytoplankton, including cyanobacteria, increases. The low density of large zooplankton also causes a decrease in the growth rate of pikeperch fry, *Stizostedion lucioperca*. As a consequence, juvenile pikeperch are not big enough to feed on fry of perch and roach when they should be switched to fish food, which drastically reduces the chances of survival of juvenile pikeperch in winter. In addition, poor growth rates of pikeperch fry result in prolonged vulnerability to cannibalistic pressure from older individuals (Frankiewicz et al., 1999). An obvious consequence of overcrowding of planktivorous fish is an increase in intra- and interspecific competition. This results in the migration of fish to the coastal zone, where, in the absence of adequate food, the growth rate of fish size decreases, causing low winter survival and, consequently, decline in the number of fish available for harvest (Zalewski et al., 1990a, 1990b).

This situation can be reversed by regulating the water level, first allowing planktivorous fish such as perch and cyprinids to spawn in the floodplain and then, by lowering the water level, causing some of the eggs to dry out and reducing the reproductive success of these species. Reducing the population of planktivorous fish allows to maintain high density of zooplankton and, through its high filtration rate, to reduce the phytoplankton biomass. High densities of zooplankton also enable rapid growth and early transition to piscivorous behavior of pikeperch fry, which increases their chances of winter survival (Frankiewicz et al., 1996). Parallel to water level regulation measures, the population of predatory fish which reduce the population of zooplanktivorous fish should be supported. For this purpose, it is necessary to follow their population dynamics, for which a precise density assessment is required, both in the coastal zone and in open water. For this purpose, fry trawls and bongo nets are used, respectively Fig. 15.2. In temperate zone reservoirs, pelagic predators such as pikeperch are among the basic tools for biomanipulation, as the second most important predator of temperate zone waters, pike, is now rare in reservoirs where water level fluctuations destroy its preferred littoral habitat.

From the point of view of trophic dynamics, such a manipulation can be described as a change in energy flow from a detritus chain, characteristic of eutrophic lakes, to a grazing chain based on consumers (zooplankton and predatory fish), and characteristic of mesotrophic lakes with high water quality. In the case of dammed reservoirs, it should be remembered that their rate of ageing as an ecosystem is quite rapid due to the accumulation of organic and inorganic matter carried by the river, and the trophy of the reservoir changes with the progress of ecological succession, affecting the nutrient content of the water and their cycles of circulation, resulting in changes in phytoplankton dynamics.

In 2006, a controlled water level adjustment procedure was carried out in the Sulejow Reservoir to regulate the density of fry. Before and during spawning, the water level was kept high to allow the fish to spawn in flooded areas, and then after spawning, the water level was lowered to eliminate excess eggs, after which the water level was returned to normal. This procedure resulted in a 50% reduction in cyanobacteria biomass in the reservoir compared with the prediction model developed on the basis of previous reservoir studies Fig. 15.3.

Improving water quality in lakes and dam reservoirs by manipulating their trophic structure based on the principles of ecohydrology allows time to develop a comprehensive catchment protection strategy. Biomanipulation and hydrobiomanipulation enhance the top-down effect in the trophic pyramid and thus strengthen the control of

FIGURE 15.2 **Collecting empirical data.** The use of a fry trawl and bongo net to assess the density of fry in the littoral and pelagial zones, respectively (Zalewski, 2020).

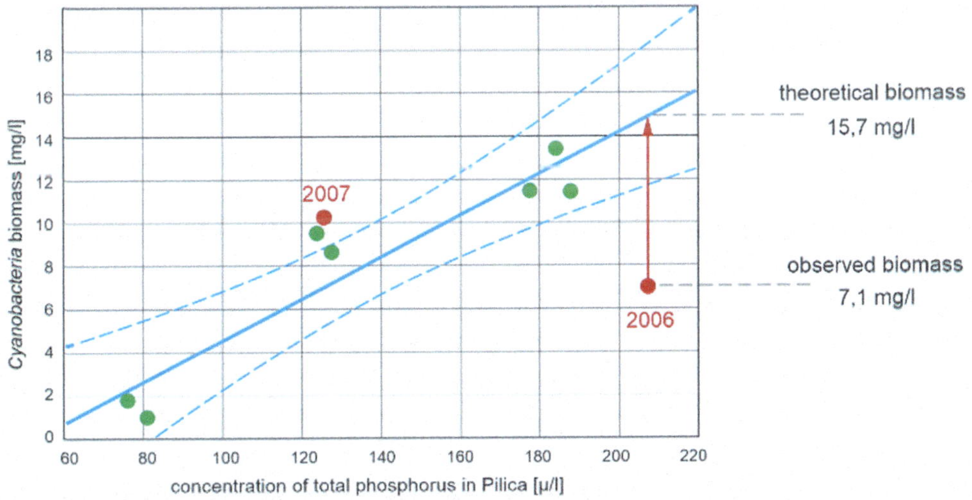

FIGURE 15.3 **Cyanobacteria biomass.** Dependence of cyanobacteria biomass on concentration of total phosphorus carried in spring and summer with Pilica River waters into the Sulejow Reservoir. In 2006, thanks to the application of hydrobiomanipulation, the potential bloom (15.7 mg/L) was reduced by half (7.1 mg/L). *Based on data from: Wagner, I., Izydorczyk, K., Kiedrzyńska, E., Mankiewicz-Boczek, J., Jurczak, T., Bednarek, A., Wojtal-Frankiewicz, A., Frankiewicz, P., Ratajski, S., Kaczkowski, Z., Zalewski, M., 2009. Ecohydrological system solutions to enhance ecosystem services: the pilica river demonstration project. Ecohydrol. Hydrobiol. 9, 13–39.*

predators over planktivores, improving water quality and minimizing the effects of the eutrophication process. According to the Intermediate Disturbances Hypothesis (Connell, 1978), moderate predation pressure on zooplankton leads to a well-balanced density of zooplankton as food for juvenile fish, resulting in faster growth rates and winter survival (Toneys and Coble, 1979). Biomanipulation is therefore an example of how, by understanding ecosystem feedback through biological processes, not only biocenosis but also water quality can be regulated in favor of WBSRCE strategies. However, it must be remembered that controlling the biotic structure of a reservoir must not result in neglecting measures to reduce pollutants reaching the water from the catchment, since the effectiveness of biomanipulation is inversely proportional to the degree of eutrophication of the reservoir. These strategies should therefore include plans to improve the efficiency of wastewater treatment plants and tools to reduce area pollution by creating ecotone zones, (see Chapter 14), e.g., using a sequential sedimentation–biofiltration system (SSBS), (see Chapter 18). Classical biomanipulation and hydrobiomanipulation serve as resilience enhancement tools, applied as complementary measures to reduce the impact of nutrient loading.

References

Benndorf, J., 1995. Possibilities and limits for controlling eutrofication by biomanipulation. Int. Rev. ges Hydrobiol. 80, 519–534.

Carpenter, S.R., Kitchell, J.F., Hodgson, J.R., 1985. BioScience 634–639.

Connell, J.H., 1978. Diversity in tropical rain forests and coral reefs. Science 199 (4335), 1302–1310.

DeMott, W., 1999. Foraging strategies and growth inhibition in five daphnids feeding on mixtures of a toxic cyanobacterium and a green alga. Freshw. Biol. 42, 263–274.

Elliott, J.A., 2010. The seasonal sensitivity of cyanobacteria and other phytoplankton to changes in flushing rate and water temperature. Glob. Change Biol. 16, 864–876.

Frankiewicz, P., Dabrowski, K., Martyniak, A., Zalewski, M., 1999. Cannibalism as a regulatory force of pikeperch, *Stizostedion lucioperca* (L.), population dynamics in the lowland Sulejow reservoir (Central Poland). Hydrobiologia 408–409, 47–55. https://doi.org/10.1007/978-94-017-2986-4_5.

Frankiewicz, P., Dabrowski, K., Zalewski, M., 1996. Mechanism of establishing bimodality in a size distribution of age-0 pikeperch, *Stizostedion lucioperca* (L.) in the Sulejöw Reservoir, Central Poland. Ann. Zool. Fenn. 33 (3–4), 321–327.

Hrbácek, J., Dvorakova, M., Korinek, V., Prochazkova, L., 1961. Demonstration of the effect of the fish stock on the species composition of zooplankton and the intensity of metabolism of the whole plankton association. Verh. - Int. Ver. Theor. Angew. Limnol. 14, 192–195.

Izydorczyk, K., Frątczak, W., Zalewski, M., 2016. Ecohydrology for reduction of cyanobacterial blooms in reservoirs: from identification of threats to development of solutions. Spilki, F.R., Scheuenstuhl, M.C.B., Tundisi, J.G. (Eds.), In: Enhancing Water Management Capacity in a Changing World: The Challenge of Increasing Global Access to Water and Sanitation Universidade Feevale, Novo Hamburgo, pp. 660–690.

Izydorczyk, K., Jurczak, T., Wojtal-Frankiewicz, A., Skowron, A., Mankiewicz-Boczek, J., Tarczyńska, M., 2008. Influence of abiotic and biotic factors on microcystin content in Microcystis aeruginosa cells in a eutrophic temperate reservoir. J. Plankton. Res. 30 (4), 393–400.

Opuszyński, K., 1987. Teoria ichtioeutrofizacji [Feedback between eutrophication processes and changes in fish assemblage. Theory of ichthyoeutrophication]. Wiadomosci Ekol. 33, 21–30.

Paine, R.T., 1969. A note on trophic complexity and community stability. Am. Nat., 103, 91–93.

Ploskey, G.R., 1985. Impacts of Terrestrial Vegetation and Preimpoundment Clearing on Reservoir Ecology and Fisheries in the United States and Canada, pp. 1–35.

Sommer, U., Gliwicz, Z.M., Lampert, W., Dunkan, A., 1986. The PEG-model of seasonal succession of planktonic events in fresh waters. Arch. Hydrobiol. 106, 433–471.

Stewart, S.D., Hamilton, D.P., Baisden, W.T., Verburg, P., Duggan, I.C., 2018. The role of mobile consumers in lake nutrient cycles: a brief review. Hydrobiologia 818, 11–29.

Toneys, M.L., Coble, D.W., 1979. Size-related, first winter mortality of freshwater fishes. Trans. Am. Fish. Soc. 108 (4), 415–419. https://doi.org/10.1577/1548-8659(1979)108[415:SFWMOF]2.0.CO;2.

Wagner, I., Izydorczyk, K., Kiedrzyńska, E., Mankiewicz-Boczek, J., Jurczak, T., Bednarek, A., Wojtal-Frankiewicz, A., Frankiewicz, P., Ratajski, S., Kaczkowski, Z., Zalewski, M., 2009. Ecohydrological system solutions to enhance ecosystem services: the pilica river demonstration project. Ecohydrol. Hydrobiol. 9, 13–39.

Wissel, B., Freier, K., Müller, B., Koop, J., Benndorf, J., 2000. Moderate planktivorous fish biomass stabilizes biomanipulation by suppressing large invertebrate predators of Daphnia. Arch. Hydrobiol. 149, 177–192.

Wojtal, A., Bogusz, D., Menshutkin, V., Izydorczyk, K., Frankiewicz, P., Wagner-Łotkowska, I., Zalewski, M., 2008. A study of daphnia-leptodora-juvenile percids interactions using a mathematical model in the biomanipulated sulejow reservoir. Ann. Limnol. - Int. J. Lim 44 (1), 69–86.

Wojtal, A., Frankiewicz, P., Izydorczyk, K., Zalewski, M., 2003. Horizontal migration of zooplankton in a littoral zone of the lowland Sulejow Reservoir (Central Poland). Hydrobiologia 506–509 (1–3), 339–346. https://doi.org/10.1023/B:HYDR.0000008627.55462.e1.

Wojtal, A., Frankiewicz, P., Wagner-Łotkowska, I., Zalewski, M., 2004. The evaluation of the role of pelagic invertebrate versus vertebrate predators on the seasonal dynamics of filtering Cladocera. Hydrobiologia 515, 123–135.

Wojtal-Frankiewicz, A., Kruk, A., Frankiewicz, P., Oleksińska, Z., Izydorczyk, K., 2015. Long-term patterns in the population dynamics of Daphnia longispina, Leptodora kindtii and cyanobacteria in a Shallow reservoir: a self-organising map (SOM) approach. PLoS One 10 (12), e0144109. https://doi.org/10.1371/journal.pone.0144109.

Zalewski, M., 2000. Ecohydrology - the scientific background to use ecosystem properties as management tools toward sustainability of water resources. Ecol. Eng. 16, 1–8.

Zalewski, M., 2014. Ecohydrology and hydrologic engineering: regulation of hydrology-biota interactions for sustainability. J. Hydrol. Eng. 20 (1). https://doi.org/10.1061/(ASCE)HE.1943-5584.0000999.

Zalewski, M., 2020. *Ekohydrologia*. Wydawnictwo Naukowe PWN, Warszawa, ISBN 978-83-01-20804-2.

Zalewski, M., Brewińska-Zaraś, B., Frankiewicz, P., 1990a. Fry communities as a biomanipulating tool in a temperate lowland reservoir. Archiv für Hydrobiologie—Beiheft Ergebnisse der Limnologie. 33, 763–774.

Zalewski, M., Brewinska-Zaraś, B., Frankiewicz, P., Kalinowski, S., 1990b. The potential for biomanipulation using fry communities in a lowland reservoir: concordance between water quality and optimal recruitment. Hydrobiologia 200 (1), 549–556. https://doi.org/10.1007/BF02530372.

Chapter 16

Application of hydroacoustic methods in ecohydrology

Malgorzata Godlewska and Maciej Zalewski
European Regional Centre for Ecohydrology of the Polish Academy of Sciences under the auspices of UNESCO, Łódź, Poland

HYDROACOUSTICS—LARGE-SCALE, TIME-EFFICIENT METHODS FOR ECOSYSTEM STATUS AND ECOHYDROLOGICAL PROCESSES DYNAMICS. *PHOTO CREDIT: M. GODLEWSKA.*

As previous chapters have demonstrated, ecohydrology can play a significant part in augmenting the ecological condition of waters (Zalewski, 2000, 2006, 2011). The first and foremost rule of ecohydrology focuses on the quantitative assessment of processes, which aim to elucidate the mechanism underpinning the interactions between biotic and abiotic factors and establish the hierarchy of how these factors impact upon one another. For ecohydrology to be effective and efficient, an adequate tool is needed for collecting both qualitative and quantitative information about various elements

Transdisciplinary Ecohydrology. https://doi.org/10.1016/B978-0-443-41596-8.01016-5

making up the aquatic environment. This applies in particular to the rapidly changing biological elements encompassing the full temporal and spatial scale, ranging from the microscale to the entire water body area.

Hydroacoustic methods are well-suited to respond to such a challenge, since they not only reveal the relationships between hydrological processes (such as changes in the water level or flow velocity, measured with a hydrographic echosounder and the Acoustic Doppler Current Profiler—ADCP) and biological elements of the environment (such as fish, macrophytes, or cyanobacteria, measured with a fishery echosounder), but they also show the "problem-solving" possibilities through evaluating the already employed methods of waterbody recultivation.

These are the only currently known methods of researching the aquatic environment, which allow large bodies of water to be measured over a short period of time and with a high resolution. In addition to speed, they also offer a whole gamut of other advantages for monitoring activities. Above all, these methods are contact-free, which means that they pose no risk to the object of the research nor the environment. They are replicable, provide quantitative data which cover the entire lake area in a continuous way (unlike point methods), are easy to automate, and can be applied from autonomous vehicles operating in the water (Goulon et al., 2021) or aerial drones (Fitzpatrick et al., 2020). Used in their traditional version, from a moving boat, they need only two personnel to conduct the monitoring (one to steer the boat and one to operate the echosounder). In 2013, hydroacoustic methods were approved as the CEN (European Committee for Standardization) standard for assessing the fish stock in deep lakes (CEN, 2014).

Ecohydrological solutions are based on the quantification and understanding of process dynamics, while hydroacoustics enables the tracing of changes in biogenic allocation in inaccessible resources such as submerged macrophytes. Conducting such research using traditional methods is a difficult and lengthy process; this also applies to sediments. The continually developed hydroacoustic methods have proven useful not only in traditional uses for fishing, but also in studying other elements of aquatic environments. This chapter provides examples of hydroacoustic methods being applied by the team of the European Regional Centre for Ecohydrology of the Polish Academy of Sciences to qualitative and quantitative measurements of fish, cyanobacterial blooms, and lake bottoms, including macrophytes. All these elements can be measured simultaneously in a given waterbody (Fig. 16.1) enabling their detailed spatiotemporal analyses and identifying interrelations between them.

1. Echosounders—principles of operation and development of hydroacoustic methods

An aquatic environment cannot easily be penetrated by sunlight or radiowaves, which are strongly absorbed by water. Mechanical (acoustic) waves are the only form of energy, which can be transmitted across large distances in the water (indeed, reaching hundreds of meters or even kilometers). In their operation, hydroacoustic devices take advantage of the phenomenon of *sound propagation* in water. If an acoustic wave on its way through water encounters an object (e.g., a fish, air bubble, bottom, macrophytes, etc.) for which the product of the sound speed times the density of the medium differs from that product for the surrounding water, part of its energy will be reflected from that object and return toward the source of the wave. The greater the difference, the stronger the reflection. Such a returning reflection of the part of the acoustic wave energy (called the "echo") has been used in the construction of hydroacoustic devices known as *echosounders*.

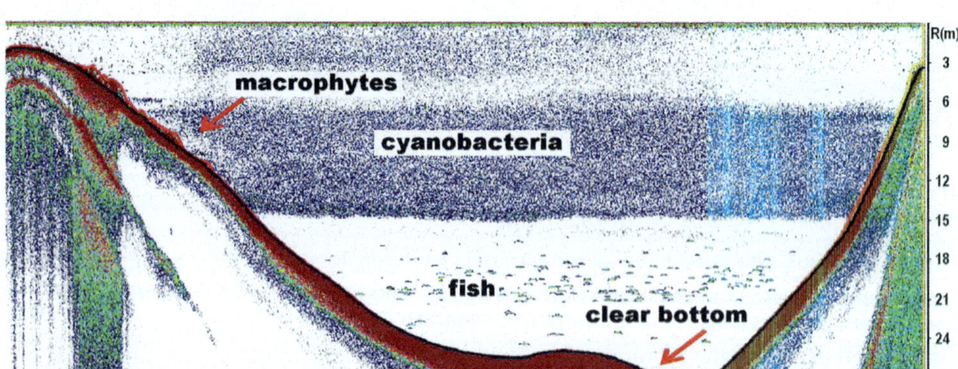

FIGURE 16.1 Echogram with indicated bottom, macrophytes, fish, and cyanobacteria. Echogram of simultaneous acoustic recording of fish, cyanobacteria, and bottom (black line) partly covered with submerged macrophytes (red line) in the Lake Czarne, Poland.

The amplitude of an acoustic wave emitted by a transducer immersed in water has two distinctive areas. The first is the so-called near field, where the sound amplitude oscillates until the distance is equal to approximately L^2/λ, where L is the maximum linear dimension of the transducer, and λ is the length of the acoustic wave. Measurements made in such a near field are not suitable for analysis precisely because of the variability of the sound amplitude (hence that area is frequently dubbed the *dead zone*). At a larger distance from the transducer, the sound amplitude is stabilized and diminishes in inverse proportion to the squared distance—we say that the wave propagates spherically. Apart from beam spreading, some acoustic energy is also lost by process of absorption, which is proportional to the distance traveled by sound. Thus, the strength of the signal during the reception of the reflected wave basically depends on the distance between the object and the transducer. To offset this phenomenon, a range gain control system (TVG—time-varied gain) is used in scientific echosounders, which compensates for both, the spherical spreading of sound and sound absorption. As a result, the signal originated from the reflected wave is directly proportional to the amount of energy reflected by the object regardless of the depth/distance from which it comes (Albers and Breazeale, 1961). This reflected energy, showing the reflecting properties of a given object, is called the target strength (TS) and is measured in logarithmic units—*decibels* (dB). By analogy with theoretical physics, often the notion of the backscattering cross section σ, expressed linearly, is used, producing the equation:

$$TS = 10 \text{ Log } \sigma$$

The frequency of an acoustic wave f, its length λ, and the speed of sound propagation in water C are associated by the standard wave relationship:

$$\lambda = C/f$$

An object immersed in water can be "seen" by an acoustic wave if its dimensions are greater than the length of the acoustic wave λ. If we assume sound velocity in water to be constant (about 1500 ms^{-1}, in fact it only slightly depends on temperature, salinity, and depth), this means that to register small objects, high acoustic frequencies need to be used. However, increase in absorption with square of frequency is the limiting factor, which determines the highest frequency possible for the object detection at a given distance. Thanks to smaller sound absorption in the freshwater as compared with salty one, the higher frequencies may be used in freshwater to achieve the same maximum range as in seawater.

The first observations on the excellent sound propagation in seawaters were made by Leonardo da Vinci as early as the 15th century. However, significant advancement of hydroacoustics did not come until the two World Wars, due to the military use of submarines, which could be detected only with hydroacoustic methods. Applications of hydroacoustics for civilian purposes appeared after World War I: studies on fish detection using hydroacoustic methods commenced in the late 1920s. It can be said, therefore, that hydroacoustics is a relatively new field in comparison with traditional methods. It was not until the 1970s that it developed into a major method for studying seas and oceans, mostly with a view to assessing their natural living resources. In the 1980s, Poland participated in BIOMASS, an international program that aimed to estimate the stock of the Antarctic krill using acoustic methods (Godlewska, 1996). Fishery echosounders came to be used in inland waters sometime later, when the devices could be miniaturized to make them suitable for small boats. Along with the development of technology, and above all due to the coupling of the echosounder with the computer and the GPS, the application potential of hydroacoustic methods rapidly increased. Even though these methods are not yet in widespread use in inland waters, they do have huge research potential and are experiencing explosive growth worldwide.

In recent years, it was observed that hydroacoustic methods prove excellent tools not only in oceanography and fishing, where they are widely used (echosounders and multibeam sonars that have a much higher degree of accuracy), but also in broadly understood ecological research of aquatic ecosystems. This is because they enable contact-free measurements of living organisms in their own environment, on the spatial scale ranging from centimeters to hundreds of meters, and with the time resolution of a fraction of a second. Such methods can be used both for assessing fish biomass for fishing purposes and for mapping entire ecosystems (characteristics of the bottom, degree of macrophyte coverage, density of living organisms at various trophic levels, i.e., having different dimensions; this can be done by a simultaneous use of several frequencies). The advantages of hydroacoustic methods primarily include the following:

1. coverage of a large area over a short period of time, which enables real macroscale, area-based measurements (as opposite to point measurements),
2. high spatiotemporal resolution of data (cm, sec),

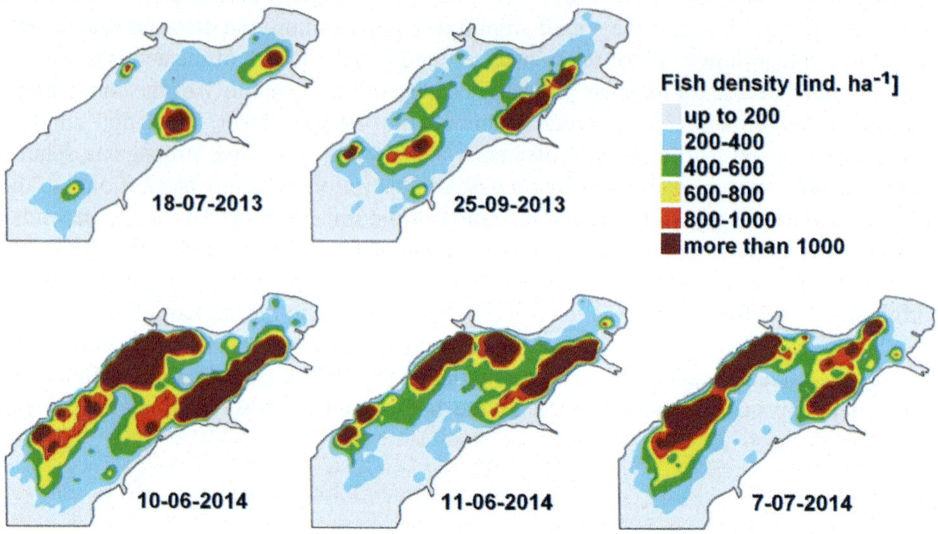

FIGURE 16.2 Fish distribution in the Sulejów Reservoir at different seasons. Distribution of fish [ind. ha^{-1}] in the Sulejów Reservoir at night, in the 2013–14 summer seasons, measured with a horizontal beam (SIMRAD EY60 echosounder, 200 kHz). 2013—a year with cyanobacterial bloom in July and September; 2014 – a year without a bloom. *From Godlewska. (2020). The use of hydroacoustic methods in ecohydrology. In M. Zalewski (Ed.), Ekohydrologia (pp. 174–182). Polish Scientific Publishers PWN.*

3. huge amount of available data (good statistics, applicability in modeling, machine learning, and AI),
4. coupling with a GPS signal receiver allows mapping and georeferencing of the ecosystem,
5. no impact on the researched object and environment (since the information is collected through remote sensing),
6. considerable savings in time, labor, and costs in comparison with traditional methods.

2. Fish detection

The publication "Fisheries Acoustics: Theory and Practice" (Simmonds and MacLennan, 2005) is the main textbook on the applications of hydroacoustics in fishery science. The swim bladder that figures in the anatomy of many fish has the product of density and sound velocity different from such product for water, which makes fish superb reflectors of sound. Acoustic measurements of fish are taken while the research boat is in motion, usually at night, when fish disperse and single individuals can be recorded. On the basis of the quotient of reflected energy in the whole water column to the energy reflected by a single, "medium-sized/average" fish, the number of fish per unit of volume or surface area is calculated; this value is known as the *fish density*. If we know the relationship between the TS expressed in dB and the fish length L expressed in cm or mm (the best-known are such relationships by Foote, 1987; Frouzova et al., 2005; Love, 1969), and the relationship between the fish length and its mass, we can estimate the fish biomass. This can be done with dedicated software systems used for postprocessing sonar data, such as Echoview, Sonar 5-Pro, BioSonic Visual Acquisition, and others. Since an acoustic transducer can have a pulse repetition rate of several impulses per second, the spatial resolution of such measurements is very high. The depth resolution in the water column depends on the length of the sounding pulse τ and equals $C\tau/2$ (C—sound velocity), which translates into a resolution of 7.5–30 cm with the most frequently used pulse lengths of 0.1–0.4 ms. A combined echosounder/GPS device enables precise mapping of the results concerning fish stock or biomass. In addition, acoustic monitoring of individual fish in the function of time helps estimate their size, speed, and direction of swimming. In waterbodies, hydroacoustics is typically used for daily, seasonal, and multiannual observation of changes in the biomass and distribution of fish in relation to various parameters of the environment (Doroszczyk et al., 2013; Godlewska, 1996; Godlewska et al., 2012, 2018; Guillard et al., 2014; Ye et al., 2013). Some examples of changes in the fish stock and distribution in the Sulejów Reservoir in 2013–14 are shown in Fig. 16.2.

Data in the worldwide literature (Draštík et al., 2017), as well as specifically Polish research (Godlewska et al., 2004), have demonstrated that the acoustically estimated fish stock shows a positive correlation with the eutrophication of the waterbody. With the transition from the mesotrophic to the eutrophic state, the number of individual fish per hectare considerably increases. In addition to the fish abundance, the fish size structure also changes, and in eutrophic waters shows a clear shift in dominance toward smaller fish. This confirms that both the fish stock and their size structure are good indicators of the ecological condition of a given ecosystem. Similarly, the vertical distribution of fish can indicate the environmental condition, because the absence of fish in deeper waters often denotes oxygen deficits in those areas. The main weakness of traditional echosounders is that they cannot distinguish between fish species. To some extent, this can be done by underwater acoustic cameras such as DIDSON or ARIS (Connolly et al., 2023; Martignac et al., 2015). The acoustic images of near-video quality obtained at short ranges allow to visually identify fish to a broad taxon, for example, to distinguish, with high confidence, salmonids from other fish taxa such as European eel, perch, pike, or roach (Gurney et al., 2014). Unfortunately, it is not an easy job; additionally, due to the high frequencies at which these acoustic cameras operate, their range is relatively small—up to 15–30 m.

3. Cyanobacteria (blue-green algae)

Intensity and extent of harmful cyanobacterial blooms have been increasing globally over the past several decades due to the increased eutrophication and climate change (Paerl and Huisman, 2009). Their mass incidence adversely affects the functioning of natural water ecosystems, waters used for recreation purposes, and potable water, and their toxicity poses a serious danger for both humans and animals, fish in particular (Carmichael, 2001; Chorus and Bartram, 1999; van Apeldoorn et al., 2007; WHO, 2003; Chorus et al., 2000). Traditionally, measurements of cyanobacteria are made through an analysis of water samples collected at a number of individual locations. Such a monitoring method is not practicable in waterbodies, which have a large surface area. Alternative satellite remote sensing may provide spatial information on phytoplankton distribution over large aquatic areas, but it only detects phytopigment densities in the near-surface water layer, at large time intervals, and is highly dependent on weather conditions (Moore et al., 2017).

Although a hydroacoustic method for cyanobacterial bloom monitoring is not fully developed as yet, during the implementation of a project "Can fish adapt to blue-green algal blooms?" at the European Regional Centre for Ecohydrology (Polish Academy of Sciences—ERCE PAS), it was found that "blooms" can successfully be acoustically recorded by the Simrad EY60, 200 kHz echosounder (Godlewska et al., 2016, 2018, 2023; Kaczkowski et al., 2017; Ostrovsky et al., 2020). What is more, the correlation between the simultaneous measurements of *Microcystis* (the most common genus of cyanobacterial blooms) conducted with an echosounder and one made with the use of a fluorometer was very high (Fig. 16.3), which gives a chance that the acoustic method may indeed find extensive use in the monitoring of how cyanobacterial blooms develop and how they move in the waterbody.

Results of preliminary analyses of simultaneous measurements of fish and cyanobacteria seem to suggest that fish are able to effectively avoid or escape the bloom area. Biomass of fish and cyanobacteria measured along the studied transects in the Sulejów Reservoir with resolution of 100 m have shown that two organisms, while having very patchy distributions, mostly mismatch in space and time. When biomass of cyanobacteria is small, the biomass of fish is high, and vice versa, when the biomass of cyanobacteria increases, the biomass of fish drops down (Godlewska et al., 2016, 2018), Fig. 16.4.

4. Characteristics of the bottom

An ecosystem's abiotic elements, including bottom-related ones such as the depth of a given waterbody (shallow waterbodies undergo eutrophication faster than deep ones), the degree of the bottom's irregularity, sediment thickness,

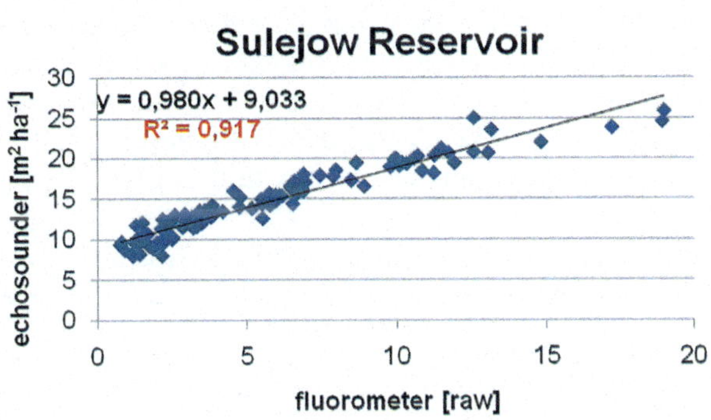

FIGURE 16.3 The correlation between cyanobacterial biomass measured by an echosounder and fluorometer. The correlation between the cyanobacterial biomass in the Sulejów Reservoir, measured with an echosounder (expressed in acoustic units, m^2ha^{-1}), and one measured with a fluorometer (relative values).

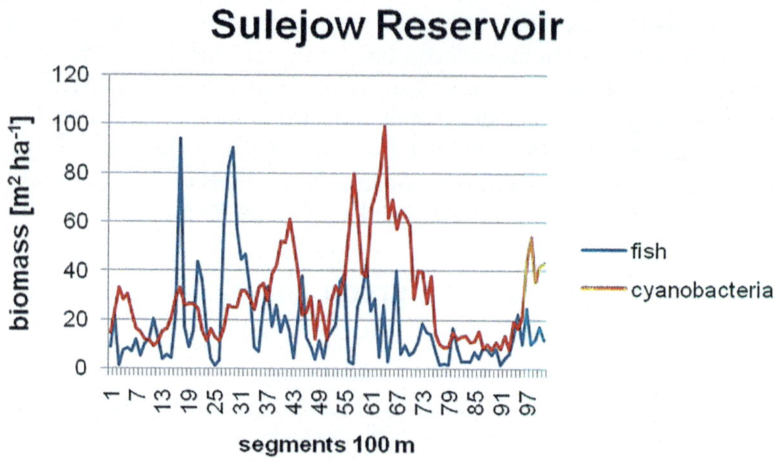

FIGURE 16.4 Comparison of fish and cyanobacteria biomass for 100 m segments. The simultaneous hydroacoustic estimates of the fish and cyanobacteria biomass in the Sulejów Reservoir along the studied transects with resolution of 100 m.

their hydration (hard, soft, hydrated), and type (sand, slime, rocky bottom, etc.), can also be measured acoustically. The vast majority of biogenic nutrients in water bodies accumulate in bottom sediments (particularly slimy ones), whereas the release of nutrients into the water is the effect of many physiochemical and biocenotic processes. The exchange of elements between the sediments and the water plays an enormous role in the functioning of the ecosystem, especially in shallow lakes in which the wave motion additionally leads to the resuspension of bottom sediments. As has been demonstrated (Horppila et al., 1998), biomanipulation in the form of intervention into the food chain in the water column alone, overlooking its impact on the benthos in the littoral and profundal zones, frequently does not produce the expected results as regard the waterbody's recultivation. The bottom plays a vital role in the circulation of biogenic substances and crucially affects forecasts relating to how the system will behave during the process of its recultivation in shallow waterbodies, at particular risk of eutrophication. Therefore, to enhance the predictability of the results of biomanipulation efforts and tailor such remedial measures to the situation of the actual waterbodies, bottom-related factors need to be considered.

The requirements for hydroacoustic equipment to be used in measuring fish and the bottom of a waterbody turn out to be largely mutually exclusive. Fishery echosounders operate at high sound frequencies (usually 120 kHz or above to ensure high resolution) and emit a narrow sound beam to be able to record individual fish. In other words, the sound beam must be sufficiently narrow to ensure that the probability of more than a single fish occurring at a given depth (or rather, at a given distance from the echosounder's transducer) is very low. Since sound absorption at the bottom is considerable and increases proportionally to the squared frequency, high-frequency echosounders are not able to penetrate bottom sediments. In addition, a narrow beam has a very small field of vision, particularly at shallow depths; in such circumstances, the reflected signal largely depends on minor irregularities in the bottom. This is the reason why lower-frequency echosounders (ca 30–50 kHz) and wider beam opening angles are used for bottom surveys (Freitas et al., 2003; T&gowski and Klusek, 1999). An extensive review of various technologies of bottom mapping and bottom sediment classification can be found in Kenny et al. (2003). One example of a specialized bottom echosounder with an acoustic bottom classifier is QTC VIEW TM. It is a device that can discriminate over 160 parameters based on the signals reflected from the bottom and can classify such signals through an analysis of its many components and cluster analysis. The large number of parameters allows subtle differences to be identified between the echo reflected from different sediment types, and even from different biotopes. Some of the results obtained with this device in the Sulejów Reservoir are shown in Fig. 16.5.

Polish researchers have significant accomplishments in the acoustic classification of the sea bottom (Łubniewski and Stepnowsk, 1997; Orłowski, 1984; Stepnowski et al., 1996; T&gowski et al., 2003; T&gowski and Łubniewski, 2000).

5. Macrophytes

Submerged macrophytes are among the key biological indicators of the quality of the water environment and play a major role in recultivation and biomanipulation processes in lakes (Babli and Whitton, 2002; Klosowski, 1985; Ozimek et al., 1990). Although a number of macrophyte indicators for environmental quality monitoring have been developed (CEN,

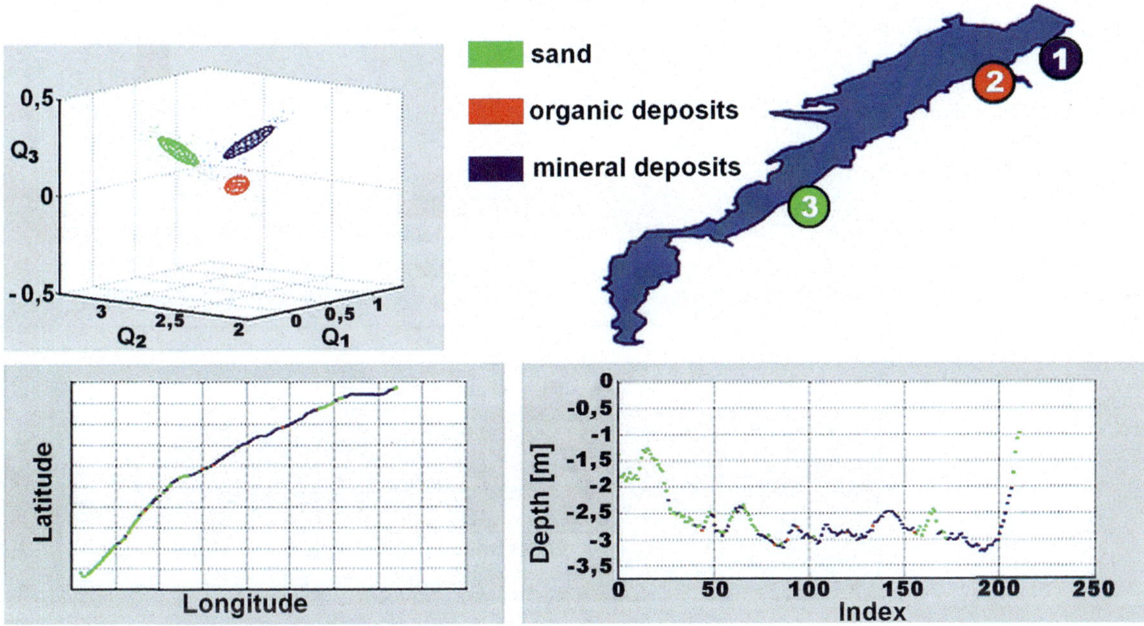

FIGURE 16.5 Acoustic classification of sediments types in the Sulejów Reservoir. Acoustic classification of sediment types in three locations of the Sulejów Reservoir, with an example of sediment classification along a selected transect (quoting the geographical coordinate) and taking into account the depth function. All results obtained with QTC VIEW TM. *From Godlewska. (2020). The use of hydroacoustic methods in ecohydrology. In M. Zalewski (Ed.), Ekohydrologia (pp. 174–182). Polish Scientific Publishers PWN.*

2003), they all draw upon strenuous phytosociological studies of selected transects, qualitative rather than quantitative and thereby hard to standardize. Furthermore, traditional research methods of underwater vegetation are extremely time- and labor-intensive, and therefore costly and restricted to relatively small areas. Alternative optical techniques (from an airplane or a satellite) provide spatial information, but in low resolution and with a quality, which largely depends on weather conditions beyond human control: clouding, surface wave action, water transparency (Madsen and Madsen, 1993; Vis et al., 2003). Hydroacoustic methods, by contrast, are practically free of such limitations and allow large areas to be surveyed in a short period of time, do not depend on the transparency of water, and, importantly, are contact-free—they neither transform nor destroy the natural environment. The assessment of such parameters as maximum and average depth of the prevalence of macrophytes, degree of bottom coverage, or the plants' height structure may prove an effective tool for the study of submerged macrophytes in the monitoring of aquatic environment (Cendrowska et al., 2014, 2015). It appears that even a partial possibility to replace traditional surveys with fast and noninvasive acoustic methods is extremely valuable in terms of both savings and effectiveness of the services in charge of environmental assessment.

Reports confirming that echosounders can be used for recording underwater vegetation first appeared as early as the 1980s, but more systematic studies of macrophytes with acoustic methods did not begin (in the United States) until the mid-1990s. This research has found that not all species yield to detection as easily. For the most part, the species that have good buoyancy thanks to their air bubbles are good sound reflectors, whereas those that form thick, crawling mats at the bottom can often be hard to distinguish from the bottom. Such freshwater species as *Myriophyllum spicatum, Hydrilla verticillata, Vallisneria* spp., or *Elodea* spp. are easily "visible," while *Chara* spp., *Nitella* spp., *Potamogeton pectinatus*, and *Ceratophyllum demersum* (Sabol et al., 2002; Sabol and Melton, 1996) present more difficulties in their detection. In addition to the detection of vegetation, an advantage of hydroacoustic methods is that they allow for both quantitative assessment of the degree of the bottom's coverage with vegetation and measurement of the plant height. Winfield (Winfield et al., 2007) demonstrated that hydroacoustic measurements were wholly consistent with the evaluation of the degree of bottom coverage made with an optical underwater camera (correlation coefficient 0.85). Macrophyte species cannot be precisely identified with the use of acoustic methods alone, similarly to fish. However, as the pilot measurement activities conducted in the Drawno National Park have found, Charophyta assemblages are easy to discriminate from assemblages with a prevalence of water milfoil or pondweed (Fig. 16.6).

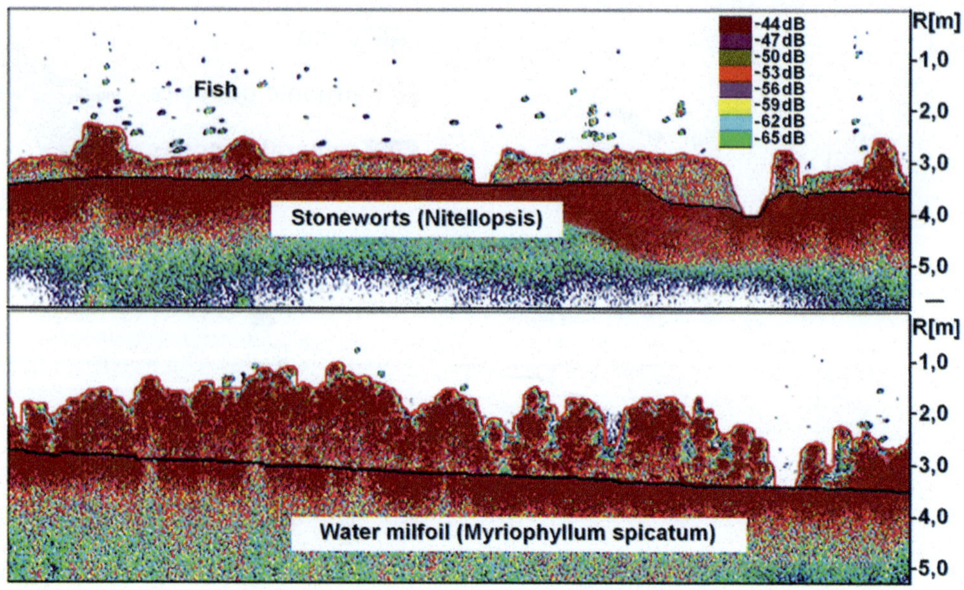

FIGURE 16.6 Echograms of various macrophyte assemblages. Echograms of various macrophyte assemblages in lakes of the Drawno National Park, Poland, obtained with SIMRAD EY60, 200 kHz, showing stoneworts (*Nitellopsis*) and milfoil (*Myriophyllum*). Black line determines the bottom; free-swimming fish are visible above the macrophytes. *From Godlewska. (2020). The use of hydroacoustic methods in ecohydrology. In M. Zalewski, Ecohydrology (pp. 174–182). Polish Scientific Publishers PWN.*

6. Restrictions of hydroacoustics in shallow waters

For obvious reasons, the various advantages of using hydroacoustics—such as encompassing large areas in a short period of time—have been central in the study of seas and oceans, the waterbodies for which these methods were developed the earliest and the most robustly. In inland waters, in turn, they have been applied mostly in the study of pelagic fish in large and deep lakes. It once seemed that hydroacoustic methods would not find any use in shallow waters due to the proximity of the medium's boundaries, which disturb the propagation of sound. Additionally, when pinging downward, the sampling volume, determined by the depth of the waterbody and the narrow beam width, is very small. However, in the 1990s, hydroacousticians developed the idea of directing the acoustic beam in shallow waters horizontally, parallel to the water surface rather than vertically downward, as in the traditional approach (Knudsen and Sægrov, 2002; Kubecka et al., 1994; Kubecka, 1996; Jan Kubecka and Wittingerova, 1998). This helped reduce the impact of the boundaries and significantly increased the sampling volume (Fig. 16.7), thus opening up possibilities for the applications of hydroacoustics in shallow waters as well.

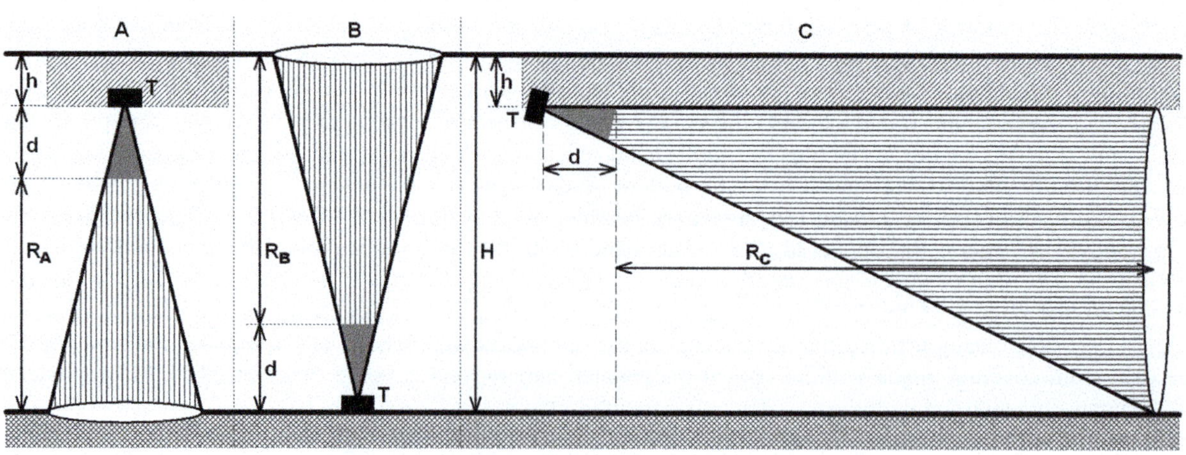

FIGURE 16.7 Comparison of sampling volume at different transducer settings. "Field of vision" in various transducer settings: (A) vertically downward, (B) vertically upward, (C) horizontally. Legend: T, acoustic transducer; H, depth of the water body; h, depth of the transducer's immersion, usually 0.5–1 m; d, transducer's dead zone, usually 0.5–4 m, depending on the width of the transducer's beam and frequency; R, transducer's maximum range determining the sampling volume V, which is proportional to R.

The sampling volume is equal to the volume of the cone formed by the beam's boundaries and is proportional to the maximum range of beam R for a given transducer setting. For a beam oriented vertically downward (variant A), that range, R_A, is equal to the depth of reservoir H less the depth of the transducer's immersion, h, and the transducer's dead zone, d:

$$R_A = H - h - d$$

For a beam oriented vertically upward (variant B), the transducer's thickness is negligible due to its small size, unless the transducer is placed on a rack, in which case its height needs to be taken into account:

$$R_B = H - d$$

For a horizontally oriented beam, h and d are negligible compared with the range, so they can be neglected (variant C):

$$R_C = H/\mathrm{tg}\theta$$

Since for a narrow beam the angle θ is small, then $\mathrm{tg}\theta \ll 1$, and $R_C \gg H$.

As is evident, the problem posed by a small sampling volume has been solved thanks to the use of a horizontal beam, but a new limitation has appeared. With a vertical beam, a single fish is always seen in the same aspect, either from above or from below. In contrast, with a horizontal beam, it can be seen in a variety of ways: in the side aspect, from the head or from the tail, and in all indirect aspects, which significantly affects the target strength (Frouzova et al., 2005), and thereby the assessment of the fish size and the biomass. To solve this problem, it was assumed that in lakes and dammed reservoirs, the direction in which the fish move in relation to the acoustic transducer is random, and therefore, each of the potential aspects is equally probable. To factor this in, the so-called deconvolution is conducted (Duncan and Kubecka, 1995)—a procedure that is typically incorporated as a function available in software (Sonar 5 Pro, for example). To carry out the procedure, we need to know the relationship between the target strength TS and the fish length L for two aspects: side (maximum TS) and the head/tail aspect (minimum TS). The difference between the maximum and minimum TS values can reach as much as 20–30 dB. Fortunately, the assumption of the randomness of fish orientation has turned out to be quite satisfactorily fulfilled (Godlewska et al., 2012; Tušer et al., 2009).

Recently, yet another deficiency of the horizontal beam has come to light (Balk et al., 2017): with an ideally smooth surface (and it has been widely acknowledged that such weather is the most favorable for surveys using the horizontal beam; Simmonds and MacLennan, 2005), a certain interference occurs between the direct signal and the signals reflected from the water surface, a process that leads to considerable fluctuations in the TS (even up to 10 dB) and an incorrect identification of the fish location. Both the amplitude of the reflected signal (TS) and the location (depth) of the fish can oscillate along with the increasing distance between the transmitter and the fish. These effects can only be minimized at a certain distance from the transducer. However, in situ research has found that, with a slight surface undulation, which is most often the case, these negative effects offset each other and can be neglected. One additional problem can be caused by a vertical temperature gradient in surface waters, which occurs mostly in the summer and results in a deflection of acoustic beams and disturbance of the spherical propagation of sound, in consequence leading to the fish size being overestimated.

The overriding function of hydroacoustic methods is to assess the fish biomass in a given body of water. It has been demonstrated, based on the comparison with the actual fish biomass measured as a result of draining the reservoir, that both in deep waters (Cadic et al., 1998) and in shallow waters (Andrzejewski et al., 2010; Godlewska et al., 2012), hydroacoustics is an effective and reliable tool for estimating the fish biomass, provided that the proper relationship

TABLE 16.1 Assessment of fish biomass in kg/hectare based on the fish catch after reservoir draining and horizontal hydroacoustic estimation using different TS/L relationships and fish directions relative to the echosounder beam (side and random).

Acoustic measurement method	Love (1969), side aspect, sea fish	Frouzova et al. (2005), side aspect, European freshwater fish	Frouzova et al. (2005), random aspect (deconvolution)	Direct measurement after draining the reservoir
SED (single echo detection)	47	118	565	537

between target strength TS and length L for a given fish species is known and a correct assumption is made on the spatial distribution of fish (i.e., whether they move in various directions randomly or along the main axis of the waterbody; Godlewska et al., 2012). Table 16.1 shows the potential significance of the differences in the estimated biomass of fish depending on the adopted TS/L regression and the assumption on the distribution of fish in relation to the axis of the transducer. The differences in the assessment of biomass can be severalfold.

In addition to the horizontal beam, another alternative way of applying the hydroacoustic method in shallow waters is to place the transducer near the bottom, facing vertically upward (Axenrot et al., 2004; Baran et al., 2017; Probst et al., 2009). This allows for the penetration of surficial waters which constitute a substantial part of shallow waterbodies (but also of deep waters); such waters tend to accumulate high numbers of water organisms and are practically overlooked, whether the beam is oriented vertically downward or horizontally (Fig. 16.7). Naturally, such a location of the transducer requires adequate safeguards to be put in place to prevent the transducer from being damaged if it hits the bottom.

As is evident, hydroacoustic methods can now also be used in shallow waters, albeit with great caution and with considerable caveats regarding the weather (as either a dead calm or strong wind practically will render this method useless), the stability of the transducer's mounting, boat stabilization, etc. Some additional knowledge of the typical spatial distribution of fish (vis-à-vis the transducer's axis) is also required, as are reliable relationships between the target strength (TS) and the fish length in the dorsal, side, and head/tail aspects.

7. Summary

In summary, hydroacoustic techniques allow for a large-scale quantification of both hydrological and biological processes, by supplying the kind of quantitative data needed to better understand how water ecosystems function and how they can be modeled. Thanks to a simultaneous measurement (in time and space) of various elements of the ecosystem, it is possible to look at the interrelationships between these elements and comprehensively evaluate the condition of the ecosystem under study, taking into account all the surveyed elements. These methods are remote, and therefore environmentally friendly, and the measurement process itself is fast and relatively inexpensive (save for the one-time expenditure on the purchase of the equipment). Unfortunately, both the measurements themselves and analysis of the data require specialist training to ensure that the hydroacoustic equipment and software are used properly and appropriately.

References

Albers, V.E., Breazeale, M.A., 1961. Underwater acoustics handbook. Phys. Today 14 (7), 51–52. https://doi.org/10.1063/1.3057657.

Andrzejewski, W., Szlakowski, J., Mastyński, J., Mazurkiewicz, J., Godlewska, M., 2010. Fish biomass and species composition in the Malta Reservoir, Poland. J. Water Land Dev. 14 (1), 67–82. https://doi.org/10.2478/v10025-011-0006-4.

Axenrot, T., Didrikas, T., Danielsson, C., Hansson, S., 2004. Diel patterns in pelagic fish behaviour and distribution observed from a stationary, bottom-mounted, and upward-facing transducer. ICES (Int. Counc. Explor. Sea) J. Mar. Sci. 61 (7), 1100–1104. https://doi.org/10.1016/j.icesjms.2004.07.006.

Babli, D.H., Whitton, B.A., 2002. Role of macrophytes in reduced summer phytoplankton abundance in the River Nene, a small nutrient-rich river in eastern England, Verhandlungen der Internationale Vereinigung főr Theoretische and Angewandte Limnologie 28, 607–610.

Balk, H., Søvegjarto, B.S., Tušer, M., Frouzová, J., Muška, M., Draštík, V., Baran, R., Kubečka, J., 2017. Surface-induced errors in target strength and position estimates during horizontal acoustic surveys. Fish. Res. 188, 149–156. https://doi.org/10.1016/j.fishres.2016.12.017.

Baran, R., Jůza, T., Tušer, M., Balk, H., Blabolil, P., Čech, M., Draštík, V., Frouzová, J., Jayasinghe, A.D., Koliada, I., Mrkvička, T., Muška, M., Ricard, D., Sajdlová, Z., Vejřík, L., Kubečka, J., 2017. A novel upward-looking hydroacoustic method for improving pelagic fish surveys. Sci. Rep. 7 (1). https://doi.org/10.1038/s41598-017-04953-6.

Cadic, N., Irz, Argillier, C., 1998. A comparison between dam draining fish-stock-taking, gillnet sampling and acoustic sampling, for fish biomass assessment on a reservoir. 1998 Proceedings of the Fourth European Conference on Underwater Acoustics 261–266.

Carmichael, W.W., 2001. Health effects of toxin-producing cyanobacteria: "The CyanoHABs". Human and Ecological Risk Assessment (HERA) 7 (5), 1393–1407. https://doi.org/10.1080/20018091095087.

CEN, 2003. TC 230/WG2/TG3/N72, Water Quality – guidance standard for the surveying of macrophytes in lakes. CEN, Brussels, p. 20.

CEN, 2014. EN15910, Water quality – guidance on estimation of fish abundance with mobile hydroacoustic methods. CEN, Brussels, pp. 1–41.

Cendrowska, M., Adamczyk, M., Godlewska, M., Puchalski, W., 2015. Using a fishery research echosounder to measure submerged macrophytes. Komunikaty Rybackie 4 (147), 17–21.

Cendrowska, M., Adamczyk, M., Puchalski, W., Godlewska, M., 2014. Acoustic mapping of submerged macrophytes in selected lakes of the Drawieński National Park. In: Papadakis, J.S., Bjørnø, L. (Eds.), Proceedings of the 2nd International Conference and Exhibition on Underwater Acoustics UA2014. 511–518. Rhodes, Greece.

Chorus, I., Bartram, J. (Eds.), 1999. Toxic Cyanobacteria in Water. A Guide to their public Health Consequencies, Monitoring and Management. Eę FN Spon, London.

Chorus, I., Falconer, I.R., Salas, H.J., Bartram, J., 2000. Health risks caused by freshwater cyanobacteria in recreational waters. J. Toxicol. 3 (4), 323–347. https://doi.org/10.1080/109374000436364.

Connolly, R.M., Jinks, K.I., Shand, A., Taylor, M.D., Gaston, T.F., Becker, A., Jinks, E.L., 2023. Out of the shadows: automatic fish detection from acoustic cameras. Aquat. Ecol. 57 (4), 833–844. https://doi.org/10.1007/s10452-022-09967-5.

Doroszczyk, L., Długoszewski, B., Kanigowska, E., Wiśniewolski, W., Godlewska, M., 2013. Spatial and temporal variability of a vendace (*Coregonus albula* (L.)) population in Lake Pluszne, Poland, based on hydroacoustics and trawl catches. Adv. Limnol. 64, 171–187. https://doi.org/10.1127/1612-166X/2013/0064-0001.

Draštík, V., Godlewska, M., Balk, H., Clabburn, P., Kubečka, J., Morrissey, E., Hateley, J., Winfield, I.J., Mrkvička, T., Guillard, J., 2017. Fish hydroacoustic survey standardization: a step forward based on comparisons of methods and systems from vertical surveys of a large deep lake. Limnol Oceanogr. Methods 15 (10), 836–846. https://doi.org/10.1002/lom3.10202.

Duncan, A., Kubecka, J., 1995. Acoustic sizes vs. real size relationships for common species of fish. National Rivers Authority Res. ę Development Note. 374.

Fitzpatrick, A., Singhvi, A., Amin, A., 2020. An Airborne sonar system for underwater remote sensing and imaging. IEEE Access 8, 189945–189959. https://doi.org/10.1109/access.2020.3031808.

Foote, K.G., 1987. Fish target strengths for use in echo integrator surveys. J. Acoust. Soc. Am. 82 (3), 981–987. https://doi.org/10.1121/1.395298.

Freitas, R., Rodrigues, A.M., Quintino, V., 2003/02/12. Benthic biotopes remote sensing using acoustics. J. Exp. Mar. Biol. Ecol. 285–286, 339–353. https://doi.org/10.1016/S0022-0981(02)00536-1.

Frouzova, J., Kubecka, J., Balk, H., Frouz, J., 2005. Target strength of some European fish species and its dependence on fish body parameters. Fish. Res. 75 (1–3), 86–96. https://doi.org/10.1016/j.fishres.2005.04.011.

Godlewska, M., 1996. Vertical migrations of krill (Euphausia superba Dana). Pol. Arch. Hydrobiol. 43 (1), 9–63.

Godlewska, M., Frouzova, J., Kubecka, J., Wiśniewolski, W., Szlakowski, J., 2012. Comparison of hydroacoustic estimates with fish census in shallow Malta Reservoir – which TS/L regression to use in horizontal beam applications? Fish. Res. 123–124, 90–97. https://doi.org/10.1016/j.fishres.2011.11.023.

Godlewska, M., Izydorczyk, K., Kaczkowski, Z., Jóźwik, A., Długoszewski, B., Ye, S., Lian, Y., Guillard, J., 2016. Do fish and blue-green algae blooms coexist in space and time? Fish. Res. 173, 93–100. https://doi.org/10.1016/j.fishres.2015.06.018.

Godlewska, M., Świerzowski, A., Winfield, I.J., 2004. Hydroacoustics as a tool for studies of fish and their habitat. Ecohydrol. Hydrobiol. 4 (4), 417–427.

Godlewska, M., Balk, H., Izydorczyk, K., Kaczkowski, Z., Mankiewicz-Boczek, J., Ye, S., 2023. Rapid in situ assessment of high-resolution spatial and temporal distribution of cyanobacterial blooms using fishery echosounder. Sci. Total Environ. 857 (Pt 2), 159492. https://doi.org/10.1016/j.scitotenv.2022.159492.

Godlewska, M., Balk, H., Kaczkowski, Z., Jurczak, T., Izydorczyk, K., Długoszewski, B., Jaskulska, A., Gągała-Borowska, I., Mankiewicz-Boczek, J., 2018. Night fish avoidance of Microcystis bloom revealed by simultaneous hydroacoustic measurements of both organisms. Fish. Res. 207, 74–84. https://doi.org/10.1016/j.fishres.2018.05.025.

Goulon, C., Le Meaux, O., Vincent-Falquet, R., Guillard, J., 2021. Hydroacoustic Autonomous boat for Remote fish detection in LakE (HARLE), an unmanned autonomous surface vehicle to monitor fish populations in lakes. Limnol Oceanogr. Methods 19 (4), 280–292. https://doi.org/10.1002/lom3.10422.

Guillard, J., Lebourges-Daussy, A., Balk, H., Colon, M., Jóźwik, A., Godlewska, M., 2014. Comparing hydroacoustic fish stock estimates in the pelagic zone of temperate deep lakes using three sound frequencies (70, 120, 200 kHz). Inland Waters 4 (4), 435–444. https://doi.org/10.5268/IW-4.4.733.

Gurney, W.S.C., Brennan, L.O., Bacon, P.J., Whelan, K.F., O'Grady, M., Dillane, E., McGinnity, P., 2014. Objectively Assigning species and Ages to salmonid length data from dual-frequency identification sonar. Trans. Am. Fish. Soc. 143 (3), 573–585. https://doi.org/10.1080/00028487.2013.862185.

Horppila, J., Peltonen, H., Malinen, T., Luokkanen, E., Kairesalo, T., 1998. Top-down or bottom-up effects by fish: issues of concern in biomanipulation of lakes. Restor. Ecol. 6 (1), 20–28. https://doi.org/10.1046/j.1526-100x.1998.00613.x.

Kaczkowski, Z., Wojtal-Frankiewicz, A., Gągała, I., Mankiewicz-Boczek, J., Jaskulska, A., Frankiewicz, P., Izydorczyk, K., Jurczak, T., Godlewska, M., 2017. Relationships among cyanobacteria, zooplankton and fish in sub-bloom conditions in the Sulejow Reservoir. J. Limnol. 76 (2), 380–396. https://doi.org/10.4081/jlimnol.2017.1521.

Kenny, A.J., Cato, I., Desprez, M., Fader, G., E Schüttenhelm, R.T., Side, J., 2003. An overview of seabed-mapping technologies in the context of marine habitat classification. ICES (Int. Counc. Explor. Sea) J. Mar. Sci. 60 (2), 411–418. https://doi.org/10.1016/s1054-3139(03)00006-7.

Klosowski, S., 1985. Habitat requirements and bioindicator value of the main communities of aquatic vegetation in North-east Poland. Pol. Arch. Hydrobiol. 32 (1), 7–29.

Knudsen, F.R., Sægrov, H., 2002. Benefits from horizontal beaming during acoustic survey: application to three Norwegian lakes. Fish. Res. 56 (2), 205–211. https://doi.org/10.1016/S0165-7836(01)00318-6.

Kubecka, J., 1996. Use of horizontal dual beam sonar for fish surveys in shallow waters. 1996 Stock Assessment in Inland Fisheries. Fishing New Books, Blackwell Science 165–178.

Kubecka, J., Duncan, A., Duncan, W.M., Sinclair, D., Butterworth, A.J., 1994. Brown trout populations of three Scottish lochs estimated by horizontal sonar and multimesh gill nets. Fish. Res. 20 (1), 29–48. https://doi.org/10.1016/0165-7836(94)90004-3.

Kubecka, J., Wittingerova, M., 1998. Horizontal beaming as a crucial component of acoustic fish stock assessment in freshwater reservoirs. Fish. Res. 35 (1–2), 99–106. https://doi.org/10.1016/s0165-7836(98)00064-2.

Love, R.H., 1969. Maximum side-aspect target strength of an individual fish. J. Acoust. Soc. Am. 46 (3B), 746–752. https://doi.org/10.1121/1.1911756.

Łubniewski, Z., Stepnowsk, A., 1997. Sea bottom typing using fractal dimensions. Hydroacoustics 1, 131–134.

Madsen, J.D., Madsen, J.D., 1993. Biomass techniques for monitoring and assessing control of aquatic vegetation. Lake Reserv. Manag. 7 (2), 141–154. https://doi.org/10.1080/07438149309354266.

Martignac, F., Daroux, A., Bagliniere, J.L., Ombredane, D., Guillard, J., 2015. The use of acoustic cameras in shallow waters: new hydroacoustic tools for monitoring migratory fish population. A review of DIDSON technology. Fish Fish. 16 (3), 486–510. https://doi.org/10.1111/faf.12071France.

Moore, T.S., Mouw, C.B., Sullivan, J.M., Twardowski, M.S., Burtner, A.M., Ciochetto, A.B., McFarland, M.N., Nayak, A.R., Paladino, D., Stockley, N. D., Johengen, T.H., Yu, A.W., Ruberg, S., Weidemann, A., 2017. Bio-optical properties of cyanobacteria blooms in western Lake Erie. Front. Mar. Sci. 4. https://doi.org/10.3389/fmars.2017.00300.

Orłowski, A., 1984. Application of multiple echoes energy measurements for evaluation of sea bottom type. Oceanologia 19, 61–78.

Ostrovsky, I., Wu, S., Lin, L., Song, L., 2020. Bloom-forming toxic cyanobacterium Microcystis: quantification and monitoring with a high-frequency echosounder. Water Res. 183, 116091. https://doi.org/10.1016/j.watres.2020.116091.

Ozimek, T., Gulati, R.D., van Donk, E., 1990. Can macrophytes be useful in biomanipulation of lakes? The Lake Zwemlust example. Hydrobiologia 200–201 (1), 399–407. https://doi.org/10.1007/BF02530357.

Paerl, H.W., Huisman, J., 2009. Climate change: a catalyst for global expansion of harmful cyanobacterial blooms. Environ. Microbiol. Rep. 1 (1), 27–37. https://doi.org/10.1111/j.1758-2229.2008.00004.x.

Probst, W.N., Thomas, G., Eckmann, R., 2009. Hydroacoustic observations of surface shoaling behaviour of young-of-the-year perch *Perca fluviatilis* (Linnaeus, 1758) with a towed upward-facing transducer. Fish. Res. 96 (2–3), 133–138. https://doi.org/10.1016/j.fishres.2008.10.009.

Sabol, B.M., Eddie Melton, R., Chamberlain, R., Doering, P., Haunert, K., 2002. Evaluation of a digital echo sounder system for detection of submersed aquatic vegetation. Estuaries 25 (1), 133–141. https://doi.org/10.1007/BF02696057.

Sabol, B.M., Melton, R.E., 1996. Development of an automated system for detection and mapping of submersed aquatic vegetation with hydroacoustic and global positioning system technologies; report 1: the submersed Aquatic Vegetation Early Warning System (SAVE-WS) – system description and users guide. In: Joint Agency Guntseville Project Aquatic Plant Management.

Simmonds, J., MacLennan, 2005. Fisheries Acoustics: Theory and Practice. Wiley-Blackwell.

Stepnowski, A., Moszyński, M., Bakiera, D., Komendarczyk, R., Burczyński, J., 1996. Visual real–time Bottom Typing System (BTS) and neural Network Experiment for sea bed classification. In: 1996 Proc. Of the 3rd European Conference on Underwater Acoustics, pp. 869–875.

T&gowski, J., Gorska, N., Klusek, Z., 2003/01. Masson SAS Poland Statistical analysis of acoustic echoes from underwater meadows in the eutrophic Puck Bay (southern Baltic Sea). Aquat. Living Resour. 16 (3), 215–221. https://doi.org/10.1016/S0990-7440(03)00015-9.

T&gowski, J., Klusek, Z., 1999. Acoustic properties of the Pomeranian Bay bottom sediments. Oceanologia 41 (3), 475–487.

T&gowski, J., Łubniewski, Z., 2000. The use of fractal properties of echo signals for acoustical classification of bottom sediments. Acoustics 86, 276–282.

Tušer, M., Kubečka, J., Frouzová, J., Jarolím, O., 2009. Fish orientation along the longitudinal profile of the Římov reservoir during daytime: consequences for horizontal acoustic surveys. Fish. Res. 96 (1), 23–29. https://doi.org/10.1016/j.fishres.2008.09.022.

van Apeldoorn, M.E., van Egmond, H.P., Speijers, G.J.A., Bakker, G.J.I., 2007. Toxins of cyanobacteria. Mol. Nutr. Food Res. 51 (1), 7–60. https://doi.org/10.1002/mnfr.200600185.

Vis, C., Hudon, C., Carignan, R., 2003. An evaluation of approaches used to determine the distribution and biomass of emergent and submerged aquatic macrophytes over large spatial scales. Aquat. Bot. 77 (3), 187–201. https://doi.org/10.1016/s0304-3770(03)00105-0.

WHO, 2003. Guidelines for Safe Recreational Water Environments Algae and Cyanobacteria in Fresh Water, Vol. 1.

Winfield, I.J., Onoufriou, C., O'Connell, M.J., Godlewska, M., Ward, R.M., Brown, A.F., Yallop, M.L., 2007/06. Assessment in two shallow lakes of a hydroacoustic system for surveying aquatic macrophytes. Hydrobiologia 584 (1), 111–119. https://doi.org/10.1007/s10750-007-0612-y.

Ye, S., Lian, Y., Godlewska, M., Liu, J., Li, Z., 2013. Day-night differences in hydroacoustic estimates of fish abundance and distribution in Lake Laojianghe, China. J. Appl. Ichthyol. 29 (6), 1423–1429. https://doi.org/10.1111/jai.12367.

Zalewski, M., 2000. Ecohydrology — the scientific background to use ecosystem properties as management tools toward sustainability of water resources. Ecol. Eng. 16 (1), 1–8. https://doi.org/10.1016/s0925-8574(00)00071-9.

Zalewski, M., 2006. Ecohydrology - an interdisciplinary tool for integrated protection and management of water bodies. River Systems 16 (4), 613–622. https://doi.org/10.1127/lr/16/2006/613.

Zalewski, M., 2011. Towards engineering harmony between water, ecosystem and society: editorial. Ecohydrol. Hydrobiol. 11 (3–4), 137–140. https://doi.org/10.2478/v10104-011-0051-2.

Zalewski, M. (Ed.), 2020. Ekohydrologia. Wydawnictwo Naukowe, PWN. ISBN 978-83-01-20804-2.

Chapter 17

Hybrid sequential biofiltration system (HSBS) for small wastewater treatment plants: An innovative ecohydrological technology

Edyta Kiedrzyńska[1, 2], Magdalena Urbaniak[2], Marcin Kiedrzyński[3] and Maciej Zalewski[1]

[1]*European Regional Centre for Ecohydrology of the Polish Academy of Sciences under the auspices of UNESCO, Łódź, Poland;* [2]*University of Lodz, UNESCO Chair on Ecohydrology and Applied Ecology, Łódź, Poland;* [3]*University of Lodz, Department of Biogeography, Paleoecology and Nature Conservation, Łódź, Poland*

PHOTO CREDIT: E. KIEDRZYNSKA.

Transdisciplinary Ecohydrology. https://doi.org/10.1016/B978-0-443-41596-8.00004-2

Municipal, domestic, and industrial wastewater discharged from wastewater treatment plants (WWTPs) located within river catchments constitutes a major source of pollution. These effluents frequently contain elevated concentrations of nutrients, particularly phosphorus (P) and nitrogen (N), as well as hazardous organic compounds including polychlorinated biphenyls (PCBs) (Chen et al., 2015; Kiedrzyńska et al., 2014a,b; Urbaniak et al., 2014; Urbaniak and Kiedrzyńska, 2015; Wang et al., 2016). Once released into the environment, these substances are transported to rivers, reservoirs, and ultimately to coastal zones, where they can exert significant ecological impacts. High nutrient loads are a primary driver of freshwater eutrophication and contribute to the proliferation of toxic cyanobacterial blooms. These blooms not only degrade water quality but also pose direct threats to aquatic organisms, domestic animals, and human health through the production of cyanotoxins. Simultaneously, municipal wastewater often contains persistent organic pollutants such as PCBs, which are toxic, bioaccumulative, and resistant to environmental degradation. These compounds originate from various industrial processes and remain in aquatic systems for extended periods, contributing to long-term contamination of sediments and food webs. One of the critical challenges in wastewater management is the disproportionately high pollution load discharged from small WWTPs. These facilities, often located in rural or semiurban areas, represent the most numerous category of treatment infrastructure in many catchments, yet they typically operate with limited resources and outdated technologies. As a result, they frequently fail to meet environmental quality standards and contribute disproportionately to point-source pollution in riverine systems (Kiedrzyńska et al., 2014a). In response to this challenge, Kiedrzyńska et al. (2017) developed and field-tested an innovative Hybrid Sequential Biofiltration System (HSBS) specifically designed to enhance pollutant removal from small-scale WWTPs, defined as those with a daily outflow of less than 100 m^3. The HSBS system was constructed at a municipal treatment plant in Rozprza, located in central Poland (Łódź Voivodeship), and operated under authentic, continuous flow conditions. The treatment system consisted of two main functional components operating in parallel: a geochemical barrier composed of filter substrates including lime, coal, and sawdust; and a biological barrier based on constructed wetlands planted with macrophytes such as *Glyceria* sp., *Acorus* sp., *Typha* sp., and *Phragmites* sp. (Fig. 17.1). The primary function of the HSBS was to reduce concentrations of suspended solids, nitrogen and phosphorus compounds, heavy metals, and PCBs in treated wastewater. The system demonstrated significant pollutant removal efficiency. Comparative analyses of inlet and outlet concentrations revealed average reductions of 16% (up to 93%) for total phosphorus (TP), 25% (up to 93%) for soluble reactive phosphorus (SRP), 15% (up to 97%) for total nitrogen (TN), and 17% (up to 98%) for nitrate nitrogen ($N\text{–}NO_3$). A 21% reduction in PCB toxicity equivalents (PCB EQ) was also observed, with the highest reduction of 43% achieved using filtration media combined with macrophyte vegetation. The biologically active barrier containing macrophytes exhibited the highest removal efficiency, with concentration reductions reaching 83% for TP, 75% for SRP, 69% for TN, 72% for $N\text{–}NO_3$, and 43% for PCB EQ. On an annual scale, the system removed approximately 0.415 kg of TP, 3.136 kg of TN, and 0.223 g of PCB EQ per square meter of biofilter surface. In total, the HSBS contributed to the removal of 25.3 kg of TP (equivalent to 18% of the total TP load from the plant), 37.1 kg SRP (26%), 191.3 kg TN (17%), 178 kg $N\text{–}NO_3$ (19%), 4.5 kg $N\text{–}NO_2$ (33%), 8.8 kg $N\text{–}NH_4^+$ (7%), and 13.6 g PCB EQ (49%) from the WWTP's annual discharge. One of the most influential factors affecting the efficiency of biofiltration systems is their physical size. Larger systems provide a longer hydraulic retention time and greater surface area for biogeochemical processes, thereby enhancing the contact between wastewater and filtering materials. However, the spatial constraints typical of small WWTPs often limit the feasibility of large-scale solutions. Therefore, compact and modular systems such as HSBS offer a valuable alternative by maximizing treatment performance within a limited footprint. The researchers also noted that the overall capacity of the HSBS could be significantly enhanced by up to two or three times, if the system were coupled with a flow-stabilizing polder or retention basin, allowing for regulated flow and optimized retention time. In conclusion, the HSBS system represents a promising, low-cost, and nature-based ecohydrological technology for improving wastewater treatment performance in small municipal facilities. Its implementation can substantially reduce the loads of nutrients and persistent organic pollutants entering surface waters from point sources, contributing to the ecological restoration of rivers and the protection of downstream aquatic ecosystems, including sensitive coastal and marine environments such as the Baltic Sea.

Building upon the outcomes and operational experience gained from the **Hybrid Sequential Biofiltration System (HSBS)** implemented in Rozprza, and guided by the **Adaptive Assessment Management** framework, a more advanced Sequential Sedimentation Biofiltration System (SSBS) configuration was subsequently designed, optimized, and constructed at a larger scale within the municipal wastewater treatment plant in **Szadek (central Poland)** (Fig. 17.2). This upgraded system integrates adaptive control of hydraulic retention times, optimized stratification of filtration substrates, and diversified macrophyte assemblages to enhance the resilience, functional stability, and long-term efficiency of nutrient and contaminant removal from wastewater. The implementation in Szadek demonstrates the scalability of the ecohydrological approach and highlights its potential for wider application in improving the performance and sustainability of **nature-based solutions** dedicated to wastewater treatment. Monitoring results indicate that the system achieves **high and**

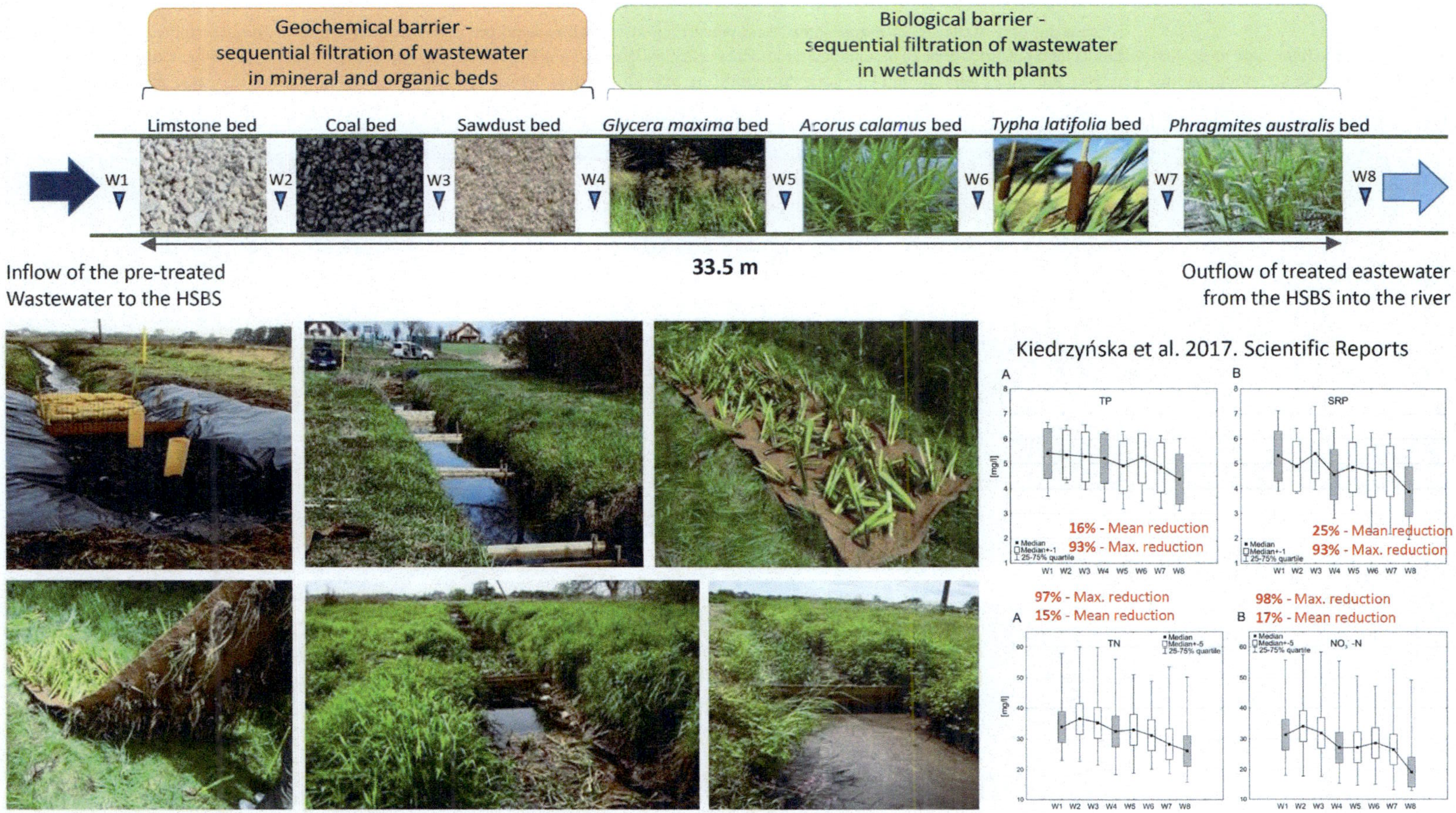

FIGURE 17.1 Construction of the hybrid sequential biofiltration system (HSBS) in a ditch carrying treated wastewater from the treatment plant to the Dąbrówka river (central Poland). Hybrid sequential biofiltration system (HSBS)—an innovative hybrid sequential biofiltration system designed for small municipal wastewater treatment plants, intended to improve effluent quality through the removal of nutrients, suspended solids, heavy metals, and persistent organic pollutants such as PCBs. *Kiedrzyńska, E., Urbaniak, M., Kiedrzyński, M., Jóźwik, A., Bednarek, A., Gągała, I., Zalewski, M., 2017. The use of a hybrid sequential biofiltration system for the improvement of nutrient removal and PCB control in municipal wastewater. Sci. Rep. 7 (1). https://doi.org/10.1038/s41598-017-05555-y.*

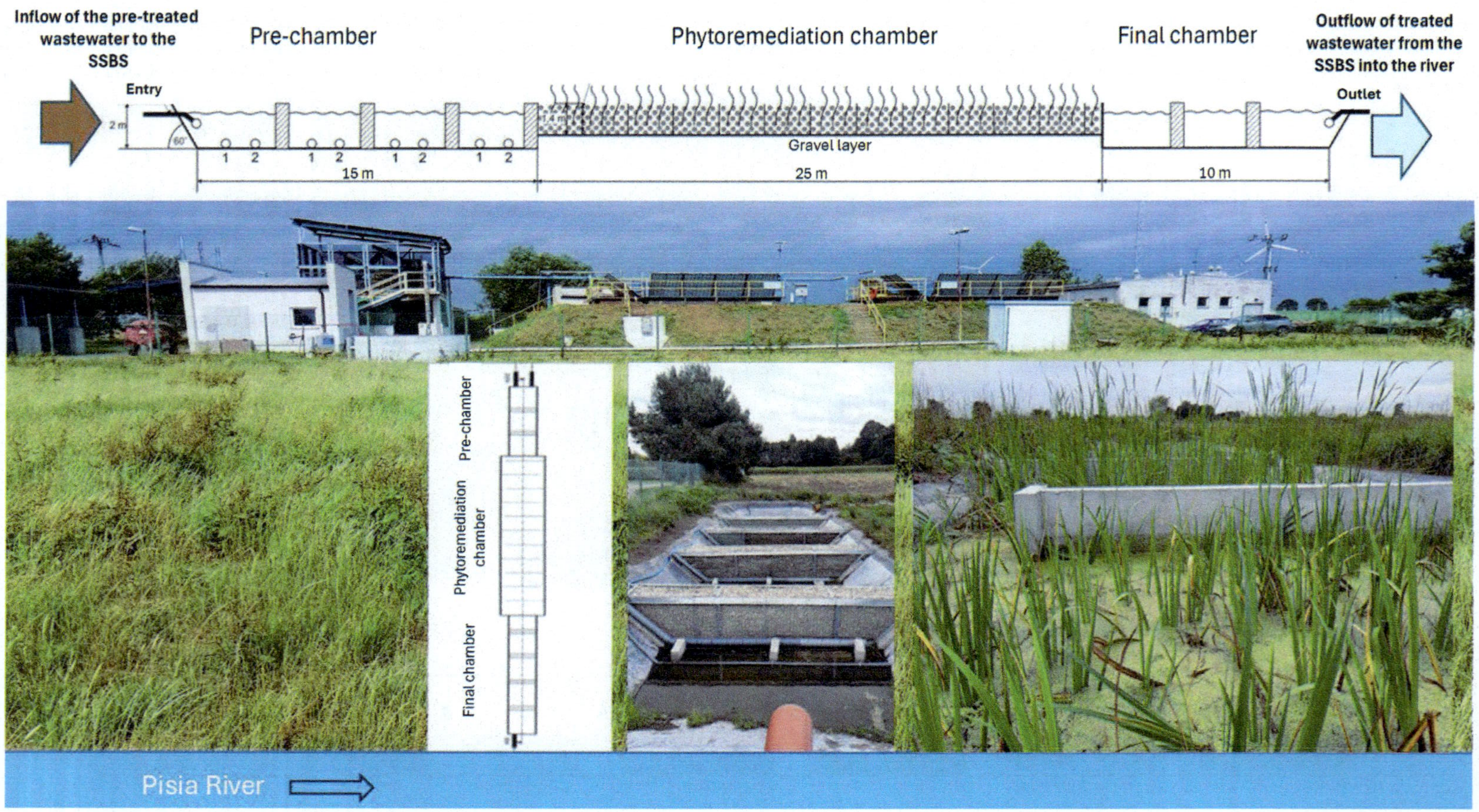

FIGURE 17.2 Advanced SSBS in central Poland—additional post-treatment system for purified, treated effluent and removing residual contaminants from municipal wastewater. Advanced sequential sedimentation biofiltration system (SSBS) constructed at the municipal wastewater treatment plant in Szadek (central Poland), functioning as an additional post-treatment unit at the outflow of the WWTP, designed to further reduce nutrient loads and other pollutants in the treated wastewater before its discharge into the river.

consistent removal efficiencies for **total suspended solids, turbidity, chemical oxygen demand (COD), biochemical oxygen demand (BOD), nitrates, sulfates,** and **total nitrogen,** while simultaneously enhancing biological activity and increasing the **purifying capacity** of effluents discharged from the treatment plant into the environment. These findings confirm that the SSBS represents a significant advancement in the development of adaptive, nature-based technologies for sustainable municipal wastewater management and for mitigating the ecological impacts of treated effluents on receiving river systems.

From an ecohydrological perspective, the SSBS contributes to reduction of impacts– supporting nutrient load reduction and the restoration of ecological balance in receiving ecosystems. Its implementation aligns with the principles of the EU Water Framework Directive, promoting the integration of technical and ecological processes to achieve good ecological status and enhance the adaptive capacity of aquatic ecosystems under changing climatic and hydrological conditions.

Acknowledgments

This research was funded by the Polish Ministry of Science and Higher Education (Project No. NN305 365738), as well as by the Polish National Science Centre under Project No. 2015/19/B/ST10/02167.

AI disclosure

During the preparation of this work the author(s) used OpenAI's ChatGPT in order to generate content. This document includes content generated and/or edited with the assistance of OpenAI's ChatGPT, used to support language refinement and structural improvements. All final content was reviewed and verified by the author. After using this tool/service, the author(s) reviewed and edited the content as needed and take(s) full responsibility for the content of the publication.

References

Chen, Y., Wen, Y., Tang, Z., Huang, J., Qi, Z., Jan, V., 2015. Effects of plant biomass on bacterial community structure in constructed wetlands used for tertiary wastewater treatment. Ecol. Eng. 84, 38–45. https://doi.org/10.1016/j.ecoleng.2015.07.013.

Kiedrzyńska, E., Jóźwik, A., Kiedrzyński, M., Zalewski, M., 2014a. Hierarchy of factors exerting an impact on nutrient load of the Baltic Sea and sustainable management of its drainage basin. Mar. Pollut. Bull. 88 (1–2), 162–173. https://doi.org/10.1016/j.marpolbul.2014.09.010.

Kiedrzyńska, E., Kiedrzyński, M., Urbaniak, M., Magnuszewski, A., Skłodowski, M., Wyrwicka, A., Zalewski, M., 2014b. Point sources of nutrient pollution in the lowland river catchment in the context of the Baltic Sea eutrophication. Ecol. Eng. 70, 337–348. https://doi.org/10.1016/j.ecoleng.2014.06.010.

Kiedrzyńska, E., Urbaniak, M., Kiedrzyński, M., Jóźwik, A., Bednarek, A., Gągała, I., Zalewski, M., 2017. The use of a hybrid sequential biofiltration system for the improvement of nutrient removal and PCB control in municipal wastewater. Sci. Rep. 7, 5477.

Urbaniak, M., Kiedrzyńska, E., 2015. Concentrations and toxic equivalency of polychlorinated biphenyls in polish wastewater treatment plant effluents. Bull. Environ. Contam. Toxicol. 95 (4), 530–535. https://doi.org/10.1007/s00128-015-1631-4.

Urbaniak, M., Kiedrzyńska, E., Kiedrzyński, M., Mendra, M., Grochowalski, A., 2014. The impact of point sources of pollution on the transport of micropollutants along the river continuum. Hydrol. Res. 45 (3), 391–410. https://doi.org/10.2166/nh.2013.242.

Wang, W., Ding, Y., Wang, Y., Song, X., Ambrose, R.F., Ullman, J.L., 2016. Intensified nitrogen removal in immobilized nitrifier enhanced constructed wetlands with external carbon addition. Bioresour. Technol. 218, 1261–1265. https://doi.org/10.1016/j.biortech.2016.06.135.

Chapter 18

Climate change, eutrophication, and cyanobacterial harmful algal blooms (CyanoHABs)

Joanna Mankiewicz-Boczek and Ilona Gagala-Borowska
European Regional Centre for Ecohydrology of the Polish Academy of Sciences under the auspices of UNESCO, Łódź, Poland

INTENSE BLOOM OF CYANOBACTERIA DOMINATED BY *MICROCYSTIS*, VISIBLE CELL DECOMPOSITION, AND RELEASE OF PHOTOSYNTHETIC PIGMENTS. MARINA ON THE SULEJOWSKI RESERVOIR, SEPTEMBER 2022. *PHOTO CREDIT: J. MANKIEWICZ-BOCZEK.*

Transdisciplinary Ecohydrology. https://doi.org/10.1016/B978-0-443-41596-8.00018-2

1. Cyanobacteria—A global threat

Cyanobacteria (blue-green algae) are one of the oldest life forms on Earth (Kaebernick and Neilan, 2001). They are classified as bacteria, but as they contain chlorophyll a, they are capable of photosynthesis, during which they convert carbon dioxide into biomass and produce oxygen. Cyanobacteria can also contain blue dye phycocyanin, red phycoerythrin, or yellow-orange carotenoids, which allow them to harness different wavelengths than algae containing only chlorophyll a, giving them a competitive advantage over other phytoplankton components in the environment (Huisman et al., 2018). Cyanobacteria are widely distributed and embrace the scum-forming genera—*Microcystis*, single-celled members of *Cyanobium* or *Synechococcus*, and filamentous species belonging to *Lyngbya, Oscillatoria, Planktothrix*, and *Phormidium*, including diazotrophic genera—*Aphanizomenon, Cylindrospermopsis, Dolichospermum*, and *Nodularia* (Paerl et al., 2018). As cosmopolitan organisms, due to their photosynthetic mode of growth, to the ability of many species to fix gaseous nitrogen into ammonia and amino acids and resistant to temperature change, cyanobacteria have the ability to adapt to changing environmental conditions and quickly colonize new areas (Codd et al., 2017; Paerl et al., 2018). In aquatic ecosystems, they are capable of existing in surface water, developing along the water column, extending over the euphotic depth and during winter overturn, below ice cover, or as benthic blooms (see Fig. 18.1).

In addition, cyanobacteria can also be symbiotes of plants and fungi. The aforementioned properties give cyanobacteria a competitive advantage over other eukaryotic phytoplankton-forming organisms, and their ability to produce secondary metabolites, often toxic to other organisms, protects them from predators (Kaebernick and Neilan, 2001).

Rapid growth of phytoplankton biomass, including cyanobacteria, in water (visible to the naked eye) is described as a **bloom**, and it is characterized by the dominance (>80%) of only one or a few species within the phytoplankton community (Meriluoto et al., 2017). This process is connected with high primary production, reduction of water transparency, and reduction of biodiversity. Therefore, the blooms caused by cyanobacteria—known as cyanobacterial harmful algal blooms (CyanoHAB)—have many serious consequences for the water body where they occur, including also the possible cyanotoxins production. **Cyanotoxins** in freshwater are represented by: microcystins (hepatotoxins) (see Fig. 18.2.A); cylindrospermopsins (cytotoxins including damage of liver and kidney); anatoxins, saxitoxins, and BMAA—β-N-methylamino-ι-alanine (neurotoxins); and LPS—lipopolysaccharides (dermatotoxins) (Merel et al., 2013; Huisman et al., 2018). Moreover, filamentous cyanobacteria exhibiting planktonic and benthic lifestyles are problematic from a risk characterization standpoint, because they contain species capable of producing multiple classes of cyanotoxins mentioned (Paerl, 2014). An additional aspect to consider in the context of the threat from CyanoHABs is the potential for aerosolized transfer of cyanotoxins and whole cyanobacterial cells. **Aerosolization of cyanotoxins** in freshwater systems may be an important factor affecting human respiratory health (Plaas and Paerl, 2021). Additionally, in the process of decomposition of cyanobacterial cells, a large amount of oxygen is absorbed, which may affect its deficiency in water during intensive CyanoHABs. Therefore, in bloom proportions, they can have **major environmental impact**, including fish kills and changes in the nutrient biogeochemistry, as well as affecting the diversity and abundance of other species (Burford et al., 2020). In turn, the mentioned changes can influence **social and economic circumstances**, for example, in the United States, the costs have been estimated to be US $2.2 billion annually, via the effects on water quality, recreation use, fishing, and property values (Burford et al., 2020; Dodds et al., 2009). Toxic CyanoHABs can influence quality of drinking water and increase the cost of its treatment (Burford et al., 2020). One a global scale, these economic costs are predicted to increase due to increasing incidences of CyanoHABs, because in the face of global climate change, cyanobacteria reach optimal growth rates at higher temperatures (>25°), which is their advantage over eukaryotic organisms in phytoplankton (Huisman et al., 2018; Paerl, 2014). This also gives them an advantage in conditions of increased fertility of aquatic ecosystems from continued eutrophication.

2. Environmental (abiotic) parameters influencing the development of CyanoHABs

Nutrients, particularly nitrogen (N) and phosphorus (P), and temperature are the main drivers of cyanobacterial dominance (Cordeiro et al., 2020; Rigosi et al., 2014). Human activity—anthropogenic pressure from domestic, agricultural, and industrial sources, including their diffuse and point character, causes increased nutrient loading to the surface water and, as a consequence, the trophic condition of a waterbody increases: in other words, **eutrophication** is observed. Considering **N**, it is important to note that if dissolved inorganic N concentrations are above 30–100 μg/L, then regardless of the season, N limitation and consequently phytoplankton abundance will not be affected (Ibelings et al., 2021; Kolzau et al., 2014; Reynolds, 2006). In study of shallow German lakes, it has been observed that a limitation of carrying capacity by nutrients could only be excluded and then **TP** and **TN** exceeding 300 and 4000 μg/L, respectively (Dolman et al., 2012; Ibelings et al., 2021). In turn, considering data from several hundred water bodies, it can be suggested that cyanobacterial

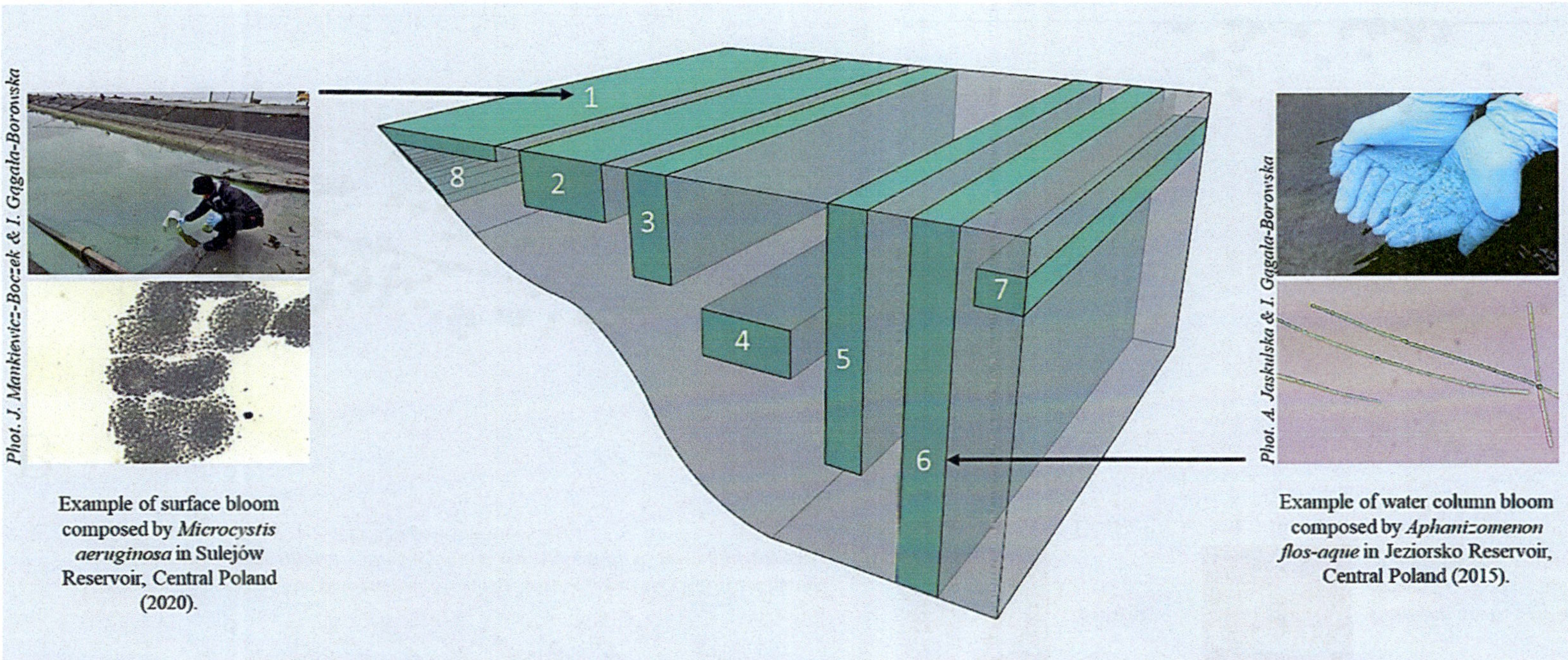

(Inspired by Salmaso et al. Basic guide to detection and monitoring of potentially toxic cyanobacteria. In: J. Meriluoto, L. Spoof and G.A. Codd [eds.] Handbook of Cyanobacterial Monitoring and Cyanotoxin Analysis. John Wiley & Sons, Ltd, The Atrium, Chichester, UK. 2017, pp 46-69).

FIGURE 18.1 Distribution of cyanobacteria in water bodies. 1 i 2—surface blooms ex. *Microcystis aeruginosa*; 3, 5, and 6—development in whole water column ex. *Aphanizomenon flos-aque*; 7—under ice; 4—metalimnetic bloom ex. *Planktothrix rubescens* (about 3–4 m deep); 8—benthic cyanobacterial blooms. *Inspired by Salmaso et al., 2017. Basic guide to detection and monitoring of potentially toxic cyanobacteria. In: Meriluoto, J., Spoof, L., Codd, G.A. (Eds.), Handbook of Cyanobacterial Monitoring and Cyanotoxin Analysis. John Wiley & Sons Ltd., pp. 46–69.*

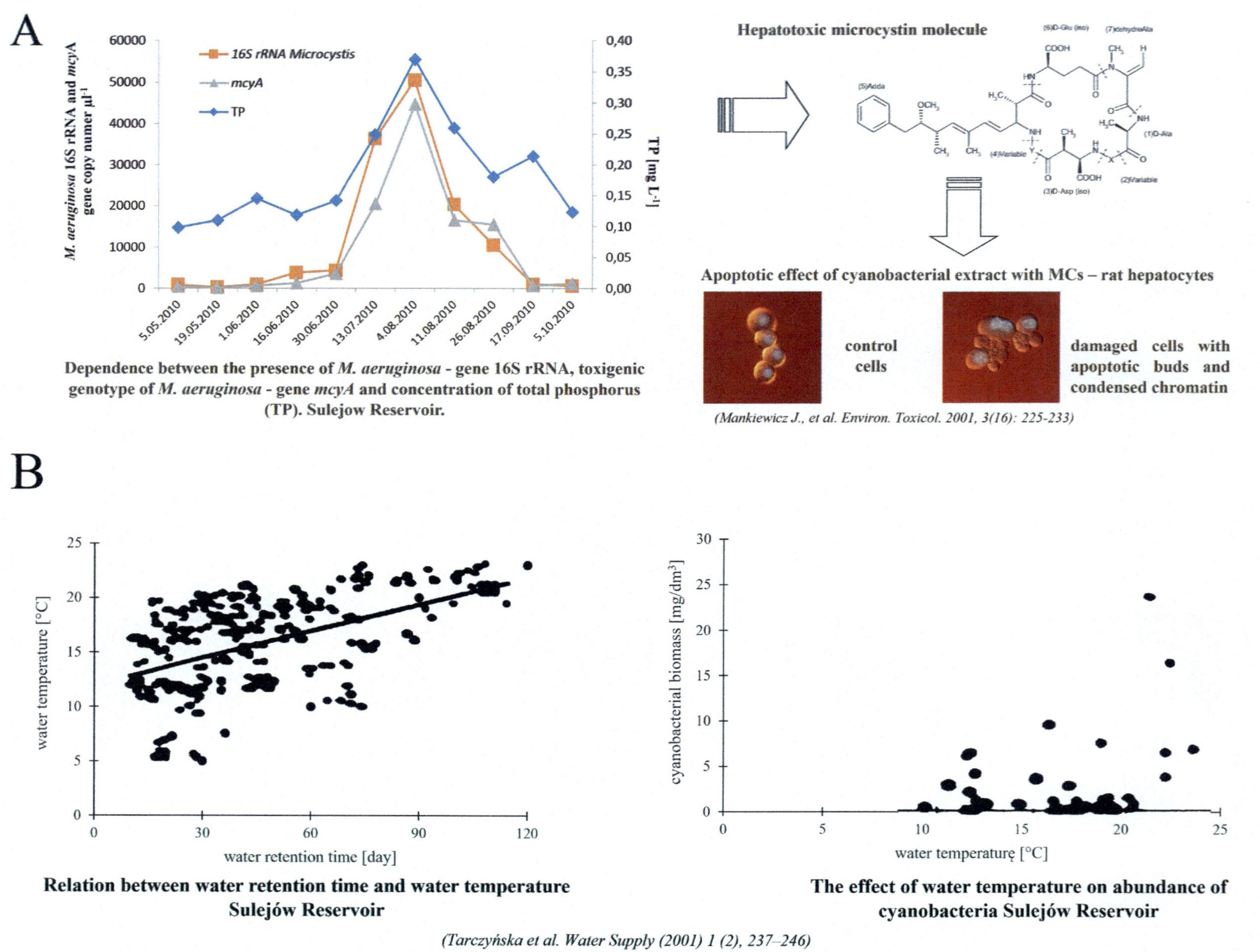

FIGURE 18.2 Parameters influencing the development and growth of CyanoHABs and consequence of hepatotoxic microcystins for mammal cells. A—total phosphorus and consequence for rat hepatocytes and DNA in an example of microcystin-producing cyanobacteria. B—temperature and water retention time. *From Tarczyńska, M., Romanowska-Duda, Z., Jurczak, T., Zalewski, M., 2001. Toxic cyanobacterial blooms in a drinking water reservoir—causes, consequences and management strategy. Water Supply 1(2), 237–246. https://doi.org/10.2166/ws.2001.0043.*

biomass rarely reaches health-relevant levels at **TP** concentrations below 10 μg/L; in the 20–100 μg/L concentration range, TP strongly determines cyanobacterial biomass levels, whereas when TP concentrations exceed 100 μg/L, then further increases in cyanobacterial biomass are rarely observed (in which case light also becomes limiting) (Ibelings et al., 2021). Additionally, Davis et al. (2009) suggested that in the face of climate change and progressive eutrophication, temperature and the presence of nutrients may cause the transformation of nontoxic cyanobacterial water blooms into toxic ones, through the expansion of cyanobacteria such as *Microcystis*. The ability of cyanobacteria to fix N, particularly in warm waters, has led in many cases to P being considered the main driver for controlling their proliferation (as an example Fig. 18.2A). On the other hand, still many of the bloom forming cyanobacteria genera are nondiazotrophic, and their proliferation is limited by N (O'Neil et al., 2012). In addition, in shallow water bodies, N can be exported more rapidly, as a result of denitrification, and consequently can be lost more rapidly than P, which can be reexchanged between sediment and water (Ibelings et al., 2021; Shatwell and Köhler, 2019). Furthermore, a reduction in N loading is essential for lakes, rivers, estuaries, and coastal waters, which are capable of assimilating more nitrogen compounds and consequently increasing their trophic status. Therefore, in preventive measures, to avoid the development of CyanoHABs, one **important step is to establish input thresholds for both N and P** (e.g., total maximum daily load; TMDL) below which CyanoHABs can be controlled in terms of size, temporal, and spatial distribution in monitored water body (Paerl, 2014).

Human **hydrologic modifications**, including water with drawls, diversions, and dams, have additionally altered the flow paths of water and nutrients entering aquatic ecosystems, which in turn influences cyanobacterial diversity, quantity, and cyanotoxins productivity (El-Shehawy et al., 2012; Paerl et al., 2018). Additionally, in Poland (temperate climate), during the summer, retention time is determined by precipitation and evapotranspiration. It has been observed that in lowland reservoirs, a long retention time (above 55–60 days) connected also with increasing of water temperature (e.g., from 10 to 25°C) and its stagnation, influenced on good conditions to development of CyanoHABs, see Fig. 18.2B (Gągała et al., 2014; Tarczyńska et al., 2001). In contrast, a short retention time (e.g., 18 days) appearing at low temperatures and additionally unstable water levels affected adversely on cyanobacteria growth (Gągała et al., 2014; Tarczyńska et al., 2001).

Given the aforementioned, climate change can be predicted to have a significant impact on catchment sustainability and CyanoHABs occurrence, including changes in their frequency, intensity, and proliferation. To date, **global warming** has been altering precipitation patterns and causing the sea level to rise, accompanied by changes in ocean and lake circulation, stratification, upwelling, wind speed, cyclone frequency, and intensity. All of those predictions play increasingly important roles in modulating CyanoHABs dynamics (Paerl, 2018). Also often associated with increasing temperature are short, intense storms that increase soil erosion and delivery of nutrients and decreased rainfall in summer or dry seasons. Consequent decreasing water levels may concentrate the nutrients already present and expose marginal sediment to mineralization and further nutrient release, favoring bigger expansion of slow growing but persistent phytoplankters like cyanobacteria (Moss, 2011). In sum, hydrologic modifications and climate change are enabling CyanoHABs to reach larger magnitudes and persist longer when accompanied by excessive nutrient loading (Paerl et al., 2018).

3. Monitoring of toxic cyanobacteria

CyanoHABs caused by the genus *Microcystis* are among the most common and most studied (as an example see Fig. 18.1). *Microcystis* can generally produce toxic secondary metabolites, microcystins (MCs), which belong to the hepatotoxin group. MCs are also among the most commonly identified cyanotoxins in surface waters worldwide (Merel et al., 2013). However, blooms can also be caused by cyanobacteria that do not produce cyanotoxins. **Whether or not toxigenic strains** with the potential to **produce cyanobacterial toxins** are present among cyanobacterial populations in a given water body **can only be investigated using molecular methods**, by detecting *mcy* gene cluster involved in the synthesis of cyanotoxin molecules, such as MCs (Mankiewicz-Boczek et al., 2006; Gągała et al., 2014). For other cyanotoxins mentioned before, such as cylindrospermopsin (*cyr* genes), anatoxin (*ana* genes), or saxitoxin (*sxt* genes), gene clusters responsible for their synthesis have also been described (Kurmayer et al., 2017). Although genetic analyses are subject to a certain margin of error related, among other factors, to cell maturity, they provide a good basis for the analysis of the generic/species structure of cyanobacteria causing a bloom and for the already mentioned assessment of the toxic potential of a CyanoHABs based on the proportion of **toxinogenic strains** in the entire cyanobacterial population (Kurmayer et al., 2017; Rinta-Kanto et al., 2005). In addition to being able to provide an early estimate of the occurrence of toxigenic cyanobacteria, these methods aim to understand how environmental factors affect cyanobacterial abundance and, consequently, the concentration of MCs (or other mentioned cyanotoxins after applied of specific gene clusters) in cyanobacterial cells and water (Davis et al., 2009; Kardinaal et al., 2007; Kurmayer et al., 2017).

The genetic monitoring mentioned before provides an example of **early warning** of the threat from toxic cyanobacteria. However, for a complete assessment of water quality exposed to eutrophication and consequent development of CyanoHABs, a **basic monitoring** should include assessment of **physical** (water temperature along a vertical profile, light availability, transparency, hydraulic retention time—reservoirs, conductivity), **chemical** (total and dissolved phosphorus, total and dissolved nitrogen, pH), and **biological variables** (chlorophyll a, quantitative determination of cyanobacteria abundance with determination of dominant cyanobacterial species or genera, and general phytoplankton description) (Meriluoto et al., 2017). Next, if cyanobacteria with the potential to produce toxins are detected, then the **cyanotoxins** should be determined qualitatively and quantitatively, together with their toxicity (biological activity) (Meriluoto et al., 2017).

References

Burford, M.A., C Carey, C., Hamilton, D.P., Huisman, J., W Paerl, H., A Wood, S., Wulff, A., 2020. Perspective: advancing the research agenda for improving understanding of cyanobacteria in a future of global change. Harmful Algae 91, 101601. https://doi.org/10.1016/j.hal.2019.04.004.

Codd, G.A., Meriluoto, J., Metcalf, J.S., 2017. Introduction: Cyanobacteria, Cyanotoxins, Their Human Impact, and Risk Management Handbook of Cyanobacterial Monitoring and Cyanotoxin Analysis. Wiley, pp. 3–8. https://doi.org/10.1002/9781119068761.ch1.

Cordeiro, R., Luz, R., Vasconcelos, V., Fonseca, A., Gonçalves, V., 2020. A critical review of cyanobacteria distribution and cyanotoxins occurrence in Atlantic Ocean Islands. Cryptogam. Algol. 41 (9), 73–89. https://doi.org/10.5252/cryptogamie-algologie2020v41a9.

Davis, T.W., Berry, D.L., Boyer, G.L., Gobler, C.J., 2009. The effects of temperature and nutrients on the growth and dynamics of toxic and non-toxic strains of Microcystis during cyanobacteria blooms. Harmful Algae 8 (5), 715–725. https://doi.org/10.1016/j.hal.2009.02.004.

Dodds, W.K., Bouska, W.W., Eitzmann, J.L., Pilger, T.J., Pitts, K.L., Riley, A.J., Schloesser, J.T., Thornbrugh, D.J., 2009. Eutrophication of U. S. freshwaters: analysis of potential economic damages. Environ. Sci. Technol. 43 (1), 12–19. https://doi.org/10.1021/es801217qUnitedStates.

Dolman, A.M., Rücker, J., Pick, F.R., Fastner, J., Rohrlack, T., Mischke, U., Wiedner, C., 2012. Cyanobacteria and cyanotoxins: the influence of nitrogen versus phosphorus. PLoS One 7 (6). https://doi.org/10.1371/journal.pone.0038757Germany.

El-Shehawy, R., Gorokhova, E., Fernández-Piñas, F., del Campo, F.F., 2012. Global warming and hepatotoxin production by cyanobacteria: what can we learn from experiments? Water Res. 46 (5), 1420–1429. https://doi.org/10.1016/j.watres.2011.11.021.

Gągała, I., Izydorczyk, K., Jurczak, T., Pawełczyk, J., Dziadek, J., Wojtal-Frankiewicz, A., Jóźwik, A., Jaskulska, A., Mankiewicz-Boczek, J., 2014. Role of environmental factors and toxic genotypes in the regulation of microcystins-producing cyanobacterial blooms. Microb. Ecol. 67 (2), 465–479. https://doi.org/10.1007/s00248-013-0303-3.

Huisman, J., Codd, G.A., Paerl, H.W., Ibelings, B.W., Verspagen, J.M.H., Visser, P.M., 2018. Cyanobacterial blooms. Nat. Rev. Microbiol. 16 (8), 471–483. https://doi.org/10.1038/s41579-018-0040-1.

Ibelings, B.W., Kurmayer, R., Azevedo, S.M.F.O., Wood, S.A., Chorus, I., Welker, M., 2021. Understanding the Occurrence of Cyanobacteria and Cyanotoxins. Informa, pp. 213–294. https://doi.org/10.1201/9781003081449-4.

Kaebernick, M., Neilan, B.A., 2001. Ecological and molecular investigations of cyanotoxin production. FEMS Microbiol. Ecol. 35 (1), 1–9. https://doi.org/10.1016/S0168-6496(00)00093-3.

Kardinaal, W.E.A., Janse, I., Kamst-van Agterveld, M., Meima, M., Snoek, J., Mur, L.R., Huisman, J., Zwart, G., Visser, P., 2007. Microcystis genotype succession in relation to microcystin concentrations in freshwater lakes. Aquat. Microb. Ecol. 48 (1), 1–12. https://doi.org/10.3354/ame048001.

Kolzau, S., Wiedner, C., Rücker, J., Köhler, J., Köhler, A., Dolman, A.M., Ban, S., 2014. Seasonal patterns of nitrogen and phosphorus limitation in four German lakes and the predictability of limitation status from ambient nutrient concentrations. PLoS One 9 (4), e96065. https://doi.org/10.1371/journal.pone.0096065.

Kurmayer, R., Sivonen, K., Wilmotte, A., Salmaso, N., 2017. Application of Molecular Tools in Monitoring Cyanobacteria and Their Potential Toxin Production Molecular Tools for the Detection and Quantification of Toxigenic Cyanobacteria. Wiley, pp. 301–333. https://doi.org/10.1002/9781119332169.ch11.

Mankiewicz-Boczek, J., Izydorczyk, K., Romanowska-Duda, Z., Jurczak, T., Stefaniak, K., Kokocinski, M., 2006. Detection and monitoring toxigenicity of cyanobacteria by application of molecular methods. Environ. Toxicol. 21 (4), 380–387. https://doi.org/10.1002/tox.20200.

Merel, S., Villarín, M.C., Chung, K., Snyder, S., 2013. Spatial and thematic distribution of research on cyanotoxins. Toxicon 76, 118–131. https://doi.org/10.1016/j.toxicon.2013.09.008.

Meriluoto, J., Spoof, L., Codd, G.A., Akçaalan, R., Albay, M., Ballot, A., Catherine, A., Fastner, J., Häggqvist, K., Horecká, M., Izydorczyk, K., Köker, L., Komárek, J., Maloufi, S., Boczek, J.M.-, Metcalf, J.S., Quesada, A., Quiblier, C., Yéprémian, C., 2017. Basic Guide to Detection and Monitoring of Potentially Toxic Cyanobacteria Handbook of Cyanobacterial Monitoring and Cyanotoxin Analysis. Wiley, pp. 46–69. https://doi.org/10.1002/9781119068761.ch6.

Moss, B., 2011. Allied attack: climate change and eutrophication. Inland Waters 1 (2), 101–105. https://doi.org/10.5268/IW-1.2.359.

O'Neil, J.M., Davis, T.W., Burford, M.A., Gobler, C.J., 2012. The rise of harmful cyanobacteria blooms: the potential roles of eutrophication and climate change. Harmful Algae 14, 313–334. https://doi.org/10.1016/j.hal.2011.10.027.

Paerl, H., 2014. Mitigating harmful cyanobacterial blooms in a human- and climatically-impacted world. Life 4 (4), 988–1012. https://doi.org/10.3390/life4040988.

Paerl, H., 2018. Mitigating toxic planktonic cyanobacterial blooms in aquatic ecosystems facing increasing anthropogenic and climatic pressures. Toxins 10 (2), 76. https://doi.org/10.3390/toxins10020076.

Paerl, H.W., Otten, T.G., Kudela, R., 2018. Mitigating the expansion of harmful algal blooms across the freshwater-to-marine continuum. Environ. Sci. Technol. 52 (10), 5519–5529. https://doi.org/10.1021/acs.est.7b05950.

Plaas, H.E., Paerl, H.W., 2021. Toxic cyanobacteria: a growing threat to water and air quality. Environ. Sci. Technol. 55 (1), 44–64. https://doi.org/10.1021/acs.est.0c06653.

Reynolds, C.S., 2006. The Ecology of Phytoplankton the Ecology of Phytoplankton. Cambridge University Press, pp. 1–535. https://doi.org/10.1017/CBO9780511542145.

Rigosi, A., Carey, C.C., Ibelings, B.W., Brookes, J.D., 2014. The interaction between climate warming and eutrophication to promote cyanobacteria is dependent on trophic state and varies among taxa. Limnol. Oceanogr. 59 (1), 99–114. https://doi.org/10.4319/lo.2014.59.1.0099.

Rinta-Kanto, J.M., Ouellette, A.J.A., Boyer, G.L., Twiss, M.R., Bridgeman, T.B., Wilhelm, S.W., 2005. Quantification of toxic Microcystis spp. during the 2003 and 2004 blooms in western Lake Erie using quantitative real-time PCR. Environ. Sci. Technol. 39 (11), 4198–4205. https://doi.org/10.1021/es048249u.

Shatwell, T., Köhler, J., 2019. Decreased nitrogen loading controls summer cyanobacterial blooms without promoting nitrogen-fixing taxa: long-term response of a shallow lake. Limnol. Oceanogr. 64, S166. https://doi.org/10.1002/lno.11002.

Tarczyńska, M., Romanowska-Duda, Z., Jurczak, T., Zalewski, M., 2001. Toxic cyanobacterial blooms in a drinking water reservoir—causes, consequences and management strategy. Water Supply 1 (2), 237–246. https://doi.org/10.2166/ws.2001.0043.

Chapter 19

Microorganisms in ecohydrological solutions for reducing cyanobacterial harmful algal bloom (CyanoHAB)

Joanna Mankiewicz-Boczek, Liliana Serwecińska and Ilona Gagala-Borowska
European Regional Centre for Ecohydrology of the Polish Academy of Sciences under the auspices of UNESCO, Łódź, Poland

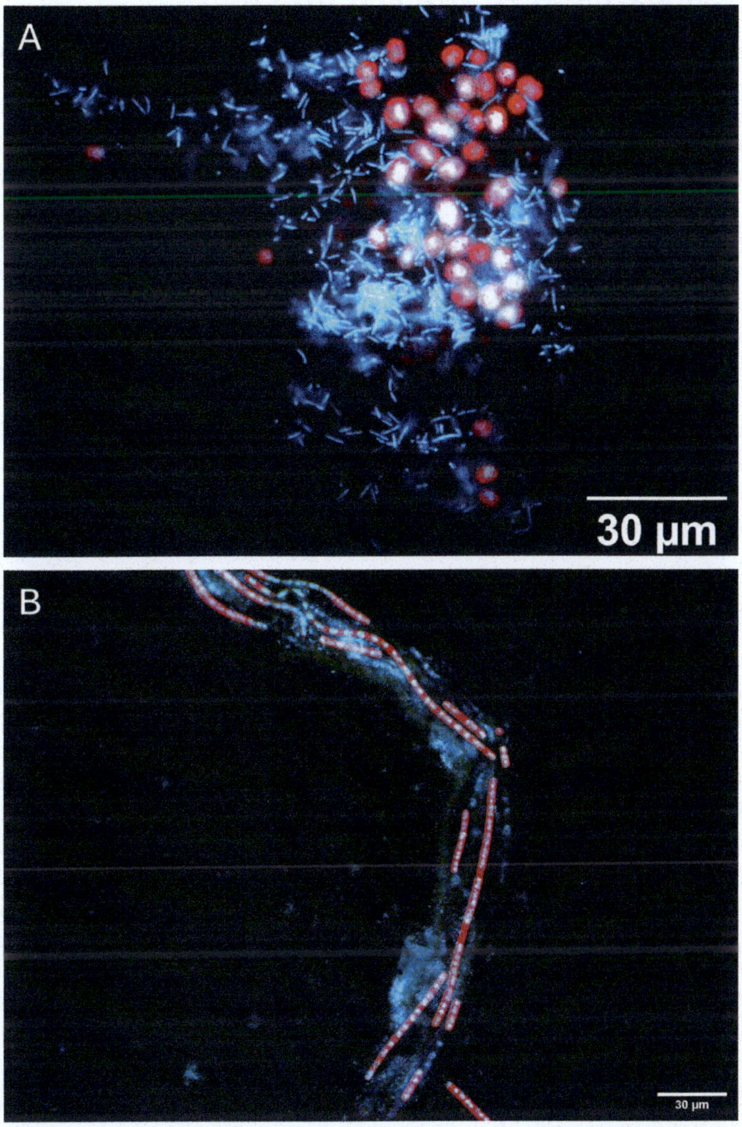

ALGICIDAL EFFECT OF BACTERIA ON CYANOBACTERIA. FLUORESCENCE MICROSCOPY IMAGE SHOWING THE INTERACTION BETWEEN A. *MICROCYSTIS AERUGINOSA* AND *MORGANELLA MORGANII*; B. *PLANKTOTHRIX AGARDHII* AND *MORGANELLA MORGANII*. RED AUTOFLUORESCENCE FROM *CHLOROPHYLL* IDENTIFIES *M. AERUGINOSA* OR *P. AGARDHII* CELLS, WHILE *M. MORGANII* CELLS ARE STAINED WITH DAPI AND OBSERVED IN BLUE. *PHOTO CREDIT: J. MORóN-LóPEZ.*

Transdisciplinary Ecohydrology. https://doi.org/10.1016/B978-0-443-41596-8.00013-3

1. Preventing the development of CyanoHAB

One of the most important factors determining the growth of cyanobacteria, as well as the appearance and duration of cyanobacterial harmful algal blooms (CyanoHABs), is the availability of nutrients, especially dissolved, easily digestible forms of nitrogen and phosphorus of anthropogenic origin (Paerl and Paul, 2012). In order to prevent the development of CyanoHAB, it is therefore important to limit their availability in the aquatic environment. Nutrients limitation methods should be twofold. On the one hand, it is important to limit the inflow of pollutants rich in N and P from the catchment area to rivers and water reservoirs, e.g., by the use of buffer zones along water bodies, as well as the construction of denitrification barriers to mitigate the runoff of pollutants into groundwater. On the other hand, it is important to eliminate the already accumulated nutrient loads from areas of rivers and reservoirs (such as river estuaries, bays, etc.) where the mixing of water is limited and leads to stagnation and increased sedimentation. The methods of action in this case are the elimination of nutrients directly from water and sediments through, among others, the use of sequential sedimentation and biofiltration systems and constructed wetlands (see other chapters in the book).

2. Regulation of cyanobacterial growth and toxicity

The different mitigation strategies and methods to eliminate the threat emerging from the occurrence of CyanoHABs are widely discussed in the literature. Up to this day a variety of chemical and physical methods have been described.

Copper sulfate, potassium permanganate, chlorine, hydrogen peroxide (H_2O_2) or ozone are broadly used **chemicals to diminish cyanobacterial blooms** (Fan et al., 2013; Miao and Tao, 2009; Qian et al., 2010; Teixeira et al., 2020). Nevertheless, using chemical methods may have negative influence on aquatic microorganisms due to the occurrence of secondary pollution from the used chemicals and the release of toxic metabolites from damaged cyanobacterial cells. As a result the accumulation of these pollutants in organisms and sediments may occur (Jancula and Marsalek, 2011), in the worst scenario leading to the imbalance of ecosystem and its collapse. Treatments in lakes with low concentrations of H_2O_2 appear to be a promising short term method for suppressing CyanoHABs. However, care should be taken in dose selection ($\leq$2 mg/L preferred) so as not to adversely affect certain zooplankton taxa in lakes. Therefore, it is advisable to carry out preliminary tests before treating lakes to determine optimal treatment concentrations that kill most cyanobacteria and minimize potential side-effects on nontarget organisms (Piel et al., 2024).

Physical method of removing cyanobacterial cells is based mainly on mixing water column using air compressor, ultrasonic damage to algal cells, pressure devices to collapse cyanobacterial gas vesicles, and UV irradiation (Ou et al., 2012; Rajasekhar et al., 2012; Teixeira et al., 2020; Visser et al., 2005). Similarly to chemical methods, physical approach may not have selective activity toward only cyanobacterial cells, require high energy level and therefore may be less effective.

Promising tool for **decrease CyanoHABs in surface water is the use of naturally occurring microorganisms**, as an example see Fig. 19.1, which may be considered as safer and more economical solutions compared to physical methods and/or the ones based on the use of chemicals (de la Cruz et al., 2011; Dziga et al., 2020).

An example of biologically active entities which are evolutionarily linked to the degradation of CyanHABs are **cyanophages**—viruses inhabiting the aquatic environment, hosted by cyanobacteria. By infecting cyanobacteria cells, cyanophages can influence the natural selection of cyanobacteria in the environment, including host population diversity, dynamics, and evolution (Avrani and Lindell, 2015; Mankiewicz-Boczek et al., 2016; Marston et al., 2012). Among the identified cyanophages there are e.g., viruses from the *Myoviridae* family that infect cyanobacteria of the species *Microcystis aeruginosa*, commonly found in water blooms worldwide (Mankiewicz-Boczek et al., 2016; Jaskulska and Mankiewicz-Boczek, 2019). The research conducted by Mankiewicz-Boczek et al. (2016) in the years 2009–2013 in the Sulejów Reservoir located on the Pilica River, showed that *Myoviridae* cyanophages appeared in the reservoir at the same time or slightly later than their hosts—the *Microcystis* cyanobacteria Fig. 19.1. Monitoring of both cyanophages and their hosts, showed that viruses may not only affect the number of cyanobacteria, but also coexist with them (Mankiewicz-Boczek et al., 2016; Yoshida et al., 2010). In the Sulejów Reservoir, a balanced coexistence of viruses and cyanobacteria was observed in the years characterized by a variable retention time between 33 and 81 days. However, when the retention time stabilized at the level of 64–71 days, the interactions of cyanophages and their hosts became more antagonistic and the increase in cyanophage abundance clearly resulted in a rapid decrease in cyanobacteria. These observations fit into the concept of the hierarchy of abiotic and biotic factors, where under stable abiotic (e.g., hydrological) conditions, the relationships between groups of organisms (biotic interactions) take on a superior character. Also, the highest amount of cyanophages in water was recorded with the mass appearance of cyanobacteria during bloom (Mankiewicz-Boczek et al., 2016). Although promising, the viral pathogens are target selective (frequently limited to a

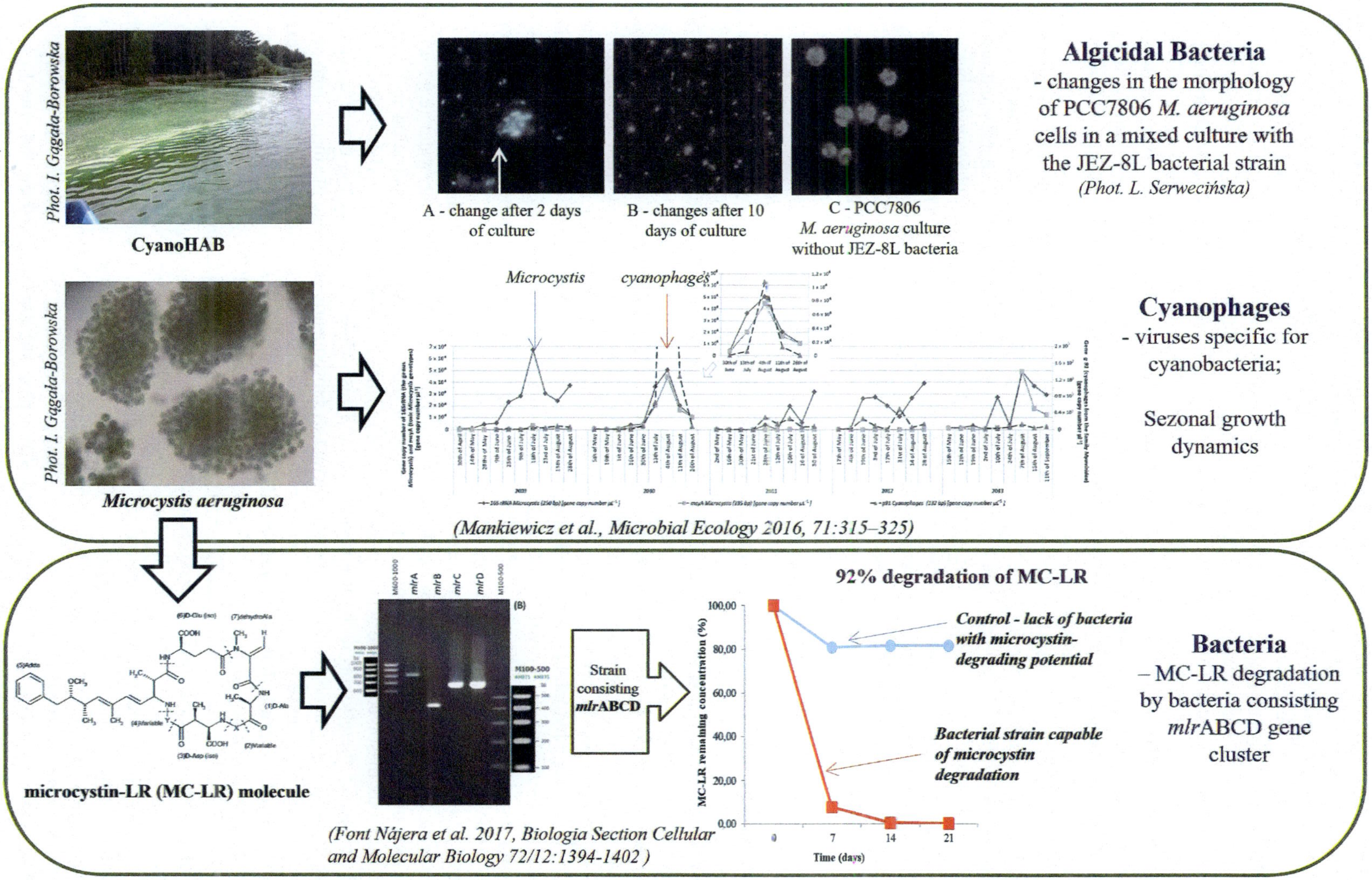

FIGURE 19.1 **Influence of bacteria and viruses on the development, occurrence and toxicity of microcystin-producing cyanobacteria.** *From: Mankiewicz-Boczek, J., Jaskulska, A., Pawełczyk, J., Gągała, I., Serwecińska, L., Dziadek, J., 2016. Cyanophages infection of* Microcystis *bloom in lowland dam reservoir of Sulejów, Poland. Microbial Ecol. 71(2), 315–325. https://doi.org/10.1007/s00248-015-0677-5; and Nájera, F., Serwecinska, L., Gągała-Borowska, I., Jurczak, T., Mankiewicz-Boczek, J., 2017. The characterization of a novel bacterial strain capable of microcystin degradation from the Jeziorsko reservoir, Poland: a preliminary study. Biol. Sect. Cell. Mol. Biol. 72/12, 1394–1402.*

single host) and their activity strictly depends on host presence in the environment (Yang et al., 2020; Yoshida et al., 2006, 2008). They could be a potential tool to reduce cyanobacterial biomass within reservoir dominated by particular host followed with strictly defined abiotic conditions. However, in a situation where the species structure of cyanobacteria changes over time, the use of a particular virus strain would be ineffective. Precisely their specificity makes cyanophages impossible to apply on a wide scale, regardless of the waterbody and the composition of CyanoHAB that prevail in it.

Therefore, bacteria tend to be more practical tool as they can grow on alternate food sources during nonbloom periods and they are less selective toward unique attachment receptors (Gumbo et al., 2008). Among others, very important is the knowledge on bacteria inhibiting growth or lyse of cyanobacterial cells and their **algicidal properties (algicidal bacteria)** (Meyer et al., 2017; Morón-López et al., 2024; Ndlela et al., 2018; Shao et al., 2013). Cyanobacterial cells are affected either as a result of direct contact or the indirect interaction via excreted diffusive lytic enzymes or extracellular products as algicidal origin (Gęgała et al., 2014; Meyer et al., 2017). Among the groups of compounds proposed as algicidal agents produced by bacteria, are for example: (1) alkaloids such as indole with its derivatives, harmine or norharmane; (2) amino acid derivatives, peptides and proteins such as argimicin A, B or C, group of the diketopiperazines, cyclo–(Gly–Pro) or hydroxamate-type siderophore; (3) terpens as saponin-type compound (Mankiewicz-Boczek and Font-Nájera, 2022; Meyer et al., 2017; Morón-López et al., 2024; Shao et al., 2013). In our study, we observed that tryptoline was the most effective algicidal compound against all species and morphotypes of cyanobacteria (represented genera of *Microcystis, Plantothrix, Aphanisomenon* and *Raphidiopsis*) but also had a significant effect on green algae (*Pseudopediastrum boryanum* and *Desmodesmus communis*). This suggests that tryptolin does not have a target effect restricted to cyanobacteria only, it also appears that filamentous cyanobacteria are more sensitive to algicidal compounds than *M. aeruginosa* strains (Morón-López et al., 2024). The algicidal bacteria isolated from freshwater or sediments described in the literature so far often represent a genus *Pseudomonas* (*i.a.* Kang et al., 2012). or *Bacillus* (*i.a.* Yu et al., 2015). Other bacteria with the ability to degrade cyanobacterial cells include *Acinetobacter* (*i.a.* Osman et al., 2017), *Aeromonas* sp. (*i.a.* Weiss et al., 2019), *Stenotrophomonas* (*i.a.* Osman et al., 2017), *Shewanella* (Li et al., 2014), *Raoultella* sp. (Su et al., 2016) and *Streptomyces amritsarensis* HG-16 (Yu et al., 2019). The algicidal bacteria may have a broad host range and affect different groups of cyanobacteria but also other eukaryotic algae (Li et al., 2014; Lin et al., 2016) or be specific only to single genus/species (Xuan et al., 2017; Yu et al., 2015). However, in our recent studies we were able to isolate from the blooms and confirm a strong algicidal character, with a selective effect only for cyanobacterial strains (no effect for green algae) for several bacterial strains: *Bacillus pumilus* (strain SU8S0818), *Morganella morganii* (strain SU7S0818), *Exiguobacterium acetylicum* (strain SU5S1020), *Chryseobacterium ureilyticum* (strain SU12S0821) and *Stenotrophomonas* sp. (strain SU6S0921) (Morón-López et al., 2024). In addition, in microcosm experiments we confirmed a promising algicidal effect for the *M. morganii* strain, by balancing phytoplankton communities toward eukaryotic dominance and reducing toxic cyanobacterial genotypes, by inhibiting the growth of *Microcystis* spp. which facilitated the growth of diatoms such as *Nitzschia* sp. (Gyrista species) (Morón-López et al., 2025) Most results however were obtained in laboratory conditions, and they still need to be supplemented (Ndlela et al., 2018).

It should be stressed that the activity of both viruses (mentioned earlier in this chapter) and algicidal bacteria strictly related to cyanobacteria cell degradation can cause cyanotoxins release from the cell into the water (e.g., microcystins intracellular concentrations are usually much higher—up to 90% of total microcystin concentration in comparison to extracellular concentrations dissolved in the water (Chorus, 2001). This in turns carries the risk of poisoning or the accumulation and transfer of these toxins along the trophic chain. Thus, while considering the use of ecohydrological biotechnology in the removal of cyanobacterial blooms, an attempt to subsequently eliminate the threat from cyanotoxins release should be included. There have been occasional literature reports about the **simultaneous ability of mentioned earlier algicidal bacteria to degrade both cyanobacterial cells and cyanotoxins**—microcystins (MCs). In example Kang et al. (2012) reported capability to degrade both—cyanobacterial cells and MCs by bacterial strain *Pseudomonas aeruginosa* UCBPP-PA14. Also, our previous studies have shown that bacterial strain *Sphingosinicella* sp. JEZ-8L (family *Sphingomonadaceae*) isolated from Jeziorsko reservoirs has a capability to degrade not only microcystin–LR molecule Fig. 19.1 (Font Nájera et al., 2017) but also is able to interact with *Microcystis aeruginosa* cells (Gęgała-Borowska et al., 2019). These bacteria were present at the same time as the toxic water bloom dominated by microcystin-producing cyanobacteria. As a consequence of this interaction disintegration of cyanobacteria cells was observed. Additionally, in year 2018, in samples collected from Sulejów Reservoir during strong *Microcystis* bloom in August and September, it was possible to obtain bacterial isolates (classified as *Stenotrophomonas* genus) potentially capable of both *Microcystis* cell lysis and MCs degradation. Also, next strains isolated from Sulejów Reservoir, homologous to *Bacillus* genus, showed the ability to degrade cyanobacterial cells. At the same time *mlr* gene cluster responsible for the synthesis of enzymes involved in degradation of MCs was detected in them. The results of this preliminary study contribute to filling the gap in the research field related to simultaneous interactions between algicidal bacteria with activity as MCs-degraders and freshwater cyanobacteria occurring in temperate climate zone. Furthermore, they indicate the potential for their future use as a tool to improve water quality and reduce dangerous CyanoHABs. Specific **bacterial strains capable of MCs molecules degradation** are very important microorganisms from the point of view of the self-purification process during cyanobacterial water bloom. The presence of bacteria capable of cyanotoxins degradation offers the possibility of a faster reduction of the negative impact of cyanobacteria and their toxins on the environment. This is because a microorganisms-driven biodegradation can occur faster than alternative pathways of cyanotoxin degradation, e.g., photodegradation (Tsuji et al., 1994) and absorption (Rivasseau et al., 1998). From literature reports, more than 40 different strains are known, representing such groups of bacteria as: Proteobacteria, Actinobacteria and Firmicutes. Of these, the *Sphingomonadaceae* family (Proteobacteria type) is the most widely represented, containing approx. 40% of all MCs-degrading bacteria (Kormas and Lymperopoulou, 2013). The first case of a bacterial strain capable of degrading a microcystin molecule was described in Australia in the Murrumbidgee River. Originally described as *Pseudomonas* sp. (Jones et al., 1994), it was later reclassified as *Sphingomonas* sp. (Bourne et al., 1996). In turn, the bacterium *Paucibacter toxinivorans*, isolated from the sediments of Lake Vihnusjarvi in Finland, was probably the first bacterium capable of degrading cyanobacterial metabolites found in Europe (Rapala et al., 2005). Once identified, MCs-degrading bacteria were incorporated in the specially designed and constructed systems for water purification (apply to operations in confined areas of water and the controlled conditions of use), separately or as an enhancement of the conventional treatment processes (Dziga et al., 2020). The systems differed in terms of design (ex. bioreactors filled with sand, beads or granular activated carbon), method of application (ex. spraying immobilized bacteria over the water) and quality of used microorganisms (naturally occurring in the form of a biofilm or inoculum of bacteria previously selected from the environment) (Dziga et al., 2020).

It should be emphasized that native bacteria with alone or both algicidal and MCs-degrading abilities perform an important ecological function, increasing the resilience of the aquatic ecosystem, especially one that receives a large stream of nutrients from area sources (e.g., farmlands) and point sources (waste discharge from urban or industrial areas)—an example of such a reservoir is Jeziorsko and Sulejów (Gęgała et al., 2014; Gqgala et al., 2012; Mankiewicz-Boczek et al., 2015) The ability to use molecular tools for observation and isolation the bacteria accompanying CyanoHABs and knowledge of their characteristics is the first step toward designing future **biotechnological solutions** aimed at regulating the occurrence of cyanobacteria and their toxins in reservoirs and lakes (Zalewski, 2020). The observations carried out in mentioned reservoirs and others referenced publications significantly broaden the knowledge of the relationship of cyanobacteria with other microorganisms and indicate the potential for their future use in **nature-based solutions** for improving water quality and reducing dangerous CyanoHABs.

References

Avrani, S., Lindell, D., 2015. Convergent evolution toward an improved growth rate and a reduced resistance range in Prochlorococcus strains resistant to phage. Proc. Natl. Acad. Sci. 112 (17). https://doi.org/10.1073/pnas.1420347112.

Bourne, D.G., Jones, G.J., Blakeley, R.L., Jones, A., Negri, A.P., Riddles, P., 1996. Enzymatic pathway for the bacterial degradation of the cyanobacterial cyclic peptide toxin microcystin LR. Appl. Environ. Microbiol. 62 (11), 4086–4094. https://doi.org/10.1128/aem.62.11.4086-4094.1996.

Chorus, I., 2001. Cyanotoxins. Occurrence, Causes, Consequences. Springer-Verlag.

de la Cruz, A.A., Antoniou, M.G., Hiskia, A., Pelaez, M., Song, W., O'Shea, K.E., He, X., Dionysiou, D.D., 2011. Can we effectively degrade microcystins?— implications on human health. Anti-Cancer Agents Med. Chem. 11 (1), 19–37. https://doi.org/10.2174/187152011794941217.

Dziga, D., Nybom, S., Gęgała-Borowska, I., Wasylewski, M., Hiskia, A.E., Triantis, T.M., Antoniou, M.G., Kaloudis, T., Dionysiou, D.D., 2020. Water Treatment for Purification from Cyanobacteria and Cyanotoxins. Wiley.

Fan, J., Ho, L., Hobson, P., Brookes, J., 2013. Evaluating the effectiveness of copper sulphate, chlorine, potassium permanganate, hydrogen peroxide and ozone on cyanobacterial cell integrity. Water Res. 47 (14), 5153–5164. https://doi.org/10.1016/j.watres.2013.05.057.

Font Nájera, A., Serwecińska, L.E., Gęgała-Borowska, I., Jurczak, T.E., Mankiewicz-Boczek, J.D., 2017. The characterization of a novel bacterial strain capable of microcystin degradation from the Jeziorsko reservoir, Poland: a preliminary study. Biologia 72 (12), 1394–1402. https://doi.org/10.1515/biolog-2017-0172.

Gęgała, I., Izydorczyk, K., Jurczak, T., Pawełczyk, J., Dziadek, J., Wojtal-Frankiewicz, A., Jóźwik, A., Jaskulska, A., Mankiewicz-Boczek, J., 2014. Role of environmental factors and toxic genotypes in the regulation of microcystins-producing cyanobacterial blooms. Microb. Ecol. 67 (2), 465–479. https://doi.org/10.1007/s00248-013-0303-3.

Gęgała-Borowska, I., Serwecińska, L., Wolska, L., Gałązowska, G., Boczek, 2019. 2019 Conference Proceedings of 11th International Conference on Toxic Cyanobacteria, Learning from the Past to Predict the Future Bacteria as a Factor Controlling the Occurrence of Microcystin-Producing Cyanobacterial Blooms.

Gągala, I., Izydorczyk, K., Jurczak, T., Mankiewicz-Boczek, J., 2012. The key parameters and early warning methods to identify presence of toxigenic blooms dominated by microcystis aeruginosa in the Jeziorsko Reservoir (Central Poland). Fresenius Environ. Bull. 21 (2), 295–303.

Gumbo, R.J., Ross, G., Cloete, E.T., 2008. Biological control of *Microcystis* dominated harmful algal blooms. Afr. J. Biotechnol. 7 (25), 4765–4773.

Jancula, D., Marsalek, B., 2011. Critical review of actually available chemical compounds for prevention and management of cyanobacterial blooms. Chemosphere 85.

Jaskulska, A., Mankiewicz-Boczek, J., 2019. Cyanophages specific to cyanobacteria from the genus *Microcystis*. Ecohydrol. Hydrobiol. 20, 83–90. https://doi.org/10.1016/j.ecohyd.2019.06.001.

Jones, G.J., Bourne, D.G., Blakeley, R.L., Doelle, H., 1994. Degradation of the cyanobacterial *hepatotoxin microcystin* by aquatic bacteria. Nat. Toxins 2 (4), 228–235. https://doi.org/10.1002/nt.2620020412.

Kang, Y.H., Park, C.S., Han, M.S., 2012. *Pseudomonas aeruginosa* UCBPP-PA14 a useful bacterium capable of lysing *Microcystis aeruginosa* cells and degrading *microcystins*. J. Appl. Phycol. 24 (6), 1517–1525. https://doi.org/10.1007/s10811-012-9812-6.

Kormas, K.A., Lymperopoulou, D.S., 2013. Cyanobacterial toxin degrading bacteria: who are they? BioMed Res. Int. 1–12. https://doi.org/10.1155/2013/463894.

Li, Z., Lin, S., Liu, X., Tan, J., Pan, J., Yang, H., 2014. A freshwater bacterial strain, Shewanella sp. Lzh-2, isolated from Lake Taihu and its two algicidal active substances, hexahydropyrrolo[1,2-a]pyrazine- 1,4-dione and 2, 3-indolinedione. Appl. Microbiol. Biotechnol. 98 (10), 4737–4748. https://doi.org/10.1007/s00253-014-5602-1.

Lin, S., Geng, M., Liu, X., Tan, J., Yang, H., 2016. On the control of *Microcystis aeruginosa* and *Synechococccus* species using an algicidal bacterium, Stenotrophomonas F6, and its algicidal compounds cyclo-(Gly-Pro) and hydroquinone. J. Appl. Phycol. 28 (1), 345–355. https://doi.org/10.1007/s10811-015-0549-x.

Mankiewicz-Boczek, J., Font-Nájera, A., 2022. Temporal and functional interrelationships between bacterioplankton communities and the development of a toxigenic *Microcystis* bloom in a lowland European reservoir. Sci. Rep. 12 (1), 19332. https://doi.org/10.1038/s41598-022-23671-2.

Mankiewicz-Boczek, J., Gągała, I., Jurczak, T., Jaskulska, A., Pawełczyk, J., Dziadek, J., 2015. Bacteria homologus to Aeromonas capable of *microcystin* degradation. Open Life Sci. 10 (1). https://doi.org/10.1515/biol-2015-0012.

Mankiewicz-Boczek, J., Jaskulska, A., Pawełczyk, J., Gągała, I., Serwecińska, L., Dziadek, J., 2016. Cyanophages infection of *Microcystis* bloom in lowland dam reservoir of Sulejów, Poland. Microb. Ecol. 71 (2), 315–325. https://doi.org/10.1007/s00248-015-0677-5.

Marston, M.F., Pierciey, F.J., Shepard, A., Gearin, G., Qi, J., Yandava, C., Schuster, S.C., Henn, M.R., Martiny, J.B.H., 2012. Rapid diversification of coevolving marine Synechococcus and a virus. Proc. Natl. Acad. Sci. U. S. A. 109 (12), 4544–4549. https://doi.org/10.1073/pnas.1120310109.

Meyer, N., Bigalke, A., Kaulfuß, A., Pohnert, G., 2017. Strategies and ecological roles of algicidal bacteria. FEMS Microbiol. Rev. 41 (6), 880–899. https://doi.org/10.1093/femsre/fux029.

Miao, H., Tao, W., 2009. The mechanisms of ozonation on cyanobacteria and its toxins removal. Separ. Purif. Technol. 66 (1), 187–193. https://doi.org/10.1016/j.seppur.2008.11.008.

Morón-López, J., Serwecińska, L., Balcerzak, Ł., Glińska, S., Mankiewicz-Boczek, J., 2024. Algicidal bacteria against cyanobacteria: practical knowledge from laboratory to application. Crit. Rev. Environ. Sci. Technol. 54 (3), 239–266. https://doi.org/10.1080/10643389.2023.2232257.

Morón-López, J., Font-Nájera, A., Kokociński, M., Jarosiewicz, P., Jurczak, T., Mankiewicz-Boczek, J., 2025. Influence of bloom stage on the effectiveness of algicidal bacteria in controlling harmful cyanobacteria: a microcosm study. Environ. Pollut. 374, 126261. https://doi.org/10.1016/j.envpol.2025.126261.

Ndlela, L.L., Oberholster, P.J., Van Wyk, J.H., Cheng, P.H., 2018. Bacteria as biological control agents of freshwater cyanobacteria: is it feasible beyond the laboratory? Appl. Microbiol. Biotechnol. 102 (23), 9911–9923. https://doi.org/10.1007/s00253-018-9391-9.

Osman, O.A., Beier, S., Grabherr, M., Bertilsson, S., 2017. Interactions of freshwater cyanobacteria with bacterial antagonists. Appl. Environ. Microbiol. 83 (7). https://doi.org/10.1128/AEM.02634-16.

Ou, H., Gao, N., Deng, Y., Qiao, J., Wang, H., 2012. Immediate and long-term impacts of UV-C irradiation on photosynthetic capacity, survival and microcystin-LR release risk of *Microcystis aeruginosa*. Water Res. 46 (4), 1241–1250. https://doi.org/10.1016/j.watres.2011.12.025.

Paerl, H.W., Paul, V.J., 2012. Climate change: links to global expansion of harmful cyanobacteria. Water Res. 46 (5), 1349–1363. https://doi.org/10.1016/j.watres.2011.08.002.

Piel, T., Sandrini, G., Weenink, E.F.J., Qin, H., Herk, M.J.v., Morales-Grooters, M.L., Schuurmans, J.M., Slot, P.C., Wijn, G., Arntz, J., Zervou, S.K., Kaloudis, T., Hiskia, A., Huisman, J., Visser, P.M., 2024. Shifts in phytoplankton and zooplankton communities in three cyanobacteria-dominated lakes after treatment with hydrogen peroxide. Harmful Algae 133. https://doi.org/10.1016/j.hal.2024.102585.

Qian, H., Yu, S., Sun, Z., Xie, X., Liu, W., Fu, Z., 2010. Effects of copper sulfate, hydrogen peroxide and N-phenyl-2-naphthylamine on oxidative stress and the expression of genes involved photosynthesis and microcystin disposition in *Microcystis aeruginosa*. Aquat. Toxicol. 99 (3), 405–412. https://doi.org/10.1016/j.aquatox.2010.05.018.

Rajasekhar, P., Fan, L., Nguyen, T., Roddick, F.A., 2012. Impact of sonication at 20kHz on *Microcystis aeruginosa*, Anabaena circinalis and Chlorella sp. Water Res. 46 (5), 1473–1481. https://doi.org/10.1016/j.watres.2011.11.017.

Rapala, J., Berg, K.A., Lyra, C., Maarit Niemi, R., Werner, M., Suomalainen, S., Paulin, L., Lahti, K., 2005. Paucibacter toxinivorans gen. nov., sp. nov., a bacterium that degrades cyclic cyanobacterial hepatotoxins microcystins and nodularin. Int. J. Syst. Evol. Microbiol. 55 (4), 1563–1568. https://doi.org/10.1099/ijs.0.63599-0.

Rivasseau, C., Martins, S., Hennion, M.-C., 1998. Determination of some physicochemical parameters of microcystins (cyanobacterial toxins) and trace level analysis in environmental samples using liquid chromatography. J. Chromatogr. A. 799 (1–2), 155–169. https://doi.org/10.1016/s0021-9673(97)01095-9.

Shao, J., Li, R., Lepo, J.E., Gu, J.D., 2013. Potential for control of harmful cyanobacterial blooms using biologically derived substances: problems and prospects. J. Environ. Manag. 125, 149–155. https://doi.org/10.1016/j.jenvman.2013.04.001.

Su, J.F., Shao, S.C., Ma, F., S Lu, J., Zhang, K., 2016. Bacteriological control by Raoultella sp. R11 on growth and toxins production of *Microcystis aeruginosa*. Chem. Eng. J. 293, 139–150.

Teixeira, R., Rosa, M., Sorlini, M.J., Biasibetti, S., Christophoridis, M., Edwards, C., 2020. Chapter 3: Removal of Cyanobacteria and Cyanotoxins by Conventional Physical-Chemical Treatment 69. Wiley.

Tsuji, K., Naito, S., Kondo, F., Ishikawa, N., Watanabe, M.F., Suzuki, M., Harada, K.-ichi, 1994. Stability of microcystins from cyanobacteria: effect of light on decomposition and isomerization. Environ. Sci. Technol. 28 (1), 173–177. https://doi.org/10.1021/es00050a024.

Visser, P.M., Ibelings, B.W., Mur, L.R., Walsby, A.E., 2005. The ecophysiology of the harmful cyanobacterium *Microcystis*. In: Huisman, J., Matthijs, H.C.P., Visser, P.M. (Eds.), Harmful Cyanobacteria. Springer, pp. 109–142. https://doi.org/10.1007/1-4020-3022-3_6.

Weiss, G., Kovalerchick, D., Hurwitz, J.L., Murik, O., De Philippis, R., Carmeli, S., Sukenik, A., Kaplan, A., 2019. Increased algicidal activity of Aeromonas veronii in response to *Microcystis aeruginosa*: interspecies crosstalk and secondary metabolites synergism. Environ. Microbiol. 21 (3), 1140–1150. https://doi.org/10.1111/1462-2920.14561.

Xuan, H., Dai, X., Li, J., Zhang, X., Yang, C., Luo, F., 2017. A Bacillus sp. strain with antagonistic activity against Fusarium graminearum kills *Microcystis aeruginosa* selectively. Sci. Total Environ. 583, 214–221. https://doi.org/10.1016/j.scitotenv.2017.01.055.

Yang, F., Jin, H., Wang, X.-Q., Li, Q., Zhang, J.-T., Cui, N., Jiang, Y.-L., Chen, Y., Wu, Q.-F., Zhou, C.-Z., Li, W.-F., 2020. Genomic analysis of Mic1 reveals a novel freshwater long-tailed cyanophage. Front. Microbiol. 11. https://doi.org/10.3389/fmicb.2020.00484.

Yoshida, M., Yoshida, T., Kashima, A., Takashima, Y., Hosoda, N., Nagasaki, K., Hiroishi, S., 2008. Ecological dynamics of the toxic bloom-forming cyanobacterium *Microcystis aeruginosa* and its cyanophages in freshwater. Appl. Environ. Microbiol. 74 (10), 3269–3273. https://doi.org/10.1128/AEM.02240-07.

Yoshida, M., Yoshida, T., Yoshida-Takashima, Y., Kashima, A., Hiroishi, S., 2010. Real-time PCR detection of host-mediated cyanophage gene transcripts during infection of a natural *Microcystis aeruginosa* population. Microb. Environ. 25 (3), 211–215. https://doi.org/10.1264/jsme2.me10117.

Yoshida, T., Takashima, Y., Tomaru, Y., Shirai, Y., Takao, Y., Hiroishi, S., Nagasaki, K., 2006. Isolation and characterization of a cyanophage infecting the toxic cyanobacterium *Microcystis aeruginosa*. Appl. Environ. Microbiol. 72 (2), 1239–1247. https://doi.org/10.1128/aem.72.2.1239-1247.2006.

Yu, J., Kong, Y., Gao, S., Miao, L., Zou, P., Xu, B., Zeng, C., Zhang, X., 2015. Bacillus amyloliquefaciens T1 as a potential control agent for cyanobacteria. J. Appl. Phycol. 27 (3), 1213–1221. https://doi.org/10.1007/s10811-014-0402-7.

Yu, Y., Zeng, Y., Li, J., Yang, C., Zhang, X., Luo, F., Dai, X., 2019. An algicidal *Streptomyces amritsarensis* strain against *Microcystis aeruginosa* strongly inhibits microcystin synthesis simultaneously. Sci. Total Environ. 650 (Pt 1), 34–43. https://doi.org/10.1016/j.scitotenv.2018.08.433.

Zalewski, M., 2020. Ekohydrologia. Wydawnictwo Naukowe PWN, Warszawa, ISBN 978-83-01-20804-2.

Chapter 20

Application of ecohydrology for the pesticides impact mitigation

Paweł Jarosiewicz[1,2]

[1]European Regional Centre for Ecohydrology of the Polish Academy of Sciences under the auspices of UNESCO, Łódź, Poland; [2]University of Lodz, Faculty of Biology and Environmental Protection, UNESCO Chair on Ecohydrology and Applied Ecology, Łódź, Poland

PESTICIDE APPLICATION IN AN APPLE ORCHARD LOCATED IN CENTRAL POLAND. *IMAGE CREDIT: JAROSIEWICZ ET AL. (2022).*

Transdisciplinary Ecohydrology. https://doi.org/10.1016/B978-0-443-41596-8.00017-0

1. Why should we be concerned of pesticides?

Population growth creates pressure on agricultural production, which in many places already uses significant areas. The increase in productivity per unit area is directly related to the optimization of the production process and the use of both mineral fertilizers and pesticides (WHO-FAO, 2019). Taking into account the currently dominant system of industrial agriculture and the limited land for expanding new crops, we can predict further intensification and, consequently, increasing loads of pollutants entering the environment. The observed warming of the climate contributes to the occurrence of more frequent, more violent, and intense rainfall with simultaneous longer periods of drought. It can lead to, inter alia, remobilization of pollutants deposited in permafrost and oceans; in addition, through disturbances in the dynamics of meteorological phenomena, it contributes to increased erosion and transport of pollutants to aquatic ecosystems (Macdonald et al., 2003; Wang et al., 2016).

Synthetic pesticides (from Latin *pestis*—plague, and *caedo*—to kill) are chemical substances used mainly to protect crops and plant products against harmful organisms. The silent threat of pesticides is their unwanted drift in the environment due to mobility or volatilization, which was underestimated for many years by decision-makers. Although the global consciousness of pesticide impact has constantly been increasing since the time of Rachel Carson's *Silent Spring* publication (Carson, 1962), our current model of agriculture is utterly addicted to the application of chemical protection. With discussion about pesticides, we must add that their application contributes to the improvement of the quality of agricultural products and meeting the needs of consumers, not to mention the stable food supply chain and disease control.

From 1990 to 2022, the worldwide agricultural use of pesticides steadily increased, **reaching 3.69 million metric tons in 2022** (FAO, 2024; WHO-FAO, 2019). The estimation of the influence of these substances on the state of the environment remains extremely difficult, mainly due to the simultaneous presence of many xenobiotic substances and metabolic processes in which reaction products may obtain different levels of toxicity. It has been shown that complex pollutants are subject to constant changes and accumulation processes, which is why they are often detected in concentrations much higher than expected (Sauvé and Desrosiers, 2014). And yet, monitoring of pesticide residues in the environment is performed by less than 30% of countries (WHO-FAO, 2019). Official environmental agency reports, if available, may not be indicative of the true scale of the pollution since they are focused rather on confirmed, high-toxicity compounds, which have mostly been banned (see. Water Framework Directive), while the market for pesticides is constantly growing and is flexible. As has been evidenced by many authors, pesticides are present in the environment in vast numbers, e.g., 173 pesticides in 3 European rivers (Busch et al., 2016). When discussing the presence of pesticides in the environment, it is important to understand the difference between what is assessed and what is actually present in the environment. The quality state of water resources is the resultant between compounds that are currently applied and move to the environment and those that remain bound in a pool that is difficult to access, even for decades, and are slowly released into the environment.

This is why **pesticides are considered part of the "novel entities" planetary boundary**, which refers to synthetic chemicals, engineered materials, and other substances introduced by humans that have the potential to disrupt Earth system processes (Persson et al., 2022). While not always listed individually, pesticides contribute to chemical pollution and biodiversity loss, and their uncontrolled use can have global-scale ecological impacts.

2. The impact of pesticides on human health, terrestrial, and aquatic ecosystems

Although the application of pesticides can be described as target-specific, due to their movement in water, wind, or vaporization, they can enter the environment and cause adverse ecological and human health effects. The occurrence of pesticides can result from a point source (e.g., drainage systems, industry) or diffuse (e.g., agriculture, urban catchments). Among diffuse pesticide pollution, we can mention baseflow seepage, surface and subsurface runoff, leaching, spray drift from the application spot, and deposition after volatilization (Fig. 20.1). Rapid transfer of pesticides could be connected with preferential flow phenomena and drainage, which force faster transfer of water with soluble pesticides, drastically decreasing their retention time. Traditionally, only agricultural catchments have been considered as a source of pesticides; however, green area management in cities, combined with sprawling urban areas with many gardens and the application of such pesticide groups as molluscicides and insecticides in households, has increased the visibility of pesticide threat in urban areas. Contamination of aquatic ecosystems can pose a serious threat to living organisms and limit drinking water resources. It is still debated how low concentrations that are found in human and animal tissues could impact the organisms' functions when exposure is chronic.

It is difficult to determine the impact of pesticides in the environment due to the prevailing importance of other abiotic elements, such as, for example, nutrient pollution, light availability, oxygen content, or disturbed hydrological processes. Nevertheless, the last 30 years of observations of pesticide toxicity and their fate in the environment revealed several negative pathways that can impact the ecosystems and human health.

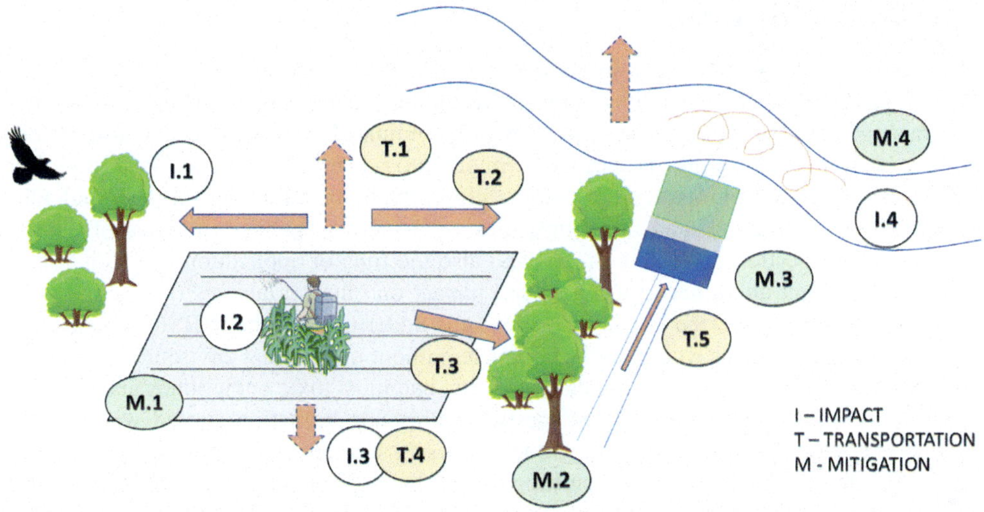

FIGURE 20.1 Impacts of pesticides. The figure presents the potential impacts of pesticides, their transportation routes, and available mitigation practices. Impacts: I.1—reduction of the biodiversity in terrestrial ecosystems, I.2—impact on human health (from farmers exposed during application to the food consumers), I.3—impact on soil microbial structure, I.4—impact on aquatic organisms. Transport: T.1—volatilization directly from application place or from water, that result later in atmospheric deposition, T.2—wind drift of the pesticide spray, T.3—surface runoff with sediments and subsurface leaching of mobile pesticides, T.4—infiltration/leaching into the soil profile and pollution of groundwater, T.5—transportation in river systems. Mitigation: M.1—better pesticide policy and more educated farmers, application of integrated pest management strategy and safe compounds, M.2—increasing the complexity of landscape with buffer zones and ecotones to limit the runoff transportation, M.3—implementation of ecohydrology and nature-based solutions in aquatic systems, M.4—enhancement of river self-purification processes with renaturalization.

3. Holistic approach toward pesticides impact reduction

Once the pesticides reach the environment, they can undergo several processes, but generally they create universal groups of extractable residues and bound residues or are metabolized by microorganisms. Important factors in determining the dynamics of pesticide transport are the n-octanol/water partition coefficient (K_{OW}) and the soil/water partition coefficient (K_{OC}), which define the equilibrium ratio of concentrations of substances in a two-phase system at a given temperature and constant pressure. These factors determine whether a substance is hydrophobic (lipophilic) or hydrophilic (lipophobic), which is important for solubility. Hydrophilic compounds are well soluble in water, and they will move much faster in the environment, for example, by surface runoff from the application site or by leaching. On the other hand, hydrophobic compounds will have a higher affinity for bioaccumulation, i.e., accumulation in the fatty tissues of organisms and transport in the food chain (biomagnification). Lipophilic compounds will also have a higher potential for adsorption to the surface of bottom sediments, soils, and organic matter found in river systems; therefore, they are often met in surface runoff, particularly in erosion-prone catchment areas. Another important feature of pesticides is their half-life (T_{50}) in water and soil. The aforementioned elements indicate strategies for reducing of pesticide presence in the environment.

In general, actions to reduce the displacement of pollutants, including pesticides, can be divided into farm management level and off-farm ecohydrological process management. Actions may include reduction of surface runoff and erosion, reduction of spray transfer range, and infiltration into groundwater (Fig. 20.1). The term "risk reduction" relates not only to the reduction of environmental impact but also to the decrease in toxicity of pesticide compounds and limiting the exposure of applicators and food consumers.

Since environmental budgets for both monitoring and management are limited, it is essential to prioritize areas for pesticide mitigation. With the development of catchment–river models and analytical studies, it has become possible to identify catchments that are more prone to pesticide pollution. In Poland, for example, catchments dominated by apple orchards with drastically reduced land /water ecotones carry significantly higher pesticide loads compared to areas with other types of agriculture or more natural river valleys (Jarosiewicz, 2021). A similar approach can be applied elsewhere through comparative studies between catchments and by analyzing pesticide consumption in specific areas. Moreover, with well-developed registration systems, it is now possible to identify precisely which pesticides are being used by farmers and to focus monitoring efforts on existing threats. At the same time, broad screening programs remain valuable, as they can help to detect illegal pesticide use or the presence of legacy compounds in sediments and soils.

4. Agriculture-level mitigation

Agriculture is the biggest emitter of pesticides to aquatic systems. Point-source inputs can be mitigated by increasing farmers' awareness regarding pesticide handling and application and encouraging them to implement loss-reducing measures of "best management practices." Yet, in catchments dominated by diffuse input, mitigation of point-source inputs alone may not be sufficient to limit the pesticide runoff into aquatic ecosystems. In terms of agricultural practice, the known methods of pollution limitation are conservation tillage, including zero tillage, mulching, application of cover crops, contour plowing, planting, and other measures, such as new precise methods of application. It is very important that the dose of the pesticide, the date of use, and also the time of application day should be selected so as to ensure effectiveness with the minimum necessary dose and impact on other organisms (respecting the principles of Integrated Pest Management). Of course, the crucial element, as is underlined by the WBSR strategy, is the education of farmers and other decision-makers. This would not be complete without mentioning environmental protection agencies or such bodies as the European Commission, which create laws that ban most toxic compounds and establish limits for their application. The evolution of EU legislation toward pesticide presence limitation started with the Water Framework Directive (Directive 2000/60/EC), which identified a group of priority substances that pose a significant risk to the aquatic environment. Currently, after 25 years, debates are shifting toward the possible impact of pollutant mixtures, as pollutants are not present individually in the environment, and cooccurrence can result in a cumulative effect and higher toxicity than expected (Sprinkle and Payne-Sturges, 2021).

Among the methods that can be used by individual farmers, proper handling of pesticides can be improved by applying certain Ecohydrological Nature-Based Solutions. A solution that mimics natural pesticide degradation processes is the biobed technology, which is usually prepared in the form of a pit filled with local biomass, for instance, a mixture of cereal straw, compost or peat, and soil. This biomass composition, when properly prepared, acts to absorb a large amount of moisture, absorb pesticide molecules, and provide an environment in which microorganisms break down the residues of pollutants.

5. Ecohydrological nature-based solutions

According to ecohydrology principles, in the face of environmental threat (e.g., pesticide presence), it is possible to use ecosystem processes as management tools (Zalewski, 2000). Following on from this idea, we can indicate the main processes that shape the self-purification potential in surface waters.

1. Physical processes such as adsorption, sedimentation, flocculation, precipitation, volatilization, dilution, and mixing.
2. Chemical processes, such as precipitation and dissolution, reducing-oxidizing reactions, coagulation, ion exchange, chemical adsorption, hydrolysis, and photolysis.
3. Biological processes, such as aerobic and anaerobic mineralization and assimilation in plant biomass.

NBS used for pesticide mitigation are, for instance, edge-of-field and riparian buffer zones, vegetated ditches, and constructed wetlands. At the molecular level, the efficiency of these NBS is controlled by the group of physical, chemical, and biological processes described above. In the following part of the chapter, we will briefly present these processes and their application in the NBS.

5.1 Adsorption process

Adsorbents are used in the water purification process. Since adsorption takes place on the surface of the solid materials used, the specific surface area, which is the total surface area per unit mass, becomes one of the most important features. For this reason, the most popular adsorbents are primarily compounds with high porosity, which determines their effectiveness. Sorption processes are the first to play a role in pesticide and environment interaction and regulate bioavailability, leaching, runoff, and reactivity. Adsorption increases the retention time, which is essential for the self-degradation of complex compounds, such as pesticides and other xenobiotics. Recently, biochar has become a potential pesticide adsorbent, which has been widely tested and applied. Biochar is a stabilized, carbon-rich by-product obtained from biomass by the process of pyrolysis. Although considered an efficient sorption material, it may have different properties due to the type of biomass and its origin. Removal of pesticides with activated carbon has been applied in drinking water treatment, where usually the column filters are used with specific filtration speed. The longer the contact time, the better the efficiency that can be achieved. This kind of filter does not have any direct application for dynamic lotic systems. Therefore, there is a growing interest in modified biopolymers and other adsorbent types (e.

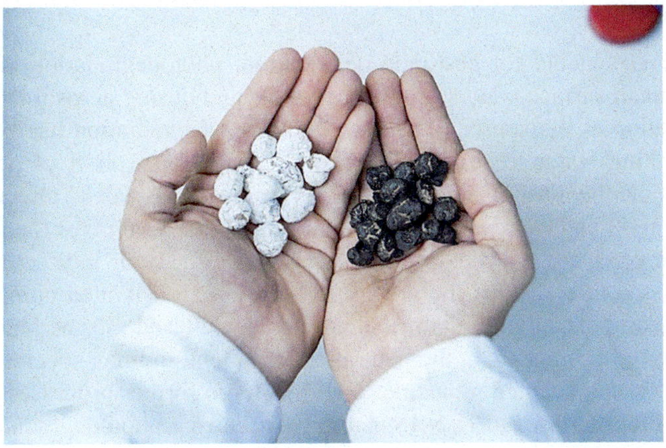

FIGURE 20.2 **BioKer.** BioKer is one of the new adsorbents developed for pesticide removal at ERCE PAS UNESCO. With the addition of biochar (right hand), it gains the ability to remove pesticides from aqueous solutions. *Image credit: P. Ławreszuk.*

g., nanoparticles) which are resistant to environmental conditions. One of these is BioKer (Jarosiewicz et al., 2022)—an adsorbent that can use the properties of biochar or active carbon on the stable ceramic core and is used to create a Permeable Reactive Barrier in flowing water conditions (Fig. 20.2).

5.2 Microbial degradation

Microbial degradation is the major process responsible for utilization of xenobiotics in the environment. The metabolic activity of a local environment that has contact with a pesticide will be different due to factors that stimulate microbial growth (e.g., temperature, water availability, and redox conditions). Nevertheless, we recognize some bacteria that are able to accelerate pesticide degradation. One of the most popular phyla of bacteria involved in pesticide metabolization are *Actinobacteria.* Other bacteria capable of pesticide biodegradation are the members of the phyla *Pseudomonas, Xantobacter,* and *Arthobacter.* All of these can effectively use pesticides as a source of carbon, metabolizing them to carbon dioxide and water. Microorganisms that can switch their metabolism to use pesticides as a source of energy are able to survive during contact with pesticides. Reduced biodiversity, after pesticide contact, entails reduced competition in the surroundings; therefore, during another drift of pesticides, microorganisms with the ability to metabolize them grow exponentially without any pressure from other microorganisms. The major mechanisms responsible for microorganisms' ability to reduce the level of pesticides in the environment are biosorption, bioaccumulation, reduction, efflux, and enzymatic reduction.

One of the particularly interesting mechanisms is the production of surface-active compounds (SACs) by microorganisms. SACs can be classified into two groups: low-molecular weight SACs named biosurfactants and high-molecular weight SACs. Biosurfactants can reduce the surface tensions, e.g., at the water–air interface, and high-molecular weight SACs reduce the interfacial tensions between immiscible liquids or at the solid–liquid interface. SACs could increase the bioavailability of contaminants of organic and inorganic origin (Franzetti et al., 2010). Although playing a positive role in microbial removal and bioremediation processes, SACs promote pollutant desorption due to their complexes (e.g., in soil or sediments) and increase mobility.

In natural ecosystems, bacteria create communities full of sophisticated interactions that enable unique features and abilities to be generated. Therefore, one of the goals of efficient EH-NbS development is to provide a suitable environment for the consortia of microorganisms able to reduce complex pollutants. Interactions among bacterial communities can be described by dedicated signals, horizontal gene transfer, and competition or cooperation. Studies comparing single strains with consortium abilities indicate that mixed cultures formed by two, three, and four strains were able to remove up to 68% of y-HCH, while single culture removal efficiency was 37% at maximum (Fuentes et al., 2014). Despite the vast number of ideas and approaches in bioremediation, only some specific techniques are really studied and developed for pesticide remediation. In most of the methods, retention time is one of the crucial factors controlling the degradation efficiency of the pesticides.

5.3 Photochemical reactions

Among the abiotic processes responsible for pesticide degradation, photodegradation is described as one of the most efficient. Naturally, it can occur in surface waters or surface runoff and during aerosol drift. Photochemical methods are based on the use of light radiation as the source of energy. Natural solar radiation has potential; however, external UV light sources are used in drinking water treatment. Radiation through absorption of photons by chemical compound generates photophysical and photochemical changes. Photochemical methods can be characterized by a simple photolysis reaction (adsorption of radiant light and transformation of the compound). It involves the homolytic cleavage of activated molecules to form degradation products, and this process can occur directly or indirectly. Within the direct pathway, a chromophore in the pesticide structure absorbs the energy to generate an activated compound. The indirect pathway is characterized by excitation of compounds other than pesticide molecules, and thus activation of chain processes that can also facilitate the transformation of pesticides. In the natural environment, indirect species include the humic acids, fluvic acids, or nitrates present in the water. Excited molecules could undergo further processes such as homolysis, heterolysis, or photoionization. From the point of view of EH-NBS, naturally occurring photodegradation processes are still not being fully exploited.

5.4 Integration of pesticides removal processes—Buffer zones and ecotones

A vegetative barrier, including a Riparian Buffer Zone, is a strip of vegetation, usually of different species composition, which creates a border between two landscape components. Multiple studies have indicated riparian vegetation zones as an effective tool for agricultural pollution buffering (Naiman et al., 1989; Izydorczyk et al., 2013). The width of buffer zones available to reduce pesticides adsorbed on sediments in surface runoff is 6 m; however, to gain the same efficiency with soluble or moderately and weakly adsorbed pesticides, the width should be at least 30 m (Osmond, 2023). The width of buffer zones is not the major factor responsible for their efficiency. Many studies have shown that other parameters such as plant composition, soil sorption capacity, and level of pollution that the buffered ecosystem is receiving may be significant. Some authors indicate that the highest efficiency can be achieved with forest-type buffer zones, rather than shrubs or grasses (Aguiar et al., 2015). Yet all authors agree that continued maintenance is extremely important for long-term high-efficiency achievement. This usually means a combination of mowing and removal of sediments. Regarding surface runoff and buffer zones, the problem starts to grow when we face a concentrated flow (preferential pathways) instead of sheet flow. Enhancement of buffer zones and ecotones is possible with level spreaders, water bards, and stiff-grass hedges. Several reviews focused exclusively on buffer strips, e.g, (Aguiar et al., 2015; Dosskey et al., 2010). Based on the ecohydrological approach, Enhanced Riparian Buffer Zones have been proposed (see Chapter 14). Their potential for stimulating microbiological growth has been identified and can be targeted in future pesticide removal case studies.

5.5 Constructed wetlands

Constructed wetlands have entailed a huge interest in pollutant removal, including pesticides. The first attempts at using wetland macrophytes for pesticide removal were conducted in the early 1970s. Recently, this knowledge has been rapidly developing, thanks to the new transdisciplinary approach of ecohydrology. Following the naturally occurring processes of water purification in a wetland environment, Constructed Wetlands are engineered systems that are human-designed. Typically, they refer to ecological engineering (Mitsch and Jørgensen, 2003). We can distinguish several categories of CWs used for pesticide (and other pollutant) removal—free water surface absence or presence, direction of flow, or macrophyte types (Vymazal and Březinová, 2015). A specific type of CW is the Sequential Sedimentation Biofiltration System developed with the ecohydrology approach (see Chapter 24). In addition to the vertical flow system with hydrophilic macrophytes, SSBS enhances the sedimentation and adsorption processes (Fig. 20.3). As previously emphasized, a longer retention time for the pollutant favors the degradation processes, either via self-degradation or biodegradation. Thus, following the abiotic–biotic regulatory concept (Zalewski and Naiman, 1985) the hydrology of CWs can play a superior role in its efficiency.

6. Conclusions

Pesticides can seriously limit the resilience of aquatic ecosystems through their toxic impact on different biotic components, biodiversity loss, and disturbances in trophic web structure. Moreover, we still do not fully recognize the

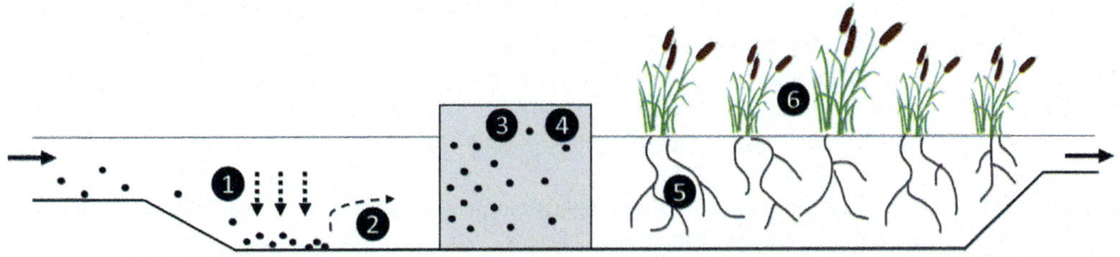

FIGURE 20.3 Sequential sedimentation biofiltration system. Sequential sedimentation biofiltration systems are particular types of constructed wetlands, developed based on understanding ecohydrological processes. Pesticide removal efficiency in SSBS is a sum of several processes. First is the sedimentation process (1) that is present due to the change in water velocity in the sedimentation chamber. Pesticides can be remobilized due to redox change or microbial processes in the bottom sediments (2). When the water flows through a permeable reactive barrier, pollutants are trapped in the barrier on the basis of chemical (3) and physical (4) adsorption. In the last zone of SSBS, rhizospheric bacteria and fungi metabolize the pesticides (5). Hydrophilic plants accumulate the pollutants in their biomass (6). The last zone of SSBS can be described by the phytoremediation process. Moreover, SSBS stimulates photolysis and creates both oxidation and reduction conditions in a relatively small area; therefore, the removal of pollutants is more complex.

potential impact of mixtures of pollutants, as they are usually applied and present in the environment as multiple-component mixtures. As the exposure to different xenobiotics in the environment can be considered as chronic, our current toxicological models are not enough to assess the real impact. Although legislation and best management practice provide tools to limit the pesticides drift into the environment, with the current model of agriculture, we are far from meeting the goal of zero pesticides in the aquatic environment. Therefore, the development and application of new Nature-Based Solutions according to the concept of ecohydrology is necessary to achieve catchment sustainability in terms of the multidimensional WBSR goal.

References

Aguiar, T.R., Bortolozo, F.R., Hansel, F.A., Rasera, K., Ferreira, M.T., 2015. Riparian buffer zones as pesticide filters of no-till crops. Environ. Sci. Pollut. Control Ser. 22 (14), 10618–10626. https://doi.org/10.1007/s11356-015-4281-5.

Busch, W., Schmidt, S., Kühne, R., Schulze, T., Krauss, M., Altenburger, R., 2016. Micropollutants in European rivers: a mode of action survey to support the development of effect-based tools for water monitoring. Environ. Toxicol. Chem. 35 (8), 1887–1899. https://doi.org/10.1002/etc.3460.

Carson, R., 1962. Silent Spring. Houghton Mifflin.

Dosskey, M.G., Vidon, P., Gurwick, N.P., Allan, C.J., Duval, T.P., Lowrance, R., 2010. The role of riparian vegetation in protecting and improving chemical water quality in streams. J. Am. Water Resour. Assoc. 46 (2), 261–277. https://doi.org/10.1111/j.1752-1688.2010.00419.x.

FAO, 2024. Pesticides use and trade – 1990–2022. FAOSTAT Analytical Briefs, No. 89. Rome. Food and Agriculture Organization of the United Nations (FAO), Rome. https://doi.org/10.4060/cd1486en.

Franzetti, A., Tamburini, E., Banat, I.M., 2010. Applications of biological surface active compounds in remediation technologies. Adv. Exp. Med. Biol. 672, 121–134. https://doi.org/10.1007/978-1-4419-5979-9_9.

Fuentes, M.S., Alvarez, A., Saez, J.M., Benimeli, C.S., Amoroso, M.J., 2014. Methoxychlor bioremediation by defined consortium of environmental Streptomyces strains. Int. J. Environ. Sci. Technol. 11 (4), 1147–1156. https://doi.org/10.1007/s13762-013-0314-0.

Izydorczyk, K., Frątczak, W., Drobniewska, A., Cichowicz, E., Michalska-Hejduk, D., Gross, R., Zalewski, M., 2013. A biogeochemical barrier to enhance a buffer zone for reducing diffuse phosphorus pollution—Preliminary results. Ecohydrol. Hydrobiol. 13 (2), 104–112.

Jarosiewicz, P., 2021. Application of the Ecohydrology Framework to Reduce Agricultural Pollution in River Systems (Doctoral Dissertation).

Jarosiewicz, P., Fazi, S., Zalewski, M., 2022. How to boost Ecohydrological Nature-Based Solutions in water quality management. Ecohydrol. Hydrobiol. 22 (2), 226–233. https://doi.org/10.1016/j.ecohyd.2021.11.005.

Macdonald, R.W., Mackay, D., Li, Y.F., Hickie, B., 2003. How will global climate change affect risks from long-range transport of persistent organic pollutants? Hum. Ecol. Risk Assess. 9 (3), 643–660. https://doi.org/10.1080/713609959.

Mitsch, W.J., Jørgensen, S.E., 2003. Ecological Engineering and Ecosystem Restoration.

Naiman, R.J., Décamps, H., Fournier, F. (Eds.), 1989. Role of land/inland water ecotones in landscape management and restoration: Proposals for collaborative research (MAB Digest No.4). UNESCO.

Osmond, D.L., 2023. Riparian Buffers and Controlled Drainage to Reduce Agricultural Nonpoint Source Pollution.

Persson, L., Carney Almroth, B.M., Collins, C.D., Cornell, S., de Wit, C.A., Diamond, M.L., Fantke, P., Hassellöv, M., MacLeod, M., Ryberg, M.W., Søgaard Jørgensen, P., Villarrubia-Gómez, P., Wang, Z., Hauschild, M.Z., 2022. Outside the safe operating space of the planetary boundary for novel entities. Environ. Sci. Technol. 56 (3), 1510–1521. https://doi.org/10.1021/acs.est.1c04158.

Sauvé, S., Desrosiers, M., 2014. A review of what is an emerging contaminant. Chem. Cent. J. 8 (1). https://doi.org/10.1186/1752-153x-8-15.

Sprinkle, R.H., Payne-Sturges, D.C., 2021. Mixture toxicity, cumulative risk, and environmental justice in United States federal policy, 1980–2016. Environ. Health 20 (1). https://doi.org/10.1186/s12940-021-00764-5.

Vymazal, J., Březinová, T., 2015. The use of constructed wetlands for removal of pesticides from agricultural runoff and drainage: a review. Environ. Int. 75, 11–20. https://doi.org/10.1016/j.envint.2014.10.026.

Wang, X.P., Sun, D.C., Yao, T.D., 2016. Climate change and global cycling of persistent organic pollutants: a critical review. Sci. China Earth Sci. 59 (10), 1899–1911. https://doi.org/10.1007/s11430-016-5073-0.

WHO-FAO, 2019. Global situation of pesticide management in agriculture and public health: Report of a 2018 WHO-FAO survey. World Health Organization and Food and Agriculture Organization of the United Nations.

Zalewski, M., 2000. Ecohydrology—the scientific background to use ecosystem properties as management tools toward sustainability of water resources. Ecol. Eng. 16 (1), 1–8. https://doi.org/10.1016/s0925-8574(00)00071-9.

Zalewski, M., Naiman, R.J., 1985. The regulation of riverine fish communities by a continuum of abiotic–biotic factors. In: Alabaster, J.S. (Ed.), Habitat modification and freshwater fisheries: Proceedings of a symposium on habitat modification and freshwater fisheries. Butterworths for FAO, Aarhus, Denmark, pp. 3–9, 23 May 1984.

Zalewski, M., Wagner, I., Frątczak, W., Mankiewicz-Boczek, J., Parniewski, P., 2012. Blue-green city for compensating global climate change. The Parliament Magazine 350, 2–3.

Chapter 21

Biological remediation methods as effective tools for micropollutants removal and protection of land–water ecosystems

Elżbieta Mierzejewska-Sinner and Magdalena Urbaniak
University of Lodz, Faculty of Biology and Environmental Protection, UNESCO Chair on Ecohydrology and Applied Ecology, Łódź, Poland

IMAGE CREDIT: M. ZALEWSKI.

Transdisciplinary Ecohydrology. https://doi.org/10.1016/B978-0-443-41596-8.00003-0

1. Introduction

xBiological remediation is one of the core components of Ecohydrological Nature-Based Solutions (NBS). It uses the natural capacities of ecosystems: microorganisms, plants and their interactions-to neutralize, transform or retain pollutants. In transdisciplinary ecohydrology, which combines hydrological, ecological and bioengineering processes at the catchment scale, bioremediation together with phytoremediation and rhizoremediation are key methods for reducing the pressure of micropollutants on aquatic ecosystems and create the foundation for Advanced Nature-based Solutions (ANbS). This chapter introduces biological remediation as a set of practical, ecosystem-based strategies that support self-purification processes, strengthen ecosystem functions and contribute to the enhancement of the environment and its resilience in landscapes increasingly affected by human activity.

As a consequence of the rapid development of modern society during the 20th and 21st century, a significant amount of micropollutants have been dispersed into the environment. Consequently, nowadays one of the major environmental problems is an intensified emission and spreading of micropollutants within the catchment area (e.g., agricultural lands, city spaces, industrial areas) and further transport and accumulation of these pollutants and their metabolites in water ecosystems (Fig. 21.1). Majority of micropollutants have been used in large scales as pesticides, herbicides, insecticides and industrial chemicals or produced as undesirable industrial byproducts. Worldwide data have identified the increased levels of micropollutants not only in agricultural lands, urban/industrial areas but also in wetlands, marine water, stormwater, rivers, groundwater and sources of drinking water (Urbaniak et al., 2012, 2014).

Ecotoxicological studies showed that micropollutants can exert a negative effect on land and aquatic organisms causing systemic, developmental and behavioral changes (Urbaniak, 2007). Micropollutants belonging to the group of Persistent Organic Pollutants (POPs), including inter alia dioxins, polychlorinated biphenyls (PCBs) as well as range of pesticides, have been recognized as substances that pose a serious risk to human health and environment. They are characterized by high persistence in the environment, high toxicity and ability to accumulate in soils and sediments, bioaccumulation in tissues and biomagnification in the trophic chains (Urbaniak, 2013).

Although most POPs are found in the environment at low concentrations, chronic exposure to the substances together with their potential for bioaccumulation and negative effects presents particular risks to the environment and to humans, as the final link of the trophic chain. The consequences of exposure to their toxic effects are changes which often manifest themselves in subsequent generations of exposed organisms. As these include adverse effects on fertility, feminization of males and masculinization of females, they consequently affect the process of reproduction and hence, population size.

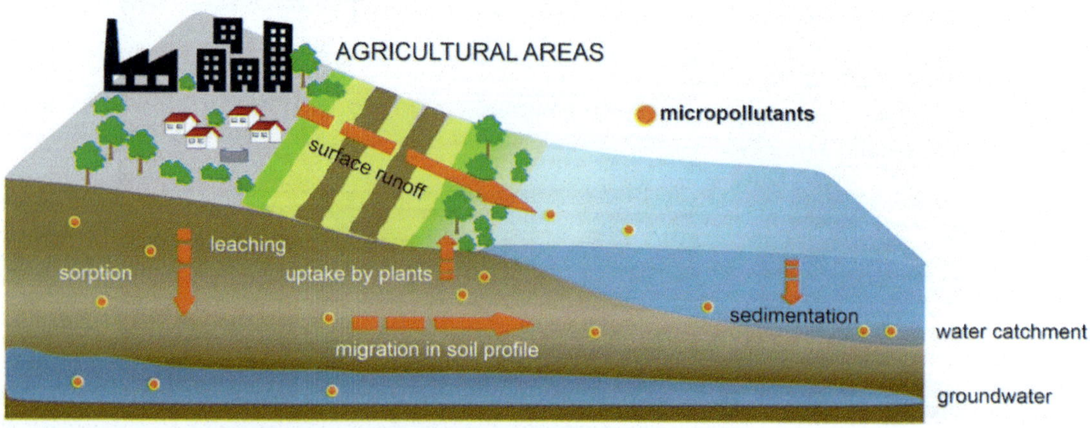

FIGURE 21.1 Micropollutants' pathways in the land–water ecosystems.

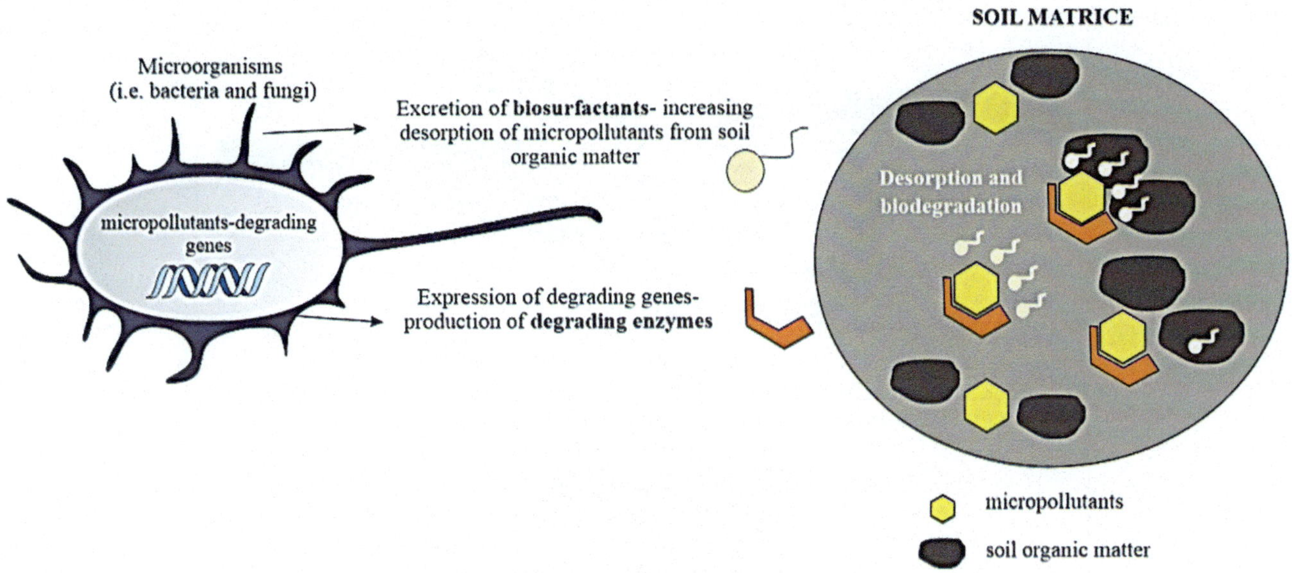

FIGURE 21.2 Bioremediation. Bioremediation of micropollutant-contaminated soil.

Furthermore, these compounds have carcinogenic, hepatotoxic, immunotoxic and neurotoxic effects (Loganathan and Matsunaga, 2015; Makles et al., 2001).

As POPs present a threat to the functioning of natural as well as anthropogenic ecosystems, seeking for tailored and effective strategies for removal of POPs has gained a special attention of scientific bodies as well as practitioners worldwide. A growing body of literature has investigated NBS such as biological remediation methods (bioremediation, phytoremediation and rhizoremediation) as effective tools for micropollutants removal and consequently prevention of their transport and accumulation in water bodies.

2. Bioremediation

Biodegradation is one of the many available methods of removing micropollutants from the environment. This process is based on using the catabolic capacity of microorganisms (bacteria and fungi) to remove micropollutants from a given environmental matrix (soil, sediment etc.) (Crawford and Crawford, 2011) (Fig. 21.2).

The process of complete conversion of complex micropollutants into simple inorganic compounds is called mineralization and a single transformation of one substance into another is called biotransformation (Crawford and Crawford, 2011). The biodegradability of some micropollutants may be altered over time with the progressive changes in the activity and structure of the microbial population and their adaptation to certain environmental conditions (Crawford and Crawford, 2011).

Microorganisms can degrade a wide range of organic pollutants, but this process might be very slow in some cases. For instance, the effectiveness of bioremediation can be influenced by physical and properties of pollutants, microbiological characteristics of soil (biomass concentration, enzymatic activity), physical and chemical parameters of the environment (water availability, pH, and temperature), sorption processes and the presence of other carbon sources (Klimiuk and Łebkowska, 2005). What is more, the rate of micropollutants removal depends on their bioavailability to microorganisms. Due to the presence of autochthonous microorganisms in contaminated matrices, that are capable of metabolizing organic pollutants, biodegradation may be accelerated by biostimulation. Biostimulation bases on introducing nutrients or other necessary components to the environment, which are necessary for the development of soil microorganisms (Klimiuk and Łebkowska, 2005). This may include by adding biogenic elements such as nitrogen or phosphorus, cometabolites and biosurfactants, which increase the bioavailability of pollutants for microorganisms.

3. Phytoremediation

An alternative method of cleaning contaminated environments from POPs is phytoremediation. It is a cost-effective and environment-friendly strategy that uses plants to transform or mineralize micropollutants to the less toxic or

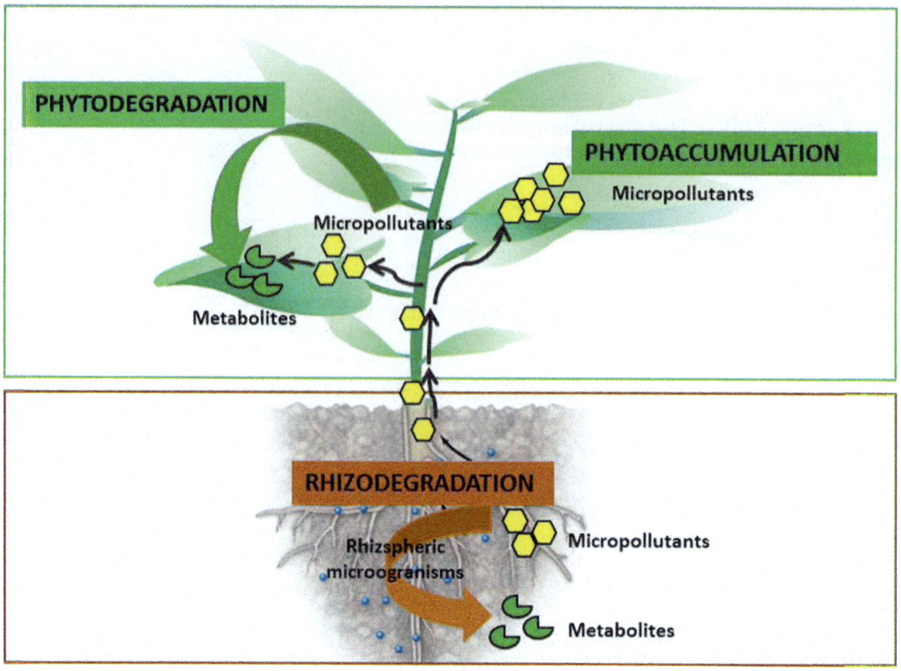

FIGURE 21.3 **Phytoremediation.** Phytoremediation of micropollutant-contaminated soil.

environmentally neutral compounds (Gerhardt et al., 2009) (Fig. 21.3). Plants produce an array of catabolic enzymes, which operate to protect the host organisms and detoxify xenobiotic compounds (Singer et al., 2003).

However, contaminants like POPs, due to their hydrophobicity and hence their strong adsorption by soil particles, are transferred to soil solution to a very small extent. Consequently, the ability of plants to uptake POPs from contaminated environment is very limited (Briggs et al., 1982; Terzaghi et al., 2018). Though, one notable exception is the Cucurbitaceae family, including zucchini (*Cucurbita pepo*) and cucumber (*Cucumis sativus*), which readily takes up highly hydrophobic compounds such as POPs from soil and translocate them to leaves and fruits (Huelster et al., 1994; Matsuo et al., 2011; White, 2002; Whitfield Åslund et al., 2010). The ability of cucurbits to remove POPs from soil matrices has also been confirmed by Urbaniak et al. (2016, 2020c), Wyrwicka et al. (2014), Wyrwicka and Urbaniak (2016).

Plants play also a crucial role in the development of soil structure and stabilization of fundamental soil ecosystem functions such as water flow (Machado, 2018). Therefore, phytoremediation not only contributes to the detoxification of the environment from micropollutants, but also has a positive influence on its functioning. Results showed that phytoremediation with *C. pepo* L. does not only effectively mitigate the ecological risk related to the presence micropollutants in the soil but also restores proper functioning of the soil ecosystem (Urbaniak et al., 2020b).

4. Rhizoremediation

Rhizoremediation is another promising approach used for micropollutants removal. This process enhances the removal of such recalcitrant compounds like POPs from the environment by exploiting the interactions between selected plants (able to grow under the presence of given pollutants), root exudates (including plant secondary metabolites, PSMs), and microorganisms (Fig. 21.3). Rhizoremediation is a naturally occurring process within the plant root zone (rhizosphere), where the growth of microorganisms and their degradative activity are stimulated by root exudates enriched by PSMs. Such plant-derived compounds can serve as substrates in cometabolism and provide energy for microbial growth, act as inducers of degradative enzymes due to their structural similarities to some micropollutants, and enhance the degree of contamination removal by increasing pollutant bioavailability in soil (Fig. 21.4).

PSMs can be also used as stimulators of POPs degradation (Haichar et al., 2008; Musilova et al., 2016; Singer et al., 2003) via promoting the growth of desirable bacteria strains: for example in mulberry (*Morus* sp.) the PSMs (morusin, morusinol and kuwanon C) have been found to support the growth of *Burkholderia* sp. LB400-a POP-degrader (Leigh et al., 2002). The PSM (L-carvone) excreted by spearmint (*Mentha spicata*) stimulates the cometabolism of selected POPs by *Arthrobacter* sp. strain B1B (Gilbert and Crowley, 1997). Similar results were obtained using PSMs carvone and

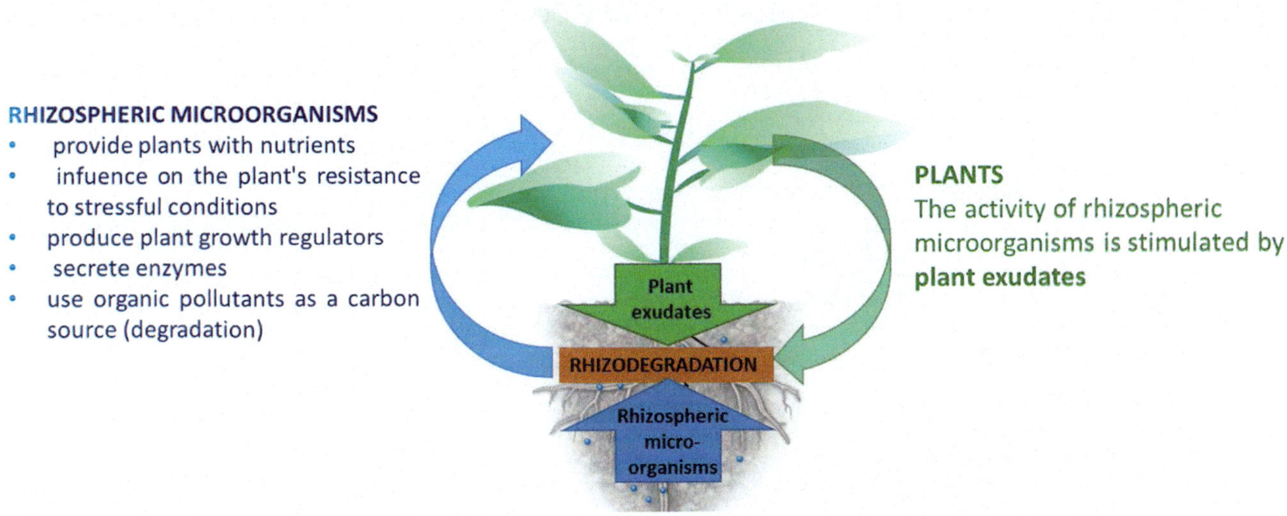

FIGURE 21.4 Phytoremediation. Integrations between plant and rhizospheric microorganisms—the processes behind rhizodegradation of micropllutants.

salicylic acid (Singer et al., 2000). In addition to acting as growth substrates for specific degradative bacteria, it has been hypothesized that PSMs may also induce detoxification mechanisms in bacteria at molecular level (Master and Mohn, 2001; Siciliano et al., 2003; Uhlik et al., 2013). Siciliano et al. (2003) report greater induction of bacterial genes involved in POP (naphthalene) degradation in the rhizosphere soil of tall fescue (*Festuca arundinaceae*) than in unplanted soil. Master and Mohn (2001) report a PSM-salicylate to have an up-regulating effect on the expression of degradative genes in the POP-degrader *Pseudomonas* sp. Cam-1. Hu et al. (2014) demonstrated the expression of degradative genes in *Ralstonia eutropha* H850 and *Pseudomonas fluorescens* P2W amended with salicylic acid.

The worldwide studies showed that the application of plants and their PSMs is a promising way to alleviate the toxic influence of POPs and in this way may contribute to protection of land and water ecosystems. Nevertheless, for effective rhizodegradation of contaminants to take place, a wide range of plants and bacterial traits is needed, involving the orchestrated interaction of a multitude of genes and enzymes. Rhizoremediation can therefore be optimized by selecting suitable plant–microbe sets, which can be achieved by combining plant and microorganisms capable of micropollutant degradation.

5. A need for land–water ecosystems quality improvement—An implementation of nature-based remediation solutions for micropollutants removal at catchment–valley–river scale

Water is one of the main mediums responsible for the export of micropollutants from catchments. Consequently, water ecosystems, due to their location at the lowest points in the landscape allow micropollutants from the catchment to be transported and accumulated therein through such processes as atmospheric deposition, runoff from urban and agricultural areas or inflow through point sources. Recently observed climate change, and related greater occurrences of prolonged summer droughts interrupted by extreme floods, additionally accelerates the transport of micropollutants into water ecosystems.

In the case of POPs the main factor determining their high accumulation in aquatic systems is their extremely low solubility in water. This characteristic allows POPs to be associated with the organic and mineral particles present in water, and hence to be transported along the river continuum, where they undergo gradual deposition and accumulation in river and especially reservoir sediments, that serve as long-term sources of micropollutants. In many cases it is necessary to remove accumulated sediments, which makes utilization of such contaminated dredged sediments a matter of concern. For example, in Europe and the US, around 200 million m^3 of sediments is dredged annually; of this amount, 2.3 to 9 million m^3 are contaminated to the extent that they require special, and often costly, handling and treatment (SedNet, 2004). At the same time, due to the beneficial properties of sediments that are rich in clay, organic matter and mineral nutrients they have been proposed to be utilized in agriculture (Baran et al., 2020; Urbaniak et al., 2020a). Consequently,

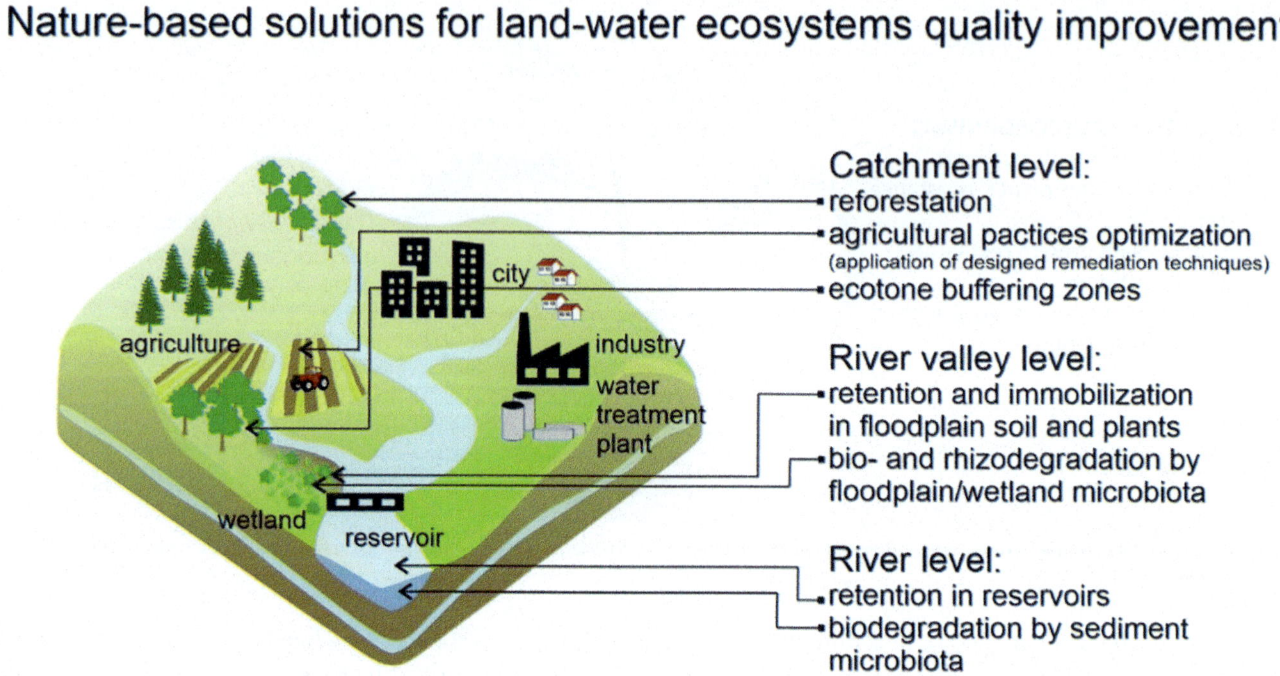

FIGURE 21.5 Remediation. Examples of nature-based remediation solutions for removal or immobilization of micropollutants at catchment–valley–river sale.

for their use in agriculture, remediation methods must be taken to limit the risk posed by substances toxic to the soil environment and ultimately, the terrestrial food chain (Siebielec et al., 2019; Terzaghi et al., 2018; Briggs et al., 1982).

Additionally, point sources of pollution related to human activity in the catchment influence negatively on the quality of water ecosystems. These include i.e., wastewater treatment plants that constantly discharge effluents into river or reservoir recipients. Wastewater treatment plants produce also sewage sludge as the end product of the treatment process. In current practice, this sewage sludge is typically stored on the area of the wastewater treatment plant or used in agriculture as organic fertilizer. Thus agriculture utilization of sewage sludge additionally contributes to an increase in micropollutants concentration in the catchment soil and their further export to aquatic ecosystems.

Concluding the above, in order to protect the water resources against POPs, it is essential to reduce the input of micropollutants from the catchment, especially from diffuse sources of pollution, which are responsible for the highest, mostly uncontrolled and unregulated, load of micropollutants.

Biological remediation methods offer an efficient tool to mitigate the risk related to the excess amount of POPs in the environmental matrices. However, for effective remediation (and protection) of contaminated land–water ecosystems, a range of actions at the level of catchment, river valley and river itself have to be implemented and harmonized between each other (Fig. 21.5).

At the catchment level the removal of micropollutants may be enhanced through their retention in the soil. This can be achieved by optimization of agricultural practices, reforestation, and creation of ecotone buffering zones. The "filtering and buffering" role of soils has been considered as efficient tool for POPs removal on a catchment scale. Moreover, mosaic landscape structure promotes bio- and rhizodegradation of POPs (Urbaniak, 2013) (Fig. 21.5).

Designed remediation methods have been applied to clean up agriculture soil fertilized with POP-contaminated sewage sludge. One of such method is to use willow (*Salix* sp.). Urbaniak et al. (2017) demonstrated that cultivation of willow on soil treated with sludge resulted in a gradual increase of humus fractions, total organic carbon content and bacterial abundance. Although the soil toxicity and POP content in sludge amended soil initially increased, these negative aspects were diminished after 10 weeks sludge application in areas planted by willow. Sludge application was also found to have a positive influence on willow, probably due to its high nutrient and carbon content. The results showed that willow plantation can be use an efficient tool toward to reducing the concentration of POPs in soil being a result of agriculture utilization of sewage sludge. It can also promote production of renewable, green energy via accelerated growth of willow plants.

Other plants frequently used for removal of POPs from soil are cucurbits. As it was mentioned in the previous section, cucurbits demonstrated unique properties, which enable them to take up POPs from soil. The ability of cucurbits to remove POPs from soil fertilized with sewage sludge has been confirmed by Urbaniak et al. (2016), who showed that cultivation of zucchini (*C. pepo* cv. Atena Polka) in POP-contaminated soil, reduced the POPs (dioxins) concentration in soil by 37%, thereby decreasing toxic equivalency by 68%. A reduction of other POPs—PCBs reached 39% after 5 weeks cultivation of cucumber (*C. sativus* cv. Cezar) (Wyrwicka et al., 2014). Study showed also that zucchini was more effective in POPs (dioxin) removal (a reduction rate of 37%) than cucumber (24%) (Urbaniak et al., 2016), but at the same time other study demonstrated that POPs toxicity on plant growth and metabolism were greater for zucchini than for cucumber (Wyrwicka and Urbaniak, 2016).

A similar approach has been adopted to remediate soil contaminated with POPs as an effect of river and reservoir sediments application. The members of the Cucurbitaceae family, including zucchini, squash and cucumber, have been used for treatment of such soil. The removal of POPs was found to reach 27% when cucumber (*C. sativus* L.) was applied (Wyrwicka et al., 2014), and 32% when zucchini (*C. pepo* L.) was used.

The above results confirmed the potential of cucurbits in remediation of soils contaminated with POPs as an effect of river/reservoir sediments agricultural utilization.

At the river valley level, a floodplain can be used as a tool to reduce the excess load of such micropollutants. Floodplain it can absorb and retain sediment and POPs such as dioxins and PCBs during stable and high-flow conditions (Kiedrzyńska et al., 2015; Skłodowski et al., 2014). Because the largest pollutant loads are usually transported by rivers during the rising water stages of floods, their redirection to floodplain areas during the initial stages of a flood would diminish the load transported along the river. The declining levels of POPs in river water occur not only via their retention and immobilization in floodplain sediments but also as an effect of bioremediation processes. Some naturally occurring plants in the floodplain zones, such as willow communities, promote the loss of POPs from the soil and sediments via rhizoremediation (Urbaniak, 2013) (Fig. 21.5). These processes were investigated in a study of Skłodowski et al. (2014). The authors demonstrated that a 40 km section of the Pilica river covered with a riparian vegetation community was associated with a 49% reduction of TEQ concentrations of dl-PCBs in sediments. These results indicate that the presence of natural floodplain terraces in river valleys significantly increases the rate of self-purification of the river and can reduce the transport of POPs along the river continuum. Similar findings can be found in the works of, among others, Barendregt (2003), Bayley (1995), Burt et al. (2022), Loe and Lamers (2003). Hence, floodplains can be considered tools for reducing the concentrations of POPs transported along the river continuum (Kiedrzyńska et al., 2008a,b, 2014, 2015; Skłodowski et al., 2014; Urbaniak et al., 2012).

At river level constructed dam reservoirs can play an increasingly important role in the improvement of river water quality. Dam reservoirs are usually constructed in the middle or downstream section of the river, thus they intersect with its continuum, and act as traps for suspended matter (Kentzer et al., 2010) and for their associated chemical toxicants such as POPs (Chi et al., 2011; Devault et al., 2009), resulting in improved downstream water quality (Kentzer et al., 2010). The water quality improvement is the effect of the limnological characteristics of the reservoirs, which affect water flow and reduce its lifting force, and as a consequence, enhance the sedimentation of coarse particulate matter and fine particles on which POPs are adsorbed. The coarser sediments tend to be found in the upper part of a reservoir, which has a riverine character, with the fine particles in the lower part of the reservoir, which is more of a limnological character. This may lead to periodical retention of up to 90% of the total amount of flowing matter, and in the case of POPs, this value can be as high as 97% (DiPinto et al., 1993). Urbaniak et al. (2014) demonstrated 29%–56% decreases in the concentrations of selected POPs in sediments along the Sulejów Reservoir (Central Poland). In addition, the total dioxins concentrations were reduced below the dam by 21% during flooding and by 24% during stable flow, while the comparative toxic equivalency concentrations were reduced by 69% at flooding and 36% during serene flow (Urbaniak et al., 2014), thus resulting in improved water quality. Study of Urbaniak et al. from 2019 showed that the Sulejów Reservoir was found to play a positive role in reducing PCDD, PCDF and dl-PCB transport within the study period, with reductions ranging from 17% to 83% for total concentrations, and 33%–79% for TEQ. These findings clearly demonstrate the potential role played by a lowland reservoir in the reduction of dioxins concentrations transported along the river continuum (Fig. 21.5).

References

Baran, A., Mierzewa-Hersztek, M., Urbaniak, M., Gondek, K., Tarnawski, M., Szara, M., Zieliński, M., 2020. An assessment of the concentrations of PCDDs/Fs in contaminated bottom sediments and their sources and ecological risk. J. Soils Sediments 20, 2588–2597.

Barendregt, A., 2003. Ecological impact of flooding in a fresh water tidal area. International conference "Towards natural flood reduction strategies". September 2003, Warsaw, Poland.

Bayley, P.B., 1995. Understanding large river: floodplain ecosystems. Bioscience 45 (3), 153–158.

Briggs, G.G., Bromilow, R.H., Evans, A.A., 1982. Relationships between lipophilicity and root uptake and translocation of non-ionised chemicals by barley. Pestic. Sci. 13, 495–504. https://doi.org/10.1002/ps.2780130506.

Burt, T.P., Bates, P.D., Stewarts, M.D., Claxton, A.J., Anderson, M.G., Price, D.A., 2022. Water table fluctuations within the floodplain of the river Severn. J. Hydrol. 262, 1–20.

Chi, K.H., Hsu, S.-C., Lin, C.-Y., Kao, S.J., Lee, T.T., 2011. Deposition fluxes of PCDD/Fs in a reservoir system in northern Taiwan. Chemosphere 83, 745–752.

Crawford, R.L., Crawford, D.L., 2011. Bioremediation: Principles and Applications. Cambridge University Press, Cambridge.

Devault, D.A., Gerino, M., Laplanche, C., Julien, F., Winterton, P., Merlin, G., 2009. Herbicide accumulation and evolution in reservoir sediments. Sci. Total Environ. 407, 2659–2665.

DiPinto, L.M., Coull, B.C., Chandler, G.T., 1993. Lethal and sublethal effect of sediment-associated PCB Aroclor 1254 on a meiobenthic copepod. Environ. Toxicol. Chem. 12, 1909–1918.

Gerhardt, K.E., Huang, X.-D., Glick, B.R., Greenberg, B.M., 2009. Phytoremediation and rhizoremediation of organic soil contaminants: Potential and challenges. Plant Sci. 176, 20–30. https://doi.org/10.1016/j.plantsci.2008.09.014.

Gilbert, E.S., Crowley, D.E., 1997. Plant compounds that induce polychlorinated biphenyl biodegradation by *Arthrobacter* sp. strain B1B. Appl. Environ. Microbiol. 63 (5), 1933–1938.

Haichar, F.Z., Marol, C., Berge, O., Rangel-Castro, J.I., Prosser, J.I., Balesdent, J., Heulin, T., Achouak, W., 2008. Plant host habitat and root exudates shape soil bacterial community structure. ISME J. 2 (12), 1221–1230. https://doi.org/10.1038/ismej.2008.80.

Hu, C., Luo, Y., Wang, X., Zhang, H., H., Li, X., 2014. PCB biodegradation and *bph*A1 gene expression induced by salicylic acid and biphenyl with *Pseudomonas fluorescence* P2W and *Ralstonia eutropha* H850 Chaohua. Pol. J. Environ. Stud. 23 (5), 1591–1598.

Huelster, A., Müller, J.F., Marschner, H., 1994. Soil-plant transfer of polychlorinated dibenzo-p-dioxins and dibenzofurans to vegetables of the cucumber family (Cucurbitaceae). Environ. Sci. Technol. 28, 1110–1115.

Kentzer, A., Dembowska, E., Giziński, A., Napiórkowski, P., 2010. Influence of the Włocławek Reservoir on hydrochemistry and plankton of a large, lowland river (the Lower Vistula River, Poland). Ecol. Eng. 36, 1747–1753.

Kiedrzyńska, E., Kiedrzyński, M., Urbaniak, M., Magnuszewski, A., Skłodowski, M., Wyrwicka, A., Zalewski, M., 2014. Point sources of nutrient pollution in the lowland river catchment in the context of the Baltic Sea eutrophication. Ecol. Eng. 70, 337–348. https://doi.org/10.1016/j.ecoleng.2014.06.010.

Kiedrzyńska, E., Kiedrzyński, M., Zalewski, M., 2008a. Flood sediment deposition and phosphorus retention in a lowland river floodplain: impact on water quality of a reservoir, Sulejów, Poland. Ecohydrol. Hydrobiol. 8 (2–4), 281–289. https://doi.org/10.2478/v10104-009-0022-z.

Kiedrzyńska, E., Kiedrzyński, M., Zalewski, M., 2015. Sustainable floodplain management for flood prevention and water quality improvement. Nat. Hazards 76 (2), 955–977. https://doi.org/10.1007/s11069-014-1529-1.

Kiedrzyńska, E., Wagner, I., Zalewski, M., 2008b. Quantification of phosphorus retention efficiency by floodplain vegetation and a management strategy for a eutrophic reservoir restoration. Ecol. Eng. 33 (1), 15–25. https://doi.org/10.1016/j.ecoleng.2007.10.010.

Klimiuk, E., Łebkowska, M., 2005. Biotechnologia w ochronie środowiska (Biotechnology in environmental protection). PWN.

Leigh, M.B, Fletcher, J.S., Fu, X., Schmitz, F.J., 2002. Root turnover: an important source of microbial substrates in rhizosphere remediation of recalcitrant contaminants. Environm. Sci. Technol. 36 (7), 1579–1583. https://doi.org/10.1021/es015702i.

Loe, R., Lamers, L., 2003. The effects of river water quality on the development of wet floodplain vegetations in the Netherlands. International conference "Towards natural flood reduction strategies". September 2003, Warsaw, Poland.

Loganathan, B.G., Matsunaga, S., 2015. PCBs, dioxins and furans: human exposure and health effects. Handbook of Toxicology of Chemical Warfare Agents. Academic Press.

Machado, et al., 2018. The importance of plants to development and maintenance of soil structure, microbial communities and ecosystem functions. Soil Tillage Res. 175, 139–149. https://doi.org/10.1016/j.still.2017.09.002.

Makles, Świętkowski, Grybowska, 2001. Niebezpieczne dioksyny (Hazardous dioxins). Arkady Sp. z o.o.

Mamindy-Pajany, Y., Geret, F., Romero, M., Hurel, C., Marmier, N., 2012. Ex situ remediation of contaminated sediments using mineral additives: assessment of pollutant bioavailability with theMicrotox solid phase test. Chemosphere 86:1112. Chemosphere 86 (1112).

Master, E.R., Mohn, W.W., 2001. Induction of bphA, encoding biphenyl dioxygenase, in two polychlorinated biphenyl-degrading bacteria, psychrotolerant Pseudomonas strain Cam-1 and mesophilic Burkholderia strain LB400. Appl. Environ. Microbiol. https://doi.org/10.1128/AEM.67.6.2669-2676.2001.

Matsuo, T., Kawaguchi, M., Ohe, N., Takahashi, M., et al., 2011. Structure-selective accumulation of polychlorinated biphenyls in *Cucurbita pepo*. J. Pestic. Sci. 36 (3), 363–369. https://doi.org/10.1584/jpestics.G11-03.

Musilova, Z., Ridl, J., Polivkova, M., Macek, T., Uhlik, O., Mackova, M., 2016. Effects of secondary plant metabolites on microbial populations: Changes in community structure and metabolic activity in contaminated environments. Int. J. Mol. Sci. 17 (8), 1205. https://doi.org/10.3390/ijms17081205.

SedNet, 2004. Contaminated Sediments in European River Basins. https://sednet.org/wpcontent/uploads/2016/03/Sednet_booklet_final_2.pdf.

Siciliano, S.D., Germida, J.J., Banks, K., Greer, C.W., 2003. Changes in microbial community composition and function during a polyaromatic hydrocarbon phytoremediation field trial. Appl. Environ. Microbiol. 69 (1), 483–489. https://doi.org/10.1128/AEM.69.1.483-489.2003.

Siebielec, S., Siebielec, S., Urbaniak, M., Smreczak, B., Grząda, E., Wyrwicka, A., Kidd, P.S., 2019. Impact of rhizobacterial inoculants on plant growth and enzyme activities in soil treated with contaminated bottom sediments. Int. J. Phytoremediation. https://doi.org/10.1080/15226514.2018.1524833.

Singer, A.C., Gilbert, E.S., Luepromchai, E., Crowley, D.E., 2000. Bioremediation of polychlorinated biphenyl-contaminated soil using carvone and surfactant-grown bacteria. Appl. Microbiol. Biotechnol. 54 (6), 838–843. https://doi.org/10.1007/s002530000472.

Singer, A.C., Smith, D., Jury, W.A., 2003. Secondary plant metabolites in phytoremediation and biotransformation. Trends Biotechnol. 21 (3), 123–130. https://doi.org/10.1016/S0167-7799(02)00041-0.

Skłodowski, M., Kiedrzyńska, E., Kiedrzyński, M., Urbaniak, M., M Zielińska, K., K Kurowski, J., Zalewski, M., 2014. The role of riparian willow communities in phosphorus accumulation and dioxin control for water quality improvement in a lowland river. Ecol. Eng. 70.

Terzaghi, E., Schievano, A., Zanardini, E., Morosini, C., Raspa, G, Borin, S., Fava, F., 2018. Modelling peak exposure of pesticides in terrestrial and aquatic ecosystems: importance of dissolved organic carbon and vertical particle movement in soil. SAR QSAR Environ. Res. 31 (1), 19–32. https://doi.org/10.1080/1062936X.2019.1686715.

Uhlik, O., Jecna, K., Leigh, M.B., Mackova, M., Macek, T., et al., 2013. Plant secondary metabolite-induced shifts in bacterial community structure and degradative ability in contaminated soil. Appl. Microbiol. Biotechnol. 97 (20), 9245–9256. https://doi.org/10.1007/s00253-012-4627-6.

Urbaniak, M., 2007. Polychlorinated biphenyls: Sources, distribution and transformation in the environment - A literature review. Acta Toxicol.

Urbaniak, M., 2013. Spatial distribution of PCDDs, PCDFs and dl-PCBs along the cascade of urban reservoirs. Hydrol. Res. https://doi.org/10.2166/nh.2012.236.

Urbaniak, M., 2013. Biodegradation of PCDDs/PCDFs and dl- PCBs. [In:]: Chamy R., Rosenkran F. (Eds). Biodegradation – Engineering and Technology. INTECH Publisher, ISBN 978-953-51-1153-5, str. 73-100 73–100.

Urbaniak, M., Baran, A., Szara, M., Mierzejewska, E., Lee, S., Takazawa, M., Kannan, K., 2020. Evaluation of ecotoxicological and chemical properties of soil amended with Hudson River (New York, USA) sediment. Environ. Sci. Pollut. Res. 27, 7388–7397.

Urbaniak, M., Kiedrzyńska, E., Kiedrzyński, M., Mendra, M., Grochowalski, A., et al., 2014. The impact of point sources of pollution on the transport of micropollutants along the river continuum. Hydrol. Res. https://doi.org/10.2166/nh.2013.242.

Urbaniak, M., Kiedrzyńska, E., Wyrwicka, A., Zieliński, M., Mierzejewska, E., Kiedrzyński, M., Kannan, K., Zalewski, M., 2019. An ecohydrological approach to the river contamination by PCDDs, PCDFs and dl-PCBs – concentrations, distribution and removal using phytoremediation techniques. Sci. Rep. 9 (19310). https://doi.org/10.1038/s41598-019-55973-3.

Urbaniak, M., Kiedrzyńska, E., Zalewski, M., 2012. The role of a lowland reservoir in the transport of micropollutants, nutrients and the suspended particulate matter along the river continuum. Hydrol. Res. 43 (4), 400–411. https://doi.org/10.2166/nh.2012.108.

Urbaniak, M., Lee, S., Takazawa, M., Mierzejewska, E., Baran, A., Kannan, K., 2020. Effects of soil amendment with PCB-contaminated sediment on the growth of two cucurbit species. Environ. Sci. Pollut. Res. 27, 8872–8884. https://doi.org/10.1007/s11356-019-06509-9.

Urbaniak, M., Wyrwicka, A., Tołoczko, W., Serwecińska, L., Zieliński, M., 2017. The effect of sewage sludge application on soil properties and willow (*Salix* sp.) cultivation. Sci. Total Environ. 586, 66–75. https://doi.org/10.1016/j.scitotenv.2017.02.012.

Urbaniak, M., Wyrwicka, A., Tołoczko, W., Serwecińska, L., Zieliński, M., 2020. Utilization of PCB-contaminated Hudson River sediment by thermal processing and phytoremediation. Sci. Total Environ. 738, 139841. https://doi.org/10.1016/j.scitotenv.2020.139841.

Urbaniak, M., Wyrwicka, A., Zieliński, M., Mankiewicz-Boczek, J., 2016. Potential for phytoremediation of PCDD/PCDF-contaminated sludge and sediments using *Cucurbitaceae* plants: a pilot study. Bull. Environ. Contam. Toxicol. https://doi.org/10.1007/s00128-016-1868-6.

Wang, L., Kwok, J.H., Tsang, D.C., Poon, C.S., 2015. Mixture design and treatment methods for recycling contaminated sediment. J. Hazard. Mater. 283, 623–632.

White, J.C, 2002. Differential bioavailability of field- weathered p,p'-DDE to plants of the *Cucurbita* and *Cucumis* genera. Chemosphere 49, 143–152. https://doi.org/10.1016/S0045-6535(02)00277-1.

Whitfield Åslund, M.L., Simpson, M.J., Simpson, A.J., 2010. Effects of amendments on the uptake and distribution of DDT in *Cucurbita pepo* ssp pepo plants. Environ. Pollut. 158 (2), 508–513. https://doi.org/10.1016/j.envpol.2009.08.030.

Wyrwicka, A., Steffani, L., Urbaniak, M., 2014. The effect of PCB-contaminated sewage sludge and sediment on metabolizm of cucumber plants (*Cucumis sativus* L.). Ecohydrol. Hydrobiol. 14 (1), 75–82. https://doi.org/10.1016/j.ecohyd.2014.01.003.

Wyrwicka, A., Urbaniak, M., 2016. The different physiological and antioxidative responses of zucchini and cucumber to sewage sludge application. PLoS One. https://doi.org/10.1371/journal.pone.0157782.

Chapter 22

Denitrification barriers for reducing nitrogen pollution in agricultural landscapes

Agnieszka Bednarek[1], Joanna Mankiewicz-Boczek[1], Liliana Serwecińska[2] and Sebastian Szklarek[2]
[1]*University of Lodz, Faculty of Biology and Environmental Protection, UNESCO Chair on Ecohydrology and Applied Ecology, Łódź, Poland;*
[2]*European Regional Centre for Ecohydrology of the Polish Academy of Sciences under the auspices of UNESCO, Łódź, Poland*

CONTRACTION OF DENITRIFICATION BARRIERS. IMAGE CREDIT: A. BEDNAREK.

Transdisciplinary Ecohydrology. https://doi.org/10.1016/B978-0-443-41596-8.00001-7

Agriculture, one of the three pillars of the economy, unlike services and industry, is the strongest dependant on natural conditions. Its impact on the environment varies around the world depending on the farming practices used. Loss of ingredients nutrients (mainly nitrates) resulting from improper management fields and crops is diffuse pollution that is difficult to limit. In turn, leachate from point sources of pollution with nitrogen compounds, including unsecured organic fertilizer storage yards or leaking building floors livestock, they often result in exceeding the permissible standards by a dozen or so times by the Nitrates Directive (Council of the European Communities, 1991); As they spread across the catchment area, they become pollution area. Nitrogen, as a result of agricultural activities, passes most often into the environment in the form of nitrogen oxides (NO_x), the most dangerous of which in the form of nitrates (NO_3^-) and ammonia compounds (NH_3) and a toxic form nitrous oxide gas (N_2O) as greenhouse gas.

To limit the negative effects of agriculture on natural disturbances the nitrogen cycle and further on the condition of the natural environment by, inter alia, destroying the layer ozone depletion, water eutrophication, land acidification and decline in diversity (Bednarek et al., 2014; Galloway et al., 2004), many guidelines have been adopted specifying the taking of direct minimizing or preventing actions these effects. Currently, the most important regulations in force in agriculture legal regulations to protect water are:

(1) The 1991 Nitrates Directive (91/676/EEC)—one of the first acts EU legislation to control pollution and improve water quality.
(2) Water Framework Directive of 2000 (2000/60/EC)—establishing community framework actions in the field of water policy.
(3) Directive (NEC) of the European Parliament and of the Council (EU) of 2016 (2016/2284) on the reduction of domestic emissions of certain types of pollutants atmospheric.

Counteracting the negative effects of excess toxic forms of nitrogen in the environment must be related not only to limiting the input nitrogen load to catchments from agricultural areas, but also with the possibility of accelerating biodegradation various forms of pollution and binding them in the so-called a pool that is difficult to access (e.g., accumulation in plant tissues). In the case of toxic forms of nitrogen, a key role the processes of denitrification and nitrification play in their transformations (Bothe et al., 2007). Due to the shortage and progressive loss of organic carbon (humus) in the soil profile, it is possible to support natural processes denitrification and nitrification in areas with increased contamination with compounds nitrogen. The pace and effectiveness of natural denitrification depends, inter alia, on the genus and availability of soil organic carbon, C/N ratio, species composition and the number of microorganisms and their activity in the transformation of compounds nitrogen (Paul and Clark, 2002; Trepel and Palmeri, 2002). Bacteria denitrification, common in the soil environment, reduce NO_3^- to N_2, using carbon compounds as electron acceptors for growth (Rivett et al., 2008). This transformation takes place in several stages, what can be written in a simplified way:

$$NO_3^- \text{ (V)} \rightarrow NO_2^- \text{ (III)} \rightarrow NO \text{ (II)} \rightarrow N_2O \text{ (I)} \rightarrow N_2 \text{ (0)}$$

Importantly, leachate from pig manure landfills is dominant NO_3^- while leachate from cattle landfills is dominated by ammonium ions (NH_4^+). The natural population of nitrifying bacteria inhabiting the soil profile is leading for the oxidation of NH_4^+ to NO_3^- and these, in turn, may succumb under appropriate conditions reduction by denitrification to the aforementioned nitrogen gas (N_2), thanks to why the nitrogen cycle is closed without negative effects on the environment (Pietrzak and Nawalny, 2009).

The above knowledge was used in the development of biotechnological solutions with the use of prepared deposits with a carbon substrate constituting the basis of horizontal and vertical denitrification barriers for compound removal nitrogen from point and area (diffuse) pollution sources agricultural sectors (Bednarek et al., 2010, 2014; Izydorczyk et al., 2015; Mankiewicz-Boczek et al., 2017). It is worth emphasizing that the proposed denitrification barriers to reduction of nitrogen compounds works best in intensively used catchments for agricultural and breeding purposes (Fig. 22.1), surrounded by point sources nitrate (e.g., organic fertilizer landfills directly next to the ground) or the coastline of surface waters) (Bednarek et al., 2014).

When constructing a deposit, optimization of conditions is a key element their work in terms of the type and availability of the organic carbon source as well enriching them with an additional pool of microorganisms participating in the transformation nitrogen compounds (e.g., as the total microbial pool obtained from barriers already working in the field and additionally breeding pure cultures of effective strains in the process of denitrification and/or nitrification, Mankiewicz-Boczek et al., 2017) for the purpose activating and accelerating the work (or increasing the efficiency) of the

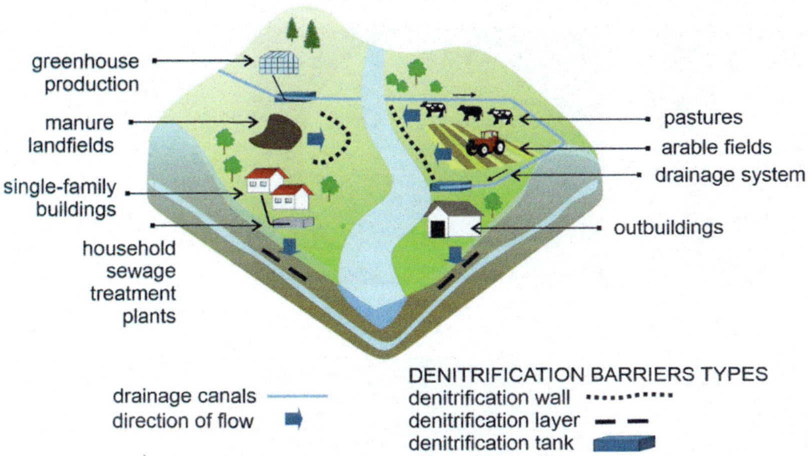

FIGURE 22.1 Possibilities of using the denitrification barriers for the removal of nitrogen compounds. *From Bednarek, A., Szklarek, S., Zalewski, M., 2014. Nitrogen pollution removal from areas of intensive farming—comparison of various denitrification biotechnologies. Ecohydrol. Hydrobiol. 14 (2), 132–141, https://doi.org/10.1016/j.ecohyd.2014.01.005.*

newly constructed deposits and increasing their resistance to environmental stress (including hydrological drought or a large load of sewage) (Figs. 22.2 and 22.3).

The proposed deposits must contain a source of organic carbon, which is the substrate for microorganisms carrying out nitrogen transformation processes (including denitrification and nitrification). Groundwater flowing through the created deposit, contaminated with nitrogen compounds from fertilizers not used by crops mineral (diffuse contamination) should be denitrified within deposits into gaseous forms before they could reach surface waters, increasing their level of eutrophication.

In contrast, organic manure plates to trap contaminants points from manure landfills may consist of deposits in a horizontal form layer under the landfill and a vertical band around the landfill, or optional from the side of the dominant outflow to groundwater. To both layers while When installing the plates in the field, an additional pool of microorganisms may be applied supporting the process of transformation of nitrogen compounds (Mankiewicz-Boczek et al., 2017, Patent PL 231278 B1: Organic Manure Plate).

Work can be an example of optimizing the composition of beds for the removal of nitrogen compounds in which the course of the denitrification process was analyzed in bioreactors filled with various carbon substrates: lignite, pine sawdust, chaff, barley straw and a mixture of brown coal and straw. The analysis of the efficiency of the process in all reactors showed that the ions NO_3^- were most efficiently removed using a straw substrate (99% reduction after 48 days of bed operation, at the starting concentration of NO_3^- over 2000 mg/L). Substrate from straw, which is an easily digestible source of carbon, promoted the growth of microorganisms and their metabolic activity. In turn, NH_4^+ ions in the same experiment were most effectively removed by microorganisms living on the carbon substrate brown (97% reduction at starting NH_4^+ over

FIGURE 22.2 Nitrogen transformation. Various types of carbon substrates for microbial stimulation nitrogen transformation. *Reproduced from: Bednarek, A., Mankiewicz-Boczek, J., Serwecińska, L., Szklarek, S., 2020. Biotechnologie ekohydrologiczne dla ograniczenia zanieczyszczenia wód w obszarach rolniczych. In: Zalewski, M. (Ed.), Ekohydrologia. Wydawnictwo Naukowe PWN, pp. 151–156.*

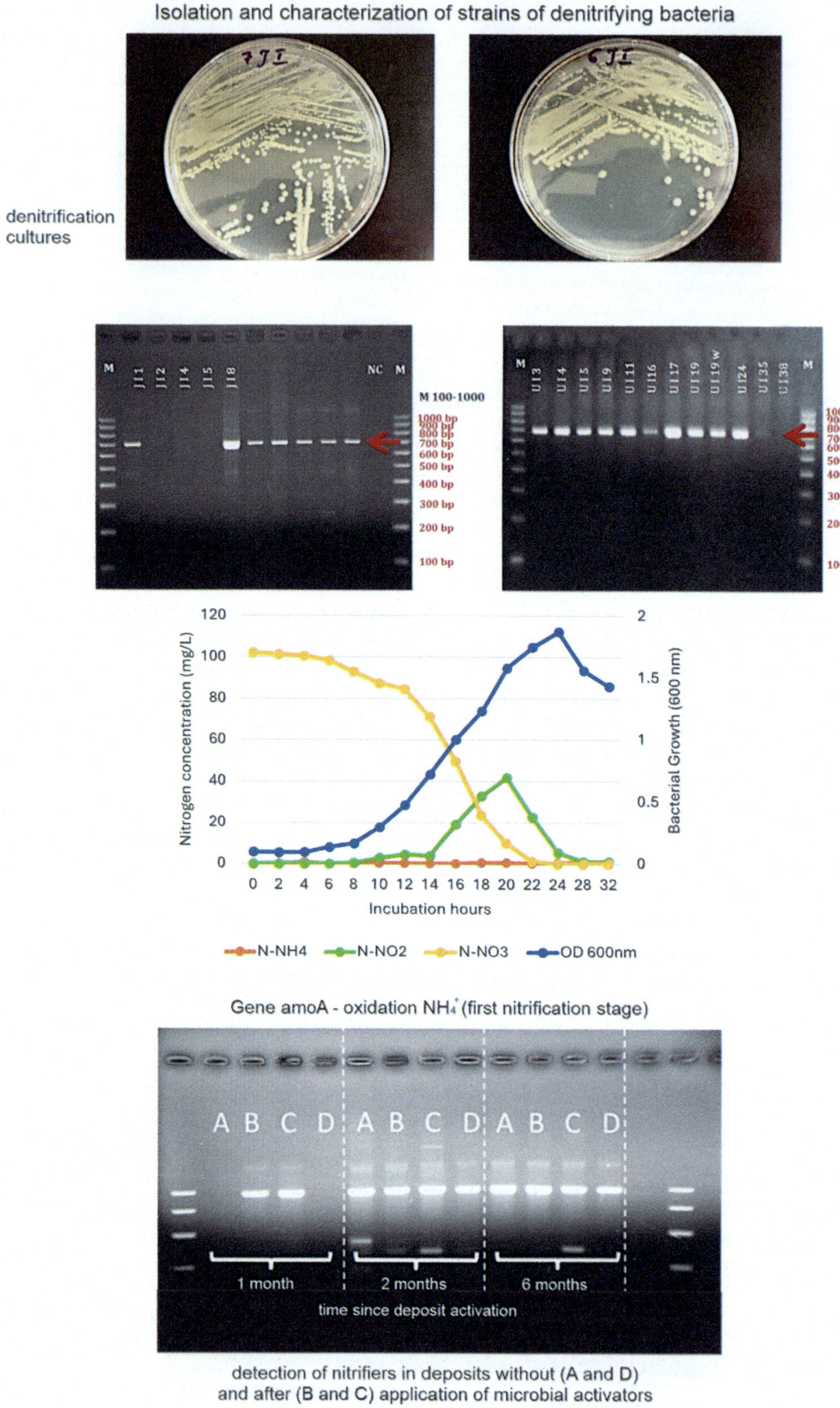

FIGURE 22.3 Development of microbiological activators—evaluation of the ability of the denitrifying strain *Pseudomonas fluorescens* 8JI, isolated from an active denitrification barrier, to transform toxic forms of nitrogen. *Based on Mankiewicz-Boczek, J., Bednarek, A., Gągała-Borowska, I., Serwecińska, L., Zaborowski, A., Kolate, E., Pawełczyk, J., Żaczek, A., Dziadek, J., Zalewski, M., 2017. The removal of nitrogen compounds from farming wastewater—the effect of different carbon substrates and different microbial activators. Ecol. Eng. 105, 341–354. https://doi.org/10.1016/j.ecoleng.2017.05.014.*

650 mg/L). Demonstrated thus, the optimal carbon substrate for newly constructed denitrification barriers there will be a mixture of brown coal and straw that will enable activation both the denitrification and nitrification processes, and thus—optimal removal various forms of nitrogen (Mankiewicz-Boczek et al., 2017).

As can be seen from the examples above, barriers to removal of compounds are described of nitrogen are highly efficient, and their construction does not require large outlays financial or construction, which makes them an ideal solution for small one's farms (point pollution); they can also be part of ecotone zones (diffuse pollution) (Fig. 22.1). It should be emphasized that the greatest efficiency of reduction of nitrogen compounds was observed in the built deposits around point sources, e.g., manure landfills (even above 95% reduction nitrogen). The proposed solution is much cheaper than concrete manure plates, and the bed itself can effectively purify the slurry effluent for a dozen or so years. Additionally, using them, for example, as an element of ecotone zones increases efficiency protection of waters on the catchment scale, especially in winter, when there is none vegetation, on the other hand, an increased runoff of nitrates with groundwater is observed. These are environmentally safe solutions as they are based on materials biodegradable.

References

Bednarek, A., Mankiewicz-Boczek, J., Serwecińska, L., Szklarek, S., 2020. Biotechnologie ekohydrologiczne dla ograniczenia zanieczyszczenia wód w obszarach rolniczych. In: Zalewski, M. (Ed.), Ekohydrologia. Wydawnictwo Naukowe PWN, pp. 151–156.

Bednarek, A., Stolarska, M., Ubraniak, M., Zalewski, M., 2010. Application of permeable reactive barrier for reduction of nitrogen load in the agricultural areas—preliminary results. Ecohydrol. Hydrobiol. 10 (2–4), 355–361. https://doi.org/10.2478/v10104-011-0007-6.

Bednarek, A., Szklarek, S., Zalewski, M., 2014. Nitrogen pollution removal from areas of intensive farming—comparison of various denitrification biotechnologies. Ecohydrol. Hydrobiol. 14 (2), 132–141. https://doi.org/10.1016/j.ecohyd.2014.01.005.

Bothe, H., Ferguson, S., Newton, W.E., 2007. Biology of the Nitrogen Cycle Biology of the Nitrogen Cycle. Elsevier, Germany. https://doi.org/10.1016/B978-0-444-52857-5.X5000-0. Available from: http://www.sciencedirect.com/science/book/9780444528575.

Council of the European Communities, 1991. Council Directive 91/676/EEC of 12 December 1991 concerning the protection of waters against pollution caused by nitrates from agricultural sources. OJEC L375, 1–8.

Galloway, J.N., Dentener, F.J., Capone, D.G., Boyer, E.W., Howarth, R.W., Seitzinger, S.P., Asner, G.P., Cleveland, C.C., Green, P.A., Holland, E.A., Karl, D.M., Michaels, A.F., Porter, J.H., Townsend, A.R., Vörösmarty, C.J., 2004. Nitrogen cycles: past, present, and future. Biogeochemistry 70 (2), 153–226. https://doi.org/10.1007/s10533-004-0370-0.

Izydorczyk, K., Michalska-Hejduk, D., Frętczak, W., Bednarek, A., Łapińska, M., Jarosiewicz, P., Kosińska, A., Zalewski, M., 2015. Strefy Buforowe I Biotechnologie Ekohydrologiczne W Ograniczaniu Zanieczyszczeń Obszarowych/Buffer Zones and Ecohydrological Biotechnologies in Reducing Diffuse Pollution [in Polish]. European Regional Centre for Ecohydrology. Polish Academy of Sciences.

Mankiewicz-Boczek, J., Bednarek, A., Gęgała-Borowska, I., Serwecińska, L., Zaborowski, A., Kolate, E., Pawełczyk, J., Żaczek, A., Dziadek, J., Zalewski, M., 2017. The removal of nitrogen compounds from farming wastewater—the effect of different carbon substrates and different microbial activators. Ecol. Eng. 105, 341–354. https://doi.org/10.1016/j.ecoleng.2017.05.014.

Paul, E.A., Clark, F.E., 2002. Mikrobiologia I Biochemia Gleb.

Pietrzak, S., Nawalny, P., 2009. Ocena stanu jakości wód będęcych pod wpływem obornika posadowionego bezpośrednio na gruncie [Assessment of the quality status of water under the influence of manure placed directly on the ground]. In: Sapek, B. (Ed.), Badania Chemiczne W Służbie Rolnictwa I Ochrony Środowiska/Chemical Research in the Service of Agriculture and Environmental Protection. Instytut Melioracji i Użytków Zielonych (IMUZ).

Rivett, M.O., Buss, S.R., Morgan, P., Smith, J.W.N., Bemment, C.D., 2008. Nitrate attenuation in groundwater: a review of biogeochemical controlling processes. Water Res. 42 (16), 4215–4232. https://doi.org/10.1016/j.watres.2008.07.020.

Trepel, M., Palmeri, L., 2002. Quantifying nitrogen retention in surface flow wetlands for environmental planning at the landscape-scale. Ecol. Eng. 19 (2), 127–140. https://doi.org/10.1016/s0925-8574(02)00038-1.

Water Framework Directive, December 22, 2000. Directive 2000/60/EC of the European Parliament and of the Council of 23 October 2000 establishing a framework for Community action in the field of water policy. OJEC 327, 1–73.

Chapter 23

Reversing lake eutrophication by sequential sedimentation biofiltration system

Agnieszka Bednarek[1], Sebastian Szklarek[2], Liliana Serwecińska[2], Arnoldo Font Nájera[2] and Joanna Mankiewicz-Boczek[2]

[1]*University of Lodz, Faculty of Biology and Environmental Protection, UNESCO Chair on Ecohydrology and Applied Ecology, Łódź, Poland;*
[2]*European Regional Centre for Ecohydrology of the Polish Academy of Sciences under the auspices of UNESCO, Łódź, Poland*

PHOTO CREDIT: A. BEDNAREK.

The main problem concerning the degradation of surface waters is the excessive influx of biogenic substances such as nitrogen or phosphorus, which penetrate the water from industrial and farming areas and are due to the ever-increasing pace of urbanization. Biogenic nutrients can flow into lakes as runoff from hardened surfaces, from point sources of pollution or as inflow via watercourses from the catchment area. This leads to their degradation, and in effect adversely affects their value as places of recreation and fishing and restricts their use for municipal purposes. The progressing eutrophication endangers shallow lakes and lakes in areas strongly transformed by human activity in particular. In the long run, gradual overgrowing and shallowing may lead to their extinction. This is why identifying the main pollution sources within the lake's catchment area should be a primary activity, to be undertaken before recultivation measures are commenced; proposing methods aimed to contain or cut off pollutants in full should come as the next step.

The innovative SED–BIO sedimentation and biofiltration system (based on earlier concept of sequential sedimentation biofiltration system—SSBS) described here is an example of a viable solution recommended for reducing the inflow of

Transdisciplinary Ecohydrology. https://doi.org/10.1016/B978-0-443-41596-8.00045-5

FIGURE 23.1 SED–BIO system. Construction diagram of the SED–BIO system, showing sections (2, 3, 3a) and 5–7, and selected stages of the system construction process: (a) pumping out the water from the pond—section A; (b) pond desludging and deepening—section A; (c) construction of filtering gabions and foundations for a mineral filter; (d) the operational engineering solution—gabions with a mineral filter; (e) lake Jelonek; (f) planted littoral vegetation; (g) "mattresses" with a denitrification bed; (h) mineral filters with floats; (i) microdictum balls with microorganic consortia.

such biogenic substances and enhancing the physical and biological parameters. The system uses combined pollutant reduction methods (engineering, physical and biological), coupled with the use of preparations containing selected microorganism consortia (Fig. 23.1i). As a prototype solution, it was used on the Struga Gnieźnieńska (SG), an urban river flowing into Lake Jelonek. The potential sources of pollutants reaching the SG river along the section between its source and outlet into Lake Jelonek include: domestic sewage, precipitation, private allotment gardens, a number of settling ponds on a former sugar plantation, stormwater drains, combined sewers, refuse from the settling pond area and the SG river, built-up areas and sealed areas, past operation of two tanneries and, periodically, stormwater drainage from the "Mieszko" sports stadium (Kupiec et al., 2017).

The SED–BIO system is an original solution developed with a view to surface water recultivation (Patent Application P.422056) and built as part of the GEKON2/03/267948/21/2016 project. The system comprises two sections: A—situated in a 0.7 ha pond through which the SG river flows, and B—sequential sedimentation biofiltration system (SSBS), constructed in a wider section of the SG river, directly preceding its point of entry into Lake Jelonek (Fig. 23.2).

FIGURE 23.2 Sedimentation biofiltration system—SED–BIO located in Gniezno, Poland (the Struga Gnieźnieńska river/lake Jelonek). *Image credit: A. Urbaniak.*

The construction of the system was preceded by the preparation of an analysis of relevant water-related legal regulations and the obtaining of the relevant permits (construction, archeological, etc.). The preparatory work involved the trimming of the branches overhanging the pond, removal of ligneous debris and rubbish around the pond, pumping the water out of the pond, and directing the current through a girdling ditch. This work was carried out in extremely difficult conditions, on peaty and boggy soils. The substratum needed to be properly stabilized before any work could commenced, and the level of water seepage from the peat had to be controlled. Comprehensive work in the pond included the deepening of the pond where the SG river entered it, i.e., in section A of the system, where intensive sedimentation took place (Fig. 23.1a–c). As a result of this process, that part of the pond became much shallower. The work in section A of

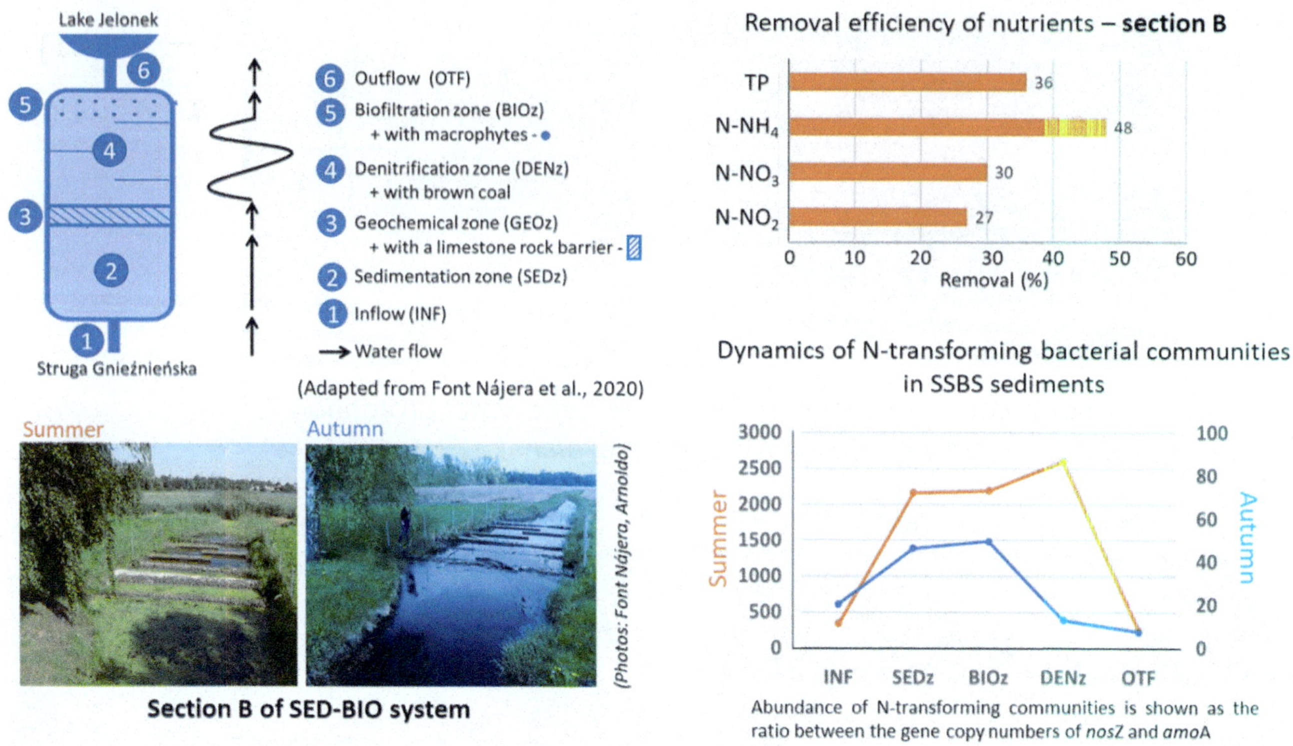

FIGURE 23.3 Section B of system (SSBS). Description of construction and performance of section B of system (SSBS) on the Struga Gnieźnienska river—reduction of nutrients for the prevention of eutrophication (in August and October 2017).

the system also included the construction of two gabions harnessing bioengineering technology, whose role was to halt suspension (Fig. 23.1c and d). During the subsequent stage of the work, mineral filters were installed between the gabions in the pond's section A; these were specially designed baskets suspended on floats absorbing phosphorus (Fig. 23.1d and h). Layered limestone rock and air-cooled, briquetted furnace, 30–70 mm fraction slag were used as the filter substrate. The baskets with the substrate were placed perpendicularly to the direction of the water flow. A bottom zone was prepared behind the gabions, with the aim of activating the denitrification processes using denitrification deposits (Fig. 23.1g), whilst plants were planted in the littoral and sublittoral zones in order to reinforce the biofiltration effect (Fig. 23.1f). In total, about 20,000 plants were planted, representing seven species with a special affinity for phosphorus.

The subsequent stage of work was performed on the section of the SG river between the pond and Lake Jelonek, i.e., on section B of the system (Figs. 23.3 and 23.4). The second part of the filter was built from mesh and stone gabions, a limestone deposit, a denitrification deposit, a system of barriers extending the watercourse and slowing down the water flow, and a plant biofilter.

FIGURE 23.4 Sequential sedimentation biofiltration system. Efficiency of the sequential sedimentation biofiltration system in reduction of dissolved phosphorus (compare point 1 and point 8) in relation to the concentration in the lake Gnieźnieńska. Stream inflow to the system (point 1). *Image credit: A. Bednarek.*

One year after the completion of the whole system, a monitoring study of the performance of section B was performed in August and October 2017 (Font Nájera et al., 2020). Samples for chemical, microbiological and molecular analysis were collected in different areas from the horizontal profile (Fig. 23.3). The results showed removal of N–NO_2 (27%), N–NO_3 (30%), N–NH_4 (48%) and TP (36%). Quantitative monitoring of the genes for denitrification (*nos*Z) and nitrification (*amo*A) revealed higher numbers of bacteria participating in the transformation of nitrogen in the system zones than in the inflow and outflow (Fig. 23.1). This difference was particularly noticeable in the summer period. The results obtained confirmed the ability to decrease the load of nitrogen and phosphorus compounds (Fig. 23.4), due to application of SSBS on SG (section B) which directly supplies the waters of Lake Jelonek. This example of the use of a natural-based solution illustrates an important technique for reducing the eutrophication process in lakes and, consequently, the limiting of harmful algal blooms, often dominated by toxic cyanobacteria.

To sum up the operation of the whole SED–BIO system, the system built on the SG river—both in section A on the pond, and in section B on the stream—helped enhance the physiochemical parameters of the water flowing into Lake Jelonek. **During the vegetation season, when all the processes activated in the system were at an optimal level, the volume of phosphates reaching the lake was reduced by approx. 52%, nitrates and nitrites by approx. 73%, and ammoniacal nitrogen was cut to marginal values in both sections, A and B**. The physical parameters, such as conductivity or pH, complied with the standards adopted for water purity class I. In addition, the water flowing through the system was better aerated than the water flowing into the system. The proposed sequential and biofiltration system had a beneficial effect on the microbiological condition of the water flowing into Lake Jelonek in such aspects as the structure of the microorganic population, its diversity, metabolic functionality of microorganisms representing a polyphyletic group of predominantly heterotrophs, not to mention the sanitary purity of the water. The system also helped enhance the quality of water in Lake Jelonek, as manifested by the first-ever year noted with the absence of blooms and oxygen deficits (Kupiec et al., 2017, 2018, 2019, 2022; Kupiec, 2020, 2021; Matłoka et al., 2020).

References

Font Nájera, A., Serwecińska, L., Szklarek, S., Mankiewicz-Boczek, J., 2020. Characterization and comparison of microbial communities in sequential sedimentation-biofiltration systems for removal of nutrients in urban rivers. J. Ecol. Eng. 149. https://doi.org/10.1016/j.ecoleng.2020.105796. Available from: www.elsevier.com/inca/publications/store/5/2/2/7/5/1.

Kupiec, J.M., 2020. Wyniki Funkcjonowania Systemu SED–BIO Zlokalizowanego na Strudze Gnieźnieńskiej w Sezonie Wegetacyjnym 2020 r, p. 19.

Kupiec, J.M., 2021. Wyniki Chemiczne Funkcjonowania Systemu SED–BIO Zlokalizowanego na Strudze Gnieźnieńskiej w Okresie X–XII 2020 r.

Kupiec, J.M., Bednarek, A., Szklarek, Mankiewicz-Boczek, J., Żeleźnik, W., Zalewski, M., 2017. Wyniki Funkcjonowania Systemu Sedymentacyjno-Biofiltracyjnego Zlokalizowanego na Strudze Gnieźnieńskiej w Kontekście Ochrony Jeziora Jelonek: Raport Roczny WOŚ, 272.13.2017.

Kupiec, J.M., Bednarek, A., Szklarek, S., Żeleźnik, W., 2018. Wyniki Funkcjonowa-Nia Systemu Sedymentacyjno-Biofiltracyjnego Zlokalizowanego na Strudze Gnież-Nieńskiej w Kontekście Ochrony Jeziora Jelonek. Raport roczny WOŚ, p. 21.

Kupiec, J.M., Bednarek, A., Szklarek, S., Żeleźnik, W., 2019. Wyniki Funkcjonowania Systemu Sedymentacyjno-Biofiltracyjnego Zlokalizowanego na Strudze Gnieźnieńskiej w Sezonie Wegetacyjnym 2019 r. Oraz Obliczenie Bilansu Ładunku Sub-stancji Biogennych za Lata 2016–2018.

Kupiec, J.M., Bednarek, A., Szklarek, S., Mankiewicz-Boczek, J., Serwecińska, L., Dąbrowska, J., 2022. Evaluation of the effectiveness of the SED–BIO system in reducing the inflow of selected physical, chemical and biological pollutants to a lake. Water 14 (2), 239. https://doi.org/10.3390/w14020239.

Matłoka, A., Ciesielski, K., Gołdyn, R., Strzelińska, S., Wekwerth, S., Kozłowska, A., Siepak, J., 2020. Wizja rekultywacji jeziora Jelonek w Gnieźnie. Przegląd Komunalny. 4, 47–51.

Chapter 24

Microorganisms as activators, regulators, and bioindicators of ecosystems

Liliana Serwecińska[1, a], Joanna Mankiewicz-Boczek[1], Agnieszka Bednarek[2] and Arnoldo Font-Nájera[1]

[1]European Regional Centre for Ecohydrology of the Polish Academy of Sciences under the auspices of UNESCO, Łódź, Poland; [2]University of Lodz, Faculty of Biology and Environmental Protection, UNESCO Chair on Ecohydrology and Applied Ecology, Łódź, Poland

DENITRIFICATION ZONE OF SEQUENTIAL SEDIMENTATION BIOFILTRATION SYSTEM (SSBS) WITH A POOL OF ACTIVE BACTERIAL STRAINS TO REDUCE NITRATES ON THE GNIEZNO STREAM. *IMAGE CREDIT: A. FONT-NÁJERA.*

a Retired.

Transdisciplinary Ecohydrology. https://doi.org/10.1016/B978-0-443-41596-8.00043-1

1. Bacterial activators and regulators in environmental compartments

Microbial activity is one of the most important environmental factors for the transformation and cycling of nutrients, thus a key element of this type of water purification system. Bacteria are almost always the most abundant organisms found in every compartment of the environment in terms of numbers. They participate in the circulation of nitrogen, phosphorus, sulfur, distribution of organic matter, energy flow and stability of ecosystems; bacterial cells are considered to be highly sensitive to their surrounding environment, they quickly respond and adapt to environmental changes due to their rapid proliferation ability and simple structure (Ryall et al., 2012). When beneficial microorganisms are absent from a polluted site or their populations have been reduced, favorable conditions can be created to facilitate the development of beneficial microflora, or specific microorganisms can be added to enhance the existing population. The efficiency and the rate of natural environmental processes are highly dependent on the structure, size, and taxonomic composition of the natural, functional populations of microorganisms inhabiting a given ecological niche. The nitrogen cycle is the best studied of the mineral cycles due to its role in the environment and the high nitrogen content in the cells of bacteria and plants; nitrogen is the fourth most common element found in bacterial and plant cells, making up approximately 12% of cell dry weight. The major processes converting nitrogen into other forms are nitrogen fixation, denitrification, nitrification, ammonification, COMMAMOX (complete ammonia oxidation) and ANNAMOX (anaerobic ammonium oxidation). All these processes are a part of the global nitrogen cycle and are microbiological processes. Human activities have disturbed the natural flow of nutrients, resulting in the arising of numerous environmental problems such as: contamination of drinking water, freshwater eutrophication, biodiversity reduction in ecosystems, stratospheric ozone depletion, acidification of soils and surface waters (Galloway et al., 2004; Bednarek et al., 2014; Erisman et al., 2015). It is commonly recognized that nitrate pollution has numerous negative effects on human health; an excess of nitrates causes, inter alia, genetic mutations, various types of cancer, birth defects, respiratory problems and methaemoglobinaemia (Bosman, 2009; Zhou, 2015; Ducci, 2018; Gao et al., 2020). The inflow of high concentration of phosphorus and nitrogen compounds contribute to the occurrence of favorable conditions for the proliferation of phytoplankton, including cyanobacteria, in bodies of water such as rivers, lakes and ponds, which affect both, aquatic ecosystems and human health due to the depletion of oxygen in the pelagic profile of the water, decreases in light transmittance, and production of cyanotoxins (Chahal et al., 2016; Codd et al., 2017; Wang et al., 2017). To address this problem, biofiltration systems have been implemented to purify of various types of inflows to rivers, lakes and ponds; these systems are innovative biotechnological solutions and are designed in accordance with the principles of ecohydrology to enhance the removal of environmental pollutants (Zalewski, 2012; Zalewski et al., 2018). One of the most studied stages in the nitrogen cycle is the reduction of nitrate, which can be conducted by dissimilatory, including denitrification, and assimilatory pathways.

Denitrification is the microbial mediated process involving four cascade reactions of the reduction: $NO_3^- \rightarrow NO_2^- \rightarrow NO \rightarrow N_2O \rightarrow N_2$, Fig. 24.1; initially it was discovered and described in heterotrophic, obligatory and facultative anaerobic bacterial strains (Tiedje, 1988; Devol, 2015). Sensitivity to oxygen is the major difference between the aerobic and anaerobic groups of bacteria. For anaerobic denitrifiers the presence of oxygen results in a shift in genes expression and inhibition of enzymes involved in denitrification, while aerobic denitrifiers carry out denitrification in the presence of oxygen (Papaspyrou et al., 2014). Many environmental prokaryotes can conduct the process of denitrification and exhibit a variety of reduction pathways for nitrogenous oxides. Among environmental microorganisms, Bacteria and Archea are of particular interest in their potential for nitrogen transformation, however, Archea has not been studied much. Most denitrifiers belong to the α-, β-, γ -, and ε- proteobacteria, however denitrification was also shown to be widespread among *Actinomycetales*, *Bacteroidetes*, *Firmicutes* and *Aquificaceae* as well as Archaea and fungi (Philippot et al., 2002; Laughlin and Stevens, 2002; Risgaard-Petersen et al., 2006; Piña-Ochoa et al., 2010). Many studies of soil, sediments and freshwater samples show, that the dominant denitrifiers in most samples were *Pseudomonas* genus (Gamble et al., 1977; Mankiewicz-Boczek et al., 2017). In the last decade denitrifying aerobic bacteria were also intensively studied due to their simultaneous nitrification ability, and their high potential for application in wastewater treatment biotechnologies. Several strains have been isolated and reported to perform simultaneous nitrification–denitrification activity (SNdN), such as *Pseudomonas*, *Bacillus*, *Acinetobacter*, *Agrobacterium*, *Alcaligenes*, *Enterobacter*, and *Klebsiella* (Holmes et al., 2019). Nitrogen-transforming bacteria have been isolated and described from such sources as water and sediments, wastewater, sewage sludge, soil, compost, manure etc. Nevertheless, it should be emphasized, that most denitrifier diversity is not covered by culturable bacteria. Nitrification in turn, is a process of two consecutive reactions $NH_4^+ \rightarrow NO_2^- \rightarrow NO_3^-$, Fig. 24.1, conducted by two groups of autotrophic bacterial strains: ammonia oxidizing bacteria (AOB)—*Nitrosomonas*, *Nitrosospira* and *Nitrosococcus* (Guo et al., 2013), and nitrite oxidizing bacteria (NOB) represented by *Nitrobacter* and *Nitrospina* (Prosser, 2005), however, the role of ammonia-oxidizing Archaea has also been elucidated in the last decade (Offre et al., 2013). Other nitrification pathways are also known: (i) direct and complete ammonium oxidation of $NH_4^+ \rightarrow$

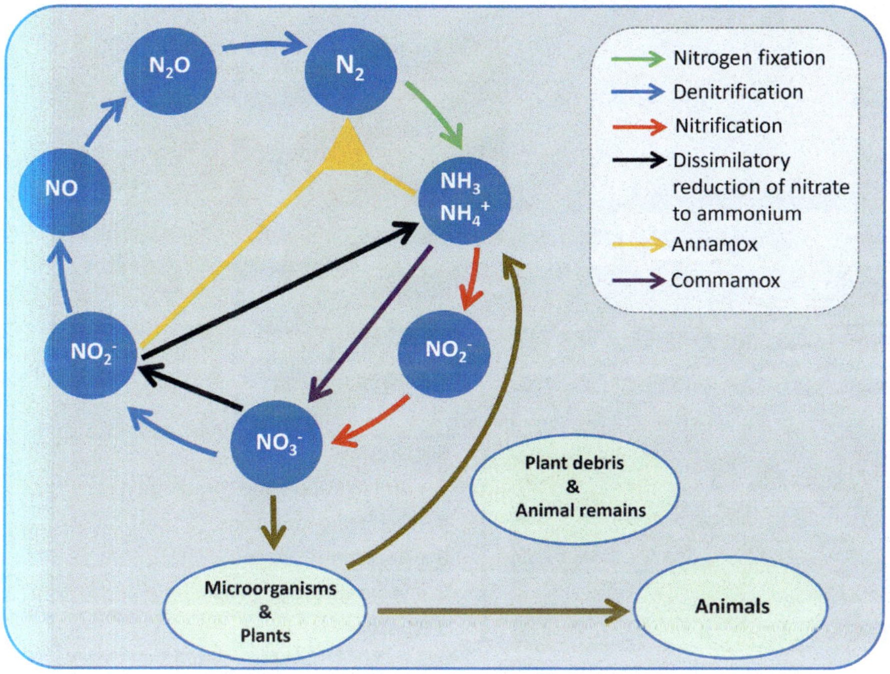

FIGURE 24.1 Nitrogen cycle. Colored lines indicate the direction of particular nitrogen transformation pathways.

NO_3 (COMMAMOX process) (Daims et al., 2015; Hu and He, 2017), (ii) nitrification via metabolism of hydroxylamine (NH_2OH), which is an intermediary product and has been described for *Nitrosomonas*, *Pseudomonas* and *Enterobacter* strains. Nitrogen transformation, different to nitrification or denitrification, has also been described as nitrogen assimilation; NO_3^- and NO_2^- are transformed to NH_4^+ and subsequently to glutamate, which is then incorporated into the process of bacterial respiration. Another pathway is ANNAMOX, where $NH_4^+$and NO_2^- are directly transformed to N_2 (Kartal and Keltjens, 2016), Fig. 24.1.

Bacterial communities inhabiting different sequential sedimentation-biofiltration systems (SSBSs), Fig. 24.2, which have been implemented for removing nutrients from urban rivers and ponds, have recently been studied with the use of microbiological and molecular techniques, Fig. 24.3. Significant positive correlations were observed between the measured concentration of NO_3^- and NH_4^+ and the abundance of bacterial key functional genes involved in nitrification and denitrification processes (*amo*A and *nos*Z genes, respectively) (Font Nájera et al., 2021a,b). Furthermore, attempts have been made to broadly characterize the microbial population of SSBSs, and to isolate the individual bacterial strains involved in nitrogen transformation in various zones of this type of system. The bacteria isolated belong to different taxonomic groups, where the phylum Proteobacteria with the families *Pseudomonadaceae*, *Comamonadaceae*, *Enterobacteriaceae* and *Oxalobacteriaceae* were dominant. Moreover, the phylum *Firmicutes* was represented by two strains of *Bacillus* spp. and the phylum *Actinobacteria* by *Kocuria* sp., Fig. 24.3A. A few strains demonstrated high efficiency in removing nitrogen compounds, achieving up to 98.9% removal of N–NH_4 and 87.7% of N–NO_3 in laboratory-controlled assays within the first 30 h of incubation, Fig. 24.3B (Font Nájera et al., 2021a,b). The size of the population of denitrifying bacteria showed a positive correlation with the concentration of NO_3^-. The highest abundance of denitrifiers was observed within the denitrification zone of one studied SSBS, with added brown coal as a carbon source for bacteria (Font Nájera et al., 2020, 2021a,b). Moreover, it was revealed that the denitrification process is more effective in summer with higher temperatures (within 20 ± 2° C) when compared to nitrification (within 14 ± 2.5° C) (Font Nájera et al., 2021a,b). The abundance of nitrifying bacteria showed a positive correlation with the concentration of NH_4^+. The highest abundances of key microbial gene involved in nitrification were noted within the geochemical zone made of limestone, where they formed a biofilm composed of a nitrifying/denitrifying bacterial complex, while dolomite used in geochemical zone of another biofiltration system was much less conducive to the bacterial setting. Furthermore, microbial community analysis, using the 16S rRNA sequencing technique, and analysis of its correlation with NO_3^- and NH_4^+ concentration, suggested that mostly bacterial families of *Comamonadaceae*, *Flavobacteriaceae*, *Crenotrichaceae* and *Rhodocyclaceae* are involved in nitrogen transformations. Study revealed that a denitrification zone, provides the best environmental

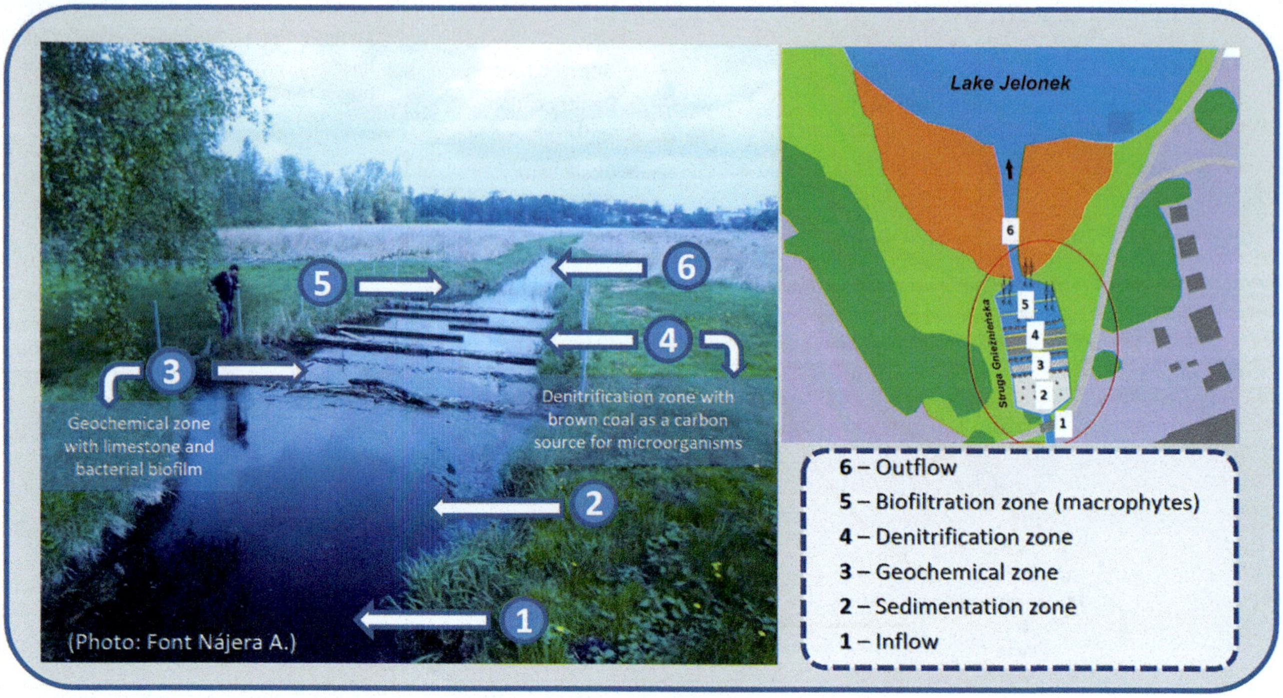

FIGURE 24.2 The structure of the part of the sequential sedimentation-biofiltration system (SSBS)—example of construction on the Struga Gnieźnieńska municipal river (project GEKON2/03/267948/21/2016). Denitrification zone with the applied pool of bacteria highly active in denitrification pathway, was installed between the biofiltration and geochemical zones. *Kupiec, J.M., Bednarek, A., Szklarek, S., Mankiewicz-Boczek, J., Serwecińska, L., Dabrowska, J., 2022. Evaluation of the effectiveness of the SED-BIO system in reducing the inflow of selected physical, chemical and biological pollutants to a lake. Water, 14, 239.*

conditions for the development and activity of these microbial communities. Moreover, their activity and abundance increased, when the temperature was higher than 10° C, and consequently it had an impact on higher removal efficiency of nutrients in SSBS (Font Nájera et al., 2021a,b). The effectiveness of some SSBS constructed in urban rivers was high, and reduction of nutrients reached up to 71.5% of total nitrogen (TN), 91.3% of NO_3^-, 55% of NO_2^-, 48% of NH_4^+ and 93% of total phosphorus (TP) (Serwecińska et al., 2017; Szklarek et al., 2018; Font Nájera et al., 2020).

Nitrate (NO_3), is also the most mobile form of nitrogen and a toxic pollutant in soil; this issue has been highlighted in the Nitrates Directive (The Council of the European Communities, EU, 1991), and this aims to reduce NO_3^- leaching from agriculture. Agricultural sources of N cause the release of nitrate (NO_3^-) to groundwater, nitrous oxide (N_2O) to the air and atmospheric emissions of ammonia (NH_3). The excess of nitrogen from crops and animal breeding can leach through the shallow groundwater and finally enter surface waters through diffuse or point discharges. A complete microbiota originated from denitrifying barriers installed in animal farm was analyzed and screened for bacteria able to reduce NO_3^-/NO_2^-; a number of culturable bacterial denitrifiers were isolated and collected for further application in nature-based solutions; their denitrifying potential differ widely and four major kinetics patterns of NO_3^- reduction with different rates and efficiencies of reaction were noted, with a minimum time of complete nitrate/nitrite reduction of 36 h, Fig. 24.3C (Mankiewicz-Boczek et al., 2017). Analysis of the 16S rRNA fragment sequence allowed most culturable denitrifying bacteria isolates to be classified as the *Pseudomonas* genus. Closer analysis of the *nos*Z gene sequence variation between those strains revealed, that they represented different environmental isolates, Fig. 24.3 DE. Such variability in closely related bacteria indicates significant bacterial diversity in the denitrifying deposits. The proportion of denitrifiers in soil represent up to 5% of the total microbial community; however, denitrification activity was confirmed in 20% of the 300 bacterial strains tested in (Mankiewicz-Boczek et al., 2017). Several of these highly metabolically active bacterial isolates are currently being used in nature-based solutions i.e., denitrifying deposits in the case of groundwater in agricultural areas, and SSBS for urban rivers, water courses, ponds and lakes. Both developed systems use microorganisms with a specific spectrum of metabolic activity, initiating beneficial biochemical processes, which in turn lead to the neutralization of nutrients, and to the restoration of the ecological status of the waters. The application of this type of ecohydrological biotechnology, which employs nonpathogenic, highly metabolically active bacterial strains or their selected compositions living in natural environments, does not cause undesirable side effects for the natural environment, their action is based on

A Detection of key functional genes involved in N-transformation pathways for bacterial strains isolated from SSBSs sediments.

Strain	16S rRNA Identification	Assimilation *nas*A $NO_3^- \rightarrow NO_2^-$	Nitrification *hao* $NH_2OH \rightarrow NO_2^-$	Denitrification *nap*A/*nar*G $NO_3^- \rightarrow NO_2^-$		*nir*S $NO_2^- \rightarrow NO$	*nor*B $NO \rightarrow N_2O$	*nos*Z $N_2O \rightarrow N_2$
Str01	*Bacillus simplex* (98.4%)	-	-	-	✓	-	-	-
Str21	*Pseudomonas mandelii* (99.6%)	✓	-	-	✓	✓	✓	✓
Bzr02	*Citrobacter freundii* (99.4%)	-	✓	✓	✓	-	-	-
Bzr07	*Pseudomonas migulae* (99.8%)	-	-	✓	-	-	-	✓
Sok01	*Hydrogenophaga taeniospiralis* (99.3%)	-	-	-	✓	-	-	✓
Sok03	*Pseudomonas guineae*(99.4%)	-	-	-	-	-	-	✓
Sok05	*Kocuria rosea* (99.1%)	-	-	-	✓	-	-	-
Sok06	*Bacillus aerius*(99.6%)	-	-	-	✓	-	-	✓
Sok20	*Janthinobacterium lividum* (99.3%)	-	-	-	✓	-	-	✓
Sok41	*Acidovorax radicis* (99.3%)	-	-	-	✓	-	-	✓

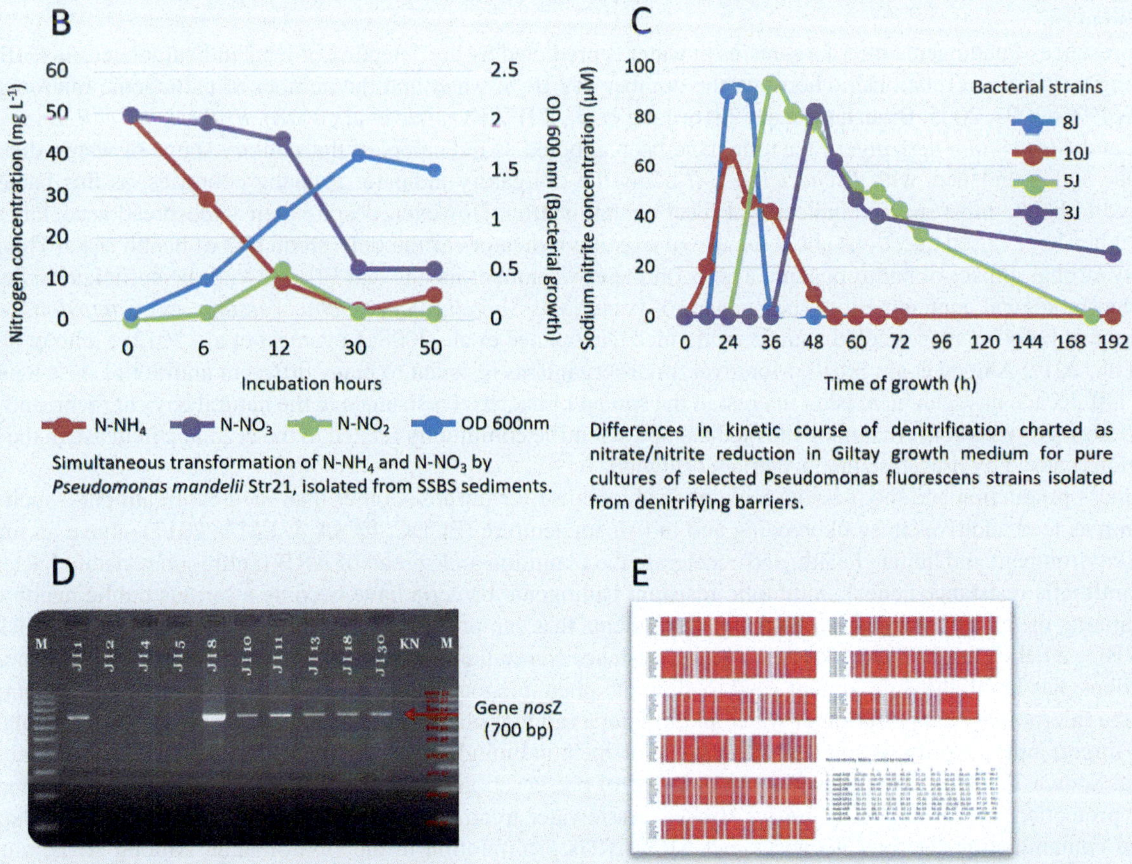

Detection of *nos*Z gene involved in denitrification pathway for different bacterial strains isolated from denitrification barriers (D), and sequence analysis of the *nos*Z gene fragments (700 bp) (E).

FIGURE 24.3 Characterization of nutrient-transforming bacterial communities in SSBS. Individual pannels are described inside the figure. *(A and B) Reproduced from: Font Nájera, A., Serwecińska, L., Mankiewicz-Boczek, J. (2021). Culturable nitrogen-transforming bacteria from sequential sedimentation biofiltration systems and their potential for nutrient removal in urban polluted rivers. Sci. Rep., 11, 7448. http://doi.org/10.1038/s41598-021-86212-3. (C) Reproduced from: Mankiewicz-Boczek, J., Bednarek, A., Gągała-Borowska, I., Serwecińska, L., Zaborowski, A., Kolate, E., Pawełczyk, J., Żaczek, A., Dziadek, J., Zalewski, M., (2017). The removal of nitrogen compounds from farming wastewater—the effect of different carbon substrates and different carbon activators. Ecol. Engine, 105, 341–354. https://doi.org/10.1016/j.ecoleng.2017.05.014.*

the stimulation and intensification of natural processes. These types of microbiological biotechnologies, like chemical methods, have a wide range of potential applications for reducing the nutrient load flowing into freshwater ecosystems. Ecohydrological solutions based on microorganisms are highly efficient, more balanced, easy to adapt to environmental processes, low-cost in terms of implementation, low-invasive and environmentally friendly. Hence, their application is important for enhancing rivers system' sustainability potential development potential WBSR and sustainable development strategies.

2. Bacterial indicators as sanitary threat detectors. Microbial source tracking and monitoring of aquatic ecosystem's contamination with pathogenic bacteria

The water environment is increasingly threatened by anthropogenic factors; thus, protection of aquatic ecosystems requires constant monitoring using the latest knowledge of a wide variety of emerging pollutants and the modern techniques available. This issue demands the continual refinement of the monitoring rules to make them comprehensive and pertinent for the target pollutants. Various physicochemical parameters, as well as hydrobiological factors and bacteriological indicators are among the common measurements utilized in water quality index (WQI) to determine the severity of water pollution.

Population growth and rapid urbanization have created a sanitation risk for surface and groundwater. Bacteriological pollution may occur as a result of the inflow from municipal and industrial sewage treatment plants, storm, urban and agricultural land runoff, as well as permanent and recreational development areas. Monitoring the sanitary status of bodies of water, is vital for the effective control of pollution by pathogenic bacteria and maintaining the microbiological safety of water resources.

The presence of pathogenic microorganisms in water is predicted by the detection of fecal indicator bacteria (FIB) due to there being a significant relationship between the number of FIB in water and the number of pathogenic microorganisms (USEPA, 1986, 2004, 2005; Bradshaw et al., 2016; Jang et al., 2017; Korajkic et al., 2018). *Escherichia coli*, *Enterococcus faecalis*, and *Clostridium perfringens* bacteria have been adopted as indicators of the sanitary status of water, due to their prevalence and abundance, with *Escherichia coli* being the obligatory indicator in many countries, as this bacterium is considered to be the most specific indicator of fecal contamination. However, despite their widespread acceptance, many recent studies have raised questions about the use of existing indicators as the only predictors of health risks. The present, extremely strong, impact of anthropogenic agents on the environment means that FIB may not be sufficient indicators of water microbiological contamination; such genera of bacteria as *Shigella*, *Salmonella*, *Yersinia* and *Campylobacter* have been responsible for many so-called aquatic epidemics (Jakopanec et al., 2008; Levantesi et al., 2012; Sidhu et al., 2013; Steele et al., 2017; Ahmed et al., 2019). Moreover, microorganisms resistant to many different antibiotics pose a particular risk; the last decade has seen increasing interest in the spread of bacterial resistance in the natural environment, and this has stemmed from growing concern among the medical and scientific community related to the dramatic increase in the number of infections caused by drug-resistant bacteria worldwide.

The mass production and the massive scale of antibiotic use for purposes other than medical treatment—such as their application as feed additives in stockbreeding and in fish aquaculture (ECDC, EFSA & EMA, 2017)—have an impact on both the environment and human health, and accelerate the evolution and spread of ARB (antibiotic resistant bacteria) and ARGs (antibiotic resistance genes). Antibiotic-resistant pathogenic bacteria have become a serious public health concern and are among the most discussed environmental pollutants that can proliferate rapidly across the globe (Czekalski et al., 2014; WHO, 2018). The European Antimicrobial Resistance Surveillance Network, a group that reports resistance trends for microbes, has highlighted the urgent need to slow the phenomenon of drug resistance in the natural environment. To prevent the microbiological pollution of water and initiate a rational plan of response to this type of environmental threat, it is first important to identify its sources with the use of specific bioindicators. A valuable tool for environmental study is Microbial Source Tracking, which responds to the need of preservation of water resources and can be used to great effect in water protection. According to numerous studies, wastewater treatment plants (WWTPs) represent a major source of microbial contamination for river networks and ARB/ARGs proliferation in the environment. Among ARB, multidrug-resistant bacteria (MDRB) pose the greatest threat and cause the most serious infections. Moreover, a number of bacterial species isolated from environment and WWTPs are of growing clinical importance, being resistant to many antimicrobials, including new generation antibiotics and pharmaceuticals of last resort used for treating bacterial infections (WHO, 2017). More worryingly, the nature of the microbial community in WWTPs promotes the exchange of antibiotic resistance between different bacterial species.

Microbial Source Tracking has been used in many studies, and research conducted on the catchment scale seems to be particularly valuable, as it provides comprehensive knowledge about the sanitary condition of water and wastewaters in

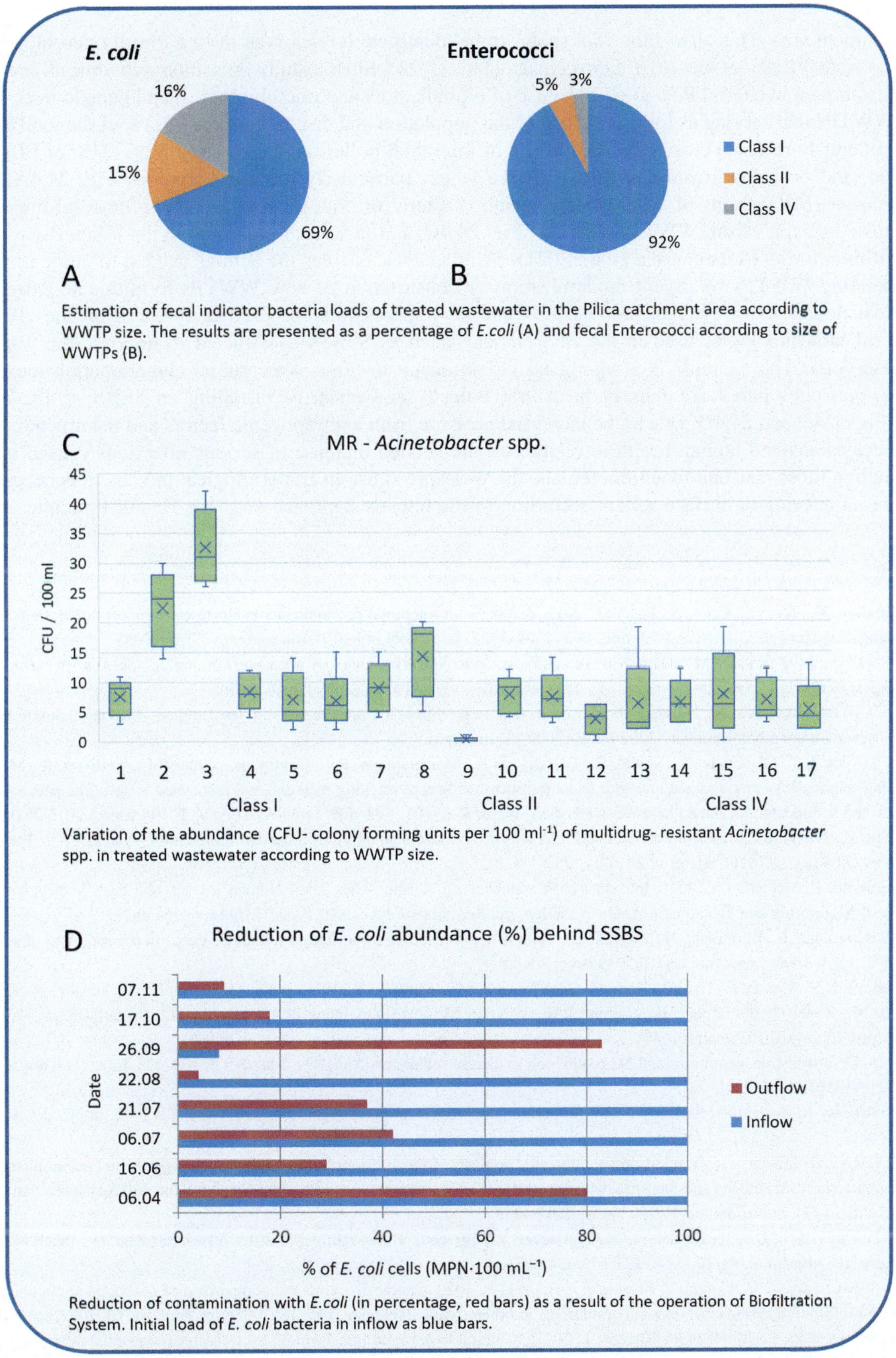

Estimation of fecal indicator bacteria loads of treated wastewater in the Pilica catchment area according to WWTP size. The results are presented as a percentage of *E.coli* (A) and fecal Enterococci according to size of WWTPs (B).

Variation of the abundance (CFU- colony forming units per 100 ml^{-1}) of multidrug- resistant *Acinetobacter* spp. in treated wastewater according to WWTP size.

Reduction of contamination with *E.coli* (in percentage, red bars) as a result of the operation of Biofiltration System. Initial load of *E. coli* bacteria in inflow as blue bars.

FIGURE 24.4 Bacterial indicators for monitoring the sanitary quality of water. Individual pannels are described inside the figure. WWTP size classes based on average treated wastewater flow (m^3/day): Class I (small) - 107; Class II (medium) - 640; Class III (large) - 4230; Class IV (very large) - 15,570. (A–C) Based on data from: Serwecińska L., Kiedrzyńska,E., Kiedrzyński, M., 2021. A cachment-scale assessment of the saniutary conditions of treated wastewater and river water based on fecal indicator bacteria and carbapenem-resistant Acinetobacter spp. Sci. Tot. Environ. 750, 142266. (D) Modifed from Kupiec, J.M., Bednarek, A., Szklarek, S., Mankiewicz-Boczek, J., Serwecińska, L., Dabrowska, J., 2022. Evaluation of the effectiveness of the SED-BIO system in reducing the inflow of selected physical, chemical and biological pollutants to a lake. Water, 14, 239. *https://doi.org/10.3390/w14020239.*

the entire catchment area. This allows the "hot spots" to be identified in regards of most harmful microbiological sources of pollution of water (Rizzo et al., 2013; Serwecińska et al., 2021). Such a study on sanitary conditions conducted in the Pilica River catchment in central Poland with the use of bioindicators and microbiological and genetic tests, revealed that the smaller WWTPs that serving as little as 0.56% of the population and discharging only 1.2% of the total wastewater in the area, continuously released extremely high levels of microbial pollution, constituting 52%–93% of FIB loads in the studied region, and pose the greatest threat to surface water, particularly to river networks, Fig. 24.4AB. Moreover, multidrug-resistant (MR) strains of *Acinetobacte*r genus, bacteria of both clinical and environmental importance, were found in all tested samples from WWTPs effluents, Fig. 24.4C, and in water samples from the Pilica River during the 2-year period of monitoring (Serwecińska et al., 2021). Such analysis enables the actions needed to guide improvement of the worst operating WWTPs by urgent modernization or construction of new WWTPs to eliminate extremely serious microbiological threats to water. Indicator bacteria can also successfully be used in research on the effectiveness of ecohydrological biotechnologies used in the environment, such as SSBSs constructed to improve the water quality in rivers and reservoirs. The example is a significant reduction up to 90% of microbial contamination measured by the number of *E. coli* cells into Lake Jelonek in central Poland, as a result of installing an SSBS on the inflow of the watercourse Figs. 24.2 and 24.4D. Due to the increased pressure from anthropogenic factors and the urgent need to protect the aquatic ecosystems and human health in relation to the sudden increase of serious infections caused by pathogenic bacteria, including those resistant to antimicrobials, the widespread use of bacteriological indicators is necessary practice in the modern monitoring of surface waters according to the holistic approach and One Health Principle.

References

Ahmed, W., Hamilton, K., Toze, S., Cook, S., Page, D., 2019. A review on microbial contaminants in stormwater runoff and outfalls: potential health risks and mitigation strategies. Sci. Total Environ. 692, 1304–1321. https://doi.org/10.1016/j.scitotenv.2019.07.055.

Bednarek, A., Szklarek, S., Zalewski, M., 2014. Nitrogen pollution removal from areas of intensive farming-comparison of various denitrification biotechnologies. Ecohydrol. Hydrobiol. 14 (2), 132–141. https://doi.org/10.1016/j.ecohyd.2014.01.005.

Bosman, C., 2009. The Hidden Dragon: Nitrate Pollution From Open-Pit Mines–A Case Study From the Limpopo Province, South Africa. Available from: https://www.imwa.info/docs/imwa_ 2009/IMWA2009_Bosman.pdf.

Bradshaw, J.K., Snyder, B.J., Oladeinde, A., Spidle, D., Berrang, M.E., Meinersmann, R.J., Oakley, B., Sidle, R.C., Sullivan, K., Molina, M., 2016. Characterizing relationships among fecal indicator bacteria, microbial source tracking markers, and associated waterborne pathogen occurrence in stream water and sediments in a mixed land use watershed. Water Res. 101, 498–509. https://doi.org/10.1016/j.watres.2016.05.014.

Chahal, M.K., Shi, Z., Flury, M., 2016. Nutrient leaching and copper speciation in compost-amended bioretention systems. Sci. Total Environ. 556, 302–309. https://doi.org/10.1016/j.scitotenv.2016.02.125.

Codd, G.A., Meriluoto, J., Metcalf, J.S., 2017. Introduction: Cyanobacteria, Cyanotoxins, Their Human Impact, and Risk Management Handbook of Cyanobacterial Monitoring and Cyanotoxin Analysis. Wiley, pp. 3–8. https://doi.org/10.1002/9781119068761.ch1.

Czekalski, N., Gascón Díez, E., Bürgmann, H., 2014. Wastewater as a point source of antibiotic-resistance genes in the sediment of a freshwater lake. ISME J. 8 (7), 1381–1390. https://doi.org/10.1038/ismej.2014.8.

Daims, H., Lebedeva, E.V., Pjevac, P., Han, P., Herbold, C., Albertsen, M., Jehmlich, N., Palatinszky, M., Vierheilig, J., Bulaev, A., Kirkegaard, R.H., Von Bergen, M., Rattei, T., Bendinger, B., Nielsen, P.H., Wagner, M., 2015. Complete nitrification by Nitrospira bacteria. Nature 528 (7583), 504–509. https://doi.org/10.1038/nature16461.

Devol, A.H., 2015. Denitrification, anammox, and N2 production in marine sediments. Ann. Rev. Mar. Sci 7, 403–423. https://doi.org/10.1146/annurev-marine-010213-135040.

Ducci, D., 2018. An easy-to-use method for assessing nitrate contamination susceptibility in groundwater. Geofluids 1–12. https://doi.org/10.1155/2018/1371825.

ECDC EFSA & EMA, 2017. Second joint report on the integrated analysis of the consumption of antimicrobial agents and occurrence of antimicrobial resistance in bacteria from humans and food-producing animals: Joint Interagency Antimicrobial Consumption and Resistance Analysis (JIACRA) Report. EFSA J. 15 (7). https://doi.org/10.2903/j.efsa.2017.4872.

Erisman, J.W., Galloway, J.N., Dise, N.B., Sutton, M.A., Bleeker, A., Grizzetti, B., Leach, A.M., 2015. Vries, nitrogen: too much of a vital resource. WWF Netherlands. https://doi.org/10.13140/RG.2.1.3664.8163.

Font Nájera, A., Serwecińska, L., Szklarek, S., Mankiewicz-Boczek, J., 2020. Characterization and comparison of microbial communities in sequential sedimentation-biofiltration systems for removal of nutrients in urban rivers. Ecol. Eng. 149, 105796. https://doi.org/10.1016/j.ecoleng.2020.105796.

Font Nájera, A., Serwecińska, L., Mankiewicz-Boczek, J., 2021a. Culturable nitrogen-transforming bacteria from sequential sedimentation biofiltration systems and their potential for nutrient removal in urban polluted rivers. Sci. Rep. 11 (1). https://doi.org/10.1038/s41598-021-86212-3.

Font Nájera, A., Serwecińska, L., Szklarek, S., Mankiewicz-Boczek, J., 2021b. Seasonal and spatial changes of N-transforming microbial communities in sequential sedimentation-biofiltration systems—influence of system design and environmental conditions. Int. Biodeterior. Biodegrad. 159, 105203. https://doi.org/10.1016/j.ibiod.2021.105203.

Galloway, J.N., Dentener, F.J., Capone, D.G., Boyer, E.W., Howarth, R.W., Seitzinger, S.P., Asner, G.P., Cleveland, C.C., Green, P.A., Holland, E.A., Karl, D.M., Michaels, A.F., Porter, J.H., Townsend, A.R., Vörösmarty, C.J., 2004. Nitrogen cycles: past, present, and future. Biogeochemistry 70 (2), 153–226. https://doi.org/10.1007/s10533-004-0370-0.

Gamble, T.N., Betlach, M.R., Tiedje, J.M., 1977. Numerically dominant denitrifying bacteria from world soils. Appl. Environ. Microbiol. 33 (4), 926–939. https://doi.org/10.1128/aem.33.4.926-939.1977.

Gao, S., Li, C., Jia, C., Zhang, H., Guan, Q., Wu, X., Wang, J., Lv, M., 2020. Health risk assessment of groundwater nitrate contamination: a case study of a typical karst hydrogeological unit in East China. Environ. Sci. Pollut. Control Ser. 27 (9), 9274–9287. https://doi.org/10.1007/s11356-019-07075-w.

Guo, J., Peng, Y., Wang, S., Ma, B., Ge, S., Wang, Z., Huang, H., Zhang, J., Zhang, L., 2013. Pathways and organisms involved in ammonia oxidation and nitrous oxide emission. Crit. Rev. Environ. Sci. Technol. 43 (21), 2213–2296. https://doi.org/10.1080/10643389.2012.672072.

Holmes, D.E., Dang, Y., Smith, J.A., 2019. Nitrogen cycling during wastewater treatment. Adv. Appl. Microbiol. 106, 113–192. https://doi.org/10.1016/bs.aambs.2018.10.003.

Hu, H.W., He, J.Z., 2017. Comammox—a newly discovered nitrification process in the terrestrial nitrogen cycle. J. Soils Sediments 17 (12), 2709–2717. https://doi.org/10.1007/s11368-017-1851-9.

Jakopanec, I., Borgen, K., Vold, L., Lund, H., Forseth, T., Hannula, R., Nygård, K., 2008. A large waterborne outbreak of campylobacteriosis in Norway: the need to focus on distribution system safety. BMC Infect. Dis. 8 (1). https://doi.org/10.1186/1471-2334-8-128.

Jang, J., Hur, H.-G., Sadowsky, M.J., Byappanahalli, M.N., Yan, T., Ishii, S., 2017. Environmental *Escherichia coli*: ecology and public health implications-a review. J. Appl. Microbiol. 123 (3), 570–581. https://doi.org/10.1111/jam.13468.

Kartal, B., Keltjens, J.T., 2016. Anammox biochemistry: a tale of heme c proteins. Trends Biochem. Sci. 41 (12), 998–1011. https://doi.org/10.1016/j.tibs.2016.08.015.

Korajkic, A., McMinn, B.R., Harwood, V.J., 2018. Relationships between microbial indicators and pathogens in recreational water settings. Int. J. Environ. Res. Publ. Health 15 (12), 2842. https://doi.org/10.3390/ijerph15122842.

Laughlin, R.J., Stevens, R.J., 2002. Evidence for fungal dominance of denitrification and codenitrification in a grassland soil. Soil Sci. Soc. Am. J. 66 (5), 1540–1548. https://doi.org/10.2136/sssaj2002.1540.

Levantesi, C., Bonadonna, L., Briancesco, R., Grohmann, E., Toze, S., Tandoi, V., 2012. Salmonella in surface and drinking water: occurrence and water-mediated transmission. Food Res. Int. 45 (2), 587–602. https://doi.org/10.1016/j.foodres.2011.06.037.

Mankiewicz-Boczek, J., Bednarek, A., Gągała-Borowska, I., Serwecińska, L., Zaborowski, A., Kolate, E., Pawełczyk, J., Żaczek, A., Dziadek, J., Zalewski, M., 2017. The removal of nitrogen compounds from farming wastewater—the effect of different carbon substrates and different microbial activators. Ecol. Eng. 105, 341–354. https://doi.org/10.1016/j.ecoleng.2017.05.014.

Offre, P., Spang, A., Schleper, C., 2013. Archaea in biogeochemical cycles. Annu. Rev. Microbiol. 67 (1), 437–457. https://doi.org/10.1146/annurev-micro-092412-155614.

Papaspyrou, S., Smith, C.J., Dong, L.F., Whitby, C., Dumbrell, A.J., Nedwell, D.B., 2014. Nitrate reduction functional genes and nitrate reduction potentials persist in deeper estuarine sediments why? PLoS One 9 (4). https://doi.org/10.1371/journal.pone.0094111.

Philippot, L., Piutti, S., Martin-Laurent, F., Hallet, S., Germon, J.C., 2002. Molecular analysis of the nitrate-reducing community from unplanted and maize-planted soils. Appl. Environ. Microbiol. 68 (12), 6121–6128. https://doi.org/10.1128/aem.68.12.6121-6128.2002.

Piña-Ochoa, E., Høgslund, S., Geslin, E., Cedhagen, T., Revsbech, N.P., Nielsen, L.P., Schweizer, M., Jorissen, F., Rysgaard, S., Risgaard-Petersen, N., 2010. Widespread occurrence of nitrate storage and denitrification among Foraminifera and Gromiida. Proc. Natl. Acad. Sci. U. S. A 107 (3), 1148–1153. https://doi.org/10.1073/pnas.0908440107Denmark.

Prosser, J.I., 2005. Nitrogen in soils—nitrification. Encyclopedia of Soils in the Environment 4, 31–39. https://doi.org/10.1016/B0-12-348530-4/00512-9.

Risgaard-Petersen, N., Langezaal, A.M., Ingvardsen, S., Schmid, M.C., Jetten, M.S.M., Op den Camp, H.J.M., Derksen, J.W.M., Piña-Ochoa, E., Eriksson, S.P., Peter Nielsen, L., Peter Revsbech, N., Cedhagen, T., van der Zwaan, G.J., 2006. Evidence for complete denitrification in a benthic foraminifer. Nature 443 (7107), 93–96. https://doi.org/10.1038/nature05070.

Rizzo, L., Manaia, C., Merlin, C., Schwartz, T., Dagot, C., Ploy, M.C., Michael, I., Fatta-Kassinos, D., 2013. Urban wastewater treatment plants as hotspots for antibiotic resistant bacteria and genes spread into the environment: a review. Sci. Total Environ. 447, 345–360. https://doi.org/10.1016/j.scitotenv.2013.01.032.

Ryall, B., Eydallin, G., Ferenci, T., 2012. Culture history and population heterogeneity as determinants of bacterial adaptation: the adaptomics of a single environmental transition. Microbiol. Mol. Biol. Rev. 76 (3), 597–625. https://doi.org/10.1128/mmbr.05028-11.

Serwecińska, L., Bednarek, A., Gągała-Borowska, I., Szklarek, S., Kupiec, J.M., Żeleźnik, W., Draczyński, Z., Krucińska, I., Mankiewicz-Boczek, J., 2017. International Symposium Ecohydrology for the Circular Economy, and Nature-Based Solutions towards Mitigation/Adaptation to Climate Change Innovative Method of Improving Water Quality on the Example of Gniezno Stream and Lake Jelonek.

Serwecińska, L., Kiedrzyńska, E., Kiedrzyński, M., 2021. A catchment-scale assessment of the sanitary condition of treated wastewater and river water based on fecal indicators and carbapenem-resistant Acinetobacter spp. Sci. Total Environ. 750, 142266. https://doi.org/10.1016/j.scitotenv.2020.142266.

Sidhu, J.P.S., Ahmed, W., Gernjak, W., Aryal, R., McCarthy, D., Palmer, A., Kolotelo, P., Toze, S., 2013. Sewage pollution in urban stormwater runoff as evident from the widespread presence of multiple microbial and chemical source tracking markers. Sci. Total Environ. 463–464, 488–496. https://doi.org/10.1016/j.scitotenv.2013.06.020.

Steele, J., Griffith, J., Noble, R., Schif, K., 2017. Tracking Human Fecal Sources in an Urban Watershed during WetWeather.

Szklarek, S., Wagner, I., Jurczak, T., Zalewski, M., 2018. Sequential sedimentation-biofiltration system for the purification of a small urban river (the Sokolowka, Lodz) supplied by stormwater. J. Environ. Manag. 205, 201–208. https://doi.org/10.1016/j.jenvman.2017.09.066.

The Council of the European Communities, EU, Directive 91/676/EEC concerning the protection of waters against pollution caused by nitrates from agricultural sources (Consolidated version 11 December 2008). Off. J. Eur. Commun. L. 375, 1991, 1. Available from: https://eur-lex.europa.eu/legal-content/EN/TXT/?uri=CELEX%3A01991L0676-20081211.

Tiedje, J.M., 1988. Ecology of Denitrification and Dissimilatory Nitrate Reduction to Ammonium Environmental Microbiology of Anaerobes, 179–244.

USEPA, 1986. Ambient Water Quality Criteria for Bacteria—1986. Available from: https://www.epa.gov/sites/default/files/2019-03/documents/ambient-wqc-bacteria-1986.pdf.

USEPA, 2005. Microbial Source Tracking Guide Document. Available from: https://nepis.epa.gov/Exe/ZyPURL.cgi?Dockey=2000D20V.TXT.

USEPA, 2004. Water Quality Standards for Coastal and Great Lakes Recreation Waters, 69 (220). Available from: https://www.govinfo.gov/content/pkg/FR-2004-11-16/pdf/04-25303.pdf.

Wang, S., Lin, X., Yu, H., Wang, Z., Xia, H., An, J., Fan, G., 2017. Nitrogen removal from urban stormwater runoff by stepped bioretention systems. Ecol. Eng. 106, 340–348. https://doi.org/10.1016/j.ecoleng.2017.05.055.

WHO Antimicrobial Resistance, 2017. Fact Sheets on Sustainable Development Goals: Health Targets. World Health Organization.

WHO World Health Organization, 2018. Food and Agricultural Organization of the United Nations, World Organisation for Animal Health. Monitoring Global Progress on Addressing Antimicrobial Resistance: Analysis Report of the Second Round of Results of AMR Country Self-Assessment Survey.

Zalewski, M., 2012. Blue-green city for compensating global climate change. Parl. Mag. 350, 2–3.

Zalewski, M., Arduino, G., Bidoglio, G., Junk, W., Cullmann, J., Uhlenbrook, S., Xia, J., de Leaniz, C.G., Rowinski, P.M., Vörösmarty, C.J., Chicharo, L., 2018. Low cost, nature-based solutions for managing aquatic resources: integrating the principles of Ecohydrology and the Circular Economy. Ecohydrol. Hydrobiol. 18 (4), 309–310. https://doi.org/10.1016/j.ecohyd.2018.12.001.

Zhou, Z., 2015. A Global Assessment of Nitrate Contamination in Groundwater. International Groundwater Resources Assessment Centre (IGRAC). Available from: https://www.researchgate.net/profile/Collins-Tay/post/Can-anybody-suggest-some-literature-on-the-contamination-and-pollution-of-groundwater-with-nitrate/attachment/5f19a8c1c82a7c0001a6029f/AS%3A916479386087427%401595517121692/download/A+Global+Assessment+of+Nitrate+Contamination+in+Groundwater.pdf.

Chapter 25

Smart blue–green city of the future

Iwona Wagner[1], Kinga Krauze[2], Renata Włodarczyk-Marciniak[2] and Maciej Zalewski[2]
[1]*University of Lodz, Faculty of Biology and Environmental Protection, UNESCO Chair on Ecohydrology and Applied Ecology, Łódź, Poland;*
[2]*European Regional Centre for Ecohydrology of the Polish Academy of Sciences under the auspices of UNESCO, Łódź, Poland*

IMAGE CREDIT: ELCHE, SPAIN. IMAGE CREDIT: M. ZALEWSKI.

1. The conditions for coexistence of the city and nature

Cities will soon become home to more than 70% of the world's human population. Thus the living conditions they provide will shape the quality of life for the majority of humanity. In recent decades, these have been increasingly compromised

Transdisciplinary Ecohydrology. https://doi.org/10.1016/B978-0-443-41596-8.00037-6

by accelerating impacts of global climate change. They superimpose on often unsustainable urbanization, enhancing urban droughts and heat islands, generating flash floods, and resulting in health risks.

At the same time, cities are major contributors to the global environmental footprint, consuming over two-thirds of the world's energy and accounting for more than 70% of global CO_2 emissions. This makes them central to both the problem and the solution. While they contribute significantly to environmental damage, cities with their social, human, and economic capital can also lead efforts toward better urban living conditions, more sustainable practices, protection of critical habitats and species, and reduction of their own global impact.

The 20th and 21st centuries have seen numerous attempts to envision the city of the future. One such concept is the *Smart City*, which has been described in the literature across multiple dimensions of urban functioning, including industry, economy, information technology, governance, transportation, education, public participation, technical infrastructure, and the environment. Various aspects of these sectoral and differently angled approaches have been brought together into one holistic concept by Giffinger et al. (2007). His report proposed a smart city ranking, based on a comprehensive catalog of indicators which have unified various sectoral perspectives into one more tangible concept. The smart elements include: *economy, people, governance, mobility, environment*, and *living*. A similar concept, differing mainly with the indicators, was proposed by Boyd Cohen, author of *the so-called Smart City Wheel*. He divides smart cities into three types, distinguished by the way they embrace technology and development in the transformation process - moving from tech-company driven, through city government driven, to, finally, citizen driven. The environment of the Smart City is described by attractive natural conditions (climate, green space etc.), pollution control, resource management and efforts toward environmental protection. This can be achieved thanks to a variety of measures, such as: energy efficient technologies, citizen awareness, smart governance, and smart mobility, among others.

As far as the *Smart City* concept looked at the city through the lens of overall indicators of technology and environmental impacts, the *Compact City (City of short distances)* concept (Dantzig and Saaty, 1973) looked closer at the city landscape and searched for specific recommendations related to the city's spatial organization. *Compact City*, driven mainly by the need to reduce energy and resources consumption and pollution, is a city with relatively high urban density and mixed uses, which facilitate better communication - both in terms of transportation (efficient-energy and shorter time) as well as social sustainability (social connections, access to services and nature). The concept has strongly influenced urban planning theory and practice, which in the early 20th century were always in strong opposition to building and population density.

The concepts discussed above highlight the fundamental importance of a clean environment and nature as integral elements of a healthy urban system. However, they do not sufficiently explore the internal dynamics and needs of natural systems themselves. To achieve the desired quality of life in cities - where natural systems face harsh conditions yet inhabitants rely heavily on ecosystem services - a deeper understanding of ecosystem management options is essential. This requires insight into ecosystem functioning, resilience, interactions with urban green infrastructure, and capacity for self-maintenance - all of which determine their ability to provide key regulatory and cultural services within highly anthropogenically stressed urban environments. A functional approach looking into the ecosystem processes and the tool for their management is rooted in ecohydrology (Zalewski, 2000a) which, in cities, seeks to couple spatial planning and urban greenery with the dynamics of the urban water cycle (Wagner and Breil, 2013). A smart blue-green city, therefore, as proposed in the title of this chapter, recognizes environmental processes as a city management tool, to improve both nature and social systems resilience. It also integrates ecosystems into city management plans, based on deeper consideration of their ecological foundations.

A city based on such an approach does not evolve under the paradigm of dominating nature (the Socio-centric–Mechanistic Paradigm, as articulated during the Industrial Age by the French philosopher La Mettrie), but rather within the Evolutionary–Ecosystemic Paradigm (Zalewski, 2021), which seeks to enhance interactions between humans and nature. Within this framework, urban ecosystemic processes - such as water and nutrient cycling, energy flow, pollination, and gene flow - aim to mirror those found in natural ecosystems as closely as possible.

2. Water in cities

Water is not only the resource, but also the foundation of growth and development. The quality of life has always been inseparably linked to water, a relationship insightfully described by Jean Delumeau (Delumeau, 2000):

The word "paradise" is an old Persian term—paridaiza - which later took the form paradeisos in Greek. It refers to the image of a "garden surrounded by walls" protecting it from the hot desert winds. An image of happiness among greenery, abundant water full of flowers and vegetation.

TABLE 25.1 **Water-oriented objectives of the city at subsequent stages of the development of the water and city relationship concept.**

Stage/Concept	Motivation for action	Urban water objectives	City type	Beneficiary
Water supply city	Provision of resources	Water supply security	Linear city	Human
Sewered city	Protection against threats	Wastewater disposal, protection of public health		Human
Drained city		Flood protection		Human
Waterways city, Environmental protection and rehabilitation	Aesthetics and quality of life	Social amenities and environmental protection for urban space quality		Human, nature
Water cycle city, Urban ecohydrology	Responding to threats resulting from environmental degradation	Mitigation of water scarcity and stormwater flooding	Regenerative city	Human, nature
Water sensitive city, Urban ecohydrology		Adaptation to climate change		Human, nature
Water smart city, Urban ecohydrology, Ecosystem-based approach	Maintaining the quality of the natural environment as fundament for provision of ecosystem services for good living	Preserving ecosystems' capacity to adapt to disturbances while maintaining their structure, functions and potential to deliver ecosystem services		Nature, human
Urban ecohydrology, Regenerative city Enhancement of urban sustainability WBSR		Mutual regulation of ecohydrological processes to improve environmental quality, quality of life, and reduce city operating costs		Nature, human

Modified from: Wagner, I. (2025). Ecohydrology and ecosystem approach in regenerative cities [in Polish: Ekohydrologia i podejście ekosystemowe w miastach regeneracyjnych]. In Regeneration of the City [Regeneracja miasta] (pp. 61–80).

Achieving such paradisiacal living conditions in a future shaped by climate change and exponential population growth (over 8 billion people now) may be extremely difficult. The dynamics and availability of water are becoming increasingly unpredictable and unstable. In particular in the urban context, water emerges as a decisive factor limiting the functioning of both natural and social systems, as well as the prospects for overall sustainability.

The relationship between cities and water has fundamentally changed over time, alongside the shifting social, political, normative and regulatory expectations that influenced decision-making in water resource management. Brown et al. (2009) identified six transitional stages in this relationship: the *Water Supply City, the Sewered City, the Drained City, the Waterways City, the Water Cycle City,* and *the Water Sensitive City* (Table 25.1). The first three historical stages reflect a paradigm in which nature was subordinated to meet human needs and to address environmental problems. In contrast, more recent stages emphasize the primacy of the environmental context in shaping urban water strategies. Cities are turning towards their rivers, which become city-forming features, both aesthetically and functionally integrated into the cityscapes (Januchta-Szostak, 2020). This is also when such approaches as urban ecohydrology (Wagner and Breil, 2013; Zalewski and Wagner, 2005) and ecosystem-based approaches, including blue–green infrastructure (Melville-Shreeve et al., 2018) and nature-based solutions (NBS) (Towards an EU Research and Innovation Policy Agenda for Nature-Based Solutions and Re-Naturing Cities, 2015) have gained prominence, reshaping the discourse on urban sustainability. Through these approaches, urban water management expands its toolbox, moving beyond purely engineering measures toward ecosystem understanding and management. Stormwater is redefined as a valuable resource, not a threat. Greenery and freshwater ecosystems (e.g., river, lake, pond, reservoir, coastal zone) functions are integrated with technological infrastructure (e.g., systems of water purification and water supply, stormwater and municipal sewage management and purification), designed in accordance with the newest technologies (smart city Gawrońska-Nowak et al., 2025). Holistic thinking about cities also involves a shift in economic and cognitive paradigms towards an "economy of values," which emphasizes solidarity and human responsibility rather than narrow economic rationality (Hausner, 2024). This approach is closely linked to the concept of a regenerative economy and regenerative cities (Markowski, 2023), aiming to close the loops for sustainable development.

Stormwater and meltwater management in the city landscape is typically the primary concern, as many cities have suffered the negative consequences of overly engineered approaches. Traditional stormwater systems were designed to remove water from the city as quickly as possible, directing it toward the nearest river, lake, reservoir or coastal waters, have lost their relevance in today's context for two key reasons. First, increasing climate unpredictability and frequency of extreme rainfalls overload drainage systems, which become unreliable. Second, stormwater drainage deprives the city of an essential source of water, necessary for ecosystem services such as water supply, microclimate stabilization, urban greenery growth and many others. The restoration of water cycling and the reuse of water in the urban landscape, advocated by urban ecohydrology, becomes therefore a tangible value. Ecosystems, which both rely on and regulate water, are increasingly seen as living infrastructures that operate alongside technological systems and co-function with technological systems to sustain the urban metabolism. However, regenerating the water cycle must become a central objective of sustainable urban planning and management.

The tools, costs, and effort required to implement such an approach vary across different parts of the city and depend largely on the degree of landscape modification arising from the intensity of urbanization.

3. Abiotic–biotic regulation in the urbanization gradient

Urbanization transforms landscapes - alters the initial terrestrial and water ecosystems mosaic by their removal, fragmentation and/or intensive maintenance. Terrestrial ecosystems are transformed by roads, housing and industrial developments, aquatic ones by traditional technical "flood control measures" (e.g., weirs, banks stabilization, channelization), severe pollution (e.g., combined sewer overflows) or even elimination (e.g., underground channelization). From the urban ecohydrology perspective (Wagner and Breil, 2013), this means not only the physical degradation of the "visible nature" but, most importantly, disruptions in ecosystems' processes. Flows of energy, water and matter - which functioned in a continuity inbetween unmanaged ecosystems are interrupted and rerouted into unsustainable, human-driven pathways. These changes intensify as urbanization accelerates, transforming the processes that regulate ecosystem functioning.

The interaction between disturbance (e.g., urbanization), biodiversity, and the efficiency of ecological processes, has been firstly introduced by Connell's intermediate disturbance hypothesis (Connell, 1978). He argued that intermediate levels of disturbances actually stimulate biodiversity and ecosystem efficiency, contrary to low or high levels. Zalewski and Naiman (1985) further developed this idea, putting forward the concept of a continuum of abiotic–biotic control of river ecosystem dynamics. This concept suggests that aquatic ecosystems are regulated through two primary mechanisms. The first is abiotic regulation, which occurs when the frequency, duration and intensity of physical disturbance is high, causing biotic processes and structure of ecosystems to be subordinated to abiotic limiting factors. The second is biotic regulation, which develops under stable and predictable environmental conditions. A similar relationship has been described in the seasonal regulation pattern of temperate rivers. During the cold season, river ecosystem dynamics is regulated by abiotic processes (e.g., rainfall, snowmelt, discharge dynamics), whereas in the growing season, biotic processes become the key drivers of ecosystem dynamics (e.g., biological self-purification, primary production, energy flow regulation through trophic cascade (Wagner and Zalewski, 2016).

This concept may also apply, under some conditions, to urban rivers. E.g., in the case of through-flowing rivers, with catchments much larger than the city, the location of the city along the river continuum (e.g., mountain, lowland, coastal city) has a decisive importance for effective water management. It determines not only risks but also opportunities for getting benefits of ecosystem services. In the case of small urban rivers, having most of their courses located within urban areas, and thus heavily modified and hydraulically overloaded, the typical abiotic–biotic control pattern becomes disrupted. Abiotic factors tend to dominate along the entire river course, reducing the role of biota in regulating river dynamics. The primary challenge of the management strategy is to stabilize the river's hydrological regime and subsequently restore its biodiversity.

Analogy can be also drawn for terrestrial ecosystems within the city, where patches of different urbanscapes form a gradient of regulatory conditions between two extremes - the suburbs and the city center. In city centers, typically dominated by gray infrastructure, ecohydrological processes are controlled primarily by abiotic factors such as water stress, elevated temperature, pollution, space constraints and the discontinuity of ecological features. Moving toward the suburban areas, biotic regulation of ecohydrological processes gradually becomes more prominent.

This continuum has fundamental practical consequences for NBS implementation and for cities management. Stronger abiotic control requires creating environmental conditions suitable for vegetation and associated fauna; this usually entails a greater investment in enabling/substitutive NBS (e.g., those with surrogate soils or reshaped river bank structures). Here, the aim of ecohydrology is not only to strengthen the biotic regulatory pathway, but also to achieve a balance between spatial and economic development and welfare of nature. Urban ecological systems under stronger biotic control exhibit greater integrity and self-regulatory potential. Here, the role of NBS is primarily to support and sustain nature (e.g., small retention in forests, mid-field ponds in agricultural areas around cities, restoration of river valleys). In water management, stabilizing the water cycle and enhancing stormwater management in the urban catchment are fundamental steps. By easing abiotic regulation, these measures reduce hydraulic and chemical pressures on rivers, stabilize their functioning, and enhance prospects for maintaining water quality and achieving effective ecological restoration.

Roe et al. (2005) identified three most common management styles: self-regulatory, bandwidth, and high-reliability, which apply well to the context of abiotic–biotic control in urbanscapes and to the management of the delivery-demand ratio for ecosystem services.

Self-regulatory management applies to the outskirts of cities, composed of relatively high-quality ecosystems, such as suburban agricultural lands, urban forests, free-flowing rivers and designated nature conservation areas. Its key management target is maintenance of as broad a spectrum of ecosystem services as possible through the preservation or active conservation of ecosystems, i.e., sustaining biodiversity, as well as water and matter cycling [Service Providing Areas—SPA (Syrbe and Walz, 2012)]. Although such areas are rather scarce in an urbanized landscape, they are nonetheless critically important due to this role of supporting service-benefiting areas (SBA).

Bandwidth management is associated with developed suburbs, mostly single-family urban housing and some more centrally located blocks of flats. There is a high demand for cultural, regulatory, and supporting services in these areas, but their delivery potential is already suppressed. The target of bandwidth management is therefore restoration that keeps the socioecological system stable within specified bandwidths of water quality, stormwater management and flood control, climate regulation, goods provision as well as an attractive and healthy city space.

High-reliability management refers to densely populated city zones (e.g., downtowns). Their landscapes are dominated by gray infrastructure, while green and blue areas are scarce, disconnected and simplified due to strong technical and maintenance interventions. Management approaches in these areas rely primarily on conventional engineering and technological measures, which can allow only a very narrow margin for error. The NBS management target is to enable nature. In these areas, the potentially high economic and societal costs exclude the use of any trial-and-error methods, while constant stress may preclude the delivery of multiple services at expected loads. To secure environmental safety and livable conditions for residents, such management is optimized for continuous peak load performance of the entire system, with only very few selected ecosystem services supported at high costs (e.g., continuous water provision) (Krauze and Wagner, 2019).

From the perspective of sustainable city management in the context of the climate crisis, cities need to develop long-term and precautionary strategies to ensure that high-reliability management zones constitute only a small proportion of the urban area. Depending on the choice of the dominating management style, the corresponding NBS may or may not be resilient to change. In some cases, their cost-efficiency may be attainable only in the long term. The relationship between the level of disturbance in urban ecosystems and the role of NBS is illustrated in Fig. 25.1 The figure shows how management targets evolve from resistance and impact avoidance in mature ecosystems (toward suburbs and natural systems) toward compensation and the establishing of novel systems in highly disturbed environments (in city centers and densely developed areas). Correspondingly, the role of NBS shifts from preserving existing system functions to enabling the creation of new ecological processes.

4. Ecosystem resilience and its management and social consequences

Recognizing the natural system as a critical (infra)structural (blue–green infrastructure - BGI) element of a city requires an understanding of its complex internal structure, the processes occurring within it, and the boundary conditions necessary to ensure its proper functioning (Wagner, 2022). In natural ecosystems, structure and function emerge from the long-term coexistence of the species with their environment and with one another, resulting in mutual adaptation. Such systems reproduce "order from order" (Schrödinger, 1998) resulting in continuous regeneration of biological structures. In cities, however, except for the relatively rare cases of ecosystems whose functioning remains close to natural conditions (e.g., urban forests or unmodified river sections), most ecosystems are transformed or heavily altered by direct and indirect human activities. Consequently, they often lose the capacity to maintain their structure and functions - that is, to reproduce "order from order" - when exposed to various disturbance factors.

An ecosystem's response to a disturbance is described by two features: *resistance* and *resilience*. Resistance refers to the ability of an ecosystem to absorb disturbance without changing the structure and functions, until it reaches a tipping point and flips to a new, completely different regime (Holling, 1973). The later the response occurs, the higher the ecosystem's resistance. Resilience marks the systems able to return to their pre-disturbance state once the disturbance ceases. The higher is the recovery capacity the higher is the resilience. In both cases if the disturbance persists, intensifies over time, and its impact is too severe, the ecosystem may not return to its former state even after the disturbance has

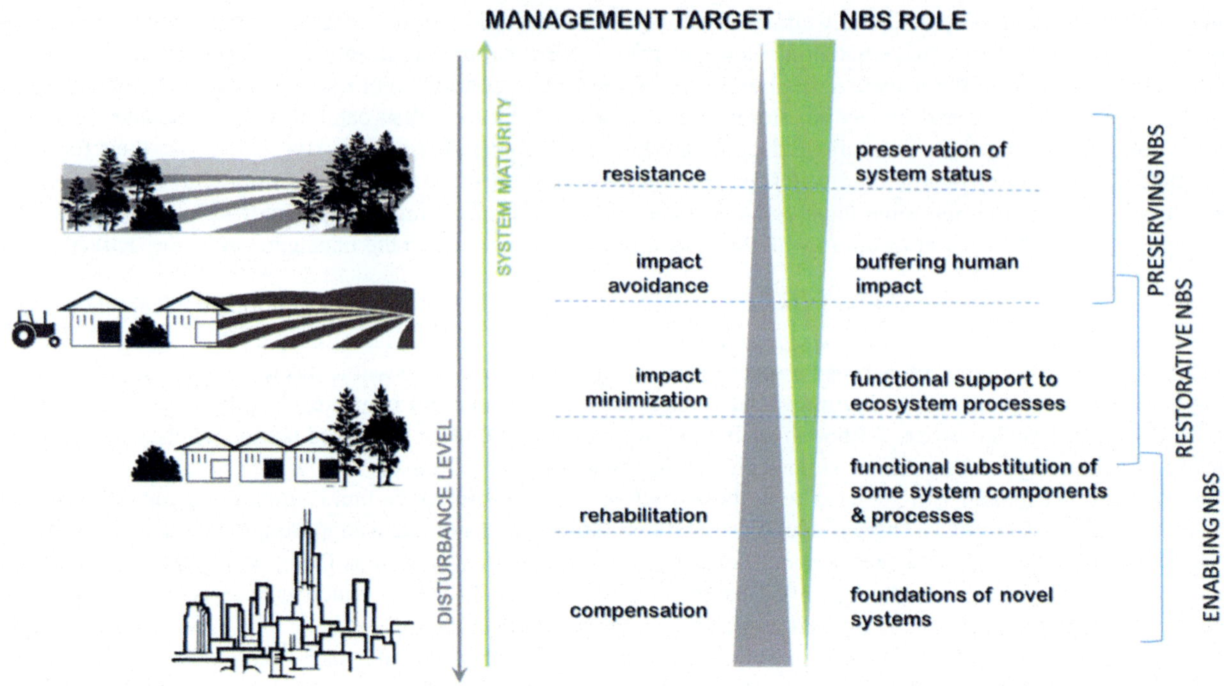

FIGURE 25.1 Conceptual framework illustrating the role of Nature-based Solutions (NBS) across gradients of urban disturbance and ecosystem maturity. Conceptual framework illustrating the role of Nature-based Solutions (NBSs) across gradients of urban disturbance and ecosystem maturity. The figure shows how management targets shift from resistance to compensation along increasing levels of human disturbance - from natural to highly urbanized environments. Correspondingly, NBS roles evolve from preserving system status and buffering human impacts (preserving NBS), through supporting and substituting ecosystem functions (restorative NBS), to building foundations for novel systems (enabling NBS). The framework highlights how different forms of NBS can maintain or restore ecosystem processes depending on the degree of urban transformation. *Modified from: Krauze, K., & Wagner, I., 2019. From classical water-ecosystem theories to nature-based solutions — Contextualizing nature-based solutions for sustainable city. Science of the Total Environment, 655, 697–706. https://doi.org/10.1016/j.scitotenv.2018.11.187.*

ended. The disturbance then drives the system to a threshold, beyond which it no longer returns to its original regime. The ecosystem thus permanently loses its initial structure and function, giving way to the dominance of a new regime, which has environmental, management and social consequences (Fig. 25.2).

Structural changes of ecosystems weaken their resilience (Holling, 1973) and limit adaptive responses to impacts. Co-occurring, especially synergistic disturbances are particularly harmful. They make ecosystems more vulnerable to factors that previously could have been absorbed without any harm. The city is an example of an area where such co-occurring pressures accumulate - disturbance factors are not only droughts, floods or heat waves, but also soil pollution and compacting, destroying of seeds banks, habitats fragmentation, noise and light pollution, introduction of alien species, reduction or elimination of green and water-dependent areas, river channelization and many others. Furthermore, many of city ecosystems are so-called *novel ecosystems* (Hobbs et al., 2009), i.e., ecosystems that develop within human-created habitats. These include various elements of urban blue–green infrastructure, such as green roofs, parks, street greenery, rain gardens, artificial rivers, constructed wetlands (e.g., SSSB described in earlier chapters). Such systems often replace earlier removed greenery, especially in central and more developed parts of the city (high-reliability management zones), or are designed to support specific ecosystem processes (e.g., water purification by SSSB). However, on contrary to the natural systems, they may lack or exhibit only limited self-regulatory potential thus require energy (finances, human effort, equipment, time) for their management. Despite this, novel ecosystems play a fundamental role in supporting gray infrastructure and improving urban quality of life. Their design should therefore emulate, as closely as possible within given constraints, the structural and functional characteristics of natural ecosystems. Moreover, their management must account for their heightened vulnerability and instability.

Harnessing the potential of ecosystems as active management tools in cities therefore depends on reinforcing the role of their biotic components. Achieving this involves three major directions in integrating ecosystems into urban management: (i) shifting from a structural to a functional understanding of ecosystems, (ii) enhancing biological performance under multiple stresses to maintain their ecological integrity, and (iii) assuming responsibility for their maintenance in order to support or enable their functioning.

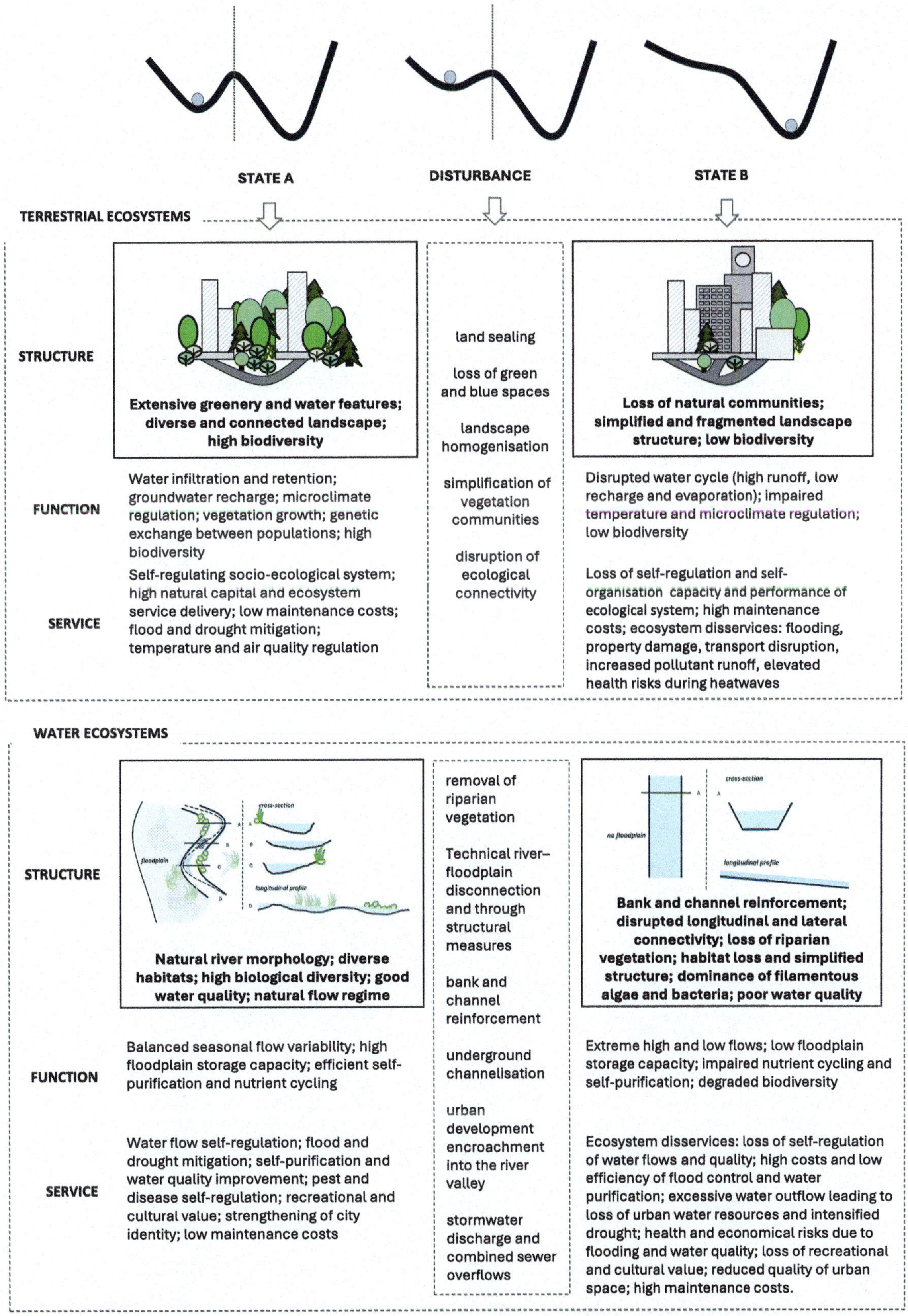

FIGURE 25.2 Conceptual model of transitions in urban terrestrial and aquatic ecosystems under unsustainable urbanization. The diagram illustrates how terrestrial (landscape) and aquatic (river) systems shift from a healthy, self-regulating state (A) to a degraded state (B) once a threshold is crossed due to disturbance from unsustainable urban development. Structural, functional, and service changes are shown for both systems, highlighting the loss of biodiversity, disruption of ecological processes and decline in ecosystem services such as water regulation, purification and climate moderation.

5. Blue-green network concept in Łódź

One of the most important tools for protecting or incorporating greenery into cities is urban planning, which establishes the rules governing the allocation and distribution of all infrastructure, including blue–green infrastructure, within urban space.

The role of different levels of planning documents is to ensure the sufficient vegetation cover, connectivity between green and blue–green areas, and preserve or restore their structure (e.g., native species, biodiversity, naturalness) and functions (e.g., water retention and alimentation, flood control, temperature stabilization). Connectivity between green areas - both within the city and with surrounding natural systems, not only enhances benefits such as city ventilation or water throughflow but also promotes biodiversity. In turn, increased biodiversity builds up resilience of urban natural systems to stress, thereby stabilize ecosystem functions.

An example of such an approach is the concept of the Blue–Green Network Concept (Zalewski et al., 2012) (BGNC, Fig. 25.3, Table 25.2) in the city of Łódź. Łódź, a city with industrial roots and a complex history of transformation, possesses unique natural potential (Kronenberg et al., 2017). As much as 60% of the city's area is green space, yet only 15% of it is formal greenery - that is, planned and legally protected. The vast majority, around 85% of green areas, is informal greenery, including ruderal areas, street greenery or brownfields (Sikorska et al., 2020). Also green space coverage is notably limited in the city center - where it accounts for only 15% - in stark contrast to the outer parts of the city, where green areas constitute 63% of the urban landscape (Łaszkiewicz et al., 2021). Informal greenery, despite representing substantial natural capital with high ecosystem potential, remains unprotected and susceptible to urban development, often perceived simply as "investment reserve areas."

BGNC was developed within the transdisciplinary EU project SWITCH and implemented into the city strategy in 2010 (Rada Miejska w Łodzi, 2010). It builds upon and extends previous studies of the natural features and green architectural elements of Łódź and its region. The concept takes advantage of the city's location at the boundary of two first-order

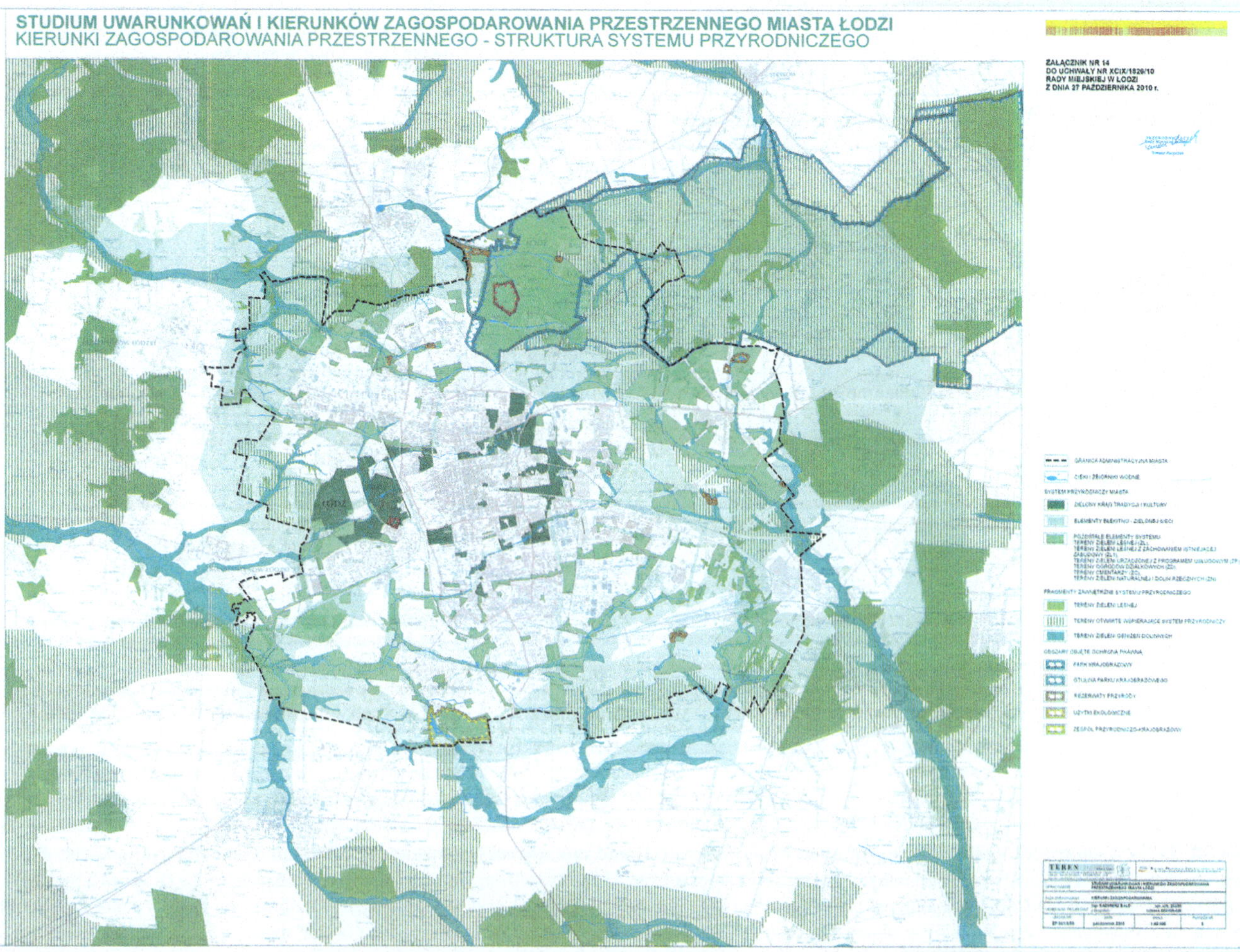

FIGURE 25.3 **The blue-green network** (Zalewski et al., 2012, Rada Miejska w Łodzi, 2010).

TABLE 25.2 Expected benefits of implementation of BGN in Lodz.

Expected benefits of implementation of BGN in Lodz		
Improvement of residents' quality of life and health	Improvement of environmental quality and reduction of management costs	Increasing city attractiveness and promoting sustainable development
• Increasing access to green areas for residents of all districts and neighborhoods, ensuring environmental justice • Maintaining quality of life under progressing climate change and urbanization • Providing opportunities for daily physical and mental regeneration through direct contact with nature, without the need for car transport • Creating an attractive network of pedestrian and cycling routes supporting sustainable mobility and recreation • Reducing the risk of developmental disorders in children, asthma, and allergic diseases • Reducing social costs associated with diseases	• Improving the microclimate, air and water quality and biodiversity • Reducing the risk and costs of flooding during periods of intense rainfall through enhanced landscape retention of stormwater • Supporting the development of urban vegetation and reducing its maintenance costs • Reducing overload of combined and stormwater sewer systems • Improving the operation and cost-efficiency of wastewater treatment plant • Limiting negative impacts on rivers during heavy rain events • Connecting green areas within and around the city to improve the integration of ecological processes (ecological corridors, biodiversity, urban ventilation) • Enhancing ecosystems resilience to anthropogenic pressures	• protection and revitalization of cultural heritage: Enriching the concept of integrated revitalization by incorporating natural elements historically linked to the city's identity alongside architectural and social components • Improving aesthetics, especially in degraded areas • Enhancing Łódź's attractiveness as a residential area • Contributing to the retention and attraction of professionals, creative individuals, and investment capital, supporting urban growth and regional competitiveness • Creating new jobs related to the blue–green network system • Improving ecological and public safety • Fostering social integration • Creating the conditions for sustainable development in line with the principles of the UN agenda 21and MGD

watersheds, originally traversed by approximately 20 small streams. It proposes using existing rivers and green areas net as the foundation of the functional, economic, and spatial organization of urban space.

The Blue–green Network (BGN) represented the first attempt to implement NBS, tailored to local challenges and capacities, to enhance stormwater purification, infiltration, and retention, thereby strengthening ecohydrological regulatory pathways within the city. Through transdisciplinary research, the challenges were identified (Wagner and Zalewski, 2009, 2012), including: dense urban development within catchments, pollution of city rivers and reservoirs, river channelization and associated ecosystems degradation, high sludge production in overloaded wastewater treatment plants, health risks arising from the degraded environment, the need for innovative urban planning, and institutional barriers limiting integrated management (Krauze and Włodarczyk-Marciniak, 2018). Integrated science also contributed to their solving, e.g., studies on the hydrological effects of urbanization provided a basis for developing water budgets and mathematical models for stormwater management; Research on nutrient transport and the distribution of organic pollutants informed new biofiltration designs and indicators for monitoring river degradation; vegetation mapping and river habitat assessments guided rehabilitation plans, including the creation of the Sokołówka River Park (Wagner and Zalewski, 2012).

Further studies optimized sewage sludge reuse for biomass production, linking wastewater management with renewable energy strategies. Research on the health impacts of urban green areas (Kupryś-Lipińska et al., 2009) supported the incorporation of the "Blue–green Network" into a city master plan, aimed at promoting a healthier urban environment. Investigations into stormwater as a driver of urban revitalization (Wagner and Zalewski, 2009) provided new directions for integrated spatial planning. Finally, institutional analyses (Krauze and Włodarczyk-Marciniak, 2018; Wagner et al., 2011) contributed to the establishment of a governance framework for Integrated Urban Water Resources Management (IUWRM), identifying decision-making mechanisms and priority actions for the Łódź water strategy.

6. Transdisciplinary science and social participation

Strengthening ecological literacy is fundamental to fostering public support for blue–green and nature-based concepts in urban environments. Also, in Łódź, the implementation and long-term success of the these approaches relies on active engagement of local communities and the integration of diverse forms of knowledge in management and planning. Such transdisciplinary collaboration bridges natural and social sciences and urban planning and governance, translating the concept into urban practice that is ecologically grounded, socially legitimate and aligned with the city's cultural and institutional context. At the same time, analyses of social preferences help identify priority areas for environmental education and reveal how blue and green solutions are valued in relation to cultural heritage urban attractiveness and the provision of livability-regulating services (Włodarczyk-Maricniak and Krauze, 2023; Włodarczyk-Marciniak et al., 2020, 2024).

For instance, studies on the preferences of residents of tenement houses in city centers have shown limited awareness of the potential of courtyards to contribute to blue-green infrastructure. The findings indicate that the concept of blue-green courtyards is largely absent from residents' consciousness. While greenery is generally welcomed, preferences lean strongly toward low-maintenance, unobtrusive forms - such as border plantings - reflecting the priority given to car parking and everyday comfort in tenement courtyards. Blue infrastructure solutions are often rejected due to concerns about safety, convenience, and disruption of established habits (Włodarczyk-Marciniak et al., 2024b). Limited environmental awareness can also have far-reaching social and spatial consequences. A poor understanding of how urbanization affects ecosystems may lead younger generations to favor suburban living, contributing to urban sprawl, the loss of green spaces and higher transport-related emissions.

Ultimately, research conducted in Łódź (Włodarczyk-Marciniak and Krauze, 2023) emphasizes how access to ecological knowledge is essential for nurturing a deeper environmental consciousness among urban residents. Education significantly influences individuals' ability to recognize and evaluate ecosystem services, particularly those provided by informal blue-green infrastructure, such as brownfields. Because such services are not as visually or tangibly apparent as esthetic or recreational benefits, their value often goes unnoticed by those without ecological training, depriving the informal greenery of local protection and stewardship. Other studies similarly emphasize that environmental understanding and awareness of ecological risks are fundamental for promoting environmentally responsible behavior (Yoon et al., 2021). Through education, individuals internalize norms and attitudes that guide daily decisions and shape collective values, reinforcing behavioral shifts toward sustainability (Geiger et al., 2019).

Yet knowledge alone is insufficient. Weak community cohesion and low levels of social trust often diminish residents' willingness to take part in collective initiatives, leading to a preference for utilitarian over communal or ecological improvements in urban areas. Under these conditions, the benefits offered by blue-green infrastructure–such as stormwater retention, mitigation of urban heat, enhanced aesthetics, and biodiversity support—are frequently overlooked or undervalued. Research shows that active and inclusive participation of residents is essential to ensure that blue–green solutions are socially accepted and sustained over time (Hospers, 2017; Lamond and Everett, 2019). When citizens are actively involved in environmental decision-making, their sense of belonging and responsibility is reinforced, civic competences are strengthened, and mutual trust between local authorities and residents grows (Hartz-Karp and Marinova, 2021; OECD, 2020; Skidmore et al., 2006).

Public participation in urban affairs, particularly those involving green spaces remains selective and largely reactive. Residents tend to mobilize when their immediate interests are affected, e.g., trees are cut or neighboring greenery redeveloped or built over. This type of involvement is largely reactive rather than proactive. It stems from both a lack of environmental literacy and a deep-rooted distrust in participatory processes, which are often perceived as superficial and ineffective - mere formalities with no real impact on decisions. To strengthen inclusiveness, diverse outreach strategies - both digital and in-person - are essential to ensure that a wider range of residents are heard and included.

Taken together, these insights point to the urgent need for integrated civic and environmental education initiatives that explicitly communicate the connections between ecosystem health, quality of life, and urban resilience. Only through accessible and inclusive educational programs can awareness be raised, trust strengthened, and acceptance of BGI-based interventions broadened. Empowering citizens with ecological knowledge and involving them in codesign processes is thus a cornerstone for cultivating a shared sense of ownership and implementing solutions that support both human well-being and the urban environment as a whole.

7. Blue–green city of the future as a socioecological system

The *smart, blue–green city of the future* requires transdisciplinary management that integrates ecological, economic, and social dimensions. Urban challenges are increasingly recognized as "wicked problems" (Head, 2010)-complex, interconnected, and resistant to simple solutions. In cities, these problems arise from spatial constraints, the concentration of assets, conflicting development priorities, and a limited understanding of the role of ecosystems in urban governance. Wicked problems emerge when solving one issue inadvertently generates another: urban expansion increases water and energy demand, while reducing the capacity of ecosystems to provide essential services such as clean air, water, and recreation. Declining living conditions fuel further sprawl, leading to economic, social, and ecological crises. Efforts to protect water and green spaces can mitigate these pressures, yet they may also challenge short-term energy and economic priorities, revealing deep interdependencies across urban systems.

Effective ecosystem-based management of cities requires not only understanding primary drivers - such as demography, economics, politics, social structures, and culture - but also a profound knowledge of ecosystem functions, responses, and resilience. Only through this comprehensive insight can governance become adaptive, conscious, and responsive to dynamic environmental and social change.

Achieving harmony between human needs and environmental capacity - grounded in compromise and cobenefits - is fundamental to the health of both cities and the planet. This transformation demands integrated planning across four interconnected domains:

- Economy: shifting from growth-driven to resource-efficient systems, embracing a circular, closed-loop model that restores the circularity of water in cities.
- Engineering: transitioning from substitution to synergy by designing infrastructures that work *with* nature, enhancing collaboration between natural and technological systems.
- Environment: moving from protection to the support and use of ecosystem processes, fostering functionality and resilience in urban nature.
- Science and Education: evolving from disciplinary silos to integration, restoring a holistic and systemic understanding of human–environment interactions.

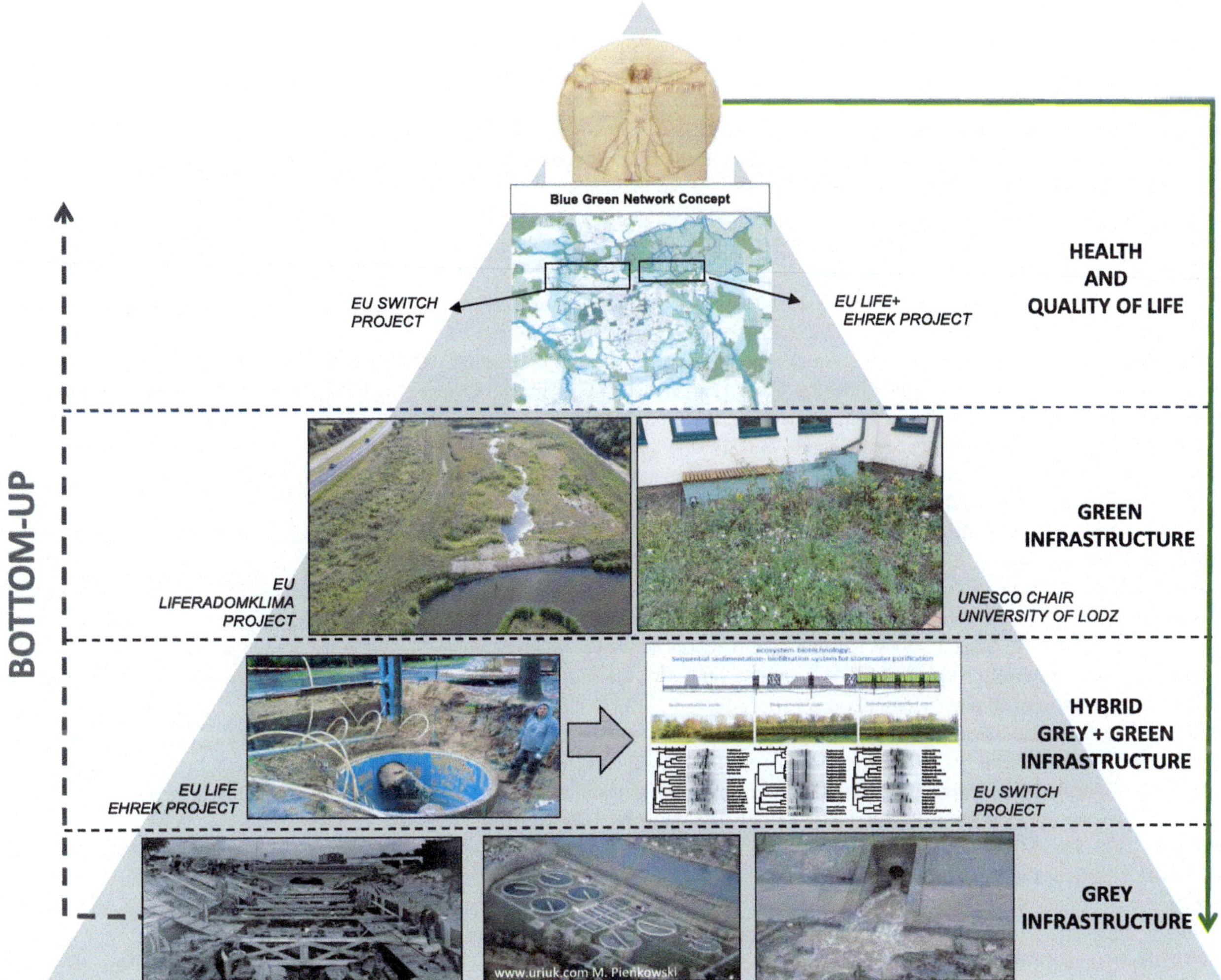

FIGURE 25.4 Smart and healthy blue-green city of the future. Zalewski, 2020b. Ekohydrologia. Wydawnictwo Naukowe PAN.

The concept of the smart, blue-green city of the future encapsulates this paradigm shift. It departs from the conventional bottom-up model, which focuses on incremental technical improvements in gray and hybrid infrastructures, toward a top-down, health-oriented framework (Fig. 25.4). This new approach begins with human health, well-being, and quality of life as central objectives of urban transformation and is driven by transdisciplinary, demand-oriented science, where social inclusion, stakeholder participation and ecological understanding guide strategic planning. The foundation of the smart, blue-green city of the future lies in the recognition and application of ecosystem mechanisms as active components of their governance. In this sense, the smart, blue-green city of the future moves beyond a purely technological paradigm toward a holistic socio-ecological framework in which ecosystem processes, governance, and human well-being are intrinsically linked.

References

Brown, R.R., Keath, N., Wong, T.H.F., 2009. Urban water management in cities: historical, current and future regimes. Water Sci. Technol. 59 (5), 847–855. https://doi.org/10.2166/wst.2009.029.

Connell, J.H., 1978. Diversity in tropical rain forests and coral reefs. Science 199 (4335), 1302–1310. https://doi.org/10.1126/science.199.4335.1302.

Dantzig, G.B., Saaty, T.L., 1973. Compact City: A plan for a liveable urban environment.

Delumeau, J., 2000. History of Paradise: The Garden of Eden in Myth and Tradition.

Gawrońska-Nowak, B., Jopek, Wagner, I., 2025. Miasto Wodne – Inteligentna Regeneracja [Water-City – Intelligent Regeneration] Regeneracja Miast, pp. 39–60. Urban regeneration.

Geiger, S.M., Geiger, M., Wilhelm, O., 2019. Environment-specific vs. general knowledge and their role in pro-environmental behavior. Front. Psychol. 10. https://doi.org/10.3389/fpsyg.2019.00718.

Giffinger, R., Fertner, C., Kramer, H., Kalasek, R., Pichler-Milanovic, N., Meijers, E., 2007. Smart Cities-Ranking of European Medium-Sized Cities [Final Report]. Available from: https://www.smart-cities.eu/download/smart_cities_final_report.pdf.

Hartz-Karp, J., Marinova, D., 2021. Using deliberative democracy for better urban decision-making through integrative thinking. Urban Sci. 5 (1). https://doi.org/10.3390/urbansci5010003.

Hausner, J., 2024. Ekonomia Wartości. Myślenie Holistyczne W Działaniu.

Head, B., 2010. Wicked Problems in Water Governance: Paradigm Changes to Promote Water Sustainability and Address Planning Uncertainty.

Hobbs, R.J., Higgs, E., Harris, J.A., 2009. Novel ecosystems: implications for conservation and restoration. Trends Ecol. Evol. 24 (11), 599–605. https://doi.org/10.1016/j.tree.2009.05.012.

Holling, C.S., 1973. Resilience and stability of ecological systems. Annu. Rev. Ecol. Systemat. 4 (1), 1–23. https://doi.org/10.1146/annurev.es.04.110173.000245.

Hospers, G.J., 2017. People, place and partnership: exploring strategies to revitalise town centres. Eur. Spatial Res. Pol. 24 (1), 65–79. https://doi.org/10.1515/esrp-2017-0004.

Januchta-Szostak, A., 2020. River-friendly Cities: An Outline of Historical changes in relations between cities and rivers and contemporary water-responsible Urbanization strategies River-Friendly Cities: An Outline of Historical Changes in Relations between Cities and Rivers and Contemporary Water-Responsible Urbanization Strategies. Peter Lang Publishing Group, Poland, pp. 1–249. https://doi.org/10.3726/b17144. Available from: https://www.peterlang.com/view/title/71818.

Krauze, K., Wagner, I., 2019. From classical water-ecosystem theories to nature-based solutions - Contextualizing nature-based solutions for sustainable city. Sci. Total Environ. 655, 697–706. https://doi.org/10.1016/j.scitotenv.2018.11.187.

Krauze, K., Włodarczyk-Marciniak, R., 2018. Defining the risk to water and natural capital in cities with risk component analysis tool (DAPSET): Case study Łódź. J. Environ. Manag. 227, 62–72. https://doi.org/10.1016/j.jenvman.2018.08.081.

Kronenberg, J., Krauze, K., Wagner, I., 2017. Focusing on ecosystem services in the multiple social-ecological transitions of Lodz. In: Frantzeskaki, N., Castán Broto, V., Coenen, L., Loorbach, D. (Eds.), Urban Sustainability Transitions. Taylor & Francis, London, pp. 331–345.

Kupryś-Lipińska, I., Elgalal, A., Kuna, P., 2009. Urban-rural differences in the prevalence of atopic diseases in the general population in Lodz Province (Poland). Postepy Dermatol. Alergol. 26 (5), 249–256. https://www.termedia.pl/Original-paper-Urban-rural-differences-in-the-prevalence-of-atopic-diseases-in-the-general-population-in-Lodz-Province-Poland-,7,13372,1,1.html.

Lamond, J., Everett, G., 2019. Sustainable Blue-Green Infrastructure: a social practice approach to understanding community preferences and stewardship. Landsc. Urban Plann. 191, 103639. https://doi.org/10.1016/j.landurbplan.2019.103639.

Łaszkiewicz, E., Kronenberg, J., Marcińczak, S., 2021. Microscale socioeconomic inequalities in green space availability in relation to residential segregation: the case study of Lodz, Poland. Cities 111, 103085. https://doi.org/10.1016/j.cities.2020.103085.

Markowski, T., 2023. Idea Miast Regeneracyjnych [The Idea of Regenerative Cities], pp. 13–26. Available from: https://hub.oees.pl/wp-content/uploads/2024/02/DP6-final-inter.pdf.

Melville-Shreeve, P., Cotterill, S., Grant, L., Mant, J., Moore, S., Ashley, R.M., 2018. State-of-the-art review on the co-benefits of green–blue infrastructure in urban areas. Urban Water J. 15 (10), 974–985. https://doi.org/10.1080/1573062X.2019.1642207.

OECD, 2020. Innovative Citizen Participation and New Democratic Institutions: Catching the Deliberative Wave. OECD Publishing, Paris. https://doi.org/10.1787/339306da-en.

Rada Miejska w Łodzi, 2010. Uchwała Nr XCIX/1826/10 Rady Miejskiej w Łodzi z dnia 27 października 2010 r., w sprawie uchwalenia Studium uwarunkowań i kierunków zagospodarowania przestrzennego miasta Łodzi. Available at: https://mpu.lodz.pl/files/mpu/public/STUDIUM/archiwum/2010/2010_SUIKZP_uchwala.pdf (Accessed 12 February 2026).

Roe, E., Schulman, P., Van Eeten, M., De Bruijne, M., 2005. High-reliability bandwidth management in large technical systems: findings and implications of two case studies. J. Publ. Adm. Res. Theor. 15 (2), 263–280. https://doi.org/10.1093/jopart/mui015.

Schrödinger, 1998. What Is Life? the Physical Aspect of the Living Cell.

Sikorska, D., Łaszkiewicz, E., Krauze, K., Sikorski, P., 2020. The role of informal green spaces in reducing inequalities in urban green space availability to children and seniors. Environ. Sci. Pol. 108, 144–154. https://doi.org/10.1016/j.envsci.2020.03.007.

Skidmore, P., Bound, Lownsbrough, H., 2006. Community Participation: Who Benefits?.

Syrbe, R.U., Walz, U., 2012. Spatial indicators for the assessment of ecosystem services: Providing, benefiting and connecting areas and landscape metrics. Ecol. Indic. 21, 80–88. https://doi.org/10.1016/j.ecolind.2012.02.013.

Towards an EU Research and Innovation Policy Agenda for Nature-Based Solutions and Re-Naturing Cities. https://doi.org/10.2777/765301, 2015.

Wagner, I., Zalewski, M., 2009. Ecohydrology as a basis for the sustainable city strategic planning: focus on Lodz, Poland. Rev. Environ. Sci. Biotechnol. 8 (3), 209–217. https://doi.org/10.1007/s11157-009-9169-8.

Wagner, I., Zalewski, M., 2012. System solutions in urban water management: The Lodz (Poland) perspective Water Sensitive Cities, pp. 231–245.

Wagner, I., Zalewski, M., 2016. Temporal changes in the abiotic/biotic drivers of selfpurification in a temperate river. Ecol. Eng. 94, 275–285. https://doi.org/10.1016/j.ecoleng.2016.05.073.

Wagner, I., Breil, P., 2013. The role of ecohydrology in creating more resilient cities. Ecohydrol. Hydrobiol. 13 (2), 113–134. https://doi.org/10.1016/j.ecohyd.2013.06.002.

Wagner, I., da Silva Wells, C., Butterworth, J., Dziegielewska-Geitz, M., 2011. Lodz: city of water. In: Butterworth, J., McIntyre, P., da Silva Wells, C. (Eds.), SWITCH in the City, Putting Urban Water Management to the Test. IRC International Water and Sanitation Centre, The Hague, The Netherlands, pp. 210–221.

Wagner, I., 2022. Formally designate blue-green infrastructure for climate adaptation. Nature 607 (657). https://doi.org/10.1038/d41586-022-02000-7.

Włodarczyk-Marciniak, R., Krauze, K., 2023. Does educational profile influence student perception of ecosystem service provision by blue-green infrastructure? Soc. Nat. Resour. 36 (7), 840–859. https://doi.org/10.1080/08941920.2023.2188502.

Włodarczyk-Marciniak, R., Krauze, K., Kretek-Kamińska, A., Krzewińska, A., 2024. Can we rely on people's choices when envisioning retrofit of semi public courtyards using blue-green infrastructure? J. Environ. Manag. 370, 123076. https://doi.org/10.1016/j.jenvman.2024.123076.

Włodarczyk-Marciniak, R, Sikorska, D, Krauze, K, 2020. Residents' awareness of the role of informal green spaces in a post-industrial city, with a focus on regulating services and urban adaptation potential. Sustain. Cities Soc. 59 (102236).

Yoon, A., Jeong, D., Chon, J., 2021. The impact of the risk perception of ocean microplastics on tourists' pro-environmental behavior intention. Sci. Total Environ. 774 (144782). https://doi.org/10.1016/j.scitotenv.2020.144782.

Zalewski, M., 2000a. Ecohydrology — the scientific background to use ecosystem properties as management tools toward sustainability of water resources. Ecol. Eng. 16 (1), 1–8. https://doi.org/10.1016/s0925-8574(00)00071-9.

Zalewski, M, 2020b. Ekohydrologia. Wydawnictwo Naukowe PWN.

Zalewski, M., 2021. Ecohydrology: An Integrative Sustainability Science. IntechOpen. https://doi.org/10.5772/intechopen.94169.

Zalewski, M., Naiman, R.J., 1985. The regulation of Riverine fish communities by a continuum of abioticbiotic factors, pp. 3–9.

Zalewski, M., Wagner, I., 2005. Ecohydrology – the use of water and ecosystem processes for healthy urban environments. In: Special issue: Aquatic Habitats in Integrated Urban Water Management. Ecohydrology and Hydrobiology 5 (4), 263–268.

Zalewski, M., Wagner, I., Fratczak, W., Mankiewicz-Boczek, J., Parniewski, P., 2012. Blue-green city for compensating global climate change. The Parliment Magazine 2–3. Available from: http://www.theparliament.com/digimag/issue350.

Chapter 26

Advanced Nature-based Solutions for sustainable and resilient city (WBSR)

Tomasz Jurczak[1], Paweł Jarosiewicz[1, 2] and Maciej Zalewski[2]

[1]*University of Lodz, Faculty of Biology and Environmental Protection, UNESCO Chair on Ecohydrology and Applied Ecology, Łódź, Poland;*
[2]*European Regional Centre for Ecohydrology of the Polish Academy of Sciences under the auspices of UNESCO, Łódź, Poland*

IMAGE CREDIT: M. ZALEWSKI.

Transdisciplinary Ecohydrology. https://doi.org/10.1016/B978-0-443-41596-8.00026-1

1. Introduction

The first example of the systemic implementation of the blue–green network concept in Łódź was the restoration of three recreational reservoirs in the upper Bzura catchment under the LIFE + EH-REK project (LIFE08 ENV/000517; Zalewski et al., 2012). The project introduced a catchment-based approach that integrated hydrotechnical infrastructure with ecohydrological Nature-Based Solutions to reduce pollution loads and improve ecological resilience. Central to the intervention was the concept of Dual Regulation, which combines hydrological control with the use of biological processes as management tools—an approach not previously applied in urban systems, where traditional engineering methods often intensified anthropopressure and accelerated surface runoff.

Despite their limited surface area, the implemented solutions proved highly effective, improving water quality, reducing nutrient inflow, and restoring recreational and ecological functions (Jurczak et al., 2019). The hybrid system, based on the sequential sedimentation-biofiltration system (SSBS), combines sediment interception, stormwater pretreatment, and biogeochemical absorption, while preserving the natural character of the landscape.

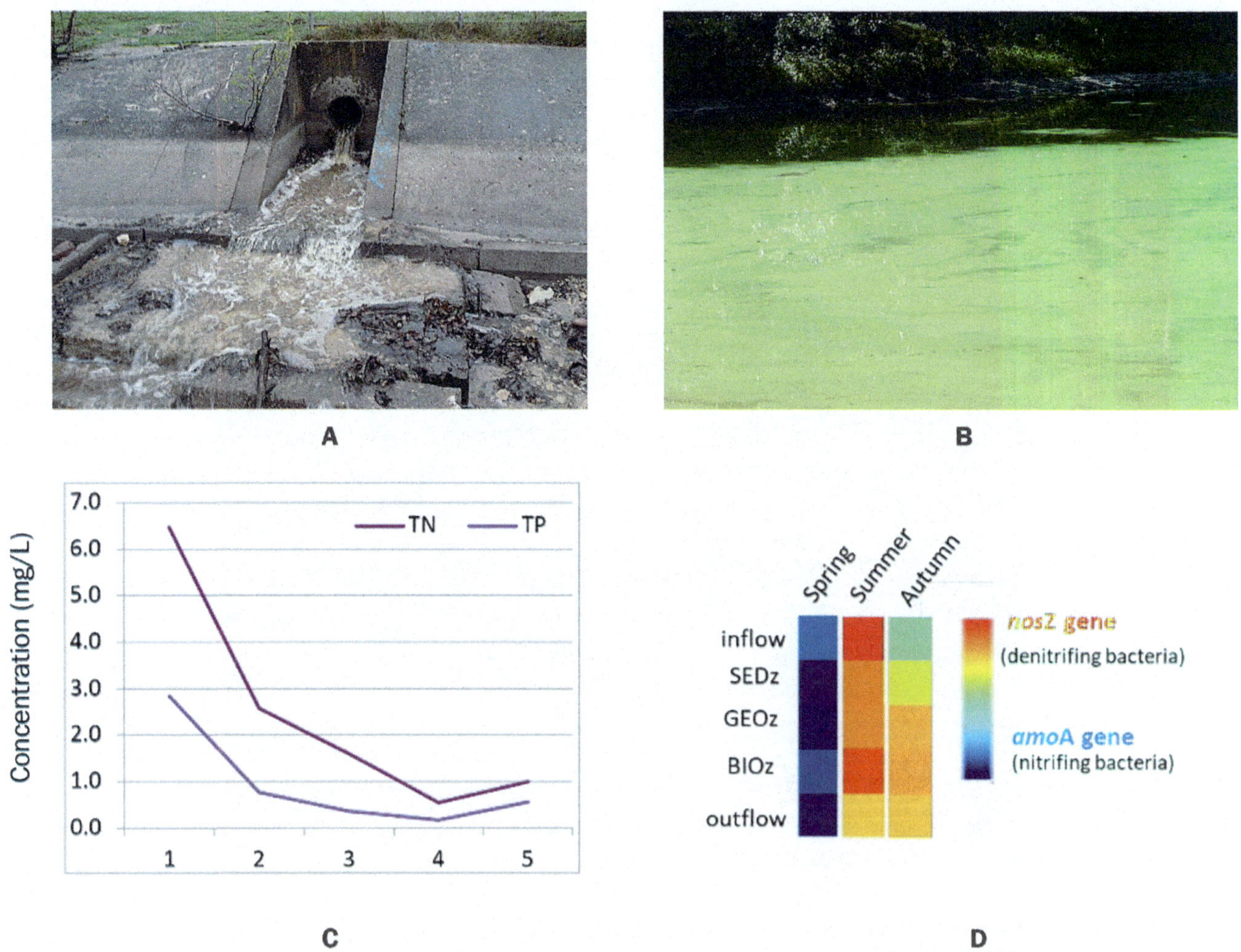

FIGURE 26.1 Urban stormwater quality and its role in generating toxic algal blooms, reducing the recreational potential of aquatic systems. (A) Urban stormwater may contain up to ten times higher concentrations of suspended mineral and organic matter, nutrients, and pollutants compared to natural streams. (B) When discharged into recreational reservoirs, it can stimulate the formation of cyanobacterial blooms. (C) The ecohydrological sequential sedimentation–biofiltration system (SSBS) hybrid system (LIFE08 ENV/PL/000517; www.arturowek.pl) demonstrates high efficiency in reducing nitrogen and phosphorus loads from stormwater originating from parking lots and roads. The system integrates separators (for hydrocarbon removal), biogeochemical barriers, phytotechnological barriers, and biogeochemical barriers, enhancing the resilience of the reservoir ecosystem (see Fig. 26.5). (1) Initial concentration in stormwater (A), mg/L; (4) Final purification stage; (5) Water quality in the receiving reservoir. (D) Seasonal analysis of microbial communities in SSBS supporting the optimization of biotechnological enhancement of system efficiency, based on data from. Font Najera, A., Serwecinska, L., Szklarek, S., Mankiewicz-Boczek, J., 2021. Seasonal and spatial changes of N-transforming microbial communities in sequential sedimentation-biofiltration systems: influence of system design and environmental conditions. Int. Biodeterior. Biodegradation 159, 105203. https://doi.org/10.1016/j.ibiod.2021.105203. *Image credit: T. Jurczak.*

T. Jurczak, et al. Ecological Engineering 131 (2019) 81–98

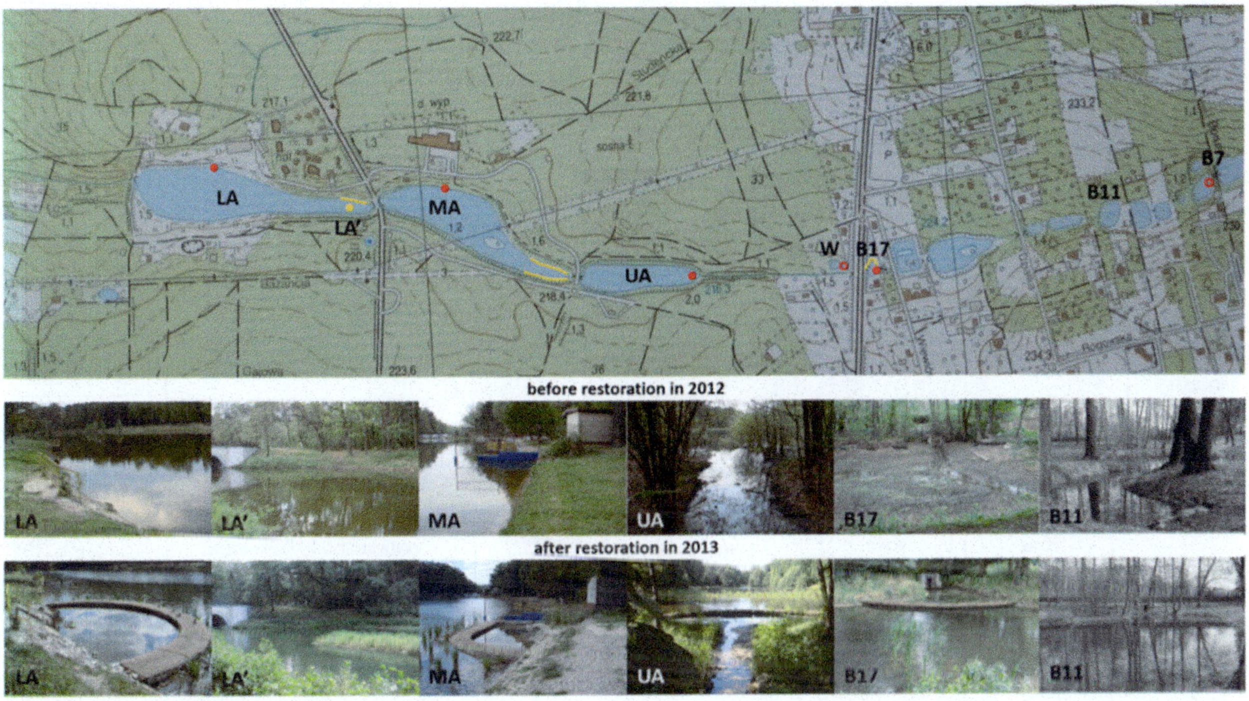

Location of the project sites in the Bzura River and the Arturówek reservoirs. Red dots – SSBSs or hybrid systems; yellow dot – the floating island; yellow lines – shoreline vegetation zones; W and B7 – locations of the work described by Jurczak et al. (2018a, 2018b). Photos depict the situation at the study sites before restoration (upper photos) and after restoration (lower photos) (photos by T. Jurczak). (For interpretation of the references to colour in this figure legend, the reader is referred to the web version of this article.)

FIGURE 26.2 Urban stormwater management. The sequence of reservoirs in the upper Bzura River, located in Arturówek. *Jurczak, T., Wagner, I., Kaczkowski, Z., Szklarek, S., Zalewski, M., 2018. Hybrid system for the purification of street stormwater runoff supplying urban recreation reservoirs. Ecol. Eng., 110, 67–77. https://doi.org/10.1016/j.ecoleng.2017.09.019.*

2. Ecohydrological hybrid solutions: SSBS in Arturówek

The Arturówek area is located in the northern part of Łódź, in the Bałuty district, adjacent to the Łagiewnicki Forest—one of the largest urban forest complexes in Europe. Both the forest and the reservoirs are of exceptional importance to the city and its residents due to their recreational and ecological value. However, strong anthropogenic pressure degraded water quality and reduced the recreational potential of the area Fig. 26.1.

The system solutions developed take into account the harmonization of hydrotechnical tools (rainwater retention and purification) Fig. 26.2 with the use of biological methods such as biotechnologies, phytotechnology (Zalewski et al., 2004), and regulation of trophic processes in water ecosystems (biomanipulation, hydrobiomanipulation see Chapter 15). An example of such a combination is the SSBS applied in Arturówek, the prototype of which was used on the Sokołówka River (see Chapter 5, Fig. 5.10). The first stages were sedimentation ponds above Wycieczkowa street. However, due to large inflow of the stormwater, sediment nutrients, and pollutants from this street, a more advanced SSBS below this street was constructed with the installation of underground separators and sedimentary chamber that are responsible for interception of stormwater (Jurczak et al., 2018). The construction of the whole system with the separators and settling tanks is placed underground, which maintains the site's ecological character and aesthetic quality—bioaesthetic of the site.

A significant quantity of the nutrient contaminants that enter the surface water is then accumulated in the sediment of the reservoirs. Due to their low solubility, longevity, and hydrophobicity, these substances easily settle on the surface of small sediment particles. Consequently, the fine fractions of the suspended matter if not intercepted at the sedimentary chamber of SSBS become a source of so-called secondary water pollution, e.g., by resuspension, bioturbation, and as result of redox potential changes, so it is important to periodically remove sediments by hydraulic and mechanical techniques from the bottom of reservoir. When a reservoir is protected by an SSBS, sediment collection becomes

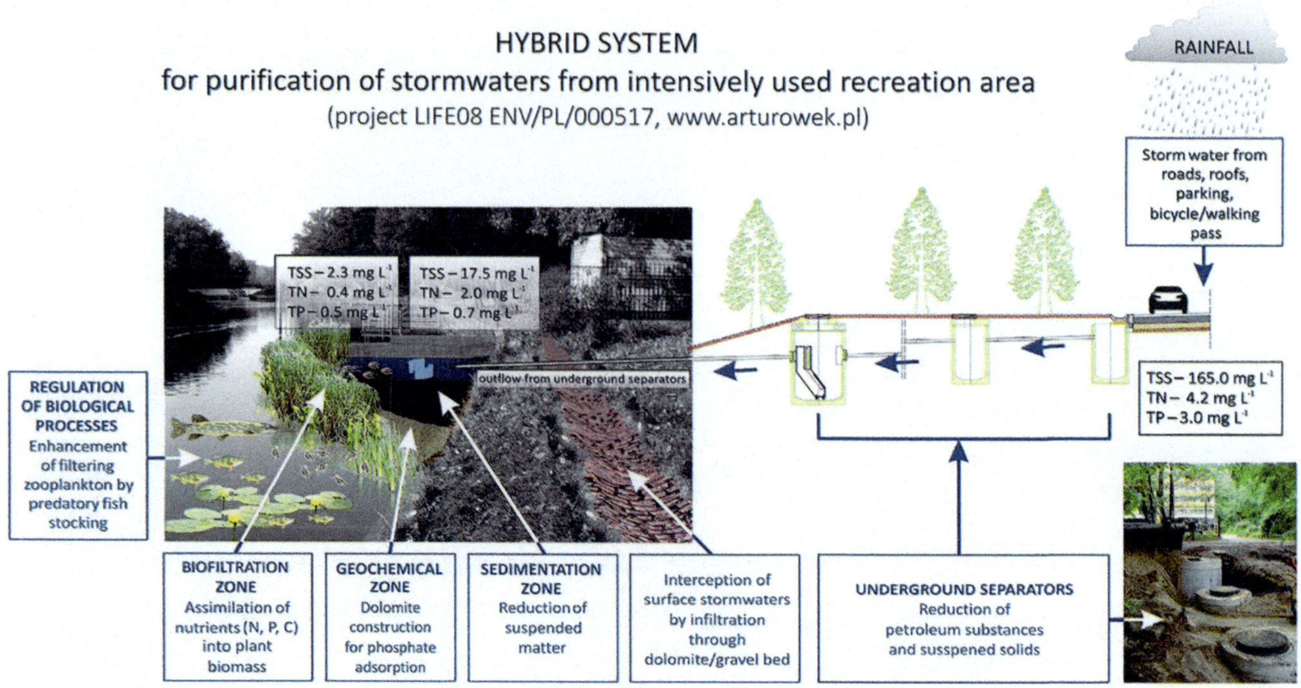

FIGURE 26.3 Restoration of urban recreational reservoirs through the construction of a hybrid system. Part 1: reduction of nutrient loading through low-cost and highly effective ecohydrological measures. An example of a hybrid system for stormwater purification, integrating hydroengineering hydrodynamic separators, biotechnologies, geochemical barriers, and phytotechnology, supported by biomanipulation to enhance the reservoir's retention capacity. *Jurczak, T., Wagner, I., Wojtal-Frankiewicz, A., Frankiewicz, P., Bednarek, A., Łapińska, M., Kaczkowski, Z., Zalewski, M., 2019. Comprehensive approach to restoring urban recreational reservoirs. Part 1—reduction of nutrient loading through low-cost and highly effective ecohydrological measures. Ecol. Eng., 131, 81–98. https://doi.org/10.1016/j.ecoleng.2019.03.006.*

considerably easier and significantly less costly, as the majority of pollutants accumulate in the SSBS sedimentation chamber rather than being dispersed throughout the reservoir.

The next important SSBS was placed in the inflow part of the upper reservoir in Arturówek (see photo at the cover page of the book). Some of applied SSBS system at shorelines of reservoirs due to limited space where continuous hybrid systems were implemented Fig. 26.3. Hydroengineering technologies—separators, geochemical barriers, and sedimentation chamber—accumulate sediments in limited space specially designed for easy and low costs dredging. Additionally, sedimentation chamber periodically can be inoculated by bacterial strains which accelerate denitrification and self-purification.

An important and often neglected source of pollution for reservoirs—including in the case of Arturówek—is the feeding of waterfowl and the intensive baiting of fish for recreational fishing. Such activities significantly increase nutrient concentrations in the water (nitrogen and phosphorus), thereby accelerating eutrophication and algal blooms. Each kilogram of phosphorus introduced into the reservoir can generate 1–2 tons of algal biomass, which may be toxic and temporarily exclude the reservoir from recreational use.

In terms of improving water quality, measures to regulate the interplay between hydrological, ecological, and biogeochemical processes can play a major role. This translates into a reduction in the load of pollutants and biochemical compounds entering reservoirs by increasing their binding in the catchment area and within the ecosystem in the form of inaccessible deposits, e.g., by the accumulation of compounds in plant biomass Fig. 26.4. In this situation, it is also essential to create buffer zones that reduce surface and ground runoff and stop the leaching of nutrients from the soil and biogeochemical barriers described in the previous chapters. In some places, vegetation strips similar to a hedge have also proved their worth, as they may hamper access to the reservoirs and limit water pollution by visitors (preventing e.g., birds from feeding, animal bathing, etc.) but also limit surface runoff and erosion at the reservoir edges. Consequently, to increase the reservoir ecosystem's retention capacity, additional littoral vegetation zones were established at the SSBS sites Fig. 26.5 (A and B).

On the reservoirs themselves, the so-called floating islands have been constructed Fig. 26.4. Mats made mainly on plastic mesh with the floating vegetation, which assimilate nutrients, create a habitat for zooplankton and by the "top down" phenomenon of the trophic pyramid reduces the development of phytoplankton responsible for the algal blooms.

FIGURE 26.4 Floating island for reduction of the internal load of nutrients (project LIFE08 ENV/PL/000517, www.arturowek.pl). *Photo credit: S. Szklarek.*

FIGURE 26.5 Hybrid system installed at lower Arturówek reservoir. The hybrid SSBS system constructed in the lower Arturówek reservoir in 2013 (A) and 3 years later (B) at the stormwater inflow point. The system treats stormwater, enhances biodiversity by creating habitat, and improves aesthetic quality, making the space more supportive of both nature and people. *Image Credit: T. Jurczak.*

The use of natural aquatic vegetation also has a positive effect on the reservoir in other respects: it provides habitats, hiding places, and breeding sites not only for various species of fish and macroinvertebrates, but also for water birds. Maintaining the biodiversity of the reservoir, on the other hand, is not only significant for enhancing WBSR but also improving its aesthetic qualities and recreational value Fig. 26.6.

The systemic approach to the restoration of reservoirs also includes social and economic aspects. Systemic approaches are aimed at achieving and maintaining a high-quality environment and ecological safety for residents, thus stimulating the creation of an attractive urban space. A good-quality environment translates into improved health and public satisfaction and creates opportunities for personal and social development for people Fig. 26.7.

The results of the EH-REK project have enabled conclusions to be drawn to achieve a balance between the river system and the reservoirs. The research and implementation work carried out under the project has also been recognized by the European Commission "Best of the best" award—of all the projects completed under the LIFE program in 2016–17. The knowledge and experience gained in the project allowed the development of a concept for the restoration of other reservoirs in Łódź (Jana Reservoir, Stefański Reservoir) but also serves as an aid in the development of strategies for the adaptation of cities to climate change.

FIGURE 26.6 The sequential sediment-biofiltration system (SSBS). The SSBS constructed at the place of the Bzura river inflow to the Bzura-7 (A), Bzura-17 (B) reservoirs, and the Upper Arturówek reservoir (C) and the prototype of this system installed on the Sokołówka River to retain and treat urban stormwater (D). *Jurczak, T., Wagner, I., Wojtal-Frankiewicz, A., Frankiewicz, P., Bednarek, A., Łapińska, M., Kaczkowski, Z., Zalewski, M., 2019. Comprehensive approach to restoring urban recreational reservoirs. Part 1 – Reduction of nutrient loading through low-cost and highly effective ecohydrological measures. Ecol. Eng. 131, 81–98. https://doi.org/10.1016/j.ecoleng.2019.03.006; Images Credit: T. Jurczak.*

The SSBS and hybrid systems integrate engineering measures with ecological biotechnologies Fig. 26.8, which increases the efficiency of stormwater purification. If maintained properly and systematically, they provide long-term prevention of eutrophication and siltation, instead of treatment of the already degraded system. The systems can be fine-tuned for intensity of impact by adjusting the size of the zones to the loads. They proved to be efficient in spite of their relatively small size and surface area; therefore, they can be used successfully in cities, where land is expensive and availability is limited. Additionally, they provide habitats for biodiversity, making the space friendlier to nature and people and resistant to anthropopressures.

3. Blue–green network implementation for a whole city scale—RadomKlima-PL project

As a continuation of the LIFE EH-REK project, the RadomKlima-PL project (LIFE14 CCA/PL/000101) was implemented to further advance the application of ecohydrology in urban environments and bring it to the level of the whole city. Its main objective was to develop and demonstrate ecohydrological nature-based solutions that enhance the resilience of urban hydrological systems to climate change. The Municipality of Radom acted as the coordinating beneficiary, while the UNESCO Chair on Ecohydrology and Applied Ecology, University of Łódź, served as a scientific partner responsible for the conceptual framework and the design of large-scale implementations. The total cost of the project amounted to PLN 24,291,746.

FIGURE 26.7 Lower Arturówek reservoir. The inflow of stormwater to the lower Arturówek reservoir before investment efforts (A) and after the construction of the hybrid system (B). The inflow of the Bzura River to the upper Arturówek reservoir, before investment activities (C) and after the SSBS construction (D). The effect of restoration and protection efforts on the water quality in Arturówek reservoirs [(E) toxic algal bloom before construction ANbS and this same section after construction of all solutions (F)] (photo: T. Jurczak). *Image credit: T. Jurczak.*

Today, Radom functions as a full-scale UNESCO Ecohydrology Demonstration Site, showcasing the application of ecohydrological principles for climate change adaptation through sustainable water management. The ecohydrological approach has been applied across urban catchments of the Mleczna, Cerekwianka, and Potok Północny Rivers, as well as within an urban reservoir system.

In addition to multiple small-scale blue–green infrastructure measures (e.g., rain gardens, green bus stops), the project implemented four major technical and ecological interventions. Each of these solutions combines hydrological regulation with ecological restoration, designed in accordance with the WBSR framework (Zalewski, 2015, 2021; Zalewski et al., 2021).

FIGURE 26.8 Hybrid system. The biofiltration section of hybrid system for third step of stormwater purification after construction (left photo) and 2 years later (right photo). *Image credit: T. Jurczak.*

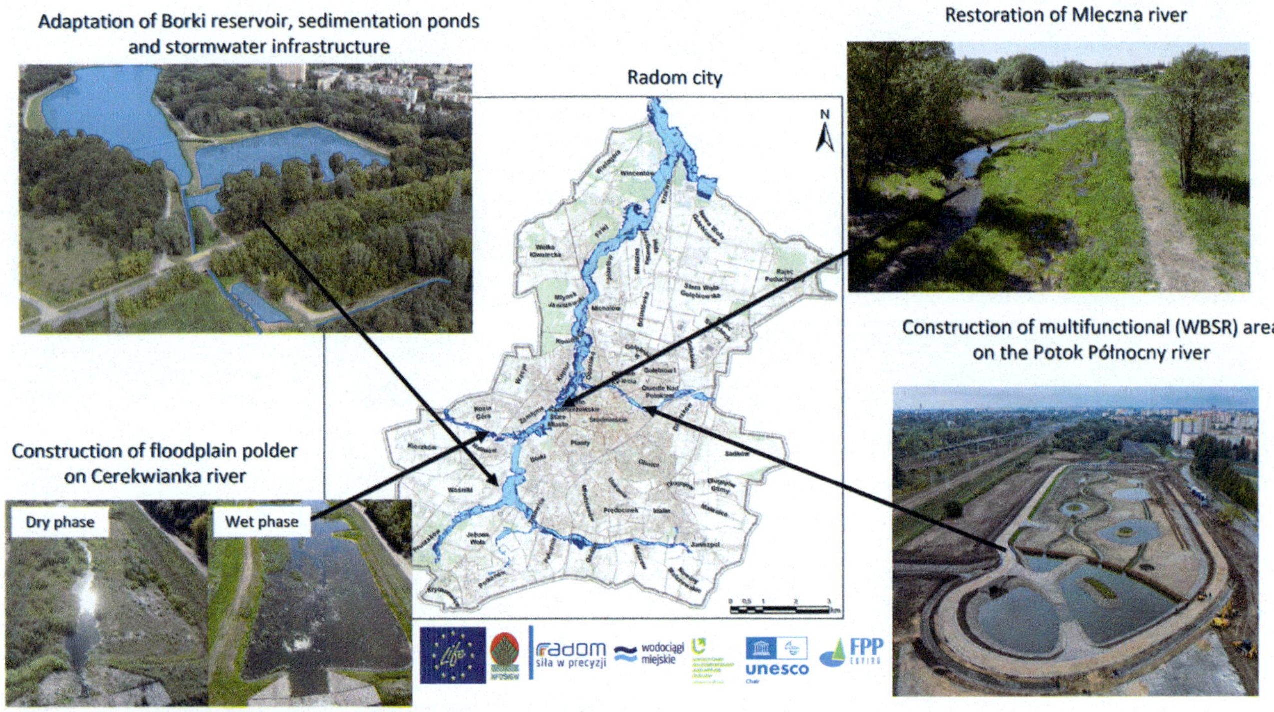

FIGURE 26.9 Integrated stormwater management. Blue–green network implementation—RadomKlima project.

4. Adaptation of the Borki Reservoir and sedimentation ponds

The Borki Reservoir and associated sedimentation ponds represent one of Radom's most important water retention and recreation areas Fig. 26.9. Over time, the system became heavily silted, reducing both retention capacity and water quality. The project aimed to adapt the reservoir and stormwater infrastructure to changing climatic conditions by enhancing retention, improving hydraulic safety, and increasing the self-purification capacity of the inflowing waters.

The adaptation involved the reconstruction of hydraulic structures and stormwater collectors, as well as the installation of a sedimentation–biofiltration system along the stormwater channel (Jarosiewicz et al., 2024).

Key works included:

- Reconstruction of the main dam and outlet structures, increasing storage volume by 10%–20%.
- Construction of a fish pass, restoring ecological connectivity.
- Dredging of accumulated sediments from approximately 70% of the reservoir's bottom area.

- Installation of the SSBSs at the sedimentation pond and stormwater inflow upstream. One of the SSBSs was innovatively supplied with aeration systems to deal with a large load of organic pollution.

As a result, the retention capacity of the Borki Reservoir increased by 24,558 m^3 (15.9%), while the sedimentation ponds' capacity increased by 12,735 m^3 (60%). The works improved water quality, restored hydraulic function, and created conditions for aquatic biodiversity recovery. Importantly, the reservoir now serves as a multifunctional ecohydrological system, integrating flood control, water purification, and recreational use.

5. Renaturalization and adaptation of the Mleczna River

The Mleczna River, which flows through the city center, was historically channelized and heavily degraded. To address the loss of retention capacity and ecological function, the project introduced a renaturalization scheme based on ecohydrological principles, aiming to restore the natural dynamics of the river while maintaining flood safety. The main goal was to recreate the natural river morphology and restore hydrological connectivity with the floodplain. By increasing the cross-section of the riverbed and creating a floodplain connection, the intervention slows flood waves and delays their passage through the lower sections of the river, protecting city center. This approach also contributes to improved water quality, groundwater recharge, and habitat diversity.

The following ecohydrological and technical works were implemented (among others):

- Remeandering of the river along a 315 m section (171 m of newly created meanders and 144 m of the preserved natural channel);
- Construction of four quasi-natural oxbow lakes for self-purification and biodiversity enhancement.
- Installation of 10 sequences of deflectors and riffle–pool formations, creating alternating fast and slow flow habitats.
- Construction of a wooden viewing platform linking the river with the Mleczna Boulevards for educational and recreational purposes.

The intervention increased the river valley retention, lowered flood peaks, and improved habitat structure. In the long term, the project supports the recovery of hydrophilic habitats, restoration of the ecological corridor, and enhancement of biodiversity (fish, birds, amphibians, and macroinvertebrates).

6. Construction of a multifunctional (WBSR) flood-control reservoir on the Potok Północny River

Another major investment involved the construction of a multi-functional retention reservoir on the Potok Północny River. This system integrates hydraulic regulation with sediment and nutrient removal processes through a sequential sedimentation–biofiltration processes in the large scale.

The facility was designed to reduce the risk of urban flooding by temporarily storing runoff during extreme precipitation events as also to improve stormwater quality before its discharge into receiving waters (W). Enhance biodiversity (B) through the creation of wetland habitats within the biofiltration zones. The system consists of a series of sedimentation chambers and vegetated filtration ponds, connected by controlled flow structures. The design allows to provide variety of ecosystem services, including flood protection and recreation (R + S). System is designed to operate in dry and wet phases.

7. Construction of a floodplain polder on the Cerekwianka River

The final component of the LIFE project was the construction of a floodplain polder on the Cerekwianka River, designed to act as a natural retention and purification system. This intervention represents a landscape-scale adaptation measure, enhancing water retention and biodiversity in the urban periphery.

The polder functions as a controlled floodplain, temporarily storing and treating runoff during high-flow events. The design was based on the creation of a mosaic of habitats and the support of natural succession processes. The investment included the formation of three zones: wet sedimentation basin, main retention area, and downstream retention area with floodwater capacity.

The Cerekwianka polder enhances catchment-scale retention, mitigates flood peaks, and supports self-purification of stormwater through natural sedimentation and biogeochemical processes. The area has become a biodiversity hotspot, providing habitats for amphibians, wetland plants, and birds. It also serves as an educational site illustrating the role of floodplains in urban ecohydrological systems.

The ecohydrological biotechnologies and systemic solutions used in the above projects not only mitigate a broad range of impacts by regulating key ecohydrological processes at different scales, from molecular to catchment, but also enhance WBSR sustainability potential of the river systems. This approach aligns with the UN's Result-Based Management (RBM) framework in its emphasis on measurable outcomes, adaptive management, and evidence-based decision-making to achieve long-term impact (*Results-Based Management Handbook: Harmonizing RBM Concepts and Approaches for Improved Development Results at Country Level.*, 2012).

AI disclosure

During the preparation of this work, the author(s) used ChatGPT. During the preparation of this work, the author(s) used OpenAI's ChatGPT. This document includes content generated and/or edited with the assistance of OpenAI's ChatGPT, used to support language refinement and structural improvements. All final content was reviewed and verified by the author. After using this tool/service, the author(s) reviewed and edited the content as needed and take(s) full responsibility for the content of the publication. After using this tool/service, the author(s) reviewed and edited the content as needed and take(s) full responsibility for the content of the publication.

References

Font Nájera, A., Serwecińska, L., Szklarek, S., Mankiewicz-Boczek, J., 2021. Seasonal and spatial changes of N-transforming microbial communities in sequential sedimentation-biofiltration systems: influence of system design and environmental conditions. Int. Biodeterior. Biodegradation 159, 105203. https://doi.org/10.1016/j.ibiod.2021.105203.

Jarosiewicz, P., Font-Najera, A., Mankiewicz-Boczek, J., Chamerska, A., Amalfitano, S., Fazi, S., Jurczak, T., 2024. Stormwater treatment in constrained urban spaces through a hybrid Sequential Sedimentation Biofiltration System. Chemosphere 367, 143696. https://doi.org/10.1016/j.chemosphere.2024.143696.

Jurczak, T., Wagner, I., Kaczkowski, Z., Szklarek, S., Zalewski, M., 2018. Hybrid system for the purification of street stormwater runoff supplying urban recreation reservoirs. Ecol. Eng. 110, 67–77. https://doi.org/10.1016/j.ecoleng.2017.09.019.

Jurczak, T., Wagner, I., Wojtal-Frankiewicz, A., Frankiewicz, P., Bednarek, A., Łapińska, M., Kaczkowski, Z., Zalewski, M., 2019. Comprehensive approach to restoring urban recreational reservoirs. Part 1 – reduction of nutrient loading through low-cost and highly effective ecohydrological measures. Ecol. Eng. 131, 81–98. https://doi.org/10.1016/j.ecoleng.2019.03.006.

Zalewski, M., 2015. Ecohydrology and hydrologic engineering: regulation of hydrology-biota interactions for sustainability. J. Hydrol. Eng. 20 (1). https://doi.org/10.1061/(ASCE)HE.1943-5584.0000999.

Zalewski, M., 2021. Ecohydrology: An Integrative Sustainability Science. IntechOpen. https://doi.org/10.5772/intechopen.94169.

Zalewski, M., Kiedrzyńska, E., Wagner, I., Izydorczyk, K., Boczek, J.M., Jurczak, T., Krauze, K., Frankiewicz, P., Godlewska, M., Wojtal-Frankiewicz, A., Łapińska, M., Urbaniak, M., Bednarek, A., Kaczkowski, Z., Gągała, I., Serwecińska, L., Szklarek, S., Włodarczyk-Marciniak, R., Font-Nájera, A., Mierzejewska, E., Połatyńska-Rudnicka, M., Belka, K., Jarosiewicz, P., 2021. Ecohydrology and adaptation to global change. Ecohydrol. Hydrobiol. 21 (3), 393–410. https://doi.org/10.1016/j.ecohyd.2021.08.001.

Zalewski, M., Wagner, I., Frątczak, W., Mankiewicz-Boczek, J., Parniewki, P., 2012. Blue-green city for compensating global climate change. Parl. Mag. 369, 36–37.

Zalewski, M., Wagner-Łotkowska, I., Robarts, R.D. (Eds.), 2004. Integrated Watershed Management: Ecohydrology and Phytotechnology Manual. UNESCO.

Further reading

United Nations Development Group (UNDG) Results-Based Management Handbook, 2012. Harmonizing RBM Concepts and Approaches for Improved Development Results at Country Level.

Chapter 27

Road salt application and freshwater eutrophication: An ecohydrological perspective

Sebastian Szklarek
European Regional Centre for Ecohydrology of the Polish Academy of Sciences under the auspices of UNESCO, Łódź, Poland

URBAN POND DURING WINTER (SOKOLOWKA RIVER, WASIAK POND IN LODZ, POLAND) *FROM SEBASTIAN SZKLAREK.*

1. Introduction

The use of road salt during the winter has become common practice in the inhabited northern hemisphere in recent decades as an inexpensive strategy to ensure road and pavement safety. However, recent findings show that such treatment has important consequences on the environment. Road salt may only be applied in the winter, but many studies show that it affects the year-round functioning of the aquatic ecosystem. Road salt is a kind of pollution, which has not received too much attention so far. Only in recent years has there been an increasing number of research focusing on the road salt impact on water bodies and connection between road salt pollution with eutrophication. Due to high concentration of salt in urban areas and their peaks in winter, the link between road salt application and flow from the urban catchment enhancing eutrophication of freshwater were investigated in the holistic ecohydrological approach.

Transdisciplinary Ecohydrology. https://doi.org/10.1016/B978-0-443-41596-8.00031-5

2. Monitoring of threat

The first step was a monitoring of threat, which informs about the threat delivered by road salt application. There were many studies informing about the chloride (Cl) concentration noted during winter and spring snow melting periods. Many of them noted Cl level above acute toxic level—few of them were presented in Table 27.1. Even if Cl concentration did not exceed the acute toxicity, the long-term road salt application resulted in increasing chlorides trends in many North America rivers (Corsi et al., 2015) and lakes (Dugan et al., 2017).

3. Caused and effect analyses

In the next step, the caused and effect analyses were made, which investigated the impact of road salt application on abiotic and biotic processes in catchment (Fig. 27.1). In the soil environment, NaCl could change the structure of the soils by cation exchange, which leads to decrease the porewater pH (Robinson et al., 2017), mobilize the organic matter and decrease the mineralization of organic N (Green et al., 2008, 2009), and decrease the retention capacity of soil (Schuler

TABLE 27.1 Chloride concentration noted in water bodies in comparison with acute toxicity level.

Chloride concentration determined for	MG CL/L	References
Acute toxicity level	640	Canada standard (Canadian Water Quality Guidelines for the Protection of Aquatic Life, 2011)
	860	US standard (Ambient Aquatic Life Water Quality Criteria for Chloride, 1988)
Highway runoff	19,135	Mayer et al. (2011)
Urban catchment stormwater outflow	1243–12,086	Szklarek et al. (2015)
Urban stream	3500	Rivett et al. (2016)
Stormwater pond	22,000	Gallagher et al. (2011)
Pore water in pond sediments	4740	Tixier et al. (2012)
Groundwater under a stormwater pond	18,540	Snodgrass et al. (2017)
Groundwater at urban river streambed	3200	Roy et al. (2015)

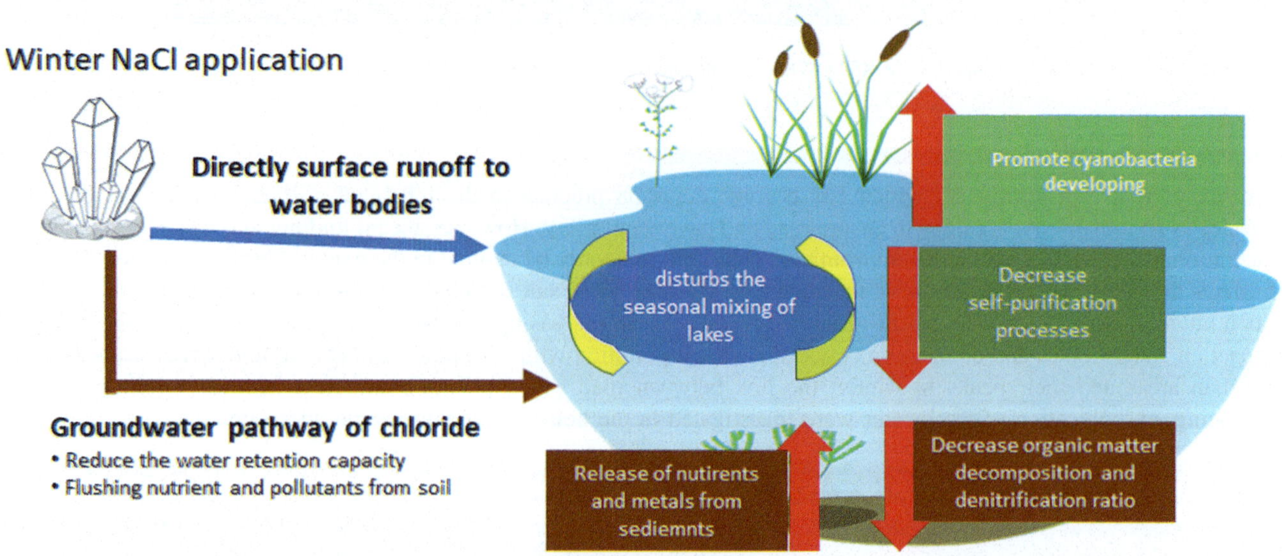

FIGURE 27.1 The impact of winter road salt application on abiotic and biotic processes in soil and water.

and Relyea, 2018). Increasing Cl concentration leads to release dissolved inorganic (DIC) and organic (DOC) carbon and release the total dissolved Kjeldahl nitrogen (TKN) (ammonium + ammonia + dissolved organic nitrogen) (Duan and Kaushal, 2015). Increasing salinization delays full or partial mixing of lakes water, which leads to anoxic condition (Kim and Koretsky, 2013).

Biotic component of freshwater ecosystems might be also affected by chloride pollution. Increasing baseflow chloride concentrations increased the number of days per year during which the chronic chloride acute criterion (230 mg Cl/L—Ambient Aquatic Life Water Quality Criteria for Chloride, 1988) is exceed (Corsi et al., 2015). In some hydrological conditions, the Cl concentration did not fall below the acute toxicity (640 mg/L—Canadian Water Quality Guidelines for the Protection of Aquatic Life, 2011) until September (Lembcke et al., 2017). Increasing Cl concentration impacts on different organisms participating in nutrients cycle and in self-purification processes. Decrease of the denitrification ratio in freshwater sediments increases the nitrates concentration in water (Lancaster et al., 2016). Increasing Cl concentration decreases the purification processes provided by macrophytes—lower reduction of heavy metal (Søberg et al., 2017) and nutrients reduction (Stoler et al., 2018) in natural and constructed biofiltration systems.

According to top-down effect, zooplankton plays an important role in controlling the phytoplankton biomass. Studies showed that the mid and high NaCl concentrations (480 and 780 mg/L, respectively) decrease the abundance of zooplankton (Jones et al., 2017). On the other hand, zooplankton community might adopt to increasing Cl concentration (Coldsnow et al., 2017), but climate change disturbances lead to unpredictable need of winter road maintenance, which might hinder the adaptation of zooplankton to regular chloride pollution. The changes in phytoplankton communities might be the result of increasing chloride concentration. Chakraborty et al. (2011) noted that the green algae where replaced by more salt-tolerant cyanobacteria (*Merismopedia* sp.). Other studies showed that freshwater cyanobacteria of Microcystis had higher salt tolerance than other phytoplankton species, especially green algal and diatoms (Tonk et al., 2007), which promote the cyanobacteria blooms in more salinity environment.

4. Solutions

The negative impact of road salt (NaCl) is significant, so there is a need to develop and optimize methods to reduce it (third step of ecohydrology approach). The use of deicing reagents other than NaCl, such as $CaCl_2$ and $MgCl_2$, is one of the solutions, but many studies showed that those reagents based on chloride salts could still have a negative impact on water environment (Lee et al., 2017; Szklarek et al., 2022). Deicing reagents based on acetates, nitrates, or formats are more environmentally friendly, but they are more expensive and less efficient than reagents based on chloride (Achkeeva et al., 2015). Optimizing the use of deicing chemicals is the other solution to reduce the negative impact of road salt. Socioeconomic models support winter maintenance strategy to decrease the used amount of road salt, which brings environmental and economic benefits (Arvidsson, 2017; Riehm and Nordin, 2012). Finally, the nature-based solution, based on halophytes, could help removing some part of chloride and sodium pollution (Guesdon et al., 2016; Suaire et al., 2016).

5. Conclusion

The impact of winter salt application on the aquatic environment is complex and wide. Therefore, there is a need to use holistic approach to investigate all possible negative impacts of road salt application and to develop system solution (fourth step of ecohydrology approach), which would help prevent the loss in water ecosystems and ecosystem services provided by them.

Acknowledgments

This study was supported by grant No. 2018/28/C/NZ8/00235, "Impact of road salt pollution in winter on zooplankton hatching success from resting eggs" funded by the National Science Center (Poland).

References

Achkeeva, R., Frolova, K., Khomyakov, D., 2015. Deicing properties of sodium, potassium, magnesium, and calcium chlorides, sodium formate and salt compositions on their basis. Theor. Found. Chem. Eng. 49 (4), 481–484. https://doi.org/10.1134/S0040579515040028.

Ambient Aquatic Life Water Quality Criteria for Chloride, 1988. https://www.epa.gov/sites/default/files/2018-08/documents/chloride-aquatic-life-criteria-1988.pdf.

Arvidsson, A.K., 2017. The Winter Model—a new way to calculate socio-economic costs depending on winter maintenance strategy. Cold Reg. Sci. Technol. 136, 30–36. https://doi.org/10.1016/j.coldregions.2017.01.005.

Canadian Water Quality Guidelines for the Protection of Aquatic Life, 2011. https://ccme.ca/en/resources/water-aquatic-life.

Chakraborty, P., Acharyya, T., Babu, R., Bandyopadhyay, D., 2011. Impact of salinity and pH on phytoplankton communities in a tropical freshwater system: an investigation with pigment analysis by HPLC. J. Environ. Monit. 13 (3), 614–620. https://doi.org/10.1039/c0em00333f.

Coldsnow, K.D., Mattes, B.M., Hintz, W.D., Relyea, R.A., 2017. Rapid evolution of tolerance to road salt in zooplankton. Environ. Pollut. 222, 367–373. https://doi.org/10.1016/j.envpol.2016.12.024.

Corsi, S.R., De Cicco, L.A., Lutz, M.A., Hirsch, R.M., 2015. River chloride trends in snow-affected urban watersheds: increasing concentrations outpace urban growth rate and are common among all seasons. Sci. Total Environ. 508, 488–497. https://doi.org/10.1016/j.scitotenv.2014.12.012.

Duan, Kaushal, 2015. Salinization alters fluxes of bioreactive elements from stream ecosystems across land use. Biogeosciences 12 (23), 7331–7347. https://doi.org/10.5194/bg-12-7331-2015.

Dugan, H.A., Bartlett, S.L., Burke, S.M., Doubek, J.P., Krivak-Tetley, F.E., Skaff, N.K., Summers, J.C., Farrell, K.J., McCullough, I.M., Morales-Williams, A.M., Roberts, D.C., Ouyang, Z., Scordo, F., Hanson, P.C., Weathers, K.C., 2017. Salting our freshwater lakes. Proc. Natl. Acad. Sci. U. S. A 114 (17), 4453–4458. https://doi.org/10.1073/pnas.1620211114.

Gallagher, M.T., Snodgrass, J.W., Ownby, D.R., Brand, A.B., Casey, R.E., 2011. Steven Lev, Watershed-scale analysis of pollutant distributions in stormwater management ponds. Urban Ecosyst. 14 (3), 469–484. https://doi.org/10.1007/s11252-011-0162-y.

Green, S.M., Machin, R., Cresser, M.S., 2008. Effect of long-term changes in soil chemistry induced by road salt applications on N-transformations in roadside soils. Environ. Pollut. 152 (1), 20–31. https://doi.org/10.1016/j.envpol.2007.06.005.

Green, S.M., Machin, R., Cresser, M.S., 2009. Does road salting induce or ameliorate DOC mobilisation from roadside soils to surface waters in the long term? Environ. Monit. Assess. 153 (1–4), 435–448. https://doi.org/10.1007/s10661-008-0369-4.

Guesdon, G., Santiago-Martín, A.de, Galvez-Cloutier, R., 2016. Phytodesalinization potential of *Typha angustifolia*, *Juncus maritimus*, and *Eleocharis palustris* for removal of de-icing salts from runoff water. Environ. Sci. Pollut. Control Ser. 23 (19), 19634–19644. https://doi.org/10.1007/s11356-016-7176-1.

Jones, D.K., Mattes, B.M., Hintz, W.D., Schuler, M.S., Stoler, A.B., Lind, L.A., Cooper, R.O., Relyea, R.A., 2017. Investigation of road salts and biotic stressors on freshwater wetland communities. Environ. Pollut. 221, 159–167. https://doi.org/10.1016/j.envpol.2016.11.060.

Kim, S.-Y., Koretsky, C., 2013. Effects of road salt deicers on sediment biogeochemistry. Biogeochemistry 112 (1–3), 343–358. https://doi.org/10.1007/s10533-012-9728-x.

Lancaster, N.A., Bushey, J.T., Tobias, C.R., Song, B., Vadas, T.M., 2016. Impact of chloride on denitrification potential in roadside wetlands. Environ. Pollut. 212, 216–223. https://doi.org/10.1016/j.envpol.2016.01.068.

Lee, B.D., Choi, Y.S., Kim, Y.G., Kim, I.S., Yang, E.I., 2017. A comparison study of performance and environmental impacts of chloride-based deicers and eco-label certified deicers in South Korea. Cold Reg. Sci. Technol. 143, 43–51. https://doi.org/10.1016/j.coldregions.2017.08.010.

Lembcke, D., Thompson, B., Read, K., Betts, A., Singaraja, D., 2017. Reducing road salt application by considering winter maintenance needs in parking lot design. J. Green Build. 12 (2), 1–12. https://doi.org/10.3992/1943-4618.12.2.1.

Mayer, R., Marsalek, P., Servos, B., Mcinnis, J., Scott, 2011. Environmental characterization of surface runoff from three highway sites in Southern Ontario, Canada: 2. Toxicology. Water Qual. Res. J. Can. 46 (2), 121–136. https://doi.org/10.2166/wqrjc.2011.036.

Riehm, M., Nordin, L., 2012. Optimization of winter road maintenance energy costs in Sweden: a critique of site specific frost warning techniques. Meteorol. Appl. 19 (4), 443–453. https://doi.org/10.1002/met.275.

Rivett, M.O., Cuthbert, M.O., Gamble, R., Connon, L.E., Pearson, A., Shepley, M.G., Davis, J., 2016. Highway deicing salt dynamic runoff to surface water and subsequent infiltration to groundwater during severe UK winters. Sci. Total Environ. 565, 324–338. https://doi.org/10.1016/j.scitotenv.2016.04.095.

Robinson, H.K., Hasenmueller, E.A., Chambers, L.G., 2017. Soil as a reservoir for road salt retention leading to its gradual release to groundwater. Appl. Geochem. 83, 72–85. https://doi.org/10.1016/j.apgeochem.2017.01.018.

Roy, J.W., McInnis, R., Bickerton, G., Gillis, P.L., 2015. Assessing potential toxicity of chloride-affected groundwater discharging to an urban stream using juvenile freshwater mussels (*Lampsilis siliquoidea*). Sci. Total Environ. 532, 309–315. https://doi.org/10.1016/j.scitotenv.2015.06.023.

Schuler, M.S., Relyea, R.A., 2018. A review of the combined threats of road salts and heavy metals to freshwater systems. Bioscience 68 (5), 327–335. https://doi.org/10.1093/biosci/biy018.

Snodgrass, J.W., Moore, J., Lev, S.M., Casey, R.E., Ownby, D.R., Flora, R.F., Izzo, G., 2017. Influence of modern stormwater management practices on transport of road salt to surface waters. Environ. Sci. Technol. 51 (8), 4165–4172. https://doi.org/10.1021/acs.est.6b03107.

Søberg, L.C., Viklander, M., Blecken, G.-T., 2017. Do salt and low temperature impair metal treatment in stormwater bioretention cells with or without a submerged zone? Sci. Total Environ. 579, 1588–1599. https://doi.org/10.1016/j.scitotenv.2016.11.179.

Stoler, A., Sudol, K., Mruzek, J., Relyea, R., 2018. Interactive effects of road salt and sediment disturbance on the productivity of seven common aquatic macrophytes. Freshw. Biol. 63 (7), 709–720. https://doi.org/10.1111/fwb.13110.

Suaire, R., Durickovic, I., Framont-Terrasse, L., Jean-Yves, L., Anne-Claire, De R., Marie-Odile, S., Phytoextraction of, N.+, Cl- by Atriplex halimus, L., Atriplex, hortensis L., 2016. A promising solution for remediation of road runoff contaminated with deicing salts. Ecol. Eng. 94, 182–189. https://doi.org/10.1016/j.ecoleng.2016.05.055.

Szklarek, S., Górecka, A., Salabert, B., Wojtal-Frankiewicz, A., 2022. Acute toxicity of seven de-icing salts on four zooplankton species– is there an “eco-friendly” alternative? Ecohydrol. Hydrobiol. 22 (4), 589–597. https://doi.org/10.1016/j.ecohyd.2022.08.005.

Szklarek, Stolarska, Wagner, Mankiewicz-Boczek, 2015. The microbiotest battery as an important component in the assessment of snowmelt toxicity in urban watercourses—preliminary studies. Environ. Monit. Assess. 187 (2). https://doi.org/10.1007/s10661-014-4252-1.

Tixier, G., Rochfort, Q., Lee, G., Marsalek, J., Lafont, M., 2012. Spatial and seasonal toxicity in a stormwater management facility: evidence obtained by adapting an integrated sediment quality assessment approach. Water Res. 46 (20), 6671–6682. https://doi.org/10.1016/j.watres.2011.12.031.

Tonk, L., Bosch, K., Visser, P.M., 2007. Jet Huisman, Salt tolerance of the harmful cyanobacterium *Microcystis aeruginosa*. Aquat. Microb. Ecol. 46 (2), 117–123. https://doi.org/10.3354/ame046117.

Chapter 28

Ecohydrological adaptation of inland waterways for enhancing multidimensional sustainability potential (WBSR)

Paweł Jarosiewicz and Maciej Zalewski
European Regional Centre for Ecohydrology of the Polish Academy of Sciences under the auspices of UNESCO, Łódź, Poland

COURTESY OF THE OFFICE FOR THE ODRA WATERWAY, SZCZECIN AND SWINOUJSCIE SEAPORTS AUTHORITY.

Transdisciplinary Ecohydrology. https://doi.org/10.1016/B978-0-443-41596-8.00046-7

1. Introduction

Ecohydrology, as presented throughout this book, is a transdisciplinary science that aims to integrate ecological and hydrological processes to achieve the sustainability of river basins. This approach contributes not only to the restoration of degraded ecosystems but also to the mitigation of environmental impacts associated with large-scale water infrastructure projects such as dams, river regulation, and inland waterway development. When implemented within such investments, ecohydrology can enhance system resilience, support biodiversity, and maintain ecosystem services, while balancing the economic and social needs that depend on water. Although any degradation of aquatic ecosystems must be avoided, sustainable management requires finding a compromise between economic stability and ecosystem health. One of the most relevant examples of this balance is found in inland waterways, which serve both as transport corridors and ecological systems. According to the OECD, there are approximately 623,000 km of navigable inland waterways worldwide, though less than 20% are actively used for freight transport. In recent years, efforts have been made to make inland navigation more sustainable, as exemplified by the FAIRway Danube Project (2023–27)—an EU cofunded initiative involving Austria, Slovakia, Hungary, Croatia, Bulgaria, and Romania, aimed at improving fairway conditions and making inland navigation more climate-resilient and environmentally responsible.

2. The Odra River as a case

The Odra River (German: Oder), Poland's second-longest river, flows for 854 km, of which approximately 742 km lie within Polish territory. Originating in the Czech Republic, it flows north through western Poland, forms part of the border with Germany, and discharges into the Baltic Sea via the Szczecin Lagoon. Historical regulation, initiated as early as the Middle Ages and intensified in the 19th–20th centuries, shortened the river's length by more than 20%, altering its natural morphology and floodplain dynamics. The upper part of Odra Basin is one of Poland's most industrialized and economically important regions, encompassing major urban centers such as Wrocław, Opole, and Szczecin. Concentration of industry, power generation, and agriculture has led to significant pressures on water quality, hydrological regimes, and ecosystem integrity.

In December 2016, Poland ratified the European Agreement on Main Inland Waterways of International Importance (AGN), under which the Odra Waterway (E30) is expected to meet the minimum Class IV navigability standard—a channel depth of at least 2.8 m and a width of 40 m to allow two vessels to pass safely.

In this context, it is important to mention that in the summer of 2022, the Odra River experienced one of the most severe environmental disasters in Central Europe in recent decades. Massive mortality of fish and aquatic organisms occurred along nearly 500 km of the river. The primary cause was a toxic bloom of the golden alga *Prymnesium parvum*, which produces prymnesins, potent toxins damaging gills and internal organs of fish and invertebrates.

The bloom was triggered by a combination of elevated salinity resulting from industrial and mining discharges, high water temperatures, low flow conditions, and nutrient enrichment. This event exposed the vulnerability of industrialized rivers to cumulative impact and underscored the urgent need for permanent integrated water quality monitoring, early warning systems, and nature-based restoration measures to reduce impacts and enhance ecological resilience to prevent similar crises in the future. Today, the Odra River should be regarded as a *novel ecosystem* (Hobbs et al., 2014), an ecological system significantly transformed by human activity and no longer retaining its natural attributes. Intensive regulation, channelization, and flood protection works have profoundly altered the river's hydromorphology, simplifying its flow patterns and reducing its self-purification capacity by cutting off floodplains from the main river channel.

3. The inland waterways debate

Poland's inland waterways network spans 3722 km, of which 3513 km are used by shipping, yet only 5.5% (206 km) meet international Class IV–V standards. Consequently, inland navigation plays a marginal role in national freight accounting for only 0.2% of total freight in 2019, compared with 0.8% in 2010.

Supporters of Odra modernization emphasize its potential to reduce road traffic, cut transport costs, and lower CO_2 emissions, aligning with the European transport policy objective of combining mobility with environmental protection. Since transport accounts for nearly 30% of CO_2 emissions in Europe (with 72% from road transport), promoting low-emission modes such as rail and inland navigation is considered beneficial.

However, from an ecohydrological perspective, river overregulation has led to simplified hydrology, reduced retention, and increased vulnerability to climate change. The construction of weirs and embankments fragments river habitats, restricts fish migration, and degrades floodplain ecosystems. According to WWF Poland (2021), the construction and maintenance of inland waterways on rivers like the Odra and Vistula would require enormous investments estimated at 213–237 billion PLN (Szymalski and Wiśniewski, 2020). Furthermore, to maintain navigability, 26–30 new dams would be required on the Odra, an intervention that would drastically alter its hydrological regime without guaranteeing year-round transport due to Odra catchment limited water resources. On the other hand, locks in the river help to maintain water in the river corridor tributaries and in consequence at the catchment during periods of droughts.

4. Ecohydrological framework for resilient inland waterways

Transcapillary ecohydrology offers a systemic framework for reconciling navigation and ecosystem protection and enhancement sustainability potential WBSR (Zalewski, 2014). By integrating hydrological, biological, and biogeochemical processes, it enables the design of inland waterways that sustain navigation while enhancing ecological resilience.

Key ecohydrological objectives for river adaptation and management include:

1. Developing innovative ecohydrological compensatory measures to mitigate navigation impacts.
2. Enhancing water retention in the catchment to support climate adaptation.
3. Improving water quality by reducing diffuse and point-source pollution, consistent with the EU Water Framework Directive.

Through deep understanding of catchment-scale processes, ecohydrology can minimize negative socio-economic impacts while strengthening the resilience of aquatic ecosystems to climate change and anthropopressure.

5. Ecohydrological solutions for sustainable inland navigation

It is essential to emphasize that ecohydrology as a scientific discipline does not support the development of inland waterways as an investment goal in itself. Large-scale river engineering projects should only be considered when their economic, social, and environmental benefits are clearly demonstrated and balanced according to the WBSR enhancement of multidimensional sustainability goal. In the absence of such justification, river restoration and the recovery of natural hydromorphological processes remain the preferred and more sustainable options for enhancing ecosystem services and climate resilience.

However, in cases where inland waterway development projects are restored, planned or implemented, ecohydrology can provide a conceptual and practical framework for mitigating their environmental impacts. In this context, the Odra Inland Waterway has served as a reference case for identifying and designing a set of ecohydrological solutions aimed at maintaining river functionality, supporting biodiversity, and enhancing ecological resilience under anthropogenic pressure. These solutions include.

5.1 Bypassing barriers and restoring connectivity

Fish migration can be supported through nature-like fishways and side channels that mimic natural stream conditions. These structures not only maintain river continuity but can also create new spawning and nursery habitats for rheophilic species. Managing water levels at weirs according to fish migration periods, based on hydroacoustic monitoring and ichthyological data, can further optimize ecological outcomes.

5.2 Increasing retention and restoring floodplain functionality

Widening or locally raising levees helps increase groundwater recharge and support wetland ecosystems. While downstream erosion can occur near dams, a cascade of low weirs may balance flow energy and sediment transport. Similar approaches under the Four Major Rivers Restoration Project (Korea) and along the Elbe River (Germany) showed positive effects on groundwater levels and floodplain wetlands (Lee et al., 2016; Matula et al., 2014). Floodplain restoration through lateral reconnection and controlled inundation should support self-purification - nutrient and pollutants transformation into biomass and enhance biodiversity and resilience to floods and droughts.

5.3 Ecofriendly shoreline design

To stabilize littoral zones, natural materials such as woody debris, boulders, and root systems should replace concrete reinforcements. Modified groyne fields and longitudinal training dams (LTDs) can diversify flow conditions and create

FIGURE 28.1 Oxbow lakes. Oxbow lakes are natural refuges of biodiversity in which they are numerous habitats, however without permanent water exchange can become highly euthrophic with the intensive algal blooms (see photo). *From: Zalewski, M. & Jarosiewicz, P. (2019). Ekohydrologiczne podstawy poprawy żeglowności i zwiększenia potencjału środowiskowego Odrzańskiej Drogi Wodnej.*

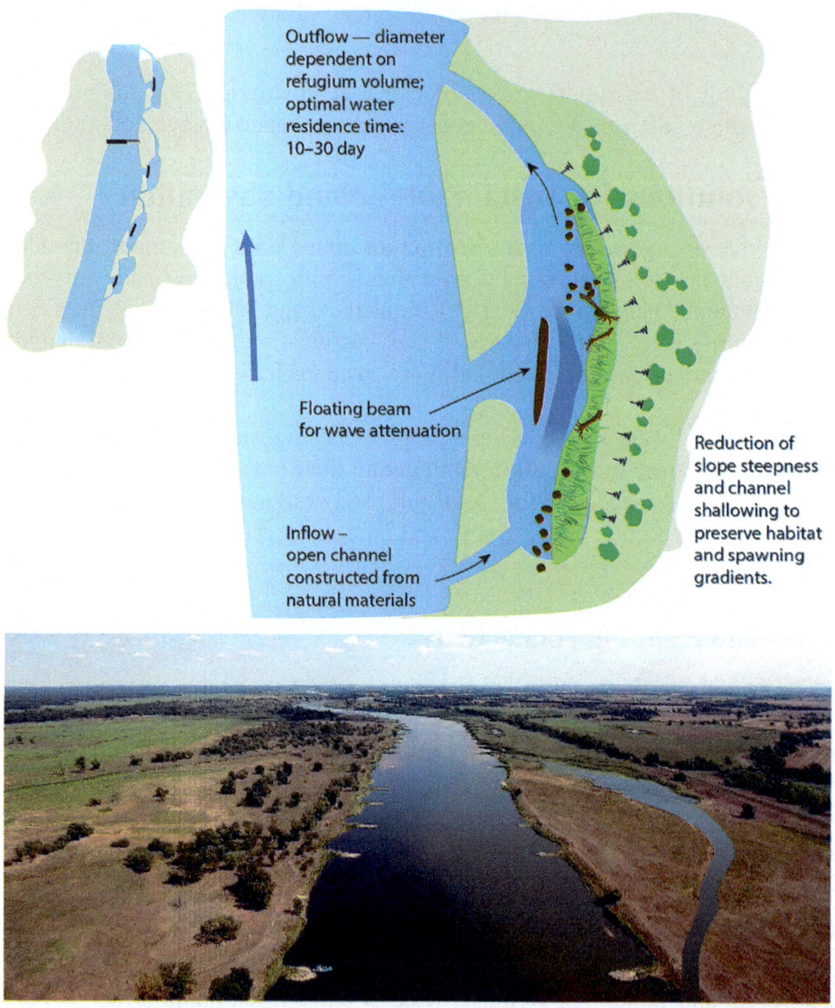

FIGURE 28.2 The constructed oxbow lakes at navigable river become natural refuges of biodiversity with numerous habitats. The periodic damming of water in the riverbed contributes to the recovery water resources in the river valley. Raising the water level affects the increase water retention in the riverbed, as well as the restoration and support of groundwater resources the functioning of terrestrial ecosystems and wetlands in the river valley. This is an example measures to adapt to climate change that enable the maintenance of reservoirs biodiversity in the river valley. *Modified from: Zalewski, M. & Jarosiewicz, P. (2019). Ekohydrologiczne podstawy poprawy żeglowności i zwiększenia potencjału środowiskowego Odrzańskiej Drogi Wodnej.*

habitats for fish and macroinvertebrates (Collas et al., 2022). LTDs implemented on the River Waal have successfully enhanced habitat diversity under heavy navigation pressure.

5.4 Reconnection of oxbow lakes

Reconnecting oxbow lakes with the main river channel helps restore flow dynamics, oxygenation, and habitat diversity. Oxbows should be designed with gradual shore slopes, variable depths, and sufficient hydraulic connectivity to ensure water exchange at least every 10–30 days. Such refugia support macrophyte diversity, mitigate eutrophication, and provide critical breeding and feeding habitats for aquatic organisms. When disconnected from the main river corridor, or with only limited (one side) hydrological exchange, these oxbows become eutrophic or hypereutrophic with intensive algal blooms Fig. 28.1; instead of supporting biodiversity, they turn into anoxic and methanogenic environments (Obolewski et al., 2018). Permanent water exchange - hydrological connectivity can therefore significantly enhance ecosystem functioning and shift the balance between respiration and primary production Fig. 28.2.

5.5 More sustainable barrier technologies

The reconstruction of groynes on the Oder River provides an example of how navigation infrastructure can be redesigned through the Transdisciplinary Ecohydrology (TEH) approach by integrating the Advanced Nature-based Solutions (ANbS) proposed in this chapter. These include floodplain extension through levee relocation, constructed oxbow lakes, and bypass channels that replace conventional fish ladders while creating diverse habitats for the migration, spawning, and juvenile rearing of rheophilic fish. Such interventions transform traditional river engineering into multifunctional infrastructure that simultaneously supports navigation, biodiversity conservation, water quality improvement, and climate resilience.

Instead of relying on repeated dredging, navigation depth can also be maintained through adaptive hydraulic structures such as movable crest weirs or inflatable rubber dams, which provide flexible water-level regulation while minimizing ecological disturbance (PIANC, 2018; Tam, 1998). When appropriately designed and operated, these structures facilitate adaptive flow management, improve sediment continuity, and support fish migration. Their operation should follow the natural hydrological regime, with gate openings and flushing flows during high-water periods used to mobilize and transport accumulated sediments, thereby maintaining channel capacity while preserving riverine processes (Kondolf, 1997; Peeters et al., 2020).

While maintaining the flow concentration required for navigation, groyne fields modified according to TEH principles create diverse ecohydrological conditions that enhance habitat heterogeneity, improve sediment dynamics, facilitate fish migration, provide spawning and refuge areas, and increase the water retention capacity of the river corridor. Collectively, these engineering and ecohydrological interventions demonstrate a transition from conventional single-purpose river regulation toward adaptive, multifunctional river management that integrates navigation with ecosystem restoration and long-term socio-ecological resilience (Fig. 28.3).

FIGURE 28.3 Reconstruction of Groynes on Section 1 from River Kilometre 581.0 to 585.7 (Świecko–Słubice/Frankfurt Reach). *Image credit: Regional Water Management Authority (PGW WodyPolskie) in Szczecin, Poland.*

6. Conclusions

The integration of ecohydrological principles into the planning and operation of inland waterways represents a systemic approach that aligns economic objectives with long-term sustainability. Rather than mitigating the environmental impacts of navigation alone, the proposed solutions enhance the capacity of river ecosystems to buffer climate change, improve biodiversity, restore sediment continuity, and strengthen ecosystem services. When adapted to local hydro-geomorphological and ecological conditions, ecohydrological interventions–including floodplain reconnection, adaptive sediment management, habitat restoration, and nature-based hydraulic infrastructure–can transform heavily modified rivers into resilient socio-ecological systems.

The Oder River illustrates both the ecological risks associated with conventional river regulation and the opportunities offered by the Transdisciplinary Ecohydrology approach. By integrating engineering with ecosystem processes through Advanced Nature-based Solutions, inland waterways can evolve from single-function navigation corridors into multi-functional river systems that simultaneously support navigation, biodiversity conservation, water quality improvement, climate adaptation, and sustainable regional development.

References

Collas, F.P.L., van Oudenhove, G.N.L., Buijse, A.D., Schoor, M.M., Leuven, R.S.E.W., 2022. Effects of longitudinal training dams on macro-invertebrate assemblages in the river Waal, 38(3). River Res. Appl., pp. 494–505

Hobbs, R.J., Higgs, E., Hall, C.M., Bridgewater, P., Chapin, F.S., Ellis, E.C., Ewel, J.J., Hallett, L.M., Harris, J., Hulvey, K.B., Jackson, S.T., Kennedy, P.L., Kueffer, C., Lach, L., Lantz, T.C., Lugo, A.E., Mascaro, J., Murphy, S.D., Nelson, C.R., Perring, M.P., Richardson, D.M., Seastedt, T.R., Standish, R.J., Starzomski, B.M., Suding, K.N., Tognetti, P.M., Yakob, L., Yung, L., 2014. Managing the whole landscape: historical, hybrid, and novel ecosystems. Front. Ecol. Environ. 12 (10), 557–564. https://doi.org/10.1890/130300.

Kondolf, G.M., 1997. PROFILE: hungry water: effects of dams and gravel mining on river channels. Environ. Manag. 21 (4), 533–551. https://doi.org/10.1007/s002679900048.

Lee, Y., Kim, H., Hong, Y., 2016. Revealed preference and effectiveness of public investment in ecological river restoration projects: an application of the count data model. Sustainability 8 (4), 353. https://doi.org/10.3390/su8040353.

Matula, S., Mekonnen, G.B., Báťková, K., Nešetřil, K., 2014. Simulations of groundwater-surface water interaction and particle movement due to the effect of weir construction in the sub-watershed of the river Labe in the town of Děčín. Environ. Monit. Assess. 186 (11), 7755–7770. https://doi.org/10.1007/s10661-014-3964-6.

Obolewski, K., Glińska-Lewczuk, K., Bąkowska, M., 2018. From isolation to connectivity: the effect of floodplain lake restoration on sediments as habitats for macroinvertebrate communities. Aquat. Sci. 80 (1). https://doi.org/10.1007/s00027-017-0556-x.

PIANC, 2018. Inflatable Structures in Hydraulic Engineering. PIANC Report No. 166. The World Association for Waterborne Transport Infrastructure (PIANC), Brussels, Belgium. ISBN 978-2-87223-254-3.

Peeters, A., Houbrechts, G., Hallot, E., Van Campenhout, J., Gob, F., Petit, F., 2020. Can coarse bedload pass through weirs? Geomorphology 359, 107131. https://doi.org/10.1016/j.geomorph.2020.107131.

Szymalski, W., Wiśniewski, J. (Eds.), 2020. Żegluga czy kolej? Perspektywy rozwoju zrównoważonego transportu w Polsce do 2050 roku. Fundacja WWF Polska.

Tam, W.M., 1998. Use of inflatable dams as agricultural weirs in Hong Kong. J. Hydraul. Eng. 124 (12), 1215–1226. https://doi.org/10.1061/(asce)0733-9429(1998)124:12(1215).

Zalewski, M., 2014. Ecohydrology and hydrologic engineering: Regulation of hydrology–biota interactions for sustainability. J. Hydrol. Eng. 19 (5), 1080–1088. https://doi.org/10.1061/(ASCE)HE.1943-5584.0000999.

Chapter 29

Lateral reservoirs for enhancing river valley sustainability and climate change adaptation

Maciej Zalewski[1], Edyta Kiedrzyńska[1, 2], Kamila Belka[1] and Paweł Jarosiewicz[1]
[1]*European Regional Centre for Ecohydrology of the Polish Academy of Sciences under the auspices of UNESCO, Łódź, Poland;* [2]*University of Lodz, Faculty of Biology and Environmental Protection, UNESCO Chair on Ecohydrology and Applied Ecology, Łódź, Poland*

GRABIA RIVER AT THE MIDDLE COURSE, AT THE PROPOSED AREA OF THE LATERAL RESERVOIR. *PHOTO CREDIT: M. ZALEWSKI.*

Transdisciplinary Ecohydrology. https://doi.org/10.1016/B978-0-443-41596-8.00047-9

1. Introduction

We are living in an era when planetary boundaries such as climate change, biodiversity loss, ocean acidification, land cover complexity ecosystems decline (deforestation) biogeochemical cycles (carbon, nitrogen, phosphorus), stratospheric ozone depletion, chemical pollution, and atmospheric aerosol loading, all moving beyond safe limits. This indicates that humanity has been approaching the Earth system's carrying capacity. At this stage of Man and Biosphere evolution Water becomes recognized by the UN (New York UN Conference, 2023) as a major factor for accelerating SDGs (United Nations, 2023). The major conclusions of this conference are: (1) Reduce fragmentation of efforts, which implicitly means also fragmentation of knowledge, (2) Use Nature-based Solutions developed on the basis integrative science ecohydrology, (3) Involve society into decision makers. The last element is the "*conditio sine qua non*" of holistic catchment management, since the widespread implementation of NbS and hybrid solutions across the catchment can only be maintained through collective societal effort. Considering the above complexity of environmental challenges, the holistic perspective (IWRM) becomes critical for sustainable catchment management, especially expanded by the idea of the enhancement of their sustainability potential Water–Biodiversity–Ecosystem Services–Resilience (WBSR) (Zalewski, 2014). The question is what in the face of progressing climate change should be the fundamental steps. No doubt it should be the increase the retentiveness of the catchment, which in turn reduces the stochasticity of hydrological processes, increase ground water recharge and reverse xerosere process—drying up ecosystems, which reduce their bio productivity and biodiversity especially in such vulnerable areas of the reduced biocomplexity as homogenized agricultural landscape.

2. Three steps toward water retentiveness in the WBSR framework

To enhance the sustainability potential of catchments and the biosphere within the WBSR framework, where water retentiveness becomes as the key element, the following three steps are necessary.

2.1 Increase catchment biomass and biocomplexity

The fundamental step to increase water retentiveness in the catchment is to increase biocomplexity and biomass by reforestation, shelter belts between fields and land water ecotones (buffer zones). Also, watering drying wetlands by excess water from intensive precipitation-river flood peaks and creating midfield ponds is very important. It is because increased catchment retention increases and stabilizes the amount of water in soil, groundwater recharge and moderate the stream run-off patterns (Falkenmark and Rockström, 2004). All those factors create the "sponge" effect, which prevents xerosere—decline of bioproductivity and biodiversity. There is an increasing amount of empirical evidence showing that forests have a large potential to increase precipitation through the "biotic pump" effect (Makarieva et al., 2013). This phenomenon is supported by findings from the European Environmental Report, which indicate that in water basins with 30% forest cover, water retention is 25% higher than in basins with only 10% forest cover. "Coniferous forests in general retain 10% more water than broad leaved forests or mixed forests." Moreover, spatial plant diversity of agricultural catchment seriously reduces pesticides and nutrients impact on ecological status of freshwater ecosystems and water quality. However, the high variability of the above regularities in different geographic regions indicate that to increase retentiveness of the catchment by reforestation it is necessary the analysis in considering the regional ecological and geomorphological conditions. The important reference point for development of transdisciplinary research on enhancement of the catchment sustainability potential WBSR, becomes the concept of ecohydrological resilience and the landscape water storage continuum during droughts (Tetzlaff et al., 2024).

2.2 Reduction of the outflow from agricultural land and forests drainage systems

The important aspect of the methodology to increase catchments retentiveness, in the face progressing climate changes, becomes the mitigation of water outflow from the drained agricultural land which e.g., in Poland occupies 6.4 million hectares, which means that the 20% of the country is drained (Marcinkowski et al., 2024). Such uncontrolled outflow from irrigated land not only reduces water quantity but also negatively affects water quality, by transferring nutrients to downstream situated freshwater and coastal ecosystems, stimulating its eutrophication and toxic algal blooms (Kiedrzyńska et al., 2014).

However, it is worth underlining that agricultural land drainage may improve soil conditions and in consequence crop yields prevent water logging, which in turn affects root system development and efficiency. Moreover, the high groundwater level reduces mycorrhizal activity, which supports nutrient uptake by plants and reduces their growth and crop yield. These important but complex processes, which are critical for enhancing catchment resilience, require advanced comparative transdisciplinary research across different types of catchments to highlight the interplay between surface ecohydrological processes and groundwater recharge in relation to geomorphology, soil types, and vegetation cover.

In urbanized areas, conventional drainage systems were historically designed to rapidly discharge stormwater from the city, typically without treatment, leading to significant river pollution. This can be positively transformed through the implementation of the Blue-Green Network Concept (Chapter 25). It provides a framework for converting the concrete city into a "sponge" by enabling stormwater purification through Sequential Sedimentation Biofiltration Systems (SSBS), followed by retention in a cascade of small reservoirs within green areas. The effect of such city-scale, systemic implementation of Ecohydrological Advanced Nature-based Solutions (ANbS) is the transformation of urban floods into groundwater recharge and increased evapotranspiration from green areas. This, in turn, improves the microclimate, becoming a key factor in urban adaptation to climate change. An important additional effect of this transformation—from "urban heat islands" to "urban oases"—is the reduction of water outflow to the oceans and a decrease in river pollution.

2.3 Lateral reservoirs for enhancement WBSR

The third important step for improving water retention in the catchment was initiated thousands of years ago with river damming (Wantzen et al., 2016). Initially, it was mainly for water supply and agriculture, and later for flood mitigation and hydropower generation. However, in recent times, dams have been considered by limnologists and environmentalists as factors degrading river continuum processes, fish migration, sediment transport, and consequently, river functioning and biodiversity (Belletti et al., 2020).

On the other hand, Stanford report (Stanford University, 2022) concluded that "analysis finds that dammed reservoirs could store more than 50% of the water needed to irrigate crops without depleting water stocks or encroaching on nature." As "more environmentally sound water storage for irrigation," the alternative for large dams they propose decentralized approach small dams, "because they lose less water due to evaporation, require less conveyance infrastructure and often create cobenefits for local communities and wildlife."

3. Natural, hydrotechnical, and lateral reservoirs at the Grabia River in Łask, Poland

It is necessary to take into consideration that small reservoirs situated in agricultural or urbanized catchments are often exposed to high nutrient pulses, which are converted into toxic algal blooms after intensive rains due to flash flooding. In such cases, most of the positive functions of these small reservoirs are drastically reduced. On the other hand, rivers of natural character (Fig. 29.1A), during periods of stable medium–low flow, possess a high self-purification potential to reduce nutrient concentrations below the critical limits for the appearance of toxic algal blooms. In parallel the resilience of reservoir depend on high biotic complexity includes littoral vegetation, which can be supported by constructed floating islands, and a high density of filtering zooplankton maintained through "top-down" control of zooplanktivorous fish by predatory fish. Additionally, resilience must be maintained through the periodic reduction of internal loads accumulated in reservoirs by removing bottom sediments, for example, through hydrosuction or siphon dredging.

The Lateral Reservoir concept was proposed as the alternative for typical hydrotechnical construction of a barrier that halts river flow and accumulates sediments, nutrients, and pollutants. Such typical dams blocking river (Fig. 29.1B) eliminate the natural free flowing section of the river with high biodiversity and fish migration. Because they accumulate sediments and nutrients, thus area downstream—floodplains become sediment starved in consequence, altering the habitats of the river valley. On the other hand, accumulation of sediments and nutrients in reservoir leads to stimulation of toxic algal blooms and diminishes most of the functions of the reservoir. As an alternative the proposition based on understanding ecohydrological processes was formulated (Fig. 29.1C). Lateral Reservoir located on one side of the floodplain. As a consequence, the wild, free flowing meandering part of the river will be maintained (Photo A). A fundamental step to maintain the good ecological status of the reservoir has to be the reduction of impact from the catchment, in above case pulses of nutrients and sediments inflow to the reservoir. To achieve that, the Automatic Clean Water Supply System was proposed based on the permanent water quality monitoring, which continuously measures turbidity, dissolved oxygen, pH, and conductivity. It should close water supply to the reservoir when suspended mineral and organic matter, nutrients concentration exceed critical values for eutrophication. It will be opened only when concentration of sediments and nutrients becomes low (Kiedrzyńska et al., 2021; Zalewski, 2020). In the case of River Grabia

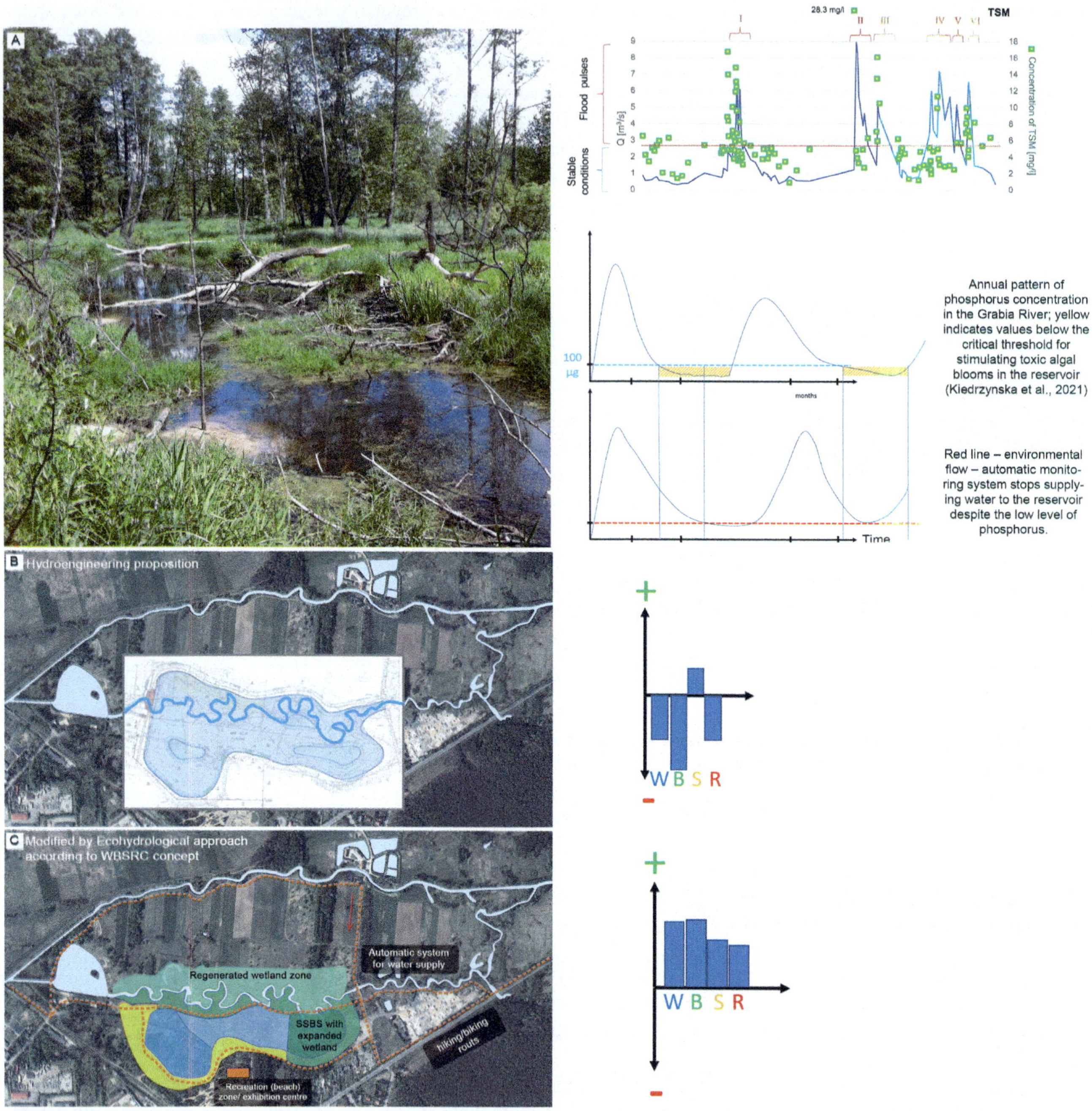

FIGURE 29.1 Lateral reservoir in Łask, Poland. (A) Natural stream, (B) Hydrotechnical proposal of a dam reservoir eliminating the natural section of the river with high biodiversity (see Photo A). In such traditional hydroengeenering construction accumulation of nutrients periodically contributed by the river from flood pulses would cause toxic cyanobacterial blooms, eliminating the reservoir as source of water supply and recreational use; (C) Lateral Reservoir—ecohydrological proposal for the location of the dam reservoir maintaining the natural part of the river on one side of the floodplain and ensuring good water quality for recreation, water supply to reservoir by authomathic monitoring system only during periods of low pollution concentration in river. Additionaly inflowing water will be purified by the Sequential Sedimentation Biofilter System (SSBS). *Based on consultating with the City of Łask (2018). Top diagram reproduced from: Kiedrzyńska, E., Belka, K., Jarosiewicz, P., Kiedrzyński, M., Zalewski, M., 2021. The enhancement of valley water retentiveness in climate change conditions. Sci. Total Environ. 798, 149191. https://doi.org/10.1016/j.scitotenv.2021.149191. (B and C) modified from: Zalewski, M. (Ed.). (2020). Ekohydrologia. Wydawnictwo Naukowe PWN. ISBN 978-83-01-20804-2.*

water quality analysis showed periodically high concentrations of nutrients, reaching 300 or even 500 µg P–PO_4/dm^{-3}. Such high concentrations of phosphorus can generate toxic cyanobacterial blooms during high periods temperatures, which would eliminate recreational use of the reservoir. To prevent such situations the supply of the reservoir by water from the river will be carried out during the periods when phosphorus concentrations in the river are below 100 µg $P\text{-}PO_4$/ dm^{-3}, this means in case of Grabia River approximately 60–90 days during a year.

4. Summary conclusions: Lateral reservoir enhances WBSR

The cascade of lateral reservoirs enhances and expands the water resources (**W**) of the river valley, increases catchment retentiveness and groundwater recharge—thereby improving biocomplexity, bioproductivity, and biodiversity (**B**). All, these processes generate numerous societal benefits (**S**), including agricultural productivity, recreation, and cultural and esthetic values. Ultimately, all these factors contribute to strengthening the overall resilience (**R**) of enhancement by mitigation of drought and desertification, thus adaptation river and its catchment to climate change.

The Lateral Reservoir Concept (LRC) is a good example of how the integration of hydrological and ecological processes understanding can be translated into measures for enhancing the sustainability potential within the WBSR framework. In this case, precise monitoring of hydrological peaking processes, specifically, the condensing and dilution stages dynamics reflected in the hydrograph, is combined with an understanding of freshwater ecosystem thresholds, such as nutrient concentrations that stimulate algal blooms. Importantly, enhancing the resilience of the river system through LRC should always follow three principles of ecohydrology beginning with monitoring of spatial, temporal hydrological process and pollution in the catchment scale. Next the distribution ecosystems in the catchment with the consideration of the protection of some of them and some (Novel) to be used as water management tools e.g., NbS land/water ecotone, purification systems SSBS, both significant for improvement of catchment retentiveness. Importantly, enhancing the WBSR of the river system through LRC should always begin with pollution reduction and the improvement of catchment retentiveness. All such transdisciplinary based systemic measures can be effectively implemented by learning-oriented governance and society involvement, both based on progress in Water/Sustainability Culture translated into Proactive Law. The Lateral Reservoir Concept is a good example of how the integration of hydrological and ecological understanding can be translated into enhancing the IWRM by the sustainability potential within the WBSR framework. In this case, precise monitoring of hydrological peaking processes, specifically, the concentration and dissolution dynamics reflected in the hydrograph, is combined with an understanding of freshwater ecosystem thresholds, such as nutrient concentrations that stimulate algal blooms.

References

Belletti, B., Garcia de Leaniz, C., Jones, J., Bizzi, S., Börger, L., Segura, G., Castelletti, A., Bund, W.van de, Aarestrup, K., Barry, J., Belka, K., Berkhuysen, A., Birnie-Gauvin, K., Bussettini, M., Carolli, M., Consuegra, S., Dopico, E., Feierfeil, T., Fernández, S., Fernandez Garrido, P., Garcia-Vazquez, E., Garrido, S., Giannico, G., Gough, P., Jepsen, N., Jones, P.E., Kemp, P., Kerr, J., King, J., Łapińska, M., Lázaro, G., Lucas, M. C., Marcello, L., Martin, P., McGinnity, P., O'Hanley, J., Amo, R.O.del, Parasiewicz, P., Pusch, M., Rincon, G., Rodriguez, C., Royte, J., Schneider, C.T., Tummers, J.S., Vallesi, S., Vowles, A., Verspoor, E., Wanningen, H., Wantzen, K.M., Wildman, L., Zalewski, M., 2020. More than one million barriers fragment Europe's rivers. Nature 588 (7838), 436–441. https://doi.org/10.1038/s41586-020-3005-2.

Falkenmark, M., Rockström, J., 2004. Balancing Water for Humans and Nature: The New Approach in Ecohydrology. Earthscan.

Kiedrzyńska, E., Belka, K., Jarosiewicz, P., Kiedrzyński, M., Zalewski, M., 2021. The enhancement of valley water retentiveness in climate change conditions. Sci. Total Environ. 799, 149427. https://doi.org/10.1016/j.scitotenv.2021.149427.

Kiedrzyńska, E., Kiedrzyński, M., Urbaniak, M., Magnuszewski, A., Skłodowski, M., Wyrwicka, A., Zalewski, M., 2014. Point sources of nutrient pollution in the lowland river catchment in the context of the Baltic Sea eutrophication. Ecol. Eng. 70, 337–348. https://doi.org/10.1016/j.ecoleng.2014.06.010.

Makarieva, A.M., Gorshkov, V.G., Li, B.L., 2013. Revisiting forest impact on atmospheric water vapor transport and precipitation. Theor. Appl. Climatol. 111 (1–2), 79–96. https://doi.org/10.1007/s00704-012-0643-9.

Marcinkowski, P., Piniewski, M., Okruszko, T., 2024. Towards sustainable agricultural water management in Poland – How to meet water demand for supplemental irrigation? Agric. Water Manage. 306.

Stanford University, 2022. Dams Could Play a Big Role in Feeding the World More Sustainably, Stanford Researchers Find. Stanford Report. https://news.stanford.edu/2022/11/14/dams-food-security/.

Tetzlaff, D., Laudon, H., Luo, S., Soulsby, C., 2024. Ecohydrological resilience and the landscape water storage continuum in droughts. Nat. Water 2 (10), 915–918. https://doi.org/10.1038/s44221-024-00300-y.

United Nations, 2023. NY UN Conference Water key driver for SDG, 22–24 March 2023. United Nations Headquarters, New York, USA. Water Conference Side Event: Transdciplinary Ecohydrology for Acceleration on SDG - Methodology of Science and Patterns of Implementaiton.

Wantzen, K.M., Ballouche, A., Longuet, I., Bao, I., Bocoum, H., Cissé, L., Chauhan, M., Girard, P., Gopal, B., Kane, A., Marchese, M.R., Nautiyal, P., Teixeira, P., Zalewski, M., 2016. River culture: an eco-social approach to mitigate the biological and cultural diversity crisis in riverscapes. Ecohydrol. Hydrobiol. 16 (1), 7–18. https://doi.org/10.1016/j.ecohyd.2015.12.003.

Zalewski, M., 2014. Ecohydrology and hydrologic engineering: regulation of hydrology–biota interactions for sustainability. J. Hydrol. Eng. 19 (5), 1080–1088. https://doi.org/10.1061/(ASCE)HE.1943-5584.0000999.

Zalewski, M., 2020. Ekohydrologia. Wydawnictwo Naukowe PWN. ISBN 978-83-01-20804-2.

Part III

Transdisciplinary ecohydrology in the world

Chapter 30

"How to make three omelettes with just one egg?"—Threefold resource-limitation of riverine habitat dynamics and possible ecohydrological solutions (the Loire River as a case study)

Karl Matthias Wantzen[1, 2, 3, 4], Stéphane Rodrigues[4] and Laurent Schmitt[5]

[1]*EUCOR Excellence Chair "Water and Sustainability", Institut Terre et Environnement de Strasbourg (ITES, CNRS/ENGEES UMR 7063), University of Strasbourg, Strasbourg, France;* [2]*Karlsruhe Institut fur Technologie (KIT), Institute for Water and Environment, Karlsruhe, Germany;* [3]*UNESCO Chair "Rivers and Heritage", University of Tours, Tours, France;* [4]*Cités, Territoires, Environnement et Sociétés (CITERES, CNRS UMR 7324) and Polytechnic School of Engineering, University of Tours, Tours, France;* [5]*Laboratoire Image, Ville, Environment (LIVE, UMR 7362), University of Strasbourg, CNRS, ENGEES, France*

"Only rarely, European rivers still have stream power to move sediments, availability of sediment material and of floodplain space where natural hydromorphological dynamics can take place (the "omelette" problem). The rewilding of the upper Moselle in France gives some hope for future redynamization projects. Image credit: Karl M. Wantzen"

Transdisciplinary Ecohydrology. https://doi.org/10.1016/B978-0-443-41596-8.00048-0

1. Introduction—the "omelette syndrome"

Climate change (CC) acts as a magnifying glass to reinforce the consequences of river mismanagement in the past, and to bring them to public awareness. Until recently, the erosion of biological and cultural diversities along rivers (Reid et al., 2019; Wantzen, 2024), their persistent heating up of the waters (Johnson et al., 2024), and the limited success in restoration (Haase et al., 2023) or in reducing the toxic load of rivers (Malaj et al., 2014) have largely been condoned (i.e., willingly taken for granted) as a collateral damage of the ever-needed (?) economical progress of human societies, and measures to mitigate or even halt the river polycrisis are still insufficient. These slowly advancing catastrophies rarely make it into the headlines of journals, they rather act as a "death by a thousand cuts" (Belletti et al., 2020), and even immanent, severe degradation menaces, such as the fate of the Pantanal (world's largest contiguous wetland with an outstanding biodiversity, people, and biocultural heritage), which are essentially threatened by the Hidrovia waterway being built (Wantzen et al., 2024), do not mobilize enough national and international audience to take action. However, in our days and in the near future, combined mismanagement- and climate change–driven mega events such as massive fish kills (as occurred in the Odra river (Köhler et al., 2024)), large human fatalities (as occurred in the Ahr and other rivers in 2021 (Wolf et al., 2024)), or the risk to shut down nuclear power plants due to lacking cooling water (Ahmad, 2021) may have a similar potential to mobilize politics and society to eventually take action in favor of nature as did the Sandoz chemical accident on the Rhine River in the 1980s (Schmitt et al., 2018; Wantzen et al., 2023b; Wantzen et al., 2022). Due to climate change, in each catchment, ecosystems generally need more water to work well and to absorb, store, and transport water. However, political decisions generally ignore even science-based minimum discharge thresholds and environmental flow requirements (Arthington et al., 2018), and consider nature's needs as of minor importance. At the same time, human water needs increase, e.g., for irrigation of previously rainfed agricultural production, for cooling overheated cities with evapotranspiration, or for hydropower production, for industrial activities, refrigeration, and e-mobility (Jones et al., 2024). Water conflicts, threatening freedom, welfare, and democracy, are immanent.

Unfortunately, the issue we present here is by no means spectacular, but it is a central problem for river restoration worldwide and an important thematics for transdisciplinary ecohydrology. And like the other thematics mentioned before, it has begun to become magnified by climate change and to contribute massively to the river polycrisis. The problem described here deals with one of the strongest human impacts on riverine biodiversity and ecosystem functions; however, with the currently lowest public attention, but it may become a highly critical issue in the context of CC mitigation and adaptation, and then, also call for action to avoid or mitigate catastrophies, especially when it comes to prolonged drought phases and the need to cope with unprecedented flood events. We call it: "*How to make three omelettes with just one egg?*" meaning that most rivers worldwide are undergoing a syndrome of triplicate, tightly interconnected limitations of essential resources: (1) *discharge* is limited and flow does not occur in natural dynamic patterns due to upstream damming, channelization, and water abstraction; therefore, sediments are hardly moved in a near-natural pattern, (2) the proper *sediments* that might become moved are not available anymore in sufficient quantity and quality (grain size composition), as they are retained by dams and stabilized banks or dredged for navigation or house construction, while island dynamics, woody debris, and riparian vegetation (all of which are of great importance for sediment dynamics) are strictly controlled (respectively, completely removed), and (3) (formerly) *dynamic floodplain space* has been largely transformed into areas for human use (cities, agriculture, industrial areas) or completely removed (in the case of islands, or gravel quarries within floodplains). The synergy of these three constraints results in that the natural hydrosedimentological dynamics of most intensively used rivers have practically ceased to exist along their entire course. And even the estuary zones, when seen from space, formerly known as "delta" (named according to their triangle-like shape, like the Greek letter), now appear more and more as an indented "beta," because sediments hardly arrive in the estuaries any more (e.g., the Ebro, Nile, or Mekong deltas, see examples in Wantzen et al. (2023)), while marine transgressions are additionally facilitated by sea-level rise (Ericson et al., 2006).

All concepts dealing with the ecological functions of rivers and their biological and cultural diversities (e.g., flood pulse (Junk et al., 1989; Junk and Wantzen, 2004; Tockner et al., 2000), ecohydrology (Zalewski, 2002), stream hydraulics and patch dynamics (Statzner and Higler, 1986; Townsend, 1989), intermediate disturbance and connectivity (Ward, 1989; Ward and Stanford, 1983), the Riverine Productivity Model and its follow-ups (Thorp et al., 2010), environmental flows (Anderson et al., 2019; Arthington et al., 2018), riverhood (Boelens et al., 2023), or the River Culture Concept (Wantzen et al., 2016) independently highlight the importance of the natural dynamics of river flow and sediment discharge. These dynamics diversify habitat structures, locally reinitiate succession processes, and decide over physical connectivity in three dimensions, creating spatial–temporal patterns that are by far more diverse than

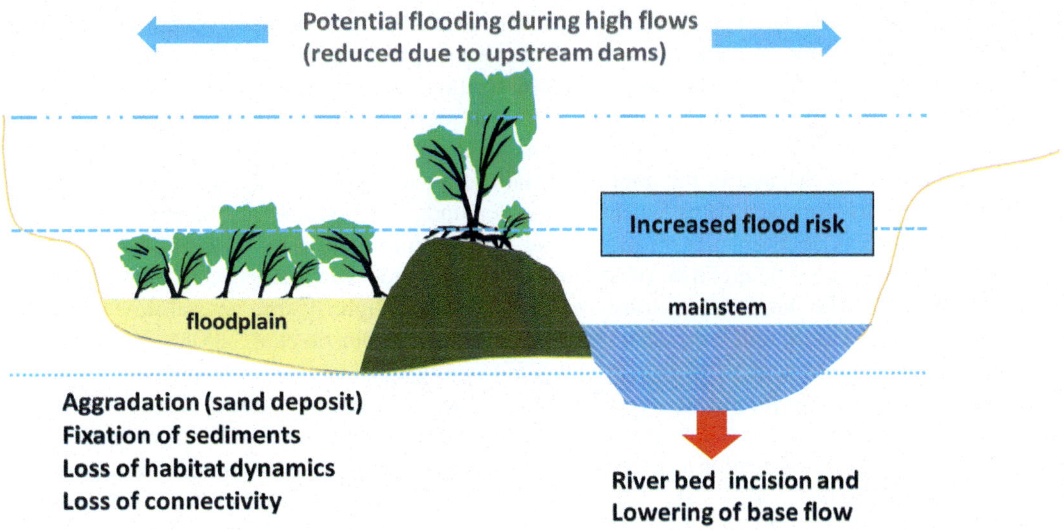

FIGURE 30.1 The "omelette syndrome". The "omelette syndrome" (i.e., the dilemma of lacking stream power plus lacking sediment supply plus lacking floodplain space) resulting in strongly reduced morphodynamics of rivers. *Graph by S. Rodrigues and Karl M. Wantzen.*

nonpulsing systems (Junk and Wantzen, 2006; Tockner et al., 1999), and, due to the cooccurrence of widely different life cycle strategies, they provide a "sequential" diversity of different strategies at the same place during different hydrological–sedimentological settings (Wagner and Schmidt, 2004; Wantzen, 2024). Moreover, sediment dislocation underwater "cleans" the interstitial pore space of the hyporheos by mobilizing fine particles and restores habitats for invertebrate and fish and reactivates biogeochemical activities by the microbial colonization of the sediment (Fischer et al., 2005; Marmonier et al., 2012). These essential processes and patterns become severely impacted when sediment movements become constrained.

Unfortunately, the "omelette syndrome" Fig. 30.1 occurs all over the world, as it is the inevitable result of regulating rivers for human use. The general consequences of this for ecological processes and biodiversity are all well-known (Grill et al., 2019; Reid et al., 2019; Tockner and Stanford, 2002); the consequences for cultural diversity, however, are much less so (Wantzen, 2024). One of the collateral, but natural effects of damming and flow reduction, is a strongly increased growth of woody vegetation on previously dynamic sediment bars or even within the stream bed, when these become exposed during low water levels. There are tipping points, at which the stream power is not sufficient any more to remove the vegetation by erosion (Becker et al., 2025; Corenblit et al., 2007). This leads to a further aggravation of the "omelette problem" described before, as erosion on the remaining mobility space of the river will promote riverbed incision, and further decoupling of the floodplain from the river (Bravard et al., 1997). Moreover, upstream riverbed incision and increased silting up of the overgrown (formerly dynamic) floodplain result in increased flood risk downstream.

Experiments to manage (or even prevent) vegetation growth on previously dynamic sediment bars, channel pattern restoration (Chadwick et al., 2025), and the potential for habitat restoration are poorly studied. Therefore, this article will focus on this phenomenon and trials to control it by river management, mostly using the Loire River as a case study.

2. Ecohydrological solution potential

The "omelette syndrome" concerns all tenets of ecohydrology (abbreviated as WBESRCE), i.e., **W**ater resources (the hydrology principle), **B**iodiversity and **E**cosystem **S**ervices (the Ecology Principle), **R**esilience to impacts, **C**ulture and **E**ducation (Elfithri et al., 2025; Zalewski, 2002) and likewise, all tenets of the River Culture Concept (Wantzen et al., 2016; Zalewski and Wantzen, 2023). The big challenge consists in harmonizing ongoing human water needs with the requirements of ecosystems to function and sustain an evolving biodiversity. A return to a pristine state or even the classical reference states is hardly possible, which is why the European Water Framework Directive has established the "ecological potential" at the place of the return to the reference status in the case of "heavily modified water bodies" (Carvalho et al., 2019).

In the case of reduced hydrosedimentological dynamics, several management options can be imagined:

1 *Pulse-like release of water and sediments from dams according to the natural flow regime.* During periods of originally high flows, e.g., springtime snowmelt or rain periods, water and stored sediments could be released from dams (Loire et al., 2019). However, stored sediments may have accumulated pollutants such as heavy metals or pesticides (Lepage et al., 2020). Water released during dam-flushing can mobilize fine sediments deposited downstream (Loire et al., 2019) and may further clog the hyporheic pore space in the downstream sections and make them inhabitable for fish fry and invertebrates (as debated in the context of dam removal (Tullos et al., 2016)). Moreover, water usually becomes released from hydropower dams when energy demands are highest (i.e., mostly during prolonged, hot, low-water phases), which is counteractive to the natural flow regime and may kill terrestrial organisms colonizing the dry-fallen floodplain, e.g., fledgelings of ground-breeding birds (Sparks et al., 1998; Wantzen, 2024). In some cases, appointments with the dam operator have succeeded in a slightly more natural flow and sediment regime, e.g., removal of filamentous algae and increase of benthos diversity (Robinson et al., 2003) and maintenance of migratory fish populations in the Dordogne (Michau et al., 2023) or Loire/Allier rivers (Wantzen et al., 2023a).

2 *Mobilization of sediments within the fluvial corridor.* Due to the "omelette syndrome" described above, mobilizable sediments are a rare resource in managed rivers, and, as with release from dams, the risk of pollutant remobilization is immanent. Technically, the mobilization can be achieved by removal of riprap embankments or dike replacement (Chardon et al., 2022; Díaz-Redondo et al., 2018; Eschbach et al., 2018; Schmitt et al., 2019), navigation groynes removal (Rodrigues and Berton, 2021), or by artificial addition of sediments (Chardon et al., 2021; Staentzel et al., 2018). Along the reach of the "Old Rhine" by-passed by the Grand Canal d'Alscae (i.e., from Kembs to Neuf-Brisach), several of such actions were tested. Gravel augmentations downstream the Kembs diversion dam diversified fluvial habitats and dynamics only over short to median temporal scales as a condition of sediment starvation occurring again after only about 5 years (Chardon et al., 2021). Consequently, a more efficient and sustainable strategy may consist in enlarging the channel downstream the gravel augmentation sites. This configuration occurred unexpectedly during the Q_{30} flood of July 2021 (Chardon et al., 2025), Fig. 30.2, when a large and continuously dynamic sediment bar was created by a flood event. Another, however planned, experiment consisted in the removal of bank protection combined with the construction of two groynes, resulting in a very weak recovery of fluvial processes, as bank erosion was limited by flow diversion and local constraints (Chardon et al., 2020, 2022), while habitat diversification was artificially enhanced due to the presence of groynes (Staentzel et al., 2019). In terms of biodiversity, the reaction of the faunal assemblages depends much on quality of the source material (especially the granulometry). In this experiment, mostly fine sediments were added, and the accumulation of these did not yield the expected increase in sensitive species. Rather, it resulted in an increase of psammophilous fauna, while these pioneer habitats were also attractive for invasive gobies and crustaceans (Staentzel et al., 2018, 2019).

3 *Flow diversification by longitudinal separation of the river in zones with different flow velocities and morphodynamics.* Canalization often goes along with by-passing downstream a diversion dam, which separates the original main channel from completely artificial sections that become used for navigation and hydropower generation. In terms of discharge, preference is given to these human uses, whereas the dotation of the remaining natural channel has a

FIGURE 30.2 Spontaneous sediment bar development. Aerial picture (taken in July 2021) of a new bar formed during a 30 years' flood in front of the upstream connection of a German flood control area at Efringen-Kirchen. Four years later, this bar is still in place while moving slightly during floods (Chardon et al., 2025). Upstream at the foreground of the image.© Regierungspräsidium Freiburg, 2021. *Image credit: Regierungspräsidium Freiburg.*

lower priority. In the case of the Rhine river, the "Grand Canal d'Alsace" on the Rhine from Kembs to Neuf-Brisach has profoundly altered the functioning of the remaining by-passed sections ("Old Rhine"/"Vieux Rhin"/"Restrhein"), including riverbed incision due to sediment deficit, bed armoring, loss of habitat dynamics, of biodiversity and of ecosystem functions (Wantzen et al., 2022). Similar patterns were found in the Rhône River, despite a different form of by-passing (Olivier et al., 2022). Several projects have been undertaken to return floodplain areas, to locally provide erodible areas in the riverbed, and increased bedload transport (see above however, the quota of remaining water to possibly redynamize the fluvial landscape is generally too low (Schmitt et al., 2018). In the Netherlands, the "Room for the River" (Rijke et al., 2012) project has taken an effort to reestablish lateral connectivity and to allow a more natural erosion/deposition regime. Here again, the lacking discharge from the upstream sections is a critical issue.

4 *Abandonment of* existing *or of planned river regulation.* Due to the incompatibility of certain water use practices with a sustainable development of a growing human population, the question arises, which of these practices can still be adapted toward a sustainability transformation, and which are incompatible, and consequently need to be abandoned. In river management planning, conservation has so far only received minor attention, and most nature reserves or places where restoration measures are taken, are "leftover" places, which would require too high investments to be used intensively (Higgins et al., 2021), while navigation continues to have a major impact on fluvial biodiversity (Jeliazkov et al., 2024; Sexton et al., 2024). However, in the context of globally needed sustainability transformation, the idea of abandonment of existing and the nonrealization of planned large-scale projects cannot be longer ignored. Many rivers are still maintained as navigation waterways despite their economical uselessness has been proven for decades. Efforts to maintain them navigable exceed the economical return by their use by far. Maintenance of these rivers as waterways often involves higher flood risks, as floodplains are usually disconnected, and results in river bed incision, thus aggravating the effects of longer droughts due to climate change. Moreover, a number of other key ecosystem services of rivers are affected that are, however, urgently needed, such as availability of drinking water resources, provisioning of habitats for threatened species (e.g., salmon or sturgeon species), water quality regulation by self-purification, or recreational opportunities. Often, the involved institutions prefer to maintain the status quo of these waterways to maintain jobs and to avoid falling into insignificance, despite the tonnage of transported goods being very low (e.g., Elbe; Odra, Weser rivers, Brunar & Pusch, communication arising) or completely absent (Lahn River, Germany). On the contrary, money saved by the abandonment of the useless navigable function could significantly boost functional restoration programs. In other cases, waterways are planned and may be installed despite that hydroclimatological and economical data clearly show the inviability of the project (as it is the case of the Tramo Norte of the Hidrovia on the Paraguay River, Brazil, Bolivia, Argentina) notwithstanding the deleterious effects on national and World Heritage, Ramsar sites, and Indigenous Reserves and their biocultural diversities (Wantzen et al., 2024). Reconsideration of such projects is urgently needed.

3. The Loire case

Despite the fact that the Loire River is often said to be "one of the last wild rivers of Europe," it has been managed for several centuries. Great initiatives by NGOs such as SOS-Loire Vivante/European Rivers Network and WWF have succeeded in keeping most of its main channel free of dams until today (Wantzen et al., 2023a). With a length of 1012 km, it is the longest national river of France, its basin covers one-fifth of its territory (Moatar et al., 2022), but its discharge is relatively low (150-year average at Montjean s/Loire: 843 m^3/s; min: 250 m^3/s (August); max: 1500 m^3/s (February)). The two main water suppliers, the Allier River and the upper Loire main course, have their catchments in the southwestern Massif Central (ca. 800m asl). Along their flow in a south-north direction, they can be considered as the hydro-sedimentological "powerhouse" of the Loire. In this steep gradient, several dams reduce the sediment transport and flatten the annual flood pulse. Near Nevers, the Loire takes a western direction until reaching the Atlantic in the wide estuary of St. Nazaire. This part of the river mostly flows in lowland areas, as a typical "sandy river" (Wantzen et al., 2014). Lateral levees built since the 12th century separate the river from a large part of its floodplain, but a certain degree of freedom for lateral fluvial morphodynamics is still maintained on a width of about 200–3000 m, mainly due to the sandy bedload, allowing the development of mobile dunes and partially to fully migrating alluvial bars (Rodrigues et al., 2015). The navigation history is very informative concerning morphodynamics: From the Middle Ages to preindustrial times, the architecture of the traditional boats has been adapted to the variable water depths and river training had a relatively limited impact on morphosedimentary dynamics, whereas during a very short phase of steamboat navigation, dikes and groynes were intensified (Wantzen et al., 2023a).

Currently, the most pressing ecohydrological problem is the ongoing riverbed incision, which has been caused by the reduced channel width, sediment retention in upstream dams, legacies from previous training works for navigation in the 19th century and massive sediment mining between 1950 and 1995 (220 Mt between Nevers and Nantes between 1949 and 1992) (Claude et al., 2012; Le Guern, 2021). Riverbed incision of the middle and downstream reaches have been documented since 1858 by Latapie et al. (2014) and Andréault (2024). The median value of channel bed variation in elevation on the reach between Nevers and Ponts-de Cé was equal to −0.3 m between 1978 and 1996; however, this trend was leveling off in the following years (incision rates of −27 mm/yr, −11 mm/yr, −0.502 mm/yr between 1978–1984, 1984–1996, and 1995–2020, respectively). Although reaches located between Nevers and Blois show a general stabilization of the river bed, reaches located downstream Blois still experience bed lowering (Andréault, 2024). This incision and the reduced sediment transport cause loss of connectivity between the floodplain and the mainstem and a decrease in habitat dynamics and biodiversity (Andréault et al., 2024). They result in increased flood risk and increasing fixation of sand banks by woody vegetation, combined with invasions of exotic plants such as the neotropical macrophyte *Ludwigia peploides*, which enhances the terrestrialization process in remaining floodplain water bodies (Thouvenot et al., 2013). Moreover, along with the impact of decreasing stream power and lateral erosion capacity on sediment deposits, sand and gravel bars begin to be overgrown by woody vegetation (diverse willow species and black poplar), which is further limiting their mobility (Figs. 30.3 and 30.4) (Rodrigues et al., 2006, 2007; Rodrigues and Gautier, 2007). The natural flow regime provides ideal conditions for these soft-wood species (Corenblit et al., 2007, 2014; Garófano-Gómez et al., 2017; Wintenberger et al., 2019), which accelerates the transformation of dynamic floodplain space into vegetation-fixed banks and islands.

To solve the dilemma of "lacking stream power *plus* lacking sediment supply", several restoration schemes were applied.

In the *BioMareau project*, reopening of lateral channels and removal of woody plants on a nonmigrating midchannel bar by caterpillars (Fig. 30.5A and B) have been made to allow sediment dynamics again. These measures have mitigated the problem only temporarily, as the softwood vegetation recovered quickly (Wintenberger et al., 2015). However, they highlighted the crucial role of sediment dynamics (erosion/deposition) on the uprooting process affecting the young seedlings of *Populus nigra* (Wintenberger et al., 2019). Three types of processes leading to mortality of seedlings during floods were identified and their contribution quantified: (1) uprooting by drag force, (2) uprooting by erosion of sediments and drag force, and (3) burial. The combination of these three causes of mortality explains more than 90% of the seedlings death on the bar studied. Still, the artificial removal of woody vegetation mimicked the erosion events on the vegetation, rejuvenated the succession process, and, in combination with a more or less natural flow regime, increased plant biodiversity, especially that of annuals (Janssen et al., 2023).

FIGURE 30.3 Alluvial bar. Alluvial bar overgrown by woody vegetation at the Loire River (center) and still-migrating bars (fore and background). Flow direction from the right to the left. *Image credit: K. M. Wantzen.*

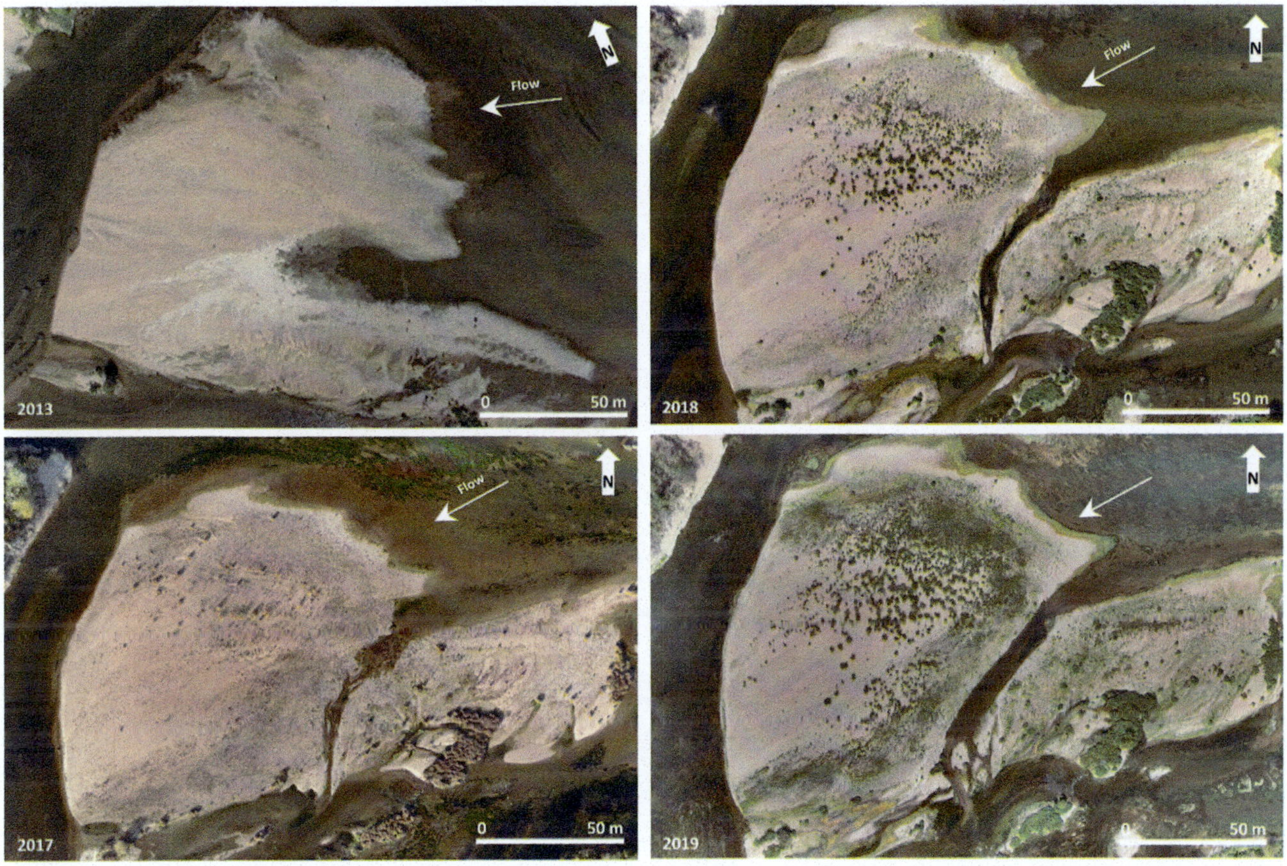

FIGURE 30.4 Regrowth after vegetation removal. Aerial views of the midchannel bar of Mareau-aux-Près in 2013 (upper left) after the vegetation removal operations carried out in 2012 (*Image credit: C.L. Wintenberger*). The three last panels of the figure show the recolonization by poplars and willows in 2017, 2018, and 2019 (upper right, lower left, lower right), respectively. *Images credit: O. Denux.*

In the Lower Loire (*R-TEMUS project*), measures to remove the vegetation overgrowing previously dynamic sediment deposits by caterpillars were combined with removal of 104 obsolete groynes on a 10 km reach near Chalonnes sur Loire in 2009 (Berton and Rodrigues, 2020; Rodrigues and Berton, 2021). Since then, this operation has been conducted on two other sites (between Montjean-sur-Loire and Ingrandes—Le Fresne-sur-Loire and between Anetz and Oudon) and combined to the construction of a low-head ground sill (at the location of a natural rocky outcrop that had been removed in 1976 to facilitate navigation) in the lowermost section of the river to decrease the water slope and the intensity of tidal erosion in the estuary. These operations, combining nature-based solutions and restoration works, are part of a larger restoration program (Contrat pour la Loire et ses Annexes, 2014) targeting the limitation of fluvial incision and flood risk and fostering the development of fluvial habitats and biodiversity. The target of this project was to give more space to the river channels, promote erosion of the sandy sediments deposited between groynes and in the side channels, and reduce river bed incision for the next 40 years. On the long run, and in the perspective of the ongoing climate change, a reappraisal of the annual discharge schemes at the hydropower dams in the upper Loire will be needed to mobilize sufficient amounts of water and sediments during flood pulses. Conflicts of interests, e.g., by groups interested in preserving the industrial heritage of the groynes (Barraud et al., 2013), or by conservationists claiming to maintain the newly established trees on the sediment banks, had to be overcome. In this context, the R-TEMUS project highlighted the efficiency of groyne removal on bed incision and sediment deficit in the main channel. Average bedload discharge upstream of the estuary was estimated to 310,000 t/yr for the period 1990–2020 (Le Guern et al., 2023). Results also documented the terrestrialization effect induced by side channel disconnection on plant communities and soil seed banks (Gaudichet et al., 2022).

(A)

(B)

FIGURE 30.5 Removal of vegetation using machines. (A) Aerial view of a side channel of the Loire River at separate as Mont-louis sur Loire that experienced removal of poplar and willow vegetation, by bar surface plowing, to reduce flood risk and remobilize sediments. *Image credit: C.L. Wintenberger.* (B) Woody vegetation removal on the nonmigrating bar of the site of Mareau au Près during summer 2012. *Image credit: D. Hemeray.*

4. Conclusions and a (hypothetical) outlook

The results presented from the Loire fall into the categories 2 and 3 of the previously cited management options. The Loire case shows that—despite limited availability of stream power (due to the dams upstream)—a certain redynamization of sediments is possible. This, however, requires the use of heavy machineries, is costly, and has only limited sustainability. The plowing of sediment surfaces with machines cuts the root systems of the willows and poplar into pieces and rather facilitates a dense regrowth. These soft-wood species have a vegetative reproduction strategy based on breaking and resprouting plant parts. Thus, the duration of the effect would need to be combined with a subsequent erosion event (e.g., by releasing substantial quantities of water from the reservoirs) to wash out the root pieces. Similarly, redigging the channels that may connect the mainstem and the lateral floodplain water bodies (french: "*annexes hydrauliques*") has often been observed to be counteracted by subsequent siltation, which is often facilitated by fast-growing aquatic macrophytes such as *Ludwigia* spp., which produce a dense network of vegetation that slows down the water velocity and increases sediment (and plant biomass) deposition. In the past, fishermen's associations often initiated and financed removal of invasive macrophytes; however, as with the willows, the regrowth of the macrophytes limits the sustainability of these measures.

The option to release water in very large quantities to facilitate natural erosion and even uprooting of woody vegetation has not (yet) been tested, as the dams in the upstream sections of Loire and Allier are strictly controlled to protect from deleterious floods, to produce hydropower, and to maintain sufficient discharge to warrant the cooling of the nuclear power plants along the Loire. In theory, such an option could be combined with measures to facilitate erosion and uprooting, e.g., by mechanically breaching the upstream sections of recently overgrown sediment bars and by increasing

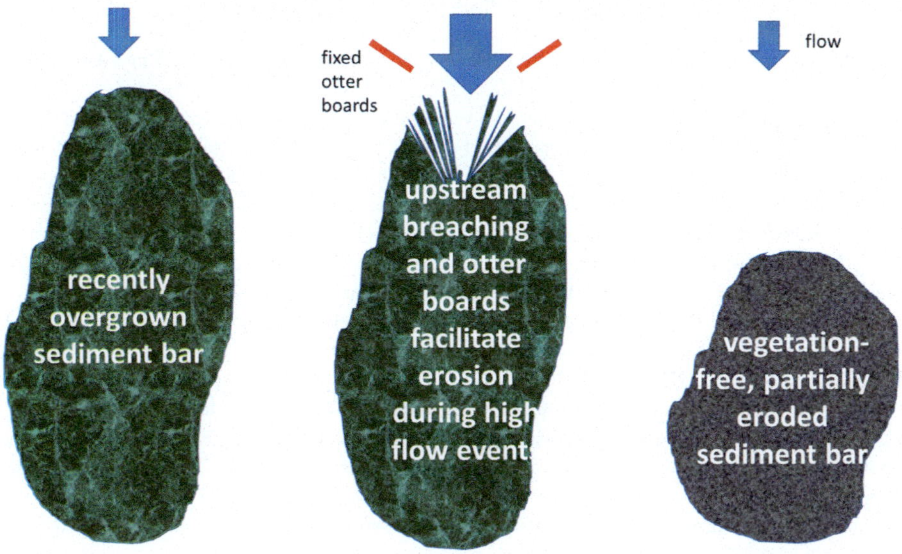

FIGURE 30.6 Hypothetical technique for the removal of softwood vegetation overgrowing previously dynamic sediment bars. From left to right: Hypothetical technique for the removal of softwood vegetation on sediment bars to remobilize sediments by combining mechanical breaching of root systems at the upstream side of the bar, combined with increased flow (release from upstream dams) and temporarily fixed otter boards to direct flow toward the breach. For this technique, mechanical removal of plants, fixation of the otter boards, and a targeted release of water to create a flash flood would need to be synchronized. *Graph by Karl M. Wantzen.*

traction forces shortly before an artificial high flow event. Additionally fixed otter boards could direct flow toward the breached zone (Fig. 30.6, from left to right). Experiments might help to find out the most efficient constellation between discharge increase and flow direction. The advantage of this method compared with the use of caterpillars to uproot softwood vegetation would be that the root system is largely maintained intact and that postmanagement sprouting from root fragments is minimized (as it happened in the case of the BioMareau project). Of course, a balance needs to be found between the removal of vegetation and the maintenance of woody habitats. The space limitation in the restoration of floodplain habitats is very critical, as the rarefaction of certain habitats contradicts the reestablishment of natural patch dynamics.

Acknowledgments

This chapter has been written under the auspices of the UNESCO Chair "Rivers and Heritage" and the EUCOR—The European Campus—Excellence Chair on "Water and Sustainability". SR received financial support by the BioMareau (M. Villard, coord., funded by the Région Centre-Val de Loire and the European Union [RCVL grant n° PRESAGE 37729 and FEDER grant n°2016-EX00604]) and the R-TEMUS2 projects (funded by the European Union [FEDER grant n. 2017-EX002207] and the Agence de l'Eau Loire-Bretagne [grant no. 2017C005]). KMW acknowledges support by the *Contrat triennal "Strasbourg, capitale européenne,"* Bürkert, EDF, Wattwiler Group, Kronenbourg, Véolia, Soprema, Heppner, Eurometropole Strasbourg, Région Grand-Est, Collectivité européenne Alsace, Ports Autonomes de Strasbourg, the Interdisciplinary Institute SWITCH at Uni Strasbourg, the Karlsruhe Institute of Technology and ENGEES. The authors acknowledge the contributions by Iris Brunar, BUND Germany and Martin Pusch, IGB Berlin, for information on the navigability of rivers.

References

Ahmad, A., 2021. Increase in frequency of nuclear power outages due to changing climate. Nat. Energy 6 (7), 755–762. https://doi.org/10.1038/s41560-021-00849-y.

Anderson, E.P., Jackson, S., Tharme, R.E., Douglas, M., Flotemersch, J.E., Zwarteveen, M., Lokgariwar, C., Montoya, M., Wali, A., Tipa, G.T., Jardine, T.D., Olden, J.D., Cheng, L., Conallin, J., Cosens, B., Dickens, C., Garrick, D., Groenfeldt, D., Kabogo, J., Arthington, A.H., 2019. Understanding rivers and their social relations: a critical step to advance environmental water management. WIREs Water 6 (6), e1381. https://doi.org/10.1002/wat2.1381.

Andréault, A., 2024. Structure verticale d'un fleuve à chenaux multiples dérivée de données LiDAR topo-bathymétriques et d'un modèle de mélange gaussien: implication pour la morphodynamique et la connectivité latérale de la Loire Tours.

Andréault, A., Rodrigues, S., Gaudichet, C., Wintenberger, C.L., 2024. Statistically derived morphological signatures of large river channels extracted from topo-bathymetric LiDAR data. Earth Surf. Process. Landf. 49 (2), 804–820.

Arthington, A.H., Bhaduri, A., Bunn, S.E., Jackson, S.E., Tharme, R.E., Tickner, D., Young, B., Acreman, M., Baker, N., Capon, S., Horne, A.C., Kendy, E., McClain, M.E., Poff, N.L., Richter, B.D., Ward, S., 2018. The Brisbane declaration and global action agenda on environmental flows (2018) [policy and practice reviews]. Front. Environ. Sci. 6 (45). https://doi.org/10.3389/fenvs.2018.00045.

Barraud, R., Carcaud, N., Davodeau, H., Montembault, D., Pordoy, C., 2013. Les épis de la Loire armoricaine, un héritage à la patrimonialité incertaine. Norois 228, 39–52. https://doi.org/10.4000/norois.4729.

Becker, I., Egger, G., Paul, M., Betz, F., 2025. Biogeomorphic tipping point in braided rivers: vegetation development at a restored Upper Drava River section (Austria). Hydrobiologia 1–17.

Belletti, B., Garcia de Leaniz, C., Jones, J., Bizzi, S., Börger, L., Segura, G., Castelletti, A., van de Bund, W., Aarestrup, K., Barry, J., Belka, K., Berkhuysen, A., Birnie-Gauvin, K., Bussettini, M., Carolli, M., Consuegra, S., Dopico, E., Feierfeil, T., Fernández, S., Zalewski, M., 2020. More than one million barriers fragment Europe's rivers. Nature 588 (7838), 436–441. https://doi.org/10.1038/s41586-020-3005-2.

Berton, J.P., Rodrigues, S., 2020. 25 ans de restauration du lit de la Loire. Entre visions hydrauliciennes, morphodynamique et écologique. *303—Paysages culturels du Val de Loire*, 2000–2020. Comparaisons, évolutions et transformations 147–154.

Boelens, R., Escobar, A., Bakker, K., Hommes, L., Swyngedouw, E., Hogenboom, B., Huijbens, E.H., Jackson, S., Vos, J., Harris, L.M., 2023. Riverhood: political ecologies of socionature commoning and translocal struggles for water justice. J. Peasant Stud. 50 (3), 1125–1156.

Bravard, J.P., Amoros, C., Pautou, G., Bornette, G., Bournaud, M., Creuzé Des Châtelliers, M., Gibert, J., Peiry, J.L., Perrin, J.F., Tachet, H., 1997. River incision in south-east France: morphological phenomena and ecological effects. Regul. Rivers Res. Manag. 13 (1), 75–90. https://doi.org/10.1002/(SICI)1099-1646(199701)13:1<75::AID-RRR444>3.0.CO;2-6.

Carvalho, L., Mackay, E.B., Cardoso, A.C., Baattrup-Pedersen, A., Birk, S., Blackstock, K.L., Borics, G., Borja, A., Feld, C.K., Ferreira, M.T., Globevnik, L., Grizzetti, B., Hendry, S., Hering, D., Kelly, M., Langaas, S., Meissner, K., Panagopoulos, Y., Penning, E., Solheim, A.L., 2019. Protecting and restoring Europe's waters: an analysis of the future development needs of the Water Framework Directive. Sci. Total Environ. 658, 1228–1238. https://doi.org/10.1016/j.scitotenv.2018.12.255.

Chadwick, A.J., Greenberg, E., Ganti, V., 2025. Single- and multithread rivers originate from (im)balance between lateral erosion and accretion. Science 389 (6756), 146–150. https://doi.org/10.1126/science.ads6567.

Chardon, V., Schmitt, L., Arnaud, F., Piégay, H., Clutier, A., 2021. Efficiency and sustainability of gravel augmentation to restore large regulated rivers: insights from three experiments on the Rhine River (France/Germany). Geomorphology 380, 107639. https://doi.org/10.1016/j.geomorph.2021.107639.

Chardon, V., Schmitt, L., Clutier, A., 2022. Bank re-erosion action to promote sediment supply and channel diversification: feedback from a restoration test on the Rhine. River Res. Appl. 38 (5), 975–987. https://doi.org/10.1002/rra.3968.

Chardon, V., Barillier, A., Loire, R., Schmitt, L., 2025. Combined gravel augmentation and flood control area to rehabilitate riverbed diversification: lesson from unexpected effects of a 30 years flood (Upper Rhine). River Res. Appl. 41 (9), 1819–1831. https://doi.org/10.1002/rra.70020.

Chardon, V., Schmitt, L., Piégay, H., Beisel, J.-N., Staentzel, C., Barillier, A., Clutier, A., 2020. Effects of transverse groynes on meso-habitat suitability for native fish species on a regulated by-passed large river: a case study along the Rhine River. Water 12, 987. https://doi.org/10.3390/w12040987.

Claude, N., Rodrigues, S., Bustillo, V., Bréhéret, J.-G., Macaire, J.-J., Jugé, P., 2012. Estimating bedload transport in a large sand–gravel bed river from direct sampling, dune tracking and empirical formulas. Geomorphology 179 (0), 40–57. https://doi.org/10.1016/j.geomorph.2012.07.030.

Corenblit, D., Tabacchi, E., Steiger, J., Gurnell, A.M., 2007. Reciprocal interactions and adjustments between fluvial landforms and vegetation dynamics in river corridors: a review of complementary approaches. Earth Sci. Rev. 84 (1–2), 56–86. Available from: http://www.sciencedirect.com/science/article/B6V62-4NX2NM7-1/2/31f73918514f08c4dee502b03ec82498.

Corenblit, D., Steiger, J., González, E., Gurnell, A., Charrier, G., Darrozes, J., Dousseau, J., Julien, F., Lambs, L., Larrue, S., 2014. The biogeomorphological life cycle of poplars during the fluvial biogeomorphological succession: a special focus on Populus nigra L. Earth Surf. Process. Landf. 39 (4), 546–563.

Díaz-Redondo, M., Egger, G., Marchamalo, M., Damm, C., de Oliveira, R.P., Schmitt, L., 2018. Targeting lateral connectivity and morphodynamics in a large river-floodplain system: the upper Rhine River. River Res. Appl. 34 (7), 734–744. https://doi.org/10.1002/rra.3287.

Elfithri, R., Zalewski, M., Arduino, G., Chicharo, L., 2025. Transdisciplinary ecohydrology for water management solutions and sustainability. Nat. Water 3 (4), 360–363. https://doi.org/10.1038/s44221-025-00395-x.

Ericson, J.P., Vörösmarty, C.J., Dingman, S.L., Ward, L.G., Meybeck, M., 2006. Effective sea-level rise and deltas: causes of change and human dimension implications. Global Planet. Change 50 (1–2), 63–82.

Eschbach, D., Schmitt, L., Imfeld, G., May, J.H., Payraudeau, S., Preusser, F., Trauerstein, M., Skupinski, G., 2018. Long-term temporal trajectories to enhance restoration efficiency and sustainability on large rivers: an interdisciplinary study. Hydrol. Earth Syst. Sci. 22 (5), 2717–2737. https://doi.org/10.5194/hess-22-2717-2018.

Fischer, H., Kloep, F., Wilcek, S., Pusch, M., 2005. A river's liver—microbial processes within the hyporheic zone of a large lowland river. Biogeochemistry 76, 349–371. Available from: \\eawag-nas\lim$\uehlinge\MyDocuments\Bibliographien\PDF\Fischer et al. 2005 Hyporheic zone.pdf.

Garófano-Gómez, V., Metz, M., Egger, G., Díaz-Redondo, M., Hortobágyi, B., Geerling, G., Corenblit, D., Steiger, J., 2017. Vegetation Succession Processes and Fluvial Dynamics of a Mobile Temperate Riparian Ecosystem: The Lower Allier River (France), 23. Groupe français de géomorphologie.

Gaudichet, C., Greulich, S., Grellier, S., Rodrigues, S., 2022. Effect of flooding gradient on soil seedbank and standing vegetation in a disconnecting side channel of the Loire River (France). Hydrobiologia 849 (6), 1383–1396. https://doi.org/10.1007/s10750-021-04785-6.

Grill, G., Lehner, B., Thieme, M., Geenen, B., Tickner, D., Antonelli, F., Babu, S., Borrelli, P., Cheng, L., Crochetiere, H., Ehalt Macedo, H., Filgueiras, R., Goichot, M., Higgins, J., Hogan, Z., Lip, B., McClain, M.E., Meng, J., Mulligan, M., Zarfl, C., 2019. Mapping the world's free-flowing rivers. Nature 569 (7755), 215–221. https://doi.org/10.1038/s41586-019-1111-9.

Haase, P., Bowler, D.E., Baker, N.J., Bonada, N., Domisch, S., Garcia Marquez, J.R., Heino, J., Hering, D., Jähnig, S.C., Schmidt-Kloiber, A., Stubbington, R., Altermatt, F., Álvarez-Cabria, M., Amatulli, G., Angeler, D.G., Archambaud-Suard, G., Jorrín, I.A., Aspin, T., Azpiroz, I., Bañares, I., Ortiz, J.B., Bodin, C.L., Bonacina, L., Bottarin, R., Cañedo-Argüelles, M., Csabai, Z., Datry, T., de Eyto, E., Dohet, A., Dörflinger, G., Drohan, E., Eikland, K.A., England, J., Eriksen, T.E., Evtimova, V., Feio, M.J., Ferréol, M., Floury, M., Forcellini, M., Forio, M.A.E., Fornaroli, R., Friberg, N., Fruget, J.F., Georgieva, G., Goethals, P., Graça, M.A.S., Graf, W., House, A., Huttunen, K.L., Jensen, T.C., Johnson, R.K., Jones, J.I., Kiesel, J., Kuglerová, L., Larrañaga, A., Leitner, P., L'Hoste, L., Lizée, M.H., Lorenz, A.W., Maire, A., Arnaiz, J.A.M., McKie, B.G., Millán, A., Monteith, D., Muotka, T., Murphy, J.F., Ozolins, D., Paavola, R., Paril, P., Peñas, F.J., Pilotto, F., Polášek, M., Rasmussen, J.J., Rubio, M., Sánchez-Fernández, D., Sandin, L., Schäfer, R.B., Scotti, A., Shen, L.Q., Skuja, A., Stoll, S., Straka, M., Timm, H., Tyufekchieva, V.G., Tziortzis, I., Uzunov, Y., van der Lee, G.H., Vannevel, R., Varadinova, E., Várbíró, G., Velle, G., Verdonschot, P.F.M., Verdonschot, R.C.M., Vidinova, Y., Wiberg-Larsen, P., Welti, E.A.R., 2023. The recovery of European freshwater biodiversity has come to a halt. Nature 620 (7974), 582–588. https://doi.org/10.1038/s41586-023-06400-1.

Higgins, J., Zablocki, J., Newsock, A., Krolopp, A., Tabas, P., Salama, M., 2021. Durable freshwater protection: a framework for establishing and maintaining long-term protection for freshwater ecosystems and the values they sustain. Sustainability 13 (4), 1950. Available from: https://www.mdpi.com/2071-1050/13/4/1950.

Janssen, P., Chevalier, R., Chantereau, M., Dupré, R., Evette, A., Hémeray, D., Mårell, A., Martin, H., Rodrigues, S., Villar, M., Greulich, S., 2023. Can vegetation clearing operations and reprofiling of bars be considered as an ecological restoration measure? Lessons from a 10-year vegetation monitoring program (Loire River, France). Restor. Ecol. 31 (3), e13704. https://doi.org/10.1111/rec.13704.

Jeliazkov, A., Martínez-Fernández, V., Altanov, V.Y., Beisel, J.N., Buijse, A.D., Consuegra, S., Felin, S., Garcia de Leaniz, C., Graf, W., He, F., Jähnig, S.C., Leitner, P., Schmidt-Kloiber, A., Sexton, A.N., Staentzel, C., Tales, E., Wantzen, K.M., Wolter, C., 2024. Effects of inland navigation and related infrastructures on freshwater ecosystem: a global map of knowledge. J. Environ. Manage. 370. https://doi.org/10.1016/j.jenvman.2024.122474.

Johnson, M.F., Albertson, L.K., Algar, A.C., Dugdale, S.J., Edwards, P., England, J., Gibbins, C., Kazama, S., Komori, D., MacColl, A.D., 2024. Rising water temperature in rivers: ecological impacts and future resilience. Wiley Interdiscip. Rev. Water 11 (4), e1724.

Johnson, M.F., Albertson, L.K., Algar, A.C., Dugdale, S.J., Edwards, P., England, J., Gibbins, C., Kazama, S., Komori, D., MacColl, A.D.C., Scholl, E.A., Wilby, R.L., de Oliveira Roque, F., Wood, P.J., 2024. Rising water temperature in rivers: ecological impacts and future resilience. Wiley Interdiscip. Rev. Water 11 (4). https://doi.org/10.1002/wat2.1724.

Jones, E.R., Bierkens, M.F.P., van Vliet, M.T.H., 2024. Current and future global water scarcity intensifies when accounting for surface water quality. Nat. Clim. Change 14 (6), 629–635. https://doi.org/10.1038/s41558-024-02007-0.

Junk, W.J., Wantzen, K.M., 2004. The flood pulse concept: new aspects, approaches, and applications—an update. In: Welcomme, R., Petr, T. (Eds.), Proceedings of the Second International Symposium on the Management of Large Rivers for Fisheries, vol. 2, pp. 117–149.

Junk, W.J., Wantzen, K.M., 2006. Flood pulsing, and the development and maintenance of biodiversity in floodplains. In: Batzer, D.P., Sharitz, R.R. (Eds.), Ecology of Freshwater and Estuarine Wetlands. University of California Press, pp. 407–435 (Reprinted from ON REQUEST (//)).

Junk, W.J., Bayley, P.B., Sparks, R.E., 1989. The flood pulse concept in river-floodplain systems. Can. Spec. Publ. Fish. Aquat. Sci. 106, 110–127.

Köhler, J., Varga, E., Spahr, S., Gessner, J., Stelzer, K., Brandt, G., Mahecha, M.D., Kraemer, G., Pusch, M., Wolter, C., 2024. Unpredicted ecosystem response to compound human impacts in a European river. Sci. Rep. 14 (1), 16445.

Latapie, A., Camenen, B., Rodrigues, S., Paquier, A., Bouchard, J.P., Moatar, F., 2014. Assessing channel response of a long river influenced by human disturbance. Catena 121, 1–12. https://doi.org/10.1016/j.catena.2014.04.017.

Le Guern, J., 2021. Morphodynamique d'un lit fluviatile sablo-graveleux: méthodes acoustiques de quantification de la charge de fond et analyse des interactions des formes du lit dans les chenaux ligériens Université de Tours (UT), Tours, FRA.

Le Guern, J., Rodrigues, S., Tassi, P., Cordier, F., Wintenberger, C.L., Jugé, P., 2023. Migrating bars influence the formation and dynamics of their peers in large sandy-gravel bed rivers. Earth Surf. Process. Landf. 48 (8), 1512–1525. https://doi.org/10.1002/esp.5563.

Lepage, H., Launay, M., Le Coz, J., Angot, H., Miege, C., Gairoard, S., Radakovitch, O., Coquery, M., 2020. Impact of dam flushing operations on sediment dynamics and quality in the upper Rhône River, France. J. Environ. Manag. 255, 109886.

Loire, R., Grosprêtre, L., Malavoi, J.-R., Ortiz, O., Piégay, H., 2019. What discharge is required to remove silt and sand downstream from a dam? An adaptive approach on the Selves River, France. Water 11 (2), 392.

Malaj, E., Von Der Ohe, P.C., Grote, M., Kühne, R., Mondy, C.P., Usseglio-Polatera, P., Brack, W., Schäfer, R.B., 2014. Organic chemicals jeopardize the health of freshwater ecosystems on the continental scale. Proceedings of the National Academy of Sciences of the United States of America 111 (26), 9549–9554. https://doi.org/10.1073/pnas.1321082111.

Marmonier, P., Wantzen, K.M., et al., 2012. The role of biodiversity in hyporheic processes: gaps in knowledge and needs for applications. Int. J. Limnol. 48, 253–266.

Michau, R., Guerri, O., Brun, A., Quetier, J., 2023. Dordogne: the first biosphere reserve covering an entire river basin. In: Wantzen, K.M. (Ed.), River Culture: Life as a Dance to the Rhythm of the Waters. UNESCO Publishing.

Moatar, F., Descy, J.-P., Rodrigues, S., Souchon, Y., Floury, M., Grosbois, C., Minaudo, C., Leitao, M., Wantzen, K.M., Bertrand, F., 2022. Chapter 7—the Loire River basin. In: Tockner, K., Zarfl, C., Robinson, C.T. (Eds.), Rivers of Europe, second ed. Elsevier, pp. 245–271. https://doi.org/10.1016/B978-0-08-102612-0.00007-9.

Olivier, J.-M., Carrel, G., Lamouroux, N., Dole-Olivier, M.-J., Malard, F., Bravard, J.-P., Piégay, H., Castella, E., Barthélemy, C., 2022. The Rhône river basin. In: Rivers of Europe. Elsevier, pp. 393–453.

Reid, A.J., Carlson, A.K., Creed, I.F., Eliason, E.J., Gell, P.A., Johnson, P.T.J., Kidd, K.A., MacCormack, T.J., Olden, J.D., Ormerod, S.J., Smol, J.P., Taylor, W.W., Tockner, K., Vermaire, J.C., Dudgeon, D., Cooke, S.J., 2019. Emerging threats and persistent conservation challenges for freshwater biodiversity. Biol. Rev. 94 (3), 849–873. https://doi.org/10.1111/brv.12480.

Rijke, J., van Herk, S., Zevenbergen, C., Ashley, R., 2012. Room for the River: delivering integrated river basin management in The Netherlands. Int. J. River Basin Manag. 10 (4), 369–382.

Robinson, C.T., Uehlinger, U., Monaghan, M.T., 2003. Effects of a multi-year experimental flood regime on macroinvertebrates downstream of a reservoir. Aquat. Sci. (submitted).

Rodrigues, S., Berton, J.-P., 2021. Freeing the bed of the Loire from its corset of groynes to give it a new lease of life. IAGF. Retrieved 10.09.2025 from. https://www.initiativesrivers.org/actualites/freeing-the-bed-of-the-loire-from-its-corset-of-groynes-to-give-it-a-new-lease-of-life/.

Rodrigues, S., Gautier, J.-N., 2007. Morphological response of alluvial channels to maintenance operations: the case of the vegetated secondary channels of the River Loire. Houille Blanche-Revue Internationale De L Eau (2), 75–84. https://doi.org/10.1051/lhb:2007022.

Rodrigues, S., Breheret, J.G., Macaire, J.J., Moatar, F., Nistoran, D., Juge, P., 2006. Flow and sediment dynamics in the vegetated secondary channels of an anabranching river: the Loire River (France). Sediment. Geol. 186 (1–2), 89–109. https://doi.org/10.1016/j.sedgeo.2005.11.011.

Rodrigues, S., Bréheret, J.G., Macaire, J.J., Greulich, S., Villar, M., 2007. In-channel woody vegetation controls on sedimentary processes and the sedimentary record within alluvial environments: a modern example of an anabranch of the River Loire, France. Sedimentology 54 (1), 223–242. https://doi.org/10.1111/j.1365-3091.2006.00832.x.

Rodrigues, S., Mosselman, E., Claude, N., Wintenberger, C., Jugé, P., 2015. Alternate bars in a sandy gravel bed river: generation, migration and interactions with superimposed dunes. Earth Surf. Process. Landf. 40 (5), 610–628.

Schmitt, L., Morris, D., Kondolf, G.M., 2018. Managing floods in large river basins in Europe: the Rhine River. In: Managing Flood Risk: Innovative Approaches from Big Floodplain Rivers and Urban Streams. Springer, pp. 75–89.

Schmitt, L., Beisel, J., Preusser, F., De Jong, C., Wantzen, K., Chardon, V., Staentzel, C., Eschbach, D., Damm, C., Rixhon, G., 2019. Sustainable management of the upper Rhine River and its alluvial plain: lessons from interdisciplinary research in France and Germany. 153–174. In: Sustainable Research in the Upper Rhine Region. Concepts and Case Studies, pp. 201–226.

Sexton, A.N., Beisel, J.-N., Staentzel, C., Wolter, C., Tales, E., Belliard, J., Buijse, A.D., Martínez Fernández, V., Wantzen, K.M., Jähnig, S.C., 2024. Inland navigation and land use interact to impact European freshwater biodiversity. Nat. Ecol. Evol. 1–11.

Sparks, R.E., Nelson, J.C., Yin, Y., 1998. Naturalization of the flood regime in regulated rivers. Bioscience 48 (9), 706–720.

Staentzel, C., Arnaud, F., Combroux, I., Schmitt, L., Trémolières, M., Grac, C., Piégay, H., Barillier, A., Chardon, V., Beisel, J.N., 2018. How do instream flow increase and gravel augmentation impact biological communities in large rivers: a case study on the Upper Rhine River. River Res. Appl. 34 (2), 153–164. https://doi.org/10.1002/rra.3237.

Staentzel, C., Combroux, I., Barillier, A., Grac, C., Chanez, E., Beisel, J.-N., 2019. Effects of a river restoration project along the Old Rhine River (France-Germany): response of macroinvertebrate communities. Ecol. Eng. 127, 114–124. https://doi.org/10.1016/j.ecoleng.2018.10.024.

Statzner, B., Higler, B., 1986. Stream hydraulics as a major determinant of benthic invertebrate zonation patterns. Freshw. Biol. 16, 127–139.

Thorp, J.H., Thoms, M.C., Delong, M.D., 2010. The Riverine Ecosystem Synthesis: Toward Conceptual Cohesiveness in River Science. Elsevier.

Thouvenot, L., Haury, J., Thiebaut, G., 2013. A success story: water primroses, aquatic plant pests. Aquat. Conserv. Mar. Freshw. Ecosyst. 23 (5), 790–803. https://doi.org/10.1002/aqc.2387.

Tockner, K., Stanford, J.A., 2002. Riverine flood plains: present state and future trends. Environ. Conserv. 29, 308–330.

Tockner, K., Schiemer, F., Baumgartner, C., Kum, G., Weigand, E., Zweimuller, I., Ward, J.V., 1999. The Danube restoration project: species diversity patterns across connectivity gradients in the floodplain system. Regul. Rivers Res. Manag. 15 (1–3), 245–258.

Tockner, K., Malard, F., Ward, J.V., 2000. An extension of the flood pulse concept. Hydrol. Process. 14 (16–17), 2861–2883.

Townsend, C.R., 1989. The patch dynamics concept of stream community ecology. J. North Am. Benthol. Soc. 8, 36–50.

Tullos, D.D., Collins, M.J., Bellmore, J.R., Bountry, J.A., Connolly, P.J., Shafroth, P.B., Wilcox, A.C., 2016. Synthesis of common management concerns associated with dam removal. JAWRA J. Am. Water Resour. Assoc. 52 (5), 1179–1206. https://doi.org/10.1111/1752-1688.12450.

Wagner, R., Schmidt, H.H., 2004. Yearly discharge patterns determine species abundance and community diversity: analysis of a 25 year record from the Breitenbach. Archiv fur Hydrobiologie 16 (4), 511–540.

Wantzen, K.M., 2024. River culture: how socio-ecological linkages to the rhythm of the waters develop, how they are lost, and how they can be regained. Geogr. J. 190 (2). https://doi.org/10.1111/geoj.12476.

Wantzen, K.M., Assine, M.L., Bortolotto, I.M., Calheiros, D.F., Campos, Z., Catella, A.C., Chiaravalotti, R.M., Collischonn, W., Couto, E.G., da Cunha, C.N., 2024. The end of an entire biome? World's largest wetland, the Pantanal, is menaced by the Hidrovia project which is uncertain to sustainably support large-scale navigation. Sci. Total Environ. 908, 167751.

Wantzen, K.M. (Ed.), 2023. River Culture: Life as a Dance to the Rhythm of the waters. UNESCO publishing, Paris, 901+XV pp., ISBN 978-92-3-100540-4. https://unesdoc.unesco.org/ark:/48223/pf0000382775.

Wantzen, K.M., Blettler, M.C.M., Marchese, M.R., Amsler, M.L., Bacchi, M., Ezcurra de Drago, I.D., Drago, E.E., 2014. Sandy rivers: a review on general ecohydrological patterns of benthic invertebrate assemblages across continents. Int. J. River Basin Manag. 12 (3), 163–174. https://doi.org/10.1080/15715124.2014.885438.

Wantzen, K.M., Aziz, B., Longuet, I., Bao, I., Bocoum, H., Cissé, L., Chauhan, M., Girard, P., Gopal, B., Kane, A., Marchese, M.R., Prakash, N., Teixeira, P., Zalewski, M., 2016. River Culture: an eco-social approach to mitigate the biological and cultural diversity crisis in riverscapes. Ecohydrol. Hydrobiol. 16 (1), 7–18. https://doi.org/10.1016/j.ecohyd.2015.12.003.

Wantzen, K.M., Uehlinger, U., Van der Velde, G., Leuven, R.S.E.W., Schmitt, L., Beisel, J.N., 2022. The Rhine River Basin Rivers of Europe. Elsevier, pp. 333–391. https://doi.org/10.1016/B978-0-08-102612-0.00010-9.

Wantzen, K.M., Berton, J.-P., Boisneau, C., Greulich, S., Marmiroli, B., Rodrigues, S., 2023a. The Loire—a cultural and environmental exception in Europe. In: Wantzen, K.M. (Ed.), River Culture: Life as a Dance to the Rhythm of the Waters. UNESCO Publishing, pp. 699–721. https://doi.org/10.54677/VNTD8327.

Wantzen, K.M., Schmitt, L., Wirbelauer, B., 2023b. The Rhine—an important biocultural axis for Europe. In: Wantzen, K.M. (Ed.), River Culture—Life as a Dance to the Rhythm of the Waters. UNESCO.

Wantzen, K.M., Assine, M.L., Bortolotto, I.M., Calheiros, D.F., Campos, Z., Catella, A.C., Chiaravalotti, R.M., Collischonn, W., Couto, E.G., da Cunha, C.N., 2024. The end of an entire biome? World's largest wetland, the Pantanal, is menaced by the Hidrovia project which is uncertain to sustainably support large-scale navigation. Sci. Total Environ. 908, 167751. https://doi.org/10.1016/j.scitotenv.2023.167751.

Ward, J.V., 1989. The four-dimensional nature of lotic ecosystems. J. North Am. Benthol. Soc. 8 (1), 2–8.

Ward, J.V., Stanford, J.A., 1983. The intermediate-disturbance hypothesis: an explanation for biotic diversity patterns in lotic ecosystems. In: S.M, B. (Ed.), Dynamics of Lotic Ecosystems. Ann Arbor Science, Ann Arbor, pp. 347–356.

Wintenberger, C.L., Rodrigues, S., Bréhéret, J.G., Villar, M., 2015. Fluvial islands: first stage of development from nonmigrating (forced) bars and woody-vegetation interactions. Geomorphology 246, 305–320. https://doi.org/10.1016/j.geomorph.2015.06.026.

Wintenberger, C.L., Rodrigues, S., Greulich, S., Bréhéret, J.G., Jugé, P., Tal, M., Dubois, A., Villar, M., 2019. Control of non-migrating bar morphodynamics on survival of Populus nigra seedlings during floods. Wetlands 39 (2), 275–290. https://doi.org/10.1007/s13157-018-1121-7.

Wolf, S., Stark, N., Holste, I., Lehmkuhl, F., Römer, W., Burghardt, L., Schüttrumpf, H., 2024. Evaluation of the high-energy flood of mid-July 2021 as a morphologic driver in the anthropogenically developed Ahr Valley, Germany, in interaction with infrastructures. Environ. Sci. Eur. 36 (1), 54. https://doi.org/10.1186/s12302-024-00860-6.

Zalewski, M., 2002. Ecohydrology—the use of ecological and hydrological processes for sustainable management of water resources. Hydrol. Sci. J. J. Sci. Hydrol. 47 (5), 823–832.

Zalewski, M., Wantzen, K.M., 2023. Ecohydrology and river culture: synergy for the enhancement of River and catchment sustainability. In: Wantzen, K.M. (Ed.), River Culture: Life as a Dance to the Rhythm of the Waters. UNESCO Publishing.

Chapter 31

Ecohydrological solutions in the catchment area of Lake Naivasha

Nic Pacini[1, 2], Silas Wanjala[3], Timothy Mwinami[4] and David M. Harper[1, 5]
[1]*University of Leicester, School of Geography, Geology and the Environment, Leicester, United Kingdom;* [2]*Department of Environmental Engineering, University of Calabria, Arcavacata di Rende, Italy;* [3]*Lake Naivasha Riparian Association, Naivasha, Kenya;* [4]*National Museums of Kenya, Department of Ornithology, Nairobi, Kenya;* [5]*Freshwater Biological Association, Cumbria, United Kingdom*

PHOTO CREDIT: N. PACINI

1. Establishing restoration priorities

Lake Naivasha is located at an altitude of 1888 m above sea level in the eastern Great Rift Valley, in Kenya, a country over 70% covered by savannas. Over the past 30 years, the lake's environmental status has gradually deteriorated due to

Transdisciplinary Ecohydrology. https://doi.org/10.1016/B978-0-443-41596-8.00027-3

population growth, deforestation, eutrophication, the introduction of invasive species, and large water fluctuations in irregular periods of drought and floods (Becht and Harper, 2002; Gherardi et al., 2011; Harper et al., 2011).

The Principles of Ecohydrology (Zalewski, 2000; Harper and Mavuti, 2004; Zalewski et al., 2008) highlight the potentially crucial role of native plant species (*Cyperus papyrus* growing in and around the lake, *Vachellia xanthophloea* in the surrounding riparian zone, riverside forest in the upper stream catchments) in supporting the resilience of local ecosystems. We designed pilot projects following these principles to rationalize the management of wetlands, to support the renewal of papyrus growth, to treat used wastewater, and to minimize erosion of soils. All these measures are illustrated by visible examples that were implemented to prevent the degradation of the lake and catchment (Harper et al., 2013; Wanjala et al., 2018).

2. Constructed wetlands

During the colonial period (around 1900–50), European farmers built approximately 900 small dams in the entire catchment for water supply and irrigation; all these structures fell into disrepair after independence (Fig. 31.1). Nowadays, these dams remain an important source of water for local communities, especially in times of drought. They are also important for preserving water in the landscape, mitigating extreme climatic conditions, and supporting biodiversity.

As part of the pilot project, fenced areas around three small dams in the Naivasha catchment were created, cutting off access to the dam water for livestock. Inside the fenced area, a mixture of native and fast-growing trees was planted to provide carbon sequestration and income for local people. The dam, damaged by neglect and past storms, was recreated with sandbags, and the water surface area returned to its former size. These actions helped to create conditions for the development of diverse aquatic vegetation, which in turn supports the deposition of silt, improving the transparency of water. At the same time, the vegetation purifies the water for potable use.

A discharge pipe was also installed in the dam at the middle of its depth, to a standpipe below connected to a water trough for livestock. The quantity and quality of water for about 3000 people from local villages around, all without a piped water supply, were thus made more reliable and cleaner. Separate access to the water was provided for animals to be watered, and a third separate bay was designated for washing clothes. An overflow outlet was created to prevent a dangerous build-up of water pressure that would threaten the dam's stability (Fig. 31.2).

These dams, subject to such rehabilitation, also increased the water retention time in the landscape, decreased outflow of nutrients downstream, and provided conditions for the development of biodiversity (Fig. 31.3).

3. Restoring the riparian zone of rivers

Throughout the Lake Naivasha Basin, removal of riparian forest from the upper catchment caused permanent damage to people and nature downstream. To help minimize erosion from cultivation and silting in the watercourses, and eventually in the lake, about 15,000 trees have been planted on the banks of the streams in the Naivasha catchment. Saplings were provided by local nurseries (Fig. 31.4).

FIGURE 31.1 Gilgil subcatchment (Naivasha catchment, Kenya). Abandoned dams from colonial times now enable conditions for the restoration of aquatic and related ecosystem services.

FIGURE 31.2 Malewa subcatchment (Naivasha catchment, Kenya). A pilot study of a rebuilt dam with a fenced area, planted with native trees. The exclusion of livestock enables submerged and floating-leaved macrophyte regrowth, which in turn causes sedimentation of silt and purification of water. Improved-quality drinking water supply is delivered by a standpipe connected to a trough downstream of the dam.

FIGURE 31.3 Artificial wetlands integrate seamlessly within the landscape.

FIGURE 31.4 Tree nurseries were encouraged in every village to support riparian and contour tree planting.

They were a mixture of native hardwood and fast-growing softwood trees that provided local farmers with income opportunities. On steep slopes, planting extended onto cultivation plots, following contours, in order to hold back soil erosion from farmland (Figs. 31.5 and 31.6).

In the lower catchment, check dams were built to hold back eroded sediment and waste from reaching the lake shore after storms (Fig. 31.7).

FIGURE 31.5 Olosho Rongai (Naivasha catchment): rows of trees planted beside crops to reduce erosion after recent deforestation.

FIGURE 31.6 Rows of endemic African olive and pencil cedar trees, along with fast-growing cypresses, were planted on steep hill ravine terraces to minimize this soil loss.

FIGURE 31.7 Gabions inserted into the banks of a temporary torrent to hold back solid waste and sediment.

FIGURE 31.8 Floating islands showing the planting of cut clumps and their successful growth, in a cell of the artificial wetland, prior to translocation to the littoral zone of the lake.

FIGURE 31.9 Fully developed artificial papyrus island.

4. Wetlands on the shores of the lake

Artificial wetlands have been constructed by flower farms in the riparian zone to collect and purify liquid waste discharged from their flower-growing operations. To support water quality in these wetlands, we planted floating islands from recycled plastic (https://www.floatingislandinternational.com) with native *Cyperus papyrus* (Fig. 31.8).

These islands were also placed in the lake shallows to try to kick-start papyrus regrowth and recreate a continuous vegetation fringe along the lakeshore (Fig. 31.9).

5. Training of farmers and flower-growers

Hundreds of farmers have been trained on sanitation, water abstraction, and sustainable agriculture to improve ecosystem services and community benefits from the environment (Harper et al., 2013) (Fig. 31.10). The project was funded by European supermarkets that are customers of the flowers grown around Lake Naivasha, particularly REWE (Germany) and COOP (Switzerland). This initiative for the sustainable use of Lake Naivasha's catchment area is the result of cooperation with Josphat Macharia and his Ndabibi Environmental Conservation Center (www.ewfandw.org).

FIGURE 31.10 Upper catchment farmers during a "Water Ambassadors" field training session.

References

Becht, R., Harper, D.M., 2002. Towards an understanding of human impact upon the hydrology of Lake Naivasha, Kenya. Hydrobiologia 488, 1–11. https://doi.org/10.1023/A:1023318007715.

Gherardi, F., Robert Britton, J., Mavuti, K.M., Pacini, N., Grey, J., Tricarico, E., Harper, D.M., 2011. A review of allodiversity in Lake Naivasha, Kenya: developing conservation actions to protect East African lakes from the negative impacts of alien species. Biol. Conserv. 144 (11), 2585–2596. https://doi.org/10.1016/j.biocon.2011.07.020.

Harper, D., Mavuti, K., Naivasha, L., 2004. Kenya: ecohydrology to guide the management of a tropical protected area. Ecohydrol. Hydrobiol. 4 (3), 287–305.

Harper, D.M., Morrison, E.H.J., Macharia, M.M., Mavuti, K.M., 2011. Caroline Upton, Lake Naivasha, Kenya: ecology, society and future. Freshw. Rev. 4 (2), 89–114. https://doi.org/10.1608/frj-4.2.149.

Harper, D.M., Pacini, N., Upton, C., Morrison, E.H., Fox, R., Kiminta, E., 2013. Water cooperation for sustainable utilization: Lake Naivasha, Kenya. Free Flow: Reaching Water Security Through Cooperation. UNESCO, Paris, pp. 256–259.

Wanjala, S., Mwinami, T., Harper, D.M., Morrison, E.H.J., Pacini, N., 2018. Ecohydrological tools for the preservation and enhancement of ecosystem services in the Naivasha Basin, Kenya. Ecohydrol. Hydrobiol. 18 (2), 155–173. https://doi.org/10.1016/j.ecohyd.2017.09.004.

Zalewski, M., 2000. Ecohydrology—the scientific background to use ecosystem properties as management tools toward sustainability of water resources. Ecol. Eng. 16 (1), 1–8. https://doi.org/10.1016/s0925-8574(00)00071-9.

Zalewski, M., Harper, D.M., Demars, B., Jolánkai, G., Crosa, G., Janauer, G.A., Pacini, N., 2008. Linking Biological and Physical Processes at the River Basin Scale: The Origins, Scientific Background and Scope of Ecohydrology. CABI Publishing, pp. 1–17. https://doi.org/10.1079/9781845930028.0001.

Chapter 32

Transdisciplinary ecohydrology for reduction of soil erosion, stream self-purification and Tilapia aquaculture development in Ethiopia

Yohannes Zerihun Negussie[1], **Maciej Zalewski**[2] **and Magdalena Urbaniak**[3]

[1]*Ministry of Water and Energy, Integrated Water Resources Management, Basin Scale Resilience Initiative for Ethiopia (BaSRINET) Project, Addis Ababa, Ethiopia;* [2]*European Regional Centre for Ecohydrology of the Polish Academy of Sciences under the auspices of UNESCO, Łódź, Poland;* [3]*University of Lodz, Faculty of Biology and Environmental Protection, UNESCO Chair on Ecohydrology and Applied Ecology, Łódź, Poland*

PHOTO CREDIT: M. ZALEWSKI.

Transdisciplinary Ecohydrology. https://doi.org/10.1016/B978-0-443-41596-8.00044-3

Ethiopia is a country with a dynamic population growth and economic development and related urbanization, intensification of agriculture and deforestation, the last one recently compensated by some reforestation efforts. About 90% of the country is mountainous, making Ethiopia one of the most biogeographically sensitive countries in the world, most susceptible to soil degradation. Deforestation and intensive agriculture are a major cause of landscape degradation and resilience to climate change decline (Amede and Nigatu, 2001; Asres and Awulachew, 2010; Belete, 2023; Berry, 2003; U.S. EPA. Selected Analytical Methods for Environmental Remediation and Recovery (SAM), 2012), which coupled with unsustainable farming and heavy rainfall drives the erosion process (Zalewski et al., 2010). Erosion, Fig. 32.1 contributes to the fertile soil degradation, transfer of nutrients and pollutants into freshwater ecosystems, and in consequence the acceleration of the process of its eutrophication (Amare, 2005; Endalew and Tollner, 2009), resulting in the dangerous toxic algal blooms (Tesfay, 2007).

Moreover, organic and inorganic matter, eluviated from the catchment area during heavy rainfall in season accelerates the siltation of rivers, lakes and reservoirs. In case of rivers this process reduces diversity and biological productivity of habitats causing decline in fish reproduction and production. In all freshwater ecosystems persistent organic pollutants e.g., dioxins, and heavy metals, pose risk for human health and the aquatic environment and biodiversity (Ongley, 1996), due to bioaccumulation in organisms (Urbaniak et al., 2010, 2012; Urbaniak and Zalewski, 2011; Zalewski et al., 2010).

The one of the major causes of the soil erosion have been the pastures overgrazing. The important element of strategy of the Ministry of Water and Energy of Ethiopia to reduce such impact become the provision of the alternative source of proteins for society. The one of the most promising appeared the development the *Tilapia nilotica* pisciculture compatible with policy of increase the water retention at the farms ponds. However, the most important problem limiting the broad implementation was the difficulties in fish seed production.

The solutions of such complex water and sustainability problems, which need systemic and innovative low costs advanced solutions, have been in the scope of Transdisciplinary Ecohydrology due to its holistic and applied perspective. The broad spectrum of research related to above environmental/sustainability problems that has been for decades conducted in European Regional Center for Ecohydrology (ERCE) Polish Academy of Sciences under UNESCO, in the framework of UNESCO Intergovernmental Hydrological Program. To develop prototype solutions into innovative universal EH NbS tools, it was necessary to test them and to makes the fine tuning into African tropical conditions. This was done in the framework of the Polish Ethiopian cooperation, financed by Polish Aid program of Ministry of Foreign Affairs of Poland, and Ministry of Water and Energy of Ethiopia.

On the basis preparatory visits and discussion, the three areas for development of and adaptation for Ethiopian conditions of the transdisciplinary ecohydrological Advanced Nature-based Solutions have been identified and implemented.

1. Reduction of soil erosion by application of biodegradable recycle geotextiles (Circular economy)
2. Reduction of stream pollutants for the restoration of drinking water reservoirs by Sequential Sedimentation Biofiltration Systems.
3. Improvements of the Tilapia Aquaculture by development the efficient techniques of fish seed for stocking

FIGURE 32.1 **Soil erosion.** Deforested and overgrazed areas in central Ethiopia. *Photo credit: M. Zalewski.*

1. Reduction of soil erosion using recycled geotextiles

Transdisciplinary Ecohydrology integrates not only water and environmental science for innovative solutions but as branch of the Sustainability Sciences has to build the link with efforts to reduce the impact for Biosphere by synergy with Circular Economy. Especially that Sustainability Science has integrative character focused on highlighting the interactions between Human and Biosphere to be used for development innovative low-cost methods and systemic solutions for harmonization of the natural resource use and society needs.

As a playground for implementation and development THE solution especially to demonstrate the importance of integration catchment scale efforts to reduce degradation of water resource by dioxins, erosion and eutrophication, the small reservoir Burkitu Reservoir, also known as Lake Assela, was selected. It was constructed in the 1970s at the Assela river valley (3 km from the town of Assela) in Ethiopia. The reservoir was built to supply the town of Assela with drinking water, it stopped fulfilling this function, as the water taken from it was found to be toxic and causes disease. Diagnostic research on this reservoir by Zalewski et al. (2010) indicate the strong contamination of the reservoir sediments with dioxins, also the appearance of the toxic algal blooms was observed (Urbaniak et al., 2012; Zalewski et al., 2010).

As a first step to reduce soil erosion, runoff loads of mineral, organic matter, nutrient from the catchment area, and siltation, the biodegradable geotextile at the pastures above the reservoir was applied Fig. 32.2A and B. As process ecological succession was progressing the plants cover was getting dense and geotextiles in few years decompose and

FIGURE 32.2 The use of geotextiles. (A) The use of biodegradable recycled geotextiles to reduce the erosion of pastures; (B) The use of geotextiles for the restoration of vegetation. In strongly eroded areas sediments collected at check dams were covered by geotextiles and different species of vegetation were planted; (C) and (D) Plantings of field vegetation to reduce the wind speed, water evaporation in the boundary strips between fields. Branches were periodically used as fodder for farm animals. *Photo credit: Yohannesa Zerihuna Negussie.*

disappear. Moreover, to broaden the catchment measures and to reduce process of agricultural landscape "homogenization" (Levia et al., 2020) and to reduce wind speed, which accelerate soil desiccation, at the boundaries between the fields, the shelterbelts—vegetative barriers was planted (Fig. 32.2C and D).

2. Reduction of stream pollutants through drinking water reservoir restoration with SSFS

In order to improve the water quality in the reservoir and thus protect the health of the population from the surrounding area, it was necessary to limit the inflow not only eroded mineral and organic substances, but also the incoming pollutants from the upper reaches of the river, mainly from animal husbandry, and from the Assela wastewater treatment plant. To reduce all eroded soils and pollutants inflowing to the reservoir, the sequential sedimentation and biofiltration was constructed (SSBS) (Zalewski et al., 2010). The construction of the system was based on of experience gathered during the development of the prototype the system on the Sokołówka River, used for purification urban stormwater in Łódź and constructed in the framework of the EU Funded project SWITCH (Zalewski, 2012).

The Lake Assela Sedimentation and Biofiltration System (SSBS) consist of three steps. First, the sedimentation chamber, in which the process of mineral and organic matter deposit ion is intensified. Accumulated sediments from this chamber were periodically used as fertilizers (Zerihun Negussie et al., 2012). Second, a geochemical barrier—gabion with high density core of gravel and calcium rock that has dual function. First, absorbs phosphorus and releases calcium to increase efficiency of biological process at phytotechnology, which was a third step. Because gabion was set on a coarse gravel foundation with lower hydraulic resistance than core of gabion it was forcing to flow of water from sedimentation chamber under the gabion directly into the Third zone—Phyto technological, where root zone of (*Chrysopogon zizanioides* fragrant vetiver). At this stage process of water purification due to slow water flow through the substratum with dense roots of plants assimilating and absorbing phosphorus and organic forms of nitrogen (NH_4; NO_3), which additionally were denitrified. The reduction of the nitrogen and phosphorus load in it, thus reducing the risk of toxic blooms in the reservoir. Moreover, this system reduces concentration of dioxins below the toxic level (Fig. 32.3). As a result, the water resources from Asella reservoir were used by society again.

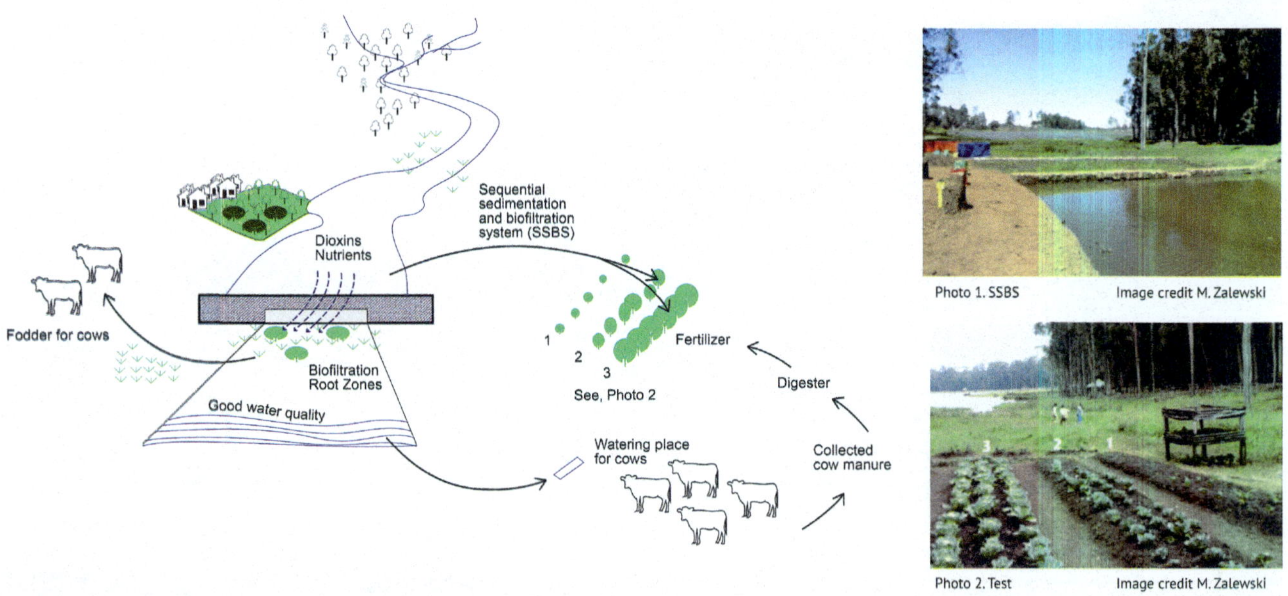

FIGURE 32.3 Test of SSBS sediments as a fertilizer for cabbage: 1. Control (no sediments applied). 2. No 1. Cabage—unfertilised land (control). No 2. Cabage—fertilised by sedniments from SSBS (periodic maintenance). No 3. Cabage—fertilised by sediments from SSBS + digested cow manure. 3. SSBS sediments combined with digestate manure. *Image credit: M. Zalewski.*

The final element of the simple system solutions was the watering points for livestock (cattle) which was constructed in secure distance from the stream and reservoir. The purpose it was the reduction of stream pollution by manure due to common cattle watering directly from stream. It was not only to prevent pollution but also to collect,

FIGURE 32.4 **Improvement of Tilapia aquaculture in Badhir dar.** Plastic bubble foil served as a nighttime cover for the experimental tilapia reproduction ponds. *Photo credit: Yohannesa Zerihuna Negussie*

digest and use digested manure as fertilizer (Circular Economy). To encourage the local society for maintenance of SSBS in high-efficiency status, which means periodical removal of the sediments from the first chamber, the three garden patches were established, Fig. 32.4. The photos indicate that the cabbage growth appears to be excellent indicator of the value of sediments, especially mixed with the collection of cattle manure which can be better that artificial fertilizers.

3. Improvement of Tilapia aquaculture through efficient reproduction techniques

The intensive development of pastoral land management has resulted in overgrazing and intensive soil erosion appears often in Africa. This stimulates development of tilapia pisciculture as alternative protein source for growing population. However, in Ethiopia, due to high-altitude conditions, fish seed production and availability was limited. Analyzing the temperature at the spawning ponds at the Ichtiological Institutes at Bahir Dar (BFALRC), the high variability of temperature during circadian period was observed and considered as potential factor restricting Tilapia reproduction efficiency. Considering it was conducted experiment—covering the ponds for night with bubble foil (Fig. 32.5). The results that are introduced indicate first, significant reduction of temperature variability (Fig. 32.6), which resulted in extremely high improvement of fish reproduction efficiency (Fig. 32.7).This low-cost solution was recently used broadly in individual farms in even small ponds for tilapia seed production as the alternative protein source.

	Temperature	
Data	Covered	Not-covered
1-Jan	22.68	17.12
2-Jan	23.2	17.83
3-Jan	24.65	19.27
4-Jan	24.03	18.9
5-Jan	24.12	19.35
6-Jan	24.3	19.26
7-Jan	24.38	19.21

FIGURE 32.5 **Temperature in ponds covered versus not covered.** The average temperature increased significantly, while diurnal temperature variability was drastically reduced.

FIGURE 32.6 **Tilapia stock production by covering the pond with bubble plastic foil.** The difference in fish reproduction in the ponds covered by bubble plastic foil versus not covered.

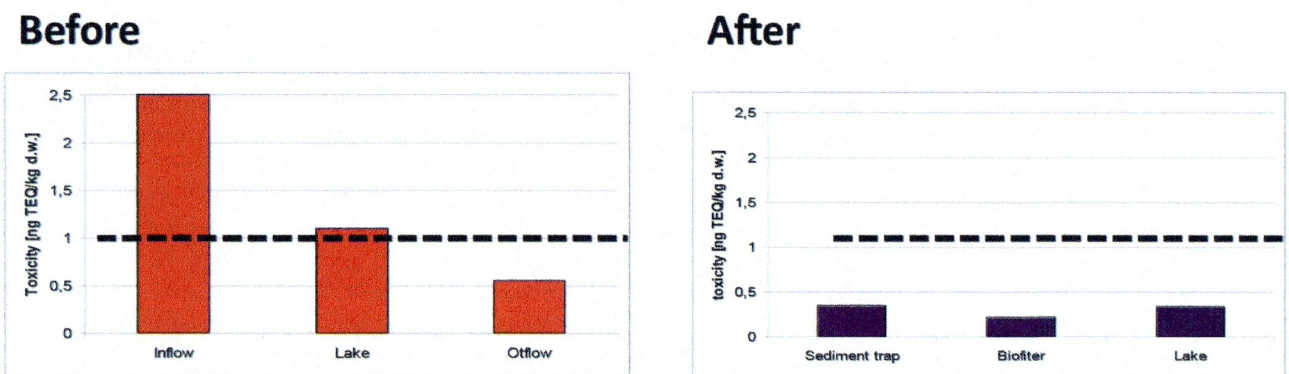

FIGURE 32.7 **Sequential sedimentaton-biofiltratoin system effectiveness in reduction of toxicity.** Before and after installation of SSBS in Assela.

4. Conclusions

All the above presented transdisciplinary ecohydrological solution supported by circular economy can significantly enhance sustainability potential of catchments at very low costs. However, this prototype solutions need further development, testing and fine tuning in catchments of different geomorphology, soils, climate, land cover and society priorities. As far as catchment scale implementation of above and related TEH solutions, will be long term with intensive society involvement process, it has to be done on diagnosis supported by principles of ecohydrology and socioeconomic and environmental foresight.

References

Amare, 2005. Study of Sediment Yield from the Watershed of Angereb Reservoir.

Amede, T., Nigatu, Y., 2001. Interaction of components of sweetpotato-maize intercropping under the semi-arid conditions of the Rift-Valley, Ethiopia. Trop. Agric. 78 (1), 1–7.

Asres, M.T., Awulachew, S.B., 2010. SWAT based runoff and sediment yield modelling: a case study of the Gumera watershed in the Blue Nile basin. Ecohydrol. Hydrobiol. 10 (2–4), 191–199. https://doi.org/10.2478/v10104-011-0020-9.

Belete, M.D., 2023. Ecohydrology-based Landscape Restoration: Theory and Practice. Routledge, p. 164.

Berry, L., 2003. Land Degradation in Ethiopia: Its Extent and Impact.

Endalew, M., Tollner, E.W., 2009. Assessment of Major Threats of Lake Tana and Strategies for Integrated Water-Use Management Proceedings of the First Annual Conference of the Ethiopian Fisheries and Aquatic Sciences Association (EFASA), pp. 174–191.

Levia, D.F., Creed, I.F., Hannah, D.M., Nanko, K., Boyer, E.W., Carlyle-Moses, D.E., van de Giesen, N., Grasso, D., Guswa, A.J., Hudson, J.E., Hudson, S.A., Iida, S., Jackson, R.B., Katul, G.G., Kumagai, T., Llorens, P., Ribeiro, F.L., Pataki, D.E., Peters, C.A., Carretero, D.S., Selker, J.S., Tetzlaff, D., Zalewski, M., Bruen, M., 2020. Homogenization of the terrestrial water cycle. Nat. Geosci. 13 (10), 656–658. https://doi.org/10.1038/s41561-020-0641-y.

Ongley, E.D., 1996. Control of Water Pollution from Agriculture (FAO Irrigation and Drainage Paper No. 55).

Tesfay, H., 2007. Spatio-temporal Variations of the Biomass and Primary Production of Phytoplankton in Koka Reservoir.

Urbaniak, M., Zalewski, M., 2011. Polychlorinated dibenzo-p-dioxins and polychlorinated dibenzofurans in sediments from two Ethiopian Rift Valley lakes. Pol. J. Environ. Stud. 20 (4), 1069–1074.

Urbaniak, M., Zerihun Negussie, Y., Zalewski, M., 2012. The ecohydrological biotechnology (SBFS) for reduction of dioxin-induced toxicity in Asella Lake, Ethiopia. Geophys. Res. Abstr. 14.

Urbaniak, M., Zieliński, M., Ligocka, D., Zalewski, M., 2010. A comparative analysis of selected persistent organic pollutants (POPs) in reservoirs of different types of anthropopression polish and ethiopian studies. Fresenius Environ. Bull. 19 (11 A), 2710–2718.

U.S. EPA. Selected Analytical Methods for Environmental Remediation and Recovery (SAM), U.S. Environmental Protection Agency, Washington, DC, EPA/600/R-12/555, 2012. Selected Analytical Methods for Environmental Remediation and Recovery (SAM), 2012.

Zalewski, M., 2012. Ecohydrology—process oriented thinking for sustainability of river basins. Ecohydrol. Hydrobiol. 12 (2), 89–92. https://doi.org/10.2478/v10104-012-0012-4.

Zalewski, M., Urbaniak, M., Zerihun Negussie, Y., 2010. Ecohydrological systemic solutions for reduction of siltation, eutophication and dioxin-induced toxicity. The pilot study of the Asella BioFarm Park lake, Ethiopia. Ecohydrol. Hydrobiol. 10 (2–4), 363–368. https://doi.org/10.2478/v10104-011-0027-2.

Zerihun Negussie, Y., Urbaniak, M., Szklarek, S., Lont, K., Gągała, I., Zalewski, M., 2012. Efficiency analysis of two sequential biofiltration systems in Poland and Ethiopia—the pilot study. Ecohydrol. Hydrobiol. 12 (4), 271–285. https://doi.org/10.2478/v10104-012-0028-9

Chapter 33

Transdisciplinary and ecohydrological solutions for integrated watershed and water resources management: Case studies from Brazil

Tundisi Jose Galizia and Takako Matsumura-Tundisi
International Institute of Ecology, São Calos, São Paulo, Brazil

IMAGE CREDIT: M. ZALEWSKI.

Transdisciplinary Ecohydrology. https://doi.org/10.1016/B978-0-443-41596-8.00040-6

1. Introduction and conceptual note

Environmental management comprises a very complex set of approaches due to the nature of ecosystems and their structure and function.

In order to develop sound environmental management, a deep understanding of the ecosystems is necessary and a prerequisite for the application of technology.

When considering management by using technical processes in a river, lake, or reservoir, it is fundamental to take into consideration the biogeographical unit where these ecosystems are located, that is, the watershed. Therefore, including the entire watershed for environmental management is a priority since aquatic ecosystems are open systems depending on the interaction with the watershed for their functioning (Likens 1992).

It is in this watershed that the forcing functions that have an important role in the functioning of aquatic ecosystems are: geomorphology, geology, climate, such as temperature, precipitation, wind, input of suspended matter, pollution from diffuse and point sources, dissolved substances, and toxins.

The proper environmental management of watersheds, water resources, aquatic ecosystems, and water quality requires a combination of environmental technologies, ecotechnologies, ecohydrology, and clean technology. These approaches were proposed, consolidated, and implemented in the last 25 years, that is, in the 21st century. In the last 10 years of the 20th century, it was concluded that the application of only engineering technologies was not sufficient to cope with the magnitude and complexity of environmental problems, such as the recovery of degraded areas and aquatic ecosystems, eutrophication, control of water quality, reduction of pollution of soil, air, and water, and conservation of biodiversity.

A whole set of new scientific approaches and technologies was needed, including interdisciplinarity and transdisciplinarity. As discussed by Jorgensen (2016), these new scientific and management developments and tools include *ecological economy*, *ecotoxicology*, *ecohydrology*, *ecological indicators*, *ecotechnologies*, *ecological modeling*, and *ecological services*.

All this knowledge and the conceptual changes were a result of many years of several research efforts by specialists worldwide (Patten et al., 1990, 1994; Jorgensen and Muller, 2000; Zalewski and Robarts, 2003; Mitsch and Jorgensen, 2004; Gopal and Masing, 1990).

Integrated Water Resources Management is therefore based on a systemic and dynamic view of the complex structure and functions of ecosystems, with an integrative transdisciplinary science that combines several scientific disciplines. There are some important developments in the introduction of these new concepts, such as:

- First, the sectoral approach is not sufficient to solve the environmental management problems. It is necessary to have an integrated biogeophysical approach, encompassing several disciplines and focused on the management of several simultaneous multiple uses of the aquatic ecosystems.
- Second, the environmental management has to be predictive and not only responsive. Therefore, modeling techniques and elaboration of predictive scenarios are fundamental and have to be introduced in the process (Tundisi 1990a).
- Third, the social environmental processes, economy, and quality of life, such as human health and education, have to be included in the diagnosis and the solutions (Tundisi 1987).
- The use of ecosystem properties and nature-based solutions enhances the management performance and provides new inputs to consolidate the natural regulations of the processes.

As discussed by Zalewski (2010), the use of ***"ecosystems properties as management tools, are complementary and harmonizes with hydro-technical solutions; the ecosystems carrying capacity has to be enhanced, by using the interplay between hydrology and biota, and the use of dual regulation to improve biodiversity, water resources and ecosystems services for society***" (sic). This is a fundamental tool to regulate complex systems, integrating scientific disciplines and social and economic issues and responding to society's demands.

In this review of demonstration of case studies from Brazil, the authors discuss the importance of the use of transdisciplinary and ecohydrological scientific concepts and advanced technologies and approaches to provide integrated management of watersheds, water resources, and water quality. The ecological, economic, and social benefits of the projects are highlighted.

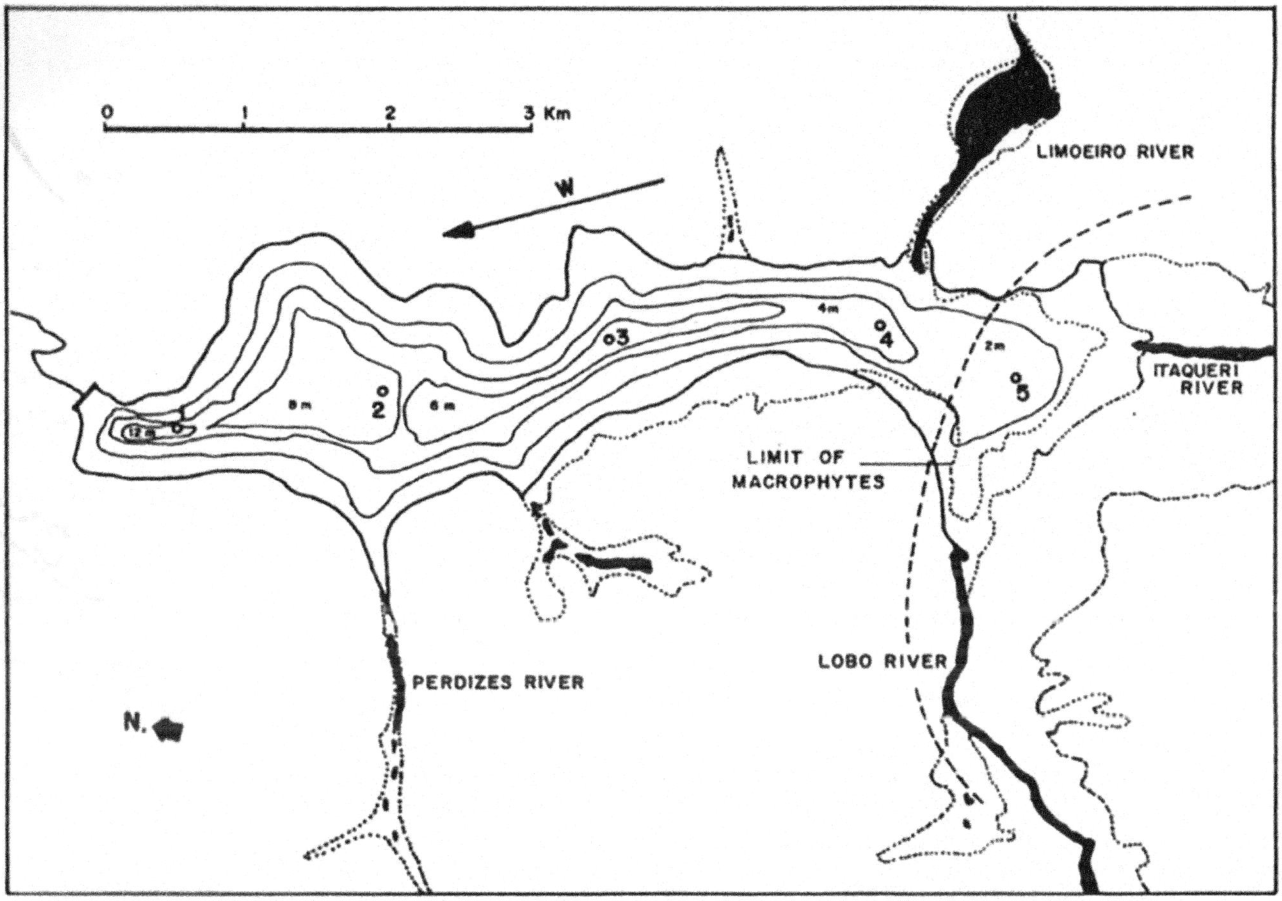

FIGURE 33.1 UHE Carlos Botelho (Lobo -Broa Reservoir) with its main characteristcs. *Tundisi J. G., Report The ecology of Lobo Broa Watershed and reservoir, Fapesp, 117pp.*

2. The sustainable management plan of the Itaqueri and Lobo river watersheds and the reservoir of the Carlos Botelho-Lobo/Broa HPP (Hydroelectric Power Plant)

The watersheds of the Itaqueri and Lobo rivers are relatively small, ecologically complex ecosystems located near the geographic center of São Paulo State, Brazil (latitude of 22 degrees 15′S and longitude of 47 degrees 49′W). These watersheds have the following general characteristics:

Area: 230 km^2

Density of drainage: 0.75 km/km^2

Declivity of the basin: 0.575 m/m

Maximum altitude: 770 m

Minimum altitude: 680 m

The reservoir of the HPP Carlos Botelho (Lobo/Broa) is located in this watershed at the confluence of the two rivers Lobo and Itaqueri. Being constructed in 1936 for hydroelectric power supply, this is a shallow ecosystem (average depth 3.00 m) with an area of 7.2 km^2 and a residence time of 20–25 days all year round.

Fig. 33.1 shows the location of the watersheds and the reservoir.

The conceptual starting point of the project to study the watersheds and the reservoir was established by Macagno (1970), who described and discussed the opportunity to promote in the region a Center of Ecological Engineering based on the integration of engineering, hydrology, biology, ecology, and limnology. Thus, the idea of a transdisciplinary approach to study the ecosystem dynamics was established at the very beginning of the project.

The research plan, organized in 1971, stated three main objectives:

- Research per se to understand ecological dynamics with a system orientation and a transdisciplinary approach.
- Scientific research and technological developments for a long-term period
- Capacity building, including faculty associates, post-doctorate fellows, students, international collaborators, and the network.

A description of the approaches and organization of research priorities follows:

As pointed out by Patten et al. (1994), every ecosystem is open: it exchanges energy, matter, and information with a dynamic process that includes biotic and abiotic factors. Lobo and Itaqueri watersheds are a spatial mosaic in which the heterogeneous distribution of habitats and communities reflects a nonuniformity of the environment. This is very well represented in the soil map of the watersheds (Fig. 33.2).

Likens (1992) discusses the importance of sustainable long-term ecological research and monitoring, consisting of the study of ecological patterns and processes for periods of more than 5 years. As stressed by this author, this long-term research is a powerful tool to focus on fundamental questions and can contribute significantly to the identification of complex environmental problems. The research in the watersheds of the Itaqueri and Lobo rivers and the Carlos Botelho (Lobo/Broa) reservoir is being completed in 2025, **54** years of continuous scientific investigation. This is why the projects, proposals, and actions of management have a strong conceptual basis (Tundisi 1990b).

With this soil composition, inputs of suspended material to rivers and the reservoir are from several sources with diverse compositions.

The hierarchical model of the watersheds includes:

a) A geographic region that incorporates climate, meteorology, geology, geomorphology, soils, and hydrology;
b) The watershed ecosystem defining the area of two sub-basins: Itaqueri and Lobo rivers.
c) The reservoir

Habitats in the ecosystems include savannah (**cerrado**, a typical vegetation of this region), *Pinus* sp. and *Eucalyptus* spp. reforestation, rivers, riparian forests, slope forests, wetlands, and reservoirs.

The driving forces of the watershed and the other components of the ecosystem are:

Driving forces: Climate, hydrology, soils, soil uses, management, and environmental disturbances

Timber Management: for *Pinus* sp. and *Eucalyptus* spp.

Soil composition

Soil biogeochemistry

Riparian forests: terrestrial/aquatic ecotones and wetlands

Impacts: fire/deforestation, which affects soil composition, soil, and biogeochemistry

The results of these several years of research on the dynamics of these components are as follows:

The ecological dynamics of the watersheds and the reservoir are influenced by the temporal climate changes as a result of the geographic position of the ecosystems and the subtropical climate (Tundisi 1977).

The response of the biological communities to the climate, soil, nutrient inputs, and global changes is the succession of species, their temporal and spatial distribution, and the rates of production/decomposition (Rodriguez and Matsumura-Tundisi 2000). The response of the reservoir to the external forcing functions, such as cold fronts and other disturbances, was in the direction of the reorganization of the system after perturbation in terms of circulation and rapid chemical changes in the biogeochemical cycles (for example, in the sediment–water interactions) or in the succession of phytoplankton and zooplankton. The feedback mechanisms forced the reservoir to be near its original status; this was enhanced by the shallowness of the reservoir, the permanent low residence time, and the function of wetlands in the upstream reservoir (Tundisi 2017). The natural wetlands at the upstream reservoir have a strong role in the control of the cycles of phosphorus, nitrogen, carbon, and the biodiversity in the whole reservoir. These wetlands are a nursery ground for fish and some species of copepods (Tundisi and Matsumura-Tundisi 2014).

The long-term ecological research demonstrated the important role of the natural environmental processes as controls of the resilience in the watersheds and the reservoir (Tundisi 1994). It was clear that the interaction and network of the hydrological, biological, biochemical, and hydrodynamic processes and the spatial heterogeneity were fundamental for the maintenance and performance of the ecosystem services (Granadeiro Rios 2003). Periotto and Tundisi (2013) conducted a

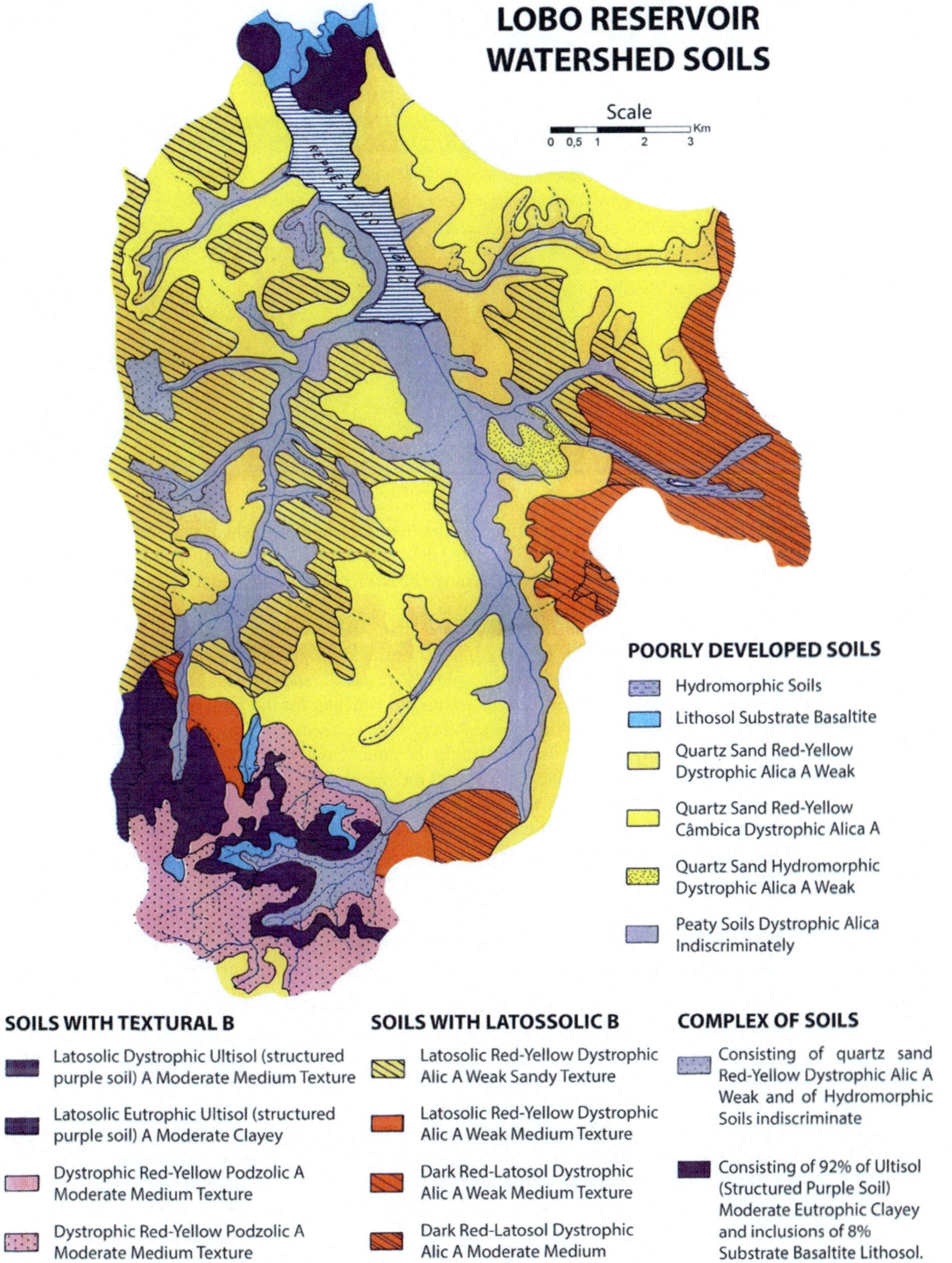

FIGURE 33.2 The soils of the watersheds of the Itaqueri and Lobo rivers. *Sobral M., Tundisi J. G., 1986. Map of Soils of the Watersheds of Itaqueri and Lobo Rivers. Federal University of São Carlos.*

FIGURE 33.3 The wetland area upstream of the reservoir of Carlos Botelho HPP (Lobo/Broa). *Photo International Institute of Ecology (2016).*

detailed study of the range and values of services provided by the Itaqueri-Lobo watersheds and the Lobo-Broa reservoir. They also measured the exergy of the ecosystem and concluded that, despite temporal changes, a high level of exergy was maintained (Tundisi et al., 2012). These authors listed 20 ecological services that produced a revenue of U$45,000 (forty-five thousand US dollars) per hectare per year.

Higher exergy levels mean higher overall buffer capacity and more stability. Being away from thermodynamic equilibrium pays off in terms of energy conservation, network interactions, and ecosystem services (Fath and Patten 1998).

The knowledge promoted by the long-term transdisciplinary research and the findings of the ecohydrological mechanisms controlling the ecosystems' responses were therefore fundamental for the proposition of a management plan, promoting the conservation of ecosystem services and the stability of the ecosystems. This was achieved by preserving the ecological infrastructure of the ecosystems, such as maintaining the aquatic and terrestrial biodiversity, conservation and restoration of riparian forests, slope forests, wetlands upstream of the reservoir, and the low retention time of the reservoir. Preserved ecosystems maintain a high diversity of ecosystem services (Constanza et al., 1997). (Fig. 33.3).

The management plan has the following additional components: ***water quality regulations at the basin level, basin infrastructure plan, basin water quality plan*** (Matsumura-Tundisi T. and Tundisi J. G. 2018; Tundisi and Matsumura-Tundisi 2023).

3. Constructed wetlands of the Bocaina municipality—Effluent of wastewater treatment with phytotechnology

The main objective of this project was to apply a phytotechnology to reduce the phosphorus and nitrogen loads of an effluent of a wastewater treatment plant. Thus, the application of phytotechnology was used in treated sewage purification systems.

A constructed wetland was installed at the municipality of Bocaina (São Paulo State) at the outlet of a sewage treatment plant.

The concept of this artificial system was developed following the design of a subsurface flow wetland model (UNEP/University of Copenhagen 2005). This consists of a constructed wetland area in which gravel is used as soil, allowing a water flow through the wetland. Rooted wetland plants with an indicated density are planted in the gravel.

The following information is necessary in order to program the removal efficiencies of organic matter:

- The water flow
- The concentrations of the organic matter in the effluent water (total nitrogen, total phosphorus, Biological Oxygen Demand (BOD))

The required removal efficiency is calculated based on a model that defines the desired concentration at the end of the wetland plant, considering the time of the water flow and the plant density (Fig. 33.4).

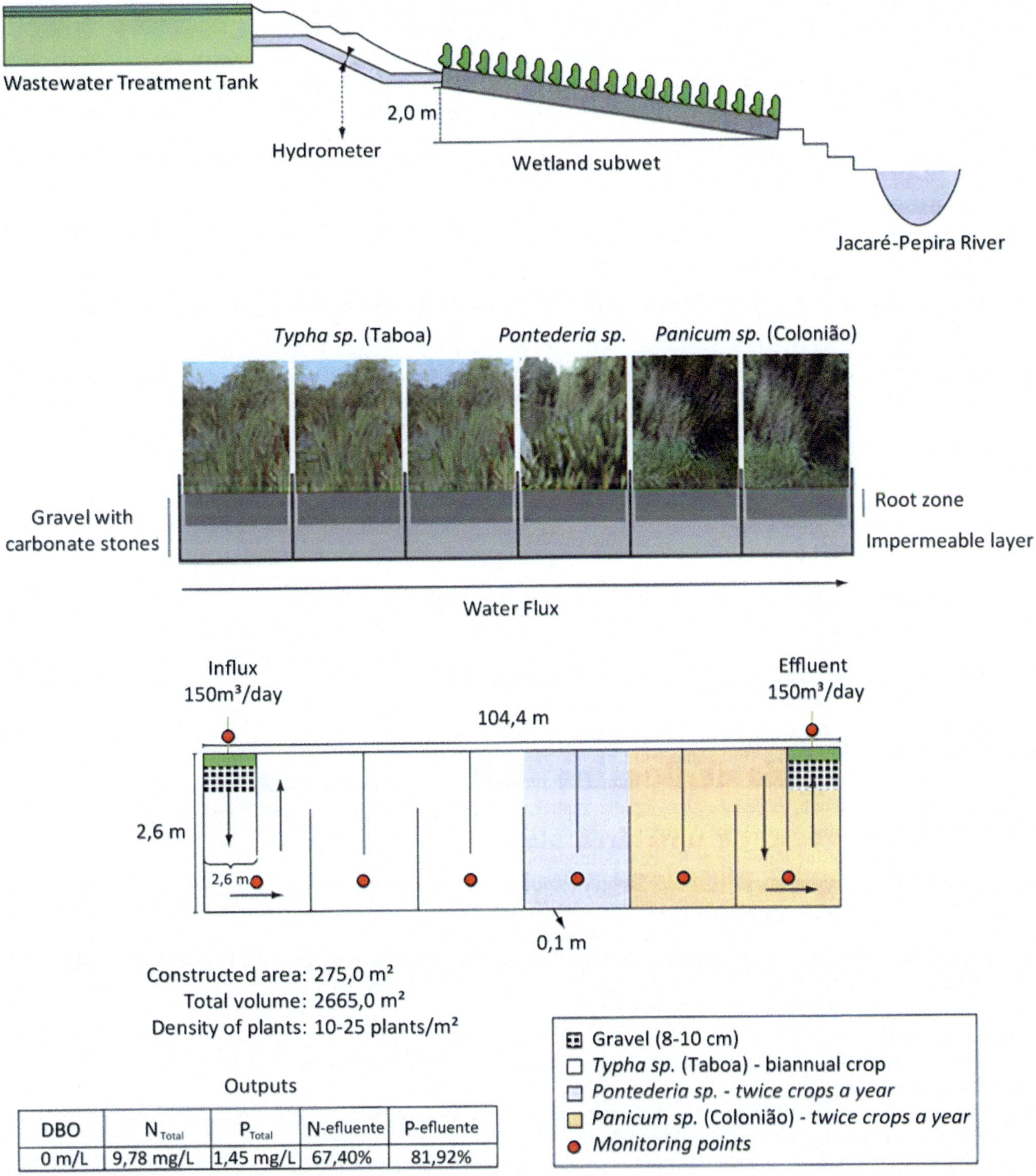

DBO	N_{Total}	P_{Total}	N-efluente	P-efluente
0 m/L	9,78 mg/L	1,45 mg/L	67,40%	81,92%

FIGURE 33.4 Details the infrastructure of the artificial wetland and other information. Constructed wetlands can be very effective in the removal of pollutants from point sources—municipal, industrial, or agricultural wastewater or nonpoint sources—mining, urban runoff, or agriculture. The main properties of wetlands include high plant productivity, large capacity to absorb sediments, and high rates of oxidation by microflora associated with plant biomass. The biological reactions involved in the performance of a constructed wetland include ***photosynthesis*, *respiration*, *fermentation*, *nitrification*, *denitrification***, and ***removal of phosphorus by microbial activity*** (Mitchel et al., 1995; Zalewski et al., 2004).

With appropriate and low wetland water velocities, suspended soils will settle from the water column within the wetland. Wetlands also remove pathogens such as bacteria, fungi, viruses, protozoa, and helminths.

Constructed wetlands are relatively inexpensive to operate and construct, provide effective wastewater treatment, provide green spaces, and enhance aquatic and terrestrial biodiversity. Some disadvantages are as follows:

Cost of gravel, imprecise design and operation criteria, possible problems with pests and mosquitoes. Adequate maintenance is needed.

The constructed wetland at Bocaina had an impact: water quality in the rivers downstream was improving, stimulating reuse for irrigation and for public supply. As a regional project, a demonstration and capacity building and education for managers and public authorities were effective (Tundisi et al., 2009).

4. Application of ecohydrology and transdisciplinary ecological approaches to water transfer in the Metropolitan Region of São Paulo: The use of a natural wetland

The Metropolitan Region of São Paulo has, in an area of 8,000 km^2, 23 shallow reservoirs (generally <30 m) that are used primarily for public water supply and support also multiple activities such as recreation, aquatic sports, fisheries, and irrigation. These artificial ecosystems were built up during the 20th century and provide water for a population of 22 million inhabitants. They are subjected to several impacts resulting from human activities in their watersheds: deforestation, point and nonpoint sources of nitrogen and phosphorus, input of toxic substances, and industrial wastewater. There is also an unknown atmospheric contribution of particulate material with C, N, P, and other elements (Tundisi et al., 2006). These reservoirs are a strategic water reserve for the Metropolitan Region of São Paulo (MRSP).

The transfer of water between the reservoirs and from some small rivers to reservoirs was adopted in periods of water scarcity or long dry periods in order to maintain an adequate supply of water in some regions of the metropolitan area. In 1992, the Board of SABESP (Company of Water Supply of São Paulo).

Determined the transference of 4 m^3/s of water from Billings reservoir to Guarapiranga reservoir, the two largest artificial ecosystems of the MRSP. In order to follow up on this resolution, a Panel of Experts[a] was established with the task to provide the best possible evaluation of this process and produce an independent report with their recommendations.

The Parelheiros River flows to the Guarapiranga reservoir, and in its final region has a large wetland with an area of 991.308 m^2. The main vegetation is rooted macrophytes: *Panicum maximum* and *Typha angustifólia*, which occupy the whole area of the wetland (Fig. 33.5a and b).

The Panel of Experts recommended continuous monitoring of the wetland area in order to evaluate its role as an efficient natural ecosystem to reduce the organic matter of the transferred water from Billings to the Guarapiranga reservoir. This monitoring was realized, and the results are presented in Tables 33.1 and 33.2. From these results, it is clear that the wetland of the Parelheiros River was fundamental in removing phosphorus, nitrogen, inorganic carbon, and sulfate. Additional experiments using water samples collected in the water from above the sediments every 2 h and incubated for 24 h were performed (Abe et al., 2006). Analysis for nitrate, nitrite, ammonium, and total dissolved nitrogen showed nitrogen removal as high as 13.43 mg N m^2 day^{-1} measured.

Processes of ammonification and denitrification are the main factors for the reduction of nitrogen, which is liberated in the atmosphere as gases dinitrogen and nitrous oxide. Seitzinger (1988) showed similar results for Lake Mendota and Lake Okeechobee (USA), Whitaker and Matvienko (1998) obtained similar results also for the reservoir of UHE Carlos Botelho—Lobo/Broa.

Intensive reduction of sulfate was measured in the Parelheiros wetland central area, probably as a result of the activity of sulfate reduction bacteria.

Point sources of wastewater at some stations explain some increases. But the overall removal process of phosphorus is demonstrated, considering the results at station 4.

(Methods for sampling and chemical analysis: Valderrama 1981; Chapman 1992).

Tropical and subtropical wetlands retain nitrogen, phosphorus, and carbon. The vegetation cover of these wetlands is composed of dense stands of rooted macrophytes that function as an effective biological filter. Several measurements in tropical or subtropical wetlands showed that they can effectively remove dissolved phosphorus, especially when there are extractable aluminum and iron in the soil. The sediments and the vegetation are major reservoirs of nutrients and help in

a The Panel of Experts was composed of Dr. J. G. Tundisi (Chairman), IIE SP Brazil. Dr. Carlos Tucci, UFRGS, Brazil. Dr. Joan Armengol, Department of Ecology, University of Barcelona, Spain. Engineer Alberto Calcagno, UNEP, Nairobi, Kenya.

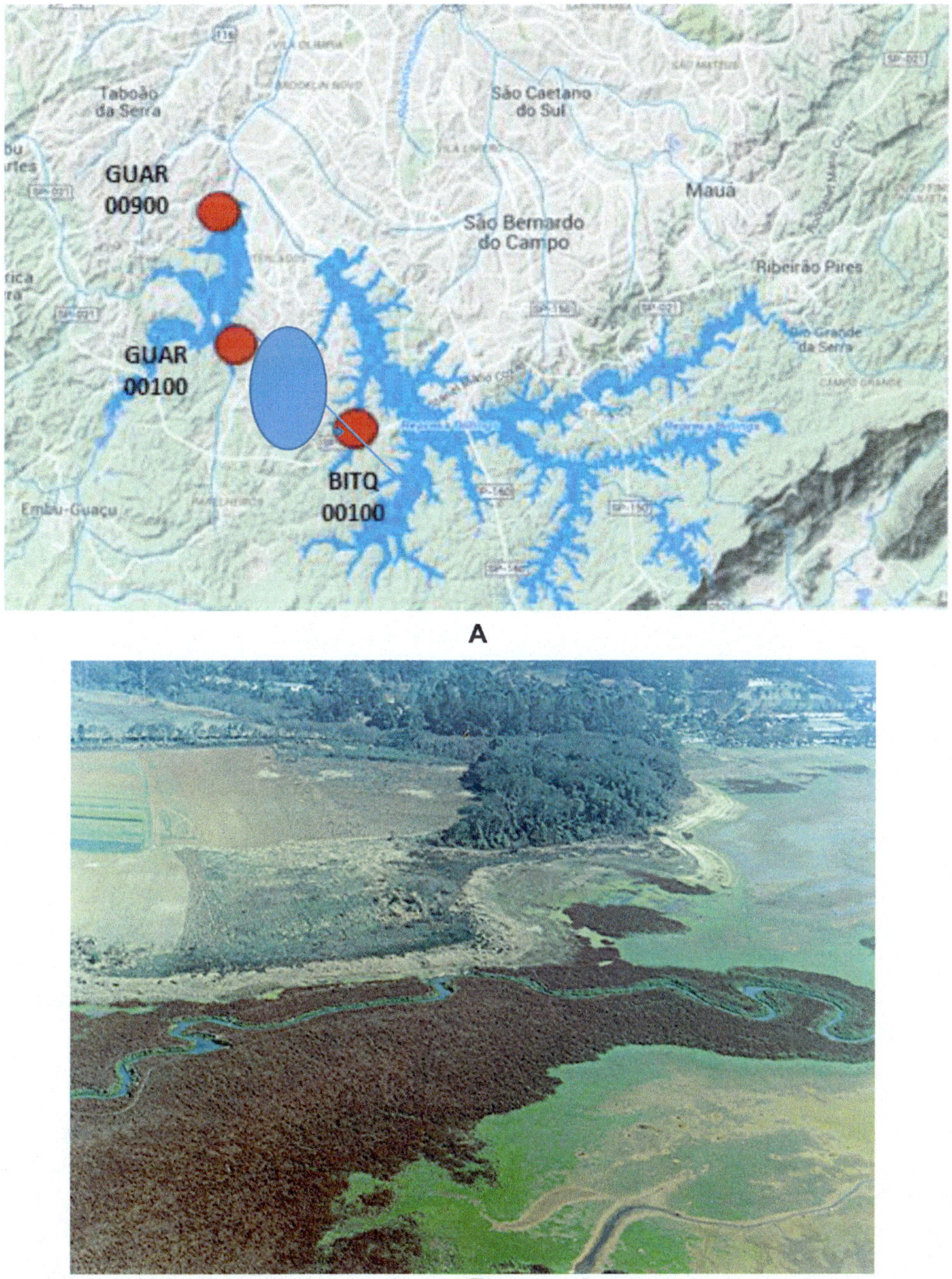

FIGURE 33.5 (a) The red dots show the area of transference of water from Billings to the Guarapiranga reservoir. The area of the wetland is shown in blue. (b) The wetland of the Parelheiros River in the Metropolitan Region of São Paulo. *CETESB., 2020. Monitoring of the Water Quality of the Parelheiros Wetland. Tables 2 and 3; International Institute of Ecology and SABESP-SP, 1992. The auditing of a water transfer in the Metropolitan Region of São Paulo. In: Report: Conclusions and Recommendations. p. 125. Photo: J.G. Tundisi, 1992.*

nutrient removal through uptake and transformations such as denitrification. Thus, the nutrient budgets are governed by nutrient uptake by plants, retention in the sediments, and losses to the atmosphere (Gopal and Masing 1990).

Wetlands associated with reservoirs are important compartments of the system, playing a significant role in the control of high levels of dissolved nutrients, organic matter, and toxic substances. In this case study, the removal capacity of the Parelheiros wetland was demonstrated. The wetland in the Metropolitan Region of São Paulo is located in a large urbanized area. Therefore, its conservation is vital for the control of the water quality being transferred from Billings

TABLE 33.1 Concentration of total dissolved phosphorus and total phosphorus in four sampling stations in the wetland area of Parelheiros in the Metropolitan Region of São Paulo.

	Total D. phosphorus (μg/l)	Total phosphorus (μg/l)
Station 1 (upstream wetland)	99.31	116.90
Station 2 (middle wetland)	13.77	48.81
Station 3 (central area wetland)	24.38	85.35
Station 4 (downstream wetland)	24.72	93.18

Source: Abe, S.D., Tundisi, J.G., Vannucci, D., Sidagis, G.C., 2006. Assessment of the nitrogen removal capacity of a floodplain in the Guarapiranga reservoir Metropolitan Region of São Paulo. In: Tundisi, J.G., Matsumura-Tundisi, T., Sidagis, T., Galli, C. (Eds.), Eutrophication in South America: Causes , Consequences Technologies for Management and Control. International Institute of Ecology, Brazilian Academy of Sciences, Interamerican Network of Academies of Sciences. pp. 241–253 CNPq.

TABLE 33.2 Concentration of total dissolved nitrogen, nitrate, sulfate, and dissolved inorganic carbon at four sampling stations in the wetland area of Parelheiros at the Metropolitan Region of São Paulo.

	T. D. Nitrogen (μg/l)	Nitrate (μg/l)	Sulfate (mg/l)	D. I. C. Carbon (μg/l)
Station 1 (Upstream W.)	870	210	5.27	12.38
Station 2 (Middle W.)	1096	177	3.41	13.48
Station 3 (Central area W.)	1538	130	3.78	15.61
Station 4 (Downstream W.)	640	129	1.32	6.80

D.I.C.,dissolved inorganic carbon.
Source: Abe, S.D., Tundisi, J.G., Vannucci, D., Sidagis, G.C., 2006. Assessment of the nitrogen removal capacity of a floodplain in the Guarapiranga reservoir Metropolitan Region of São Paulo. In: Tundisi, J.G., Matsumura- Tundisi, T., Sidagis, T., Galli, C. (Eds.), Eutrophication in South America: Causes , Consequences Technologies for Management and Control. International Institute of Ecology, Brazilian Academy of Sciences, Interamerican Network of Academies of Sciences. pp. 241–253 CNPq.

Reservoir to Guarapiranga. This reservoir supplies drinking water for five million people of the Metropolitan Region of São Paulo; therefore, the transference of 4 m^3/s helps to sustain the quality of the water available, especially in periods of water scarcity.

This water transfer project, which used a wetland, was the first large-scale project developed in the MRSP and in Brazil. This technology offered several benefits: (i) the non-extensive use of pipelines and pumping stations, as recommended by the Panel of Experts, resulted in an infrastructure and energy consumption savings of US$1,000,000—one million American dollars; (ii) the removal of organic matter and nutrients by the wetland resulted in the transfer of less contaminated water to the receiving reservoir, which prevented eutrophication; and (iii) the wetland enhances biodiversity and can be used for observation and teaching. The project also demonstrated an advanced technology.

5. Conclusions

These three case studies showed the importance of the use of transdisciplinary and ecohydrological approaches to solve problems of integrated watershed management and integrated management of water resources. The implementation of the projects was at different scales, which was useful as a demonstration of the application of technologies and concepts.

The use of advanced techniques for management has several ecological and environmental consequences and benefits due to the reduction of impacts, maintenance, and conservation of ecological processes and ecosystem services and better control of water quality, reducing pollution and costs for water treatment used for public supply (Tundisi and Matsumura-Tundisi 2010). Also, economic benefits are expected due to the reduction of costs of projects with nature-based solutions and maintenance expenses.

Conservation and improvement of aquatic and terrestrial biodiversity is another important result of the application of these technologies. Therefore, natural ecohydrological processes enhance the exergy of ecosystems and the biodiversity of natural capital. The resulting increase in ecological services mirrors growth in money flows, as discussed by Patten 2019.

The growth of ecological services is a consequence of the increasing complexity of networks, which is one of the results of the dual regulation process of hydrology and biology. The increased complexity of the network also results in more genetic information and plays a significant role in ecosystem stability (Jorgensen and Svirezhev 2004).

But besides important achievements in environmental quality and sustainability, which include a better human quality of life, it is relevant to consider the effects of the applications of these technologies and scientific approaches on the regional economies.

For the watersheds of the Lobo/Itaqueri rivers and the Carlos Botelho-Lobo/Broa reservoir, the improvement and maintenance of a watershed and reservoir with good water quality for tourism, recreation, aquatic sports, camping, sports fisheries, and enhanced and stimulated the regional economy based on the ecological services provided by the watersheds and the reservoir. The watersheds and the reservoirs receive annually 300,000 people for tourism and recreation.

The conservation of the slope forests was also fundamental to maintain the water quality of the rivers flowing downstream through these regions and the biodiversity of the vegetation and animal life.

The artificial wetland constructed in the municipality of Bocaina promoted the improvement of the water quality discharge at low cost and the reduction of phosphorus load to the rivers near the town. The use of water for irrigation and public supply downstream with a lower cost of treatment was stimulated. It also improved the local biodiversity.

The third case study, the use of Parelheiros natural wetland in the Metropolitan Region of São Paulo, proved to be a very good example of ecohydrological technology applied on a larger scale, since this is a natural ecosystem in a metropolitan region with multiple impacts of urbanization.

This wetland significantly reduced the nitrogen, phosphorus, and carbon concentrations in the discharged water to the receiving reservoir. This reduced the costs of water treatment for public supply and prevented further eutrophication of the waters of the receiving reservoir.

Another consequence is that the water transfer system did not use expensive engineering technology such as pipelines, pumped stations, and energy. The water was transferred by flowing freely from one reservoir to the other. The application of these technologies therefore had a social, economic, and environmental impact (IIE/SABESP, 1992).

Since this was the first time it was implemented at this scale in a metropolitan region in Brazil, it was also a demonstration with an educational impact and a capacity-building effect on engineers, managers, and the water resources administration authorities.

Wetlands associated with reservoirs are important compartments of the system, playing a significant role in the control of high levels of dissolved nutrients, organic matter, and toxic substances. In this case study, the removal capacity of the Parelheiros wetland was demonstrated. The wetland in the Metropolitan Region of São Paulo is located in a large urbanized area. Therefore, its conservation is vital for the control of the water quality being transferred from Billings Reservoir to Guarapiranga. This reservoir supplies drinking water for five million people of the Metropolitan Region of São Paulo; therefore, it is a strategic water resource.

The use of a wetland in this water transfer was the first large-scale nature-based solution project developed in the MRSP and in Brazil. The economic benefits of this technology were demonstrated; the removal of organic matter and nutrients by the use of a wetland was another important benefit, as it was transferring water with lower contamination to the receiving reservoir, preventing eutrophication. This wetland is an area that enhances biodiversity; therefore, this is an important benefit considering its role as a buffer ecosystem in a densely populated area. Once the project was established and completed, schools constructed observation platforms for birds and used the wetland as a teaching site for biology and ecology (Patten and Fath, 2018).

The project was also a demonstration of an advanced and new technology for engineers, managers, and water resources administrators; therefore, it had a capacity-building component.

Acknowledgments

The National Research Council of Brazil (CNPq), the São Paulo State Foundation for Support of Research (FAPESP), and the Organization of American States (OEA) have provided financial support for research in the watersheds of the Itaqueri/Lobo rivers and the HPP Carlos Botelho (Lobo/Broa) reservoir for many years. Additional financial support was provided by the ILEC—International Lake Environment Committee (Shiga, Japan), the United Nations University (Tokyo, Japan), and the University of São Paulo's School of Engineering in São Carlos and the Federal University of São Carlos.

The constructed wetland area in Bocaina was supported by the United Nations Environmental Programme (UNEP—IETC—Japan) and the Municipality of Bocaina.

The Panel of Experts on water resources and the field visits, logistics report, and consultancy were supported by a contract between SABESP (the São Paulo State water management company) and the International Institute of Ecology in São Carlos.

Mr. Felipe Blanco and Ms. Rogerio Almeida assisted with the illustrations and word processing for different versions of the manuscript.

References

Abe, S.D., Tundisi, J.G., Vannucci, D., Sidagis, G.C., 2006. Assessment of the nitrogen removal capacity of a floodplain in the Guarapiranga reservoir Metropolitan Region of São Paulo. In: Tundisi, J.G., Matsumura- Tundisi, T., Sidagis, T., Galli, C. (Eds.), Eutrophication in South America: Causes , Consequences Technologies for Management and Control. International Institute of Ecology, Brazilian Academy of Sciences, Interamerican Network of Academies of Sciences, pp. 241–253. CNPq.

Chapman, D. (Ed.), 1992. Water Quality Assessments. UNEP., UNESCO,WHO, p. 585.

Constanza, R., et al., 1997. The value of the words ecosystems services and natural capital. Nature 387, 253–260.

Fath, B.D., Patten, B.C., 1998. Network synergism emergency of positive relation in ecological systems. Ecol. Model. 107 (2–3), 122–143.

Gopal, B., Masing, V., 1990. Biology and ecology. In: Paten, B.C., et al. (Eds.), Wetlands and Continental Water Bodies. Vol 1 Natural and Human Relationships. SPB Publishing, The Hague, The Netherlands, pp. 91–239.

Granadeiro, R.C.M.M., 2003. Modelagem da circulação em um reservatório tropical polimiticco. Tese de Doutorado UFSCAR . 58pp.

International Institute of Ecology and SABESP-SP, 1992. The auditing of a water transfer in the Metropolitan Region of São Paulo. In: Report: Conclusions and Recommendations, p. 125.

Jorgensen, S.E., 2016. Ecohydrology as an important concept and tool in environmental management. Ecohydrol. Hydrobiol. 16 (1), 4–6.

Jorgensen, Svirezhev, Y.M., 2004. Towards a Thermodynamic Theory for Ecological Systems. Elsevier, Amsterdam, p. 366.

Jorgensen, Muller, F., 2000. In: Handbook of Ecosystem Theories and Management. CRC Press LLC Lewis Publishers, p. 583.

Likens, G., 1992. The ecosystem approach : its use and abuse. In: Kinne, O. (Ed.), Excellence in Ecology, vol. 3. Ecology Institute, Oldendorf Luhe Germany, p. 166.

Macagno, E., 1970. A Regional Center for Ecological Engineering. University of São Paulo, School of Engineering at São Carlos, p. 34. Report to FAPESP.

Matsumura-Tundisi, Tundisi, J.G., 2018. Water Resources Management. Editora Scienza, São Carlos, Brazil, p. 248pp.

Mitisch, W., Jorgensen, J., 2004. Ecological Engineering and Ecosystem Restoration. John Wiley and Sons Inc, p. 409.

Mitchel, D.S., Chick, A.J., Rasin, G.W., 1995. The use of wetlands for water pollution control in Australia : an ecological perspective. Water Sci. Technol. 32, 365–373.

Patten, B.C., Fath, B.D., 2018. Notes from an introductory course on field systems ecology. Ecol. Model. 368, 33–40.

Patten, B.C., Jorgensen, S.E., Dumont, H., Gopal, B., Kvet, J., Svirezhev, Y.M., Koryakov, P.P., Lofler, H., Tundisi, J.G. (Eds.), 1990. Wetlands and Continental Water Bodies Volume 1 Natural and Human Relationships. SPB Academic Publishing,, The Netherlands the Hague, p. 759.

Patten, B.C., Jorgensen, S.E., Dumont, H., Gopal, B., Kvet, J., Svirezhev, Y.M., Koryakov, P.P., Lofler, H., Tundisi, J.G. (Eds.), 1994. Wetlands and Continental Waterbodies .Vol 2. Case Studies. SPB Academic Publishing, the Netherlands The Hague, p. 732.

Periotto, N.A., Tundisi, J.G., 2013. Ecosystem services of the UHE Carlos Botelho (Lobo/Broa): a new approach for the management and planning of dams multiple uses. Braz. J. Biol. 73 (3), 471–482.

Rodrigues, M.P., Matsumura-Tundisi, T., 2000. Variations in the density of species composition and species dominance of rotifers at a shallow tropical reservoir (Broa reservoir SP. Brazil) in a short scale of time. Braz. J. Biol. 60 (1), 1–9.

Seitzinger, S.P., 1998. Denitrification in freshwater and coastal marine ecosystems. In: Ecological and Geochemical Significance, 33, pp. 702–724.

Tundisi, J.G., 1977. Primary Production, Phytoplankton Standing Stock and Ecological Processes at Lobo/Broa Reservoir São Paulo State , Brazil. DsC Thesis University of São Pauulo, p. 245.

Tundisi, J.G., 1987. Local community involvement in environmental planning and management: the Lobo/Broa reservoir case study. Reg. Dev. Dialog. 8, 135–155.

Tundisi, J.G., 1990a. Perspectives for ecological modelling of tropical and subtropical reservoirs in South America. Ecol. Model. 52, pp7–20.

Tundisi, J.G., 1990b. Ecology and development: perspectives for better society. In: Kawanebe, H., Ogushi, T., Higashi, M. (Eds.), Ecology for Tomorrow. Physiology and Ecology, pp. 93–130. Japan Vol. 27 Special Number.

Tundisi, J.G., 1994. Ecosystem model for the Lobo(Broa) reservoir ,Brazil. In: Patten, B.,C., et al. (Eds.), Wetlands and Shallow Continental Water Bodies, vol 2, pp. 217–223 (Case Studies).

Tundisi, J.G., 2017. Long Term Ecological Research in the Watersheds of Itaqueri and Lobo Rivers and in the Carlos Botelho (Lobo/Broa), Reservoir. Itirapina SP Brazil, p. 70 (Report). Fapesp.

Tundisi, J.G., Matsumura-Tundisi, t., 2010. Potential impacts of alterations on the forest law in Brazil in the water resources. Biota Neotropica 10 (4), 68–73.

Tundisi, J.G., Matsumura-Tundisi, T., 2014. The ecology of Carlos Botelho (Lobo/Broa) reservoir and its watersheds . São Paulo state, Brazil. Freshwater Rev. 6 (2), 75–91.

Tundisi, J.G., Matsumura-Tundisi, T., Periotto, N.A., 2012. Exergy and ecological services in reservoirs. Braz. J. Biol. 72 (4), 979.

Tundisi, J.G., Abe, S.D., Matsumura- Tundisi, T., Tundisi, J.E.M., Vannucci, D., 2006. Reservoirs of the metropolitan region of São Paulo: consequences of eutrophication and perspectives for recovery and management. In: Matsumura, T.J.G., Tundisi, T., Sidagis Galli, C. (Eds.), Eutrophication in South America : Causes, Consequences , Technologies for Management and Control . International Institute of Ecology, Brazilian Academy of Sciences. Interamerican Network of Academies of Science, CNPq, pp. 161–182.

Tundisi, J.G., et al., 2009. A Manual for the Management of Watersheds. International Institute of Ecology, São Carlos Municipality of São Paulo/ Secretary of Environment, p. 134.

Tundisi, J.G., Matsumura-Tundisi, T., 2023. Systems ecology of watersheds of Itaqueri and Lobo rivers and Lobo-Broa reservoir (Itirapina sp) : "a new wine for a old bottle". Braz. J. Biol. 83, 1–9.

UNEP-IETC, 2005. Handbook of Phytotechnology for Water Quality Improvement and Wetland Management through Modelling Applications. University of Copenhagen, Olentangy River Wetlands, p. 239.

Valderrama, J.C., 1981. The simultaneous analysis of total nitrogen and total phosphorus in natural waters. Mar. Chem. 10, 109–122.

Whitaker, V., Matvienko, B., 1998. The denitrification potential and hydrological conditions in the wetlands of the Lobo Reservoir. Verh. Internat. Verin. Limnol. 26, 11377–11380.

Zalewski, M., 2010. Ecohydrology for Implementation of the EU Water Framework Directive. Publication of the Institute of Civil Engineers, p. 9.

Zalewski, M., Robarts, R., 2003. Ecohydrology – a New Paradigm for Integrated Water Resources Management. SIL NEWS, pp. 1–5.

Zalewski, M., Wagner-Lotkowska, I., Robarts, R.D. (Eds.), 2004. Integrated Watershed Management – Ecohydrology & Phytotechnology – Manual, UNESCO-IHP, UNESCO-ROSTE, UNEP-DTIE-IETC, Venice, Osaka, Warsaw and Łódź.

Further reading

Patten, B.C., 1994. Systems ecology of Okefenokee swamp. (USA). In: Patten, B.C., et al. (Eds.), Wetlands and Shallow Continental Water bodiesVol 2: Case Studies, pp. 129–180.

Tundisi, J.G., 2025. Reservoirs: Ecology, Limnology, Multiple Uses. Management.Editora Scienza, São Carlos. Brazil, p. 240 (in press).

Tundisi, J.G., Matsumura- Tundisi, T., 2022. Represas Artificiais: Ecologia, Limnologia , Usos Múltiplos, Gerenciameento. Editora Scienza. São Carlos, p. 426.

Zalewski, M., Santiago-Fandino, V., Neate, J., 2003. Energy ,water plant interactions, "green feedback". In: As a Mechanism for Environmental Management and Control Through the Application of Phytotecnology and Ecohydrology. Hydrological Processes, 17, pp. 2753–2767.

Chapter 34

Ecohydrological approaches to assessing environmental flow in rivers: Focus on hydrological methods

Mohammad Mahdi Dorafshan[1], Saeid Eslamian[2], Mohammad Hossein Golmohammadi[3], Ali Shokri[4] and Maciej Zalewski[5]

[1]Department of Civil Engineering, Isfahan University of Technology, Isfahan, Iran; [2]Department of Water Science and Engineering, College of Agriculture, Isfahan University of Technology, Isfahan, Iran; [3]Department of Civil Engineering, Isfahan University of Technology, Isfahan, Iran; [4]School of Engineering, Waikato University, Hamilton, New Zealand; [5]European Regional Centre for Ecohydrology of the Polish Academy of Sciences under the auspices of UNESCO, Łódź, Poland

IMAGE CREDIT: M. ZALEWSKI.

Transdisciplinary Ecohydrology. https://doi.org/10.1016/B978-0-443-41596-8.00028-5

1. Introduction

In the past few decades, water resource management and ecological development have faced significant challenges due to climate change and human activities (Dorafshan and Eslamian, 2023; Wang et al., 2024a,b; Dorafshan et al., 2024; Dorafshan et al., 2025b). As a result of human activities (such as the construction of water conservancy projects and changes in land use), rivers are impacted significantly, resulting in considerable changes in hydrology (Wang et al., 2012; Lv et al., 2021). Because altered river flows are increasingly harming river ecosystems' health (Kuriqi et al., 2020; Parasiewicz et al., 2023), and as public concern about this damage has risen, scientists have developed environmental flow assessments to identify the water needed to safeguard these ecosystems (Suwal et al., 2020).

The concept of environmental flows emerged alongside the rapid expansion of modern dam construction in the first half of the 20th century (World Commission on Dams, 2000). As the number of dams increased, their adverse effects on river ecosystems became a growing concern (Richter et al., 1996). In response, advancements in scientific understanding, along with the development of technical tools and evolving societal and conservation expectations, led to improved dam design and operation to mitigate ecological harm (Griggs et al., 2013; Poff and Matthews, 2013). Prior to the 1980s, environmental flow determinations were characterized by a reductionist paradigm, emphasizing the maintenance of a minimum instream flow. This approach was largely implemented downstream of major dams, predominantly in the United States and Western Europe, with the primary objective of sustaining a limited number of aquatic taxa, typically a singular commercially or recreationally significant fish species (Hoque et al., 2024, 2025). At that time, ecological science introduced the significance of hydrological variations in preserving ecosystem structure and function, as well as the role of each aquatic species within environmental interactions and conditions (Acreman and Dunbar, 2004; Chen et al., 2025a,b). By the early 1990s, discussions surrounding river flows expanded into the broader issue of sustainable water resource management.

A significant milestone was reached at the 10th Brisbane River Symposium in 2007, where more than 800 engineers, researchers, water resource managers, economists, and policymakers from 57 countries gathered. The symposium led to the adoption of the Brisbane Declaration on environmental flow principles (Arthington et al., 2018). This declaration is valued because it links ecological sustainability to social well-being. As a result, environmental flows evolved from a narrowly focused conservation program into a broader ecological–social sustainability framework, applicable even within governance structures (Acreman et al., 2000; Cui et al., 2020).

The Brisbane Declaration articulates environmental flows as encompassing the volume, quality, and temporal dynamics of water regimes necessary to preserve freshwater ecosystems and support the livelihoods and welfare of dependent human populations. Over time, research on river flows extended beyond fish populations to encompass other in-stream water uses. Studies now address habitat biodiversity, additional biodiversity components, water quality, and ecosystem functions (Stalnaker and Arnette, 1976; Bhuyan and Sharma, 2025).

After the Millennium Ecosystem Assessment highlighted the significance of ecosystem "goods and services" (Millennium Ecosystem Assessment, 2005), the discussion around environmental flows expanded to include flow requirements for human well-being and the sociocultural needs of downstream communities. Consequently, questions arose regarding the desirable and acceptable conditions of river ecosystems and the benefits of allocating water to them. The outcome of these discussions was the recognition that acceptable river conditions should be determined through a "social–political consensus reached via public consultation." However, implementing and enforcing this decision requires strong political will and support. Given the vast diversity of rivers and their varying social, economic, and cultural contexts, environmental flow assessments have evolved into numerous methodologies.

Currently, more than 200 different approaches exist for determining the water requirements of river ecosystems (Overton et al., 2014; Wałęga et al., 2025). Various approaches developed to determine the environmental flow of rivers possess different defensibility based on complexity and the data used (Chen et al., 2025a,b). Various researchers have classified approaches for assessing environmental flows. For example, Dyson et al. (2003) have classified them into four approaches (look-up table, desktop, functional analysis, and habitat modeling), Brown and King (2003) have classified them into two approaches (prescriptive and interactive methods), and Tharme (2003) have classified them into four approaches (hydrological, hydraulic rating, habitat simulation, and holistic methods). According to the classification of Tharme (2003), hydrological and hydraulic approaches often assume a direct correlation between reduced water availability and habitat degradation in rivers. In contrast, habitat simulation models prioritize an "optimal" flow regime to maintain ecological balance (Jowett, 1997; Kuriqi et al., 2020). Holistic environmental flow approaches offer a more comprehensive approach, aiming to replicate the natural flow patterns of the river (Williams et al., 2019). Each of these methods has its own advantages and disadvantages (Efstratiadis et al., 2014; Leone et al., 2023). Hydrological approaches rely solely on time series data of rivers, while holistic approaches require extensive data (Karakoyun et al., 2018). Hydrological-based methodologies represent the most straightforward and globally prevalent techniques for environmental flow estimation (Zhao et al., 2021; Wałęga et al., 2025).

This chapter aims to explore the concepts of ecohydrology and hydroecology, define environmental flow and its various interpretations, present different classification systems for assessing environmental flows, introduce environmental flow assessment approaches, highlighting their advantages and disadvantages, and focus specifically on common hydrological methods used to determine the environmental flow of rivers.

2. The concept of ecohydrology and hydroecology

Since hydrological regimes are principally determined by runoff data (Ma et al., 2014; Dorafshan et al., 2025a,d), maintaining the ecological health and biodiversity of rivers requires a thorough understanding of the natural hydrological regime (Cui et al., 2020; Dorafshan et al., 2025c). Ecohydrology is a subdiscipline of hydrology that focuses on the ecological aspects of the hydrological cycle and the use of ecosystem properties as management tools through "Dual Regulation" (Zalewski, 2000, 2014). In contrast, hydroecology studies water in relation to organisms inhabiting wetlands, ponds, and similar ecosystems, examining how plant and animal populations interact with water–soil systems, water levels, and groundwater dynamics. Ecohydrology emphasises interactions between hydrology and ecosystem processes to develop innovative Advanced Nature-based Solutions (ANbS) that enhance the catchment sustainability potential of WBSR (Zalewski, 2014), whereas hydroecology focuses on the role of hydrology in shaping the ecological status and functioning of streams, ponds, wetlands, and estuarine ecosystems. In essence, ecohydrology integrates hydrology within the broader framework of catchment sustainability, while hydroecology emphasises ecological processes for ecosystem conservation (Rosner, 2003).

3. Definitions of environmental flow

There are various definitions of environmental flows, some of which include the following:

(1) Environmental flows designate the water regime allocated to rivers, wetlands, or coastal zones with the express purpose of safeguarding ecosystems and their associated benefits, particularly in scenarios characterized by competing water demands and regulated flow regimes (Hudson, 2002).
(2) Flows specifically allocated and determined to sustain habitats and ecosystem processes (Arthington et al., 2000).
(3) The portion of a river's original flow regime that must continue downstream and into floodplains to preserve certain valuable ecological characteristics (Tharme, 2003).
(4) A prescribed distribution of water, both in terms of time and location, that is primarily released into a river to manage its health and maintain the integrity of ecosystems dependent on its flow (Hirji and Panella, 2003).

The primary distinction between the terms "environmental water requirements (EWR)" and "environmental flows" lies in their focus. The former is more concerned with temporal variations in flow and is recognized as an ecologically desirable hydrological regime. The total annual volume of water designated as the environmental flow regime is referred to as the "environmental water requirement" and is expressed in volumetric units. In the case of wetlands, the term "environmental water requirement" is more commonly used (Gleick, 1993; Gillilan and Brown, 2013). Additionally, the process of determining environmental flows is known as "environmental flow assessment" (Gleick, 1993; Karakoyun et al., 2018).

Ecologists maintain that a fundamental criterion for environmental flow determination should encompass the preservation of both the spatial and temporal dynamics of river discharge. This emphasis stems from the recognition that flow variability exerts a significant influence on the structural and functional diversity of riverine systems and their associated floodplains, consequently impacting species richness within these ecosystems. Therefore, environmental flows should not merely specify the required water volume but should also address when and how this water should be released into the river (Smakhtin and Anputhas, 2006; Sanz et al., 2025).

Several methodologies exist for determining environmental flows. These methods vary significantly in their objectives, the type of input data required, and the accuracy of their output results (Shu-feng et al., 2024). Different approaches serve diverse purposes, ranging from comprehensive water resource planning to the detailed scheduling of reservoir release management (Wang et al., 2023). The choice of an appropriate method depends on several factors, including Wallingford (2003).

(1) The high costs and complexity of collecting extensive data.
(2) The river classification, encompassing perennial, seasonal, high base flow, and flood-prone types.

(3) The ecological significance of the river system.
(4) The consequential effects of water resource development projects on riverine ecosystems. Environmental flow assessment methods have predominantly been developed and applied in industrialized nations. A noticeable gap in knowledge and experience regarding environmental flows exists within current water resource management approaches in nearly all developing countries. National and international environmental regulations rarely incorporate explicit considerations for environmental flows (Li et al., 2024). Only a few countries—such as Australia, South Africa, and the United Kingdom—have fully integrated this concept into their water resource management frameworks. In many Latin American, African, and Asian countries, either no clear legislation exists on environmental flows, or there are no widely accepted methodologies for assessing them. Additionally, some assessment methods developed elsewhere remain either unknown or seldom implemented in these regions (Hughes and Smakhtin, 2009; Li et al., 2024).

4. Classifications for environmental flow

The study of environmental water requirements (EWR) was initially conducted in the United States by the US Fish and Wildlife Service between 1940 and 1970. This research eventually led to the official recognition of environmental flows in 1970 as a result of the National Environmental Policy Act (1969) and the Water Resources Planning Act (1956) (Tennant, 1976; Spiegel, 1988). The majority of early efforts in this field primarily focused on rivers. Initially, the concept was centered around improving fish habitats and ensuring their migration routes. However, over time, the focus expanded to encompass broader ecological concerns, such as the preservation of ecosystem cycles. In cases where wetlands were directly linked to river ecosystems, efforts were made to integrate them into environmental water allocations for rivers (Bounds and Lyons, 1979; Belmar et al., 2013; Wang et al., 2024a,b).

Determining environmental water requirements involves various methodologies, depending on the spatial scale of the study, available data, assessment timeframe, and technical and financial capacities (Yu et al., 2022). These methodologies range from rapid assessments designed for large-scale planning to more detailed approaches for individual rivers with protected species. Globally, approximately 207 methodologies for assessing environmental water requirements have been identified across 44 countries (Linnansaari et al., 2012; Wałęga et al., 2025).

The application of these methods generally falls into two main levels (Waters, 1976; Tharme, 2003):

(1) Comprehensive level: This level primarily employs habitat simulation methods and holistic approaches to determine environmental water requirements.
(2) Preliminary identification level: At this stage, hydrological methods are predominantly used. Approximately 30% of global environmental flow assessment methodologies are classified within this category, establishing it as the most frequently implemented approach worldwide.

In general, approaches to determining environmental flows have been classified by various organizations:

(1) International Union for Conservation of Nature and Natural Resources (IUCN) (Dyson et al., 2003).
(2) Word Bank (Brown and King, 2003).
(3) International Water Management Institute (IWMI) (Tharme, 2003).

4.1 International Union for Conservation of Nature and Natural Resources

The International Union for Conservation of Nature and Natural Resources (IUCN) classifies all environmental flow assessment methods into three broad categories: methods, approaches, and frameworks (Dyson et al., 2003; Karakoyun et al., 2018). For instance, the Expert Panel approach is considered a methodology under which multiple specific methods can be applied. Concurrently, flow management frameworks offer a more comprehensive strategic approach to environmental flow assessment. These frameworks typically integrate a specific approach along with a set of methods for determining environmental flows (Gillilan and Brown, 2013).

Methods developed in different countries around the world to define environmental flow allocations are divided into four categories (Dyson et al., 2003): (1) lookup tables, (2) desktop analysis, (3) functional analysis, and (4) hydraulic habitat modeling. Each of these methods may vary in their reliance on expert input and in the scope of river system components they assess. Some methods require extensive expert involvement, while others rely on minimal expertise. Similarly, some methods assess the entire river system, while others focus on specific segments. Consequently, expert involvement and the comprehensiveness of each method (i.e., whether it considers all parts of the river system) are key distinguishing features of different environmental flow assessment approaches (Acreman and Dunbar, 2004).

4.2 Word Bank

The World Bank categorizes environmental flow assessment methods into two main perspectives: the prescriptive approach and the interactive approach (Brown and King, 2003).

4.2.1 Prescriptive approach

Prescriptive approaches typically focus on a specific environmental target and, in the end, recommend a fixed environmental flow value. If the prescribed flow cannot be provided, alternative solutions are generally not considered. As a result, these approaches lack flexibility in water allocation negotiations. They are best suited for situations where objectives are clear-cut and relatively simple. Prescriptive approaches are divided into four main categories (Brown and King, 2003): hydrological index methods, hydraulic rating methods, expert panels, and holistic approaches.

Among these, hydrological index methods are among the oldest, most well-established, rapid, and easy-to-apply approaches. These methods rely on historical flow records, but they give minimal consideration to the natural conditions and aquatic biodiversity of the river. Some of the most common prescriptive methods include the Tennant (Montana) method, flow duration curve (FDC) analysis, and the wetted perimeter method (Cavendish and Duncan, 1986; Brown, 2003; Zhao et al., 2021).

4.2.2 Interactive approach

In contrast, interactive approaches focus on the relationship between river flow alterations and multiple ecological characteristics. The outcomes of these methods are usually presented as different scenarios, allowing for a more flexible decision-making process (Brown and King, 2003). However, they require more time and data compared to prescriptive approaches.

Instead of prescribing a single environmental flow value, interactive approaches provide a range of flow regimes, each with specific ecological implications. Notable methods in this category include the range of variability approach (RVA), which identifies an optimal range for environmental flows and selects key parameters based on conservation goals. Other important methods include the in-stream flow incremental methodology (IFIM) and downstream response to imposed flow transformation (DRIFT), both of which propose multiple scenarios for environmental flow management (Cavendish and Duncan, 1986; Brown, 2003).

4.3 International Water Management Institute

The International Water Management Institute (IWMI) classifies these methods into four distinct categories: hydrological methods, hydraulic methods, habitat simulation methods, and holistic methods (Stalnaker and Arnette, 1976; IWMI, 2004). This classification is considered the most logical because it is based on the necessary biophysical data rather than methodological characteristics, which may frequently change or overlap (Korsgaard, 2007).

4.3.1 Hydrological approach

The simplest approach of environmental flow utilizes hydrological data, commonly long-term monthly or daily historical statistical records, to generate environmental flow recommendations (Wałęga et al., 2025). A fixed percentage of flow or other historical hydrological indices—often referred to as low flow—is selected to represent environmental flow requirements necessary to maintain a key ecosystem characteristic at a predetermined, acceptable level. These recommended flows are typically prescribed over an annual, seasonal, or often shorter (monthly) timeframe (Karakoyun et al., 2018). Most of these methods are entirely dependent on flow data to determine environmental water requirements. Moreover, even more complex and intricate methods, particularly holistic approaches, extensively incorporate such hydrological indices as part of a broader suite of tools (King et al., 1999).

Some hydrology-based methods are region-specific or designed for particular geographic areas. For instance, the basic flow method was originally developed for use in New England, the United States, although it has since been applied in other parts of North America (Reiser et al., 1989; Ringler and Nguyen, 2004; Zhao et al., 2021). Generally, the most commonly used hydrological method is the Tennant (Montana) method (Tennant, 1976), which currently ranks as the second most widely applied environmental flow assessment method in North America (Reiser et al., 1989). The hydrological approach includes methods such as 7Q10, 7Q2, sustainability boundary approach, RVA, presumptive standard, and median monthly flow (Q50) (Caissie and El-Jabi, 1995; Belzile et al., 1997; Richter et al., 1997, 2012; Richter, 2010).

4.3.2 Hydraulic approach

Initially, these methods were developed in the United States to determine the in-stream flow requirements for fish populations. Over time, they have been enhanced alongside habitat simulation methods or incorporated into holistic approaches. These methods rely on time-series data and cross-sectional profiles of critical and significant river channels (Stucchi and Bocchiola, 2024). By modeling river hydraulics as a function of flow, these approaches generate a correlation between hydraulic variables—including wetted perimeter, depth, and velocity—and the resulting habitat suitability for the species of interest. The environmental flow is then defined as the flow that ensures optimal habitat conditions (Li and Zhang, 2023). If this flow is not maintained, a significant decline in the target species' population is expected (Korsgaard, 2007). One of the most widely used hydraulic methods globally—and the third most common method in North America—is the wetted perimeter method (Gippel and Stewardson, 1998). This method examines the general relationship between river discharge and wetted perimeter (Kashaigili et al., 2005; Gillilan and Brown, 2013). The hydraulic approach includes methods such as flowing perimeter method and R-2 cross (Espegren, 1996; Gippel and Stewardson, 1998).

4.3.3 Habitat simulation approach

These methods have been extensively applied and utilized a combination of hydrological, hydraulic, and ecological data. They establish a direct link between river flow, habitat availability, and habitat suitability for target species within a simulation model (Chen et al., 2023). Consequently, habitat conditions are explicitly connected to the specific requirements of the target species (Korsgaard, 2007). These methods estimate environmental water requirements based on various relationships between flow and physical habitats. To quantify the relationship between flow and selected variables (such as depth, wetted perimeter, etc.), which serve as indicators of habitat conditions, field studies are required. These approaches rely on defining habitat suitability curves for species, modeling the quantity and quality of habitats, and implementing necessary flow modifications based on the desired water regime (Swales and Harris, 1995; Tharme and Smakhtin, 2003).

These methods are particularly flexible in accommodating different flow scenarios and enjoy a high degree of scientific acceptance. However, they require extensive datasets and are primarily applied to protect target species within ecosystems (Tharme, 2003).

The IFIM falls within this category. This method was introduced in the late 1970s by the Environmental Flow Working Group of the US Fish and Wildlife Service (USFWS) in Colorado and has been recognized by certain environmental lawyers as the most scientifically and legally robust approach for assessing environmental flow percentages (Tharme, 2003; Karakoyun et al., 2016). Another widely used and significant habitat simulation method is the Physical Habitat Simulation System (PHABSIM) (Bovee, 1982).

A notable limitation of PHABSIM is that it was specifically developed for fish habitat conditions and does not account for other animal and plant species (Bovee, 1982; Bovee et al., 1998; Karakoyun et al., 2018). The habitat simulation approach includes methods such as RYHABSIM (River hYdraulic and HABitat SIMulation), River2D, MesoHABSIM, EVHA (EValuation de HAbitat), MesoCASIMIR, CASIMIR (Computer-Aided SIMulation of habitat In Regulated streams), and generalized habitat models (e.g., STATHAB) (Jowett, 1989; Ginot, 1995; Jorde and Bratrich, 2000; Steffler and Blackburn, 2002; Parasiewicz, 2001; Eisner et al., 2005; Lamouroux and Jowett, 2005).

4.3.4 Holistic approach

These methods provide frameworks that incorporate hydrological, hydraulic, and habitat simulation models. They are the only methods that explicitly adopt a holistic, ecosystem-based approach to determining environmental flows in rivers (St-Hilaire et al., 2021). The holistic approach assumes that if the key components of the natural flow regime are maintained in the altered flow regime, plant and animal species will persist, and the ecological diversity of the system will be preserved (Arthington et al., 2007).

All holistic methods must address two fundamental questions. First, which flows are beneficial? In relation to this, what temporal and quantitative flow patterns are required to sustain aquatic life, plant species, and river ecosystems? Second, what are the potential ecological consequences of flow alterations? Addressing both questions necessitates an understanding—and preferably a clear definition—of the interrelationships among hydrology, geomorphology, water quality, and ecological systems (Armitage and Petts, 1992; Knights, 2002; Williams et al., 2019).

Six key characteristics of flow have been identified as being geomorphologically and ecologically relevant, as well as highly sensitive to modifications in the flow regime caused by water abstraction, diversion, groundwater extraction, hydropower production, and land-use changes within the basin (Richter et al., 1997; Arthington et al., 2007). These key flow characteristics include the following (Tessman, 1980; Matthews et al., 2000; Ren et al., 2021):

(1) The duration for which specific flows persist
(2) The magnitude (volume) of river flows at any given time
(3) The frequency of specific flow events, such as floods
(4) The timing of particular flow events
(5) The rate of change in flow conditions, such as rising and falling floodwaters
(6) The seasonality, variability, and predictability of the overall flow regime

These characteristics of river flow regimes must be maintained or restored to ensure the preservation of aquatic biodiversity, ecosystem processes, and the long-term evolutionary potential of freshwater ecosystems (Olden and Poff, 2003). It is also well recognized that these flows support the ecological goods and services that rivers provide to society (Naiman et al., 2002; King et al., 2003). Transdisciplinary ecohydrology goes a step further by defining an ambitious goal—the parallel enhancement of catchment WBSR (Water, Bioproductivity/Biodiversity, Services for society, and Resilience)—to be achieved through the "Dual Regulation" of hydrology–biota interactions, from molecular biology to catchment and biosphere-scale processes (Zalewski, 2021).

Holistic environmental flow assessment methods are widely used in evaluating environmental flows. The building block methodology (BBM) falls within this category. This method follows a "bottom-up" approach. A bottom-up approach does not start with a predetermined flow but instead builds a modified flow regime to meet identified environmental needs (Bovee, 1995). This approach adds flow components to a baseline of zero flow to construct an adjusted flow regime (Arthington et al., 2004).

The developed flow stress or response (FSR) method utilizes relationships between low-flow (and high-flow) conditions and their corresponding ecological stress factors to generate a stress index time series linked to the flow regime. This method also follows a bottom-up approach (O'Keeffe et al., 2017). Derived from BBM and similar environmental flow assessment methods, the downstream response to imposed flow transformation (DRIFT) method was introduced. DRIFT comprises four modules—biophysical, social, economic, and scenario development. This method focuses on identifying the consequences of river flow reductions from natural conditions through a series of flows linked to a specific set of biophysical functions, hydrological characteristics, and hydraulic properties, based on the degree of system degradation. Although scenario-driven, DRIFT provides significant flexibility for comparative evaluation of multiple proposed flow regimes (King et al., 2003). Furthermore, it establishes links between social outcomes for livelihood-dependent users, ecological and geomorphological conditions, and economic applications, considering the costs of damage reduction or compensation. This methodology has been implemented in projects across southern Africa (Tharme, 2003).

DRIFT follows a "top-down" approach. A top-down approach in environmental flow assessment begins with the natural or existing flow regime as a reference point and examines the effects of modifying or removing different components of that regime (Belmar et al., 2013; Pauls et al., 2016). The core principle of this approach revolves around quantifying the permissible extent of river flow regime modification that can be tolerated before substantial or irreversible ecological damage manifests in the aquatic ecosystem (Arthington et al., 2004).

Among environmental flow assessment methods, BBM and DRIFT are the only ones that explicitly consider the socioeconomic dimensions of environmental flows (Knights, 2002). The holistic approach includes methods such as CASIMIR (Computer Aided SIMulation of habitat In Regulated streams), MesoHABSIM, and MesoCASIMIR, River2D (Jorde and Bratrich, 2000; Steffler and Blackburn, 2002; Parasiewicz, 2001; Eisner et al., 2005).

5. Advantages and disadvantages of flow assessment approaches

Table 34.1 provides a comprehensive evaluation of the strengths and weaknesses associated with different approaches to environmental flow determination.

6. Hydrological approach in determining the environmental flow of rivers

To determine EWR, various methods are employed depending on the spatial scale of the study, the availability of data, the temporal resolution of the assessment, and technical and financial capacities. The most important hydrological methods include Tennant, Tessman, FDC, low flow indices, Smakhtin, FDC Shifting, desktop reserve model (DRM), and RVA.

TABLE 34.1 Advantages and disadvantages of flow assessment approaches.

Environmental flow assessment approaches	Advantages	Disadvantages	Considerations	References
Hydrological approach	- Flow-based statistical methods are cost-effective and quick, requiring only historical flow data. - They are useful for preliminary identification and assessment phases and can be improved with expert judgment.	- The inherent variability and dynamic nature of the hydrological regime are not sufficiently captured by these approaches. - The long-term effects of maintaining low flows rarely mirror the short-term impacts reflected in historical flow statistics. - There is no ecological data to validate these methods, making their reliability weak and difficult to justify.	- Such methods are only suitable for an initial level of assessment, particularly in cases where no negotiation is involved in the decision-making process. - They should be applied cautiously in regions with hydrological regimes significantly different from the original contexts in which these methods were developed.	Poff et al. (1997), Bradford (2008), Alfredsen et al. (2012), Ge et al. (2018), Zhou et al. (2023), Wałęga et al. (2025)
Hydraulic approach	- These methods incorporate ecological data from habitats while remaining relatively simple and cost-effective. - They offer considerable flexibility, as they can utilize existing data and are suitable for preliminary basin assessments.	- These methods are predicated on the overly simplistic assumption that a singular hydraulic variable or a limited set of variables can sufficiently characterize the environmental flow needs of a particular species and its associated activities. - The reliance on simplified cross-sectional models, coupled with the inherent characteristics of the flow-hydraulic parameter relationship, introduces uncertainties regarding the precision of the resulting data. - Key relationships with the hydrological regime are often overlooked in environmental flow assessments, leading to results that are rarely spatially or temporally precise.	- Since these methods focus on in-channel habitats, they cannot easily be applied to broader ecosystem components such as riparian vegetation or other biotic elements beyond the watercourse. .	Jowett (1997), Gippel and Stewardson (1998), Shang (2008), Fonstad and Marcus (2010), Shu-Feng et al. (2024), Stucchi and Bocchiola (2024)
Habitat simulation approach	- These methods can assess the impacts of increasing flow variations on physical habitats. - They are effective for evaluating different	- Their primary focus is on key indicator species or target species, which can be a limitation when the goal is to protect the	- These methods are not easily adaptable for evaluating certain river ecosystem components, such as riparian vegetation.	Hatfield et al. (2003), Armstrong (2010), Lamouroux et al. (2010), Li et al. (2024)

TABLE 34.1 Advantages and disadvantages of flow assessment approaches.—cont'd

Environmental flow assessment approaches	Advantages	Disadvantages	Considerations	References
	environmental flow scenarios concerning various species, life stages, and ecological communities. - Widely accepted within the scientific community, they are well-supported and defensible.	overall river ecosystem. - Selecting the appropriate target species can be challenging. - In many countries, there may be insufficient knowledge of riverine biota, limiting the method's applicability.		
Holistic approach	- They provide a comprehensive assessment of all river ecosystem components and maintain strong links to the natural hydrological regime. - Riverine processes are considered within multiple scientific disciplines. - They offer significant flexibility in incorporating available data and provide high reliability in generating responsive environmental flow recommendations.	- These methods are costly, time-intensive, and labor-intensive.	- There is a potential for inconsistencies in findings when different scientists conduct the assessments.	Arthington (1998), Tharme (2003), Poff and Zimmerman (2010), Lumbroso et al. (2011), Williams et al. (2019)

The aforementioned methods are widely used worldwide because (1) they rely on historical streamflow data (monthly and daily), requiring fewer data inputs; (2) they are effective for initial identification and assessment phases; and (3) they demand less time and lower costs. Other approaches for estimating environmental flows require precise ecological data, making them significantly more expensive and time-consuming. Moreover, inconsistencies in the obtained information are more likely in these methods (Wałęga et al., 2025). For instance, hydraulic methods consider only a single river cross section, and incorporating multiple sections increases both costs and time (Karakoyun et al., 2018). Similarly, habitat simulation methods focus exclusively on key species. The following is a detailed explanation of each hydrological approach.

6.1 Tennant method

In 1976, Donald Tennant proposed a methodology for estimating environmental flow requirements for fish, widely recognized as the "Montana method" or, more frequently, the "Tennant method" (Tennant, 1976). Employing a percentage-based approach relative to the average annual flow (AAF), this method assesses fish habitat quality. Tennant's analysis of 58 cross sections across 11 rivers in Montana, Nebraska, and Wyoming yielded the conclusion that 10% of the AAF constitutes the minimum flow necessary for short-term fish survival. Furthermore, 30% of the AAF is deemed

TABLE 34.2 Tennant's proposed method for environmental flow for fish, wildlife, and recreation (Tennant, 1976).

Description of flows	Recommended base flow regimes	
	October–March	April–September
Flushing or maximum	200%*AFF*	200%*AFF*
Optimum range	60 – 100%*AFF*	60 – 100%*AFF*
Outstanding	40%*AFF*	60%*AFF*
Excellent	30%*AFF*	50%*AFF*
Good	20%*AFF*	40%*AFF*
Fair or degrading	10%*AFF*	30%*AFF*
Poor or minimum	10%*AFF*	10%*AFF*
Severe degradation	< 10%*AFF*	< 10%*AFF*

adequate for maintaining reasonably favorable survival conditions, while 60% of the AAF is considered optimal for habitat provision (Tennant, 1976; Khan et al., 2024).

Tennant presented a simple, standardized approach that could be applied even with minimal data (Uday Kumar and Jayakumar, 2021). This technique relies solely on the AAF of the river, allowing specialists to estimate the required flow using a percentage of the AAF without needing additional on-site data collection, as shown in Table 34.2.

Since 1976, several modifications have been made to the Tennant method, including adjustments for spring runoff, equations to account for existing flow modifications, and revisions to incorporate monthly low-flow levels.

When applying the Tennant method, the following points should be considered (Yu et al., 2022; Khan et al., 2024; Zhang et al., 2024).

(1) This method does not account for flow fluctuations and seasonal effects.
(2) It is more suitable for large, permanent streams, where flow variability is less compared to seasonal streams.
(3) The river channel geometry is not considered in this method.

The comparable studies have been carried out in the Tennant method (Mann et al., 2006; Wałęga et al., 2015; Mostafavi and Yasi, 2015; Ahmadpari et al., 2019; Karimi et al., 2021; Zargari et al., 2023).

6.2 Tessman method

In 1980, Tessman adapted the Tennant method by incorporating seasonal flow considerations. This methodology integrates mean monthly flow (MMF) and mean annual flow (MAF) to establish the minimum monthly flow requirement (Tessman, 1980; Wałęga et al., 2015). The method follows these steps, which are illustrated in a flowchart in Fig. 34.1.

- If MMF > MAF, then (0.4 × MMF) is taken as the minimum monthly flow.
- If 0.4 × MAF > MMF, then (0.4 × MAF) is considered the minimum monthly flow.
- If MMF is less than 0.4 × MAF, MMF itself is used as the minimum monthly flow.

The comparable studies have been carried out in the Tessman method (Mostafavi and Yasi, 2015; Wałęga et al., 2015; Ahmadpari et al., 2019; Zargari et al., 2023).

6.3 Flow duration curve method

One of the key characteristics of water flow used in assessing river flow variability and fluctuations from an environmental perspective is the FDC (Chouaib et al., 2019). This curve serves as one of the most effective tools for visualizing the full range of river flows, spanning from drought conditions to flood events (Fouad and Loáiciga, 2020). The FDC effectively depicts the correlation between flow magnitude and its corresponding frequency of occurrence (Olden and Poff, 2003; Ni et al., 2022). Various hydrological low-flow indices are derived from these curves. Typically, flows falling within the

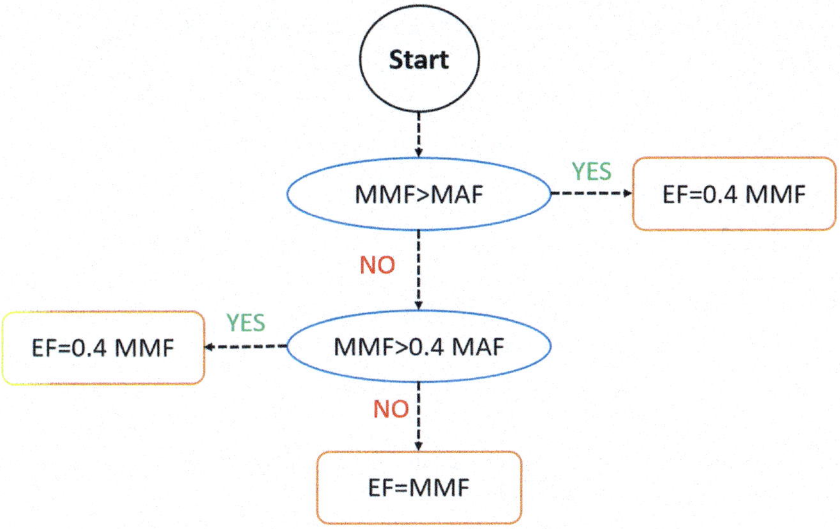

FIGURE 34.1 Flowchart for determining environmental flow using the Tessman method (EF, environmental flow).

70%–99% exceedance range (Q70 to Q99) are widely regarded as low flows. Among them, Q90 and Q95 are frequently used as standard hydrological low-flow indices, while Q50, the median monthly flow, is another commonly used flow duration index, particularly in summer months (Pyrce, 2004; Młyński et al., 2020).

6.3.1 Capabilities, applications, and limitations of FDCs

Unlike flow frequency analysis, FDCs provide limited information on extreme flow variability—whether in the form of exceptionally low flows (severe droughts) or extreme floods. However, their advantage over other variability measures lies in their ability to capture a relatively broad range of flow variability over a given timeframe (e.g., daily). This makes them highly valuable for ecological considerations in flood and flow management, particularly when assessing the impact of large reservoir structures or implementing flood control strategies for downstream river regimes. Their main limitation, however, is that while they provide insight into the relative frequency of different flow magnitudes over a period, they do not reveal how these flows are sequentially distributed within the time series (Pyrce, 2004).

6.3.2 Drawing the FDC

To construct an FDC, streamflow data is arranged in descending order, meaning the highest flow value is assigned the first rank. These flows are then plotted against their corresponding probability (P). The probability of occurrence for each flow magnitude is calculated using Eq. (34.1):

$$P = \frac{m}{n+1} \tag{34.1}$$

where m represents the rank of a given flow in the ordered time series, and n denotes the total number of observations.

The comparable studies have been carried out in the FDC method (Gao et al., 2009; Linnansaari et al., 2012; Wałęga et al., 2015; Wang et al., 2017; Verma et al., 2017; Liu et al., 2021; Karimi et al., 2021; Zhu et al., 2023; Zargari et al., 2023).

6.4 Hydrological low flow indices (7Q10 and 7Q2)

Pyrce (2004) identified two of the most widely used individual low-flow indices: the 7-day minimum flow with a 10-year return period (7Q10) and the 7-day minimum flow with a 2-year return period (7Q2). These indices are commonly employed in hydrological studies. To calculate EWR using this approach, specialized software such as HydroOffice and SMADA (ver 2) can be utilized. Estimating the 7-day minimum flow with return periods of 10 and 2 years requires daily streamflow records. The 7-day minimum flows for different years are computed using the software, and then statistical distribution functions within SMADA are applied to analyze these flow statistics.

TABLE 34.3 Estimation of maximum environmental flow requirement in the Smakhtin method (Smakhtin, 2004).

Environmental low-flow requirement (Q90)	Environmental high-flow requirement (HFR)	Description
$Q90 < 10\%MAR$	$HFR = 20\%MAR$	In basins with flood-based variable regimes occurred during wet seasons
$10\%MAR < Q90 < 20\%MAR$	$HFR = 15\%MAR$	–
$20\%MAR < Q90 < 30\%MAR$	$HFR = 7\%MAR$	–
$Q90 > 30\%MAR$	$HFR = 0$	In the constant regime-based basins with constant annual flows wherein the minimal flow is required as the principal element

The 7Q10 flow is a widely used single-index flow metric. Several reasons justify its application in regulatory frameworks, including the following (Caissie et al., 2007; Zhu et al., 2023).

(1) Protecting or regulating water quality in response to wastewater discharges.
(2) Preserving aquatic habitats during drought conditions.
(3) Serving as a long-established criterion for aquatic life protection.

The primary use of 7Q10 is in river water quality standards, particularly for pollution regulation. However, its applications have extended beyond this scope. Concerns have been raised regarding the suitability of 7Q10 as an environmental flow indicator (Boner and Furland, 1982). Caissie and El-Jabi (1995) cautioned that relying on 7Q10 significantly underestimates environmental flows and could lead to harmful biological consequences. Consequently, its use for environmental purposes is not recommended.

The comparable studies have been carried out in the low flow indices method (Annear et al., 2002; Watt et al., 2007; Caissie et al., 2015; Karimi et al., 2021).

6.5 Smakhtin method

Smakhtin (2004) employed this method to examine 128 river basins across different parts of the world, aiming to assess global river utilization while considering EWR. This approach is a hydrological method primarily used for preliminary assessments on a large spatial scale, such as national and international levels. Within this framework, EWR is defined as a combination of the environmental low-flow requirement (LFR) and the environmental high-flow requirement (HFR). LFR represents the minimum water necessary for fish and other aquatic organisms throughout the year, whereas HFR accounts for flood events and their impact on river morphology and riparian vegetation.

To maintain a "fairly good" river condition under this method, LFR should be set equal to Q90, which denotes the flow level exceeded 90% of the time within a year. Meanwhile, the HFR value depends on environmental management objectives and the river's flow regime.

Smakhtin (2004) categorized river basins into four classes and, based on this classification, expressed HFR as a percentage of the mean annual runoff (MAR), as shown in Table 34.3. If a river has a highly variable flow where Q90 is less than 10% of MAR, HFR is set at 20% of MAR. Conversely, in rivers with stable flows, where Q90 exceeds 30% of MAR, HFR is considered zero. For rivers where Q90 falls between 10%–20% and 20%–30% of MAR, HFR is assigned as 15% and 7% of MAR, respectively. The total annual EWR is obtained by summing the LFR and HFR components (Smakhtin, 2004).

The comparable studies have been carried out in the Smakhtin method (Mahmood et al., 2020; Karimi et al., 2021; Zargari et al., 2023).

6.6 Flow duration curve shifting

Smakhtin and Anputhas (2006) introduced this method to evaluate environmental flows in river systems. This approach, which offers a hydrological regime to sustain rivers in an ecologically favorable state, is known as "FDC shifting". It consists of four main steps:

(1) Modeling existing hydrological regimes.
(2) Establishing environmental management categories.
(3) Developing environmental flow duration curves.
(4) Producing monthly environmental flow time series.

6.6.1 Modeling existing hydrological regimes

The first step involves constructing the natural flow duration curve for the targeted river reach using monthly flow data. In this method, the probability axis of the flow duration curve is created by displaying the 17% exceedance probability (0.01, 0.1, 1, 5, 10, 20, 30, 40, 50, 60, 70, 80, 90, 95, 99, 99.9, and 99.99). These points ensure sufficient representation of the entire range of flows, facilitating subsequent steps (Smakhtin and Anputhas, 2006).

6.6.2 Establishing environmental management categories

Smakhtin and Anputhas (2006) proposed an environmental flow assessment methodology that employs monthly river discharge data to generate environmental flow duration curves for each defined environmental management class. This process is based on a comparative analysis of both natural and existing flow duration curves. The fundamental objective of implementing environmental flows within this framework is to sustain ecosystems in their designated states, variously termed "preferred future condition," "environmental management class," "ecological management category," or "level of environmental protection." A higher environmental management class necessitates a greater volume of water to sustain the ecosystem. Ideally, these classes should be grounded in empirical relationships between flow levels and ecological conditions, incorporating identifiable thresholds. The classification of a river within a particular environmental management class is often determined by expert judgment. This method employs six environmental management classes, detailed in Table 34.4.

TABLE 34.4 Environmental management classes (EMC) (Smakhtin and Anputhas, 2006).

Environmental management classes	Ecology definition	Management perspective
A (natural)	Natural rivers with slight changes in the instream and riparian habitat	The protection of rivers and basins; reserves and national parks; prohibition of new water projects, such as dams' construction, or diversions
B (slightly modified)	Rivers with minor modifications or ecological importance with considerable intact biodiversity and habitats regardless of the development of water resources and basins	Permitted projects to develop the water supply and irrigation
C (moderately modified)	The habitats and dynamics of the biota destructed despite the intactness of basic ecosystem functions; loss or decrease of a number of sensitive species; presence of alien species	Implementation of multiple projects, such as dams' construction,dDiversions, as well as habitat and water quality development with the purpose of the socioeconomic development
D (largely modified)	Modification of natural habitat, biota and basic ecosystem functions; inadequate species richness and intolerant species presence; prevalent alien species	Significant evident interventions, such as dams, diversions, transfers, habitat modification, and water quality degradation with the purpose of developing the basin and water resources
E (seriously modified)	Decreased diversity and accessibility of habitats; inadequate species richness; the tolerant species-limited survival rate; destruction of indigenous species; the alien species-invaded ecosystem	The dense human population and highly exploited water resources; this is not required to be accepted as a management perspective due to the fact that managerial strategies are with the purpose of restoring the flow patterns and moving a river into a higher management category
F (critically modified)	Critical modifications; the ecosystem modification based on the total loss of natural habitat and biota; complete destruction of basic ecosystem functions with irreversible changes as the worst outcomes	Not accepted in the management perspective; the managerial strategies are with the purpose of restoring the flow patterns or river habitats and, if possible, moving a river into a higher management category

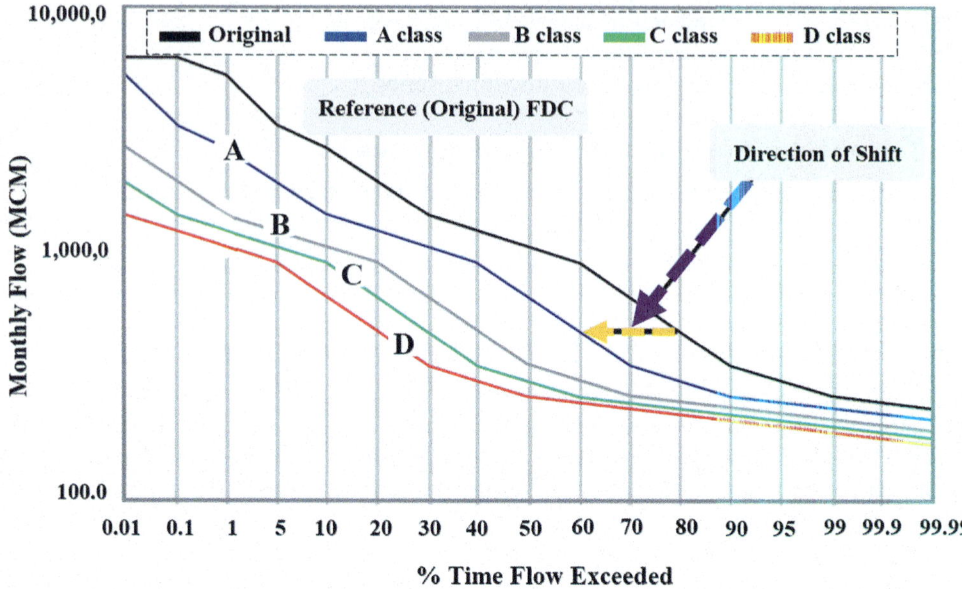

FIGURE 34.2 Transverse shifting is utilized to estimate environmental flow continuity curves for specified environmental management classifications. *Drawn by Smakhtin, V.U., Anputhas, M., 2006. An Assessment of Environmental Flow Requirements of Indian River Basins, vol 107. IWMI Research Report. Spiegel, M.R., 1988. Theory and Problems of Statistics. McGraw-Hill Publishing Company, McGraw-Hill Publishing Company.*

6.6.3 Generation of environmental FDC

After plotting the natural FDC, the next step involves using changes (shifts) along the probability axis to calculate the EFDC for each management class. A shift in the natural FDC means that a flow, which previously occurred 99.99% of the time, now occurs 99.9% of the time. The key issue in this method is the appropriate use of cross-sectional shifts in each environmental management class. Determining the number of cross-sectional shifts in the FDC for each class is challenging without an understanding of the relationship between ecological characteristics and flow changes in rivers with different hydrological regimes. In Fig. 34.2, a single cross-sectional shift is used to extract the EFDC for rivers in Class A, while two leftward shifts are applied for Class B rivers, and similarly, three and four shifts are applied for Class C and D rivers, respectively. Each shift to the left means that, although some variability is lost with each shift, the overall flow variability pattern is maintained. This loss is due to reduced certainty about monthly flows, meaning uniform flows will occur less frequently. The total environmental flow is expressed as the mean annual environmental runoff (MAER), which is reduced.

6.6.4 Generation of environmental flow time series

The EFDC for each class briefly outlines the acceptable environmental flow regime for that class. By using spatial interpolation, the EFDC can be converted into time-series monthly environmental flow data. For this purpose, the method proposed by Hughes and Smakhtin (1996) is used (Hughes and Smakhtin, 2009). This method allows the generation of time series for sites lacking data, using sites with available data. For each month, a percentage is identified on the natural FDC, and then the monthly flow value is read off the EFDC at the same percentage. The steps of this process are shown in Fig. 34.3.

The MAR is computed from the natural monthly flow time series of the river, while the MAER is analogously determined using the generated environmental monthly flow time series. Then, by dividing the MAER by the MAR (MAER/MAR), the percentage of MAR that should be considered as environmental flow for each management class can be determined.

The comparable studies have been carried out in the FDC shifting method (Mostafavi and Yasi, 2015; Ahmadpari et al., 2019; Daechini et al., 2021; Zargari et al., 2023).

6.7 Desktop reserve model

Recognizing the constraints of limited ecological data, various environmental flow assessment techniques have been developed, predicated solely on hydrological indices derived from historical streamflow datasets. Although it has been established that many environmental characteristics, including water temperature, water quality, and turbidity, affect the

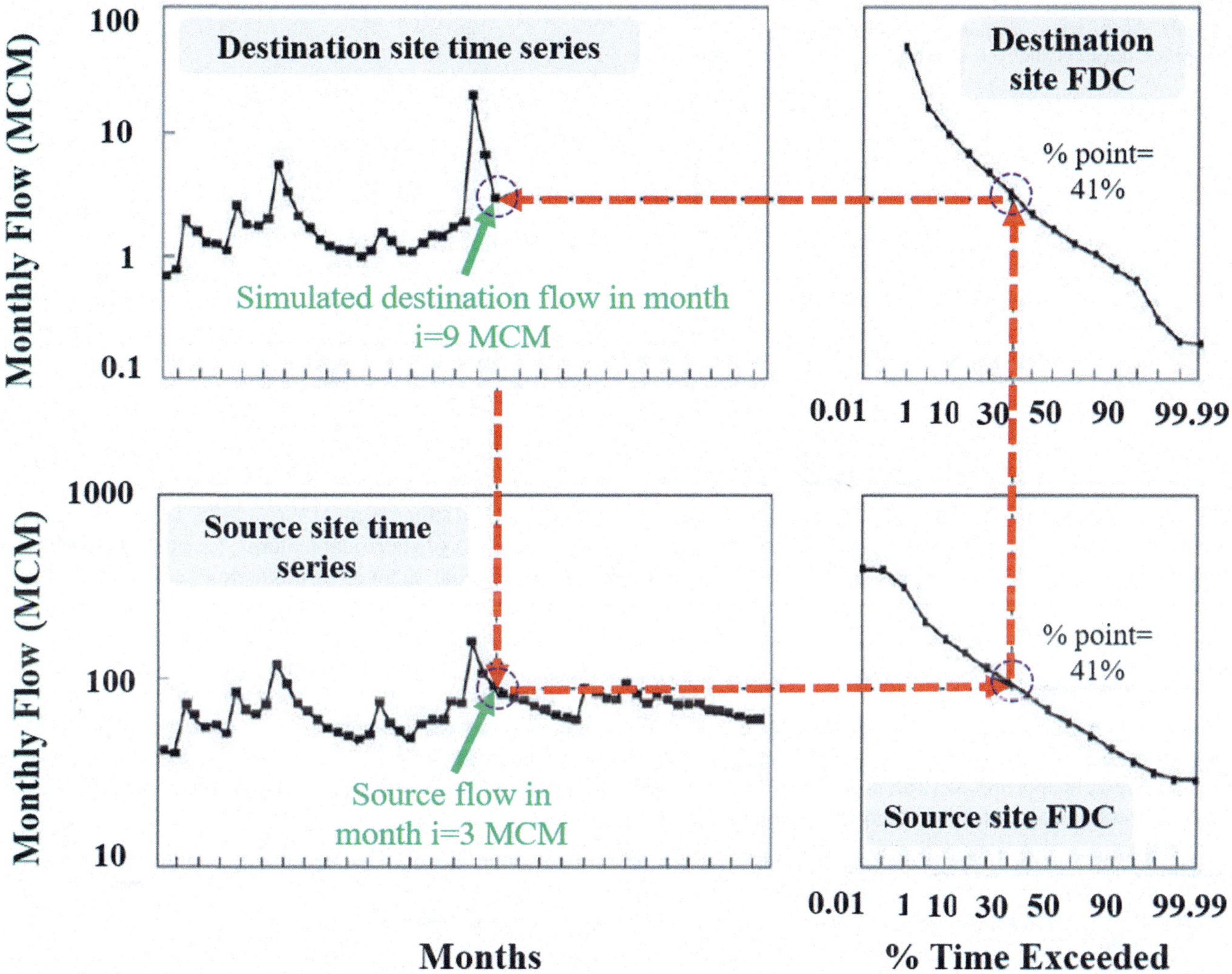

FIGURE 34.3 An illustration of how to use spatial interpolation to generate a comprehensive monthly EF time series from the environmental FDC that has been established (Smakhtin and Anputhas, 2006).

aquatic ecosystem (Abedi-Koupai et al., 2020, 2022; Dorafshan et al., 2023), the common assumption in these methods is that the flow regime is the main driving force (Richter et al., 1996). The DRM is one of these methods, which can calculate the ecological flow requirements in situations where a rapid assessment is needed and the available data is limited. This model was developed for observed conditions in South Africa (Hughes and Münster, 2000). In South Africa, rivers are classified according to their ecological status, and subsequently, the flow requirements are also classified. This classification system shows that, while some rivers are ecologically important, due to socioeconomic development needs, not all rivers can remain in conditions close to their natural state. Therefore, four possible "Environmental Management Classes" (A–D) are defined. Class A includes natural and undisturbed rivers, Class B includes modified but largely natural rivers, Class C includes moderately modified rivers, and Class D includes highly modified rivers with significant damage to habitat, biota, and ecosystem function (DWAF, 1997). In these classifications, transitional categories (e.g., A/B and B/C) are also used to expand the range of possible environmental flows, and this classification system is utilized in the DRM model to calculate flow requirements based on it. More water is allocated to the higher class (i.e., Class D) to maintain the ecosystem, and greater flow variability is preserved.

The comparable studies have been carried out in the DRM method (Palau and Alcazar, 1996; Mostafavi and Yasi, 2015; Verma et al., 2023).

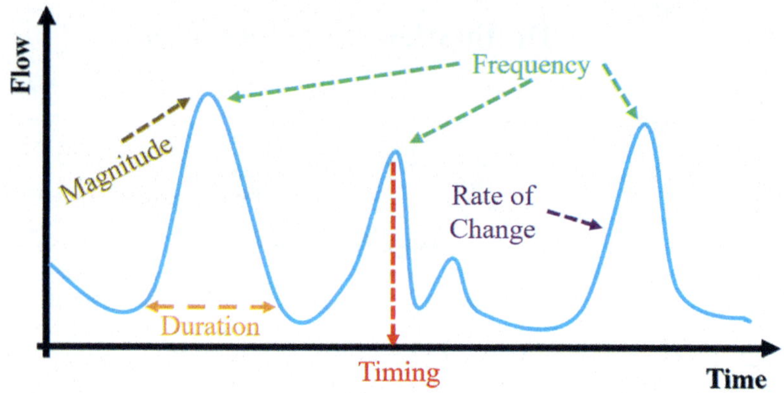

FIGURE 34.4 RVA method: Magnitude, timing, frequency, duration, and rate of change.

6.8 Range of variability approach

A complex hydrological technique, known as the RVA, was developed by Richter et al. (1996, 1997). This method is the most preferred among the hydrological indices. Its goal is to generate a series of statistical features, focusing on ecological aspects of the flow regime by highlighting the important role of hydrological changes in protecting ecosystems (Huang et al., 2017). This method is designed for use in regulated rivers where the conservation of natural and indigenous ecosystem functions and biodiversity is a primary management goal (Wang et al., 2023). To facilitate a connection between river management goals and prevailing ecological paradigms, the RVA was developed. This method uses 33 hydrological parameters in terms of magnitude, time, frequency, duration, and rate of change (Fig. 34.4). Of these parameters, 17 are related to the magnitude, time, duration, frequency, and morphology of the river, while the rest are central indices of the intensity of changes. A minimum of 20 years of data for periods before and after flow regime changes and trend analysis is appropriate (Zuo and Liang, 2015). The RVA method incorporates an adaptive management perspective. This method consists of the following six main steps (Richter et al., 1997).

Step 1: Define the natural spectrum of hydrological variability through the application of ecologically relevant hydrological indices, specifically the indicators of hydrologic alteration (IHA).
Step 2: Determine management objectives for each IHA parameter.
Step 3: Implement the management flow objectives, designated as the RVA, to formulate management protocols that guarantee target flow regimes in the majority of years.
Step 4: Implement a monitoring program to assess the ecological effects of the new management system.
Step 5: Determine the actual flow changes in the river using the same hydrological parameters and compare them with the RVA objectives.
Step 6: Repeat the first five steps by incorporating recent management results and any ecological research outcomes that lead to a revision of the management system or RVA objectives.

The comparable studies have been carried out in the RVA method (Hu et al., 2008; Mwedzi et al., 2016; Ge et al., 2018; Pirnia et al., 2019; Zhou et al., 2023).

7. Conclusions

According to the First Hydrological Principle of transdisciplinary ecohydrology, the development, optimisation, and fine-tuning of Advanced Nature-based Solutions (ANbS) to local conditions—based on "Dual Regulation" for the enhancement of catchment sustainability potential—must be grounded in a thorough environmental flow analysis. Therefore, this chapter presents a review of this methodology. Various methods developed for determining the environmental flow of rivers have different levels of defensibility based on their complexity and the data used. These methods range from simple hydrological approaches that rely solely on river time series data to comprehensive methods that require extensive data. In general, to select an appropriate method, one must consider the design phase and the available temporal and informational resources. Therefore, a series of methods can be considered at two levels: initial recognition and preliminary planning, as well as at the level of detailed ecosystem studies. For example, at the initial recognition level, hydrological methods are primarily used, while at the comprehensive level and in the detailed studies stage for determining environmental water needs, comprehensive methods are mainly employed. Methods based on hydrological approaches are the simplest and most widely used methods for estimating environmental flow worldwide. These methods primarily rely on historical river flow statistics. In these methods, environmental flow is determined as a percentage of the average annual flow of the river. Among the most common of these methods is the Tennant method. Based on various studies conducted, the main weaknesses and obstacles of existing environmental flow assessment methods include the following:

(1) Socioeconomic aspects are often overlooked.
(2) The link between flow and the performance of ecosystem components is often assumed. This uncertainty often arises in opposition to the allocation of environmental flows.
(3) The focus is on the needs of riverine ecosystems, while the needs of floodplains, riverbanks, floodplains, estuaries, and deltas are often ignored.
(4) Validation is difficult and requires long-term monitoring using verifiable indicators.
(5) Habitat simulation methods and comprehensive approaches heavily rely on expert judgments.
(6) None of the methods have been tested with high accuracy. Large-scale experiments are needed.

References

Abedi-Koupai, J., Jamalian, M.A., Dorafshan, M.M., 2020. Improving Isfahan landfill leachate quality by phytoremediation using vetiver and Phragmites plants in green space irrigation. J. Water Wastewater. 31 (3), 101–111. https://doi.org/10.22093/wwj.2019.186145.2867.

Abedi-Koupai, J., Mahdi Dorafshan, M., Gohari, A., 2022. Investigation of bioremediation of Quinoa plant for desalination of unconventional waters. J. Water Soil Sci. 26 (3), 329–342.

Acreman, M.C., Dunbar, M.J., 2004. Defining environmental river flow requirements—a review. Hydrol. Earth Syst. Sci. 8 (5), 861–876. https://doi.org/10.5194/hess-8-861-2004.

Acreman, M., Farquharson, F.A.K., McCartney, M.P., Sullivan, C., Campbell, K., Hodgson, N., Morton, J., Smith, D., Birley, M., Knott, D., Lazenby, J., 2000. Managed Flood Releases from Reservoirs: Issues and Guidance. Report to DFID and the World Commission on Dams. Centre for Ecology and Hydrology.

Ahmadpari, H., Ladez, B.R., sadat Shokoohi, E., Biglou, M.kheiry G., 2019. Evaluation of environmental flows in Dalfard river using hydrological methods. Int. J. Eng. Technol. 11 (3), 530–539. https://doi.org/10.21817/ijet/2019/v11i3/191103036.

Alfredsen, K., Harby, A., Linnansaari, T., Ugedal, O., 2012. Development of an inflow-controlled environmental flow regime for a Norwegian river. River Res. Appl. 28 (6), 731–739. https://doi.org/10.1002/rra.1550.

Annear, T., Chisholm, I., Beecher, H., Locke, A., Aarrestad, P., Burkardt, N., Coomer, C., Estes, C., Hunt, J., Jacobson, R., Jobsis, G., 2002. Instream flows for riverine resource stewardship, Revised edition. Instream Flow Council.

Armitage, P.D., Petts, G.E., 1992. Biotic score and prediction to assess the effects of water abstractions on river macroinvertebrates for conservation purposes. Aquat. Conserv. Mar. Freshw. Ecosyst. 2 (1), 1–17. https://doi.org/10.1002/aqc.3270020102.

Armstrong, J., 2010. Variation in Habitat Quality for Drift-Feeding Atlantic Salmon and Brown Trout in Relation to Local Water Velocity and River Discharge Salmonid Fisheries: Freshwater Habitat Management. Wiley-Blackwell, United Kingdom, pp. 1–27. https://doi.org/10.1002/9781444323337.ch1. Available from: http://onlinelibrary.wiley.com/book/10.1002/9781444323337.

Arthington, A.H., 1998. Comparative Evaluation of Environmental Flow Assessment Techniques: Review of Methods, vol 27.

Arthington, A., Kennard, M., Mackay, S., Zalucki, J., Brizga, S., Choy, S., McCosker, R., Ruffini, J., 2000. Environmental Flow Requirements of the Brisbane River Downstream from Wivenhoe Dam.

Arthington, A.H., Tharme, R., Brizga, S.O., Pusey, B.J., Kennard, M.J., 2004. Proceedings of the Second International Symposium on the Management of Large Rivers for Fisheries, FAO Regional Office for Asia and the Pacific Environmental Flow Assessment with Emphasis on Holistic Methodologies.

Arthington, A., Baran, E., Brown, C., Dugan, Halls, A., King, J.M., Minte-Vera, C., Tharme, R., Welcomme, R., 2007. Water Requirements of Floodplain Rivers and Fisheries: Existing Decision Support Tools and Pathways for Development. Comprehensive Assessment of Water Management in Agriculture Research Report, 17.

Arthington, A.H., Bhaduri, A., Bunn, S.E., Jackson, S.E., Tharme, R.E., Tickner, D., Young, B., Acreman, M., Baker, N., Capon, S., Horne, A.C., Kendy, E., McClain, M.E., Poff, N.L.R., Richter, B.D., Ward, S., 2018. The Brisbane declaration and global action agenda on environmental flows. Front. Environ. Sci. 6. https://doi.org/10.3389/fenvs.2018.00045. Available from: https://www.frontiersin.org/articles/10.3389/fenvs.2018.00045/full.

Belmar, O., Bruno, D., Martínez-Capel, F., Barquín, J., Velasco, J., 2013. Effects of flow regime alteration on fluvial habitats and riparian quality in a semiarid Mediterranean basin. Ecol. Indic. 30, 52–64. https://doi.org/10.1016/j.ecolind.2013.01.042.

Belzile, L., Bérubé, P., Hoang, V.D., Leclerc, M., 1997. Méthode écohydrologique de détermination des débits réservés pour la protection des habitats du poisson dans les rivières du Québec.

Bhuyan, J.M., Sharma, P., 2025. Analysing Water Quality and Aquatic Vegetation Dynamics in a Proposed Bird Sanctuary: A Case Study of Satajaan Beel, North Lakhimpur, Assam. Research Square, India. https://doi.org/10.21203/rs.3.rs-4760803/v1. Available from: https://www.researchsquare.com/browse.

Boner, M.C., Furland, L.P., 1982. Seasonal treatment and variable effluent quality based on assimilative capacity. J. Water Pollut. Control Fed. 54 (10), 1408–1416.

Bounds, R.L., Lyons, B.W., 1979. Existing reservoir and stream management: Statewide minimum streamflow recommendations. Texas Parks and Wildlife Department. Federal Aid Project F-30-R-4 Austin, TX.

Bovee, K.D., 1982. A Guide to Stream Habitat Analysis Using the Instream Flow Incremental Methodology, vol. 1. Office of Biological Services, Fish and Wildlife Service, US Department of the Interior.

Bovee, K.D., 1995. A Comprehensive Overview of the Instream Flow Incremental Methodology. National Biological Service.

Bovee, K.D., Lamb, B.L., Bartholow, J.M., Stalnaker, C.B., Taylor, J., Henriksen, J., 1998. Stream Habitat Analysis Using the Instream Flow Incremental Methodology.

Bradford, A., 2008. An ecological flow assessment framework: building a bridge to implementation in Canada. Can. Water Resour. J. 33 (3), 215–232. https://doi.org/10.4296/cwrj3303215.

Brown, C., 2003. Environmental Flows: Concepts and Methods. Water Resource and Environment Technical Note C-1. World Bank.

Brown, C.A., King, J.A., 2003. Water Resources and Environment Technical Note C. 1. Environmental Flows: Concepts and Methods. World Bank, Washington, DC.

Caissie, D., El-Jabi, N., 1995. Comparison and regionalization of hydrologically based instream flow techniques in Atlantic Canada. Can. J. Civ. Eng. 22 (2), 235–246. https://doi.org/10.1139/l95-033.

Caissie, D., El-Jabi, N., Hébert, C., 2007. Comparison of hydrologically based instream flow methods using a resampling technique. Can. J. Civ. Eng. 34 (1), 66–74. https://doi.org/10.1139/L06-095.

Caissie, J., Caissie, D., El-Jabi, N., 2015. Hydrologically based environmental flow methods applied to rivers in the Maritime Provinces (Canada). River Res. Appl. 31 (6), 651–662. https://doi.org/10.1002/rra.2772.

Cavendish, M.G., Duncan, M.I., 1986. Use of the instream flow incremental methodology: a tool for negotiation. Environ. Impact Assess. Rev. 6 (4), 347–363. https://doi.org/10.1016/0195-9255(86)90027-2.

Chen, D., Yang, Z., Zeng, Q., Wang, W., Yan, L., Zhang, P., Li, X., Peng, H., Wang, H., 2023. Analysis of the suitable ecological flow of benthic animals in the lower reaches of Xiangjiaba Reservoir in the upper reaches of the Yangtze River based on the physical habitat model. J. Hydrol. 625, 130132. https://doi.org/10.1016/j.jhydrol.2023.130132.

Chen, X., Li, Q., Jia, Z., Xiao, R., Cheng, Z., Peng, Y., 2025a. Study on the evolution of ecological flow in river and its guarantee degree during different hydrological periods. J. Clean. Prod. 490, 144761.

Chen, Y., Nie, H., Liu, G., Xiao, J., Qiu, H., Xu, B., Chen, H., Huang, S., Xie, H., 2025b. An analysis of the current situation of ecological flow release from large- and medium-sized reservoirs in the southeastern River Basins of China. Water 17 (3), 451. https://doi.org/10.3390/w17030451.

Chouaib, W., Alila, Y., Caldwell, P.V., 2019. On the use of mean monthly runoff to predict the flow–duration curve in ungauged catchments. Hydrol. Sci. J. 64 (13), 1573–1587. https://doi.org/10.1080/02626667.2019.1657233.

Cui, T., Tian, F., Yang, T., Wen, J., Khan, M.Y.A., 2020. Development of a comprehensive framework for assessing the impacts of climate change and dam construction on flow regimes. J. Hydrol. 590. https://doi.org/10.1016/j.jhydrol.2020.125358. Available from: www.elsevier.com/inca/publications/store/5/0/3/3/4/3.

Daechini, F., Vafakhah, M., Moosavi, V., 2021. Silabi, Estimation of environmental flow indicators in the downstream of golestan and Voshmgir dams. J. Ecohydrol. 8 (3), 677–690.

Dorafshan, M.M., Eslamian, S., 2023. Modeling Human Dimensions to Reduce the Disaster Risk: A Socio-Hydrological Approach Disaster Risk Reduction for Resilience: Disaster Socio-Hydrological Resilience and Sustainability. Springer International Publishing, Iran, pp. 3–24. https://doi.org/10.1007/978-3-031-43177-7_1.

Dorafshan, M.M., Abedi-Koupai, J., Eslamian, S., Amiri, M.J., Grass, V., Chrysopogon zizanoides, L., 2023. A hyper-accumulator crop for bioremediation of unconventional water. Sustainability 15 (4), 3529. https://doi.org/10.3390/su15043529.

Dorafshan, M.M., Golmohammadi, M.H., Eslamian, S., 2024. Impacts of Climate Change on the Quantity and Quality of Surface Water Resources Handbook of Climate Change Impacts on River Basin Management: Fundamentals and Impacts. CRC Press, Iran, pp. 356–369. https://doi.org/10.1201/9781003473336-29. Available from: https://www.routledge.com/Handbook-of-Climate-Change-Impacts-on-River-Basin-Management-Fundamentals-and-Impacts/Eslamian-Huda-Rather-Eslamian/p/book/9781003473336.

Dorafshan, M.M., Golmohammadi, M.H., Asghari, K., De Michele, C., 2025a. A novel fuzzified Markov chain approach to model monthly river discharge. Water Resour. Manag. 39 (4), 1931–1951. https://doi.org/10.1007/s11269-024-04053-x.

Dorafshan, M.M., Golmohammadi, M.H., Michele, Safavi, H.R., Eslamian, S., 2025b. Geo-engineering solutions for urban climate adaptation. In: Handbook of Nature-Based Drought Solutions. Elsevier, pp. 573–586. https://doi.org/10.1016/B978-0-443-29993-3.00027-8.

Dorafshan, M.M., Golmohammadi, M.H., De Michele, C., 2025c. A health index for surface water resources during drought: an automated fuzzy-based performance criteria approach. Appl. Water Sci. 15 (8). https://doi.org/10.1007/s13201-025-02543-w. Available from: https://www.springer.com/journal/13201.

Dorafshan, M.M., Golmohammadi, M.H., De Michele, C., Eslamian, S., 2025d. Risk analysis of inflow to the Zayandehrud Dam under historical and future scenarios using reliability, resiliency, and vulnerability indicators: a comparative study of the runoff risk index and runoff risk ranking methods. Water Reuse 15 (1), 1–23. https://doi.org/10.2166/wrd.2024.070.

DWAF, 1997. White Paper on a National Water Policy for South Africa. Department of Water A¥ airs and Forestry.

Dyson, M., Bergkamp, G., Scanlon, J., 2003. Flow: The Essentials of Environmental Flows. IUCN, Switzerland and Cambridge, UK, pp. 20–87. Gland.

Efstratiadis, A., Tegos, A., Varveris, A., Koutsoyiannis, D., 2014. Assessment of environmental flows under limited data availability: case study of the Acheloos River, Greece. Hydrol. Sci. J. 59 (3–4), 731–750. https://doi.org/10.1080/02626667.2013.804625.

Eisner, A., Young, C., Schneider, M., Kopecki, I., May 2005. Proceedings of Final COST 626 Meeting in Silkeborg 65 95 MesoCASiMiR: New Mapping Method and Comparison with Other Current Approaches.

Espegren, G.D., 1996. Development of Instream Flow Recommendations in Colorado Using R2CROSS: Denver. CO. Water Conservation Board.

Fonstad, M.A., Andrew Marcus, W., 2010. High resolution, basin extent observations and implications for understanding river form and process. Earth Surf. Process. Landf. 35 (6), 680–698. https://doi.org/10.1002/esp.1969.

Fouad, G., Loáiciga, H.A., 2020. Independent variable selection for regression modeling of the flow duration curve for ungauged basins in the United States. J. Hydrol. 587, 124975. https://doi.org/10.1016/j.jhydrol.2020.124975.

Gao, Y., Vogel, R.M., Kroll, C.N., Poff, N.L.R., Olden, J.D., 2009. Development of representative indicators of hydrologic alteration. J. Hydrol. 374 (1–2), 136–147. https://doi.org/10.1016/j.jhydrol.2009.06.009.

Ge, J., Peng, W., Huang, W., Qu, X., Singh, S.K., 2018. Quantitative assessment of flow regime alteration using a revised range of variability methods. Water 10 (5), 597. https://doi.org/10.3390/w10050597.

Gillilan, D.M., Brown, T.C., 2013. Instream Flow Protection: Seeking a Balance in Western Water Use. Island Press.

Ginot, V., 1995. EVHA, un logiciel d'évaluation de l'habitat du poisson sous Windows. Bull. Fr. Peche Piscic. 303–308. https://doi.org/10.1051/kmae:1995034 (337-338-339).

Gippel, C.J., Stewardson, M.J., 1998. Use of wetted perimeter in defining minimum environmental flows. Regul. Rivers Res. Manag. 14 (1), 53–67. https://doi.org/10.1002/(SICI)1099-1646(199801/02)14:1<53::AID-RRR476>3.3.CO;2-Q.

Gleick, P.H., 1993. Water in Crisis. Oxford University Press, p. 100.

Griggs, D., Stafford-Smith, M., Gaffney, O., Rockström, J., Öhman, M.C., Shyamsundar, P., Steffen, W., Glaser, G., Kanie, N., Noble, I., 2013. Sustainable development goals for people and planet. Nature 495 (7441), 305–307. https://doi.org/10.1038/495305a.

Hatfield, T., Lewis, A., Ohlson, D., Bradford, M., 2003. Development of Instream Flow Thresholds as Guidelines for Reviewing Proposed Water Uses. British Columbia Ministry of Sustainable Resource Management.

Hirji, R., Panella, T., 2003. Evolving policy reforms and experiences for addressing downstream impacts in World Bank water resources projects. River Res. Appl. 19 (5–6), 667–681. https://doi.org/10.1002/rra.754.

Hoque, M.M., Islam, A., Pal, S.C., Sarkar, B., Hossain, Z.M., 2024. Assessment of fish habitat suitability in the wake of elevated pollution of a tropical river in India. Environ. Sci. Pollut. Control Ser. 1–15.

Hoque, I., Pal, S., Hossain, 2025. Assessment of fish habitat suitability in the wake of elevated pollution of a tropical river in India. Environ. Sci. Pollut. Res. Int.

Hu, W.W., Wang, G.x., Deng, W., Li, S.n., 2008. The influence of dams on ecohydrological conditions in the Huaihe River basin, China. Ecol. Eng. 33 (3–4), 233–241. https://doi.org/10.1016/j.ecoleng.2008.04.003.

Huang, F., Li, F., Zhang, N., Chen, Q., Qian, B., Guo, L., Xia, Z., 2017. A histogram comparison approach for assessing hydrologic regime alteration. River Res. Appl. 33 (5), 809–822. https://doi.org/10.1002/rra.3130.

Hudson, H.R., 2002. Linking the Physical Form and Processes of Rivers with Ecological Response. IAHS-AISH Publication, pp. 121–139, 276.

Hughes, D.A., Münster, F., 2000. Hydrological Information and Techniques to Support the Determination of the Water Quantity Component of the Ecological Reserve for Rivers.

Hughes, D.A., Smakhtin, V., 1996. Daily flow time series patching or extension: a spatial interpolation approach based on flow duration curves. Hydrol. Sci. J. 41 (6), 851–871.

Hughes, D.A., Smakhtin, V., 2009. Daily flow time series patching or extension: a spatial interpolation approach based on flow duration curves. Hydrol. Sci. J. 41 (6), 851–871. https://doi.org/10.1080/02626669609491555.

IWMI, 2004. Environmental Flows. Environmental Perspectives on River Basin Management in Asia, vol. 1. International Water Management Institute.

Jorde, K., Bratrich, C., 2000. Ecological evaluation of instream flow regulations based on temporal and spatial variability of bottom shear stress and hydraulic habitat. Qual. Ecohydraul.

Jowett, I.G., 1989. River Hydraulic and Habitat Simulation. RHYHABSIM Computer Manual, 222.

Jowett, I.G., 1997. Instream flow methods: a comparison of approaches. Regul. Rivers Res. Manag. 13 (2), 115–127. https://doi.org/10.1002/(sici)1099-1646(199703)13:2<115::aid-rrr440>3.0.co;2-6.

Karakoyun, Y., Yumurtaci, Z., Dönmez, A.H., 2016. Environmental flow assessment for energy generation sustainability employing different hydraulic evaluation methods: Çambaşi hydropower plant case study in Turkey. Clean Technol. Environ. Policy 18 (2), 583–591. https://doi.org/10.1007/s10098-015-1023-9.

Karakoyun, Y., Dönmez, A.H., Yumurtacı, Z., 2018. Comparison of environmental flow assessment methods with a case study on a runoff river–type hydropower plant using hydrological methods. Environ. Monit. Assess. 190 (12). https://doi.org/10.1007/s10661-018-7107-3. Available from: www.wkap.nl/journalhome.htm/0167-6369.

Karimi, S., Salarijazi, M., Ghorbani, K., Heydari, M., 2021. Comparative assessment of environmental flow using hydrological methods of low flow indexes, Smakhtin, Tennant and flow duration curve. Iran Acta Geophysica 69 (1), 285–293. https://doi.org/10.1007/s11600-021-00539-z.

Kashaigili, J.J., Kadigi, R.M.J., Lankford, B.A., Mahoo, H.F., Mashauri, D.A., 2005. Environmental flows allocation in river basins: exploring allocation challenges and options in the Great Ruaha River catchment in Tanzania. Phys. Chem. Earth 30 (11–16), 689–697. https://doi.org/10.1016/j.pce.2005.08.009.

Khan, I., Zakwan, M., Pulikkal, A.K., Lalthazula, R., 2024. Environmental flow assessment for the Musi river, India. Environ. Dev. Sustain. https://doi.org/10.1007/s10668-024-04927-5. Available from: https://www.springer.com/journal/10668.

King, J., Tharme, R.E., Brown, C., 1999. Definition and Implementation of Instream Flows Thematic Report for the World Commission on Dams. Southern Waters Ecological Research and Consulting, Cape Town, SA.

King, J., Brown, C., Sabet, H., 2003. A scenario-based holistic approach to environmental flow assessment for rivers. River Res. Appl. 19 (5–6), 619–639. https://doi.org/10.1002/rra.709.

Knights, P., 2002. Proceedings of International Conference: Dialog on Water, Food and Environment Environmental Flows: Lessons from an Australian Experience.

Korsgaard, L., 2007. Environmental Flows in Integrated Water Resources Management: Linking Flows, Services and Values.

Kuriqi, A., Pinheiro, A.N., Sordo-Ward, A., Garrote, L., 2020. Water-energy-ecosystem nexus: balancing competing interests at a run-of-river hydropower plant coupling a hydrologic–ecohydraulic approach. Energy Convers. Manag. 223, 113267. https://doi.org/10.1016/j.enconman.2020.113267.

Lamouroux, N., Jowett, I.G., 2005. Generalized instream habitat models. Can. J. Fish. Aquat. Sci. 62 (1), 7–14. https://doi.org/10.1139/f04-163.

Lamouroux, N., Mérigoux, S., Capra, H., Dolédec, S., Jowett, I.G., Statzner, B., 2010. The generality of abundance-environment relationships in microhabitats: a comment on Lancaster and Downes (2009). River Res. Appl. 26 (7), 915–920. https://doi.org/10.1002/rra.1366.

Leone, M., Gentile, F., Lo Porto, A., Ricci, G.F., De Girolamo, A.M., 2023. Ecological flow in southern Europe: status and trends in non-perennial rivers. J. Environ. Manag. 342. https://doi.org/10.1016/j.jenvman.2023.118097. Available from: https://www.sciencedirect.com/journal/journal-of-environmental-management.

Li, B., Zhang, X., 2023. Investigation on the influence of spur dikes on the weighted usable area for the two native species in a city-crossing river. J. Hydrol. 620, 129428. https://doi.org/10.1016/j.jhydrol.2023.129428.

Li, X., Zhang, Q., Diao, Y., Shi, Y., Li, S., Yao, C., Su, R., Guo, S., 2024. Ecological flow considering hydrological season and habitat suitability for a variety of fish. Ecol. Model. 489, 110625. https://doi.org/10.1016/j.ecolmodel.2024.110625.

Linnansaari, T., Monk, W.A., Baird, D.J., Curry, R.A., 2012. Review of approaches and methods to assess Environmental Flows across Canada and internationally. DFO Can. Sci. Advis. Sec. Res. Doc. 39.

Liu, L., Wei, L., Xu, Y., Xin, X., Xiao, C., 2021. Projection of climate change impacts on ecological flow in the Yellow River basin. Adv. Water Sci. 32 (6), 824–833. https://doi.org/10.14042/j.cnki.32.1309.2021.06.002.

Lumbroso, D.M., Sakamoto, D., Johnstone, W.M., Tagg, A.F., Lence, B.J., 2011. Development of a life safety model to estimate the risk posed to people by dam failures and floods. Dams Reservoirs 21 (1), 31–43. https://doi.org/10.1680/dare.2011.21.1.31.

Lv, X., Liu, S., Li, S., Ni, Y., Qin, T., Zhang, Q., 2021. Quantitative estimation on contribution of climate changes and watershed characteristic changes to decreasing streamflow in a typical basin of yellow river. Front. Earth Sci. 9. https://doi.org/10.3389/feart.2021.752425. Available from. https://www.frontiersin.org/journals/earth-science.

Ma, Z.Z., Wang, Z.J., Xia, T., Gippel, C.J., Speed, R., 2014. Hydrograph-based hydrologic alteration assessment and its application to the Yellow River. J. Environ. Informat. 23 (1), 1–13. https://doi.org/10.3808/jei.201400252.

Mahmood, R., Shaofeng, J.I.A., Lv, A., Zhu, W., 2020. A preliminary assessment of environmental flow in the three rivers' source region, Qinghai Tibetan Plateau, China and suggestions. Ecol. Eng. 144, 105709. https://doi.org/10.1016/j.ecoleng.2019.105709.

Mann, J., Fassnacht, S.R., Merritt, D., Rathburn, S., 2006. Instream Flow Methodologies: An Evaluation of the Tennant Method for Higher Gradient Streams in the National Forest System Lands in the Western US.

Matthews, E., Payne, R., Rohweder, M., Murray, S., 2000. Pilot Analysis of Global Ecosystems: Forest Ecosystems. World Resources Institute.

Millennium Ecosystem Assessment, 2005. Ecosystems and human well-being: Synthesis. A report of the Millennium Ecosystem Assessment, vol. 5, 2005. Island Press, Washington, DC.

Młyński, D., Operacz, A., Wałęga, A., 2020. Sensitivity of methods for calculating environmental flows based on hydrological characteristics of watercourses regarding the hydropower potential of rivers. J. Clean. Prod. 250, 119527. https://doi.org/10.1016/j.jclepro.2019.119527.

Mostafavi, S., Yasi, M., 2015. Evaluation of environmental flows in rivers using hydrological methods (case study: the Barandozchi River-Urmia Lake basin). Water Soil 29 (5), 1219–1231.

Mwedzi, T., Katiyo, L., Mugabe, F.T., Bere, T., Bangira, C., Mangadze, T., Kupika, O.L., 2016. A spatial assessment of stream-flow characteristics and hydrologic alterations, post dam construction in the Manyame catchment. WaterSA 42 (2), 194–202. https://doi.org/10.4314/wsa.v42i2.03.

Naiman, R.J., Bunn, S.E., Nilsson, C., Petts, G.E., Pinay, G., Thompson, L.C., 2002. Legitimizing fluvial ecosystems as users of water: an overview. Environ. Manag. 30 (4), 455–467. https://doi.org/10.1007/s00267-002-2734-3.

Ni, X., Dong, Z., Xie, W., Wu, S., Chen, M., Yao, H., Jia, W., 2022. A practical approach for environmental flow calculation to support ecosystem management in Wujiang River, China. Int. J. Environ. Res. Publ. Health 19 (18), 11615. https://doi.org/10.3390/ijerph191811615.

O'Keeffe, J., Hughes, D., Tharme, R., 2017. Linking ecological responses to altered flows, for use in environmental flow assessments: the Flow stressor—response method. SIL Proc. 28 (1), 84–92. https://doi.org/10.1080/03680770.2001.11902550.

Olden, J.D., Poff, N.L., 2003. Redundancy and the choice of hydrologic indices for characterizing streamflow regimes. River Res. Appl. 19 (2), 101–121. https://doi.org/10.1002/rra.700.

Overton, I.C., Smith, D.M., Dalton, J., Barchiesi, S., Acreman, M.C., Stromberg, J.C., Kirby, J.M., 2014. Implementing environmental flows in integrated water resources management and the ecosystem approach. Hydrol. Sci. J. 59 (3–4), 860–877. https://doi.org/10.1080/02626667.2014.897408.

Palau, Alcazar, J., 1996. Jun Ecohydraulics 2000. Proceedings of the 2nd International Symposium on Habitat Hydraulics. INRS-Eau the Basic Flow: An Alternative Approach to Calculate Minimum Environmental Instream Flows.

Parasiewicz, P., 2001. MesoHABSIM: a concept for application of instream flow models in river restoration planning. Fisheries 26 (9), 6–13. https://doi.org/10.1577/1548-8446(2001)026<0006:m>2.0.co;2.

Parasiewicz, P., Belka, K., Łapińska, M., Ławniczak, K., Prus, P., Adamczyk, M., Buras, P., Szlakowski, J., Kaczkowski, Z., Krauze, K., O'Keeffe, J., Suska, K., Ligięza, J., Melcher, A., O'Hanley, J., Birnie-Gauvin, K., Aarestrup, K., Jones, P.E., Jones, J., Wiśniewolski, W., 2023. Over 200,000 kilometers of free-flowing river habitat in Europe is altered due to impoundments. Nat. Commun. 14, 6289. https://doi.org/10.1038/s41467-023-40922-6.

Pauls, M.A., Wurbs, R.A., Wre, H.D., 2016. Environmental flow attainment metrics for water allocation modeling. J. Water Resour. Plann. Manag. 142 (8), 04016018. https://doi.org/10.1061/(ASCE)WR.1943-5452.0000652.

Pirnia, A., Golshan, M., Darabi, H., Adamowski, J., Rozbeh, S., 2019. Using the Mann–Kendall test and double mass curve method to explore stream flow changes in response to climate and human activities. J. Water Clim. Change 10 (4), 725–742. https://doi.org/10.2166/wcc.2018.162.

Poff, N.L.R., Matthews, J.H., 2013. Environmental flows in the Anthropocence: past progress and future prospects. Curr. Opin. Environ. Sustain. 5 (6), 667–675. https://doi.org/10.1016/j.cosust.2013.11.006.

Poff, N.L., Zimmerman, J.K.H., 2010. Ecological responses to altered flow regimes: a literature review to inform the science and management of environmental flows. Freshw. Biol. 55 (1), 194–205. https://doi.org/10.1111/j.1365-2427.2009.02272.x.

Poff, N.L.R., Allan, J.D., Bain, M.B., Karr, J.R., Prestegaard, K.L., Richter, B.D., Sparks, R.E., Stromberg, J.C., 1997. The natural flow regime: a paradigm for river conservation and restoration. Bioscience 47 (11), 769–784. https://doi.org/10.2307/1313099.

Pyrce, R., 2004. Hydrological Low Flow Indices and Their Uses.

Reiser, D.W., Wesche, T.A., Estes, C., 1989. Status of instream flow legislation and practices in North America. Fisheries 14 (2), 22–29. https://doi.org/10.1577/1548-8446(1989)014<0022:SOIFLA>2.0.CO;2.

Ren, M., Liu, Y., Fu, X., Ding, L., Wang, G., Kan, G., Zhao, L., Zhu, Y., 2021. The changeable degree assessment of designed flood protection condition for designed unit of inter-basin water transfer project based on the entropy weight method and fuzzy comprehensive evaluation model. IOP Conf. Ser. Earth Environ. Sci. 826.

Richter, B.D., 2010. Re-thinking environmental flows: from allocations and reserves to sustainability boundaries. River Res. Appl. 26 (8), 1052–1063. https://doi.org/10.1002/rra.1320.

Richter, B.D., Baumgartner, J.V., Powell, J., Braun, D.P., 1996. A method for assessing hydrologic alteration within ecosystems. Conserv. Biol. 10 (4), 1163–1174. https://doi.org/10.1046/j.1523-1739.1996.10041163.x.

Richter, B.D., Baumgartner, J.V., Wigington, R., Braun, D.P., 1997. How much water does a river need? Freshw. Biol. 37 (1), 231–249. https://doi.org/10.1046/j.1365-2427.1997.00153.x.

Richter, B.D., Davis, M.M., Apse, C., Konrad, C., 2012. A presumptive standard for environmental flow protection. River Res. Appl. 28 (8), 1312–1321. https://doi.org/10.1002/rra.1511.

Ringler, C., Nguyen, V.H., 2004. Water allocation policies for the Dong Nai River Basin in Vietnam: an integrated perspective. SSRN Electron. J. 1556–5068. https://doi.org/10.2139/ssrn.644441.

Rosner, A., 2003. The Institute of Rural and Agricultural Development, Polish Academy of Sciences. Eastern European Countryside, pp. 185–189.

Sanz, D.B., Iribas, B.L., Calvo, P.C., 2025. Proposal for analysis of minimum ecological flow regimes based on the achievement of technical and environmental objectives: Tagus River basin case study (Spain). Environ. Earth Sci. 84 (1). https://doi.org/10.1007/s12665-024-12024-z. Available from: https://www.springer.com/journal/12665.

Shang, S., 2008. A multiple criteria decision-making approach to estimate minimum environmental flows based on wetted perimeter. River Res. Appl. 24 (1), 54–67. https://doi.org/10.1002/rra.1047.

Shu-feng, L.I.U., Zhi-hua, W.A.N.G., Shuai, G.U.A.N., Yan-hu, H.E., 2024. An improved hydraulic method for calculating ecological flow considering the changes of rainfall and temperature within a year. Ecol. Indic. 167. https://doi.org/10.1016/j.ecolind.2024.112677. Available from: https://www.sciencedirect.com/science/journal/1470160X.

Smakhtin, V., 2004. Taking into Account Environmental Water Requirements in Global-Scale Water Resources Assessments, vol. 2. International Water Management Institute.

Smakhtin, V.U., Anputhas, M., 2006. An Assessment of Environmental Flow Requirements of Indian River Basins, vol 107. IWMI Research Report.

Spiegel, M.R., 1988. Theory and Problems of Statistics. McGraw-Hill Publishing Company, McGraw-Hill Publishing Company.

St-Hilaire, A., Ferchichi, H., Berthot, L., Caissie, D., 2021. The fate of stationary tools for environmental flow determination in a context of climate change. Water 13 (9), 1203. https://doi.org/10.3390/w13091203.

Stalnaker, C.B., Arnette, J.L., 1976. Methodologies for Determining Instream Flows for Fish and Other Aquatic Life. Methodologies for the Determination of Stream Resource Flow Requirements. US Fish and Wildlife Service, pp. 89–138.

Steffler, P., Blackburn, J., 2002. Two-Dimensional Depth Averaged Model of River Hydrodynamics and Fish Habitat. River2D User's Manual. University of Alberta, Edmonton, Canada.

Stucchi, L., Bocchiola, D., 2024. A multi-criteria analysis for Environmental Flow Assessment based on hydraulic modelling. MethodsX 13, 102874. https://doi.org/10.1016/j.mex.2024.102874.

Suwal, N., Kuriqi, A., Huang, X., Delgado, J., Młyński, D., Walega, A., 2020. Environmental flows assessment in Nepal: the case of Kaligandaki river. Sustainability 12 (21), 8766. https://doi.org/10.3390/su12218766.

Swales, S., Harris, J.H., 1995. The Expert Panel Assessment Method (EPAM): A New Tool for Determining Environmental Flows in Regulated Rivers. The Ecological Basis for River Management, pp. 125–146.

Tennant, D.L., 1976. Instream flow regimens for fish, wildlife, recreation and related environmental resources. Fisheries 1 (4), 6–10. https://doi.org/10.1577/1548-8446(1976)001<0006:IFRFFW>2.0.CO;2.

Tessman, S.A., 1980. Environmental Assessment, Technical Appendix E. Environmental Use Sector Reconnaissance Elements of the Western Dakotas Region of South Dakota Study.

Tharme, R.E., 2003. A global perspective on environmental flow assessment: emerging trends in the development and application of environmental flow methodologies for rivers. River Res. Appl. 19 (5–6), 397–441. https://doi.org/10.1002/rra.736.

Tharme, R.E., Smakhtin, V.U., 2003. Proceedings of the First Southeast Asia Water Forum, Environmental Flow Assessment in Asia: Capitalizing on Existing Momentum, 2, pp. 301–313.

Uday Kumar, A., Jayakumar, K.V., 2021. Modelling of environmental flow requirements using hydraulic and habitation models. Ecol. Indic. 121, 107046. https://doi.org/10.1016/j.ecolind.2020.107046.

Verma, R.K., Murthy, S., Verma, S., Mishra, S.K., 2017. Design flow duration curves for environmental flows estimation in Damodar River Basin, India. Appl. Water Sci. 7 (3), 1283–1293. https://doi.org/10.1007/s13201-016-0486-0.

Verma, R.K., Pandey, A., Verma, S., Mishra, S.K., 2023. A review of environmental flow assessment studies in India with implementation enabling factors and constraints. Ecohydrol. Hydrobiol. 23 (4), 662–677. https://doi.org/10.1016/j.ecohyd.2023.06.006.

Wałęga, A., Młyński, D., Kokoszka, R., Miernik, W., 2015. Possibilities of applying hydrological methods for determining environmental flows in select catchments of the upper Dunajec basin. Pol. J. Environ. Stud. 24 (6), 2663–2676. https://doi.org/10.15244/PJOES/59294.

Wałęga, A., Kędzior, R., Skalski, T., Młyński, D., 2025. A Global hydrological index describes ecological conditions in rivers: a new approach to environmental flow calculation reflecting macroinvertebrate requirements. Ecol. Indic. 170, 113082. https://doi.org/10.1016/j.ecolind.2025.113082.

Wallingford, H.R., 2003. Handbook for the assessment of catchment water demand and use. Darren Lumbroso 248. HR Wallingford, Howbery Park, Wallingford, Oxon, OX10 8BA, UK.

Wang, S., Yan, M., Yan, Y., Shi, C., He, L., 2012. Contributions of climate change and human activities to the changes in runoff increment in different sections of the Yellow River. Quat. Int. 282, 66–77. https://doi.org/10.1016/j.quaint.2012.07.011.

Wang, Y., Wang, D., Lewis, Q.W., Wu, J., Huang, F., 2017. A framework to assess the cumulative impacts of dams on hydrological regime: a case study of the Yangtze River. Hydrol. Process. 31 (17), 3045–3055. https://doi.org/10.1002/hyp.11239. Available from: http://onlinelibrary.wiley.com/journal/10.1002/(ISSN)1099-1085.

Wang, H., Yan, Y., Huang, L., He, N., Guo, W., 2023. Ecological flow research and optimal selection of the most ecologically relevant hydrologic indicators at the Jingjiang River's three outlets. J. Water Clim. Change 14 (11), 4099–4120. https://doi.org/10.2166/wcc.2023.284.

Wang, D., Jia, Y., Niu, C., Yan, X., Hao, C., 2024a. A multiple criteria decision-making approach for water allocation of environmental flows considering the value trade-offs-A case study of Fen River in China. Sci. Total Environ. 912, 169588.

Wang, H., Liu, J., Klaar, M., Chen, A., Gudmundsson, L., Holden, J., 2024b. Anthropogenic climate change has influenced global river flow seasonality. Science 383 (6686), 1009–1014. https://doi.org/10.1126/science.adi9501. Available from: https://www.science.org/doi/epdf/10.1126/science.adi9501.

Waters, B.F., 1976. Proceedings of the Symposium and Specialty Conference on Instream Flow Needs A Methodology for Evaluating the Effects of Different Streamflows on Salmonid Habitat.

Watt, S.P., Watt, E., Anderson, B., 2007. A Methodology for Environmental Protection of Ontario Watercourses with Respect to the Permit to Take Water Program. Queen's University, Canada, p. 124. Master Thesis.

Williams, J.G., Moyle, P.B., Webb, J.A., Kondolf, G.M., 2019. Environmental Flow Assessment: Methods and Applications Environmental Flow Assessment: Methods and Applications. Wiley, United States, pp. 1–220. Available from: https://www.wiley.com/en-us/Environmental+Flow+Assessment%3A+Methods+and+Applications-p-9781119217381.

World Commission on Dams, 2000. Dams and development: A new framework for decision-making. The report of the World Commission on Dams. Earthscan Publications Ltd., London.

Yu, J., Chen, W., Fang, Z., Li, D., Li, Z., Yang, Le, 2022. Zoning study on influence intensity of Cao'e river sluice on river network drainage based on 2D numerical model. J. Phys., Conf. Ser. 2271 (1), 012028. https://doi.org/10.1088/1742-6596/2271/1/012028.

Zalewski, M., 2000. Ecohydrology—The scientific background to use ecosystem properties as management tools toward sustainability of water resources. Ecol. Eng. 16 (1), 1–8. https://doi.org/10.1016/S0925-8574(00)00071-9.

Zalewski, M., 2014. Ecohydrology and hydrologic engineering: Regulation of hydrology–biota interactions for sustainability. J. Hydrol. Eng. 19 (5), 1080–1088. https://doi.org/10.1061/(ASCE)HE.1943-5584.0000999.

Zalewski, M. (2021). Ecohydrology: An integrative sustainability science. In T. V. V. Hromadka II & P. Rao (Eds.), Hydrology. IntechOpen. https://doi.org/10.5772/intechopen.94169

Zargari, A., Salarijazi, M., Ghorbani, K., Ahmad Dehghani, A., 2023. Effect of dam construction on changes in river's environmental flow (case study: Gorganrood river in the south of the Caspian Sea). Appl. Water Sci. 13 (11). https://doi.org/10.1007/s13201-023-02011-3. Available from: https://www.springer.com/journal/13201.

Zhang, X., Yu, J., Wang, L., Zhang, R., 2024. Determination of river ecological flow thresholds and development of early warning programs based on coupled multiple hydrological methods. Water 16 (14), 1986. https://doi.org/10.3390/w16141986.

Zhao, F., Pang, A., Li, C., Zheng, X., Wang, X., Yi, Y., 2021. Progress in ecological water requirements in the mainstream and estuary ecosystem of the Yellow River Basin. Sheng Tai Xue Bao 41 (15), 6289–6301. https://doi.org/10.5846/stxb202004110861.

Zhou, F., Zhang, W., Liu, X., Jiang, A., Chen, G., Li, A., Peng, H., 2023. Evaluating effects of terraces on flow regimes in a hilly and mountainous basin. Water 15 (22), 3980. https://doi.org/10.3390/w15223980.

Zhu, S., Dong, Z., Wu, S., Fu, G., Li, Z., Meng, J., Shao, Y., Zhang, K., 2023. Evaluation of reservoir-induced ecological flow with consideration of ecological responses to multiple hydrological alterations: a case study in the Feiyun River Basin. Ecol. Indic. 154, 110859. https://doi.org/10.1016/j.ecolind.2023.110859.

Zuo, Q., Liang, S., 2015. Effects of dams on river flow regime based on IHA/RVA. Proc. Int. Assoc. Hydrol. Sci. 368, 275–280. https://doi.org/10.5194/piahs-368-275-2015.

Further reading

Aid Project F 30-R-4. Performance Report, Texas Parks and Wildlife Department, Austin, Texas, 28pp.

Arthington, A.H., Pusey, B.J., 1993. In-stream flow management in Australia: methods, deficiencies and future directions. Aust. Biol. 6 (1), 52–60.

Arthington, A.H., Brizga, S.O., Kennard, M.J., 1998. Comparative Evaluation of Environmental Flow Assessment Techniques: Best Practice Framework. LWRRDC Occasional Paper 25- 98. Canberra, Land and Water Research and.

Bartschi, D.K., 1976. A habitat-discharge method of determining instream flows for aquatic habitat. In: Symposium on Instream Flow Needs. American Fisheries Society, Bethseda, Maryland, pp. 285–294.

Behmanesh, J., Mostafavi, S., Zamanzad Ghavidel, S., 2017. Use of soft calculations at estimation and prediction of environmental flow discharge (Case study: Khorkhoreh Chay river). J. Civ. Environ. Eng. 47 (88), 9–22.

Development Corporation, 26 pp.

Dorafshan, M.M., Golmohammadi, M.H., De Michele, C., Eslamian, S., 2024b. Risk analysis of inflow to the Zayandehrud Dam under historical and future scenarios using reliability, resiliency, and vulnerability indicators: a comparative study of the runoff risk index and runoff risk ranking methods. Water Reuse jwrd2024070.

Dorafshan, M.M., Golmohammadi, M.H., Asghari, K., De Michele, C., 2024a. A novel fuzzified Markov chain approach to model monthly river discharge. Water Resour. Manag. 1–21.

Dunbar, M.J., Alfredsen, K., Harby, A., 2012. Hydraulic-habitat modelling for setting environmental river flow needs for salmonids. Fish. Manag. Ecol. 19 (6), 500–517.

Gippel, C.J., 2006. Environmental flows: managing hydrological environments. In: Anderson, M.G. (Ed.), Encyclopedia of Hydrological Sciences. John Wiley & Sons, Chichester, UK.

Goguen, G., Caissie, D., El-Jabi, N., 2020. Uncertainties associated with environmental flow metrics. River Res. Appl. 36 (9), 1879–1890.

Gopal, B.R.I.J., 2013. Methodologies for the assessment of environmental flows. In: Environmental Flows: An Introduction for Water Resources Managers, pp. 129–182.

Hughes, D.A., Hannart, P., 2003. A desktop model used to provide an initial estimate of the ecological instream flow requirements of rivers in South Africa. J. Hydrol. 270 (3–4), 167–181.

Li, Y.Y., Chang, J.X., Tu, H., 2014. Impact of controlling cascade reservoir joint operation on hydrologic regimes in the lower Yellow River. Resour. Sci. 36 (6), 1183–1190.

Liu, Z., Zhang, Z., Ma, H., Gao, J., Su, Z., Wang, H., Li, G., Ding, X., 2024. Optimal scheduling study of cascade hydropower stations to ensure ecological flow requirements—a case study of the Gorge Section of the Yongding River, China. Sci. Total Environ. 953, 175977.

Naderi, M., Zakerinia, M., Salarijazi, M., 2020. Ecosystems protecting of Gharasoo river based on regulation of environmental flow regime using hydrological methods. Irrigation and Water Engineering 11 (1), 118–136.

O'Keeffe, J., Hughes, D., Tharme, R., 2002. Linking ecological responses to altered flows, for use in environmental flow assessments: the Flow Stressor—response method. Internationale Vereinigung für theoretische und angewandte Limnologie: Verh Proc. Trav. SIL 28 (1), 84–92.

Parker, G.W., Armstrong, D.S., Richards, T.A., 2004. Comparison of Methods for Determining Streamflow Requirements for Aquatic Habitat Protection at Selected Sites on the Assabet and Charles Rivers, Eastern Massachusetts, 2000-02 (No. 2004). DIANE Publishing.

Pastor, A.V., Ludwig, F., Biemans, H., Hoff, H., Kabat, P., 2014. Accounting for environmental flow requirements in global water assessments. Hydrol. Earth Syst. Sci. 18 (12), 5041–5059.

Pauls, M.A., Wurbs, R.A., 2016. Environmental flow attainment metrics for water allocation modeling. J. Water Resour. Plann. Manag. 142 (8), 04016018.

Ringler, C., Huy, N.V., 2004. Water allocation policies for the Dong Nai River Basin in Vietnam: an integrated perspective. Int. Food Policy Res. Inst.

Rosenberg, D.M., McCully, P., Pringle, C.M., 2000. Global-scale environmental effects of hydrological alterations: introduction. Bioscience 50 (9), 746–751.

Shakeri Zare, H., Karam, A., Saffari, A., Kiyani, T., 2020. Assessment of environmental flow needs of Harirud border riverbed after construction and dewatering of Salma dam in Afghanistan (by hydrological methods). J. Geogr. Environ. Hazards 9 (2), 207–224.

Smakhtin, V.U., 2001. Low flow hydrology: a review. J. Hydrol. 240 (3–4), 147–186.

Smakhtin, V., Revenga, C., Döll, P., 2004. A pilot global assessment of environmental water requirements and scarcity. Water Int. 29 (3), 307–317.

Smakhtin, V.U., Shilpakar, R.L., Hughes, D.A., 2006. Hydrology-based assessment of environmental flows: an example from Nepal. Hydrol. Sci. J. 51 (2), 207–222.

The Brisbane Declaration, 2007. September. The Brisbane Declaration: environmental flows are essential for freshwater ecosystem health and human well-being. In: 10th International River Symposium, pp. 3–6. Brisbane, Australia.

Chapter 35

Integrating ecohydrology and cultural heritage: Lessons from the Paltas–Catacocha demonstration site

Marco Albarracín[1, 2]

[1]UNESCO Scientific Advisory Committee (SAC) on Ecohydrology, Quito, Ecuador; [2]WasiLab - Center for Sustainability Sciences, Pontificia Universidad Católica del Ecuador (PUCE), Quito, Ecuador

1. Introduction

Water management has been a persistent challenge throughout human history, leading many ancestral cultures to develop innovative, place-based, and bioculturally rooted strategies that remain relevant today (Berkes, 2017). These traditional

Transdisciplinary Ecohydrology. https://doi.org/10.1016/B978-0-443-41596-8.00042-X

practices, deeply intertwined with local ecosystems and cultural worldviews, provide important perspectives for contemporary approaches to water governance—particularly when interpreted through the lens of nature-based solutions (Berkes, 2017). Ecohydrology provides a holistic framework grounded in the principle of dual regulation. This principle emphasizes reciprocal interactions between water and biota. This approach addresses immediate management while strengthening the long-term ecological resilience of watersheds (Zalewski, 2000, 2021; Zalewski et al., 1997).

Building upon this, the WBSR + CE ecohydrological framework—encompassing Water, Biodiversity, Ecosystem Services, and Resilience, along with Culture and Education—advocates for an integrated approach in which scientific and cultural perspectives interact to foster sustainable and inclusive water resource management (Zalewski, 2021). Within this context, cultural heritage emerges not merely as a legacy of the past but as a living body of knowledge capable of enriching ecohydrological design and implementation.

The integration underscores the relevance of ancestral ecological knowledge into modern water management strategies is particularly relevant in the Andean region, where indigenous communities have historically developed adaptive systems aligned with local hydrological and climatic conditions (Berkes, 2017; Elfithri et al., 2025). These perspectives highlight how the Inca worldview positioned water at the spiritual and functional center of society, linking it to both environmental stewardship and social order (Boelens, 2014, 2015). Techniques such as *amunas* in Peru (Ochoa-Tocachi et al., 2019) and *cochas* in Ecuador (Albarracín et al., 2021) exemplify these systems, continuing to support agricultural productivity and water security.

These practices are part of a broader set of water sowing and harvesting (WS&H) strategies, which promote water infiltration and aquifer recharge in upper watershed zones (Albarracín et al., 2021; Herrera-Franco et al., 2024; Martos-Rosillo et al., 2021; Ochoa-Tocachi et al., 2019). While similar to managed aquifer recharge (MAR) in function—defined as the intentional recharge of groundwater to enhance availability and ecosystem benefits (Sprenger et al., 2017)—WS&H distinguishes itself through its cultural and historical depth. The terminology itself underscores the intergenerational knowledge and experiential foundations of these methods.

A notable example is the *acequias de careo* of Spain's Sierra Nevada, in use since the 11th century, which channels glacial meltwater to irrigate one of Europe's driest regions. These community-managed canals have been preserved through local stewardship and recent conservation efforts, gaining recognition for their hydrological and ecological significance (Jódar et al., 2017, 2022; Martos-Rosillo et al., 2019). Today, they are included in UNESCO-IHP's Ecohydrology Demonstration Site network.

This chapter focuses on the Paltas–Catacocha ecohydrology demonstration site in southern Ecuador as a primary case study. Located within the Binational Biosphere Reserve Bosques de Paz—recognized by UNESCO's Man and the Biosphere (MAB) Programme—it explores how ancestral symbols, sacred landscapes, and WS&H practices are being revitalized through a community-based ecohydrological approach. While grounded in local traditions, this case reflects a broader model increasingly adopted across UNESCO's Ecohydrology Demonstration Sites—such as the *amunas* and *bofedales* in Peru (Ochoa-Tocachi et al., 2019; Monge-Salazar et al., 2022), the *albarradas* and *cochas* in Ecuador (Albarracín et al., 2021; Herrera-Franco et al., 2024), and the *acequias de careo* in Spain's Sierra Nevada (Jódar et al., 2022; Martos-Rosillo et al., 2021)—demonstrating the potential of ancestral water infrastructures to strengthen hydrological resilience and inclusive watershed governance (Arduino, 2021; Elfithri et al., 2025).

2. Ecohydrology as a transdisciplinary framework for watershed management

Ecohydrology has emerged as a science-based yet culturally sensitive approach to sustainable water resource management. It integrates hydrological, biological, and social dimensions within a coherent framework aimed at enhancing ecosystem resilience and fostering participatory governance (Zalewski, 2019; Arduino, 2021). As conceptualized by Professor Maciej Zalewski and further developed through the UNESCO Intergovernmental Hydrological Programme (IHP), ecohydrology is not only a technical paradigm but also a transdisciplinary strategy that aligns ecological processes with cultural values and local agency (Zalewski, 2019; Arduino, 2021; Elfithri et al., 2025).

At its core, ecohydrology rests upon the principle of dual regulation, which proposes the concurrent modulation of water and biotic dynamics to stabilize and strengthen ecosystem functionality (Zalewski, 2019). This duality is operationalized through the ecohydrological implementation cycle, composed of three interlinked principles:

- **Hydrological diagnosis (hydrological principle):** A quantitative understanding of water flows, infiltration patterns, and retention capacities across scales. This principle involves diagnosing problems at multiple scales—from the watershed to molecular level.

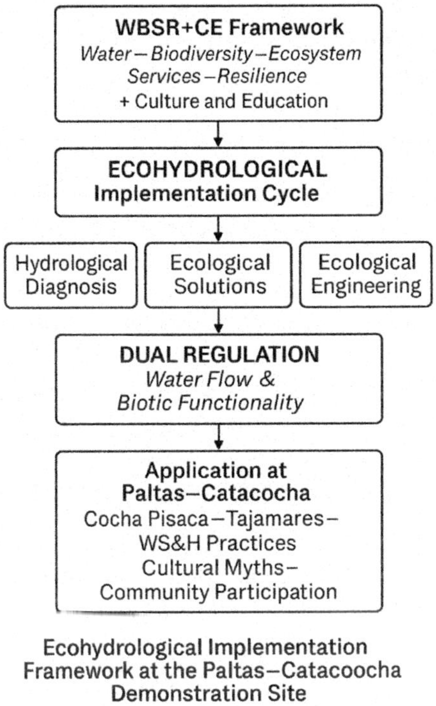

FIGURE 35.1 Ecohydrological implementation framework at the Paltas–Catacocha demonstration site. This diagram illustrates the integration of the WBSR + CE framework (water, biodiversity, ecosystem services, and resilience + culture and education), the ecohydrological implementation cycle (hydrological diagnosis, ecological solutions, and ecological engineering), and the principle of dual regulation (simultaneous regulation of water and biotic processes). These components converge in the applied strategies of the Paltas–Catacocha site, including water sowing and harvesting (WS&H), community participation, and the revitalization of cultural values.

- **Ecological solutions (ecological principle):** The use of biotic processes—such as vegetation cover, wetlands, or biota—to regulate hydrology and enhance ecosystem services.
- **Ecological engineering (ecohydrological principle):** The implementation of low-cost, nature-based interventions that are continuously evaluated and adapted, ideally through community involvement.

These principles are embedded within the broader WBSR + CE framework (Arduino, 2021; Zalewski, 2014) which expands ecohydrological planning to include the following:

- **Water, biodiversity, ecosystem services, and resilience (WBSR)** as environmental pillars, and
- **Culture and education (CE)** as social dimensions essential for sustainability and replicability.

This integrated structure—*dual regulation + ecohydrological principles + WBSR + CE*—constitutes a replicable model for the implementation of demonstration sites, adaptable to diverse ecological and cultural contexts (Elfithri et al., 2025) (see Fig. 35.1).

2.1 Culture and education: Foundational components for replication

Unlike conventional approaches to water management that treat cultural heritage as a passive context, ecohydrology recognizes culture as an active driver of system design and long-term resilience. Cultural practices, symbolic landscapes, and traditional knowledge systems often contain embedded ecohydrological logic—such as the strategic location of sacred water sources, collective water sowing rituals, or local rules governing irrigation (Boelens, 2014; Berkes, 2017).

The culture and education components of the WBSR + CE framework ensure that each intervention is context-sensitive, socially acceptable and capable of mobilizing local stewardship (Zalewski, 2021). Furthermore, by incorporating educational initiatives—from community workshops to school-based programs—ecohydrological strategies become transgenerational, fostering resilience not only in ecosystems but in knowledge systems as well (Berkes, 2017).

These dimensions are particularly relevant for the development of ecohydrology demonstration sites, where local narratives, community engagement, and adaptive learning form the foundation for both technical effectiveness and sociocultural resonance (Elfithri et al., 2025).

2.2 Toward a global replicable model

By combining scientific rigor with cultural responsiveness, ecohydrology offers a flexible yet robust model for watershed management. Its foundational components—dual regulation, ecohydrological principles, and the WBSR + CE framework—enable the design of interventions that are ecologically functional, socially embedded, and scalable across geographies (Arduino, 2021; Elfithri et al., 2025; Zalewski, 2014).

In the following section, we examine how this conceptual structure has been operationalized through UNESCO IHP's global network of ecohydrology demonstration sites. We then explore the case of Paltas–Catacocha, a pioneering example in Latin America where the ecohydrological approach has been harmonized with ancestral practices, community engagement, and cultural revitalization.

3. Ecohydrology demonstration sites and WS&H practices

The UNESCO Intergovernmental Hydrological Programme (IHP) has promoted the development of Ecohydrology Demonstration Sites as platforms for testing, adapting, and showcasing nature-based strategies for integrated water and ecosystem management. These sites are designed to reflect the transdisciplinary principles of ecohydrology by integrating scientific research, cultural context, and community engagement. Their role goes beyond technical implementation; they serve as "living laboratories" where theoretical frameworks are translated into tangible, measurable environmental and social outcomes (Arduino, 2021; Elfithri et al., 2025).

Each demonstration site embodies the core ecohydrological components of dual regulation, the implementation cycle (hydrological diagnosis, ecological solutions, and ecological engineering), and the WBSR + CE framework, all adapted to the specific ecological and cultural features of the watershed. Through this adaptation, the sites become both context-sensitive and replicable, offering strategic lessons for sustainable water governance across diverse regions (Zalewski, 2021).

A growing number of these sites—from urban wetland restoration in Poland to high-altitude irrigation systems in Spain and water retention projects in Ecuador—illustrate how ecohydrological approaches can be tailored to different hydrosocial realities. One of the most culturally embedded strategies employed across these landscapes is WS&H, a practice with deep roots in indigenous water cultures of the Andes, Mediterranean, and parts of Asia (Jódar et al., 2022; Ochoa-Tocachi et al., 2019).

WS&H practices are based on the principle of capturing and retaining rainwater in upper catchments to enhance infiltration, recharge aquifers, and maintain streamflow during dry periods. Techniques include infiltration ponds and canals, microdams, subsurface galleries, and the recovery of ancestral water bodies such as *cochas* in Ecuador and Peru, or *amunas* in the Andean highlands (Albarracín et al., 2021; Boelens, 2014; Ochoa-Tocachi et al., 2019). These strategies not only align with the ecological objectives of ecohydrology but also reinforce traditional environmental stewardship and intergenerational knowledge transmission.

Within the IHP ecohydrology framework, WS&H can be interpreted as a culturally embedded form of ecological engineering—decentralized, low-cost, and intimately linked to local worldviews and values. When reintroduced or enhanced through scientific guidance and community collaboration, these practices offer scalable and resilient solutions to modern water challenges in both rural and periurban settings (Arduino, 2021; Berkes, 2017).

As of 2024, several UNESCO-IHP ecohydrology demonstration sites have adopted WS&H approaches as core components of their watershed strategies. These include the following:

- *The Paltas–Catacocha*, Ecuador (Albarracín et al., 2018, 2021)
- *The Acequias de Careo* in the Sierra Nevada of Spain (Jódar et al., 2022)
- *The Amunas* of Carhuayumac, near Lima, Peru (Ochoa-Tocachi et al., 2019)
- *The Bofedales* of Cusco, in the southern Peruvian Andes (Monge-Salazar et al., 2022)
- *The Albarradas* of Manglaralto, along Ecuador's Pacific coast (Herrera-Franco et al., 2024)

Despite their distinct ecological and cultural settings, these sites demonstrate how WS&H techniques can be revitalized as biocultural innovations—strengthening hydrological resilience while empowering local governance and identity.

The following section explores how the Paltas–Catacocha demonstration site in southern Ecuador has implemented this integration by combining ecohydrological science with ancestral WS&H knowledge, community mobilization, and symbolic landscape recovery.

4. The ecohydrology demonstration site Paltas Catacocha, Ecuador

The Paltas–Catacocha demonstration site, located in the dry Andes of southern Ecuador, offers a compelling example of how ecohydrology can integrate scientific innovation with ancestral knowledge and symbolic landscapes (Albarracín et al., 2021; Elfithri et al., 2025). As one of the designated ecohydrology demonstration sites under UNESCO's Intergovernmental Hydrological Programme (IHP), it showcases how water sowing and harvesting (WS&H) practices—rooted in pre-Incan traditions—can be revitalized through the principles of dual regulation and participatory ecological engineering (Albarracín et al., 2021; Zalewski, 2021).

This section analyzes the site through five interrelated dimensions: cultural foundations, hydrological diagnosis, ecological solutions, social participation, and measurable outcomes. Together, these elements demonstrate how ecohydrology can function not only as a technical framework but also as a culturally embedded practice of watershed restoration.

4.1 Cultural and hydrological foundations

The cultural legacy of the Palta people pre-Incan inhabitants of the southern Andes—is preserved through both oral traditions and material symbols (Albarracín et al., 2021; Ramón, 2008, 2015, 2018). Unlike the Incas, who used *khipus* (knotted strings) to register agricultural and accounting data (Urton, 2003), the Paltas encoded hydrological and territorial knowledge through myths, sacred sites, and carved stones (Ramón, 2008, 2015, 2018). Although many of these cultural elements were suppressed during colonial and republican periods, vestiges of this ancestral knowledge have persisted in the collective memory of communities such as Catacocha.

One such element is the myth of Torito Cango, preserved in the oral tradition of the region (Albalá, 1995; Albarracín et al., 2018, 2021). According to this narrative, the goddess of Pisaca—an *Apu* or sacred mountain—fell in love with the hill of Cangonamá and bore a son named Torito Cango. When the boy cried, Pisaca, moved by emotion, produced rain that nourished the fields. When Torito Cango was taken away, drought settled over the land; but upon his return, the rains came back as a blessing from the gods.

This narrative played a critical role in mobilizing the community for the ecohydrological restoration project *Sembrando Agua para la Vida*. The phrase "El regreso del Torito Cango" ("The Return of Torito Cango") was adopted as a unifying cultural slogan, linking hydrological recovery with spiritual and social reawakening. This worldview reflects a deep-rooted Andean belief that water flows not merely from natural cycles, but from sacred relationships and affective reciprocity between humans, deities, and the landscape (Boelens, 2014).

The return of the Torito Cango thus symbolized the reestablishment of cosmic and ecological harmony, allowing the goddess to weep once again and fertilize the land. As noted by Ramón (2015) and Albalá (1995), this narrative served as a central cultural axis for the revival of ancestral wetlands and the reactivation of traditional water knowledge at the Paltas–Catacocha Ecohydrology Demonstration Site.

Another important symbol is the hill of Pisaca, which contains *Cocha Pisaca*, a seasonal wetland that functions not only as an aquifer recharge site but also as a *huaca*—a sacred space in Andean cosmovision where natural elements such as water are believed to embody the divine. This dual hydrosymbolic function reinforces the ethical obligation to protect such features as part of a culturally grounded watershed management strategy (Boelens, 2014).

A significant expression of this cultural memory is the "Snake Rock" (*Roca de la Serpiente*), located in Quilanga, Loja. According to Ramón (2015), this carved stone displays curvilinear motifs and basin-shaped depressions, interpreted as microwatersheds and *cochas*. The stylized serpent figure (*conza*) engraved on the rock evokes water-related symbolism also found in other pre-Hispanic cultures of the Americas (Ramón, 2015), such as *Amaru* among the Incas (Boelens, 2014) and *Kukulkán* among the Itza Maya. These serpentine forms—linked to water, fertility, and sacred geography—suggest that the Palta people encoded hydrological cosmology into a symbolic map.

By recognizing this legacy, the Paltas–Catacocha site affirms the ecohydrological principle that effective water management must be rooted in cultural context. This ancestral knowledge was transmitted orally and materially inscribed in the landscape. The Snake Rock, documented through photography and interpreted as a conceptual watershed model, highlights how cultural memory, cosmology, and hydrological reasoning were intertwined in the pre-Hispanic Andes (see Fig. 35.2). This carved stone, known as *Roca de la Serpiente* or Snake Rock, has been interpreted as a symbolic

FIGURE 35.2 Cultural–hydrological interpretation of *Roca de la Serpiente*, Quilanga, Ecuador. Photograph of *Roca de la Serpiente* (snake rock) in Quilanga, Ecuador (left), alongside its conceptual interpretation as a watershed model (right). The carved surface displays petroglyphs, including depressions interpreted as *cochas* (water-harvesting ponds), mountain peaks (*apus*), and the *conza*—a stylized serpent deity of the *Palta* culture. Together, these elements symbolize the integration of hydrological knowledge and sacred geography within the Andean landscape.

hydrological diagram of the watershed, featuring depressions that represent *cochas*, sacred peaks (*apus*), and a stylized serpent associated with the deity *Conza* of the Palta culture.

4.2 Hydrological diagnosis and problem identification

Historically, the territory of the Paltas was structured around a complex and efficient ancestral system of water regulation based on artificial wetlands (*cochas*) and watershed logic (Ramón, 2008, 2015, 2018). However, during the colonial period, many of these *cochas* were deliberately drained. One notable case is the largest lagoon in the region, which gave rise to the very name of the town: "Catacocha," from *Cata* (great) and *Cocha* (lagoon). Oral history and archeological evidence suggest that a large water body once existed where the present-day town square, municipal government, and church are now located (Ramón, 2015, pers. comm.). This hydrosocial transformation disrupted the ecological and spiritual fabric of the region and dismantled the water's central role in the landscape.

Another significant case is *Cocha Pisaca*, located on the slopes of the sacred mountain of the same name (*Apu Pisaca*). In the 20th century, a private landowner drained the lagoon to convert it into grazing pasture. Community testimonies recount how fish (*tilapia*) were introduced for personal consumption, and the wetland was later dried to expand cattle ranching (Albarracín et al., 2021). This process undermined the ancestral water retention system, severely reducing aquifer recharge, soil moisture, and baseflow in downstream springs.

By the year 2000, the hydrological crisis in *Catacocha* had reached a critical point. In some neighborhoods, the municipal water supply was limited to just 2 hours per day—or even as little as 15 minutes. This collapse coincided with a markedly unimodal rainfall regime, as confirmed by historical meteorological data (1990–2015) from the nearest weather station—*Catacocha* station (INAMHI, code M-515). Precipitation occurs mainly between December and May, with a peak in March and a minimum in August. Average monthly values during the period 1990–2015 ranged from 261 mm in March to only 2 mm in August. These seasonal extremes, combined with recurrent droughts and the absence of functional water retention structures, pushed the region into a state of severe water stress and made hydrological restoration an urgent necessity (Albarracín et al., 2021).

The degraded state of the *Cocha Pisaca*, as seen in Fig. 35.3, reflects this collapse. The wetland, once central to aquifer recharge and local cosmology, had become a dry pasture, stripped of both its ecological and spiritual function.

4.3 Ecological solution and dual regulation

In response to the hydrological and cultural collapse diagnosed in Catacocha, a comprehensive ecohydrological restoration strategy was launched in 2005 under the leadership of the municipal government. The project *Sembrando Agua*

FIGURE 35.3 Historical condition of the desiccated *Cocha Pisaca* before restoration (Catacocha, Ecuador). This photo shows the degraded state of *Cocha Pisaca* prior to the implementation of the ecohydrological restoration project "Sembrando Agua para la Vida." The area had been converted into pastureland, following the artificial drainage of the lagoon by a landowner. The image reflects the loss of water retention capacity and vegetation, highlighting the urgency of restoring ancestral hydrological structures. *Photo: Municipal Archive of Paltas.*

para la Vida ("Sowing Water for Life") was implemented with the support of the Fundación COMUNIDEC and the NGO Nature and Culture International (NCI). A key initial action was the purchase of approximately 400 ha of land—including the site of the desiccated *Cocha Pisaca*. This purchase, led by the mayor and supported by institutional and community partners, marked the beginning of a new phase of integrated watershed management (Albarracín et al., 2021). The Demonstration Site of Paltas–Catacocha is located within the Transboundary Biosphere Reserve *Bosques de Paz*, a binational conservation area recognized by UNESCO and shared by Ecuador and Peru, which reinforces its ecological and strategic significance in the region.

The intervention was guided by the ecohydrological principle of dual regulation, in which both hydrological and ecological functions were restored (Zalewski, 2000, 2021). One of the principal measures was the reconstruction of *Cocha Pisaca* (see Fig. 35.4), which had been desiccated decades earlier for cattle grazing. With the removal of cattle and the restoration of natural depressions, the cocha was reestablished as a functional water body with enhanced recharge capacity (Albarracín et al., 2018, 2021; Ramón, 2015).

In addition to the main lagoon, a network of smaller human-made ponds—known locally as *albarradas*—was constructed. Between 2005 and 2013, a total of 28 *albarradas* were built, collectively offering a storage capacity of 182,482 m^3, with an average of 6,517 m^3 per pond. The largest—locally called "Laguna Pisaca"—can hold approximately

FIGURE 35.4 Aerial view of Cocha Pisaca and Pisaca hill. Aerial photograph of *Cocha Pisaca*, the largest artificial pond (*albarrada*) restored under the ecohydrological initiative *Sembrando Agua para la Vida*, located at the base of Pisaca Hill. The lagoon plays a key role in seasonal water storage and aquifer recharge, while also holding ecological and cultural significance within the Paltas–Catacocha demonstration site. *Courtesy of César Aguirre Torres/Nature and Culture International (NCI). Permission granted for academic use.*

FIGURE 35.5 ***Tajamares*** **constructed as part of the ecohydrological intervention.** Composite image depicting three *tajamares* (small earthen or stone check dams) built within the Catacocha watershed. These structures were collaboratively designed and built with local communities as part of the *Sembrando Agua para la Vida* project, aiming to enhance water retention, reduce runoff velocity, and restore microhabitats. *Source: Municipal Archive of Paltas. Courtesy of the Municipal Archive of Paltas. Permission granted for academic use.*

78,422 m^3, while the smallest stores just 143 m^3 (Albarracín et al., 2021). Seventeen of these structures are located within the Pisaca Reserve, a conservation area owned by NCI, and represent 82% of the total capacity (149,794 m^3). The remaining ponds lie on private lands owned by local farmers, illustrating a hybrid model of communal and private stewardship. These artificial wetlands regulate runoff, promote aquifer recharge, and foster biodiversity while supporting agroecological transition (Albarracín et al., 2021).

Composite image depicting three *tajamares* (small earthen or stone check dams) built within the Catacocha watershed. These structures were collaboratively designed and built with local communities as part of the *Sembrando Agua para la Vida* project, aiming to enhance water retention, reduce runoff velocity, and restore microhabitat (Albarracín et al., 2021; Herrera-Franco et al., 2024; Martos-Rosillo et al., 2021). This initiative exemplifies biocultural comanagement, in which ecological engineering was grounded in traditional practices and cultural meaning (see Fig. 35.5).

Reforestation efforts were undertaken to reinforce the landscape's hydrological buffering capacity. Native tree species were prioritized to accelerate ecosystem recovery, stabilize soils, and restore evapotranspiration cycles. These actions enhanced infiltration and aquifer recharge and helped revitalize the spiritual dimension of the landscape, reconnecting communities with sacred water bodies and ceremonial sites (Albarracín et al., 2021).

From a dual regulation standpoint, the intervention successfully enhanced both the physical management of water (through storage, slowing, and infiltration) and the ecological integrity of the system (via biodiversity restoration and social cogovernance). Crucially, this approach went beyond technical solutions—it mobilized ancestral knowledge, symbolic narratives, and communal rituals to ensure legitimacy and resilience.

4.4 Cultural narratives and symbolic landscapes

In the Andes, hydrological systems are not merely hydraulic structures—they are deeply embedded within cultural narratives and sacred geographies (Boelens, 2014; Urton, 2013). Among the Palta people of southern Ecuador, water was historically governed through a hydrosocial and cosmological worldview, in which mountains, rivers, and wetlands were seen as sentient beings, forming a network of reciprocal relationships between nature and society (Ramón, 2015, 2018; Albarracín et al., 2021).

The myth of the water spirit Torito Cango exemplifies this perspective. As described earlier, this narrative links water abundance to emotional reciprocity between a sacred mountain (Cerro *Pisaca*) and her son (*Torito Cango*), whose presence or absence determined the arrival of rain (Albalá, 1995; Ramón, 2015). Such myths do not merely explain hydroclimatic dynamics—they shape collective memory, moral responsibility, and the legitimacy of water management practices.

Tangible material symbols further support these narratives. The Roca de la Serpiente (Serpernt or snake Rock) in Quilanga, Loja, serves as a carved geosymbolic representation of the watershed. Its surface features a stylized serpent (*conza*), representing the Palta water deity, alongside carved depressions interpreted as *cochas* or water-harvesting ponds. According to Ramón (2015), this stone functioned as a ceremonial marker and as a pedagogical tool for transmitting water management knowledge across generations.

Recent interpretations propose that the stone itself constitutes a three-dimensional symbolic representation of a Palta watershed. The rock's topology reflects the actual hydrological landscape, including subwatershed divisions and water pathways (Albarracín et al., 2018, 2021). This conceptual alignment between symbolic art and physical terrain reinforces the notion that Palta water governance was both scientific and spiritual—an early form of ecohydrological encoding rooted in Andean ontology.

The ecohydrological restoration of Cocha Pisaca and the wider landscape was therefore not merely a technical effort, but a profoundly cultural act of reconnection. Community members often referred to these interventions as "the return of sacred water bodies"—an expression of symbolic and emotional healing. As documented in participatory workshops and oral histories, the physical reappearance of water in formerly desiccated areas was interpreted as the revival of ancestral guardianship and a reaffirmation of divine reciprocity (Galo Ramón, per comm).

Such symbolic landscapes carry powerful implications for sustainability. They foster a sense of territorial rootedness and intergenerational continuity, ensuring that environmental stewardship is not enforced externally, but arises from within a shared ethical cosmological worldview (Berkes, 2017; Boelens, 2014; Ochoa-Tocachi et al., 2019; Ramón, 2008; Šprajc and Sánchez Nava, 2013). In Paltas–Catacocha, the convergence of hydrological restoration and cultural revival illustrates a form of ecohydrocultural governance that transcends technocratic models—one in which watersheds are managed not only as physical systems but as living and meaning-laden entities.

4.5 Social participation and ancestral governance

The success of the Paltas–Catacocha Ecohydrology Demonstration Site is inseparable from the deep-rooted participation of local communities and the reactivation of ancestral governance mechanisms. From the outset, the project "Sembrando Agua para la Vida" was conceived as a collaborative effort among the municipal government, civil society organizations, and the rural population of Paltas. Community members participated as active co-creators of the interventions rather than passive beneficiaries, contributing manual labor, ecological knowledge, and symbolic legitimacy at every stage (Albarracín et al., 2018, 2021).

Traditional forms of organization, such as *mingas* (communal labor systems), were revitalized to facilitate the construction of *albarradas* (water-harvesting ponds), *tajamares* (earthen dikes), and reforestation plots. These practices, rooted in Andean traditions of reciprocity and communal responsibility, became a cornerstone of implementation (Boelens, 2014). By building on these cultural frameworks, the project strengthened territorial ownership and reduced reliance on external actors.

Institutional coordination was also a key factor. The involvement of the local government, particularly the leadership of the mayor during the project's initial phase, ensured alignment between community priorities and public investment. The Fundación COMUNIDEC provided technical assistance and served as a bridge between ancestral practices and scientific methods. Meanwhile, the NGO Nature and Culture International (NCI) played a central role in securing land for conservation and establishing long-term protection strategies within the core area of the Transboundary Biosphere Reserve "Bosques de Paz" (Albarracín et al., 2018, 2021).

Ancestral governance principles, such as respect for the *Apus* (sacred mountains) and recognition of the *huacas* (sacred water sites), were reintroduced into local planning. Decision-making spaces incorporated ancestral knowledge, oral traditions, territorial memory, and the spiritual dimension of water, reinforcing the idea that watershed management involves technical ethical and relational dimensions (Berkes, 2017; Boelens, 2014, 2015).

This participatory and culturally grounded governance model ensured the project's sustainability. Rather than relying solely on external incentives or enforcement, stewardship of water infrastructure and reforested areas became part of a broader narrative of identity, belonging, and ecological responsibility. As a result, the Paltas–Catacocha site, located within the Bosques de Paz Transboundary Biosphere Reserve, exemplifies how social participation, when informed by ancestral governance, can serve as a catalyst for resilient ecohydrological transformations.

4.6 Measurable results and landscape resilience

The ecohydrological intervention carried out in the Paltas–Catacocha Demonstration Site has yielded significant improvements in landscape functionality, water regulation, and ecological integrity. These results are evident in field observations and local testimonies anc can also be quantified through geospatial analysis and hydrological indicators.

A key metric for evaluating the success of the intervention has been the change in vegetation cover, which serves as a proxy for soil moisture, ecosystem productivity, and landscape stability. Remote sensing analyses comparing pre-intervention and postrestoration periods demonstrate a clear increase in vegetative cover and ecosystem health across the watershed. This is particularly evident in areas surrounding Cocha Pisaca and within the Pisaca Reserve, where reforestation and wetland recovery have been most intensive.

Figure 35.6 presents a Soil-Adjusted Vegetation Index (SAVI) map illustrating the spatial distribution of vegetative vigor in the restored watershed. Higher index values (green tones) correspond to areas of dense vegetation and sustained soil moisture, while lower values (brown shades) reflect more degraded or recently intervened areas. The map confirms that restored wetlands and reforested slopes have significantly increased their ecohydrological function (Albarracín, 2022).

These biophysical changes are reinforced by hydrological improvements reported by the local population. Formerly dry springs now flow seasonally, and downstream irrigation channels exhibit prolonged water availability. The construction of 28 albarradas and over 200 tajamares has allowed for greater infiltration, reduced peak runoff, and enhanced aquifer recharge, contributing to a more balanced hydrological regime (Albarracín et al., 2021).

Beyond biophysical and ecological indicators, local testimonies provide compelling evidence of improved water availability. Residents of Catacocha report that the public water supply, which had previously delivered water for as little as 15 min to 2 h per day, now reaches many neighborhoods for extended periods of 6–8 h daily—even during the dry season. Although no formal study has yet quantified this improvement, the widespread community observations strongly suggest that aquifer recharge has been successfully restored through the ecohydrological interventions.

At the socioecological level, the project has strengthened community resilience by securing water for domestic, agricultural, and ecological use, while also reinforcing cultural heritage and territorial belonging. This convergence of biophysical and cultural restoration exemplifies the long-term vision of ecohydrology: developing adaptive landscapes capable of withstanding climatic variability and anthropogenic pressures through integrated, place-based solutions.

5. Concluding reflections and cultural horizons

The ecohydrological restoration of Cocha Pisaca inside the Paltas–Catacocha Demonstration Site exemplifies the convergence of ancestral knowledge and scientific methodology to rehabilitate degraded watersheds and revitalize biocultural systems. This experience reinforces the principle that watershed management is more than a technical endeavor; it is a relational practice grounded in memory, symbolism, and social organization.

A central insight from this study is the role of WS&H techniques as time-tested strategies to buffer climatic extremes and enhance landscape resilience. Practices such as *cochas*, *bofedales*, *albarradas*, *tajamares*, and *amunas*—developed by pre-Incan cultures across the Andes—demonstrate how indigenous infrastructure can function as nature-based solutions (NbS) for aquifer recharge and hydrological regulation (Albarracín et al., 2021; Herrera-Franco et al., 2024; Martos-Rosillo et al., 2021; Monge-Salazar et al., 2022; Ochoa-Tocachi et al., 2019; Ramón, 2018). Recent studies from entirely different ecological contexts—such as the Sierra Nevada range in Spain, have shown that these ancestral systems can delay runoff, increase dry-season flows, and provide ecosystem services critical for adaptation to climate change (Jódar et al., 2017, 2022; Martos-Rosillo et al., 2019, 2021).

These techniques have gained institutional recognition and are now integrated into UNESCO's Ecohydrology Demonstration Sites network, including examples such as the restored *cochas* and *albarradas* in Ecuador, the *amunas* and *bofedales* in the Andes of Peru, and the *acequias de careo* in Spain (Arduino, 2021; Elfithri et al., 2025). Their success lies in effective water management as well as in their capacity to preserve and reactivate cultural identity, traditional governance, and intergenerational knowledge systems.

Within this broader context, the symbolic and spiritual dimensions of water should remain central to the interpretation of these systems. Elements like the snake rock of Quilanga, engraved with hydrological motifs, and the myth of Torito Cango, linked to rainfall and fertility, highlight the Andean understanding of water as a sentient and reciprocal being. In

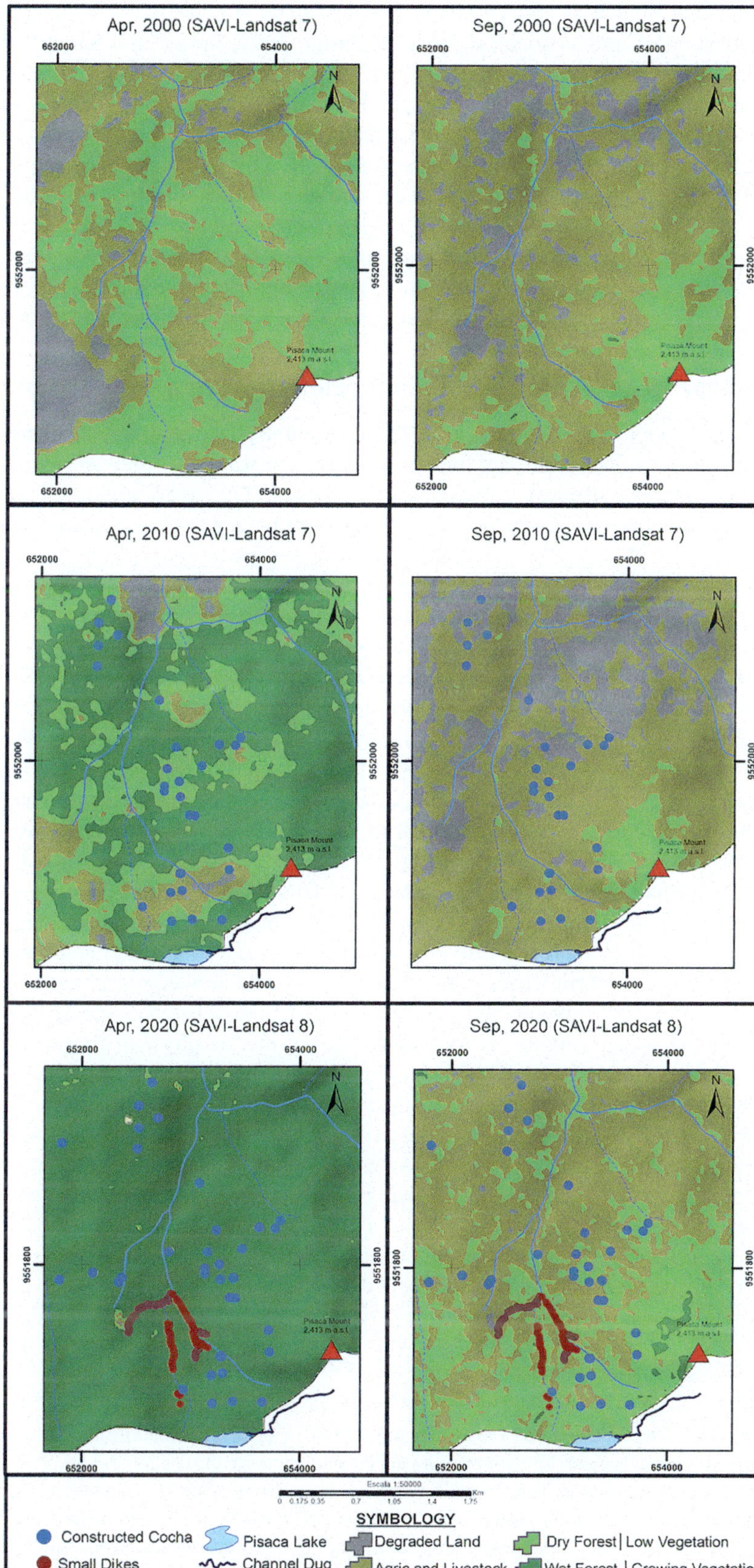

FIGURE 35.6 Soil-adjusted Vegetation Index (SAVI) map of the Palta–Catacocha ecohydrology demonstration site. SAVI showing vegetative cover before and after ecohydrological interventions in the Palta–Catacocha demonstration site, located within the Bosques de Paz Transboundary Biosphere Reserve. Areas around Cocha Pisaca and the Pisaca Reserve exhibit a marked increase in vegetation density and restored ecohydrological function. *Source: Adapted from Albarracín, 2022. Enfoque ecohidrolgico para restaurar cochas ancestrales del cerro Pisaca: Siembra y Cosecha de Agua en Catacocha, Loja, Ecuador. In Siembra y Cosecha de Agua en Iberoamérica, 109–126.*

other traditions, serpentine figures such as *Amaru* (Inca) and *Kukulkán* (of the Itza Maya) embody similar notions of cosmic movement and hydrological order (Boelens, 2014; Milbrath, 1999; Šprajc and Sánchez Nava, 2013). While the *Conza* associated with the Snake Rock has not yet been studied within this symbolic-hydrological framework, future research could examine its relevance as a regional archetype for water governance.

The integration of ecological infrastructure, sacred landscapes, and community-led restoration offers a transferable and adaptable model for other biocultural territories facing water insecurity. This calls for transdisciplinary methodologies that combine ecohydrological analysis, ethnographic knowledge, and codesigned solutions to sustain both ecosystems and cultural worldviews.

Ultimately, the Paltas–Catacocha case confirms that the path toward water sustainability is not a return to the past, but a living dialogue between ancestral memory and innovation. By treating rivers, wetlands, and mountains not merely as resources but as relational beings, we unlock new possibilities for resilient, inclusive, and equitable futures.

5.1 The serpent archetype in ancestral hydrocosmologies: Toward a transdisciplinary ecohydrology

Recent contributions such as *River Culture* (Wantzen, 2023) have emphasized that water governance must transcend its technical and ecological dimensions. Rivers are not merely hydrological features—they are living biocultural landscapes, shaped by the rhythms of human civilizations, rituals, and identities. This view aligns closely with Andean, Mesoamerican, and other indigenous cosmologies, which have long portrayed water as a sacred being that flows between worlds.

In these ancestral frameworks, water is not an object, but a relational and sentient entity, part of a wider hydrocosmic system that connects the celestial, terrestrial, and subterranean realms. The Andean concept of the *Chakana*, the tripartite structure of *Hanan Pacha—Kay Pacha—Uku Pacha*, and the serpent *Amaru* as a celestial and subterranean water bearer illustrate a sophisticated understanding of water cycles encoded in cosmology (Boelens, 2014; Urton, 2013). Similarly, in Mesoamerica, the feathered serpent Kukulkán descends during the spring equinox in Chichén Itzá, marking the beginning of the rainy season and embodying the link between celestial rhythms and agricultural cycles (Šprajc and Sánchez Nava, 2013).

These serpents—*Amaru, Kukulkán*, and the *Conza* of Quilanga—function as hydrocosmic archetypes. They represent the flow of water from the stars to the soil, and from myth to governance. In fact, modern science has begun to explore similar questions: hypotheses about water arriving on Earth via icy comets echo ancestral beliefs about the descent of water from the sky through divine serpents—sometimes even envisioned as the tail of a comet.

This convergence invites the development of a truly transdisciplinary ecohydrology, one that embraces multiple ontologies of water. The WBSR + CE approach provides a replicable model for integrated water resource management while also offering a philosophical platform where scientific and ancestral knowledge can meet. In this vision, water is no longer merely regulated—it is revered, remembered, and restored, as both substance and story.

Ultimately, the figure of the serpent invites us to read rivers not just as physical flows, but as living narratives of meaning and connection. By integrating symbols, science, and spirituality, ecohydrology can evolve into a science of relational and inclusive water, capable of guiding more sustainable and inclusive futures.

This reflection does not seek to close the discussion, but to open new paths for research and dialog. Understanding how ancestral cultures conceived, inhabited, and represented water—not just as a resource, but as a living, sacred, and sovereign manifestation—can offer crucial insights for addressing today's water security challenges. Perhaps, by deciphering the hidden meanings behind ancient infrastructures such as the Peruvian *amunas*, the *cochas* of the Ecuadorian Andes, or the Great Pyramid of Chichén Itzá, we may recover other ways of knowing and governing water: ways that reconnect humanity with the rhythms of the planet and restore water's dignity as *crystalline mother, divine serpent, and source of life.*

Acknowledgments

I am deeply grateful to Professors Maciej Zalewski and Luis Chícharo, as well as to Drs. Rahmah Elfithri, Giuseppe Arduino, Miguel Doria, Sergio Martos-Rosillo, and Galo Ramón Valarezo, for their encouragement and guidance throughout the development of this work. I also wish to thank my colleagues Jorge González, Eduardo Toral, Saskia Flores, Fabián Rodas, Carlos Iñiguez, Paul Carrión, Gricelda Herrera and Santiago Calero for their valuable support and collaboration. Special thanks are extended to the coordinators of the UNESCO Ecohydrology Demonstration Sites in Latin America and the Caribbean, whose efforts and shared experiences have inspired and enriched the reflections presented here. I also acknowledge the members of the LAC Working Group and the Scientific Advisory Committee (SAC) on Ecohydrology of UNESCO, whose technical insights and institutional commitment continue to shape the advancement of the ecohydrological approach in the region. I am especially thankful to the inhabitants of Paltas and Catacocha, whose knowledge, trust, and active participation have been instrumental to the success of ecohydrological actions on the ground. Their commitment to restoring ancestral practices and caring for the territory serves as a living example of resilience and cultural continuity. Finally, I am likewise grateful to the Ecuadorian universities: Pontificia Universidad Católica del Ecuador (PUCE), *Universidad Politécnica Salesiana (UPS), Escuela Superior Politécnica del Litoral (ESPOL)*, and *Universidad Técnica Particular de Loja (UTPL)* for their continued support in the implementation of ecohydrology initiatives in Ecuador

AI disclosure

This chapter was developed with the assistance of generative artificial intelligence using OpenAI's ChatGPT. The tool was used to support language refinement, structural editing, and the drafting of specific sections such as workflow articulation, outcomes, and recommendations. All content was reviewed and curated by the author to ensure academic accuracy and contextual appropriateness. After using this tool/service, the author(s) reviewed and edited the content as needed and take(s) full responsibility for the content of the publication.

References

Albala, L., 1995. Paltas: Leyendas y Tradiciones. Casa de la Cultura Ecuatoriana.

Albarracín, M., 2022. Enfoque ecohidrológico para restaurar cochas ancestrales del Cerro Pisaca: Siembra y Cosecha de Agua en Catacocha, Loja, Ecuador. Siembra y Cosecha de Agua en Iberoamérica. Agencia Española de Cooperación Internacional para el Desarrollo, pp. 109–126.

Albarracín, M., Gaona, J., Chicharo, L., Zalewski, M., 2018. Ecohydrology and its implementation in Ecuador. Original in Spanish. EDILOJA, Loja, Ecuador.

Albarracín, M., Ramón, G., González, J., Iñiguez-Armijos, C., Zakaluk, T., Martos-Rosillo, S., 2021. The ecohydrological approach in water sowing and harvesting systems: the case of the Paltas Catacocha ecohydrology demonstration site, Ecuador. Ecohydrol. Hydrobiol. 21 (3), 454–466. https://doi.org/10.1016/j.ecohyd.2021.07.007.

Arduino, G., 2021. Ecohydrology: engineering harmony for a sustainable world, theme 5 of the intergovernmental hydrological Programme (IHP) phase VIII (2014-2021). Ecohydrol. Hydrobiol. 21 (3), 380–386. https://doi.org/10.1016/j.ecohyd.2021.08.009.

Berkes, F., 2017. Sacred Ecology. Routledge.

Boelens, R., 2015. Water, Power and Identity: The Cultural Politics of Water in the Andes. Taylor and Francis Inc., Netherlands, pp. 1–365. Available from: http://www.taylorandfrancis.com/books/details/9780415719186.

Boelens, R., 2014. Cultural politics and the hydrosocial cycle: water, power and identity in the Andean highlands. Geoforum 57, 234–247. https://doi.org/10.1016/j.geoforum.2013.02.008.

Elfithri, R., Zalewski, M., Arduino, G., Chicharo, L., 2025. Transdisciplinary ecohydrology for water management solutions and sustainability. Nat. Water 3 (4), 360–363. https://doi.org/10.1038/s44221-025-00395-x.

Herrera-Franco, G., Morante-Carballo, F., Bravo-Montero, L., Valencia-Robles, J., Aguilar-Aguilar, M., Martos-Rosillo, S., Carrión-Mero, P., 2024. Water sowing and harvesting (WS&H) for sustainable management in Ecuador: a review. Heritage 7 (7), 3696–3718. https://doi.org/10.3390/heritage7070175.

Jódar, J., Cabrera, J.A., Martos-Rosillo, S., Ruiz-Constán, A., González-Ramón, A., Lambán, L.J., Herrera, C., Custodio, E., 2017. Groundwater discharge in high-mountain watersheds: a valuable resource for downstream semi-arid zones. Sci. Total Environ. 593–594, 760–772. https://doi.org/10.1016/j.scitotenv.2017.03.190.

Jódar, J., Martos-Rosillo, S., Custodio, E., Mateos, L., Cabello, J., Casas, J., Salinas-Bonillo, M.J., Martín-Civantos, J.M., González-Ramón, A., Zakaluk, T., Herrera-Lameli, C., Urrutia, J., Lambán, L.J., 2022. The recharge channels of the Sierra Nevada range (Spain) and the Peruvian Andes as ancient nature-based solutions for the ecological transition. Water (Switzerland) 14 (19). https://doi.org/10.3390/w14193130.

Martos-Rosillo, S., Durán, A., Castro, M., Vélez, J.J., Herrera, G., Martín-Civantos, J.M., Mateos, L., Durán, J.J., Jódar, J., Gutiérrez, C., Hermoza, R. M., Peña, F., 2021. Ancestral Techniques of Water Sowing and Harvesting in Ibero-America: Examples of Hydrogeoethical Systems Advances in Science, Technology and Innovation. Spain Springer Nature, Spain, pp. 489–492. https://doi.org/10.1007/978-3-030-59320-9_104. Available from: https://www.springer.com/series/15883.

Martos-Rosillo, S., Ruiz-Constán, A., González-Ramón, A., Mediavilla, R., Martín-Civantos, J.M., Martínez-Moreno, F.J., Jódar, J., Marín-Lechado, C., Medialdea, A., Galindo-Zaldívar, J., Pedrera, A., Durán, J.J., 2019. The oldest managed aquifer recharge system in Europe: new insights from the Espino recharge channel (Sierra Nevada, southern Spain). J. Hydrol. 578, 124047. https://doi.org/10.1016/j.jhydrol.2019.124047.

Milbrath, S., 1999. In: Gods of the Maya: Astronomy in Art, Folklore, and Calendars. University of Texas Press, Austin.

Monge-Salazar, M.J., Tovar, C., Cuadros-Adriazola, J., Baiker, J.R., Montesinos-Tubée, D.B., Bonnesoeur, V., Antiporta, J., Román-Dañobeytia, F., Fuentealba, B., Ochoa-Tocachi, B.F., Buytaert, W., 2022. Ecohydrology and ecosystem services of a natural and an artificial bofedal wetland in the central Andes. Sci. Total Environ. 838. https://doi.org/10.1016/j.scitotenv.2022.155968.

Ochoa-Tocachi, B.F., Bardales, J.D., Antiporta, J., Pérez, K., Acosta, L., Mao, F., Zulkafli, Z., Gil-Ríos, J., Angulo, O., Grainger, S., Gammie, G., De Bièvre, B., Buytaert, W., 2019. Potential contributions of pre-Inca infiltration infrastructure to Andean water security. Nat. Sustain. 2 (7), 584–593. https://doi.org/10.1038/s41893-019-0307-1.

Ramón, G., 2008. La Historia Aborigen y Colonial, vol. 1. Casa de la Cultura Ecuadoriana, Loja, Ecuador.

Ramón, G., 2015. Quilanga: Historia y perspectiva. Casa de la Cultura Ecuatoriana Benjamín Carrión. Núcleo de Loja.

Ramón, G., 2018. Recuperación de saberes ancestrales de los Paltas para el manejo del agua en Catacocha. In Ecohidrología y su Implementación en el Ecuador. EDIJOJA, pp. 134–135.

Šprajc, I., Sánchez Nava, P.F., 2013. Astronomy in the architecture of Chichen Itza: a reevaluation. Estud. Cult. Maya 41 (1), 31–60. https://doi.org/10.1016/s0185-2574(13)71376-5.

Sprenger, C., Hartog, N., Hernández, M., Vilanova, E., Grützmacher, G., Scheibler, F., Hannappel, S., 2017. Inventaire des sites de gestion des aquifères par recharge en Europe: développement historique, situation actuelle et perspectives. Hydrogeol. J. 25 (6), 1909–1922. https://doi.org/10.1007/s10040-017-1554-8.

Urton, G., 2003. In: Signs of the Inka Khipu: Binary Coding in the Anden Knotted-String Records. University of Texas Press, Austin.

Urton, G., 2013. At the Crossroads of the Earth and the Sky: An Andean Cosmology. University of Texas Press.

Wantzen, K.M., 2023. In: River Culture: Life as a Dance to the Rhythm of the Waters. Springer, Cham.

Zalewski, M., 2021. Ecosystem biotechnologies for the enhancement of ecohydrological potential of the catchments – water, biodiversity, ecosystem services, resilience, culture and education. IOP Conf. Ser. Earth Environ. Sci. 1, 012031. https://doi.org/10.1088/1755-1315/789/1/012031.

Zalewski, M., Janauer, G.A., Jolankai, G., 1997. Ecohydrology. A New Paradigm for the Sustainable Use of Aquatic Resources.

Zalewski, M., 2000. Ecohydrology—the scientific background to use ecosystem properties as management tools toward sustainability of water resources. Ecol. Eng. 16 (1), 1–8. https://doi.org/10.1016/s0925-8574(00)00071-9.

Zalewski, M., 2014. Ecohydrology and hydrologic engineering: Regulation of hydrology–biota interactions for sustainability. J. Hydrol. Eng. 19 (5), 1080–1088. https://doi.org/10.1061/(ASCE)HE.1943-5584.0000999.

Zalewski, M., 2019. Ecohydrology as a framework for enhancing the catchment sustainability potential. In: Ecohydrology and its Implementation in Ecuador. EDILOJA, LOJA, pp. 49–57 (Chapter 5).

Chapter 36

Enhancing sustainable water resources management and ecosystem services at the Saguling reservoir, West Java (ecohydrology demonstration site)

Ignasius D.A. Sutapa[1,2], Apip[1,2], M. Fakhrudin[1,2], Reynard P.S.[3] and Maria Yustiningsih[4]

[1]*Asia Pacific Centre for Ecohydrology (APCE) – UNESCO C2C, Bogor Regency – West Java, Indonesia;* [2]*Research Centre for Limnology and Water Resources, National Research and Innovation Agency, Bogor Regency – West Java, Indonesia;* [3]*Industrial Engineering Study Program, Faculty of Engineering, University of Indonesia, Depok City – West Java, Indonesia;* [4]*Biology Education, Faculty of Teacher Training and Education (FKIP), University of Timor, Kefamenanu – TTU Regency, East Nusa Tenggara, Indonesia*

PHOTO CREDIT: I. SUTAPA

Transdisciplinary Ecohydrology. https://doi.org/10.1016/B978-0-443-41596-8.00036-4

1. Introduction

Indonesia, as an archipelagic nation with an extensive network of rivers, lakes, and wetlands, relies heavily on its water resources for domestic, agricultural, industrial, and ecological needs. Among these, the Citarum River basin holds particular significance (Kurniawan et al., 2024). It is the longest river basin in West Java, with an area of approximately 6822 km^2, supporting around 25 million residents, supplying about 80% of raw water for Jakarta, irrigating large agricultural areas, and powering three cascade reservoirs (Saguling, Cirata, Jatiluhur) with a combined capacity of 2600 MW (Abdillah et al., 2024). However, the basin faces severe environmental challenges. Rapid urbanization, industrial activity, and inadequate waste management infrastructure have resulted in severe pollution, particularly in the upper basin (Hadfield et al., 2024). Flooding and erosion risks are heightened by land-use transformation and climate pressures, while restoration efforts remain constrained by fragmented governance across 13 municipalities (Hadfield et al., 2024; Novalia et al., 2024). The cumulative impact of these pressures has resulted in deteriorated water quality, increased sedimentation, loss of aquatic biodiversity, and diminished ecosystem services. Recognizing these challenges, the Indonesian government implemented integrated programs such as Citarum Harum, spanning from 2018 to 2024, aiming to revitalize the watershed through coordinated multi-stakeholder action (Abdillah et al., 2024). Nevertheless, implementing such comprehensive strategies remains challenging due to limited community engagement, inadequate data, and the need for practical, scalable solutions tailored to local conditions (McKenna and Hanrahan, 2024).

Ecohydrology emerges as a promising paradigm that integrates ecological principles with hydrological processes to achieve sustainable water and ecosystem management (Zalewski and Robarts, 2003; Zalewski et al., 2009). Defined broadly, ecohydrology examines the interactions and feedback mechanisms between terrestrial, aquatic, and human systems, highlighting the importance of ecosystem integrity in regulating hydrological functions such as water quality, sediment transport, flood control, and habitat provision (Firoozi Akbar and Firoozi Asghar, 2024; Sutapa et al., 2023). In the context of the Citarum basin, ecohydrology offers a pathway to address complex environmental issues through nature-based solutions (NBS). For instance, restoring riparian vegetation can stabilize riverbanks, reduce sediment runoff, and improve water filtration—principles supported by models of ecological and hydrological integration (Firoozi Akbar and Firoozi Asghar, 2024). Similarly, implementing green infrastructure, including wetlands, vegetated buffer zones, and phytoremediation ponds, can mitigate pollution loads, decrease erosion rates, and enhance groundwater recharge (Granata and Nunno, 2025). These measures not only contribute to environmental betterment but also foster social resilience and community participation, essential components for long-term sustainability.

Numerous international case studies reinforce the potential of ecohydrological approaches. Sarandon et al. (2009), demonstrated how adjustments to land use policy at Lacar Lake in Argentina improved ecosystem health and water management. Pawitan and Haryani (2011), highlighted Indonesia's need for integrating ecohydrological principles into water sustainability strategies that consider local sociocultural contexts. These examples underscore that ecohydrology can serve as an effective framework for balancing ecological conservation with human development goals, particularly in tropical developing countries such as Indonesia. Furthermore, the adoption of ecohydrological practices aligns with global commitments, such as the Sustainable Development Goals (SDGs), specifically Goals 6 (Clean Water and Sanitation), 13 (Climate Action), and 15 (Life on Land) (Cao et al., 2025). By adopting nature-based solutions within a comprehensive management plan, stakeholders can improve resilience against climate variability, enhance biodiversity, and secure water for present and future generations (Marijuan et al., 2024; Sutapa and Fakhrudin, 2021).

The objective of this study is to establish and operationalize an Ecohydrological Demonstration Site (EDS) at the Saguling Reservoir, situated within the Citarum River basin, with the overarching goal of promoting integrated, sustainable water resource management practices rooted in ecohydrological principles (Budiman et al., 2023; Sutapa et al., 2021). This demonstration area serves as a living laboratory, showcasing how ecological restoration, through riparian planting, phytoremediation, and green infrastructure can mitigate water pollution, control erosion, and enhance overall ecosystem health.

2. Study area, context, and approach

2.1 Geographical and hydrological overview

The Saguling Reservoir, situated in West Java, Indonesia, is a vital component of the Citarum River Basin, which spans approximately 11,000 square kilometers. The reservoir is positioned at coordinates 107°26′E to 107°95′E longitude and 6°73′S to 7°25′S latitude, with an elevation ranging from 600 to 3000 m above sea level (Wardhani et al., 2018) Fig. 36.1. It serves multiple functions, including hydroelectric power generation, flood control, irrigation, and water supply for both local communities and industrial activities. Hydrologically, the Citarum River is characterized by high seasonal variability,

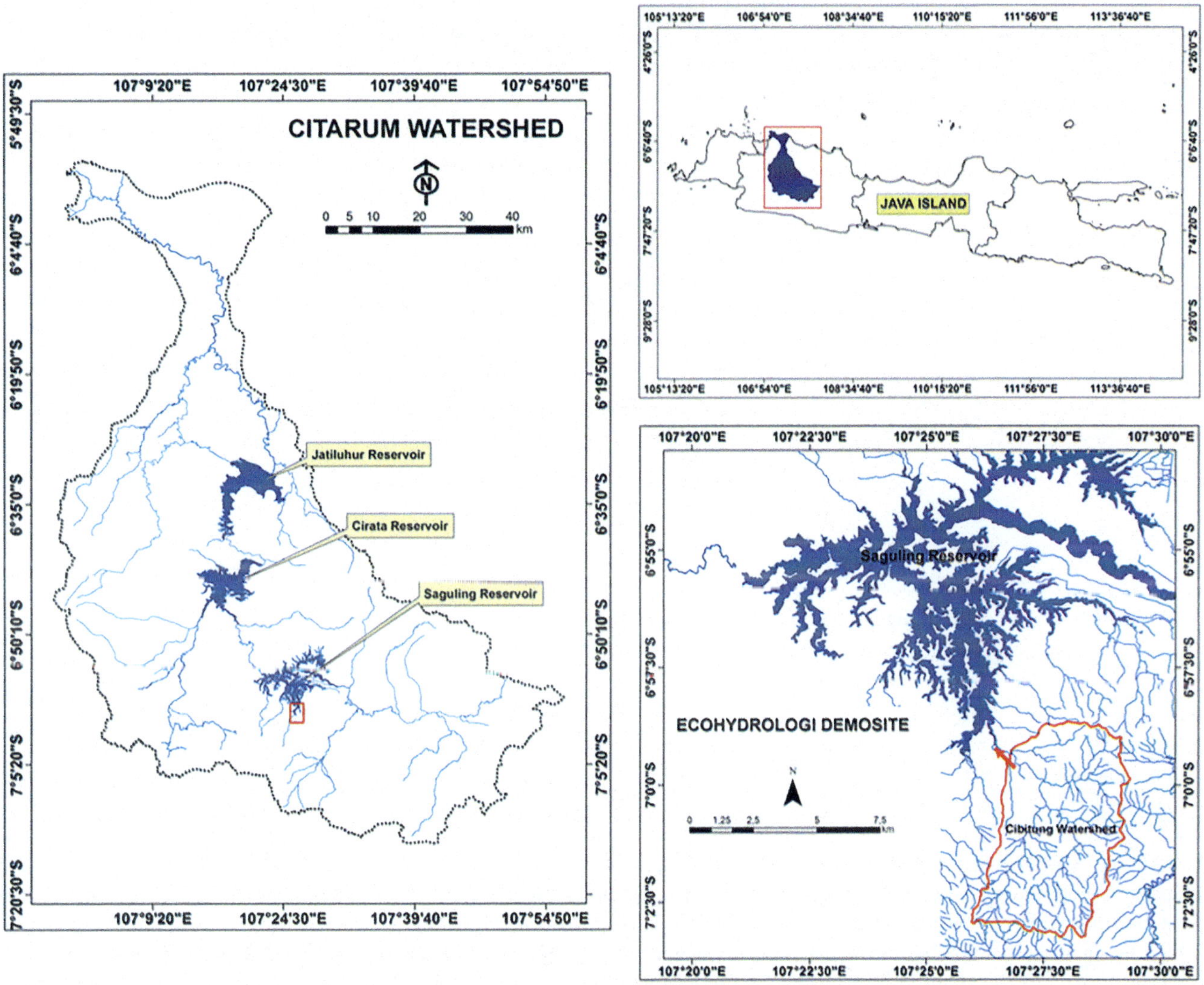

FIGURE 36.1 **Loction** Map of Saguling reservoir and Citarum river catchment. *Based on Sutapa, I.D.A., Fakhrudin, M., 2021. Ecohydrology: a management tool for aquatic ecosystem and SDGs implementation.* IOP Conf. Ser. Earth Environ. Sci. *789 (1), 012032. https://doi.org/10.1088/1755-1315/789/1/012032.*

with significant rainfall during the monsoon period (November to March) and diminished flows during the dry season (April to October) (Dimyati et al., 2024). The basin's hydrology is heavily influenced by rapid land-use changes, urbanization, and climate variability, posing significant challenges for sustainable water resource management. The reservoir's catchment area is characterized by diverse topography, ranging from mountainous uplands to floodplains, which influences local hydrodynamics and sediment transport processes. The basin's annual rainfall averages around 2500 mm, supporting abundant water flows but also contributing to frequent flooding and erosion issues (Rixson et al., 2024). Figure 36.2A represents the reservoir volume trend over 15 years at Saguling, Cirata and Jatiluhur dams, while Fig. 36.3 b shows the Hydrograph Unit Synthesis (SUH) of Saguling reservoir based on the report of (Empung et al., 2023).

2.2 Socioeconomic, environmental conditions and challenges

The socioeconomic fabric of the Saguling catchment is primarily based on agriculture, industry, and local livelihoods that are closely linked to water resources. Local communities rely heavily on irrigated rice paddies, small-scale fisheries, and water-dependent industries, such as textile manufacturing and power plants (Pueppke et al., 2020). Recent socioeconomic assessments indicate that the community's average monthly income ranges from IDR 1,500,000 to IDR 3,500,000, with a significant portion of this income dedicated to water expenses, including irrigation, domestic use, and industrial processes (Table 36.1).

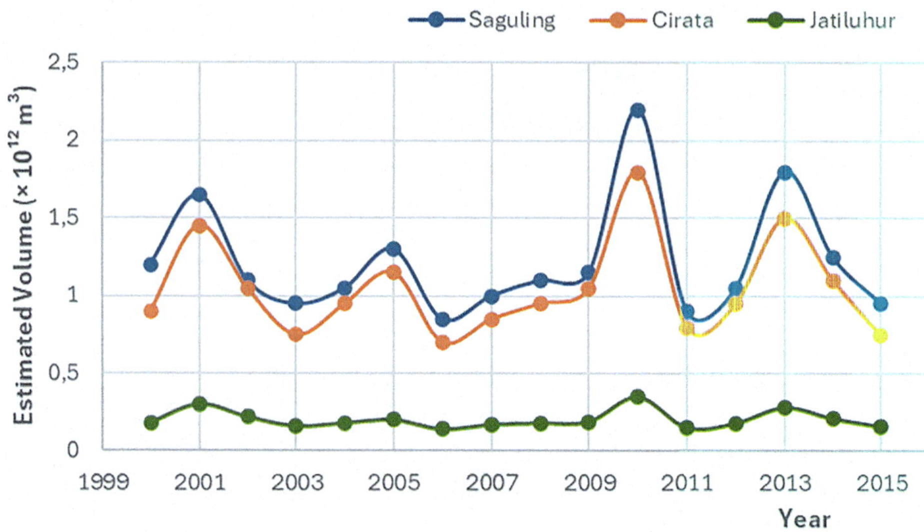

FIGURE 36.2 Fifteen-Year Trends in Reservoir Volume at Saguling, Cirata, and Jatiluhur Dams (*This graph was developed and constructed using data processed from the sources reported in* Empung et al. (2023)): *The Saguling Reservoir consistently shows the highest storage volume throughout the 2000–2015 period; Major peaks are visible around 2010 and 2013, indicating periods of significantly increased water storage; Cirata Reservoir follows a similar fluctuation pattern to Saguling, suggesting hydrological linkage within the same river system; Jatiluhur Reservoir remains comparatively stable with lower overall storage volumes.*

TABLE 36.1 Land cover changes in Saguling catchment (2010–20).

Land cover type	2010 (%)	2020 (%)
Forest	65	58
Agriculture	25	30
Urban/Development	10	12

Community participation and awareness regarding sustainable water use vary, often influenced by education levels, sociocultural traditions, and understanding of ecohydrological principles (Santos et al., 2023). Despite its ecological significance, the area faces multiple environmental pressures, including deforestation, sedimentation, water pollution, and climate variability, which threaten the sustainability of local ecosystems and the livelihoods of its inhabitants (Saoum et al., 2024).

Land cover in the Saguling catchment is predominantly characterized by forest (approximately 60%), agricultural lands (paddy fields, seasonal crops, and plantations), and built-up areas, including industrial zones and settlements (Nahib et al., 2021). Over the past decade, rapid land conversion driven by urbanization and agricultural expansion has significantly altered natural land cover. The forested areas, primarily located in upland regions, play a critical role in stabilizing slopes, controlling erosion, and maintaining hydrological cycles. Conversely, extensive deforestation has led to increased surface runoff, sediment yield, and decreased water quality downstream. Satellite imagery analysis reveals a decline in forest cover from 65% to 58% over a 10-year period, accompanied by an increase in cultivated and urban areas, as shown in Table 36.2. These land cover dynamics have intensified surface runoff and sedimentation issues, which in turn impact

TABLE 36.2 Socioeconomic indicators of local communities in Saguling catchment.

Indicator	Value
Average monthly income	IDR 2,500,000
Water expense (% of income)	10%–15%
Education level	Primary to secondary
Main livelihoods	Agriculture, industry
Community water awareness	Moderate

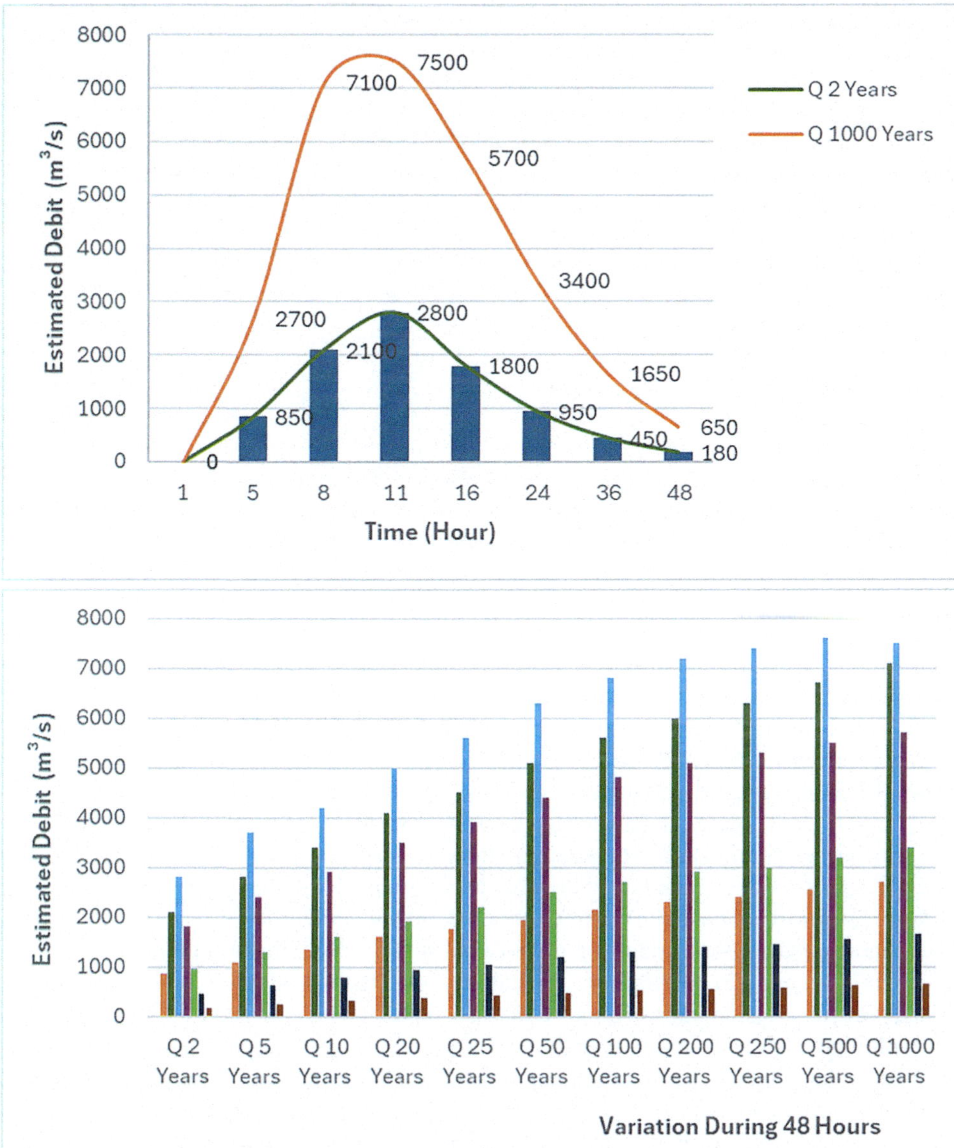

FIGURE 36.3 Synthesis of Unit Hydrographs (SUH) for Saguling Reservoir (*These graphs were developed and constructed using data processed from the sources reported in Empung et al. (2023)): All hydrograph series exhibit a similar hydrological response pattern with a rapid rising limb, a distinct peak flood at approximately Hour 11, and a long recession period; Increasing return periods (Q2 → Q1000) correspond to substantially higher peak discharges and streamflow magnitudes; The hydrograph shapes indicate that the Saguling watershed responds rapidly to rainfall events, followed by gradual runoff attenuation over time.*

reservoir capacity and ecosystem health. The need for integrated land management and ecohydrological intervention is critical to mitigate these pressures (Shaik and Fan, 2025).

The Saguling–Citarum system faces numerous ecological and hydrological challenges that compromise its sustainability, including (Ali et al., 2025; Rixson et al., 2024): (1) Sedimentation and Erosion: The basin's steep terrain and deforestation contribute to high rates of soil erosion, leading to sediment deposition within the reservoir. Continuous sedimentation reduces storage capacity and affects hydroelectric operations; (2) Water Pollution: Industrial discharges, domestic wastewater, and agricultural runoff introduce nutrients, heavy metals, and organic pollutants, impairing water quality and threatening aquatic biodiversity. Elevated levels of turbidity and chemical contaminants have been reported, affecting downstream ecosystems; (3) Flooding and Droughts: Seasonal rainfall disparities cause frequent floods during monsoon periods and water shortages during dry months. Urban expansion and land use change exacerbate these extremes, challenging flood control and drought mitigation efforts; (4) Climate Variability: Climate change has intensified rainfall variability and increased the frequency of extreme weather events, impacting hydrological regimes and ecological resilience; (5) Biodiversity Loss: Habitat fragmentation, pollution, and land use change threaten the rich biodiversity of

TABLE 36.3 Key features of the demonstration area.

Feature	Specifications	Purpose
Reforestation zones	Native species planting on denuded slopes	Stabilize soils, enhance transpiration, improve biodiversity
Buffer strips	Vegetative buffers along waterways	Reduce runoff sedimentation, filter pollutants
Constructed wetlands	Vegetated shallow ponds	Treat wastewater, support habitat, retain water
Infiltration basins	Permeable basins at strategic points	Promote groundwater recharge, reduce surface runoff
Riparian zones	Natural vegetation along streams	Erosion control, habitat connectivity
Water harvesting structures	Rainwater harvesting ponds	Supplement water supply, reduce runoff velocity

the basin, including endemic fish species and riparian flora, reducing ecological functions; (6) Limited Community Engagement: A lack of community awareness and participation hampers effective ecosystem management and sustainable water use practices, necessitating participatory approaches integrated into intervention strategies.

2.3 Approach methodology

This section delineates the comprehensive approach employed to develop, implement, and evaluate the EDS at the Saguling Reservoir within the Citarum River catchment, West Java, Indonesia. The methodology integrates site selection principles, community participation, technical interventions, monitoring systems, data analysis, and modeling tools, all aimed at enhancing sustainable water resource management and ecosystem services.

The strategic development of the EDS requires meticulous site selection based on ecological vulnerability, hydrological significance, socioeconomic factors, and logistical feasibility (UNESCO EDS, 2024). The criteria include: (1) Hydrological importance: Areas contributing significantly to sedimentation, erosion, or flood risks in the Saguling Reservoir; (2) Ecological sensitivity: Regions with degraded or vulnerable ecosystems, particularly in riparian zones and wetlands, (3) Community accessibility: Sites with active local communities engaged in agriculture or fisheries, facilitating participatory actions; (4) Land use characteristics: Surrounded by mixed land uses such as forest, agriculture, and settlements, enabling integrative interventions; and (5) Feasibility: Availability of existing infrastructure, feasibility of installing monitoring equipment, and security considerations. Using GIS-based spatial analysis (see Fig. 36.1), potential sites are mapped to identify zones with high erosion potential, critical ecological functions, and potential for community engagement. The selected site, located in the Cibitung River catchment area, meets these criteria, given its existing land degradation concerns and community interest. Table 36.3 summarizes the key features of the ecological design inside the demonstration area, incorporating diverse elements tailored to hydrological and ecological objectives.

2.3.1 Implementation strategies

The implementation strategies outlined in this study emphasize an integrated approach combining technical interventions with advanced monitoring systems to promote sustainable water resource management and ecosystem revitalization within the Saguling reservoir catchment. On the technical front, soil conservation measures such as terracing and the installation of contour bunds are employed on agricultural slopes to effectively mitigate erosion and enhance land stability. Complemented by the construction of rainwater harvesting ponds, these structures aim to facilitate groundwater recharge and reduce surface runoff, thus alleviating pressure on downstream ecosystems. According to Majumdar and Avishek (2023), riparian buffers comprising native grasses, shrubs, and trees are strategically planted along riverbanks to stabilize stream banks, filter pollutants, and improve water quality. Extensive reforestation with indigenous species restores degraded slopes, fostering infiltration, reducing sediment load, and enhancing biodiversity conservation. Additionally, biodegradable erosion control mats are deployed on vulnerable slopes to accelerate vegetation establishment and further prevent soil loss. To support adaptive management, the project integrates a comprehensive ecological and hydrological monitoring system, deploying multi-parameter sensors to continuously measure key water quality parameters, including turbidity, pH, dissolved oxygen, and nutrients. Satellite imagery and UAV-based remote sensing enable monitoring of land use changes, vegetative cover, and sediment plumes, while river gauges and meteorological stations provide real-time data on flow and rainfall patterns. The development of a centralized data management platform ensures online accessibility of comprehensive datasets, facilitating remote analysis, timely decision-making, and adaptive strategies to address emerging hydrological and ecological challenges effectively.

2.3.2 Data collection and analysis

The data collection and analysis framework employed in this study incorporates a comprehensive suite of hydrological, water quality, ecological, and socioeconomic assessments to evaluate the effectiveness of ecohydrological interventions in the Saguling reservoir catchment (Maranata et al., 2024). Hydrological measurements include continuous flow discharge monitoring via ultrasonic flow meters at critical cross-sections, complemented by sediment load quantification through sediment samplers and turbidity sensors to assess sediment transport and deposition dynamics; rainfall data are obtained from strategically positioned weather stations and supplemented with regional meteorological agency records to ensure data robustness. Water quality parameters are monitored weekly using portable checkers that measure in situ variables, including temperature, pH, dissolved oxygen, conductivity, total dissolved solids, and turbidity. Additionally, laboratory analyses evaluate chemical pollutants such as phosphate, nitrite, nitrate, and organic loads to understand contaminant sources and the success of mitigation efforts. Ecological indicators encompass biodiversity surveys of aquatic and terrestrial species during seasonal periods, remote sensing-based assessments of vegetation health and reforestation success, and habitat quality evaluations through biotope mapping to determine ecosystem integrity (Zalewski, 2015). Socioeconomic impacts are gauged through structured questionnaires, focus group discussions, and baseline as well as follow-up surveys, with statistical tools applied to analyze trends, correlations, and measure the effectiveness of interventions, thereby supporting evidence-based adaptive management decisions.

2.3.3 Modeling and simulation

The modeling and simulation efforts at the Saguling ecohydrological demonstration site utilize a physically based Distributed Rainfall–Runoff–Sediment Model (DRRSM), designed to explicitly capture the complex interactions among hydrological, erosion, and ecological processes (Sutapa et al., 2021). This model integrates modules for rainfall interception, surface and subsurface runoff, soil detachment and transport, sediment deposition, and water quality parameters, enabling comprehensive scenario analysis. Calibration employed observed data from 2009–10, collected during the demonstration site's monitoring activities, focusing on discharge, sediment load, and water quality indicators; subsequent validation tested the model against 2010 datasets, with performance evaluated using statistical metrics such as the Nash-Sutcliffe efficiency coefficient (NSE), root mean square error (RMSE), and coefficient of determination (R^2). Various simulation scenarios were developed to evaluate the impact of interventions, including reforestation, buffer strips, and water harvesting structures, under baseline and climate change conditions—altered precipitation and temperature patterns—aiming to optimize management practices. The outputs from these simulations provide critical insights into sediment reduction, water quality enhancements, and structural placement, forming an evidence-based foundation for adaptive decision-making and sustainable watershed management.

3. Results and findings

3.1 Baseline conditions

The establishment of the EDS at Saguling Reservoir in the Citarum River Catchment aimed to provide a comprehensive understanding of pre-intervention environmental conditions. Initial assessments revealed several critical baseline indicators concerning water quality, flow regimes, and ecological status that framed subsequent intervention strategies.

Analysis of baseline water samples collected from the Cibitung River, the primary inlet to Saguling Reservoir, indicated elevated turbidity levels averaging 120 NTU, and can reach more than 3000 NTU in certain conditions (Sutapa et al., 2021), significantly exceeding the standard water quality threshold for aquatic life and human use (WHO, 2017). Parameters such as Total Dissolved Solids (TDS), pH, dissolved oxygen (DO), and nutrient concentrations also indicated eutrophication risks, with phosphate levels at approximately 0.15 mg/L and nitrite levels near 0.08 mg/L. These readings reflect substantial anthropogenic impacts, primarily resulting from upstream land-use activities such as agriculture and deforestation. Moreover, hydrological data recorded over a year showed distinct seasonality aligned with monsoon cycles, with peak flows during November–March and reduced flows during the dry season (June–September). Notably, flow variability was significant, with peak discharges reaching up to 500 m^3/s during storm events, leading to erosion and sediment mobilization. The sediment load was high, averaging 200 mg/L, contributing to siltation in the reservoir. Meanwhile, the aquatic ecosystem was characterized by low biodiversity, with surveys indicating a dominance of

TABLE 36.4 The key baseline conditions of water quality.

Parameter	Value	Descriptive condition
Turbidity	~120 NTU	High, impair photosynthesis
TDS	~500 mg/L	Elevated, suggests pollution
pH	6.4–6.8	Slightly acidic
Dissolved oxygen	3.2–4.5 mg/L	Low, stress for aquatic life
Phosphate	~0.15 mg/L	High, eutrophication potential
Nitrite	~0.08 mg/L	Elevated, pollution indicator
Sediment load	~200 mg/L	Significant erosion source

pollution-tolerant species, such as Lemna minor and Pistia, in shallow water zones. Macroinvertebrate diversity was limited, with key indicator species absent, signaling degraded habitat conditions. Riparian vegetation consisted mainly of monoculture crops and secondary forest patches, providing limited natural buffering. Table 36.4 summarizes the baseline conditions of water quality.

3.2 Ecological and hydrological improvements post-intervention

Following the implementation of ecohydrology-based interventions, including reforestation, rainwater harvesting, phytoremediation, and community engagement, measurable improvements have been observed across hydrological and ecological parameters. The reforestation of degraded slopes with native vegetation improved land cover, significantly enhancing soil permeability. Rainwater infiltration capacity increased from 10 mm/h at baseline to approximately 25 mm/h after the intervention, resulting in reduced surface runoff. Consequently, peak discharge levels decreased by approximately 20%, as monitored during consecutive extreme weather events. This augmented infiltration not only mitigates erosion but also recharges groundwater stocks, which is crucial for maintaining base flows during dry seasons. The cumulative effect has been a more stabilized flow regime, reducing the risk of flash floods and sediment mobilization.

Phytoremediation systems utilizing endemic aquatic plants, such as Lemna, Landoltia, Salvinia, Pistia, and Eichhornia, have demonstrated a remarkable ability to reduce turbidity and nutrient loads. Water turbidity within the riparian zones decreased by up to 65%, reaching levels below 10 NTU Fig. 36.4, approaching national water quality standards (PPRI No. 22, 2021). Moreover, the reintroduction of native vegetation and habitat structures fostered an increase in biodiversity. Macroinvertebrate diversity indices improved from an index value of 0.3 (poor) to over 0.6 (moderate), indicating enhanced habitat complexity. Bird and fish species surveys recorded the appearance of previously absent indicator species such as Ichthyocentris and Ardea alba.

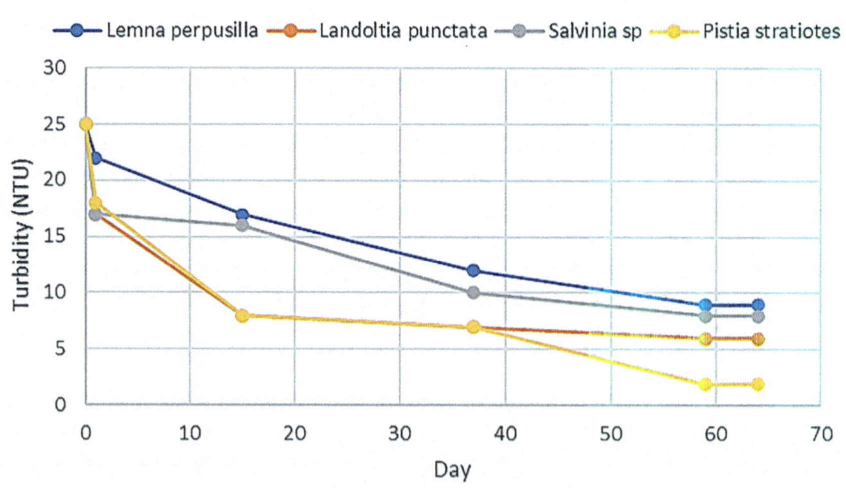

FIGURE 36.4 Reduction of turbidity inside the phytoremediation pond using aquatic plants Lemna, Landoltia, Salvinia, Pistia, and Eichhornia.

TABLE 36.5 Ecosystem service metrics before and after interventions.

Service	Baseline	Post-intervention	Improvement (%)
Water turbidity (NTU)	120	45	62.5
Nutrient load (mg/L PO_4^{3-})	0.15	0.07	53.3
Peak discharge (m^3/s)	500	375	25
Fish biodiversity index	0.2	0.5	150

TABLE 36.6 Key outcomes post-intervention.

Parameter	Baseline	Post-intervention	Change/Impact
Turbidity (NTU)	~120	45	Reduced by 62.5%
Nutrients (mg/L PO_4^{3-})	0.15	0.07	Halved, improved water quality
Flood peaks (m^3/s)	500	375	Decreased by 25%
Biodiversity index	0.2	0.5	150% improvement
Community participation	0%	75% Involved	Significant engagement

3.3 Ecosystem services enhancement

A quantitative assessment of ecosystem services at Saguling reveals substantial gains attributable to the ecohydrological measures. The phytoremediation system, integrated within the riparian zones, was responsible for approximately a 45% reduction in sediment load and a 50% decrease in nutrient concentrations, thus improving downstream water quality and reducing treatment costs for local communities (Mazilamani et al., 2024; Sutapa et al., 2021). It enhanced natural filtration and sediment retention capacities, effectively prolonging the lifespan of the Saguling Reservoir. Meanwhile, restoring native vegetation along riparian buffers increased habitat diversity, providing niches for aquatic and terrestrial fauna. Bird species richness increased by 30%, with potential benefits for ecological balance and ecotourism. The enhanced habitat complexity supports pollinators and fish spawning grounds, which are crucial for local livelihoods Table 36.5.

3.4 Community and stakeholder feedback

The success of ecohydrological activities heavily depends on local community participation and stakeholder acceptance. Surveys and participatory assessments reflected positive responses towards the project. Approximately 75% of community members expressed awareness of the project benefits, with 60% actively involved in reforestation, maintenance, and monitoring activities (Irianti et al., 2023; Sutapa et al., 2021). Focus group discussions revealed increased awareness of water conservation and soil erosion issues.

Households reported an increase in income through ecotourism, sustainable agriculture, and the availability of clean water for domestic use. The improvement of water quality resulted in a 15% reduction in health expenses associated with waterborne diseases across participating villages. Furthermore, the collaboration among government agencies, local communities, NGOs, and academic institutions facilitated knowledge transfer, capacity building, and shared responsibility. The participatory approach fostered a sense of ownership, ensuring the sustainability of ecohydrological practices. Table 36.6 compiles the key outcomes following the intervention.

4. Discussion

This section discusses the multifaceted aspects of establishing and operating the EDS at the Saguling Reservoir, emphasizing its contribution to sustainable water resource management and ecosystem services in the Citarum River catchment. The analysis encompasses the effectiveness of the demonstration, lessons learned, challenges faced, policy implications, and the potential for scaling and replicating the approach in other regions.

FIGURE 36.5 Overview of the phytoremediation pond of Saguling ecohydrological demonstration (Sutapa et al., 2021). *Based on Sutapa, I.D.A., Fakhrudin, M., 2021. Ecohydrology: a management tool for aquatic ecosystem and SDGs implementation.* IOP Conf. Ser. Earth Environ. Sci. *789 (1), 012032. https://doi.org/10.1088/1755-1315/789/1/012032.*

4.1 Effectiveness of the demonstration area

The Saguling EDS has demonstrated significant potential in illustrating how ecohydrological principles can be effectively integrated into regional water management strategies. The core success lies in the development of a comprehensive online monitoring system that captures real-time data on river water quality, flow, and weather conditions, thus facilitating timely decision-making and adaptive management (Sutapa et al., 2021). Such real-time data collection is crucial for understanding the dynamics of sediment transport and water pollution, especially in high-erosion landscapes like the Cibitung watershed, where land use is predominantly agricultural and forest cover is limited. The ecohydrological modelling system deployed at Saguling has shown promising calibration and validation results, with high NSE values indicating robust performance in predicting sediment yield and hydrological responses (Sutapa et al., 2021). The ability to simulate sediment transport and water quality trends under various land-use scenarios provides a valuable tool for planning erosion control measures and evaluating the impact of interventions, such as phytoremediation with aquatic plants. Empirical evidence from the planted phytoremediation ponds revealed substantial reductions in water turbidity, reaching up to 92% with Eichhornia and Pistia, signifying their efficacy in trapping sediments and absorbing nutrients.

The integration of community participation, particularly in socioeconomic and cultural contexts, has enhanced the initiative's effectiveness. Community engagement activities, such as surveys, focus group discussions, and interviews, have fostered a local understanding of the benefits of ecohydrology, leading to an increased willingness to participate in and sustain related activities. This participative approach underscores the importance of local knowledge and social capital in implementing ecohydrological measures effectively. Seamless alignment of scientific tools with sociocultural realities has reinforced the demonstration's role as an educational and capacity-building platform. The demonstration area serves as a real-world laboratory for stakeholders to observe and learn ecohydrological processes, enabling replication and adaptation across different geographic and socioeconomic settings. The success of these integrated efforts supports ecohydrology as a viable strategy for achieving multiple water and environmental objectives, including flood mitigation, sedimentation control, and habitat restoration. Fig. 36.5 shows the phytoremediation pond of the Saguling ecohydrological demonstration.

4.2 Lessons learned and the best practices

The Saguling EDS project offers valuable lessons and best practices vital for implementing sustainable water resource management in tropical landscapes. Foremost, adopting a holistic approach that integrates hydrological monitoring, ecological restoration, such as phytoremediation, and active social engagement creates synergistic benefits, ultimately bolstering the resilience of water resources against environmental pressures (Sutapa et al., 2021). Data-driven decision making, supported by real-time online monitoring systems and advanced modeling, enables policymakers and stakeholders to formulate proactive strategies rather than reactive responses to sedimentation, water quality degradation, or hydrological fluctuations. Community-centered implementation emerges as a cornerstone, emphasizing the importance of

TABLE 36.7 Summary of best practices and lessons learned at Saguling EDS.

Practice	Description	Outcome
Integrated monitoring	Real-time hydrometeorological data collection	Improved responsiveness and planning
Phytoremediation in ponds	Use of native aquatic plants to reduce turbidity and nutrients	Significant water quality improvement
Participative community engagement	Local involvement via surveys, discussions, and education	Increased community buy-in and sustainability
Modelling and scenario	Analysis simulation of sedimentation and land use impacts	Informed management and policy formulation

involving local residents through participatory decision-making, capacity-building workshops, and respecting indigenous knowledge to foster a sense of ownership and ensure the long-term sustainability of interventions. Additionally, the scalability and adaptability of decentralized measures, such as planting aquatic macrophytes and installing sediment traps, are demonstrated to be cost-effective and easily tailored to specific sites. Importantly, adopting an adaptive management framework that incorporates continuous monitoring and feedback loops enables flexible modifications aligned with climate variability, land-use dynamics, and evolving socioeconomic conditions, thereby ensuring the project's resilience and sustainability over time (Table 36.7).

4.3 Challenges and limitations

Despite the promising results of the Saguling EDS, several inherent challenges and limitations hinder its comprehensive success and scalability. Technically, the reliance on real-time monitoring systems presents vulnerabilities, including device maintenance issues, power interruptions, and data transmission disruptions that can compromise data integrity and ongoing analysis. Additionally, the complexities inherent in ecological and hydrological modelling, such as uncertainties during extreme weather events or anomalous land-use changes, complicate prediction accuracy and decision-making processes. The site-specific nature of hydrological and ecological processes further limits the direct application or scaling of models to broader geographic regions, necessitating localized calibration and adaptation. Socioeconomically, sustaining active community participation remains resource-intensive, requiring continuous incentives, education, and capacity-building efforts, which pose logistical and financial challenges. Upstream land-use practices, including deforestation and intensive agriculture, exacerbate sedimentation issues, undermining mitigation efforts and highlighting the need for integrated landscape management. Policy-related constraints include fragmented governance, with inadequate coordination among water, land, and environmental regulatory frameworks, creating gaps in holistic resource management. Furthermore, securing consistent funding for long-term maintenance, monitoring, scaling initiatives, and capacity development remains elusive, which threatens the project's sustainability and replication. Addressing these multifaceted challenges requires integrated strategies that encompass technological resilience, stakeholder engagement, policy coherence, and sustained financial investment, all of which are essential for unlocking the full potential of ecohydrological interventions in tropical settings (Table 36.8).

TABLE 36.8 Summary of challenges at the Saguling EDS.

Category	Specific challenges
Technical	Data gaps, equipment maintenance, and model uncertainties
Socioeconomic	Community engagement, sustainability, and land management conflicts
Policy	Fragmented policies, funding constraints

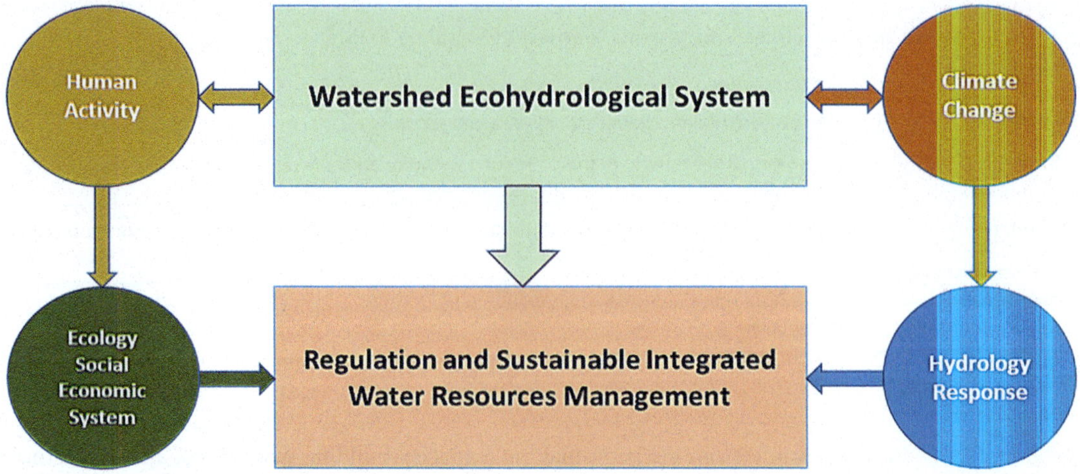

FIGURE 36.6 Policy framework integrating ecohydrology principles into water resource management. *Adopted from Cao, Z., Wang, S., Luo, P., Xie, D., Zhu, W., 2022. Watershed Ecohydrological Processes in a Changing Environment: Opportunities and Challenges. Water 14, 1502. doi:10.3390/w14091502.*

4.4 Policy implications and recommendations

The findings from the Saguling Ecohydrological Demonstration Site (EDS) highlight the critical importance of integrating ecohydrology into the broader water resource management policy landscape. Policymakers are encouraged to establish institutional mechanisms that promote cross-sectoral coordination among agencies responsible for water, agriculture, forestry, and urban development, thereby ensuring cohesive planning and implementation. Establishing dedicated platforms for data sharing, joint planning, and policy harmonization can facilitate more comprehensive approaches to water and land management, addressing the interconnected challenges of sedimentation, erosion, and water quality. Furthermore, developing a legal and regulatory framework that incentivizes sustainable land use practices, such as establishing buffer zones, promoting afforestation, and protecting wetlands, can reinforce positive behavioural shifts among stakeholders, encouraging adoption of eco-friendly interventions that support ecosystem resilience Fig. 36.6. To sustain such initiatives, financial mechanisms, including grants, subsidies, and schemes such as Payment for Ecosystem Services (PES), should be promoted to motivate community participation and cover the costs associated with ecohydrological practices. These financial incentives are vital for scaling up successful pilot projects and ensuring their long-term viability. Additionally, capacity building is a cornerstone for policy effectiveness; therefore, targeted training programs should be prioritized for local officials, community leaders, engineers, and practitioners to deepen their understanding of ecohydrological principles, introduce advanced monitoring technologies, and foster adaptive management strategies. Finally, establishing a robust framework for ongoing monitoring, evaluation, and adaptive policymaking—grounded in scientific evidence from demonstration sites—will enable continuous improvements, ensuring that policies remain responsive to environmental changes and stakeholder needs, ultimately advancing sustainable water management in tropical landscapes.

4.5 Scalability and replicability of the approach

The scalability and replicability of the Saguling EDS are underpinned by its modular design, which utilizes adaptable, low-cost infrastructure that can be easily integrated into diverse tropical environments. Components such as community-managed phytoremediation ponds, sediment traps, and mobile monitoring units allow for flexible deployment tailored to local ecological and socioeconomic conditions. The comprehensive documentation of protocols, lessons learned, and operational strategies not only aids in knowledge transfer but also serves as a valuable blueprint for implementing similar systems elsewhere, taking into account unique local contexts. Furthermore, fostering strong stakeholder engagement through active community participation, local leadership, and multi-stakeholder partnerships is essential in scaling up interventions, ensuring local ownership and sustainability of ecohydrological initiatives. An enabling policy environment that recognizes the importance of ecohydrology significantly enhances the potential for wider adoption; support from national policies and international cooperation platforms, such as UNESCO, can facilitate knowledge exchange and capacity building across regions.

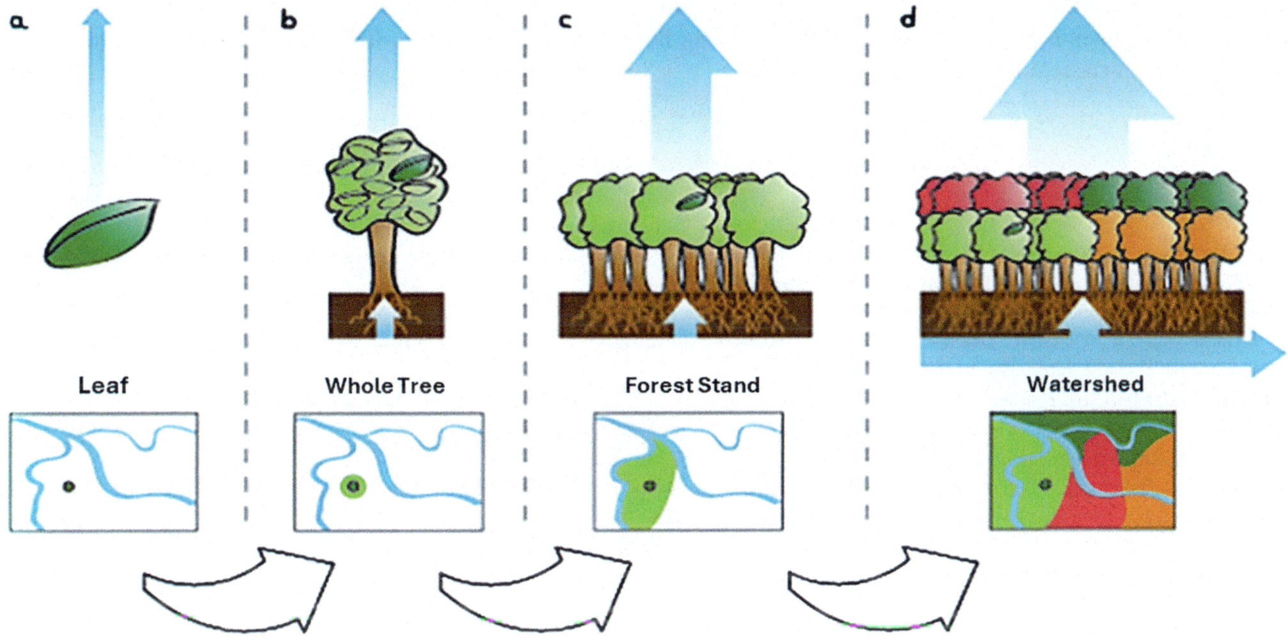

FIGURE 36.7 Conceptual framework inspiration for scaling ecohydrological demonstration sites across tropical catchments (Asbjornsen et al., 2011). *Based on Asbjornsen, H., Goldsmith, G. R., Alvarado-Barrientos, M. S., Rebel, K., Van Osch, F. P., Rietkerk, M., et al., 2011. Ecohydrological advances and applications in plant-water relations research: a review. J. Plant Ecol. 4 (1–2), 3–22. https://doi.org/10.1093/jpe/rtr005.*

Technological advancements further bolster the extensibility of this approach. Innovations in remote sensing, geographic information systems (GIS), and mobile data collection tools enable efficient and large-scale monitoring with minimal incremental costs, thereby increasing the feasibility of expanding these interventions over broader spatial scales. Collectively, with appropriate adaptation, the Saguling model holds significant promise for fostering sustainable water resource management across tropical and developing regions worldwide. Fig. 36.7 illustrates the inspiration model for the conceptual framework of scaling ecohydrological demonstration sites across tropical catchments (Asbjornsen et al., 2011).

5. Conclusions

The establishment of the Ecohydrological Demonstration Site (EDS) at the Saguling Reservoir in the Citarum River Catchment represents a significant stride toward sustainable water resource management and ecosystem conservation in West Java, Indonesia. The key achievements of this initiative include the successful implementation of an integrated online monitoring system capable of real-time assessment of water quality, quantity, and environmental conditions, facilitating more informed and timely decision-making. Additionally, the development and calibration of a distributed rainfall-runoff-sediment model have enabled accurate prediction of sediment yields and hydrological responses, which are essential for mitigating erosion, reducing sedimentation in reservoirs, and maintaining water quality. The application of ecohydrological principles, particularly phytoremediation using aquatic plants, has demonstrated promising results in decreasing water turbidity and trapping pollutants, thereby enhancing ecosystem health. Importantly, the project has emphasized community engagement, incorporating socioeconomic, cultural, and traditional knowledge to ensure local participation and ownership, which are vital for long-term sustainability. The impacts of these efforts extend beyond local environmental benefits; they contribute valuable insights into ecohydrological approaches as effective, nature-based solutions for addressing complex water-related challenges posed by rapid urbanization, climate variability, and land-use changes. The integration of technological tools such as remote sensing, GIS, and online monitoring systems exemplifies how modern innovations can complement traditional ecological knowledge to create resilient and adaptive water management strategies.

Looking forward, future research should focus on expanding the ecological modules within the modeling system, enhancing predictive capabilities under diverse climate scenarios, and exploring the socioeconomic aspects of community-based water management. Policy actions should prioritize the adoption of ecohydrological principles in regional water governance, incorporating community participation, environmental education, and sustainable land-use planning. Strengthening institutional support for ecohydrological projects and scaling successful models across other vulnerable catchments will be instrumental in achieving holistic and sustainable water and ecosystem management for Indonesia's tropical regions.

References

Abdillah, A., Widianingsih, I., Buchari, R.A., Nurasa, H., 2024. Collaborative strategies for sustainable management of the Citarum watershed in Indonesia: a Quintuple Helix approach. Discov. Sustain. 5 (1), 347. https://doi.org/10.1007/s43621-024-00571-1.

Ali, M.H., Biswas, P., Hassan, S.M.Q., Islam, M.A., 2025. Climate change and its impacts on hydrological regimes over the Bengal delta. Results Eng. 25, 103861. https://doi.org/10.1016/j.rineng.2024.103861.

Asbjornsen, H., Goldsmith, G.R., Alvarado-Barrientos, M.S., Rebel, K., Van Osch, F.P., Rietkerk, M., Chen, J., Gotsch, S., Tobon, C., Geissert, D.R., Gomez-Tagle, A., Vache, K., Dawson, T.E., 2011. Ecohydrological advances and applications in plant-water relations research: a review. J. Plant Ecol. 4 (1–2), 3–22. https://doi.org/10.1093/jpe/rtr005.

Budiman, F., Ismardi, A., Hardinah, T., Muhammad, R., Nurwijayadi, H.A., Nurhidayat, A., Sasto, I.H.S., Sutapa, I.D.A., 2023. Strengthening oil pollution monitoring system in aquatic environment through development of IoT-based oil-water separator device. Ecohydrol. Hydrobiol. 24 (3), 617–623. https://doi.org/10.1016/j.ecohyd.2023.06.002.

Cao, S., Li, X., Jiang, J., Lu, L., Du, M., Wen, S., 2025. Investigating the influence of wetland area change on environmental sustainable development goals: evidence from 31 provinces of China. Environ. Sustain. Indic. 27, 100776. https://doi.org/10.1016/j.indic.2025.100776.

Cao, Z., Wang, S., Luo, P., Xie, D., Zhu, W., 2022. Watershed Ecohydrological Processes in a Changing Environment: Opportunities and Challenges. Water 14, 1502. https://doi.org/10.3390/w14091502.

Dimyati, M., Rustanto, A., Ash Shidiq, I.P., Indratmoko, S., Siswanto, Dimyati, R.D., Nurlambang, T., Zubair, A., Fakhruddin, A., Siddiq, A., Adhanto, D.H., Maulidina, K., Auni, R., 2024. Spatiotemporal relation of satellite-based meteorological to agricultural drought in the downstream Citarum watershed, Indonesia. Environ. Sustain. Indic. 22, 100339. https://doi.org/10.1016/j.indic.2024.100339.

Empung, Dharma, I.G.B.S., Yekti, M.I., Putera, I.G.A.A., 2023. Navigating environmental risks: exploring the effects of flood hydrographs and probable maximum floods in a series cascade reservoir system-a case study. Civ. Eng. Archit. 11 (6), 3763–3774. https://doi.org/10.13189/cea.2023.110638.

Firoozi Akbar, A., Firoozi Asghar, A., 2024. Water erosion processes: mechanisms, impact, and management strategies. Results Eng. 24 (2024), 103237. https://doi.org/10.1016/j.rineng.2024.103237.

Granata, F., Nunno, F.D., 2025. Pathways for hydrological resilience: strategies for adaptation in a changing climate. Earth Syst. Environ. 10, 203–231. https://doi.org/10.1007/s41748-024-00567-x.

Hadfield, P., Prescott, M., Holden, J., Novalia, W., Suwarso, R., Marthanty, D.R., Priadi, C., Kirana, K.H., Endyana, C., Hardesty, B.D., Dzakwan Taufik, F., Zurbrügg, C., Josey, B., Astuti, N., Wong, T., Ramirez-Lovering, D., Raven, R., Senapathi, V., Living Lab, C., 2024. Co-creating visions for sustainable river revitalisation. PLoS Water 3 (8), e0000200. https://doi.org/10.1371/journal.pwat.0000200.

Irianti, S., Yunianto, A., Dharmayanti, I., Hidayangsih, P.S., Zahra, Z., Mbarep, D.P.P., Sutapa, I.D.A., 2023. Implementation of drinking water quality surveillance and household water management practices in selected provinces in Indonesia. IOP Conf. Ser. Earth Environ. Sci. 1201 (1), 012106. https://doi.org/10.1088/1755-1315/1201/1/012106.

Kurniawan, T.A., Bandala, E.R., Othman, M.H.D., Goh, H.H., Anouzla, A., Chew, K.W., Aziz, F., Al-Hazmi, H.E., Khoir, A.N., 2024. Implications of climate change on water quality and sanitation in climate hotspot locations: a case study in Indonesia. Water Supply 24 (2), 517–542. https://doi.org/10.2166/ws.2024.008.

Majumdar, A., Avishek, K., 2023. Riparian zone assessment and management: an integrated review using geospatial technology. Water Air Soil Pollut. 234 (5). https://doi.org/10.1007/s11270-023-06329-1.

Maranata, D., Asmaranto, R., Sudaryanti, S., 2024. Water quality identification and analysis of Saguling reservoir, West Bandung Regency. J. Presipitasi 21 (2), 433–445.

Marijuan, R., Díez, B., Peláez-Sánchez, S., Sánchez, C., Iglesias, J., Şirin, B., Baba, A., Gündüz, O., Sánchez, R., 2024. Evaluating the impact of nature-based solutions on the provision of water-related and water-dependant ecosystem services. Nat. Based Solut. 6, 100194. https://doi.org/10.1016/j.nbsj.2024.100194.

Mazilamani, L.S., Annammala, K.V., Martin, P., Qi, L.Y., Sugumaran, D., Ling, Y.E., Reynard, P.S., Nurhidayat, A., Syawal, M.S., Sutapa, I.D.A., 2024. Comprehensive assessment of water and sediment quality to support sustainable management practices and mitigate potential risks of trace metal pollution in the Johor River Basin, South Peninsular Malaysia. J. Ecohydrol. Hydrobiol. 25, 315–329. https://doi.org/10.1016/j.ecohyd.2024.04.009.

McKenna, F., Hanrahan, J., 2024. Meaningful community engagement through advanced indicator systems for sustainable destination planning. Environ. Sustain. Indic. 22, 100392. https://doi.org/10.1016/j.indic.2024.100392.

Nahib, I., Ambarwulan, W., Rahadiati, A., Munajati, S.L., Prihanto, Y., Suryanta, J., Turmudi, T., Nuswantoro, A.C., 2021. Assessment of the impacts of climate and LULC changes on the water yield in the Citarum River Basin, West Java Province, Indonesia. Sustainability 13, 3919. https://doi.org/10.3390/su13073919.

Novalia, W., Rogers, B., Yosafat, Y., Suwarso, R., 2024. Enhancing local capacity for river revitalisation: municipal enablers in the Upper Citarum. APN GCR Final Report. Asia-Pacific Network for Global Change Research, p. XYZ.

Pawitan, H., Haryani, G.S., 2011. Water resources, sustainability and societal livelihoods in Indonesia. Ecohydrol. Hydrobiol. 11 (3–4), 231–243. https://doi.org/10.2478/v10104-011-0050-3.

PP No 22, 2021. Peraturan Pemerintah Republik Indonesia Nomor 22 Tahun 2021 Tentang Penyelenggaraan Perlindungan dan Pengeloiaan Lingkungan Hidup.

Pueppke, S.G., Nurtazin, S., Ou, W., 2020. Water and land as shared resources for agriculture and aquaculture: insights from Asia. Water 12, 2787. https://doi.org/10.3390/w12102787.

Rixson, L., Wenbiao, D., Aliyanta, B., Godbold, D., Lubis, A.A., Nugraha, E.D., 2024. Assessment of soil erosion in the upper Citarum watershed for sustainability of the Saguling reservoir: unmixing model approach. Environ. Monit. Assess. 196, 615. https://doi.org/10.1007/s10661-024-12750-8.

Santos, E., Carvalho, M., Martins, S., 2023. Sustainable water management: understanding the socioeconomic and cultural dimensions. Sustainability 15, 13074. https://doi.org/10.3390/su151713074.

Saoum, M.R., Kumar, Sarkar, S.K., 2024. Monitoring mangrove forest change and its impacts on the environment. Ecol. Indic. 159, 111666. https://doi.org/10.1016/j.ecolind.2024.111666.

Sarandón, R., Gaviño Novillo, M., Muschong, D., Guerrero Borges, V., 2009. Lacar Lake demonstration project for ecohydrology: improving land use policy at Lacar Lake watershed based on an ecohydrological approach (San Martín de los Andes – Neuquén – R. Argentina). Ecohydrol. Hydrobiol. 9 (1), 125–134. https://doi.org/10.2478/v10104-009-0041-9.

Shaik, E., Fan, D., 2025. Changes in runoff and sediment discharge and their influencing factors in the lower Brahmaputra River, Bangladesh. J. Hydrol. 662 (Part A), 133924. https://doi.org/10.1016/j.jhydrol.2025.133924.

Sutapa, I.D.A., Apip, Fakhrudin, M., Yogaswara, H., 2021. Implementation of ecohydrology to support sustainable water resources management in tropical region, Indonesia. Ecohydrol. Hydrobiol. 21 (3), 501–515. https://doi.org/10.1016/j.ecohyd.2021.08.010.

Sutapa, I.D.A., Fakhrudin, M., 2021. Ecohydrology: a management tool for aquatic ecosystem and SDGs implementation. IOP Conf. Ser. Earth Environ. Sci. 789 (1), 012032. https://doi.org/10.1088/1755-1315/789/1/012032.

Sutapa, I.D.A., Mbarep, D.P.P., Hasibuan, H.S., Zalewski, M., 2023. Ecohydrology approach to strengthen public green open space management towards comfortable common space and playground in Kalijodo area – Jakarta Province, Indonesia. J. Ecohydrol. Hydrobiol. 23 (4), 518–531. https://doi.org/10.1016/j.ecohyd.2023.04.005.

UNESCO EDS, 2024. Call for New Ecohydrology Demonstration Sites for the Year 2024. https://www.unesco.org/en/articles/call-new-ecohydrology-demonstration-sites-year-2024.

Wardhani, E., Notodarmojo, S., Roosmini, D., 2018. Assessment of heavy metal contamination in Saguling reservoir water West Java Province Indonesia. E3S Web Confer. 73, 06009. https://doi.org/10.1051/e3sconf/2018730.

WHO, 2017. Water Quality and Health - Review of Turbidity: Information for Regulators and Water Suppliers. https://iris.who.int/bitstream/handle/10665/254631/WHO-FWC-WSH-17.01-eng.pdf.

Zalewski, M., 2015. Ecohydrology and hydrologic engineering: regulation of hydrology-biota interactions for sustainability. J. Hydrol. Eng. 20 (1). https://doi.org/10.1061/(ASCE)HE.1943-5584.0000999.

Zalewski, M., Robarts, R., 2003. Ecohydrology- a new paradigm for integrated water resources management. SILnews 40, 1–5.

Zalewski, M., Harper, D., Wagner, I., 2009. Ecohydrology – why demonstration projects throughout the world? Ecohydrol. Hydrobiol. 9 (1), 3–11. https://doi.org/10.2478/v10104-009-0043-7.

Chapter 37

Reimagining urban green spaces: Integrating functionality and aesthetic experience

Izabela A. Levy
European Regional Centre for Ecohydrology of the Polish Academy of Sciences under the auspices of UNESCO, Łódź, Poland

WALLER CREEK RENDERING. IMAGE CREDIT: MICHAEL VAN VALKENBURGH ASSOCIATES INC. (MVVA).

Transdisciplinary Ecohydrology. https://doi.org/10.1016/B978-0-443-41596-8.00034-0

1. Aesthetic experience as a bridge between people, nature, and culture

Aesthetic experience plays a crucial role in shaping how people perceive and relate to their surroundings. Individuals from diverse social and cultural backgrounds can recognise and appreciate the aesthetic qualities of urban green spaces. These values extend far beyond the visual appreciation of vegetation, eliciting emotional, sensory, and psychological responses that resonate across different audiences (Gobster et al., 2007). Human perception of landscape is inherently intuitive and deeply personal, shaped by lived experience yet rooted in a shared human affinity for connection with the natural world (Kellert and Wilson, 1993).

This innate affinity suggests that contact with nature is not merely a cultural preference but a fundamental human need. Increasingly, this understanding is reflected in international policy frameworks. In 2021, the United Nations Human Rights Council (UNHRC) affirmed the right to a clean, healthy, and sustainable environment (United Nations Human Rights, 2021). From a psychological perspective, this connection has measurable benefits for human well-being. Interacting with aesthetically rich and functional landscapes supports cognitive clarity, emotional stability, and social cohesion (Kaplan and Kaplan, 1989; Ulrich, 1984). Engaging with landscapes through multiple sensory and experiential dimensions enhances cognitive clarity, supports psychological restoration, and creates opportunities for reflection and moments of wonder—states shown to reduce stress and strengthen mental health (Bratman et al., 2019). When ecohydrological processes become perceptible through spatial design, these benefits are further amplified. Urban green spaces can function not only as recreational environments but also as settings that support attention recovery and emotional resilience (Gonçalves et al., 2021; Marselle et al., 2014). They contribute not only to physical and mental well-being—often the primary focus of policy discussions—but also to a deeper human need for connection with living systems. In this sense, beauty is not merely aesthetic ornamentation detached from function. Rather, it provides experiential access to underlying ecohydrological processes, reinforcing the sense of belonging and stewardship that underpins long-term environmental care (Kwon et al., 2021).

Such emotional and perceptual connections to landscape are increasingly important in the context of global environmental change. Climate change, however, is not solely a physical or technical challenge; it is also a deeply social and cultural one. Its effects are visible in shifting weather patterns, ecological succession in degraded landscapes, and the intensification of urban flooding, heat island effects, and droughts now experienced globally. Effective adaptation depends not only on strategic decision-making but also on public engagement, behavioral change, and collective investment in long-term resilience (Adger, 2003; IPCC, 2022). In this context, aesthetic and emotional engagement with landscapes plays a vital role in shaping collective responses to environmental change, cultivating attachment, ecological awareness, and stewardship that support the transition toward more sustainable and resilient socio-ecological systems.

Addressing these challenges requires approaches that recognise the interconnected dynamics of water, ecosystems, and society. The integration of hydrological and ecological processes is fundamental to developing systemic responses to environmental pressures (Zalewski, 2000). Transdisciplinary ecohydrology emphasises that effective environmental management must move beyond fragmented, sectoral solutions toward strategies that acknowledge feedback mechanisms operating within complex socio-eco-hydrological systems. Landscapes that integrate hydrological regulation and ecological functioning with aesthetic and cultural meaning can help foster a broader culture of sustainability—one that recognises the interdependence of natural processes and human societies across scales, from the catchment to the biosphere. Within this perspective, aesthetic experience serves as a critical mediating layer, translating ecohydrological knowledge into lived experience and embedding scientific understanding within cultural practice. The Culture & Education (C+E) pillar emphasizes that sustainable design must engage not only hydrological and ecological processes but also the sensory, emotional, and symbolic dimensions of human experience. An eco-hydro-cultural approach translates this integration into spatial practice, shaping environments in which hydrological regulation, ecological vitality, and aesthetic expression become inseparable dimensions of the same system. Such environments can function as living laboratories for cultivating ecohydrological awareness, social responsibility, and environmental stewardship, while supporting community participation in the co-creation of resilient urban landscapes. Through the analysis of the Waller Creek project, this chapter demonstrates how aesthetic experience can deepen and enrich the relationship between people and the natural environment. Aesthetic appreciation is closely linked to perceptions of ecohydrological vitality and environmental health: landscapes perceived as biodiverse and functioning ecosystems tend to evoke positive emotional responses, encourage care, and strengthen environmental attachment (Gobster et al., 2007; Gonçalves et al., 2021; Hartig et al., 2014; Kaplan and Kaplan, 1989).

Although the restoration of ecohydrological systems such as Waller Creek required substantial environmental intervention and complex engineering, the significance of the project extends beyond its technical achievements. Its value also lies in its cultural resonance, social meaning, and the quality of aesthetic experience it offers to the public. By restoring ecological functions and hydrological processes, the project contributes to increased community resilience to water-related hazards, as ecosystem restoration can reduce vulnerability to environmental risks (Paudel et al., 2023). In this sense, landscape design becomes not only a technical intervention but also a medium through which ecohydrological processes are made visible and meaningful to society.

2. From knowledge to practice: The innovation process

Although the importance of urban green spaces has been widely documented by researchers, public health professionals, and environmental planners (Chiesura, 2004; Kabisch et al., 2015), their integration into urban development remains inconsistent. Despite substantial evidence demonstrating their contributions to environmental quality, public health, and social well-being (World Health Organization & UN-Habitat, 2016; Hartig et al., 2014; Tzoulas et al., 2007), urban greenery is often treated primarily as an ornamental element rather than as essential ecohydrological infrastructure. Such a perception overlooks the critical role that green spaces play in sustaining ecological processes, supporting biodiversity, and enhancing human well-being, thereby limiting their potential to function as resilient components of urban systems.

In practice, the development of multifunctional blue-green infrastructure is frequently constrained by practical and institutional limitations, including insufficient financial resources, a shortage of trained specialists, and the absence of dedicated implementation frameworks or experienced contractors capable of delivering complex projects (Haase et al., 2017; Niemelä et al., 2010; Pauleit et al., 2019). These challenges contribute to a persistent gap between scientific knowledge, policy aspirations, and practical implementation in urban planning and design (Ahern, 2013; Benedict and McMahon, 2006). The realization of innovative landscape projects therefore requires the coordinated integration of interdisciplinary expertise, technical capacity, and supportive institutional structures, see Fig. 37.1.

Many of these barriers ultimately reflect limitations in existing governance structures. Addressing complex urban sustainability challenges therefore requires decision-making frameworks capable of integrating interdisciplinary

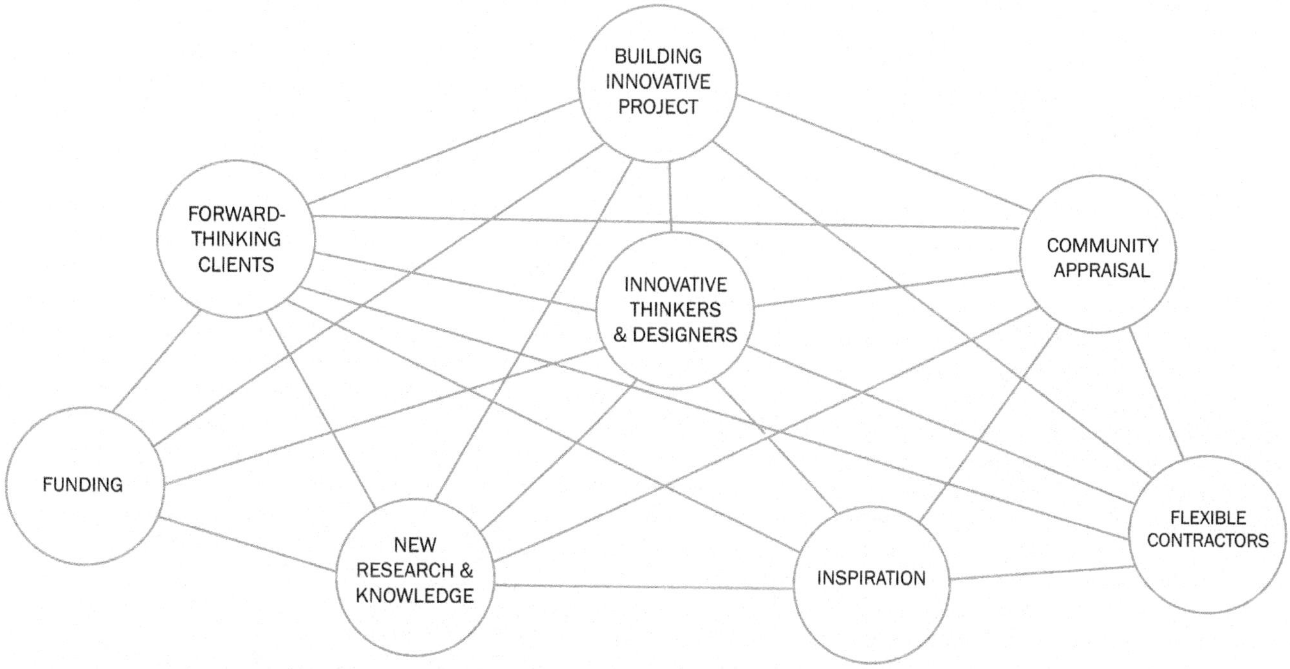

FIGURE 37.1 The transdisciplinary path to innovation. The diagram presents the interconnected system required to build an innovative project. At its core are *Innovative Thinkers and Designers*, who act as mediators between multiple actors and resources. Their work is supported by *New Research and Knowledge*, which provides the conceptual and scientific foundation, and by *Inspiration*, which fuels creativity and vision. These core processes are sustained by *Funding* and reinforced through collaboration with *Forward-Thinking Clients*—decision-makers open to experimentation and long-term perspectives. The success of innovation further depends on *Flexible Contractors* capable of translating visionary ideas into practical, adaptive solutions, and on *Community Appraisal*, which ensures social understanding and legitimacy. Together, these interlinked components form a dynamic feedback system that integrates knowledge, creativity, resources, and stakeholder engagement. The non-linear structure of the diagram emphasizes that innovation does not follow a linear sequence but rather an iterative process of exchange, adaptation, and cocreation among all participants. Like pieces of a puzzle, each actor and process contributes indispensably to bringing visionary projects to life.

knowledge and coordinating action across multiple institutions and stakeholders. The capacity for meaningful agency and transformative action within today's political landscape, particularly in the realms of urban sustainability and environmental governance, requires a paradigm shift. Traditional top-down models of decision-making have proven increasingly inadequate for addressing complex systemic challenges such as climate change, social inequality, and biodiversity loss (Meadows, 1999; Ostrom, 2010). Effective implementation of visionary ideas demands not only institutional innovation but also inclusive, participatory governance that empowers communities, decentralizes authority, and integrates knowledge across disciplines and sectors (Bulkeley and Betsill, 2005; Fischer, 2000; Healey, 2007).

The creation of a culture of sustainability (Chapter 42), thus constitutes a prerequisite for fostering the transformative change needed to translate existing knowledge and methodologies into effective real-world practice. As argued throughout this book, such cultural transformation provides the social foundation for an evolutionary–ecosystemic paradigm that transcends the sociocentric–mechanistic worldview (Zalewski, 2017), which has traditionally defined human–environment relations. This paradigm calls for a profound redefinition of humanity's role—from a dominant force to a collaborative partner in the shared process of sustaining life (Boulangeat et al., 2022). Grounded in the dynamic biogeochemical transformations of ecosystems (Bianchi, 2021), the evolutionary–ecosystemic approach provides both a scientific and an ethical framework for this transition (Costanza et al., 2007).

3. Transdisciplinary thinking in practice: The case of Waller Creek

Initiated in response to recurrent flooding in downtown Austin, the project combines large-scale engineering interventions with ecological restoration. A 1.6-kilometer flood-control tunnel constructed beneath Waller Creek diverts stormwater flow underground before releasing it into Lady Bird Lake, significantly reducing flood risk along the corridor. At the same time, the degraded urban stream was restored and reintegrated into the surrounding city fabric through landscape interventions developed by Michael Van Valkenburgh Associates (MVVA). Across its 3.9-km length, the Waller Creek corridor was redesigned as a living system—one that functions simultaneously as infrastructure, ecological habitat, and civic space, see Fig. 37.2. The result is not only a technically resilient blue–green network woven into the urban fabric, but also a landscape where ecohydrological performance and aesthetic experience reinforce one another. The Waller Creek project demonstrates how a transdisciplinary approach can integrate hydrological engineering, ecological restoration, and landscape design within a single framework. By combining expertise from hydrology, ecology, soil science, civil engineering, and planting design, the project aligns flood-control infrastructure with habitat enhancement and public space development, illustrating how landscape performance and spatial experience can be addressed simultaneously.

The planning and design process closely aligns with the principles of transdisciplinary ecohydrology, see chapter 7. According to the initial stage, the first principle of transdisciplinary ecohydrology—the *Hydrological Principle*—site

FIGURE 37.2 **Michael Van Valkenburgh Associates, Inc., *Waller Creek Framework Plan*.** Waller Creek corridor site plan shows the scale of the project: 3.9 km of the river corridor. *Taken from: Michael Van Valkenburgh Associates Inc. (MVVA). (2013). Waller Creek Corridor Framework Plan. City of Austin, Texas.*

analysis phase involves a comprehensive assessment of the hydrological cycle, including the evaluation of water flows, water quality analyses, and hydrological modeling using the Soil and Water Assessment Tool (SWAT), among other analytical approaches. By integrating these methodologies, the hydrological assessment of Waller Creek generated a comprehensive understanding of watershed dynamics, thereby informing management and restoration strategies.

The second principle of transdisciplinary ecohydrology—the *Ecological Principle*—focuses on analyzing the structure and spatial distribution of the ecosystems in the catchment with a focus on "novel ecosystems" (Hobbs et al., 2006) and identifying opportunities for intervention through the application of Nature-based Solutions (NbS). In the case of Waller Creek, this principle was applied through detailed mapping of the creek corridor's ecological status and the planning of targeted interventions Fig. 37.4. These ranged from restoring native vegetation and improving habitat connectivity to enhancing biodiversity.

The third principle of transdisciplinary ecohydrology—*the Ecological Engineering Principle*—emphasizes the use of ecosystem properties as management tools to regulate water-biota interactions (Zalewski, 2014). In the Waller Creek project, the regulation of integrated ecohydrological processes at the river basin scale—supported by blue-green infrastructure elements such as constructed wetlands and riparian buffers—improves water quality, reduces flood risk, and creates spaces where the public can experience and engage with urban nature. A series of diagrams were created to explain the project's design goals and the ecohydrological processes to be restored Fig. 37.3.

The Before and After visualizations transform scientific understanding into accessible imagery that communicates the design vision, see Fig. 37.5. In this approach, restored ecohydrological processes are carefully integrated with urban space, allowing people to experience natural dynamics within the city environment.

This project demonstrates that well-designed restoration and landscape enhancement projects can generate both economic and social benefits, see Fig. 37.6. The Waller Creek project transformed a degraded and flood-prone urban corridor into a vibrant public landscape that attracts investment, tourism, and community activity. By integrating transdisciplinary landscape sciences, and high quality public space design, the project not only reduced long-term maintenance and flood-related costs but also significantly increased surrounding property values and urban development potential.

4. Balancing functionality and aesthetics: Resilience as lived experience

The Waller Creek project demonstrates that when ecological succession, ecohydrological function, and design intention are aligned, landscapes become adaptive systems—capable of evolving while maintaining identity and purpose. It operates simultaneously as infrastructure, habitat, and civic space. Its ecohydrological processes are not hidden beneath engineering solutions but revealed and interpreted through spatial form. In doing so, the project transforms environmental performance into public experience. This case challenges the common assumption that landscape restoration is primarily a financial burden. Well-designed hybrid landscapes that stabilise hydrological and biogeochemical cycles—reflected in improvements in water quantity, water quality, and biodiversity—can generate environmental resilience, economic value, and social benefits. Beauty, in this context, is not superficial embellishment; it is a mediator between scientific knowledge and collective action. The future of urban landscape architecture lies in this convergence. When ecohydrological function and aesthetic meaning are integrated, cities reconnect with the processes that sustain them. Resilience becomes not only quantifiable in system models but tangible in everyday experience.

As cities confront accelerating climate instability, biodiversity loss, and widening social inequalities, the deliberate integration of ecohydrological processes and aesthetic experience becomes increasingly essential. Urban landscape architecture must evolve from merely mitigating environmental damage to actively creating conditions that sustain life-supporting systems. In this transition, urban green spaces emerge not as decorative amenities but as fundamental ecological infrastructure—forming the structural foundation of resilient and sustainable urban futures.

AI disclosure

During the preparation of this work the author(s) used OpenAI's ChatGPT in order to generate the content. This document includes content generated and/or edited with the assistance of OpenAI's ChatGPT, used to support language refinement and structural improvements. All final content was reviewed and verified by the author. After using this tool/service, the author(s) reviewed and edited the content as needed and take(s) full responsibility for the content of the publication.

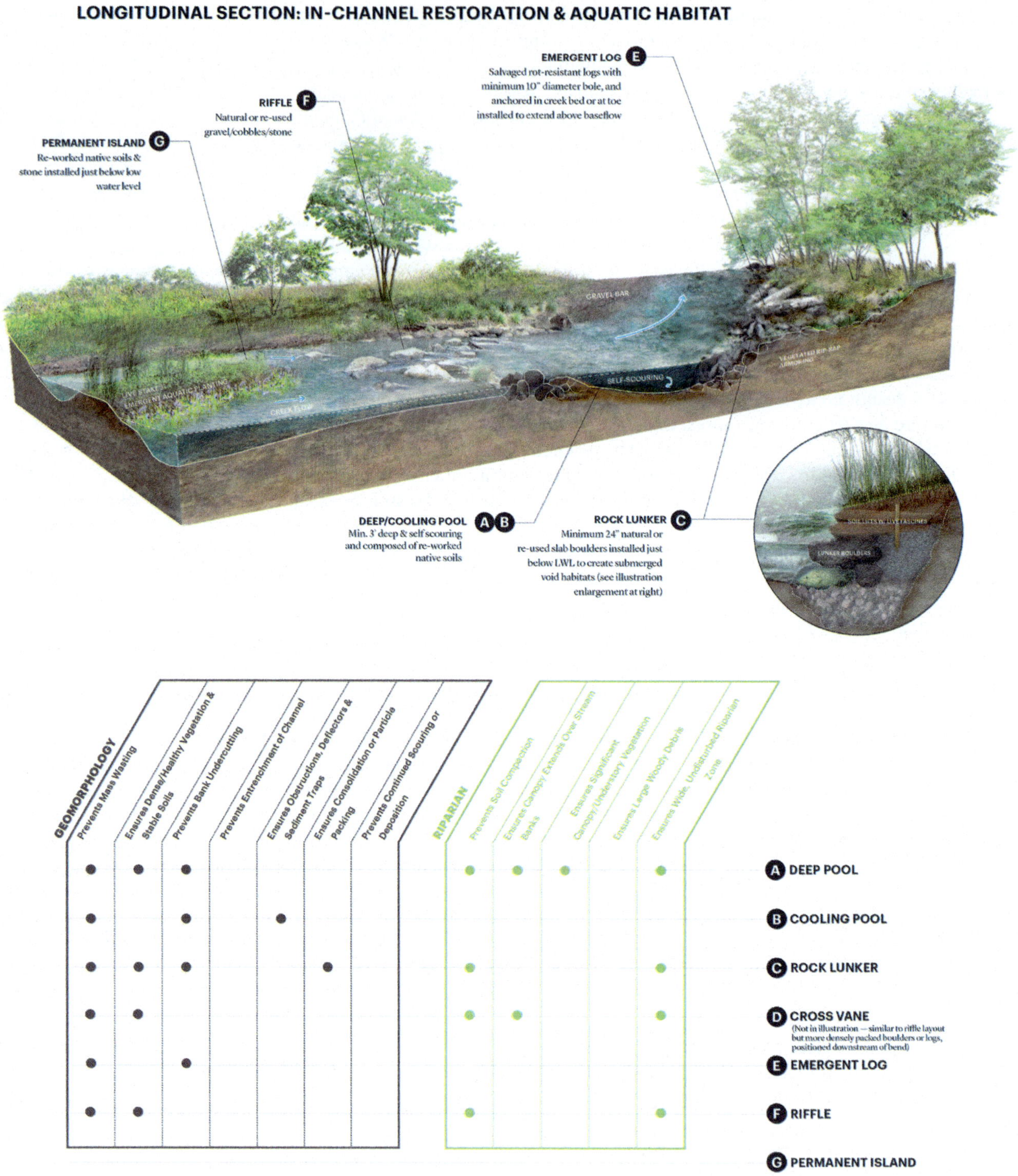

FIGURE 37.3 Longitudinal section: In-channel restoration and aquatic habitat. *Taken from: Michael Van Valkenburgh Associates Inc. (MVVA). (2013). Waller Creek Corridor Framework Plan. City of Austin, Texas.*

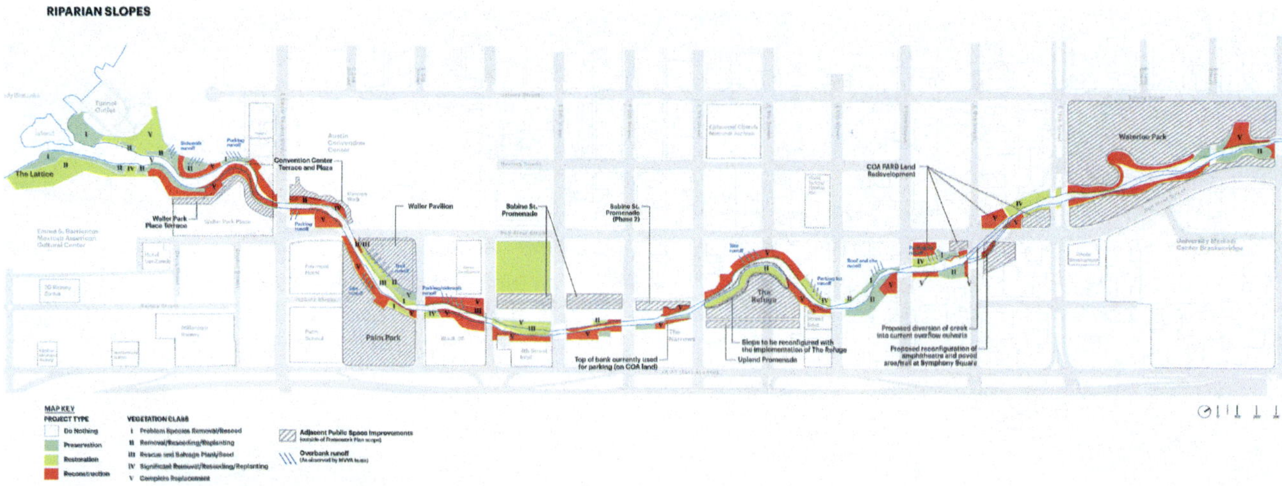

FIGURE 37.4 **Intervention strategies.** This plan describes Waller Creek intervention strategies: do nothing, preserve, restore and reconstruct. *Taken from: Michael Van Valkenburgh Associates Inc. (MVVA). (2013). Waller Creek Corridor Framework Plan. City of Austin, Texas.*

FIGURE 37.5 **Before and After Waller Creek Corridor.** *Taken from: Michael Van Valkenburgh Associates Inc. (MVVA). (2013). Waller Creek Corridor Framework Plan. City of Austin, Texas.*

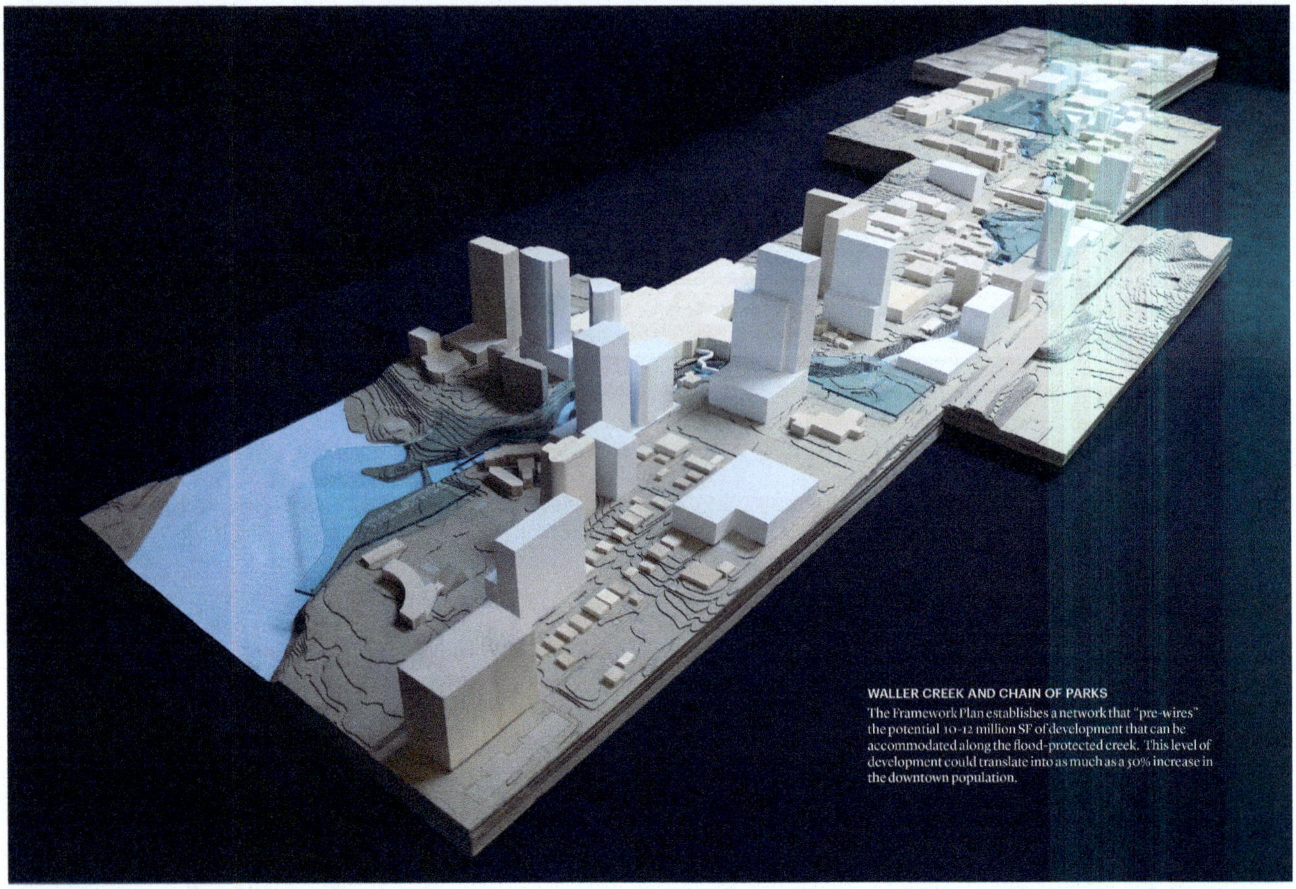

FIGURE 37.6 Proposed developments. Waller Creek and chain of parks–the framework plan establishes a network that "pre-wires" the potential 10–12 million SF of development that can be accommodated along the flood-protected Creek. This level of development could translate into as much as a 50% increase in the downtown population (Michael Van Valkenburgh Associates, Inc. MVVA, 2013). The proposed developments are shown in white on the model. *Taken from: Michael Van Valkenburgh Associates Inc. (MVVA). (2013). Waller Creek Corridor Framework Plan. City of Austin, Texas.*

References

Adger, W.N., 2003. Social capital, collective action, and adaptation to climate change. Econ. Geogr. 79 (4), 387–404. https://doi.org/10.1111/j.1944-8287.2003.tb00220.x.

Ahern, J., 2013. Urban landscape sustainability and resilience: the promise and challenges of integrating ecology with urban planning and design. Landsc. Ecol. 28 (6), 1203–1212. https://doi.org/10.1007/s10980-012-9799-z.

Benedict, M.A., McMahon, E.T., 2006. Green Infrastructure: Linking Landscapes and Communities. Island Press.

Bianchi, T.S., 2021. The evolution of biogeochemistry: revisited. Biogeochemistry 154 (2), 141–181. https://doi.org/10.1007/s10533-020-00708-0.

Boulangeat, I., Allain, S., Crouzat, E., Girard, S., Granjou, C., Poirier, C., Ruault, J.F., Paillet, Y., Arpin, I., 2022. From human-nature dualism towards more integration in socio-ecosystems studies. In: Human-nature Interactions: Exploring Nature's Values Across Landscapes. Springer International Publishing, pp. 37–49. https://doi.org/10.1007/978-3-031-01980-7_4.

Bratman, G.N., Anderson, C.B., Berman, M.G., Cochran, B., De Vries, S., Flanders, J., Folke, C., Frumkin, H., Gross, J.J., Hartig, T., Kahn, P.H., Kuo, M., Lawler, J.J., Levin, P.S., Lindahl, T., Meyer-Lindenberg, A., Ouyang, Z., Roe, J., Scarlett, L., Daily, G.C., 2019. Nature and mental health: An ecosystem service perspective. Sci. Adv. 5 (7), eaax0903. https://doi.org/10.1126/sciadv.aax0903.

Bulkeley, H., Betsill, M.M., 2005. Rethinking sustainable cities: multilevel governance and the 'urban' politics of climate change. Environ. Polit. 14 (1), 42–63. https://doi.org/10.1080/0964401042000310178.

Chiesura, A., 2004. The role of urban parks for the sustainable city. Landsc. Urban Plann. 68 (1), 129–138. https://doi.org/10.1016/j.landurbplan.2003.08.003.

Costanza, R., Graumlich, L.J., Steffen, W., 2007. Sustainability or collapse: What can we learn from integrating the history of humans and the rest of nature? Ambio J. Human Environ. 36 (7), 522–527. https://doi.org/10.1579/0044-7447(2007)36[522:SOCWCW]2.0.CO;2.

Fischer, F., 2000. Citizens, Experts, and the Environment: The Politics of Local Knowledge. Duke University Press. https://doi.org/10.2307/j.ctv11smwd9.

Gobster, P.H., Nassauer, J.I., Daniel, T.C., Fry, G., 2007. The shared landscape: what does aesthetics have to do with ecology? Landsc. Ecol. 22 (7), 959–972. https://doi.org/10.1007/s10980-007-9110-x.

Gonçalves, P., Grilo, F., Mendes, R.C., Vierikko, K., Elands, B., Marques, T.A., Santos-Reis, M., 2021. What's biodiversity got to do with it? Perceptions of biodiversity and restorativeness in urban parks. Ecol. Soc. 26 (3). https://doi.org/10.5751/ES-12598-260325.

Haase, D., Frantzeskaki, N., Elmqvist, T., 2017. Ecosystem services in urban landscapes: practical applications and governance implications. Ambio 46 (7), 591–602.

Hartig, T., Mitchell, R., de Vries, S., Frumkin, H., 2014. Nature and health. Annu. Rev. Publ. Health 35 (1), 207–228. https://doi.org/10.1146/annurev-publhealth-032013-182443.

Healey, P., 2007. Urban Complexity and Spatial Strategies: Towards a Relational Planning for Our Times. Routledge.

Hobbs, R.J., Arico, S., Aronson, J., Baron, J.S., Bridgewater, P., Cramer, V.A., Epstein, P.R., Ewel, J.J., Klink, C.A., Lugo, A.E., Norton, D., Ojima, D., Richardson, D.M., Sanderson, E.W., Valladares, F., Vilà, M., Zamora, R., Zobel, M., 2006. Novel ecosystems: Theoretical and management aspects of the new ecological world order. Global Ecol. Biogeogr. 15 (1), 1–7. https://doi.org/10.1111/j.1466-822x.2006.00212.x.

Intergovernmental Panel on Climate Change (IPCC), 2022. Climate Change 2022: Impacts, Adaptation and Vulnerability. Contribution of Working Group II to the Sixth Assessment Report of the Intergovernmental. Panel on Climate Change Cambridge University Press IPCC.

Kabisch, N., Qureshi, S., Haase, D., 2015. Human–environment interactions in urban green spaces — a systematic review of contemporary issues and prospects for future research. Environ. Impact Assess. Rev. 50, 25–34. https://doi.org/10.1016/j.eiar.2014.08.007.

Kaplan, R., Kaplan, S., 1989. The experience of nature: a psychological perspective. In: Cambridge University Press.

Kellert, S.R., Wilson, E.O., 1993. The Biophilia Hypothesis. Island Press.

Kwon, O.H., Hong, I., Yang, J., Wohn, D.Y., Jung, W.S., Cha, M., 2021. Urban green space and happiness in developed countries. EPJ Data Sci. 10 (1), 28. https://doi.org/10.1140/epjds/s13688-021-00278-7.

Marselle, M.R., Irvine, K.N., Warber, S.L., 2014. Examining group walks in nature and multiple aspects of well-being: A large-scale study. Ecopsychology 6 (3), 134–147. https://doi.org/10.1089/eco.2014.0027.

Meadows, D.H., 1999. In: Leverage Points: Places to Intervene in a System. The Sustainability Institute.

Michael Van Valkenburgh Associates, Inc. (MVVA), 2013. In: Waller Creek Corridor Framework Plan. City of Austin.

Niemelä, J., Saarela, S.-R., Söderman, T., Kopperoinen, L., Yli-Pelkonen, V., Väre, S., Johan Kotze, D., 2010. Using the ecosystem services approach for better planning and conservation of urban green spaces: a Finland case study. Biodivers. Conserv. 19 (11), 3225–3243. https://doi.org/10.1007/s10531-010-9888-8.

Ostrom, E., 2010. Polycentric systems for coping with collective action and global environmental change. Glob. Environ. Change 20 (4), 550–557. https://doi.org/10.1016/j.gloenvcha.2010.07.004.

Paudel, P.K., Lamichhane, A., Acharya, K.P., Bastola, R., 2023. Ecosystem restoration reduces community vulnerability to water-induced disasters: Need to rethink Chure conservation in Nepal. Int. J. Disaster Risk Reduct. 90, 103647. https://doi.org/10.1016/j.ijdrr.2023.103647.

Pauleit, S., Ambrose-Oji, B., Andersson, E., Anton, B., Buijs, A., Haase, D., Elands, B., Hansen, R., Kowarik, I., Kronenberg, J., Mattijssen, T., Olafsson, A.S., Rall, E., van der Jagt, A.P.N., van den Bosch, C.K., 2019. Advancing urban green infrastructure in Europe: Outcomes and reflections from the GREEN SURGE project. Urban For. Urban Green. 40, 4–16. https://doi.org/10.1016/j.ufug.2018.10.006.

Tzoulas, K., Korpela, K., Venn, S., Yli-Pelkonen, V., Kaźmierczak, A., Niemela, J., James, P., 2007. Promoting ecosystem and human health in urban areas using Green Infrastructure: a literature review. Landsc. Urban Plann. 81 (3), 167–178. https://doi.org/10.1016/j.landurbplan.2007.02.001.

Ulrich, R.S., 1984. View through a window may influence recovery from surgery. Science 224 (4647), 420–421. https://doi.org/10.1126/science.6143402.

United Nations Human Rights Council. (2021, October 8). The human right to a clean, healthy and sustainable environment (Res. A/HRC/RES/48/13). United Nations.

World Health Organization & UN-Habitat, 2016. Global report on urban health: Equitable, healthier cities for sustainable development. WHO.

Zalewski, M., 2000. Ecohydrology—The scientific background to use ecosystem properties as management tools toward sustainability of water resources. Ecol. Eng. 16 (1), 1–8. https://doi.org/10.1016/S0925-8574(00)00071-9.

Zalewski, M., 2014. Ecohydrology and hydrologic engineering: Regulation of hydrology–biota interactions for sustainability. J. Hydrol. Eng. 19 (8), 04014020. https://doi.org/10.1061/(ASCE)HE.1943-5584.0000999.

Zalewski, M., 2017. Ecohydrology measures for sustainable catchment management: An outline. In: Radecki-Pawlik, A., Pagliara, S., Hradecky, J. (Eds.), Open channel hydraulics, river hydraulic structures and fluvial geomorphology: For engineers, geomorphologists and physical geographers. CRC Press.

Chapter 38

Ecohydrology based landscape restoration

Mulugeta Dadi Belete
Hawassa University, Faculty of Biosystems and Water Resources Engineering, Hawassa, Sidama Region, Ethiopia

PHOTO CREDIT: M. DADI

1. Introduction

Landscape restoration plays a vital role in preserving ecological integrity, securing water resources, and sustaining livelihoods, especially in areas prone to land degradation and climate variability (Molden, 2007). However, the

Transdisciplinary Ecohydrology. https://doi.org/10.1016/B978-0-443-41596-8.00029-7

contemporary conventional landscape restoration practices, while often effective in certain contexts, exhibit several fundamental limitations that hinder their ability to achieve long-term sustainability. These shortcomings can be addressed through the integration of sustainability sciences such as ecohydrology, which combines ecological and hydrological principles to restore ecosystem function and resilience. The following points outline these key limitations and the potential for ecohydrology to provide more effective solutions (Belete, 2022):

- ***Limited integration of biota and hydrology***: Traditional methods emphasize physical structures like soil bunds or terracing, often neglecting the role of vegetation, soil microbes, and hydrological interactions. Ecohydrology leverages these natural synergies to improve water retention, infiltration, nutrient cycling, and overall ecosystem health, making landscapes more resilient to climate and land use changes.
- ***Loss of productive land***: Conventional measures such as fanya juu and soil bunds can occupy 8%–30% of farmland (Tenge et al., 2005; Teshome et al., 2013), reducing land available for cultivation and affecting food security. Ecohydrological approaches are designed to coexist with land use, enhancing both productivity and ecological function without sacrificing arable land.
- ***Over-engineering***: Heavy structures like gabions and concrete barriers are often inflexible and may lack community acceptance due to their artificial appearance. Ecohydrology favors nature-based solutions that are adaptive and culturally acceptable.
- ***Narrow erosion focus***: Traditional restoration often focuses solely on erosion control, overlooking broader processes like water quality regulation, nutrient cycling, and biodiversity support. Ecohydrology broadens the scope by restoring water pathways, enhancing infiltration and groundwater recharge, and supporting ecosystem services like soil fertility and carbon storage.
- ***Underutilization of biological measures***: In many projects, vegetation-based interventions are treated as secondary rather than essential. Ecohydrology positions biological components, such as plants and microbes, as coregulators of hydrological functions, integrating them into the core of restoration strategies.
- ***Lack of multifunctionality***: Conventional interventions often aim to solve a single issue, such as erosion or flooding. In contrast, ecohydrology promotes multifunctional solutions that deliver cobenefits: increased agricultural productivity, improved water quality, biodiversity conservation, and climate regulation through carbon sequestration.
- ***Misconception of passive recovery***: There is often an assumption that degraded landscapes can recover naturally over time. However, severely degraded systems typically need active interventions. Ecohydrology applies proactive, nature-based techniques to accelerate recovery, rebuild ecological function, and restore resilience.
- ***Soil carbon loss from disturbance***: Structural interventions like bunding and terracing can disturb soil and release stored organic carbon, contributing to CO_2 emissions (Lal, 2004). Ecohydrology minimizes such disturbances, emphasizing vegetation and natural stabilization processes to preserve or enhance soil organic carbon.

The above shortcomings of the conventional landscape restoration practices together with the unprecedented pressure on the Earth's ecosystems due to human activities have become increasingly untenable in this Anthropocene Era where the global ecological footprint has reached 1.7 global hectares (gha) per person, implying that humanity is consuming 1.7 times more resources than the planet can regenerate (WWF, 2010). This ecological overshoot threatens planetary health and underscores the urgency of adopting sustainable land management strategies (Steffen et al., 2015).

It is in response to the above pushing factor and the presence of ecohydrological strategy as pulling factor, a fairly new approach known as "Ecohydrology-based landscape restoration" (Belete, 2022) has emerged as a promising alternative. To overcome these limitations, it is imperative to renovate the conventional landscape restoration approaches by embedding the principles of ecohydrology, which emphasize the use of ecosystem properties as management tools. Ecohydrology provides a framework for harnessing nature's regulatory functions to enhance landscape resilience, ensuring that interventions work with, rather than against, natural processes (Zalewski, 2000; Zalewski et al., 2009).

This paper presents an applied synthesis of three case studies from the Ethiopian Rift Valley Lakes Basin, demonstrating the practical implementation of ecohydrological principles in landscape restoration.

2. Methodology

2.1 The proposed ecohydrological systemic solution (EHSS) to optimize and sustain hydrological and ecological processes in the landscape

The general question posed in the formulation of the concept of ecohydrology concerns how to regulate the biological processes of freshwater ecosystems using hydrology and—vice versa—how to use biotic ecosystem properties as a tool in

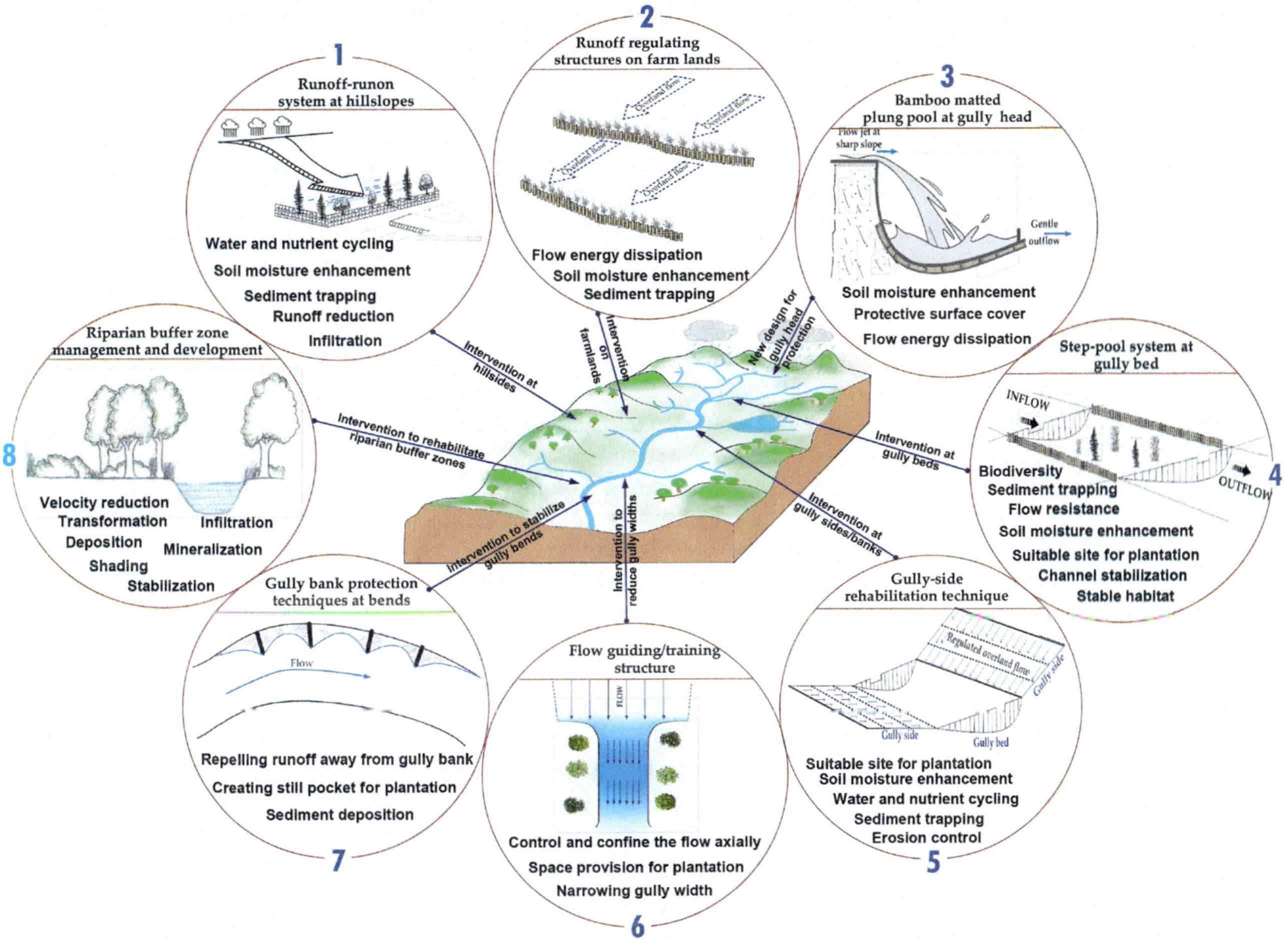

FIGURE 38.1 **Ecosystem services.** Key components of the blended semigray and green infrastructure and their ecosystem services. *From Belete, M. D., 2022.* Ecohydrology-based Landscape Restoration: Theory and Practice. *Taylor and Francis, Ethiopia, pp. 1–180. https://doi.org/10.4324/9781003309130.*

water management. As shown in Fig. 38.1, the proposed EHSS considers the entire landscape as a system where the interactions between land, water, and ecosystems are recognized for landscape restoration. It offers an integrative, nature-based approach to reversing land degradation by strategically applying the three core principles of ecohydrology: hydrological regulation, enhancement of ecosystem resilience, and ecological engineering (Zalewski, 2000). By restoring the biophysical functionality of degraded landscapes, the system is designed to reinforce the synergy between water dynamics and biological processes to ensure long-term sustainability and productivity.

The system considers the upstream–downstream continuum by dealing with the hydrological regulation (*1st principle of EH*) and ecological feedback (*2nd principle of EH*) following the typical sequence of land use and ecological zones in agricultural and seminatural watersheds: hillslopes → farmland → gully network → riparian buffer zone. The sequence represents how water and sediment move through a watershed from the headwaters (hillslopes) to the riparian zones.

At the hillslope level, the runoff–run-on system plays a foundational role by redistributing water, nutrients, and organic matter from bare zones to vegetated patches. This mechanism enhances infiltration and moisture retention, supports plant growth, and initiates a self-sustaining ecosystem cycle. Its design is rooted in the source–sink and patch–interpatch ecological models (Ludwig et al., 1997), where water and nutrient fluxes are regulated to enhance biomass production and prevent further degradation. This component reflects the first ecohydrology principle—hydrological regulation—while contributing to ecological resilience and engineering through self-reinforcing feedback loops.

In agricultural landscapes, an overland flow regulating system has been introduced to manage surface runoff and minimize sheet erosion. Unlike conventional terraces that occupy substantial land area, this system employs minimal earthwork structures paired with vegetation to intercept runoff, improve infiltration, and gradually transition into vegetative barriers. Over time, the planted grasses take over the structural role, stabilizing soils and sustaining ecohydrological functionality (Nyssen et al., 2007). This illustrates dual regulation by combining physical design with ecological succession and highlights the second and third ecohydrology principles in practice.

To address advanced forms of land degradation, the solution extends into gullied landscapes by applying a comprehensive and ecohydrologically informed strategy. At the gully head, bamboo-matted plunge pools reduce water velocity and erosion potential, enabling vegetation establishment. Along gully beds, step-pool formations slow down flow, trap sediments, and support habitat regeneration (Chin et al., 2008). At bends, groynes stabilize eroding banks, while runoff–run-on techniques reinforce gully sides with redistributed water and nutrients that promote vegetative growth. This approach incorporates Lacey's regime theory (Lacey, 1934), especially in reconfiguring excessively wide gully segments into stable, multifunctional eco-corridors. The strategy exemplifies how ecohydrology can be used to regulate flow, restore soil stability, and enhance ecosystem function across critical landscape units.

Further downstream, riparian buffer zones act as the final defense against aquatic ecosystem degradation. These vegetated strips intercept overland and subsurface flows, reducing nutrient and sediment loads before they enter water bodies. They also provide essential ecological functions, including habitat connectivity, erosion control, microclimate regulation, and pollutant filtration (Lowrance et al., 1995; Naiman and Décamps, 1997). Functioning as biogeochemical filters, these buffers reduce land-to-water pollution transfer and enhance the resilience of adjacent aquatic ecosystems.

Together, these eight interconnected components form a comprehensive, adaptive system that operates across ecological scales and landscape types. Each aligns with one or more principles of ecohydrology, but their integration magnifies their collective impact. The system not only rehabilitates degraded landscapes but also builds long-term ecological memory, adaptive capacity, and socio-environmental resilience (Zalewski, 2008). It is a scalable, cost-effective model of restoration rooted in the dynamic interaction of water and life, offering a pathway toward sustainable land and water management in vulnerable regions.

3. Description of the experimental watershed

The Lake Hawassa subbasin, located within the Central Rift Valley of Ethiopia, spans approximately 1400 km^2 and features a complex and heterogeneous landscape. The basin's topography varies from highland plateaus and rolling hills to lowland plains, with elevations ranging between 1500 and 2100 m above sea level. The Lake Hawassa, a key hydrological feature, is a terminal lake with no surface outlet, making it highly susceptible to fluctuations in water balance, sediment influx, and nutrient loading due to anthropogenic and climatic influences.

Hydrologically, the basin experiences a bimodal rainfall pattern, with annual precipitation ranging from 800 mm to 1200 mm. The climate varies significantly depending on altitude, with temperatures fluctuating between 10°C and 30°C.

The dominant land uses in the subbasin include smallholder agriculture, commercial farming, urban settlements, and industrial activities. However, rapid population growth, deforestation, land degradation, and unsustainable land-use practices have intensified environmental stressors. Key challenges include accelerated soil erosion, declining water quality, wetland degradation, and habitat fragmentation. These stressors not only compromise the ecological integrity of the Lake Hawassa but also threaten the livelihoods of local communities dependent on ecosystem services.

3.1 Set-up the demo sites

As shown in Table 38.1, this paper synthesizes results of three ecohydrological interventions emanated from the systemic approach shown in Fig. 1. The interventions include:

- *Experimental site 1:* Rehabilitation of a former mining site behaving like a gully system (component #4) at Wondo Genet District of Lake Hawassa subbasin and
- *Experimental site 2:* Restoration of a highly degraded hillslope (component #1) at Boricha District of the Lake Hawassa subbasin
- *Experimental site 3:* Application of nature-based solutions on erosion-prone farmland (component #2) at Shalla District of the Lake Hawassa subbasin

TABLE 38.1 The three experimental sites in Lake Hawassa Subbasin of Ethiopian Rift Valley Lakes Basin.

Before	Descriprion	After
	**Experimental site 1:** **Problem description:** This site is located in Wondo Genet district within the lake Hawassa watershed and covers an area of 12 ha. It was previously a mining site used for extracting selected materials for asphalt road construction in the district. The excavation extended deep into the subsoil and was left untreated. **Objective:** To regulate runoff through a water-spreading system using locally available wooden materials.	
	**Experimental site 2:** **Problem description:** This dysfunctional landscape (left) is located in Boricha district within the lake Hawassa watershed and covers an area of 12 ha. **Objective:** To enhance the ecohydrological functionality of the landscape by increasing its capacity to retain vital resources by introducing ecological patch-interpatch sequences	
	**Experimental site 3:** **Problem description:** This experimental site is located in Shalla wereda (district) of the Oromiyaa Regional State within the lake Hawassa subbasin. The site represents a typical case of erosion-prone farmlands. **Objective:** To reduce soil erosion by regulating overland flow through the implementation of ecohydrologically oriented green- (semi) gray infrastructure serving as bunding system with slim design and minimum soil disturbance.	

Photos credit: M. B. Belete.

TABLE 38.2 Results of landscape organization analysis (at experimental site #1).

Landscape organization	Number of patches per 10 m =	*Passive restoration*	0.32
		Active restoration	1.4
	Landscape organization index[a]=	*Passive restoration*	0.16
		Active restoration	0.94
	Average inter-patch length =	*Passive restoration*	22 m
		Active restoration	2 m

[a]*The landscape organization index (LOI) is the proportion of the length of all patches to the total length of the gradsect, thus the percentage of the gradsect that consisted of patches. These parameters reflect the ability of a system to either conserve or lose resources (Tongway and Hindley, 2004). It is calculated as = Length of patches/length of transects. A totally bare transect would have an index of 0 (zero) or if it was all patch) the index would be 1.*

4. Results and discussion

4.1 Setting the foundation for ecosystem resilience through hydrological regulation *(the first principle of ecohydrology)*

4.1.1 Landscape organization index (LOI) and patch connectivity as indicators for regulation of hydrological connectivity in the landscape

The Landscape Organization Index (LOI) is a quantitative measure that describes the spatial arrangement and connectivity of resource-conserving patches (vegetation, organic matter, stable soil) within a landscape. It reflects the degree to which a landscape retains or loses resources (water, nutrients, and sediments) by influencing hydrological connectivity (Tongway and Ludwig, 2010).

As shown in Table 38.2 (Belete, 2022), compared the results of landscape organization analysis between ecohydrologically treated (active restoration) and untreated sites. Results of experimental site #1 show that the active restoration site is found to have a landscape organization index value of 0.94 (=94% of the gradsect is consisting of resources conserving patches) as compared to 0.16 (= only 16% of the gradsect is consisting of resources conserving patches) in the case of the passive restoration site.

Implication to landscape restoration:

- Higher LOI (0.94 in active restoration sites) implies that the landscape is well-structured, with continuous vegetation patches that slow down runoff, enhance water infiltration, and promote stable hydrological connectivity. It generally suggests functional source-sink dynamics, where vegetated patches act as sinks, capturing water and nutrients.
- On the other hand, lower LOI (0.16 in passive restoration sites) implies that the landscape is fragmented, dominated by bare soil and erosion-prone areas, leading to rapid water loss (uncontrolled runoff and nutrient loss), high erosion, and nutrient depletion.

4.1.2 Effect on enhancement of soil stability

Soil stability is a critical parameter in Landscape Functionality Analysis (LFA), as it determines the soil's ability to withstand erosive forces and maintain structural integrity under varying environmental conditions.

Belete (2022) reported that pre-intervention @experimental site #2 (baseline condition) indicated an LFA-derived soil stability index of 30.6% implying highly susceptibility of the landscape to erosion, sediment loss, and overall ecosystem degradation. Whereas, the post-intervention record showed a value of 47.2%, demonstrating substantial improvements.

4.1.3 Enhancement of the ability of soil to absorb water as indicated by infiltration capacity parameters of landscape functionality analysis (LFA)

According to Belete (2022), the ecohydrological intervention significantly influenced microtopography (surface roughness), leading to notable improvements in soil infiltration potential.

- *Pre-intervention*: before implementing restoration measures, the soil exhibited a low infiltration potential of 10.4% **@experimental site #1**, as assessed using Landscape Functionality Analysis (LFA). This indicated a severely limited ability of the soil to absorb water.

- *Post-intervention:* following the application of ecohydrological techniques, infiltration capacity rose to 25.1%, more than doubling its initial value. This improvement reflects enhanced soil porosity, deeper root penetration, and increased organic matter content, all of which contribute to greater water retention and improved hydrological function.

4.1.4 Effect on the enhancement of soil moisture

Belete (2022) took 10 randomly selected samples (Table 38.3) along the landscape (experimental site #1) reported the impact of ecohydrological intervention to significantly improve water availability for biota, as evidenced by the observed increase in soil moisture ($p(x \leq T) = 0.001$).

4.1.5 Ecological response to hydrological regulation (the second principle of ecohydrology)

The second principle of ecohydrology emphasizes how biological systems respond to hydrological regulation, particularly in terms of nutrient cycling, vegetation dynamics, and ecosystem functionality. In this case, the results show how hydrological improvements lead to enhanced nutrient retention and cycling within the ecosystem, ultimately increasing productivity and resilience.

4.1.6 Effect on nutrient cycling potential

Before ecohydrological interventions, the landscape exhibited poor nutrient cycling efficiency, with a recorded potential of 10.5% @experimental site #1as measured using LFA technique (at experimental site #1) (Belete, 2022). This low value indicates a degraded ecosystem where essential biogeochemical processes, such as decomposition, mineralization, and nutrient retention, were significantly impaired. Following the implementation of ecohydrological interventions, nutrient cycling potential increased significantly to 25.6%, indicating a marked improvement in ecosystem functionality.

4.1.7 Effect on species richness

In the passive restoration treatment, species richness was recorded at eight species, indicating a relatively low level of biodiversity. This suggests that the environmental conditions were not yet optimal for supporting a diverse range of plant species. On the other hand, after the ecohydrological intervention, species richness increased to 18 species, representing a significant improvement in biodiversity.

4.1.8 Effect on plant species diversity

In the passive restoration site, plant species diversity was recorded at 0.44, a relatively low value indicating limited species variation within the ecosystem. The low diversity index reflects an ecosystem that is still struggling to recover, with weak species interactions and limited resilience to environmental changes. Following active restoration efforts, plant species diversity increased significantly to 1.22, nearly three times higher than in passive restoration. This jump in diversity suggests that the ecosystem became more complex and functionally dynamic.

4.1.9 Advancing the third principle of ecohydrology in the context of "landscape–ecosystem services—stewardship nexus

The third principle of ecohydrology emphasizes the integration of ecological and societal processes to ensure sustainable use of freshwater and terrestrial ecosystems. This principle recognizes that ecological restoration efforts must not only address biophysical functionality but also support human well-being and long-term environmental stewardship. In this context, the landscape–ecosystem services–stewardship nexus provides a conceptual and practical framework for aligning ecological integrity with socioeconomic benefits. Landscapes are more than physical spaces-they are dynamic socio-ecological systems that generate vital ecosystem services such as water regulation, soil fertility, climate moderation, and biodiversity support. By embedding the principles of stewardship—understood as the responsible and participatory management of natural resources—into restoration efforts, ecohydrology offers pathways for coproducing ecological resilience and community livelihoods.

Belete et al. (2024) reported that through field studies, 11 key ecosystem services Fig. 38.2 were identified as direct outcomes of the ecohydrological intervention:

The findings suggest that ecohydrology, through its emphasis on the interactions between hydrological processes and biological systems, leads to a range of positive outcomes. By integrating ecological principles into hydrological practices,

TABLE 38.3 Results of soil moisture (%) at passive and active restoration sites (Belete, 2022).

	Samp1	Samp2	Samp3	Samp4	Samp5	Samp6	Samp7	Samp8	Samp9	Samp10	Samp11	Ave.	Std.
Active restoration	11.08	3.66	21.99	14.1	8.58	13.12	7.06	5.82	10.8	14.44	8.3	10.81	5.06
Passive restoration	7.8	5.37	7.38	5.74	7.18	7.86	4.49	–	–	–	–	6.55	1.33

FIGURE 38.2 *The diverse ecosystem services acknowledged by the community (stewards) in positively influencing the socioeconomic system.*

ecohydrology addresses the limitations of traditional approaches that rely heavily on structural solutions, such as terraces and check dams.

5. Conclusion and recommendations

5.1 Conclusion

This study demonstrates the pivotal role of ecohydrological principles in enhancing ecosystem resilience, particularly within degraded landscapes. The application of the first principle of ecohydrology—hydrological regulation—was shown to significantly improve landscape functionality, as evidenced by marked increases in the LOI, soil stability, infiltration capacity, and soil moisture content. These hydrological improvements laid the foundation for robust ecological responses.

The second principle—biological response to hydrological change—revealed that restored hydrological conditions led to measurable gains in nutrient cycling, species richness, and plant species diversity, indicating improved ecosystem productivity and biodiversity. Such changes are critical for long-term ecological resilience and functional recovery of degraded ecosystems.

Finally, the third principle—integration of ecological and societal processes—was advanced through the landscape–ecosystem services–stewardship nexus. The ecohydrological interventions delivered not only biophysical benefits but also catalyzed community recognition of 11 key ecosystem services, strengthening the socioecological fabric of the landscape. These findings highlight the potential of ecohydrology as a transdisciplinary tool for uniting enhanced sustainability potential of cachment WBSR with human well-being.

5.2 Recommendations

Based on the findings of this study, it is recommended that ecohydrological interventions be scaled up and replicated in other degraded landscapes with similar ecological and socioeconomic conditions. The demonstrated improvements in hydrological regulation, soil functionality, and biodiversity indicate that ecohydrology provides an effective nature-based solution for landscape restoration. Moreover, the success of restoration efforts was closely linked to the involvement and recognition of local communities as stewards of the land. Therefore, it is crucial to promote community-based stewardship approaches by enhancing local participation, providing capacity-building opportunities, and ensuring that ecological interventions align with community values and livelihoods. Embedding the principles of stewardship into restoration planning strengthens the sustainability of ecosystem services and fosters long-term ecological and social resilience.

Furthermore, ecohydrology should be mainstreamed into watershed governance frameworks and restoration policies to reduce dependency on conventional, often costly, structural solutions such as terraces and check dams. Incorporating ecohydrological principles into policy can enhance the effectiveness of restoration strategies, especially in climate-vulnerable and resource-constrained areas. Finally, interdisciplinary research and knowledge exchange should be encouraged to expand the understanding of ecohydrological dynamics and their implications for ecosystem service delivery, biodiversity conservation, and sustainable development. Collaboration among hydrologists, ecologists, social scientists, and policymakers will be essential to unlocking the full potential of ecohydrology in addressing today's complex environmental challenges.

References

Belete, M.D., 2022. Ecohydrology-based Landscape Restoration: Theory and Practice. Taylor and Francis, Ethiopia, pp. 1–180. https://doi.org/10.4324/9781003309130.

Belete, M., Richards, N., Gehrels, A., 2024. Ecosystem services provision through nature-based solutions: a sustainable development pathway to environmental stewardship as evidenced in the protecting Lake Hawassa partnership in Ethiopia. Socio-Ecol. Pract. Res. 6, 331–345. https://doi.org/10.1007/s42532-024-00193-x.

Chin, A., Anderson, S., Collison, A., Ellis-Sugai, B.J., 2008. Linking theory and practice for restoration of step-pool streams. Environ. Manag. 43 (4), 645–661. https://doi.org/10.1007/s00267-008-9171-x.

Lacey, G., 1934. Uniform flow in alluvial rivers and canals (includes appendices). In: Minutes of the Proceedings of the Institution of Civil Engineers, 237, pp. 421–453. https://doi.org/10.1680/imotp.1934.15351.

Lal, R., 2004. Soil carbon sequestration to mitigate climate change. Geoderma 123 (1–2), 1–22. https://doi.org/10.1016/j.geoderma.2004.01.032.

Lowrance, R.R., Altier, L.S., Newbold, J.D., Schnabel, R.R., Groffman, P.M., Denver, J.M., Todd, A.H., 1995. Water quality functions of Riparian forest buffer systems in the Chesapeake Bay watershed. Environ. Manage. 21, 687.

Ludwig, J., Tongway, D., Freudenberger, D., Noble, J., Hodgkinson, K., 1997. Landscape Ecology, Function and Management: Principles from Australia's Rangelands. CSIRO Publishing.

Molden, D., 2007. Water for Food, Water for Life: A Comprehensive Assessment of Water Management in Agriculture. Earthscan.

Naiman, R.J., Décamps, H., 1997. The ecology of interfaces: Riparian zones. Annu. Rev. Ecol. Systemat. 28, 621–658. https://doi.org/10.1146/annurev.ecolsys.28.1.621.

Nyssen, J., Poesen, J., Gebremichael, D., Vancampenhout, K., D'aes, M., Yihdego, G., Govers, G., Leirs, H., Moeyersons, J., Naudts, J., Haregeweyn, N., Haile, M., Deckers, J., 2007. Interdisciplinary on-site evaluation of stone bunds to control soil erosion on cropland in Northern Ethiopia. Soil Tillage Res. 94 (1), 151–163. https://doi.org/10.1016/j.still.2006.07.011.

Steffen, W., Richardson, K., Rockström, J., Cornell, S.E., Fetzer, I., Bennett, E.M., Biggs, R., Carpenter, S.R., de Vries, W., de Wit, C.A., Folke, C., Gerten, D., Heinke, J., Mace, G.M., Persson, L.M., Ramanathan, V., Reyers, B., Sörlin, S., 2015. Planetary boundaries: guiding human development on a changing planet. Science 347 (6223). https://doi.org/10.1126/science.1259855.

Tenge, A.J., De graaff, J., Hella, J.P., 2005. Financial efficiency of major soil and water conservation measures in West Usambara highlands, Tanzania. Appl. Geogr. 25 (4), 348–366. https://doi.org/10.1016/j.apgeog.2005.08.003.

Teshome, A., Rolker, D., de Graaff, J., 2013. Financial viability of soil and water conservation technologies in northwestern Ethiopian highlands. Appl. Geogr. 37 (1), 139–149. https://doi.org/10.1016/j.apgeog.2012.11.007.

Tongway, D.J., Hindley, N.L., 2004. Landscape Function Analysis: Procedure for Monitoring and Assessing Landscapes-With Special Reference to Minesites and Rangelands. CSIRO Sustainable Ecosystesm, Canberra.

Tongway, D.J., Ludwig, J.A., 2010. Restoring Disturbed Landscapes: Putting Principles into Practice. Island Press.

WWF International, 2010. Living Planet Report 2010: Biodiversity, Biocapacity and Development. WWF, Gland, Switzerland.

Zalewski, M., 2000. Ecohydrology — the scientific background to use ecosystem properties as management tools toward sustainability of water resources. Ecol. Eng. 16 (1), 1–8. https://doi.org/10.1016/s0925-8574(00)00071-9.

Zalewski, M., 2008. Ecohydrology of urban aquatic ecosystems for healthy cities. In: Aquatic Habitats in Sustainable Urban Water Management: Science, Policy and Practice, p. 95.

Zalewski, M., Harper, D.M., Wagner, I., 2009. Ecohydrology—Why demonstration projects throughout the world? Ecohydrol. Hydrobiol. 9 (1), 3–11. https://doi.org/10.2478/v10104-009-0043-7.

Chapter 39

Theoretical basis for implementing the principles of ecohydrology in a tropical catchment

Nic Pacini[1, 2] and David M. Harper[1, 3]

[1]*University of Leicester, School of Geography, Geology and the Environment, Leicester, United Kingdom;* [2]*Department of Environmental Engineering, University of Calabria, Arcavacata di Rende, Italy;* [3]*Freshwater Biological Association, Milnthorpe, Cumbria, United Kingdom*

PHOTO CREDIT: NIC PACINI.

Transdisciplinary Ecohydrology. https://doi.org/10.1016/B978-0-443-41596-8.00039-X

1. Restoring biotic regulation

The prevailing neo-Darwinian paradigm typically tends to conceptualize ecosystems as passive recipients of disturbances induced by anthropogenic activities, which can adapt to these external changes. This is particularly clear from the limited understanding that most people have of how relevant plants and bacteria are in ecosystems, as highlighted for the first time by Wandersee and Schlusserl (1999), coining the expression "plant blindness" to epitomize this phenomenon. By contrast, the innovative contribution of ecohydrology in emphasizing the regulatory roles of plants and bacteria as critical components driving the main ecosystem processes aligns this discipline with other modern ecological regulatory frameworks such as the Gaia hypothesis (Lovelock and Margulis, 1974) and biotic regulation (Gorshkov et al., 2000).

In the Lake Naivasha Catchment, Kenya (see Chapter 31), the principles of ecohydrology (Zalewski, 2000; Harper and Mavuti, 2004; Zalewski et al., 2008) enable us to interpret the role of native plant communities in stabilizing nutrient fluxes, water discharge and levels, and the catchment's thermal regime. In Naivasha, these are: Papyrus—*Cyperus papyrus*—swamp grows within and at the lakeshore; Acacia—*Vachellia xanthophloea*—woodland in areas surrounding the lake, with mixed papyrus-dominated, species-rich grassland between the two; and native Red Cedar—*Juniperus procera*—forest in the upper catchment.

We have tried to restore these plant communities to support the resilience of local ecosystems. We implemented pilot demonstration projects to rationalize the management of wetlands, to support the renewal of native plant communities, to treat wastewater, and to reverse the degradation of the catchment area (Wanjala et al., 2018). These pilot studies looked to develop suitable wetland management methods that local communities could apply more extensively after practical teachings that were integrated within the project. Solutions described below focus on ecohydrological methods. Success will eventually depend on the adoption of these tools by the catchment communities, fully supported by environmental policies implemented by the Kenyan Government and underpinned by multinational NGOs.

2. Anthropogenic impact in the Lake Naivasha catchment

Over the past century, anthropogenic impacts upon Lake Naivasha have turned it into a "novel ecosystem" (Hobbs et al., 2006; Harper et al., 2024)—a system characterized by a combination of species abundances, dominated by invasive aliens never observed before in its history, leading to the establishment of a new ecological dynamic equilibrium.

The major drivers of ecosystem change were not direct anthropogenic pollution or water abstractions, as might have been expected, but two highly influential processes that developed in parallel:

1) Timber extraction from surrounding mountain forests, deforestation of stream riparian zones, and clearing of the lakeshore papyrus swamp have radically affected the structure of the catchment's vegetation (Pacini et al., 2024) and hence ground temperature and rainfall patterns.
2) Spread of directly (and/or indirectly) introduced alien invasive plants (Harper et al., 1995; Adams et al., 2002; Giberti et al., 2022), fishes (Gherardi et al., 2011; Pacini et al., 2021), crustaceans (Foster and Harper, 1999), and mollusks (Albrecht et al., 2025) have entirely restructured aquatic trophic chains and revolutionized human use of the lake's resources.

A new composition of plant assemblages has developed through change in the dominant life form and in the physical structure of ecosystems (i.e., from forest to pasture, from wooded savannah to dry bushland, from riparian wetland to farmland, and from submerged plant community in the lake littoral to floating aliens). These have modified the temperature regime of the whole catchment because of a considerable decrease in evapotranspiration. Degradation of the Small Water Cycle of local evaporation–condensation–precipitation, *sensu* Kravčík et al. (2008), weakened biotic control over hydrological processes. Land-to-water fluxes became more intense with an increase in upstream erosion and in downstream sedimentation rates, at the same time as nutrient accumulation in receiving aquatic ecosystems and a rise in lake water levels.

These recent changes in the catchment followed certain expectations. Lake Naivasha was predicted to develop eutrophic conditions (Harper et al., 1993), and it did, as shown by the increase in planktonic primary production (Hubble and Harper, 2002; Pacini et al., 2021), intensified cyanobacterial blooms (Harper, 2006; Krienitz et al., 2013; Zepernick et al., 2024), frequent oxygen supersaturation (Keyombe and Waithaka, 2017), silicate depletion (Pacini et al., 2021), and occasional fish kills (Njiru et al., 2017).

Dissolved nitrogen and phosphorus did not significantly increase within the water column, however, because constantly warm temperatures favored rapid uptake within plant biomass; the half-life of a molecule of phosphate in Lake Naivasha is a few seconds (Peters and McIntyre, 1976). Increased nutrient loading did not cause direct effects on the aquatic fauna. However, an increase in temperature and in the ratio of dissolved phosphate:silicate was likely to have significantly tipped the balance in favor of a more frequent cyanobacterial dominance of the lake phytoplankton (Huisman et al., 2018; Wang et al., 2021). Indirect effects on the aquatic fauna that could be related to cyanobacterial biomass increase included sudden supersaturation/deoxygenation cycles causing temporary asphyxia, potential build-up of toxic ammonia (NH_3) as a consequence of photosynthesis-driven high pH (Wang and Leung, 2015; Pacini et al., 2019), and increased risk of poisoning through ingestion of cyanobacterial toxins (Krienitz et al., 2013; Zepernick et al., 2024).

Predictions from hydrological modeling (Becht and Harper, 2002) did not match the drastic variations in water levels observed during the last 15 years, however. Traditional catchment hydrology (Odongo et al., 2015) failed to perceive the significant increase in precipitation that has occurred since 2010, which was eventually resolved by integrating satellite recordings (Herrnegger et al., 2021). Byrne et al. (2024) have now produced evidence of a highly contrasted spatial pattern of rain-prone and dry catchments across eastern Africa, with a short distance between each other. This outcome does not show a region affected by excessive drainage and overuse of aquatic resources, as was initially predicted from hydrological modeling. Rather, it shows a degradation of local biotic control over precipitation, i.e., a disconnection between local Small Water Cycle dynamics and the Large Water Cycle (Kravčík et al., 2008). In other words, while water vapor generated by evaporation from the Indian Ocean is the main source of moisture sustaining East African ecosystems, catchment forests once represented key elements in this Water Cycle as "rain acceptors." Sharp increases in vapor pressure generated by their intense transpiration tended to promote local condensation and atmospheric depression. Forests trigger moist air circulation. In this way, according to the *biotic pump* theory of Makarieva and Gorshkov (2007), the *evaporative force* exerted by forests drew distant humid air masses inwards that ensured frequent moderate rainfall occurred over the entire catchment. In this way, the *biotic pump* is an atmospheric extension of the ecohydrology concept, reaching beyond the catchment scale.

The current state of the Naivasha catchment, however, is that ongoing deforestation has lowered precipitation, which is now focused upon the few remaining high-altitude forests, where rains occur because of orographically determined cooler temperatures. This is causing an uneven distribution of rainfall, with increased soil erosion in some catchments, which causes floods and raised lake water levels, while other catchments nearby are still dry. This results in a highly uneven distribution of rainfall over the region.

3. Ecohydrological solutions

Designing ecohydrological solutions for this novel Naivasha ecosystem thus implies developing strategies with three parallel objectives.

3.1 Wise use of the opportunities offered by the new Lake Naivasha ecosystem for a sustainable delivery of ecosystem services to the local population

In particular, research targeting fish biology, fish population distribution, and relative densities has concluded that fishing pressure on alien invasive carp and catfish should be unimpeded to prevent the potential impact on the aquatic ecosystem by these species, which are able to affect both plankton and benthos at different stages of their ontogenic cycles (Pacini et al., 2021). Their flesh is both healthy and tasty. Specific training in fish processing methods could enhance income for local fishing communities. Crayfish allowed the development of a dedicated fishery in the past, but its numbers have more recently significantly decreased. Newcomer species—most recently two species of catfish (*Clarias gariepinus* and *Synodontis victoriae*; the former far more plentiful)—could be exploited optimally, as well as the recently introduced Nile perch *Lates niloticus*. The last two introductions, *S. victoriae* and *L. niloticus*, appeared to come unexpectedly from government-sponsored tilapia fingerlings introduced from fish farms near Kibos, on Lake Victoria's shore (Wanjala, pers. comm.). Tilapiine species fetch a higher market price, so their populations would benefit from the identification and protection of shoreline breeding sites and a targeted decrease in the density of carp and catfish through intensive fishing pressure. The Nile perch population does not seem to be expanding, probably because of insufficient water transparency.

3.2 Restoration of its previous ecological state, as far as it is possible

The dual regulation principle of ecohydrology guides strategies to optimize the immediate benefits of sustainable development through the restoration of the Small Water Cycle. To retain biotic control, ecosystems must preserve their structural integrity, including plant species assemblage composition and populations of key species with appropriate age distributions. Ecosystem restoration should come as a secondary priority to protection—in particular, the protection of old-growth forests, as their unique ecological properties can be impossible to reproduce once lost. Reforestation is always to be preferred to afforestation (planting trees where there was no forest before). The latter can increase tree cover, but it has much lower chances of success; it may not replicate the original forest ecosystem's full ecological functions, and it can sometimes disrupt existing ecosystems or hydrological cycles.

In Naivasha, we planted trees in the upper catchment by integrating them within the edges of agricultural plots on cleared forests as a strategy to avoid vandalism and incentivize farmer participation in supporting restoration efforts. The use of a mixed-species approach, combining endemic slow-growing hardwoods with fast-growing exotic species, provided farmers with the opportunity to harvest timber through progressive thinning of the tree lines. Additionally, establishing tree rows along the contours effectively mitigated soil erosion while preserving space for cultivating crops. This multipurpose silvicultural system enhanced local biodiversity and contributed to the moderation of the local microclimate. Farmers who adhered to our initiative by promising to look after the young trees for at least 3 years received a number of saplings delivered at the doorstep. Successive monitoring trips confirmed that about 50% of the trees survived until the next year, while the remainder were destroyed by cattle trampling and by goat grazing.

Papyrus restoration was conducted along the lakeshore by transplanting young papyrus shoots, approximately 4 weeks old, into floating mats constructed from recycled biodegradable materials (see Chapter 31). This method effectively addressed two primary threats to papyrus regeneration: damage caused by cattle and submersion-related mortality, enhancing the likelihood of progressively regenerating a riparian papyrus fringe along the lakeshore. It proved to be successful only within restricted access compounds and artificial lakeshore wetlands managed by flower farm attendants, however. An initiative aiming at promoting the wise use of papyrus as a source of charcoal (which should have generated an interest in preserving papyrus as a raw material) was abandoned following a dramatic reduction of lake fringing papyrus due to water level rise. This has resulted in high usage of the remaining narrow width of riparian land, so it currently does not offer options for sustainable use.

3.3 Development of new nature-based solutions for improved water storage and water treatment in the catchment

3.3.1 Artificial wetlands for flower farm effluent recycling

Until recently, the hinterland of Lake Naivasha hosted the largest and most intensively managed cut-flower production hub in the world. Because of the risk of storm damage, most production has relied on plants grown in large (>10 ha) greenhouses, irrigated with lake water or with water taken from the shallow lake aquifer. Water use was intensive, and the effluents were often released directly into the lake. Starting from around 2005, however, the largest flower farms invested in biological water treatment using digestors and wetlands discharging effluents into small (ca. 300–500 m^2) artificial lagoons with dense plant mats, which developed a diverse fauna. Water analysis showed that a substantial portion of nutrients and contaminants was effectively retained by the wetlands, which could then be used as a source for irrigation (after microfiltration through 200 μm filters) in a closed-circuit water recycling mode (see also Atuga and Jembe, 2022).

In 2009, water levels were at their lowest for 2 decades, capturing international attention. The UK Prince of Wales' International Sustainability Unit (2010) was invited to undertake a fact-finding mission to the lake by the Kenyan prime minister. It suggested there should be a single coordinating body for decisions about the lake, to be called Imarisha (Swahili for "strengthen") Naivasha, which was gazetted in May 2011. It produced a 5-year plan for achieving sustainability from 2012 to 2017, building upon three previous lake management plans. It completed, in partnership with the Water Resource Users Associations (WRUAs), the first survey of all water users throughout the lake and catchment areas (including groundwater), which showed a high proportion were unlicensed. All abstractors voluntarily agreed to a "traffic-light" system of abstraction against water-level change. It also established a 3-year Water Allocation Plan 2010–12 to assist this voluntary agreement. Although the Water Allocation Plan's lifeline ended in 2012 and has not been reviewed to date, the plan is still used to regulate abstraction, although, at the time of writing, high lake levels have made voluntary restrictions on abstraction unnecessary.

3.3.2 Rainwater harvesting and water storage pan-dams for irrigation supply

Water availability is an increasingly critical issue in the Naivasha catchment, where approximately one-third of the runoff occurs during intense storms. For the majority of the year, there is a significant evapotranspiration deficit, characterized by potential evaporation greatly exceeding precipitation. Drought stress is a source of sometimes violent clashes between local communities and environmentally unfriendly exploitation of water sources. To address these challenges and support local livelihoods, we piloted water storage facilities to cover community needs. Local schools and a few private homes were provided with rainwater harvesting facilities. A prototype pan-dam, insulated with a clay lining, was constructed to supply irrigation to a community during the dry season and facilitate sustainable agricultural production under water scarcity, benefitting a cooperative of women farmers representing diverse ethnic groups. These initiatives promoted water retention and relieved ecosystems from water abstraction.

3.3.3 Instream wetlands for flood control and drinking water supply

More than 90% of the river discharge converging into Lake Naivasha drains the Nyandarwa Range and the Bahati Highlands across a 2000 m altitudinal range. Stream morphologies include frequent waterfalls, riffles, pools, and partially silted instream shallow wetlands. The restoration and enhancement of these wetlands, illustrated in detail in Chapter 31, enhances biodiversity, improves water quality by increasing water retention time, and provides an opportunity for biofiltration. It also represents a precious source of water for people and cattle.

4. Ecohydrology and society: How to make ecohydrological solutions work?

The implementation of ecohydrological tools necessarily impacts on the competition for space and for resources between economic sectors. Without the active mobilization of Social Capital by raising ecosystem service awareness among stakeholders and promoting their participation in decision-making, it may be impossible to create sufficient momentum for the protection of Natural Capital; neither it can be possible to evaluate and compare alternative management options. Social ecohydrology includes three essential components (Zalewski et al., 2021).

4.1 Awareness of key cause-and-effect linkages underlying sustainability

For a correct evaluation of ecosystem services, awareness is key; conversely, the lack of awareness is cited as one of the main causes of uncertainty and undervaluation that prevent reaching a correct estimate of the true value of ecosystem services (Costanza et al., 1997). Factors that influence environmental awareness are: income (Üstün and Celep, 2007), access to basic services, social relations, knowledge of local species, experience in the direct use of local resources, and stakeholders' attitude, often conditioned by misconceptions (Ibrahim et al., 2023). At Lake Naivasha, ecosystem service awareness would create a perspective of challenges and opportunities offered by interaction with the catchment's natural resources, such as (1) pristine forests in highlands affected by deforestation, (2) riparian vegetation around the lake being severely degraded, (3) desertification-prone rangelands providing grazing resources, and (4) lake ecosystem including pelagic, littoral and riparian zones, threatened by pollution and habitat destruction. Sensitization and awareness building need to be implemented with continuity, because the identity and the number of resource users change rapidly in an economic environment so full of manual labor opportunities.

4.2 Participation in the protection of values

Active participation by all stakeholders in both ecosystem services evaluation and ecosystem restoration is essential to ensure that shared objectives can be established. Without this, evaluation projects may lead to unrealistic conclusions, and restoration projects fail to achieve durable outcomes. Government-facilitated participation is essential to support community-based natural resource management. In Naivasha, local institutions are often not connected to the grassroots level.

4.3 Trade-offs in resource use to maximize overall benefits

Ecosystem restoration, as well as the simple preservation of the status quo, necessarily requires bold decisions motivated by strategic choices (for example, enhanced resilience, sustainability, and recognition of overall ecosystem service value). Almost by definition, these decisions will impact the socio-economic sphere and alter the cost/benefit balance for some stakeholders. A redistribution of resources will occur, and its impact should not be underestimated. The lack of a solution that would represent a short-term "win" for all sectors of society should not become a justification for inaction. At Lake Naivasha, ecosystem restoration clashes with the vision promoted by short-term profit-seeking initiatives, often sought by vulnerable groups of the population, and by economic sectors that perceive themselves as disconnected from a legacy of natural resources. At the same time, a rapid land-ownership transition is evident, with large-scale farmers selling out to new, incoming hoteliers and resort developers.

5. Knowledge, attitudes, and practices in the Naivasha catchment

No formal KAP (knowledge, attitudes, and practices) surveys, *sensu* Cleland (1973), have been carried out in Naivasha, but we recognize the relevance of this theoretical framework for assessing public awareness of environmental change and public motivation in opposing resource degradation.

5.1 Historical heritage

KAP characterizing the lake basin community are closely dependent on human history and on the origin of today's residents. The portion of the Rift Valley comprising the Lake Naivasha catchment was never permanently settled until the arrival of Europeans; it served as a "Commons" for Maasai herders who visited it during the dry seasons (Waller, 2024) to water their cattle. The reason for this is, first of all, edaphic; i.e., caused by the infertile alkaline soils that characterize the sedimentary plains around the lake. This characteristic strongly discouraged small-scale farmers, who are more typical of the settlements on the Rift Valley slopes.

Immediately after annexation to the "East African Protectorate of Naivasha, the Western Highlands and the Mau" (1902), dispossessing the Maasai of their grazing rights became a priority of the colonial administration to unlock the catchment's farming potential (Waller, 2024). Historical accounts of shifting land tenure (Kuiper and Gravesen, 2024) and early conservation initiatives by newly immigrated European settlers (Harper and Styles, 2024) illustrate the peculiar socio-ecological condition that the Lake Naivasha catchment found itself in throughout the 20th century. Local development history explains why, despite its inherent biodiversity, the lake and its surroundings remained devoid of formal nature protection measures by the Government of Kenya (a few online sources still insist that Lake Naivasha is a National Park, but it is not), despite international recognition and nomination as a *wetland of international importance designated under the Ramsar Convention on Wetlands* in 1995.

5.2 Environmental legacy and society

The present socio-ecological status of the catchment defies known interpretative frameworks but fits the *shifting baseline syndrome* (Pauly, 1995), i.e., steady degradation proceeds almost unnoticed because younger generations cannot imagine what the status of the environment was in the past. The key principles of wetland management established by the *Ramsar Convention on Wetlands*, which insist on the integration of local culture and practices, are also missing. No ancestral human community, and no tribe, could truly claim to have developed an ancestral sense of connection to the Naivasha catchment landscape, other than the Maasai, to whom the wetland was an important dry-season refuge for generations. The Maasai today represent a small minority.

During the past 50 years, environmental degradation has progressed rapidly and unabated in the Naivasha catchment, and yet few have noticed it. The *shifting baseline syndrome* is thus applicable to the Naivasha catchment because only a minor proportion of today's residents have settled there for more than one generation; new arrivals see the baseline as it was on their arrival. This situation underlines the lack of relevance, in the case of Lake Naivasha, of the *Guiding principles for taking into account the cultural values of wetlands for the effective management of sites* (Resolution VIII.19 of the Ramsar Convention contracting parties, Valencia, 2002), which is universally recognized as an important component of wetland management guidelines.

The lack of affection toward their surroundings implicitly downgrades the value of cultural ecosystem services provided by the Lake Naivasha catchment, which include esthetic, spiritual, recreational, educational, and other nonmaterial benefits (TEEB, 2010). Out of the cited ones, only the recreational value could be considered a significant contribution. Its value, however, is highly dependent on the interaction between Human, Social, and Built Capital (i.e., transport and accommodation infrastructure) rather than Natural Capital per se, yet such infrastructures represent a significant pressure that potentially degrades the lake riparian zone. In particular, public beaches and private hotels have wiped out much of the papyrus fringe to open the view over the water. Although intuitively recreational services at the lake may be considered to be related to biodiversity and ecosystem health, this linkage is not always evident. Boat rides may carry on independently from the water quality, and lakeshore restaurants serving imported meat and fish will be in business despite the degradation of the lakeshore environment supporting local wildlife and their illegal destruction (fish, gazelles, and hippopotami).

This tragedy of the Naivasha commons is exacerbated by the sharp polarization that persists within the resident community, leading to contrasting perceptions of social, economic, and cultural values. Specifically tailored strategies are necessary to effectively raise awareness (He and Zhang, 2024). While affluent landowners congregate in the Lake Naivasha Residents Association (LNRA), laborers and self-employed traders attend church groups and trade unions. Members of this latter group do not own land, and they typically live in rented accommodation. The large community, of many tens of thousands of individuals, is mostly young people in their 20s and 30s, with a high female/male ratio and very few older age groups. Average education level, a factor that tends to be highly correlated to environmental awareness (Ibrahim et al., 2023), does not reach the standard 8-4-4 school leaving certificate. The community is internally divided, as its members tend to originate from various parts of Kenya, characterized by different natural resource availability and inhabited by different tribes. Thus, most of the people who live along the lakeshores today do not have direct experience in the wise use and management of the local natural resources found on-site. It is imaginable that their life expectations and demands reflect contrasting values in a scenario of mutual mistrust that generates anxiety, land and water conflict (Müller-Mahn and Gemälich, 2024), and risk of violent clashes at critical times, such as during severe drought (as in Mai Mahiu in 2005–09) or during postelection violence (in Naivasha townships in early 2008). This cultural setting prevents residents from developing a sense of belonging and shared responsibility toward their immediate surroundings. Consequently, they tend not to show a direct interest in the uses of the lake's resources. They are transient communities that live within the catchment because of employment, but are ready to leave when better conditions arise elsewhere.

5.3 Future outlook

The future of a sizable part of the lakeshore environment depends, therefore, on choices that are made by private large-scale landowners and investors. Many of them have established nature conservancies; these could be said to feel "connected to nature" and to be motivated to preserve the lake's natural beauty and ecosystem services. Many other landowners, however, perceive the lake surroundings as a great business opportunity through the production of cut flowers and vegetables for European markets and/or the development of upper-class accommodation resorts. Presently, access to the lake is highly restricted by landowners' private fences, and only three official public beaches allow visitors to freely reach the lake edge.

The Naivasha Catchment contributes a good fraction of the national income. Both the county and the national government are torn between the interests that represent numerically important working-class operators (large numbers of voters), employed in hotels and in flower farms, and, on the other hand, the few but wealthy landowners whose investment maintains the economy of the region. The Naivasha case suggests an increasing contrast between the actual value of the natural resources present within the catchment and local people's perception; this is particularly striking when considered at a larger scale. While deforestation and land-use intensification are degrading increasing portions of the country, the relative value of Lake Naivasha keeps rising. It is increasingly evident that the catchment is a priority of national interest and that local residents should not see it as their exclusive prerogative.

6. Tools to integrate society within an ecohydrological framework

6.1 Education

An initiative for the sustainable use of Lake Naivasha's catchment area is the result of a cooperation established with the Naivasha Environmental Conservation Center (NECC) of Ndabibi, led by Josphat Macharia (www.ewfandw.org). Hundreds of farmers were trained on sanitation, water abstraction and use, sustainable agriculture, wise use of ecosystem services, and greater awareness of the community benefits derived from the catchment resources and from its regulatory services (Harper et al., 2013).

Through a collaboration with ENKI s.r.o. (Czech Republic) and Czech national broadcasting, in 2012, students from local universities were trained in developing documentary films on subjects selected from around the catchment area. Some of these were broadcast at the national level; others were shown during public meetings and in buses transporting flower farm workers to and from their place of work.

6.2 Representation

WRUAs, introduced by the Water Act in 2006, represent an important "point of entry" to develop a competent interlocution with grassroots resource users' organizations. WRUA's voice and impact should be supported with adequate funding and training. In 2010–13, WRUAs were approached and involved through meetings, whole catchment workshops, and (in some cases) on-site restoration works and reforestation projects. Upper and lower catchment WRUAs were involved in a Payment for Ecosystem Services (PES) scheme that recognized shared interests in water quality and quantity management.

Local youth were invited to join a large community-based organization totaling >200 members (the Naivasha Basin Sustainability Initiative, NBSI) (Wanjala et al., 2018). A broader organization, called Lake Naivasha Umbrella WRUA, handles catchment-wide projects such as River Health Assessment based on chemical and biological indicators (river invertebrates), with the help of KenGen (Kenyan national energy operator), the Kenya Marine Fisheries Research Institute (KMFRI), and the Water Resources Authority (WRA), after citizen scientists were trained by WRA.

6.3 Paying for sustainability

A PES program has existed for about a decade, whereby several of the large horticultural growers make a payment that helps farmers in the steep-sided upper Malewa catchment undertake sediment control measures (Wanjala et al., 2018). It is too small to make a difference beyond the individual farms or to be measured as lowered sediment/nutrient runoff into rivers, but several alternative mechanisms have been suggested to raise funds that might provide an adequate quantity and reliability of resources that would enable catchment scale, measurable improvement in ecosystem health and services to people at some future date.

Mekonnen et al. (2012) suggested that a water-sustainability agreement should be made between major companies growing flowers that included a premium on the final bouquet of flowers in the supermarket that bought them in the Northern Hemisphere. They felt that this would support a Water Fund that could be used, in their example, to implement the technology that would optimize the sustainable use of water in the flower-growing companies and minimize their "blue water" footprint (Hoekstra and Chapagain, 2006) and thus improve watershed management. Mulatu et al. (2015) suggested that instead of funding direct activities in the upper catchment, horticultural enterprises at the lake would be willing to pay directly into a Water Fund for ecosystem restoration. Cook et al. (2013) went further and examined each major water user—horticultural and agricultural companies, tourist enterprises, water service companies, and geothermal generation. They evaluated whether a small levy—on top of what is already paid to WRA—could be charged to support a "Lake Naivasha Basin Sustainable Development Fund" (a Water Fund, as in the other two studies). They found that a modest levy on commercial users of 0.5-1 Kenya shilling per cubic meter a day and of 1% of room charge for the tourist industry could raise between 70 and 150 million Kenya shillings (between US\$666 $\times$ 10^3 and 1.3 $\times$ 10^6) per annum. This was the first calculation of the possible revenue that could be generated for ecosystem support and would be greater than any previous ecological funds spent in the Naivasha basin, either donated or from the government. It would, in theory, pay for substantial natural resource restoration if the raising of such an income and its use could be without corruption. More recently, following the Water (Amendment) Bill of 2023, the Ministry of Water established that a small portion (1%–2%) of the water use fees paid by customers of water supply services should be reinvested in catchment protection. However, Water Users' Associations often lack the capacity to develop adequate proposals to be submitted to the WRA.

6.4 Government (local + national + international)

A larger portion of the wealth generated through the use of Lake Naivasha resources could thus be reinvested back in the catchment for the benefit of its residents and for sustaining its Natural Capital. Tangible improvement of life standards has been achieved through road improvements, upgrading roads and road-ditch networks in low-income residential areas, developing new facilities for the local KMFRI, and overseeing lake water quality and fisheries. While this happens, the Kenyan Government is evidently struggling to ensure decent living standards (accessibility to water, to medical services, and to education) in informal settlements and to enforce a minimal regulatory network (security, planning permits, and nature conservation). Local institutions are weak, and water users, as well as squatters and illegal developers, evade restrictions. Government projects are supplemented by corporate social responsibility initiatives financed by large-scale flower farmers (Müller-Mahn and Gemälich, 2024) or by large-scale buyers, such as the Swiss COOP and German REWE supermarkets. Some of these projects are educational, environmental, and health-related. In 2018, a Sustainable Agrifood Systems Strategies (SASS) emerged from the initiatives led by NBSI in collaboration with Italian universities and involved coordinating rural farmers with the county government and the national government in a joint effort to boost local vegetable production for the benefit of school and hospital canteens.

Nevertheless, 10 years on from the creation of the many demonstration projects within this ecohydrology demonstration site, there is no clear evidence that anybody—government or NGO—has progressed beyond management plans and discussion meetings. The powerful horticultural industry and its international customers (who paid for many of the Pilot Projects described here through grants to the authors) continue to do just that, because photographs (such as the ones in Chapter 31) can fill their annual reports. The new county governments fail to sustain Naivasha, as clearly shown by Dittmann and Ogolla (2024), and the national government has "mega-plans" that take absolutely no account of Naivasha's unique natural resources (Kioko and Wanjala, 2024). In the meantime, the economic scenario is evolving with a shift of lakeside land property toward hoteliers, while horticultural companies are progressively relocating abroad (e.g., Ethiopia) in search of cheaper labor and lower taxation. These new investors have no knowledge about Naivasha's ecohydrology and are not concerned about long-term sustainability issues.

7. Conclusion

Ecohydrology principles offer the potential to be adopted as a general paradigm to interpret ecosystem functioning and evaluate the performance of ecosystem services, as well as to design impact mitigation measures and/or measures for ecosystem service enhancement.

The Lake Naivasha DS illustrates potential implementation measures as well as social and economic challenges that hamper sound ecohydrological solutions. The focus internationally is given to climate change issues, limited to direct CO_2 emissions, which are low in the Naivasha catchment. This has moved the public's attention from severe deforestation, which has direct and far worse ecohydrological consequences on local ecosystem services and on ecological equilibria.

References

Adams, C.S., Boar, R.R., Hubble, D.S., Gikungu, M., Harper, D.M., Hickley, P., Tarras-Wahlberg, N., 2002. The Dynamics and Ecology of Exotic Tropical Species in Floating Plant Mats: Lake Naivasha, Kenya, vol 488. Springer Science and Business Media LLC, pp. 115–122. https://doi.org/10.1007/978-94-017-2031-1_11.

Albrecht, C., Kochey Kipyegon, J., Junginger, A., Clewing, C., 2025. Returners and new arrivals after the crash: intermediate hosts and global invaders dominate gastropod fauna of Lake Naivasha, Kenya. Diversity 17 (4), 265. https://doi.org/10.3390/d17040265.

Atuga, G., Jembe, T., 2022. Constructed Wetlands' Application for Flower Farms Wastewater Treatment in Developing Countries: Case Study in Kenya. Elsevier BV, pp. 257–269. https://doi.org/10.1016/b978-0-12-821664-4.00004-2.

Becht, R., Harper, D.M., 2002. Towards an understanding of human impact upon the hydrology of Lake Naivasha, Kenya. Hydrobiologia 488, 1–11. https://doi.org/10.1023/A:1023318007715.

Byrne, A., Norris, K., Chadwick, M.A., Avery, S., Olaka, L., Tebbs, E.J., 2024. Rising lake levels in central East Africa are driven by increasing rainfall and land-use intensification. J. Hydrol.: Reg. Stud. 56. https://doi.org/10.1016/j.ejrh.2024.101999.

Cleland, J., 1973. A critique of KAP studies and some suggestions for their improvement. Stud. Fam. Plann. 4 (2), 42. https://doi.org/10.2307/1964829.

Cook, P., Contor, B.A., Harper, D.M., Vereecke, B., 2013. The Value of Water – an Economic Basis and Mechanism to Assess Voluntary Contributions to a Proposed "Lake Naivasha Basin Sustainable Development Fund.

Costanza, R., d'Arge, R., de Groot, R., Farber, S., Grasso, M., Hannon, B., Limburg, K., Naeem, S., O'Neill, R.V., Paruelo, J., Raskin, R.G., Sutton, P., van den Belt, M., 1997. The value of the world's ecosystem services and natural capital. Nature 387 (6630), 253–260. https://doi.org/10.1038/387253a0.

Dittmann, J., Ogolla, A.F., 2024. In: Agricultural Intensification and Resource Conflicts: Social Dynamics in the Lake Naivasha Basin Devolution of Governance and the Politics of Fishery at Lake Naivasha, Kenya, 34. The Netherlands Africa-Europe Group for Interdisciplinary Studies, pp. 279–304. https://doi.org/10.1163/9789004695429_013.

Foster, J., Harper, D., 1999. The alien Louisianan red swamp crayfish *Procambarus clarkii* Girard in Lake Naivasha. Freshw. Crayfish 15, 195–202.

Gherardi, F., Britton, R.J., Mavuti, K.M., Pacini, N., Grey, J., Tricarico, E., Harper, D.M., 2011. A review of allodiversity in Lake Naivasha, Kenya: developing conservation actions to protect East African lakes from the negative impacts of alien species. Biol. Conserv. 144 (11), 2585–2596. https://doi.org/10.1016/j.biocon.2011.07.020.

Giberti, G.S., Gumiero, B., Kiprotich, A.K., Methu, S.W., Harper, D.M., Pacini, N., 2022. Longitudinal vegetation turnover in an eastern Rift Valley riparian corridor. Afr. J. Ecol. 60 (1), 27–42. https://doi.org/10.1111/aje.12920.

Gorshkov, V.G., Gorshkov, V.V., Makarieva, A.M., 2000. Biotic regulation of the environment: key issue of global change. Ecol. Stud. 146. https://doi.org/10.1007/978-3-662-04474-4.

Harper, D.M., 2006. The future of Lake Naivasha. Swara 2, 2.

Harper, D.M., Phillips, G., Chilvers, A., Kitaka, N., Mavuti, K., 1993. Eutrophication prognosis for Lake Naivasha, Kenya. SIL Proc 25 (2), 861–865. https://doi.org/10.1080/03680770.1992.11900268.

Harper, D.M., Adams, C., Mavuti, K., 1995. The aquatic plant communities of the Lake Naivasha wetland, Kenya: pattern, dynamics and conservation. Wetl. Ecol. Manag. 3 (2), 111–123. https://doi.org/10.1007/BF00177693.

Harper, D.M., Mavuti, K., 2004. Lake Naivasha, Kenya: ecohydrology to guide the management of a tropical protected area. Ecohydrol. Hydrobiol. 4 (3), 287–305.

Harper, D.M., Pacini, N., Upton, C., Morrison, E.H.J., Fox, R., Kiminta, E., 2013. Reaching water security through cooperation. In: Economic Development and Water UNESCO Water Cooperation for Sustainable Utilization: Lake Naivasha, Kenya.

Harper, D., Styles, M.A., 2024. Agricultural intensification and resource conflicts: social dynamics in the Lake Naivasha. In: Basin Environmental Conservation and Sustainability: Rhetoric and Reality, pp. 40–65. https://doi.org/10.1163/9789004695429_004.

Harper, D.M., Pacini, N., Morrison, E., Kimani, D., Mwinami, T., Upton, C., 2024. Agricultural intensification and resource conflicts: social dynamics in the Lake Naivasha Basin. In: The Unique Natural Resources of Lake Naivasha; Can They Survive?, 34, pp. 93–127. https://doi.org/10.1163/9789004695429_006.

He, W.-J., Zhang, K., 2024. The moderating effect of cultural orientation on the relationship between growth mindset and learning self-efficacy: a dimension-specific pattern. Behav. Sci. 14 (12), 1155. https://doi.org/10.3390/bs14121155.

Herrnegger, M., Stecher, G., Schwatke, C., Olang, L., 2021. Hydroclimatic analysis of rising water levels in the Great Rift Valley Lakes of Kenya. J. Hydrol.: Reg. Stud. 36, 100857. https://doi.org/10.1016/j.ejrh.2021.100857.

Hobbs, R.J., Arico, S., Aronson, J., Baron, J.S., Bridgewater, P., Cramer, V.A., Epstein, P.R., Ewel, J.J., Klink, C.A., Lugo, A.E., Norton, D., Ojima, D., Richardson, D.M., Sanderson, E.W., Valladares, F., Vila, M., Zamora, R., Zobel, M., 2006. Novel ecosystems: theoretical and management aspects of the new ecological world order. Global Ecol. Biogeogr. 15 (1), 1–7. https://doi.org/10.1111/j.1466-822X.2006.00212.x.

Hoekstra, A.Y., Chapagain, A.K., 2006. Water footprints of nations: water use by people as a function of their consumption pattern. Water Resour. Manag. 21 (1), 35–48. https://doi.org/10.1007/s11269-006-9039-x.

Hubble, D.S., Harper, D.M., 2002. Nutrient Control of Phytoplankton Production in Lake Naivasha, Kenya, vol 488. Springer Science and Business Media LLC, pp. 99–105. https://doi.org/10.1007/978-94-017-2031-1_9.

Huisman, J., Codd, G.A., Paerl, H.W., Ibelings, B.W., Verspagen, J.M.H., Visser, P.M., 2018. Cyanobacterial blooms. Nat. Rev. Microbiol. 16 (8), 471–483. https://doi.org/10.1038/s41579-018-0040-1.

Ibrahim, M.S.N., Assim, M.I.S.A., Johari, S., Mohammad, S.K.W., Afandi, S.H.M., Hassan, S., 2023. Public awareness on biodiversity conservation and well-being: case of Gunung Mulu National Park, Sarawak. Geojournal 88 (3), 3471–3496. https://doi.org/10.1007/s10708-022-10818-x.

Keyombe, J.L., Waithaka, E., 2017. Analysis of some aspects of water quality of Lake Naivasha. Int. J. Chem. Life Sci. 6 (1), 2001. https://doi.org/10.21746/ijcls.2017.1.1.

Kioko, E.M., Wanjala, S., 2024. Megaprojects and the Transformation of Lake Naivasha Basin: Aspirations, Uncertainties and Social-Ecological Implications, vol 34. Brill Academic Publishers, undefined Africa-Europe Group for Interdisciplinary Studies, pp. 353–374. https://doi.org/10.1163/9789004695429_016.

Kravčik, Pokorny, J., Kohutiar, J., Kovač, M., Toth, E., 2008. Water for the Recovery of the Climate – A New Water Paradigm. People and Water NGO, Kosice (Slovakia), pp. 1–122. Available from: http://www.waterparadigm.org. Available from:

Krienitz, L., Dadheech, P.K., Fastner, J., Kotut, K., 2013. The rise of potentially toxin producing cyanobacteria in Lake Naivasha, great African Rift Valley, Kenya. Harmful Algae 27, 42–51. https://doi.org/10.1016/j.hal.2013.04.005.

Kuiper, G., Gravesen, M.L., 2024. Shifting Land Tenure Arrangements: Private Land-Buying Companies and the Shape of Naivasha's Anthropogenic Landscape, vol 34. Brill Academic Publishers, undefined Africa-Europe Group for Interdisciplinary Studies, pp. 66–89. https://doi.org/10.1163/9789004695429_005.

Lovelock, J.E., Margulis, L., 1974. Atmospheric homeostasis by and for the biosphere: the gaia hypothesis. Tellus 26 (1–2), 2–10. https://doi.org/10.3402/tellusa.v26i1-2.9731.

Makarieva, A.M., Gorshkov, V.G., 2007. Biotic pump of atmospheric moisture as driver of the hydrological cycle on land. Hydrol. Earth Syst. Sci. 11 (2), 1013–1033. https://doi.org/10.5194/hess-11-1013-2007.

Mekonnen, M.M., Hoekstra, A.Y., Becht, R., 2012. Mitigating the water footprint of export cut flowers from the Lake Naivasha Basin, Kenya. Water Resour. Manag. 26 (13), 3725–3742. https://doi.org/10.1007/s11269-012-0099-9.

Mulatu, D.W., van Oel, P.R., van der Veen, A., 2015. Firms' willingness to invest in a water fund to improve water-related ecosystem services in the Lake Naivasha basin, Kenya. Water Int. 40 (3), 463–482. https://doi.org/10.1080/02508060.2015.1050580.

Muller-Mahn, D., Gemahlich, A., 2024. The Cut-Flower Industry in the Social-Ecological System of Lake Naivasha: Production Networks and Marketisation, vol 34. Brill Academic Publishers, pp. 159–175. https://doi.org/10.1163/9789004695429_008.

Njiru, J., Waithaka, E., Aloo, P.A., 2017. An overview of the current status of Lake Naivasha fishery: challenges and management strategies. Open Fish Sci. J. 10 (1), 1–11. https://doi.org/10.2174/1874401x01710010001.

Odongo, V.O., van der Tol, C., van Oel, P.R., Meins, F.M., Becht, R., Onyando, J., Su, Z., 2015. Characterisation of hydroclimatological trends and variability in the Lake Naivasha basin, Kenya. Hydrol. Process. 29 (15), 3276–3293. https://doi.org/10.1002/hyp.10443.

Pacini, N., Baxa, M., Kosik, M., Grey, J., Lepšova-Skacelova, O., Mbogo, D.K., Mwinami, T., Přikryl, I., Pokorny, J., Darlington, J.P.E.C., Hesslerova, P., Harper, D.M., 2021. Ecohydrological costs and benefits of common carp, the dominant species in a `novel' tropical lake ecosystem. Ecohydrol. Hydrobiol. 21 (3), 467–489. https://doi.org/10.1016/j.ecohyd.2021.08.011.

Pacini, N., Giberti, G., Harper, D.M., 2024. Agricultural intensification and resource conflicts: social dynamics in the Lake Naivasha Basin. In: Linking Hydrology to Biodiversity, Ecosystem Services and Climate within the Lake Naivasha Catchment, 34. Africa-Europe Group for Interdisciplinary Studies, pp. 128–155. https://doi.org/10.1163/9789004695429_007.

Pacini, N., Pechar, L., Harper, D.M., 2019. Chemical determinands of freshwater ecosystem functioning. In: Hughes, J.M.R. (Ed.), Techniques in Ecology and Conservation. Oxford University Press (OUP), Oxford. https://doi.org/10.1093/oso/9780198766384.003.0005.

Pauly, D., 1995. Anecdotes and the shifting baseline syndrome of fisheries. Trends Ecol. Evol. 10 (10), 430. https://doi.org/10.1016/s0169-5347(00)89171-5.

Peters, R.H., MacIntyre, S., 1976. Orthophosphate turnover in East African lakes. Oecologia 25 (4), 313–319. https://doi.org/10.1007/BF00345604.

TEEB, 2010. The Economics of Ecosystems and Biodiversity: Ecological and Economic Foundations.

Ustun, C., 2007. The connection between environmental awareness and socio-economic and cultural structure. WIT Trans. Ecol. Environ. 102, 623–631. https://doi.org/10.2495/SDP070602.

Waller, R., 2024. Agricultural intensification and resource conflicts: social dynamics in the Lake Naivasha Basin. In: Arrivals and Departures: The Maa-Speakers and Their Successors in Naivasha-Nakuru, pp. 15–39. https://doi.org/10.1163/9789004695429_003.

Wandersee, J.H., Schussler, E.E., 1999. Preventing plant blindness. Am. Biol. Teach. 61 (2), 82–86. https://doi.org/10.2307/4450628.

Wang, Z., Akbar, S., Sun, Y., Gu, L., Zhang, L., Lyu, K., Huang, Y., Yang, Z., 2021. Cyanobacterial dominance and succession: factors, mechanisms, predictions, and managements. J. Environ. Manag. 297, 113281. https://doi.org/10.1016/j.jenvman.2021.113281.

Wang, Z., Leung, K.M.Y., 2015. Effects of unionised ammonia on tropical freshwater organisms: implications on temperate-to-tropic extrapolation and water quality guidelines. Environ. Pollut. 205, 240–249. https://doi.org/10.1016/j.envpol.2015.05.045.

Wanjala, S., Mwinami, T., Harper, D.M., Morrison, E.H.J., Pacini, N., 2018. Ecohydrological tools for the preservation and enhancement of ecosystem services in the Naivasha Basin, Kenya. Ecohydrol. Hydrobiol. 18 (2), 155–173. https://doi.org/10.1016/j.ecohyd.2017.09.004.

Zalewski, M., Harper, D.M., Demars, B., Jolankai, G., Crosa, G., Janauer, G.A., Pacini, N., 2008. Linking Biological and Physical Processes at the River Basin Scale: The Origins, Scientific Background and Scope of Ecohydrology. CABI Publishing, pp. 1–17. https://doi.org/10.1079/9781845930028.0001.

Zalewski, M., 2000. Ecohydrology — the scientific background to use ecosystem properties as management tools toward sustainability of water resources. Ecol. Eng. 16 (1), 1–8. https://doi.org/10.1016/s0925-8574(00)00071-9.

Zalewski, M., Kiedrzyńska, E., Wagner, I., Izydorczyk, K., Mankiewicz Boczek, J., Jurczak, T., Krauze, K., Frankiewicz, P., Godlewska, M., Wojtal-Frankiewicz, A., Łapińska, M., Urbaniak, M., Bednarek, A., Kaczkowski, Z., Gągała, I., Serwecińska, L., Szklarek, S., Włodarczyk-Marciniak, R., Font-Najera, A., Mierzejewska, E., Połatyńska-Rudnicka, M., Belka, K., Jarosiewicz, P., 2021. Ecohydrology and adaptation to global change. Ecohydrol. Hydrobiol. 21 (3), 393–410. https://doi.org/10.1016/j.ecohyd.2021.08.001.

Zepernick, B.N., Hart, L.N., Chase, E.E., Natwora, K.E., Obuya, J.A., Olokotum, M., Houghton, K.A., Kiledal, E.A., Achieng, D., Barker, K.B., Basweti, G.M., Beal, M., Brown, K.M., Byrne, A., Drouillard, K.G., Getabu, A., Kiteresi, L.I., Lawrence, T., Lomeo, D., Miruka, J.B., Mohney, S., Njiru, J., Okutoyi, P., Omondi, R., Otieno, D., Owino, O.A., Owoko, W., Owuor, B., Shitandi, A., Stoll, J., Swaleh, M., Tebbs, E., Varga, E., Wagner, R.S., Sheik, C.S., Sherman, D.H., Dick, G.J., Wilhelm, S.W., Sitoki, L., Otiso, K.M., McKay, R.M.L., Bullerjahn, G.S., 2024. Molecular investigation of harmful cyanobacteria reveals hidden risks and niche partitioning in Kenyan Lakes. Harmful Algae 140. https://doi.org/10.1016/j.hal.2024.102757.

Chapter 40

Hydronexialism—A didactic concept to teach complex problem-solving within the framework of transdisciplinary ecohydrology

Karl Matthias Wantzen[1, 2, 3, 4]

[1]*EUCOR Excellence Chair "Water and Sustainability", Institut Terre et Environnement de Strasbourg (ITES, CNRS/ENGEES UMR 7063), University of Strasbourg, Strasbourg, France;* [2]*Karlsruhe Institut fur Technologie (KIT), Institute for Water and Environment, Karlsruhe, Germany;* [3]*UNESCO Chair "Rivers and Heritage", University of Tours, Tours, France;* [4]*Cites, Territoires, Environnement et Societes (CITERES, CNRS UMR 7324) and Polytechnic School of Engineering, University of Tours, Tours, France*

UNNUMBERED FIGURE 40.1 PHOTO TO BREAK THE CHAPTERS' STRUCTURE *PHOTO CREDIT: K. M. WANTZEN.*

Transdisciplinary Ecohydrology. https://doi.org/10.1016/B978-0-443-41596-8.00041-8

1. Introduction

Water is an essential resource for life. All biological species and processes depend on it. Inland freshwater hydrosystems (lakes, rivers, headwaters, wetlands etc, in the following short: hydrosystems) are hot spots of biological speciation and for human cultural diversification. Development of ideas, agriculture, urbanism … they all have their origins in hydrosystems and their fertile plains, which are the result of a rhythm of the waters between inundations and dry or low-water periods. Being perfect conveyor belts for dispersal of biocultural diversity, rivers and large lakes have a double importance for creation and distribution of strategies adapted to the rhythm of the waters (Wantzen, 2024; Wantzen et al., 2016).

Water resources (the hydrology principle), **B**iodiversity and Ecosystem **S**ervices (the Ecology Principle), **R**esilience to impacts, **C**ulture and **E**ducation (WBSRCE) are the pillars of the Ecohydrology philosophy (Elfithri et al., 2025; Zalewski, 2002, 2014, 2021) underline the importance of Transdisciplinarity in Ecohydrology. The River Culture Concept, with its five tenets on (i) acknowledging the value of hydrological rhythms, (ii) adapting human use practices to the natural flow regime, by learning from and improving on (iii) traditional ecological knowledge or (iv) biological adaptive strategies (landscape bionics), and (v) considering the hydrographical basin as the basic unit for all territorial and land/water use policies ("bassin de responsabilité") places the relationships between humans and their environment into the foreground, which can be summarized as "learn from the river, live with the river, love the river" (Wantzen et al., 2016). In a recent UNESCO book publication (Wantzen, 2023) "River Culture—life as a Dance to the Rhythm of the Waters," Zalewski and Wantzen (2023) provided a synthesis of both concepts.

To face the dilemma of increased water needs and reduced water availability, driven by inadequate catchment management and climate change (CC), considerable changes in water use practices will be needed but this is hampered by behavioral inertia and lacking socio-cultural connectivity with nature. Within the framework of Transdisciplinary Ecohydrology, the River Culture approach tackles this problem by systematically analyzing bionic and cultural strategies concerning their usefulness for physical application in CC mitigation but also by employing environmental psychology to engage cultural connectivity as a tool to facilitate social acceptance and coconstruction of Nature-based Solutions.

During cultural development, the relationship of humans with hydrosystems has undergone a strong change from a respectful attitude in early societies (often forced by the unability to control the dimensions of floods and droughts) to a resource-exploitation oriented one in industrial and postindustrial societies. Taking care for the water-providing hydrosystems has been neglected for centuries. As a result, the water crisis is one of the severest among a number of current global environmental crises. Quantity and quality of the water are equally menaced, and, along with the degradation of hydrosystem structures, so are their ecological and social functions, and their biological and cultural diversities (Harper et al., 2021; Reid et al., 2019; Wantzen, 2023). Climate change is exacerbating the existing problems, by increasing the water demand during pronounced dry and hot periods, and by increasing flood risk (and water pollution, by releasing substances deposited in the dry-fallen floodplain) during excessive precipitation events (Folke et al., 2021).

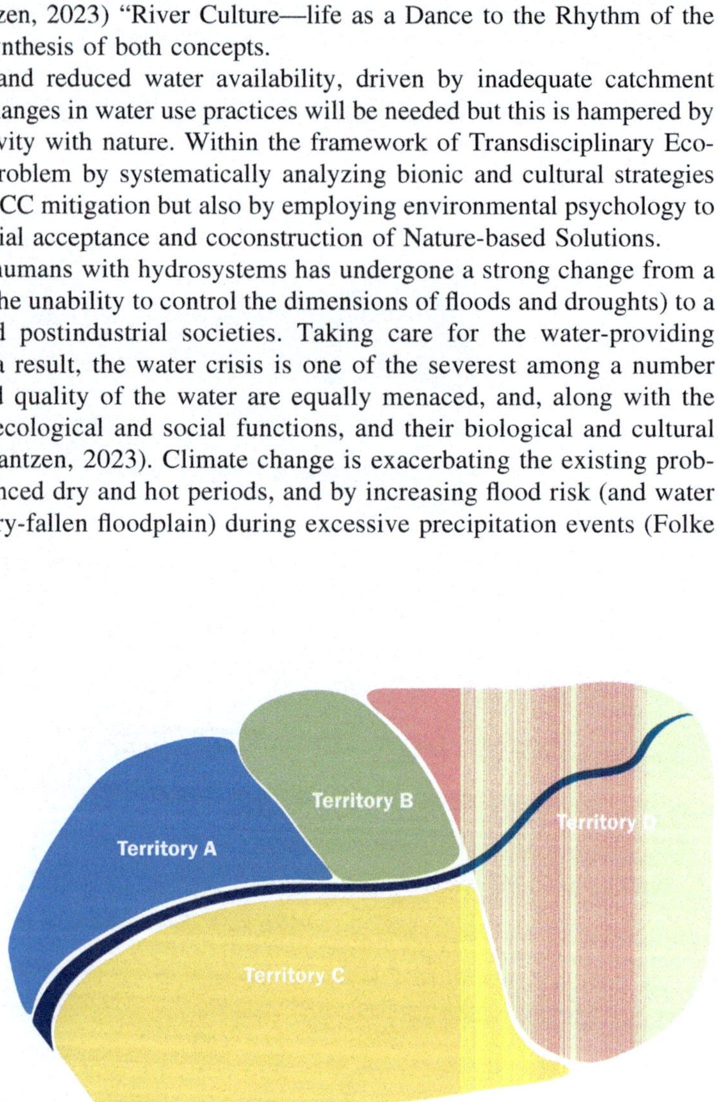

FIGURE 40.1 The "territorial disadvantage" of river systems. Rivers have an elongate shape and flow in the lowest part of the landscape (hydrographic basin). Human territories have the tendency to have geographical barriers as their borders, such as mountain ranges, marine shores, or banks of large rivers and lakes. Consequently, rivers are "cut into pieces" by human territories, unless the catchment is considered as the basic shape to the territory ("Bassin de responsabilité") (Wantzen et al., 2016).

This chapter will not discuss the causes for the global water crisis in detail, rather, it will deliver a didactical approach how to train young experts to codevelop adequate, local, case-wise solutions. Specifically, it hinges on the concept of Transdisciplinary Ecohydrology (Elfithri et al., 2025; Zalewski, 2014), by reinforcing its C + E (Culture and Education) elements.

To do so, this paper will first describe relevant thematics to be tackled, then deliver a definition of hydronexialism, and describe the didactic concept for a Master Course training young talents for problem solving in the water context.

1.1 Possible causes for inadequate hydrosystem management

Intellectual path dependency. The way how humans are using water has diversified from direct use for drinking and households to irrigation and many kinds of industrial processes, cooling, hydropower, navigation etc. Governmental, administrative and social structures have evenly been split up with this diversification and hardly match with the complexity of the problems. Rather, they are mostly oriented toward individual, specific use forms or a restricted number of uses. Claudia Pahl-Wostl, one of the leading researchers in hydropolicies, termed the expression "*intellectual path dependency*" (Pahl-Wostl et al., 2020). Path dependency in industrial processes describes the dependency on a once-taken decision to invest into a certain production line, e.g., into very expensive machines to produce a given good. Once a change is needed (e.g., as the machine or process is too resource-consuming), it is difficult to change toward a more appropriate technique, as the material goods are immobilizing the capital and hamper its reattribution. In the intellectual context, path dependency describes the way how humans are trained in universities and other institutions, which is "formatting" the way how people are reasoning and are making decisions. The continuing misunderstanding between scientists from the humanities and natural sciences is one vivid example for this intellectual path dependency, likewise, scientists whose axiology depends on the reputation (often: impact factor of publications) of their colleagues can hardly understand why "the people" would not implement a solution that they have elaborated over years. Lacking trust and communication are among the greatest hurdles in jointly finding or in implementing solutions.

The geographical problem of rivers. Humans are terrestrial organisms who organize their *terri*-tories (name!) mostly according to geographical landmarks. This puts hydrosystems, especially river systems with their elongate shape, in a disadvantage: While the faces of the physical landscapes are defined by orogenic and subsequent erosion/deposition processes within hydrographical basins (catchments) that are centered by a hydrosystem (e.g., the thalweg of rivers, the deepest point of lakes), human territories generally seek for geographical barriers as their borders. If rivers and lakes are big enough, they go beyond the borders of territories. These are very often at their banks but hardly place them in the center (as it is the case within the physical landscape Fig. 40.1). An extreme example for this longitudinal cutting of a river by territories is the Rhine between the Middle Ages and the Congress of Vienna (1813), which caused serious problems in terms of river management (Wantzen et al., 2023b).

The spatial dimensions of a river, lake or wetland are also crucial here. While it is relatively easy to include a smaller catchment into a management scheme, a territory, or a nature reserve, with growing size (breadth, depth), rivers become cultural and biogeographical borders (Kondolf and Pinto, 2017). Resulting from this thematical and geographical mismatch between the nature of hydrosystems and the ways how humans are using them, a huge number of problems has arisen, which can be summarized as:

1. *Lateral divide.* In larger rivers, left bank territories are managed in a different way than right bank territories, e.g., in the Upper Rhine region belonging to Germany, France and Switzerland (Wantzen et al., 2023a). Despite decade-long efforts to harmonize these approaches via the Upper Rhine (Schmitt et al., 2019) there are still many open questions, occupying the Upper Rhine Counsel and the International Commission for the Protection of the Rhine.
2. *Longitudinal divide.* Headwater areas are often (mostly) treated by other administrations than the middle and lower sections of hydrosystems. Hong Kong is an example, where these institutions have different interests (Cao et al., 2024). This may lead to conflicts, especially when it comes to dam construction, pollution control or flood/drought risk management.

Ignorance of nature's needs. What makes things worse for hydrosystems, is the way how humans make decisions. Anthropocentrism has established an inherently unequal human–nonhuman relationship (Strang, 2019, 2024). Moreover, humans have a lack of understanding for nonlinear proceses and consequantily, are unable to assess ecosystem carrying capacities and tipping points (Krause et al., 2015). Only a small part of the human population is actually willing to accept factual statements as a driver for changing their behavior. Decisions on the use of water and on the management of hydrosystems are largely driven by geopolitical or economic motivations, and the question about how much water (and in which quality) needs to remain to maintain essential ecosystem functions is rarely asked (Sehring et al., 2022). In this context, water and hydrosystems are rarely considered as subjects that need to be respected or preserved for their own sake, rather, they are assets that can be maximally cite Ambio.

Evolution: time–space displacement of selective feedback mechanisms hampers adequate decision-making. The biological evolutionary process, in a nutshell, can be described as that organisms become equipped with new qualities (by genetic recombination), which are then tested to be adequate (or not) in the environment at a given place and time (selection). An inadequate strategy (e.g., butterflies having brightly colored wings in a dark-colored environment) will be eliminated (negative selection) by the environment. In our example, the butterflies that can be distinguished from their background are preferentially eaten by birds and have a lower chance to propagate, whereas better adapted genetic assemblages will endure and flourish (Luz, 2019). Cultural evolution diversifies under similar selective procedures, however, it is multi-lateral and independent of parenthood and time-consuming reproductive processes (Creanza et al., 2017).

Prior to globalization and modern/postmodern times, human strategies that were not-adapted to the local limitations of natural resources failed and died off (e.g., Easter Island, Angkor Wat ...). Modern human strategies have circumvented this selective feedback mechanism in a way that the consequences of our activities are either displaced into other areas (e.g., consumption of water-demanding vegetables in central-northern Europe causing water-scarcity in Spain and Northern Africa), or into future times (e.g., the greenhouse gas emissions or the concentrations of toxic pesticides or heavy metals in soils and sediments that will severely impact future generations). Consumers and decision-makers rarely feel the consequences of their activities directly on their lives and—despite increasingly severe legal regulations—only attribute them minor concern. For example, pesticides that have been banned decades ago from agricultural production due to their well-documented toxicity for humans are still amply used in house paintings, despite that they have strongly negative effects on the environment (Imfeld et al., 2013).

Only exceptionally, there is a direct feedback between environmental impact and consequences to human health, which may then launch a set of activities to mitigate the situation. In the case of the Sandoz accident, at 1. November 1986 in Switzerland, polluting the entire Rhine down to The Netherlands, including massive fish kills, the riparian states launched a series of activities, which can be seen as the predecessor of the European Water Framework Directive launched in 2000 (European Parliament, 2000; Plum and Schulte-Wülwer-Leidig, 2014; Wantzen et al., 2023a) The International Commission for the Protection of the Rhine has been celebrating its 75th anniversary in July 2025 but prior to the Sandoz accident, the interest of the riparian states in common activities to significantly improve the environmental quality of the Rhine were rather meager. The Sandoz case can be interpreted with evolutionary patterns in human behavior. While humans are very well adapted to react adequately, and cooperate in groups, to face immediate threats to human lives (e.g., when being attacked, under fire, reacting to a environmental catastrophe like the Sandoz

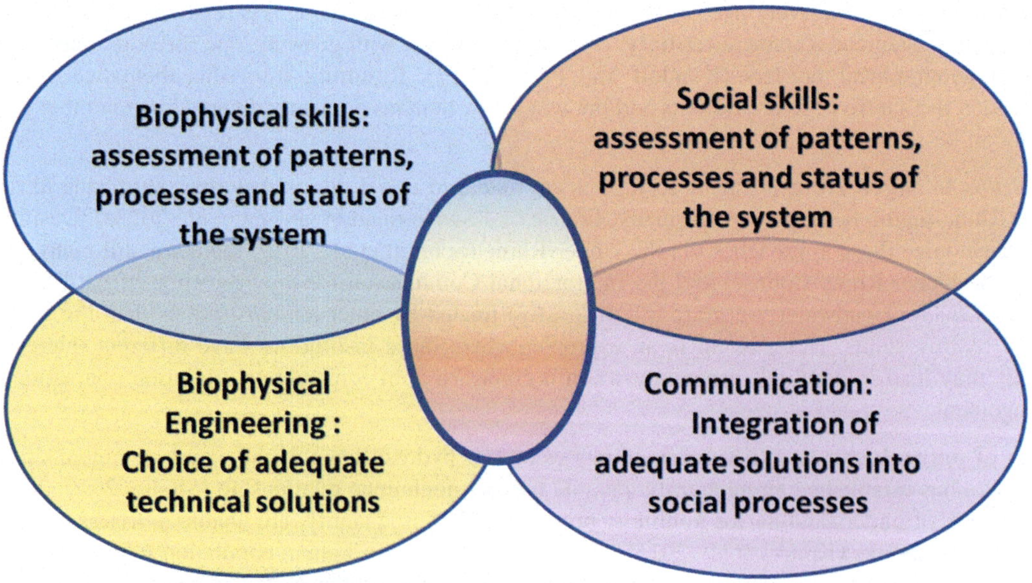

FIGURE 40.2 The concept of hydronexialism. In analogy to the wings of a butterfly, the concept of hydronexialism teaching is composed by four elements in two sides: the left side describes the physical-biological-chemical "problemsphere" (ecosystemic approach), how it can be analyzed (upper wing) and how technical solutions can be attributed (lower wing), the right side deals with the "social sphere" (humanities approach), how social structures can be analyzed (upper wing) and how holistic solutions can be communicated (lower wing). These four elements are centered around the holistic, interactive and reiterative approach of hydronexialsm (the "body" of the butterfly).

accident (Harari, 2015)), our capacities to fancy future or remote dilemmata and to act accordingly, are very limited (Folke et al., 2021).

Another consequence from biological evolution and social development is the fact that humans have the tendency to consider the most common procedure as "normal" and therefore automatically good, and have great difficulties in adopting different attitudes, even if they were more efficient. The current tendency of "looking away" or even being repelled by news on environmental challenges. "Anxiety, worry, and grief in a time of environmental and climate crisis" (Ojala et al., 2021), is another behavioral constraint when it comes to action-taking in favor of nature.

This list is by no means complete. Rather it is intended to give an impression on why it is so difficult to handle and manage inland waters adequately and why hydro-nexialism is so much needed. Most of the issues mentioned above are interconnected or even interdependent. Solving these complex problems is specifically difficult as the specialists need much time to learn their business and can always just oversee a small portion of the problem.

2. What is hydronexialism? A definition

For many years, ecologists have been proposing "holistic" approaches for ecosystem assessment, which also alludes to the interdependences of the individual players. However, due to the above-mentioned reasons, in ecosystem management, hydrosystem management is still too often stuck to a set of single-sidednessess. The term "nexialist" has been introduced by a science fiction writer A. E. Vogt in "The Voyage of the Space Beagle" (1950), describing a person who can combine very different sources of knowledge and who can motivate very different characters to collaborate, and it has found its way into pedagogics (Love, 2008). Etymologically, the term derives from the latin verb "*nectere*", which means "to bind together, to combine". A "nexus" is, originally, a nod. Since then, the term has gained increasing importance, describing persons that are able to see connections between disciplines, are skilled in conflict resolution, get people to solve complex problems, employing a project design and an adaptive reorganization of teams.

We use the term nexialism in this connotation and define *hydronexialism (and hydronexialists) as an approach (and persons) to employ nexalisistic skills (*i.e., *combining very different sources of knowledge and motivating very different characters to collaborate) for solving complex problems in the context of hydrosystems.*

2.1 Elements of hydronexialism—the butterfly metaphor

The thematics dealt with by hydronexialism can be metaphorized with the four wings of a butterfly (Fig. 40.2). On its left side, the biological, physical and technological thematics are addressed, including the problem-understanding on the upper left "wing" and problem-solving aspects on the lower left one. This side aims at identifying and combining the best possible management practices for problem solving, in an analogy to the "Key Type Measures" presented by the European Water Framework Directive (Hering et al., 2010). Here it is very important to exploit existing technical knowledge better before redundantly studying problems for which solutions had already been developed (see below, "Going beyond the DPSIR approach").

However, the left side of the butterfly is not sufficient to settle problems. Quite often, well-thought technical solutions are planned and executed, without sufficiently involving the local population. Omission of this coconstruction phase can have negative consequences o the medium to long term. A sad example for this is the dam removal in the Sélune River (flowing to the Baie de Mont-St. Michel, Normandy, France). Two dams were removed in order to re-establish passability for migratory Atlantic Salmon. All details of before-after design of the environmental studies, the complex engineering of sediment release following the removal, and the recovery of the denudated sediment surfaces were excellent (Piscart et al., 2024). Problems arose from the social side. Citizens who had become acquainted with the dam lake for decades, and had integrated it into their leisure activities (mostly by constructing anglers huts in which they spent their week ends) opposed to the project and claimed that their interests were not sufficiently considered in the project. As a matter of fact, their anglers' huts became obsolete; rather than being at the lakeside as before, they are now several meters above and dozens of meters apart from the water body, while the previous lake sediments are now being overgrown by vegetation, making access to the stream difficult. Salmon is protected, so even if the fish will run upriver, they must not be caught, contrary to the carps and pikes that the anglers fished and released regularly with great pleasure. As a consequence, a general protest movement arose (Germaine and Gonin, 2024).

This example shows how important the knowledge of the societal structures and their interdependencies are (Morandi et al., 2021). The French expression "*jeu d'acteurs*" is describing these interactions very well. The axiology of values of different social groups is very complex and displays a spatial and temporal dynamics (Murphy et al., 2022), which can be analogized with the dynamic patch mosaic of different succession stages in floodplains (Pringle et al.,

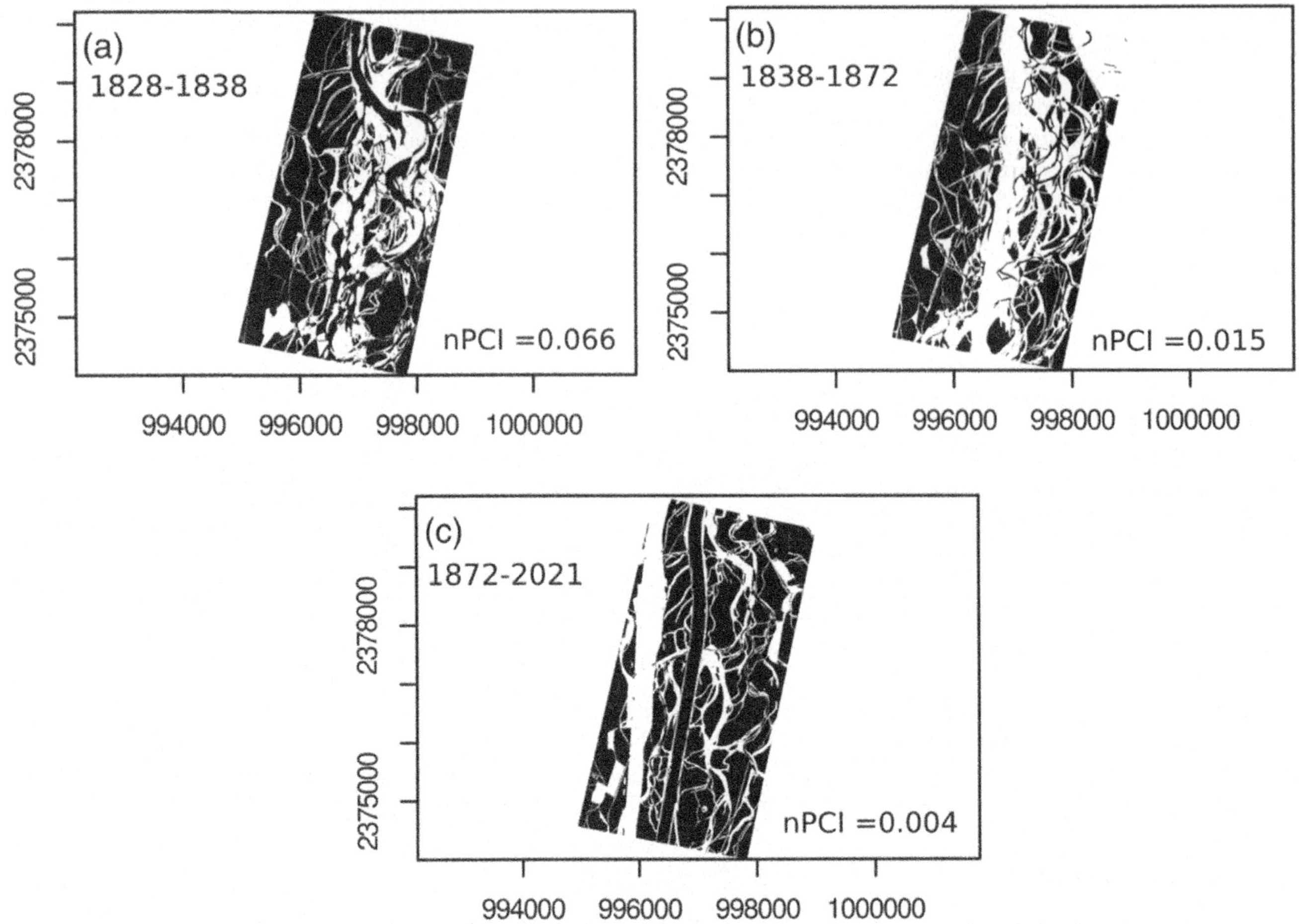

FIGURE 40.3 Ecological habitat patch dynamics and successions. Example of the upper Rhine, from (Chardon et al., 2022) are a result of physical driving forces (erosion/sedimentation) interacting with biological successions, which are space-dependent and can be rejuvenated. *From Chardon, V., Herrault, P.A., Staentzel, C., Skupinski, G., Finance, O., Wantzen, K.M., et al., 2022. Using transition matrices to assess the spatio-temporal land cover and ecotone changes in fluvial landscapes from historical planimetric data.* Earth Surf. Process. Landf. *47 (11), 2647–2659. doi: 10.1002/esp.5437.*

1988; Tockner and Stanford, 2002). In the case of floodplain landscape topography Fig. 40.3, the morphology and grain size structure of the sediment are changed by erosion and sedimentation, resulting in a dynamic mosaic of physical habitats, which, due to different environmental drivers, develops a patchy set of vegetation and other biota over time (successional stages, flood-time dependent vegetation zones), which can become rejuvenated by hydrological forces, depending on the flow regime (see Wantzen and Rodrigues, this volume). In the case of societal value-scapes Fig. 40.4, the driving forces are historical developments (e.g., middle age urbanism according to professional guilds, or constructions adapted to risk-reduction (e.g., *intra/extra muros*)), whose structures become changed by societal dynamics (e.g., immigration, gentrification, abandonment of industrial activities etc.). However, like morphological landscapes, valuescapes interact with the spatial setting (e.g., eco-social disparities between flood risk-prone and safe parts of a city, Hein, 2025; Wantzen et al., 2019).

Thus, the stakeholder analysis (corresponding to the environmental impact assessement on the upper left wing) and the corresponding communication techniques (in analogy to the environmental engineering on the lower left side of the butterfly graph) are equally important, however much less often considered in the environmental literature or restoration-oriented courses. There is still a great difficulty in convincing natural scientists or engineers about the necessity to integrate social studies into their projects. This can be partially explained by the intellectual path-dependency described above. In the mindscape of natural scientists, the power of convincing ideas has the highest values. Researchers are in a permanent competition for providing the better solution to the problem, which is translated into the impact factor of the journal, in which the results are published or the financial volume of a grant that a group of scientists receives. In this mindscape, the power of narratives or social trends is mostly tuned out. A further problem is the degree of sophistication

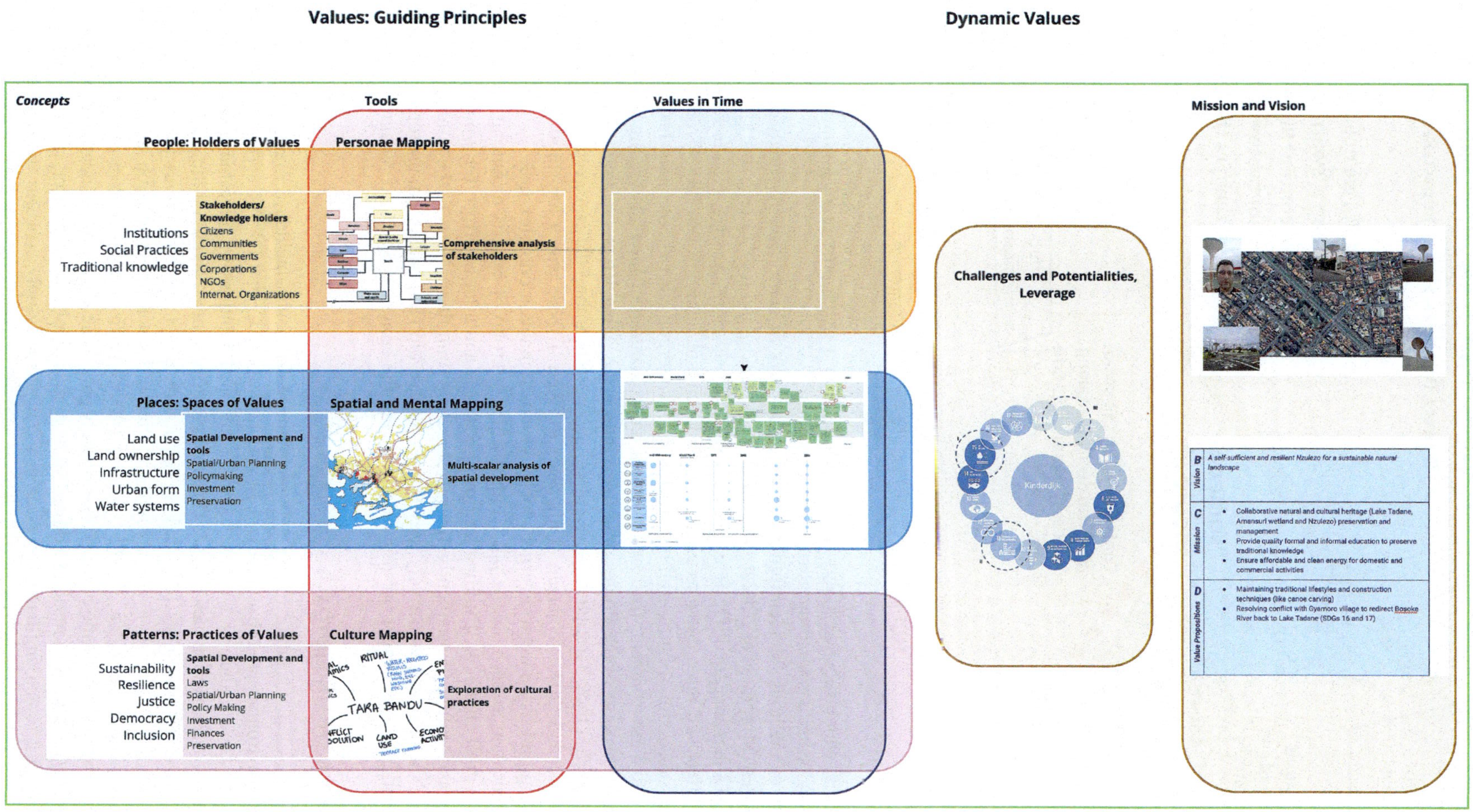

FIGURE 40.4 **Valuescapes.** Contrary to physical patch dynamics in morphological landscapes (Fig. 43.3) valuescapes are driven by historical developments and societal drivers. *From Hein, C., 2025. Water systems design: connecting and developing methods for the value case approach.* Blue Papers *4 (1), 88–99. https://doi.org/10.58981/bluepapers.2025.1.06.*

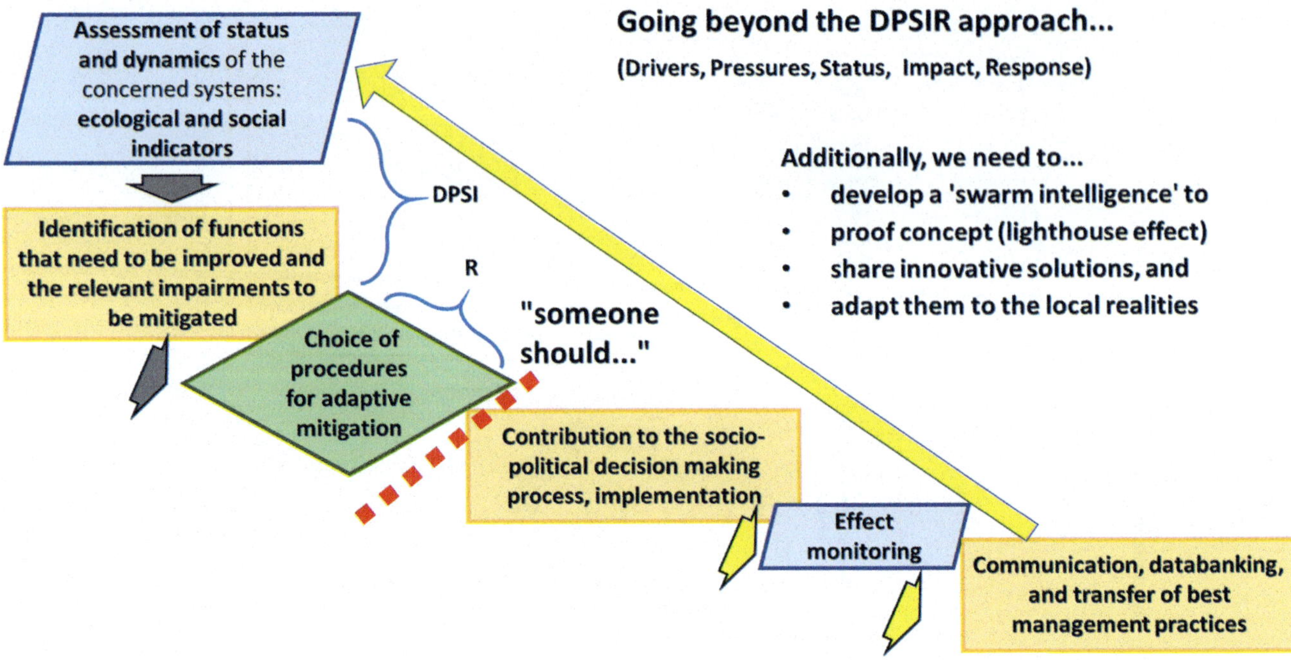

FIGURE 40.5 Going beyond the DPSIR approach.

of environmental studies, which requires a very high skillfulness in a given area (e.g., groundwater hydraulics), which hardly allows the natural scientists to study also social thematics.

There is a plethora of methods for identifying the different types of actors and for analyzing their value-scapes in different times and places (Murphy et al., 2022; Pahl-Wostl et al., 2020) and transferring this knowledge into sustainable action. When talking to site managers of Designated Areas (e.g., World Heritage sites, Nature Reserves, Biospheres), the approach is less theoretical and more down-to-earthed. In independent interviews in three French sites, the bottom-line message was that the successful solution-finding process had undergone a tedious, long-lasting process of stepwise enlargement of their social environment, trust-building and harmonization of interests among the different stakeholders (Yousry et al., 2022). In the case of the Poutès dam lowering on the Allier River (an important salmon breeding river, main tributary to the Loire), it took 3 decades to streamline the interests among the conservation, tourism, agriculture and hydropower interest (Michaud, 2021; Wantzen et al., 2023a). Even if there was a legislation that could be applied or reinforced, finding compromises between divergent lobby groups remains a complex task.

Lastly, the communication as such, taking place either within a group of experts or among the expert teams and the stakeholders, remains a great challenge. Building up trust takes time and a lot of effort, for which scientists often are not available as they have other duties. Consequently, many researchers limit their efforts to the publishable results of a study, and are not involved in the project, respectively they do not receive the feed-back coming from the stakeholders during and after the implementation (and cannot use this for improving the approach, see next paragraph).

2.2 Going beyond the DPSIR approach

Many studies aimed to develop a solution for a socio-environmental problem include a multi-disciplinary analysis of the technical problem (e.g., environmental impact analysis, biomonitoring, BACI and DPSIR approaches (England et al., 2021; Wantzen et al., 2019)) and the finding of the corresponding technical solutions to solve the problem (e.g., environmental engineering, restoration and conservation techniques). There are plentiful courses and scientific publications dealing with these thematics. Often enough, scientific studies (financed by application-oriented programmes) provide a scientifically sound problem analysis but do not even provide a suggestion for solving it. Rather, the conclusion section of these papers can be summarized in the half-phrase: "*someone should* ..." (Fig. 40.5). Even if the problem-solving approach was finally implemented, the monitoring of the outcome of the project is rarely part of the study, for financial or time reasons (but there are exceptions, see Lorenz et al., 2018). As a result, we do not know in many cases, if the projects had reached the pre-established targets or not. Moreover, if a monitoring has taken place at all, it generally only concerns the technical part of the project, whereas the hurdles that were to overcome concerning societal constraints are not studied at the same moment. Hands-on experiences of project managers or site managers are still an under-exploited source of knowledge (Yousry et al., 2022). Data management by private consultancies is even worse: The data they provide to private or public bodies are at the same time their capital (especially when it concerns the results of time-consuming and/or expensive biodiversity data), so they hardly deposit these data in GBIF-compatible databases.

Thus, the beyond-DPSIR approach includes the monitoring of the social *and* ecological effects of the developed project (including the detailed analysis of technical and societal hurdles and how they were overcome), the contribution of these date into publically available databanks, respecting the FAIR principles, adequate communication of the results (with tailored communication strategies for other stakeholders, the general public and scientists) and the feedback-loop to integrate this knowledge into future projects. To enable this additional effort, additional funding and a prolongation of the project period (to include monitoring) is required.

3. Teaching hydronexialism

3.1 The beginnings

Since 2023, in the framework of EUCOR—the European Campus, the EUCOR Excellence Chair on "Water and Sustainability" and under the auspices of the UNESCO Chair on "Rivers and Heritage, River Culture" and UNESCO's IHP-Ecohydrology branch, a program on training MSc. Students in "Continental Water Sustainability" (CWS, switch.unistra.fr/formation/master-cws/) has been developed. A steering committee from five universities across the Upper Rhine (Strasbourg, KIT Karlsruhe, Freiburg, Basel and Mulhouse) has contributed to it, integrating the experiences and expertises from their existing MSc and engineering courses in the field of hydrosystem management and UNESCO's approaches with the Ecohydrology/IHP (Intergovernmental Program for Hydrology (http://ecohydrology-ihp.org)), online courses and IHE (Institute of Water Education, Delft, https://www.un-ihe.org). It also integrated experiences from the 40 year-existing Master (now: Engineer's) course on hydrosystem management at University/PolyTech Tours (IMACOF/DAE-IMAGE) in France.

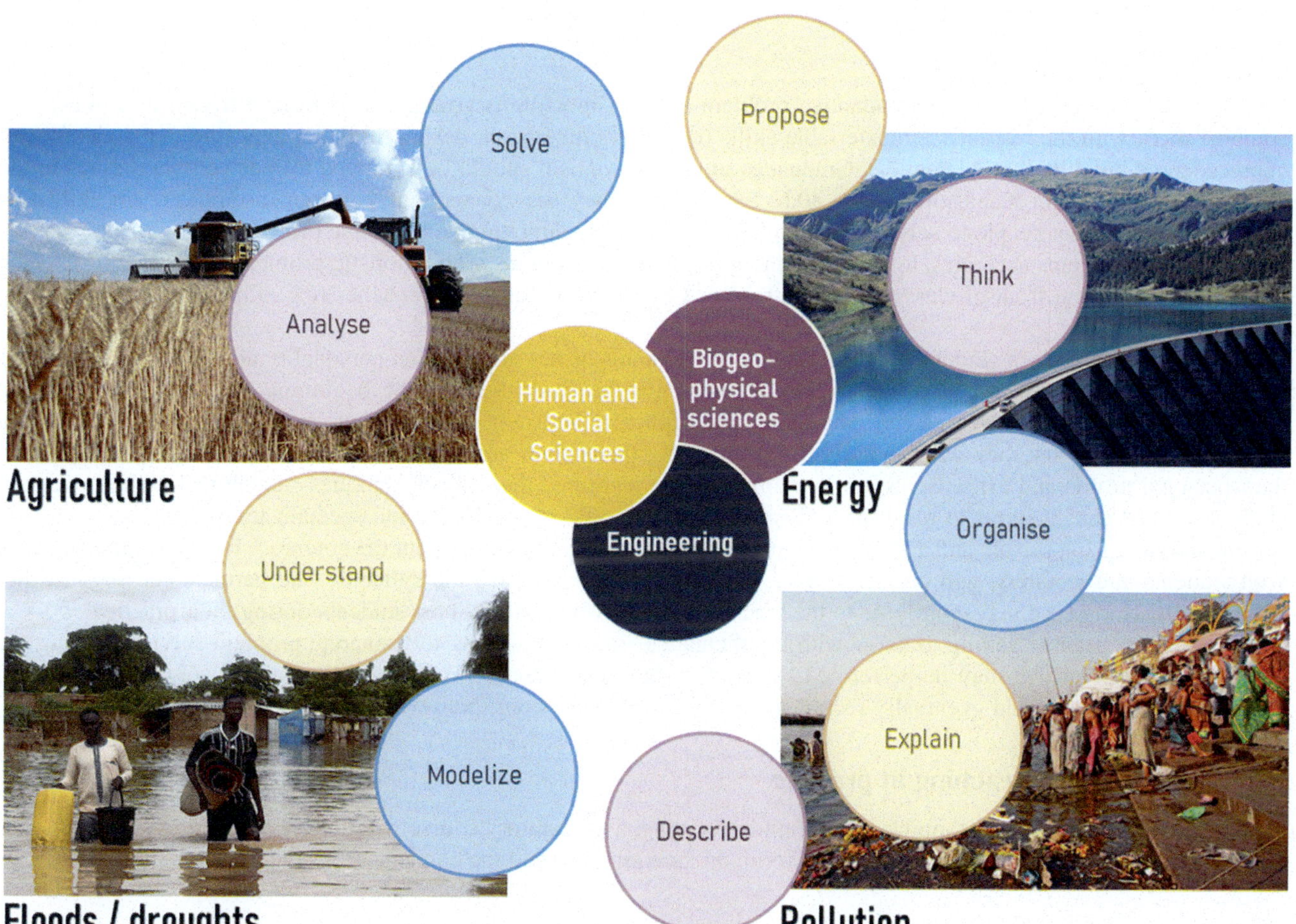

FIGURE 40.6 Thematics to be dealt with by hydronexialism. *Graph by V. Marchant.*

As the overarching goal of the new MSc. course is to train students in interdisciplinary problem-solving, the contents of the course will develop from different disciplinary background and practitioners' knowledge in problem-oriented units. The 4 C's (critical thinking, collaboration, communication, creativity) are supported in all educational units.

In the beginning, this project faced several challenges. Interdisciplinarity is key, however most lecturers insisted on delivering profound training on their individual thematics, which would have been incompatible with a holistic course for time reasons. The advantages resulting from multi-, inter- and transdisciplinary approach of this course are paid at the price that students coming from different disciplines will have different degrees of preparedness for participating in the advanced Master-lectures contributing to our course. Thus, the syllabus had to be harmonized in a way that all relevant thematics are covered in the required depth of knowledge. To overcome the dilemma resulting from the thematical diversity versus required in-depth knowledge of the graduates in each of the different disciplines, we opted for giving an additional focus on fast learning, data-acquisition, evaluation and interpretation techniques and to enable students to inform themselves efficiently in small teams, e.g., during hackathon meetings. Inter- and transdisciplinary courses, often including practitioners as lecturers, and scientists with long-term experience in holistic ecosystem analysis and management, help to provide students with the "big picture" needed to understand problems from different angles. Special care is taken for IA-supported information generation and a critical evaluation of their outcomes by skillful human minds and comparisons with peer-reviewed articles and reliable Geographical Information Systems.

Originally, it was planned to keep the Master course open for bachelors of all disciplines, but due to the required compatibility with existing courses, this idea had to be given up. Currently, students from all types of natural sciences are admissible, provided they have a good basis in environmental and aquatic sciences. Another great challenge was the transboundary character. EUCOR—the European Campus (https://www.eucor-uni.org/), is a GECT organization unifying five universities from three nations, attended by about 120,000 students. Each of the three countries has its own rules and standards, and each of the participating universities and Engineering Schools (e.g., the ENGEES French National School on Water and Environmental Engineering), which had never been syntonized before. The European ECTS mark system has been a great asset but the dates and lengths of semesters, the way how diploma are issued etc. vary greatly between schools and nations. Specific conventions were needed to make the course compatible at these levels and to provide a double degree.

3.2 The general concept

As the focus is on collaborative and adaptive problem-solving, most units arrange skills from different disciplines and combine them with the respective transferrable skills (e.g., communication, design thinking, exposing the students with unprecedented "real life" problematics). Graduades in hydronexialism should be equipped to face complex issues in the context of hydrosystem management Fig. 40.6. Specifically, they see connections between disciplines, are skilled in conflict resolution, get people to solve complex problems, thereby employing an evolutive project design and an adaptive reorganization of teams according to each upcoming problem. This begins with a holistic problem-understanding, which includes that classical front-of-class teaching is replaced by modern didactic approaches (e.g., flipped classroom, group work) when possible.

At the same time, each student will be trained to develop his or her professional personality and personal skills. This will be achieved through a course unit which accompanies the students from the beginning, in which they develop entrepreneurship for their own career, employing the Authentic Learning concept (Lombardi and Oblinger, 2007). In practice, this means that they will regularly discuss (among themselves and with tutors/experts) to identify and sharpen their personal profile and expertise. Specific training on the identified thematics/techniques—combined with the general skills described above—aims to result in a "T-shaped" skill profile of generality and specialization.

Practitioners from private and and/or public partnerships are engaged throughout the course. A board of experts, along with students and lecturers, will critically revise and permanently improve the syllabus. The partners can suggest thematics for teaching (and may contribute to the teaching themselves) and will host students during their practical work at different levels, from a "getting to know with the thematics" in the beginning to "concrete problem-solving" during the MSc. thesis work (which may be performed by a small team of students). Partners will profit by solutions developed during the MSc. theses and eventually recruit graduates (see Conclusions, below).

3.3 Hydronexialism teaching in practice

To bring enrolled students to a similar level of understanding, a preparatory course will deliver an overview the thematics of the MSc course. It presents, in a condensed form, the overall goals and the contents of individual units they will follow in the next 2 years. Most of all, it will help students to approach, understand and solve problems including highly different angles, ranging from the ecosystem perspective to the individual stakeholders and their respective constraints. Moreover, students will exercise team work, solution-oriented reasoning and design-thinking from the first moment onwards. Applying very different pedagogical approaches, from incentive talks to interactive group work, and representative field visits, students will learn more about themselves and their role in solving water-related problems.

The key word of the **first semester** is 'to consolidate', from different perspectives. Lectures and courses with a focus on the biophysical and engineering aspects of hydrosystems will help all students to acquire knowledge and technical skills necessary to fill the gaps from their respective past studies (especially considering the social-sciences/natural-sciences complementarity). The students should consolidate their knowledge and learn how to break down problems into sections that can be solved by an individuum with specific tasks. The "consolidation" also has a social component. Students assemble teams that can tackle the problems by a close and interactive cooperation. They will encounter difficulties when they experience that the expertise of the team will not suffice to solve the problem, and learn how to get the knowledge and the skills that are needed to overcome these hurdles.

The main objective of the **second semester** is to undergo the transformation from the discipline-oriented approach to the inter- and transdisciplinary working style. Students are now acquainted with each other, with most of the lecturers, know their own (and their peers') strengths and weaknesses, and have already learned some first elements of nexialism. Now, more and more integrative and practical courses are on schedule. Students will from now on work almost exclusively in teams. This includes a substantial change in the format of the pedagogics. Frontal lectures will have increasingly less importance, and will be replaced by group work, and direct exchange with stakeholders.

Semester 3 will prepare the students for their MSc thesis. It is an orientation phase, helping the students to chose the thematic issue they would like to dedicate their thesis to. One by one, "classical" water conflict situations will be played through by variable teams and lecturers. In addition to the expertises covered by the CWS student team members, the students will learn to find and integrate "help from outside the team", i.e., experts from other universities, experienced stakeholders etc. and to combine this knowledge for finding adequate solutions to the diverse problems. This is a reiterative process, as the first proposal may not fully meet the requirements for an environmentally sound and socially acceptable solution. Economical and social feasibility are strong selective parameters, so that the finally proposed solution may not always be the ideal one, but the best available under the current socio-economical-ecological setting. Moreover, the communication between the team and the target audience (politicians, financing agencies, environmental protection agencies, national and international bodies etc.) will be a key skill to be learned by the students in this phase: without adequate communication, even the best solution will not reach its target audience, and will not be implemented in the end.

This will be followed by the MSc. thesis in the **4th semester**, which will in most cases take place in the institutions of stakeholders or in the laboratories of the universities, but in all cases in tight contact with the team of supervisors from both sides. There will be two different elements to be evaluated in the theses. Of course, the performance by the individual student in terms of scientific rigor, adequacy of the selected methods, quality of the results and the knowledgeable and clear discussion of the results in the context of solving a specific part of the overall problem will be crucial. Ideally, a good deal of this work has already been presented and evaluated at the end of Semester 3. Constructive critiques by the

FIGURE 40.7 The EUCOR/UNESCO transboundary fall school 2024: 'Transdisciplinary approaches to the sustainability of water use: case studies from the Upper Rhine'. *Photo credit: Maxime Lantz.*

supervisors will be integrated into the introductory part. The acquired data during the practical work phase (which will mostly take place in form of an internship) and their analysis/interpretation will make up the bulk of the evaluated individual part. But only the team can win! Supervisors, students and hosting institutions will be encouraged to develop collaborative solutions in small teams of students. In this case, the added value delivered by the team, beyond the individual performance, will be assessed. When the theses will have been finished, they will be presented to the concerned public (stakeholders etc), who will be invited to provide their own evaluation of the solution. Their evaluation will also be considered (along with the academic jury of the project supervisors at the universities) to form the final mark. A special program for students opting for research and further PhD studies will be provided, and these students will receive specific support (networking, laboratory facilities, conferences) by their supervisors and programs such as the "FabLab" programm at the University of Strasbourg.

4. Outlook

Hydronexialism is a novel concept and the above-described CWS Master Course is an experimental approach to teach complex problem solving in the context of hydrosystems. By integrating knowledge from most diverse (natural and human) sciences, traditional ecological knowledge, engineering and stakeholders' experiences, it contributes to Transdisciplinary Ecohydrology in a practical way—by training solution-oriented, polyvalent hydrosystem managers.

A "*comité de perfectionnement*", composed by experienced researchers and stakeholders will warrant a permanent improvement of the contents and the pedagogical methods. Seminar series and summer schools Fig. 40.7 may serve as test drives for implementing new thematics. The job perspectives of the graduates are expected to be excellent. In the European "water business", about 50% of the positions will be renewed within the next 5 years, and due to the current changes (climate, technology), new types of jobs will appear, they may also accompany spin-offs and branching off within their enterprise or institution (intrapreneurship). Effort will be taken to enact the responsibility of the Global North (GN), which disposes of a large wealth of long-term experiences with water-related environmental problems and their potential solutions. Whenever possible, fellowships for students from the Global South (GS) will be provided, same for those from Ukraine, which urgently needs an "Environmental Marshall Plan" to further develop and restructure an environmental monitoring, decision-making and enactment program. However, the GN–GS exchange is not a one-way road. Much can be learned from the GS concerning creativity, (still existing) human–nature relationships, and traditional ecological knowledge (Wantzen et al., 2019).

References

Cao, Y., Chen, W.Y., Wantzen, K.M., 2024. 'One River, Two Systems': Hong Kong's river management. Environ. Manag. 73 (1), 81–101. https://doi.org/10.1007/s00267-023-01923-9.

Chardon, V., Herrault, P.A., Staentzel, C., Skupinski, G., Finance, O., Wantzen, K.M., Schmitt, L., 2022. Using transition matrices to assess the spatio-temporal land cover and ecotone changes in fluvial landscapes from historical planimetric data. Earth Surf. Process. Landf. 47 (11), 2647–2659. https://doi.org/10.1002/esp.5437.

Creanza, N., Kolodny, O., Feldman, M.W., 2017. Cultural evolutionary theory: how culture evolves and why it matters. Proc. Natl. Acad. Sci. U. S. A 114 (30), 7782–7789. https://doi.org/10.1073/pnas.1620732114.

Directive 2000/60/EC of the European Parliament and of the Council of 23 October 2000 establishing a framework for Community action in the field of water policy. Off. J. Eur. Communities - Legislation 327, 2000, 1–73.

Elfithri, R., Zalewski, M., Arduino, G., Chicharo, L., 2025. Transdisciplinary ecohydrology for water management solutions and sustainability. Nat. Water 3 (4), 360–363. https://doi.org/10.1038/s44221-025-00395-x.

England, J., Angelopoulos, N., Cooksley, S., Dodd, J., Gill, A., Gilvear, D., Johnson, M., Naura, M., O'Hare, M., Tree, A., Wheeldon, J., Wilkes, M.A., 2021. Best practices for monitoring and assessing the ecological response to river restoration. Water 13 (23), 3352. https://doi.org/10.3390/w13233352.

Folke, C., Polasky, S., Rockström, J., Galaz, V., Westley, F., Lamont, M., Scheffer, M., Österblom, H., Carpenter, S.R., Chapin, F.S., Seto, K.C., Weber, E.U., Crona, B.I., Daily, G.C., Dasgupta, P., Gaffney, O., Gordon, L.J., Hoff, H., Levin, S.A., Lubchenco, J., Steffen, W., Walker, B.H., 2021. Our future in the Anthropocene biosphere. Ambio 50 (4), 834–869. https://doi.org/10.1007/s13280-021-01544-8.

Germaine, M.A., Gonin, A., 2024. Experiencing the landscape: landscape agency in a multifunctional valley after dam removal on the Sélune River, France. Ecol. Soc. 29 (1). https://doi.org/10.5751/ES-14868-290123.

Harari, Y.N., 2015. Sapiens: A Brief History of Humankind. Harper.

Harper, M., Mejbel, H.S., Longert, D., Abell, R., Beard, T.D., Bennett, J.R., Carlson, S.M., Darwall, W., Dell, A., Domisch, S., Dudgeon, D., Freyhof, J., Harrison, I., Hughes, K.A., Jähnig, S.C., Jeschke, J.M., Lansdown, R., Lintermans, M., Lynch, A.J., Meredith, H.M.R., Molur, S., Olden, J.D., Ormerod, S.J., Patricio, H., Reid, A.J., Schmidt-Kloiber, A., Thieme, M., Tickner, D., Turak, E., Weyl, O.L.F., Cooke, S.J., 2021.

Twenty-five essential research questions to inform the protection and restoration of freshwater biodiversity. Aquat. Conserv. Mar. Freshw. Ecosyst. 31 (9), 2632–2653. https://doi.org/10.1002/aqc.3634.

Hein, C., 2025. Water systems design: connecting and developing methods for the value case approach. Blue Papers 4 (1), 88–99. https://doi.org/10.58981/bluepapers.2025.1.06.

Hering, D., Borja, A., Carstensen, J., Carvalho, L., Elliott, M., Feld, C.K., Heiskanen, A.S., Johnson, R.K., Moe, J., Pont, D., Solheim, A.L., de Bund, W. v., 2010. The European Water Framework Directive at the age of 10: a critical review of the achievements with recommendations for the future. Sci. Total Environ. 408 (19), 4007–4019. https://doi.org/10.1016/j.scitotenv.2010.05.031.

Imfeld, G., Lefrancq, M., Maillard, E., Payraudeau, S., 2013. Transport and attenuation of dissolved glyphosate and AMPA in a stormwater wetland. Chemosphere 90 (4), 1333–1339. https://doi.org/10.1016/j.chemosphere.2012.04.054.

Kondolf, G.M., Pinto, P.J., 2017. The social connectivity of urban rivers. Geomorphology 277, 182–196. https://doi.org/10.1016/j.geomorph.2016.09.028.

Krause, S., Lewandowski, J., Dahm, C.N., Tockner, K., 2015. Frontiers in real-time ecohydrology - a paradigm shift in understanding complex environmental systems. Ecohydrology 8 (4), 529–537. https://doi.org/10.1002/eco.1646.

Lombardi, M.M., Oblinger, D.G., 2007. Authentic learning for the 21st century: an overview. Educ. Learn. Init. 1, 1–12.

Lorenz, A.W., Haase, P., Januschke, K., Sundermann, A., Hering, D., 2018. Revisiting restored river reaches – assessing change of aquatic and riparian communities after five years. Sci. Total Environ. 613–614, 1185–1195. https://doi.org/10.1016/j.scitotenv.2017.09.188.

Love, J., 2008. 17: meeting the challenges of integrative learning: the Nexia concept. To Improve Acad. 26 (1), 263–274. https://doi.org/10.1002/j.2334-4822.2008.tb00513.x.

Luz, L.D., 2019. In: GEOMORFOLOGIA DA PLANÍCIE DE INUNDAÇÃO E HIDRODINÂMICA DA CONFLUÊNCIA PARAGUAI-CUIABÁ: PANTANAL MATO-GROSSENSE, BRASIL (Floodplain Geomorphology and Hydrodynamics of the Paraguay-Cuiabá Confluence: Pantanal Wetlands).

Michaud, C., 2021. In: Barrage de Poutès : EDF ne fait plus obstacle à la migration des saumons Le Parisien.

Morandi, B., Cottet, M., Piégay, H., Styler-Barry, B., 2021. How to Better Involve Stakeholders in River Restoration Projects River Restoration: Political, Social, and Economic Perspectives. Wiley, pp. 147–168. https://doi.org/10.1002/9781119410010.ch7.

Murphy, B.M., Russell, K.L., Mould, S., Vietz, G., Nelson, P.A., 2022. Managing urban riverscapes: an assessment framework to integrate social-ecological values and physical processes. J. Environ. Manag. 322. https://doi.org/10.1016/j.jenvman.2022.115862.

Ojala, M., Cunsolo, A., Ogunbode, C.A., Middleton, J., 2021. Anxiety, worry, and grief in a time of environmental and climate crisis: a narrative review. Annu. Rev. Environ. Resour. 46, 35–58. https://doi.org/10.1146/annurev-environ-012220-022716.

Pahl-Wostl, C., Knieper, C., Lukat, E., Meergans, F., Schoderer, M., Schütze, N., Schweigatz, D., Dombrowsky, I., Lenschow, A., Stein, U., Thiel, A., Tröltzsch, J., Vidaurre, R., 2020. Enhancing the capacity of water governance to deal with complex management challenges: a framework of analysis. Environ. Sci. Pol. 107, 23–35. https://doi.org/10.1016/j.envsci.2020.02.011.

Piscart, C., Dézerald, O., Pellan, L., Le Bris, N., Rodríguez-Pérez, H., Beauverger, T., Huteau, D., Roussel, J.M., 2024. Persistent disconnect between flow restoration and restoration of river ecosystem functions after the removal of a large dam on the Sélune River. Front. Environ. Sci. 12, 1250810. https://doi.org/10.3389/fenvs.2024.1250810.

Plum, N., Schulte-Wülwer-Leidig, A., 2014. From a sewer into a living river: the Rhine between Sandoz and Salmon. Hydrobiologia 729 (1), 95–106. https://doi.org/10.1007/s10750-012-1433-1.

Pringle, C.M., Naiman, R.J., Bretschko, G., Karr, J.R., Oswood, M.W., Webster, J.R., Welcomme, R.L., Winterbourn, M.J., 1988. Patch dynamics in lotic systems: the stream as a mosaic. J. North Am. Benthol. Soc. 7 (4), 503–524. https://doi.org/10.2307/1467303.

Reid, A.J., Carlson, A.K., Creed, I.F., Eliason, E.J., Gell, P.A., Johnson, P.T.J., Kidd, K.A., MacCormack, T.J., Olden, J.D., Ormerod, S.J., Smol, J.P., Taylor, W.W., Tockner, K., Vermaire, J.C., Dudgeon, D., Cooke, S.J., 2019. Emerging threats and persistent conservation challenges for freshwater biodiversity. Biol. Rev. 94 (3), 849–873. https://doi.org/10.1111/brv.12480.

Schmitt, L., Beisel, J., Preusser, F., De, J., Wantzen, C., Chardon, K., Staentzel, V., Eschbach, C., Damm, D., Rixhon, C., 2019. Sustainable management of the upper Rhine river and its Alluvial plain: lessons from interdisciplinary research in France and Germany. In: Sustainable Research in the Upper Rhine Region. Concepts and Case Studies, pp. 201–226.

Sehring, J., Schmeier, S., Horst, R.t., Offutt, A., Sharipova, B., 2022. Diving into water diplomacy - exploring the emergence of a concept. Diplomatica 4 (2), 200–221. https://doi.org/10.1163/25891774-bja10082.

Strang, V., 2019. The Rights of the River: Water, Culture and Ecological Justice Conservation: Integrating Social and Ecological Justice. Springer International Publishing, pp. 105–119. https://doi.org/10.1007/978-3-030-13905-6_8.

Strang, V., 2024. Listening to the river: representing nonhuman needs and interests in debates about water governance and management. River Res. Appl. 9, 1698–1709. https://doi.org/10.1002/rra.4137.

Tockner, K., Stanford, J.A., 2002. Riverine flood plains: present state and future trends. Environ. Conserv. 29 (3), 308–330. https://doi.org/10.1017/S037689290200022X.

Wantzen, K.M., 2023. River Culture – Life as a Dance to the Rhythm of the Waters. UNESCO publishing.

Wantzen, K.M., 2024. River culture: how socio-ecological linkages to the rhythm of the waters develop, how they are lost, and how they can be regained. Geogr. J. 190 (2). https://doi.org/10.1111/geoj.12476.

Wantzen, K.M., Aziz, B., Longuet, I., Bao, I., Bocoum, H., Cissé, L., Chauhan, M., Girard, P., Gopal, B., Kane, A., Marchese, M.R., Prakash, N., Teixeira, P., Zalewski, M., 2016. River Culture: an eco-social approach to mitigate the biological and cultural diversity crisis in riverscapes. Ecohydrol. Hydrobiol. 16 (1), 7–18. https://doi.org/10.1016/j.ecohyd.2015.12.003.

Wantzen, K.M., Alves, C.B.M., Badiane, S.D., Bala, R., Blettler, M., Callisto, M., Cao, Y., Kolb, M., Kondolf, G.M., Leite, M.F., Macedo, D.R., Mahdi, O., Neves, M., Peralta, M.E., Rotgé, V., Rueda-Delgado, G., Scharager, A., Serra-Llobet, A., Yengué, J.L., Zingraff-Hamed, A., 2019. Urban stream and wetland restoration in the global south-a DPSIR analysis. Sustainability 11 (18), 4975. https://doi.org/10.3390/su11184975.

Wantzen, K.M., Berton, J.-P., Boisneau, C., Greulich, S., Marmiroli, B., Rodrigues, S., 2023a. The Loire – A Cultural and Environmental Exception in Europe. UNESCO, pp. 699–721. https://doi.org/10.54677/vntd8327.

Wantzen, K.M., Schmitt, L., Wirbelauer, E., 2023b. The Rhine – an Important Biocultural Axis for Europe. UNESCO, pp. 561–587. https://doi.org/10.54677/kohd7709.

Yousry, L., Cao, Y., Marmiroli, B., Guerri, O., Delaunay, G., Riquet, O., Wantzen, K.M., 2022. A socio-ecological approach to conserve and manage riverscapes in designated areas: cases of the Loire river valley and Dordogne basin, France. Sustainability 14 (24), 16677. https://doi.org/10.3390/su142416677.

Zalewski, M., 2002. Ecohydrology - the use of ecological and hydrological processes for sustainable management of water resources. Hydrol. Sci. J. 47 (5), 823–832. https://doi.org/10.1080/02626660209492986.

Zalewski, M., 2014. Ecohydrology and hydrologic engineering: regulation of hydrology–biota interactions for sustainability. J. Hydrol. Eng. 19 (5), 1080–1088. https://doi.org/10.1061/(ASCE)HE.1943-5584.0000999.

Zalewski, M., 2021. Ecohydrology: an integrative sustainability science. In: Hromodka, T.V., Rao, P. (Eds.), Hydrology. IntechOpen. https://doi.org/10.5772/intechopen.94169.

Zalewski, M., Wantzen, K.M., 2023. Ecohydrology and river culture: synergies for the enhancement of river and catchment sustainability. UNESCO, pp. 807–819. https://doi.org/10.54677/dtuq3662.

Part IV

Transdisciplinary ecohydrology: A way forward

Chapter 41

Creating a culture of sustainability through transdisciplinary ecohydrology

Izabela A. Levy and Maciej Zalewski
European Regional Centre for Ecohydrology of the Polish Academy of Sciences under the auspices of UNESCO, Łódź, Poland

CREATING OPPORTUNITIES FOR PEOPLE TO EXPERIENCE THE WALLER CREEK CORRIDOR, AUSTIN, TX. *IMAGE CREDIT: MICHAEL VAN VALKENBURGH ASSOCIATES INC., WALLER CREEK.*

Transdisciplinary Ecohydrology. https://doi.org/10.1016/B978-0-443-41596-8.00035-2

1. Toward a culture of sustainability

The ongoing degradation of the biosphere, combined with population growth and rising consumption, and a widespread lack of understanding of life-supporting processes that sustain water quality, availability, and ecosystem vitality, has created a situation where indifference is no longer an option. Despite a general decline in global poverty (Poverty, Prosperity and Planet 2024), achieving SDGs by 2030 seems still unattainable. There are also some positive developments, e.g., discussions on the circular economy have nearly tripled over the past 5 years and reached megatrend status; however, in reality, global circularity has declined by 21% (The Circularity Gap Report 2024). In other words, simply talking more about possible solutions does not necessarily bring us closer to implementing environmentally conscious decisions. Why is that? One reason lies in the persistent gap between knowledge and action: while awareness of environmental problems has grown, translating this awareness into concrete policies and behavioral changes remains difficult. Discussions often stay at the level of abstract visions or broad strategies, without addressing the institutional, cultural, and economic barriers that hinder real-world application. Moreover, decision-making processes are frequently shaped by short-term political and financial interests, which overshadow long-term ecological considerations. Social inertia also plays a role—people and organizations may resist change because it requires new habits, investments, or a shift in values. This chapter explains why building a culture of sustainability enhancement is essential for making positive changes that will benefit future generations. It shows that lasting change depends not only on technology or policy, but also on shared values and everyday choices that put care for the environment first.

The approach of transdisciplinary ecohydrology (TEH) is both holistic and multifaceted. This book explores its theoretical foundations as well as its practical applications. Within this framework, a key pillar is the development of Culture and Education, discussed in this chapter, integrated through the WBSR + CE + LPG framework, which integrates water, biodiversity, ecosystem services, and resilience (WBSR) with culture and education (CE) and law policy governance (LPG). Together, these elements provide a foundation for building a sustainable society that is both environmentally resilient and socially engaged. TEH enhances the biosphere's capacity to sustain life in the face of escalating human and environmental pressures. With global population growth and intensifying consumption, enhancing life-supporting processes becomes a crucial strategy for long-term sustainability. Achieving this, however, requires both systemic and individual action: roughly 50% of the reduction in the biosphere's impact depends on strategic decision-makers who formulate and implement large-scale measures, while the remaining 50% relies on a culture of sustainability shaped by each individual's awareness and knowledge. For this reason, fostering a culture of sustainability must be emphasized as a cornerstone of lasting environmental change. This is why raising awareness of the environmental crisis is essential, and new ways of integrating this awareness into citizens' everyday lives must take priority.

Culture, as a fundamental component of human behavior, holds the power to drive positive change. Culture refers to the collection of attitudes, values, beliefs, and behaviors that are shared within a group, but different for each individual, and passed down across generations (Matsumoto, 1996). Culture forms a foundational element of social systems and plays a critical role in shaping perception, behavior, and decision-making. It is embedded in language, social norms, and collective practices, and is influenced by both biological predispositions and contextual factors. As a reflection of shared understanding within a society, community, or organization, culture defines how individuals interpret, experience, and respond to challenges. Creating a culture of sustainability therefore requires a conscious reorientation of collective values and worldviews toward long-term enhancement of the sustainability potential of the biosphere.

Culture has long been recognized as a dynamic and integral component of human identity, social cohesion, and sustainable development. According to UNESCO (2003), culture forms the foundation of people's identity, offering individuals and communities a sense of continuity, belonging, dignity, and meaning. It encompasses cultural capital, including knowledge systems, practices, values, and artistic expressions, that support social resilience and innovation. Cultural diversity, in this context, serves as a vital driver of sustainable development, contributing to the social, economic, and environmental dimensions of the sustainable development goals (SDGs), particularly by fostering inclusive growth and intercultural dialogue (Sustainable Development Goals, 2015).

Within this broader perspective, the interrelation between cultural and ecological systems becomes essential to achieving sustainability. Piotr Korzeniowski (2012) emphasizes that ecological security constitutes a key institution of environmental law, serving as a comprehensive framework that integrates legal, social, and ecological dimensions of sustainability. He proposes, in this context, the application of a systemic concept of human needs, which recognizes that environmental protection and social development are inseparably linked to the fulfillment of fundamental human needs, including safety, belonging, growth, and harmony with the environment (Korzeniowski, 2013). Additionally, from this perspective, ecohydrological security acquires a cultural dimension, reflecting shared values and collective responsibility for the continuity of life and the quality of the environment. Therefore cultural vitality and ecohydrological security together form the essential basis for societal resilience in the face of environmental and social change.

Historically, culture has never been static. Rather, it evolves through an ongoing interplay between human societies and their environments. This coevolutionary relationship, defined by mutual feedback between ecological conditions and cultural adaptation, shapes distinct ways of perceiving and managing landscapes and resources, and is defined as the ecological culture (Sutton and Anderson, 2004). The ecological culture is the study of how human societies adapt to their environments, emphasizing the interplay between culture and the natural world (Lapka et al., 2012). While the uniqueness of any culture thus emerges as the cumulative result of long-term adaptive processes shaped by both natural environment constraints and social innovation.

Cultural landscapes are the combined works of nature and humans, illustrating the evolution of societies and settlements shaped by natural conditions and ongoing social, economic, and cultural influences. They reveal the deep connection between people and the environment. Culture, rooted in shared values and the pursuit of positive change, aligns closely with the principles of sustainability and resilience. In this context, degraded landscapes can be transformed into spaces of renewal, where cultural practices support the transition to healthier, more sustainable environments. Recognizing this potential not only sheds light on past transformations but also emphasizes the role of culture in fostering environmental responsibility. By cultivating a sense of place, identity, and stewardship, cultural frameworks can drive collective efforts to enhance the sustainability potential of the biosphere (Zalewski et al., 2021) On the other hand, a "lack of culture" in this context can be understood as a lack of sustainability education and a limited awareness of the vital connections between human well-being and the natural environment. It reflects a failure to recognize how deeply we depend on nature and ecosystem services, such as clean water, air, fertile soil, and climate regulation, for our survival and quality of life.

For millennia, nature has been a source of awe, inspiration, and curiosity. Today, however, it has largely been pushed out of sight for many citizens and forgotten by the broader public. Still, people often seek out natural settings, especially during holidays, to rest and recharge, revealing an enduring need for connection with the natural world. Fortunately, there are ways to reintegrate functional, biodiverse ecosystems into both new novel environments, restoring nature not only for its own sake but also to improve our well-being and quality of life. This book offers a roadmap for enhancing the functionality of the biosphere and reversing its degradation, while simultaneously enriching the aesthetic value. Deepening our understanding of TEH approaches enables the creation of multifunctional landscapes where functionality is seamlessly integrated with aesthetics. Furthermore, culture nurtures our appreciation of art and the aesthetic dimensions of nature, yet it also demands active effort and practice. Through the practice of nature appreciation, we experience aesthetics that transcend the self and connect us with the wider world, opening pathways toward environmental stewardship. By making the human need for connection with nature more visible through scientific research or artistic work, we create the potential to reintroduce this bond into collective understanding, recovering what has been lost and revealing how deeply inspiring it can be. For example, photography can be both documentary and evocative, serving as a powerful catalyst for the change we seek (Domański, 2025). In this way, raising environmental awareness can transform culture into a driving force for a more desirable future, embedding the very notion of a culture of sustainability.

2. Educating for a sustainable future

TEH offers concrete, integrative methodologies for managing water resources and mitigating human-induced environmental pressures. At the same time, the river culture concept provides a biocultural framework that recognizes rivers as both ecological and cultural entities. It emphasizes the need to incorporate traditional knowledge, material and immaterial cultural practices, and organismic strategies into river basin management, particularly in transboundary contexts (Wantzen et al., 2016).

Both TEH and river culture share a foundational respect for natural processes and promote the maintenance of flow regimes and ecosystem conditions that support resilience and sustainability. As anthropogenic pressures on riverine systems intensify, these complementary frameworks converge in the water, biodiversity, ecosystem services, resilience, plus culture, and education (WBSR + CE) goals, which call for a multidimensional approach to sustainable catchment management. Learning-oriented governance reinforces and institutionalizes a flexible and adaptive approach to decision-making. It integrates multiple sources of knowledge, stakeholder engagement, and public participation to ensure that governance systems are continuously improved through reflection and feedback. By embracing learning as a core principle, such governance remains responsive to the complex and evolving challenges of contemporary socioecological systems, enabling policies and institutions to adapt in the face of uncertainty and change (Schwella, 2014).

Often, universities that focus on technological solutions tend to overlook the importance of sustainability as a foundational element of culture (Adams et al., 2018). It is hard to promote a culture of sustainability when people cannot agree on whether it matters or how to pursue it. We hope this book demonstrates the necessity of environmental change and cultural transformation, both of which call for a paradigm shift. New educational models should guide a culture of sustainability enhancement by moving beyond the traditional sociocentric–mechanistic worldview and adopting an evolutionary–ecosystemic paradigm (see Chapter 1). The sociocentric-mechanistic perspective views the biosphere has been assumed to provide unlimited resources and to absorb unlimited quantities of pollution and waste. Within this outdated framework, energy and material inputs are predominantly derived from nonrenewable resources such as fossil fuels, which are exploited with little regard for long-term consequences. In contrast, the evolutionary-ecosystemic paradigm places emphasis on the sustainable use of renewable resources and the principles of a circular economy. The term *evolutionary* highlights the recognition that ecosystems and the biosphere are dynamic, continually changing entities. In the Anthropocene, an epoch characterized by significant human-induced transformations of natural processes, individual and collective decisions play a crucial role in determining the trajectory of Earth's environmental future. This paradigm shift necessitates a holistic approach to education, one that fosters systems thinking and interdisciplinary communication. This shift must be supported by TEH, which emphasizes the dual regulation of ecological and hydrological processes, the integration of diverse scientific disciplines, and the implementation of nature-based solutions tailored to specific socioecological contexts (Zalewski, 2000).

This paradigm shift fundamentally transforms the discourse, conceptual paradigms, and intervention strategies applied to water-related challenges. It encourages the implementation of ecohydrological advanced nature based solutions (ANbS), which regulate water-biota interplay to improve water quality, enhance biodiversity, and increase ecosystem resilience. These strategies rely on cultural participation and indigenous knowledge (e.g., the Hehenusa water-sharing system developed at a UNESCO Ecohydrology Demonstration Site in Indonesia), making the integration of education Fig. 41.1 community engagement, and ecological science not only desirable but essential.

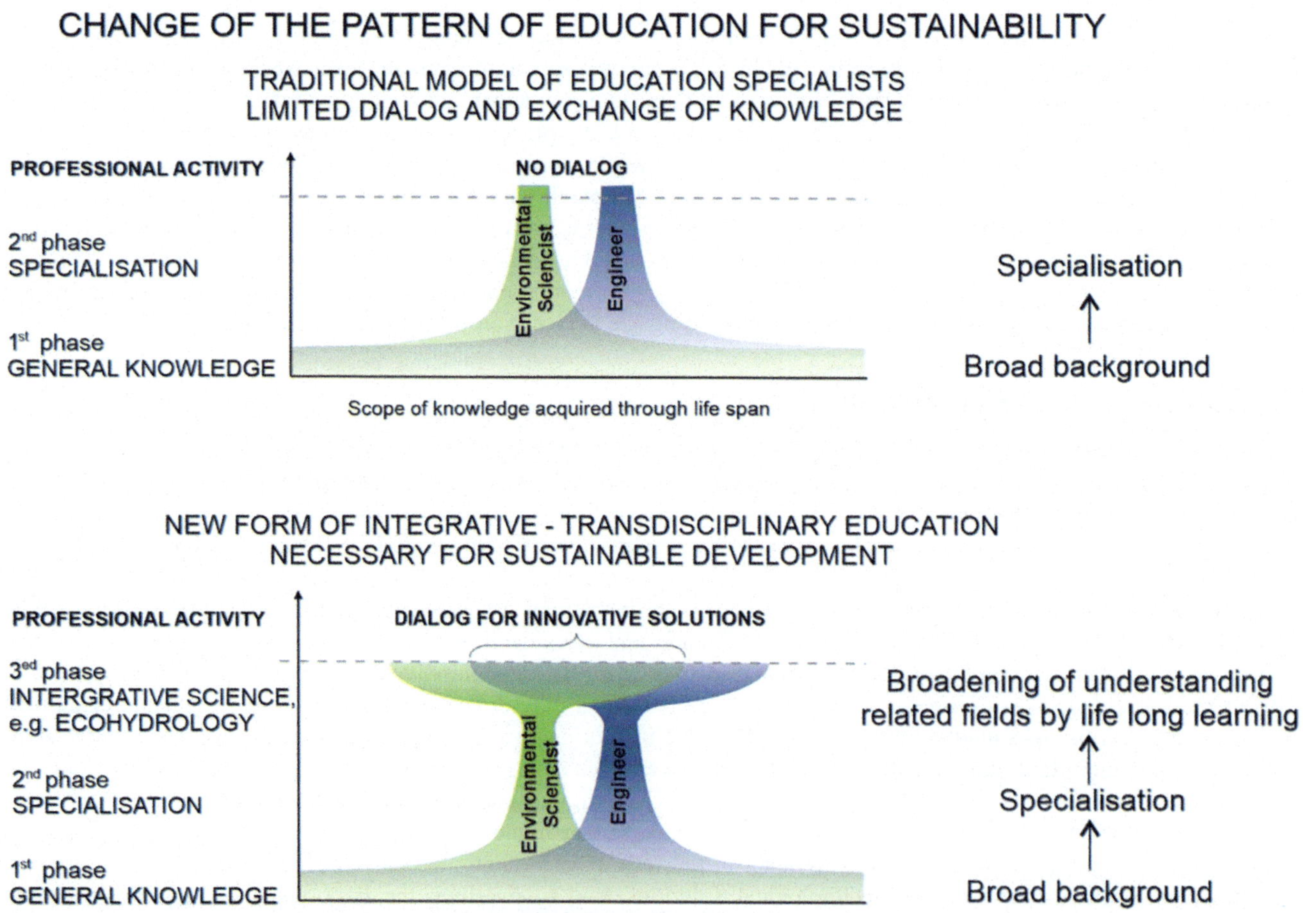

FIGURE 41.1 Education and sustainability. Change in the pattern of education for sustainability. *From Zalewski, M., 2014. Ecohydrology and hydrologic engineering: Regulation of hydrology–biota interactions for sustainability. J. Hydrol. Eng.*

3. The role of sustainability education in shaping culture

Understanding the historical role of culture is essential for shaping a sustainable future on what Buckminster Fuller metaphorically called "Spaceship Earth (Fuller, 1969)." Throughout human history, environmental conditions, particularly those providing fertile and regenerative landscapes such as the floodplains of the Nile, Tigris, and Euphrates, have enabled the development of complex societies. These water-biota permanent interactions provided not only food security but also the ecological stability necessary to support the growth of intellectual, artistic, and architectural pursuits (Konarzewski, 2005).

In the contemporary era, however, the interaction between humans and the environment has grown exponentially more complex. This complexity is driven by population growth, rising consumption, technological advancement, and escalating emissions of pollutants. The result is a significant increase in the global ecological footprint, which now exceeds 170% of the Earth's regenerative capacity (Footprint Data Foundation, 2023). This rapid intensification of human impact suggests that linear, reductionist solutions are increasingly inadequate. Despite technological advances that seem capable of addressing all kinds of challenges, access to clean water, sufficient food, and natural resources remains far from becoming reduced to "technological products." Addressing these challenges requires a new generation of sustainability experts trained in systems thinking, transdisciplinary collaboration, and culturally grounded approaches. Starting with education for sustainable development (ESD) holds the potential to drive systemic change in how people approach sustainability, not as a slogan, but as a shared responsibility that must permeate culture. ESD necessitates an integrated approach to learning, one that connects curriculum, institutional practices, and community engagement. Within this framework, whole institution approaches (WIAs) emphasize the embedding of sustainability not only in teaching but also in governance, operations, and campus culture, fostering systemic change across the entire educational environment (Holst, 2023).

AI disclosure

During the preparation of this work the author(s) used OpenAI's ChatGPT. This document includes content generated and/or edited with the assistance of OpenAI's ChatGPT, used to support language refinement and structural improvements. All final content was reviewed and verified by the author. After using this tool/service, the author(s) reviewed and edited the content as needed and take(s) full responsibility for the content of the publication.

References

Adams, R., Martin, S., Boom, K., 2018. University culture and sustainability: designing and implementing an enabling framework. J. Clean. Prod. 171, 434–445. https://doi.org/10.1016/j.jclepro.2017.10.032.

Domański, M., 2025. Riverscape: Ujawniając Rzeki = Riverscape: Revealing Rivers. Galeria Amcor, Łódź.

Footprint Data Foundation, 2023. York University Ecological Footprint Initiative, and Global Footprint Network. National Footprint and Biocapacity Accounts.

Fuller, 1969. Operating Manual for Spaceship Earth. Southern Illinois University Press.

Holst, J., 2023. Towards coherence on sustainability in education: a systematic review of Whole Institution Approaches. Sustain. Sci. 18 (1015–1030).

Konarzewski, M., 2005. Na Początku Był Głód. Warszawa. Prószyński i S-ka.

Korzeniowski, P. 2013. Bezpieczeństwo Ekologiczne Jako Instytucja Prawa Ochrony Środowiska.

Korzeniowski, P., 2012. About the Need to Standardize the Norms of Law of the Environment Protection. Acta Universitatis Lodziensis Folia Iuridica 71, 35–54.

Lapka, M., Vavra, J., Sokolickova, Z., 2012. Cultural ecology: contemporary understanding of the relationship between humans and the environment. Tájökológiai Lapok 5 (2), 12–24. https://doi.org/10.2478/v10285-012-0050-z.

Matsumoto, D., 1996. Culture and Psychology. Brooks/Cole, Pacific Grove, CA.

Poverty Prosperity and Planet 2024: Pathways Out of the Polycrisis, 2024. World Bank Group.

Schwella, E., 2014. Knowledge based governance, governance as learning: the leadership implications. Int. J. Leadersh. Public Serv. 10 (2), 84–90. https://doi.org/10.1108/ijlps-05-2014-0004.

Sustainable Development Begins with Education: How Education Can Contribute to the Proposed Post-2015 Sustainable Development Goals, 2015. United Nations Educational, Scientific and Cultural Organization.

Sutton, M.Q., Anderson, E.N. 2004. Introduction to Cultural Ecology.

The Circularity Gap Report 2024. https://www.circularity-gap.world/2024#download.

UNESCO, 2003. Convention for the Safeguarding of the Intangible Cultural Heritage. UNESCO, Paris, France.

Wantzen, K.M., Ballouche, A., Longuet, I., Bao, I., Bocoum, H., Cissé, L., Chauhan, M., Girard, P., Gopal, B., Kane, A., Marchese, M.R., Nautiyal, P., Teixeira, P., Zalewski, M., 2016. River culture: an eco-social approach to mitigate the biological and cultural diversity crisis in riverscapes. Ecohydrol. Hydrobiol. 16 (1), 7–18. https://doi.org/10.1016/j.ecohyd.2015.12.003.

Zalewski, M., 2000. Ecohydrology—the scientific background to use ecosystem properties as management tools toward sustainability of water resources. Ecol. Eng. 16 (1), 1–8. https://doi.org/10.1016/s0925-8574(00)00071-9.

Zalewski, M., 2014. Ecohydrology and hydrologic engineering: Regulation of hydrology–biota interactions for sustainability. J. Hydrol. Eng.

Zalewski, M., Kiedrzyńska, E., Wagner, I., 2021. Ecohydrology and adaptation to global change. Ecohydrol. Hydrobiol. 21 (2).

Chapter 42

Transdisciplinary ecohydrology: A summary

Maciej Zalewski
European Regional Centre for Ecohydrology of the Polish Academy of Sciences under the auspices of UNESCO, Łodź, Poland

SELF-PORTRAIT OF LEONARDO DA VINCI, C.1517–18. *FROM: HTTPS://PL.WIKIQUOTE.ORG/WIKI/LEONARDO_DA_VINCI#/MEDIA/PLIK: LEONARDO_SELF.JPG*

Transdisciplinary Ecohydrology. https://doi.org/10.1016/B978-0-443-41596-8.00030-3

Leonardo da Vinci represents the ideal synthesis of scientific exploration and artistic perception, much like how ecohydrology integrates hydrology, ecology, and social dimensions. His fluid, systemic thinking, and problem-solving approach serve as a timeless inspiration for developing a culture of sustainability that embraces scientific rigor, ecological sensitivity, and creative innovation, which are core tenets of transdisciplinary ecohydrology's vision for a resilient and adaptive future.

1. UNESCO water vision and strategic framework for sustainable development

1.1 Vision

Enhancing water resources and ecosystems resilience for achieving the sustainable development goals (SDGs) and the European Green Deal.

This vision acknowledges water as a critical driver of sustainable development, ecological balance, and social well-being. The enhancement of water resources is positioned as a central strategy to support biodiversity, human health, climate adaptation, and long-term planetary resilience UN 2023 "WATER Game Changer for Acceleration of SDGs" (UNESCO, 2023).

1.2 Mission—W4All (water security for the planet)

Accelerating the Implementation of the SDGs and the Green Deal through water-centered action. The mission commits to mobilizing science, policy, and community-level engagement to address water-related challenges in a rapidly changing world. The one of the important large scale funding program for science and innovations is **Water4All partnership–European Commission**, Water security for the planet (European Commission, Water4All Partnership, 2022).

1.3 Strategy

Overcoming fragmentation by evolving from a sociocentric–mechanistic to an evolutionary–ecosystemic paradigm by using the principles of ecohydrology as framework for assessment and management for synergy.

Current development strategies often remain confined within fragmented sectoral approaches grounded in a sociocentric/mechanistic worldview. To overcome this limitation, the United Nations Educational, Scientific and Cultural Organization (UNESCO) advocates for the integration of the evolutionary–ecosystemic paradigm, wherein ecohydrological processes—from molecular to catchment scale—serve as the scientific basis for implementing advanced nature-based solutions (ANbS), hydroengineering, and hybrid interventions.

The catchment is promoted as the universal operational framework, capable of integrating cross-sectoral efforts across water management, biodiversity conservation, agriculture, and industry. This approach:

- Reduces redundancy and fragmentation across initiatives, and
- Facilitates integration with IWRM and EU Water Framework Directive (WFD) principles.

The catchment sustainability potential WBSR: Water–biodiversity–ecosystem services–resilience should be analyzed according to the three principles of ecohydrology:

1 **The hydrological principle focuses on quantifying hydrological processes and identifying their impacts to determine the optimal area for WBSR.** It quantifies and optimizes hydrological processes within the river basin to understand its water flow dynamics, interception capacity, infiltration rates, hydrological connectivity, groundwater recharge and discharge, evapotranspiration, surface runoff, sediment transport, drought and flood cycles, pollutant presence, nutrient cycling, and other anthropogenic pressures such as land-use changes, deforestation, urbanization, and climate-induced hydrological alterations such as land-use changes, deforestation, and climate-induced hydrological alterations.
2 **The ecological principle analyses the distribution of ecosystems at the catchment scale,** with special attention to ecosystem to be protected (e.g. UNESCO Biosphere Reserve, EU Nature Restoration Law) and novel ecosystems, human-modified for the enhancement of WBSR.
3 **Ecological Engineering principle** applies ANbS in combination with dual regulation, hydrological engineering to develop integrated, systemic approaches, based on IWRM, that enhance water resources, bioproductivity, biocomplexity, biodiversity, services for society, to enhance the carrying capacity of river basins and the biosphere Fig. 42.1.

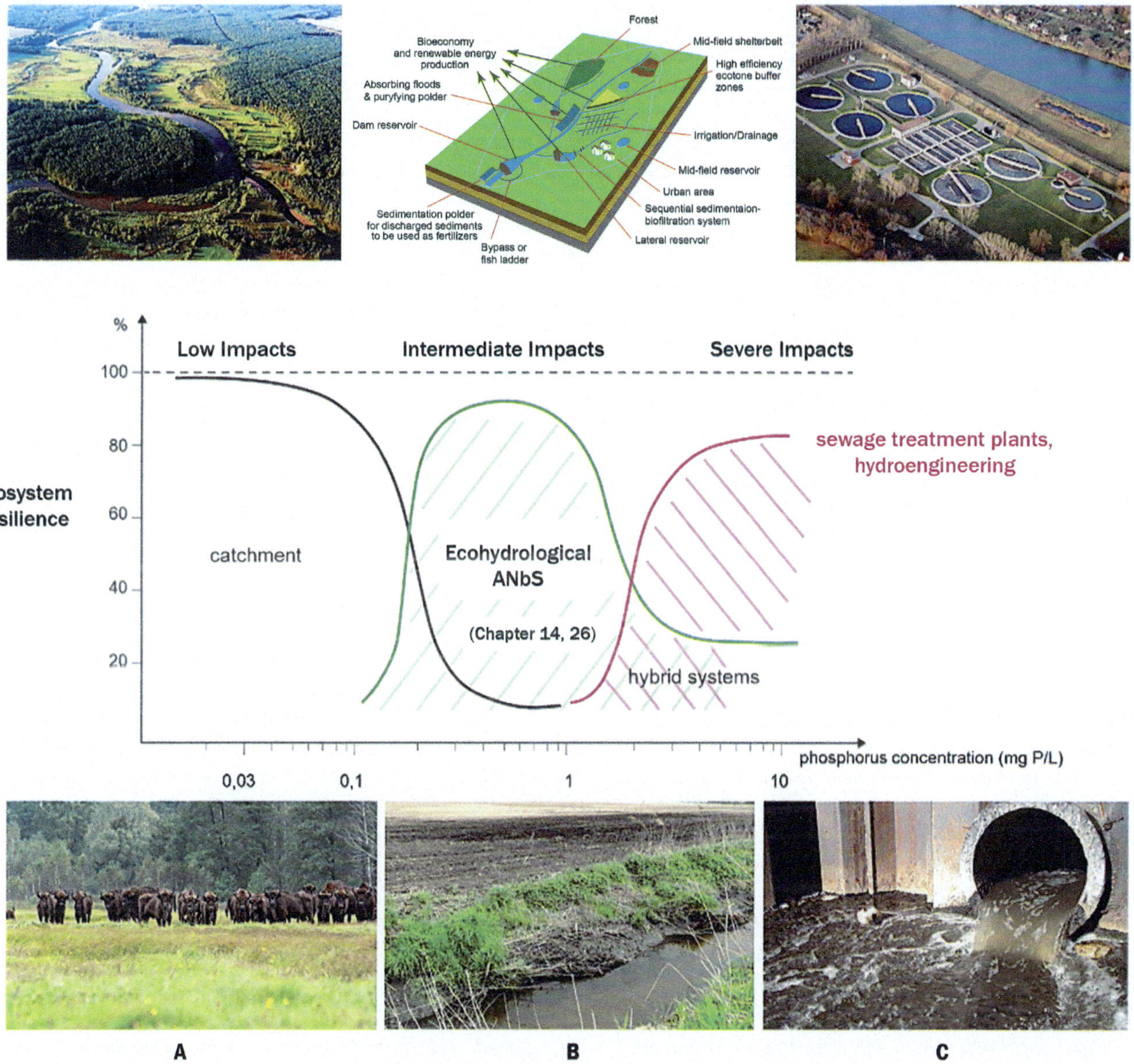

FIGURE 42.1 Ecohydrological Advanced Nature-based Solutions as major tools for mitigation of intermediate impacts. Key methodology for reduction of nonsource pollution from agricultural landscape, urban stormwaters and supporting element for hydroengineering technologies–hybrid systems. *Modified from Zalewski, M. (Ed.), 2020. Ekohydrologia. Wydawnictwo Naukowe PWN, Warszawa. ISBN:978-83-01-20804-2.*

These principles provide a foundational framework for developing, integrating, multiscale, and hybrid systemic solutions to enhance the WBSR framework.

1.4 Tactics

Fostering International Scientific and Policy Cooperation.

UNESCO promotes the translation of scientific knowledge into practical, implementable solutions. Cooperation between major global actors, including the European Commission (EC), UNESCO-Intergovernmental Hydrological Programme (IHP), FAO, WMO, and WHO, and scientific associations as IAHR is essential for achieving alignment between human systems and biosphere dynamics.

1.5 Operational level

UNESCO's IHP has been instrumental in promoting the application of ecohydrology (http://ecohydrology-ihp.org). A key initiative in this endeavor is the establishment of more than 60 ecohydrology demonstration sites worldwide Fig. 42.2, which serve as practical examples of implementing ecohydrological principles to address environmental and societal challenges. UNESCO has been launching a global call for submissions of new ecohydrology demonstration sites. These sites are designed to implement ecohydrological strategies in real-world settings, aiming to improve water quality, enhance biodiversity, and increase resilience to climate change. They also focus on integrating cultural and educational aspects, as well as proactive law, policy, and governance related to water management. The ecohydrology web platform provides an interactive map and extensive resources: ecohydrology-ihp.org.

In a broader context, attaining the SDG implies recognizing the complex relation between humans and the biosphere, and in particular the role of water as a key driver of bioproductivity, biodiversity, and nutrient cycles, all fundamental life-supporting processes, thus the urgent need to harmonize the demand with enhanced water resources Box 42.1.

1.5.1 Fitness check of the Water Framework Directive

27. (...) a new, holistic dimension of the WFD to be developed in which flood prevention should be integrated with drought prevention management and measures <u>to enhance catchment sustainability potential</u> (from an ecohydrological perspective consisting of five elements: water, biodiversity, resilience to climate change, ecosystem services for society, and other impacts, specifically culture and education—WBSR + CE)

"34. (...) the Common Agriculture Policy, the Nitrates Directive, and the Plant Protection Product Regulation should be approximated with the WFD, aiming at the reduction of nonsource pollution (nitrogen and phosphorus), which has recently been generating 20%–50% of the nutrient load to lakes, reservoirs and coastal zones. In regions where industrialized livestock farming is concentrated, it seems difficult to achieve this objective if the CAP and national policies do not commit—in accordance with the objective of the Green Deal, and the biodiversity and "farm to fork" strategies—to a significant reduction in this form of farming. Constructing highly effective land-water ecotone zones, consisting of denitrification and geochemical barriers, would also contribute to this. Emphasizes that increasing the complexity of the agricultural landscape (land-water ecotones (LWE), tree rows (TR) and shelterbelts (SB)) reduces water loss from the soil caused by strong winds but also prevents the loss of organic matter and carbon from the soil. Thus such methods (LWE, TR, and SB), as well as other measures to increase water retentiveness in the landscape, should be incorporated into the bioeconomy strategy. This could considerably increase carbon capture and storage, which is one of the priorities of the Green Deal"

1.5.2 Strategic plan for the Ninth Phase of the Intergovernmental Hydrological Programme (UNESCO IHP-IX 2022–2029)

The most important challenge for water management is how to increase water resources quantity and quality, and in parallel, increase biodiversity, ecosystem services for society, and resilience to impacts (WBSR). The answer is through a holistic approach which incorporates the innovative nature-based and ecohydrological solutions methods (NBS) and catchment scale systemic solutions, based on understanding the water ecosystems' interplay, as both recipients and producers of water, as well as promoting society involvement through culture and education of water and sustainability (CE). Culture is deeply embodied in societal behavior patterns, defining perception and actions, efficiency and effectiveness.

1.5.3 European partnership water security for the planet "Water4All," under Horizon Europe

In support of the European Green Deal and EU water-related policies, the main objective of Water4All is to enable water security for all in the long term, by boosting systemic transformations and changes across the water-research pipeline, fostering the matchmaking between problem owners and solution providers.

2. Ecohydrology: Summary and way forward

2.1 Water is life. Water—Ecosystem interplay is sustainability

Water is a key driver in strategies aimed at reversing the degradation of Biosphere Sustainability, particularly in the context of the United Nations SDGs and the increasing impacts of climate change.

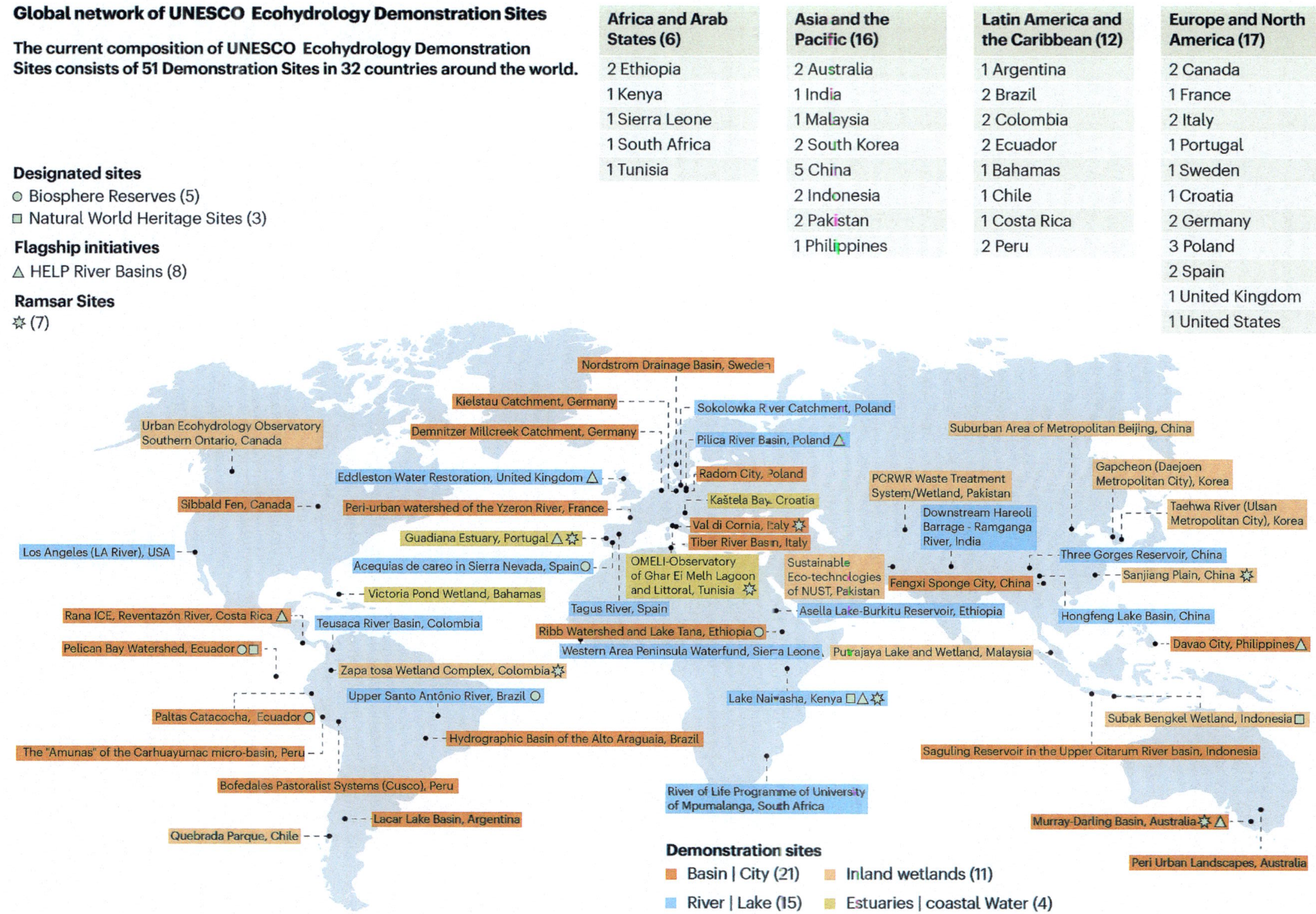

FIGURE 42.2 Global network of UNESCO ecohydrology demonstration sites. UNESCO ecohydrology demonstration sites around the world illustrate the impact of these efforts. *Reproduced from Elfithri, R., Zalewski, M., Arduino, G., Chicharo, L., 2025. Transdisciplinary ecohydrology for water management solutions and sustainability. Nature Water 3, 360–363. https://doi.org/10.1038/s44221-025-00395-x*

BOX 42.1 Translating the concept of TEH WBSR + CE to policy

The Special Side Event
New York UN Water Conference "WATER Game Changer for Acceleration of SDGs" Figs. 42.3. The conclusions of Nature Water 2023 1.301 "The water crisis does not stop in New York" can be summarised as follows:

1 **Governance:**
- A radical reshaping of water governance is essential to move away from current fragmentation.
- Coordinated climate and water actions are imperative.
- Reducing subsidies for water and agriculture that lead to overconsumption and aquifer depletion is necessary.

These points underscore the urgency for integrated policies and reforms to ensure sustainable water management.

2 **Ecohydrology:**
- A better understanding and utilization of the relationships between hydrological and ecological processes is crucial.
- Ecohydrological ANbS play a key role in enhancing biodiversity, reducing risks associated with hydro-climatic extremes, and ensuring water-food-energy security and carbon cycling.

Implementing ANbS can provide sustainable and cost-effective strategies to address water-related challenges.

Translating the concept of TEH WBSR + CE to policy
The European Commission's Committee of the Regions (CoR) is calling for a paradigm shift in water management, from a mechanistic, human-centered approach to an evolutionary, ecosystem-based strategy (Zalewski, 2000). This transition is essential to enhance water resilience, restore ecosystems, and integrate ecohydrological solutions into policy frameworks. The CoR's Fitness Check report reflects this perspective, reinforcing the need for holistic, transdisciplinary science-based water governance at the EU level Figs. 42.4 and 42.5.

Ecohydrology and earth system sustainability: WHY, WHAT, HOW, WHO
WHY Water is a key driver for SDGs as it determines bioproductivity and biodiversity, when biocomplexity of catchment and ecosystems is maintained, restored, and enhanced. This, in turn, improves ecosystem services for society and resilience to climate change (WBSR).

WHAT must be done to accelerate the SDGs beyond merely reducing exploitative anthropogenic pressures (degradation of ecosystems and emission of pollution)? Enhancement of river systems and their catchment sustainability potential (WBSR) through a "Dual Regulation" framework to compensate and reverse the declining carrying capacity of the Earth system, in the face of the growing consumption, and increasing population.

HOW
- Extension of the paradigm: Sociocentric–mechanistic by evolutionary–ecosystemic perspective.
- Water–biota interplay as fundamental factor for hightailing life-supporting processes within a catchment as a framework for a holistic approach: This involves the quantification of hydrological, ecohydrological, social, and economic processes, the assessment of ecological status, and the development and implementation of innovative technologies, biotechnologies and nature-based solutions to create synergies for sustainability, supported by IA necessary for integration of Big Data and according to transdisciplinary ecohydrology (TEH) framework.
- Translation and interpretation of science into solutions across the scales (see Chapter 1)

WHO Society's involvement through the development of the sustainability culture and education implemented and enforced by a new proactive sustainability law, policy, and governance based on evolutionary–ecosystemic paradigm.

A fundamental step in establishing holistic strategies and systemic solutions is a deep understanding of the hydrological mesocycle, water–biota interactions, and the hierarchy of regulatory drivers. This must be reinforced by scientific knowledge based on broad scope of environmental and hydroengineering sciences, while also considering economic, legal, and societal interactions.

Therefore to achieve harmony between the environment and society, it is essential to develop an integrative sustainability science, with transdisciplinary ecohydrology (TEH) serving as both the starting point and framework for a hydrocentric approach.

A crucial step in this process is translating accumulated scientific information into actionable knowledge, understanding the processes, feedback mechanisms, and hierarchical regulatory structures within these systems. This knowledge must then be transformed into wisdom: the ability to apply ANbS and hydroengineering technologies within systemic solutions e.g. IWRM, that not only mitigate human impacts but also enhance carrying and adaptive capacity of catchment and biosphere. This approach strengthens society's involvement and ability to adapt to an evolutionary, ecosystem-based paradigm and fosters the development of relevant technologies, ultimately reshaping priorities and transforming societal needs.

FIGURE 42.3 **A. The photograph presents the special session "Transdisciplinary Ecohydrology for acceleration of SDG–methodology of science and patterns of implementation" during the New York UN Water Conference, 2023.** "WATER Game Changer for Acceleration of SDGs." *Photo credit: K. Izydorczyk.* **B. UN 2023 Water Conference, New York.** The photograph presents the New York UN Water Conference, "WATER Game Changer for Acceleration of SDGs" 2023. *Photo credit: M. Zalewski.*

FIGURE 42.4 **The stakeholder meeting.** The stakeholder meeting held on February 2, 2020, in Brussels, organized by the Committee of the Regions (CoR), was a crucial event aimed at proposing an upgrade to the EU Water Framework Directive (WFD) to address the increasing pressures on water resources due to climate change, pollution, and unsustainable water use. "Alfonso Alcolea Martinez, European Commission," "Piotr Calbecki, Rapporteur," and "Maciej Zalewski, Expert." *Photo credit: Aliona Fornea.*

To accelerate the achievement of biosphere sustainability (UN SDGs), the following ecohydrological steps are necessary, based on the TEH introduced in this book:

1 ***Proactive Education of society—coping with uncertainty in the changing world, and especially climate.*** There is an urgent need for shaping societal attitudes and understanding of sustainability: understanding the biosphere as a

FIGURE 42.5 Scientific Advisory Committee IHP 9th Plenary Speakers and Panelists. The special event on ecohydrology during 50th Anniversary of UNESCO IHP 2025. Left Bottom row: Anny Chavez, Costarica, Shahbaz Khan, Director, UNESCO Beijing Office, Maciej Zalewski, ERCE PAS UNESCO, Poland, Elfithri Rahmah, UNESCO, AmbassadorMariusz Lewicki, Permanent Delegation of the Republic of Poland to UNESCO, Oula Amrouni, Tunesia. Left Upper Row: Omime O. Yaya, Regional Centre for Integrated River Basin Management (RC-IRBM) under auspices UNESCO, Nigeria, Stefano Fazi, Italy, Zhe Li China, Paweł Jarosiewicz, ERCE PAS, Poland, Marco Albarracin, Ecuador, Luis Chícharo, Portugal, Giuseppe Arduino former Cheif of Section Ecohydrology UNESCO IHP, Benjamin Lopez, CEO Water4All-partnership, European Union, Johannes Cullmann, Federal Institute of Hydrology, Germany, Michele Mossa IAHR, Antonio Nobre, Stefan Krauze, UK, David Hannah, UK, Remy Rozainy, Malaysia. *Photo credit: UNESCO.*

dynamic system where water, carbon, phosphorus, and nitrogen cycling serve as the primary drivers of sustainability, and that these are influenced by each of us and in turn determine the amount and quality of the ecosystem services we need. This approach requires proactive education, which will stimulate a shift from a sociocentric–mechanistic to an evolutionary–ecosystemic paradigm based on three assumptions:

- the unity of humankind and nature
- consciousness that happiness is not correlated with consumption, but first and foremost, a fair and good relationship with other people, and with functioning in a healthy environment
- our positive mental status is primarily defined by the quality of our environment. Such beliefs can largely stimulate social capital—confidence and cooperation.

2 ***Social capital as an important factor for* the *translation of knowledge into wisdom and of innovations in catchment management.*** The philosophy of the exchange of ideas and openness for controversial opinions has been broadening the holistic perception of the problems to be solved and generating the most efficient innovations. It is worth noting that there appear to be many examples that both encourage team spirit but also respect for seniority and leadership toward achievement of strategic goals, first and foremost reversing degradation of biosphere, synergies of both should also amplify the translation of science into technology.

3 ***Socio-economic Foresight of the catchment as a tool for creating a desirable future.*** Action without vision and strategies has typically resulted in a waste of human potential and resources. Hence, the primary tool used for the development of responsible vision and strategy in achieving SDG UN should be the foresight methodology, which should consider the circular economy, i.e., reduction of impact and bioeconomy production of commodities from renewable resources, and the enhancement of sustainability potential water–biodiversity–ecosystem services–resilience + culture and education (WBSR + CE), where the water mesocycle has to be used as a framework for assessment, planning, and management.

2.1.1 Why is ecohydrology, within the framework of UNESCO-IHP, becoming a "rocket science" for achieving the sustainable development goals, and how AI will support it?

1 **LIFE-SUPPORTING PROCESESS**—highlighted by integration hydrology and ecology in a broad scope of environmental conditions.

2 **WATER FOR ECOSYSTEM AND ECOSYSTEMS FOR WATER**—retaining, recirculating water and biogenic substances, purifying water

3 **NEW PARADIGM**—Sociocentric–Mechanistic + Evolutionary-Ecosystemic; (decision makers and hydroengineers need to open the "Black Box" of water–biota interplay).

4 **DUAL REGULATION background for the ANbS**—Integration Hydrology + Ecology + Hydroengineering + Molecular biology + Biotechnologies + Phytotechnology
5 **COST-EFFICIENT tools (ANbS) for integrated water resources management (IWRM)** (synergy with circular economy and bioeconomy)
6 **ENHANCEMENT** of catchment Sustainability Potential Water-Biodiversity, Bioproductivity and Biodiversity-Ecosystem Services—Resilience (WBSR) supported by AI.
7 **HARMONIZATION** of society's needs with enhanced ecosystem potential (WBSR) by Culture, Education + Law, Policy, Governance—(WBSR + CE + LPG),
8 **CATCHMENT FORESIGHT for** shaping a desirable and sustainable vision of the future **by the long-term enhancement of WBSR by implementing transdisciplinary ecohydrology**.

2.1.2 The future trends and evolution in transdisciplinary ecohydrology

1 Understanding the evolution of hydrology-biota, particularly through the lens of climate change and paleohydrology, as a reference point for climate adaptation strategies and the water–biota interplay from the molecular level up to the catchment and biosphere scale as a framework for ANbS and systemic solutions.
2 Understanding anthropogenic modifications of hydrological, ecological, and ecohydrological processes.
3 Developing systemic solutions to harmonize societal priorities with enhanced catchment sustainability potential (WBSR).
4 Developing a methodology for holistic catchment management (HCM) by integrating IWRM with ecohydrology, foresight, and multidisciplinary scientific and AI support.
5 Implementing HCM through adaptive assessment management (AAM) using AI as a tool.
6 Development of ecohydrology demonstration sites and systemic solutions for catchment management, integrating holistic frameworks such as IWRM, the capital of knowledge, such as intergovernmental and international organizations as UNESCO-IHP, UNECE Water Convention, EU Water Framework Directive, the Nile Basin Initiative, Mekong River Commission, Amazon Basin TEI, Large Rivers UNESCO-IHP; to support sustainability and resilience in water management.
7 Strengthening regional water-climate-sustainability strategies, including Synergy for Africa, and other continents, which possess complimentary resources and human capital. All highly exposed to climate change and demographic challenges. This includes leveraging water-ecosystem-economy feedback mechanisms to balance ecological integrity and economic growth, in alignment with OECD, FAO, UNEP, and the EU Green Deal.
8 AI in water management can act as a bridge between principles (e.g., sustainability, resilience, and equity) and algorithms (e.g., prediction, optimization, detection), enabling smart, adaptive, and integrated solutions (AI-DSS).

2.1.3 Why transdisciplinary ecohydrology has to become the "Rocket Science"?

It is more and more understood that humanity, however, as a matter of fact, each of us is at the forefront of the decision to change Anthropocene into Ecocene or continuing "business as usual" approach by staying in the recent "comfort zone," the growing consumption and impacts on water resources and the environment. This will lead soon to the global tensions and conflicts and a broad scope of dramatic consequences; the alternative is to create a sustainable, desirable future (Costanza and Kubiszewski, 2014).

The reduction of pollutant emissions and stopping climate change at this stage is fundamental, but not enough. There is an urgent need to enhance the biosphere's sustainability potential, which depends on the integrity of the life-supporting processes driven by water.

It becomes broadly accepted that the WATER is key for shaping the safe, sustainable future in the face of climate change, dynamic demographic conditions in some regions, necessity poverty eradication, nonrenewable resources, and resilience decline (e.g., "biotic pump" degradation). Such complex, complicated, and growing negative interactions analogous to "quantum entanglement" relations between man and biosphere, urgently need solutions based on a holistic approach built on advanced transdisciplinary science, which integrates hydrologists, ecologists, hydroengineers, molecular biologists, biotechnologists, economists, social science specialist and many others. Such a broad scope of knowledge and wisdom to be integrated need scientific framework based on advanced methodology of science to break through the "Babel Tower" effect. This book proposes "hydrocentric" approach as a fundamental step to these efforts.

Moreover, by shoving in the global process's perspective, the water-sustainability challenges on different continents, the examples of the experiments which led to the discovery of emerging properties of ecosystems fundamental for consciousness and understanding the diversity of interlink dependencies between water and ecosystems + dual regulation." This, in turn, was a background for the formulation of the evolutionary ecosystemic perspective, with the key concept of the enhancement of sustainability potential (WBSR + CE + LPG), which will be more and more instrumental in the face of progressing climate and impacts.

With the plethora of sustainability challenges for humanity, its dynamic is specific for every catchment, which has to be a basic unit for global actions. With limited time, there is an urgent need to release the creative potential of the water and sustainability scientists, who will be step by step translating accumulated knowledge into wisdom, with growing support of AI.

The ideas presented here possess several dimensions, which I believe should accelerate progress in achieving a sustainable, desirable future. This book is not a review; rather, its main aim is to present a way of thinking and to guide readers toward understanding complex processes and envisioning holistic solutions. The first methodology of science is built on Kuhn, Popper, Feyerabend, and Lakatos perspective, expressed in the theoretical models of processes (Abiotic Biotic Regulatory Concept [ABRC], holistic reductionistic interplay; deductive-inductive approach). Next, the special emphasis on ecosystem dynamics and metabolism is exemplified by the results of field experiments of which working hypotheses were formulated on the basis of proposed models. Both were crucial for the development of innovative low-cost/high-efficiency ANbS and systemic solutions.

Finally very important—the chapters of invited authors who in the framework of UNESCO Intergovernmental hydrological program, participate in development the ideas introduced in this book representing the different specializations of water sciences, who demonstrated in their chapters how creative thinking related to TEH, generate ideas and knowledge on the basis the field and laboratory experiments and, how they are translated to wisdom—innovative solutions.

I believe that such an untypical formula of this book, with numerous graphs and photos, was proposed because it is addressed to a broad audience of scientists, students, decision makers, NGO who intend to use their knowledge and passion for leading the efforts toward achieving a sustainable, desirable future. It was somehow intended as the analogy for modern art, not only easy to absorb the message and aesthetics, but also encourages and inspires observers/readers into creative thinking toward the development of innovations for a fast transition from Anthropocene to Ecocene-the forward-looking term for emerging era focused on enhancement of the Biosphere sustainability potential, towards the human-nature harmony.

AI disclosure

During the preparation of this work the author(s) used OpenAI's ChatGPT. This document includes content generated and/or edited with the assistance of OpenAI's ChatGPT, used to support language refinement and structural improvements. All final content was reviewed and verified by the author. After using this tool/service, the author(s) reviewed and edited the content as needed and take(s) full responsibility for the content of the publication.

References

Costanza, R., Kubiszewski, I., 2014. In: Costanza, R., Kubiszewski, I. (Eds.), Creating a Sustainable and Desirable Future: Insights from 45 Global Thought Leaders. World Scientific.

European Commission, Water4All Partnership, 2022. Water4All: Water security for the planet – A European partnership under Horizon Europe. European Union.

UNESCO, 2023. A Game-Changer for UN 2023 Water Conference: Intergovernmental Science-Policy Platform for Water Sustainability. UNESCO.

UNESCO, 2024. Ecohydrology for Water Security: Call for New Ecohydrology Demonstration Sites. UNESCO.

Zalewski, M., 2000. Ecohydrology—the scientific background to use ecosystem properties as management tools toward sustainability of water resources. Ecol. Eng. 16 (1), 1–8. https://doi.org/10.1016/s0925-8574(00)00071-9.

Zalewski, M. (Ed.), 2020. Ekohydrologia. Wydawnictwo Naukowe PWN, Warszawa. ISBN:978-83-01-20804-2.

Further readings

European Commission, 2000. Directive 2000/60/EC of the European Parliament and of the Council establishing a framework for Community action in the field of water policy (Water Framework Directive). Off. J. Eur. Comm. L 327, 1–73.

European Commission, 2019. The European Green Deal (COM(2019) 640 final). European Commission.

European Commission, 2020a. Fitness check of the Water Framework Directive, Groundwater Directive and Floods Directive (SWD(2019) 439 final). European Commission.

European Commission, 2020b. A Farm to Fork Strategy for a fair, healthy and environmentally-friendly food system (COM(2020) 381 final). European Commission.

European Commission, 2022. Proposal for a regulation on nature restoration (COM(2022) 304 final). European Commission.

European Parliament & Council of the European Union, 1991. Directive 91/676/EEC concerning the protection of waters against pollution caused by nitrates from agricultural sources (Nitrates Directive). Off. J. Eur. Comm. 375.
OECD, 2015. OECD Principles on Water Governance. OECD Publishing.
UNESCO, 2022. Strategic Plan of the Ninth Phase of the Intergovernmental Hydrological Programme IHP-IX, 2022–2029. UNESCO.
United Nations, 2015. Transforming Our World: The 2030 Agenda for Sustainable Development. United Nations.

Part V

Glossary and bibliography

Ecohydrology and UNESCO's Intergovernmental Hydrological Program—past, present, and the future

Giuseppe Arduino

Erasmus Masters of Applied Ecohydrology (MAEH), UNESCO Chair in Ecohydrology: Water for Ecosystems and Societies University of Algarve, Faculty of Sciences and Technologies, Gambelas, Portugal

Water security represents one of the most pressing challenges of the 21st century, as societies worldwide face growing pressures from climate change, biodiversity loss, urbanization, and pollution. The UNESCO Intergovernmental Hydrological Program (IHP), established in 1975, has played a pivotal role in advancing international scientific cooperation in water sciences and became the first and only intergovernmental freshwater initiative institutionalized within the United Nations system.

Over the years, IHP has evolved through a series of phases, beginning with the International Hydrological Decade in 1965 and moving through successive thematic cycles that addressed experimental basins, flood management, vulnerable environments, water systems under stress, and, ultimately, the broader concept of water security in a changing world, Fig. 1.

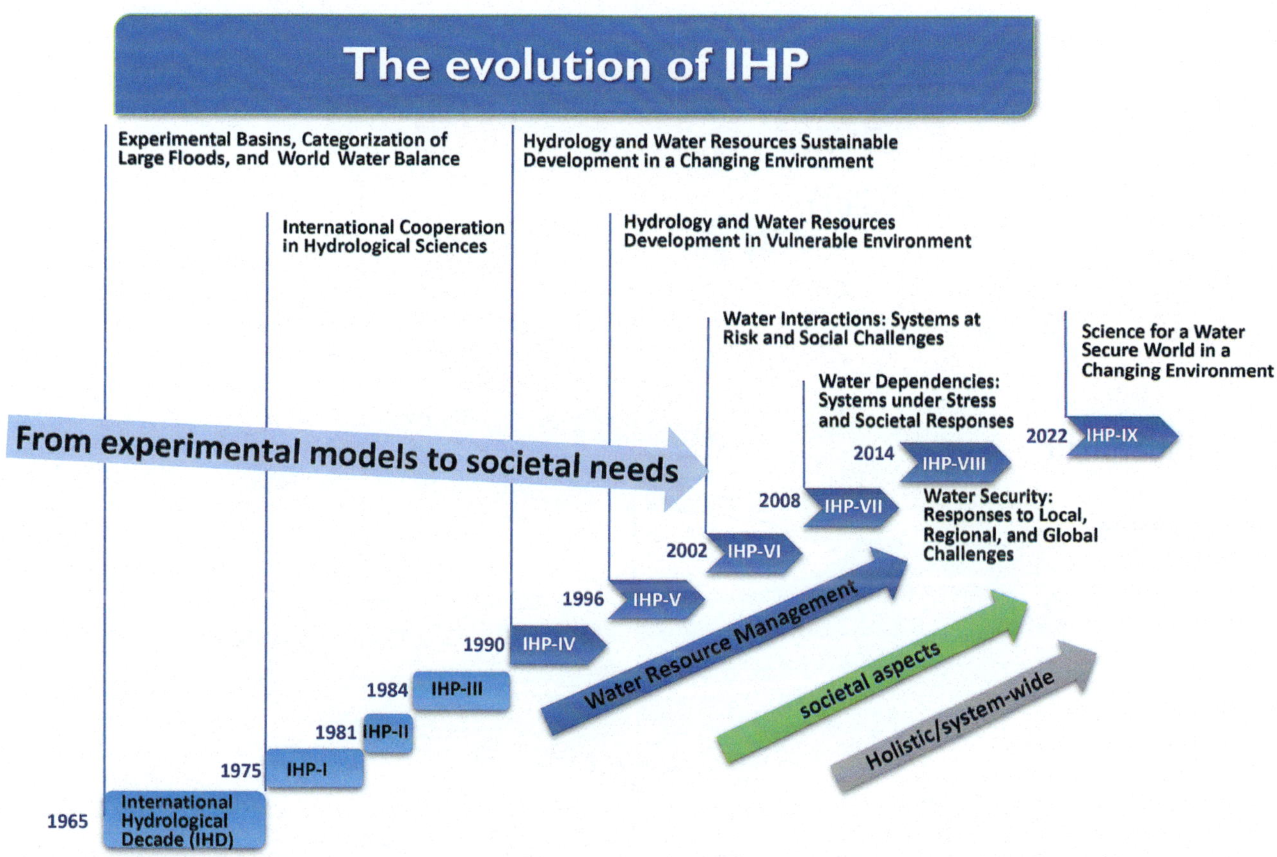

FIGURE 1 The evolution of UNESCO's intergovernmental hydrological program (IHP). *Modified from Jimenez-Cisneros, B., 2015. Responding to the challenges of water security: the Eighth Phase of the International Hydrological Programme, 2014–21. Proc. Int. Assoc. Hydrol. Sci. 366, 10–19. https://doi.org/10.5194/piahs-366-10-2015.*

Within this evolving framework, ecohydrology has emerged as a central concept. Introduced in the fifth phase of IHP (1996–2001) (Arduino, 2021), ecohydrology was initially conceived through a collaboration between experts in the Man and Biosphere (MAB) program and hydrological sciences. The focus was on understanding the role of the hydrological cycle in ecosystems and the interplay between abiotic and biotic processes in regulating water quality and buffering extreme events. Over time, ecohydrology has transitioned from being a purely scientific endeavor to a solution-oriented science that addresses the societal, environmental, and educational needs characterizing our present times.

This development was further reinforced in the subsequent IHP phases. The sixth phase (2002–07) introduced the concept of *Dual Regulation* (Zalewski, 2006, based on earlier research Zalewski and Naiman, 1985; Zalewski et al., 1990), highlighting the mutual regulation between hydrology and the biota. The seventh phase (2008–13) emphasized ecohydrology for sustainability, and it was during this time that international programs such as the Erasmus Mundus Master Course in Ecohydrology were launched, training hundreds of students from across the world. Later, in the eighth and ninth phases, ecohydrology became embedded in the broader themes of water security, resilience, and integrated approaches to global challenges. Today, ecohydrology is strategically promoted within the UNESCO Water Family and across a network of demonstration sites, as well as within UNESCO's designated sites, such as biosphere reserves, geoparks, and world heritage areas. These long-term projects engage local stakeholders to develop cost-effective, nature-based solutions for water management, food security, and disaster risk reduction.

Ecohydrology contributes to the United Nations Sustainable Development Goals (SDGs). Water (Goal 6) and biodiversity (Goals 14 and 15) are seen as fundamental building blocks for achieving sustainable development, Fig. 2. The ecohydrology framework expressed through the WBSR + CE model (Water, Biodiversity, Ecosystem Services,

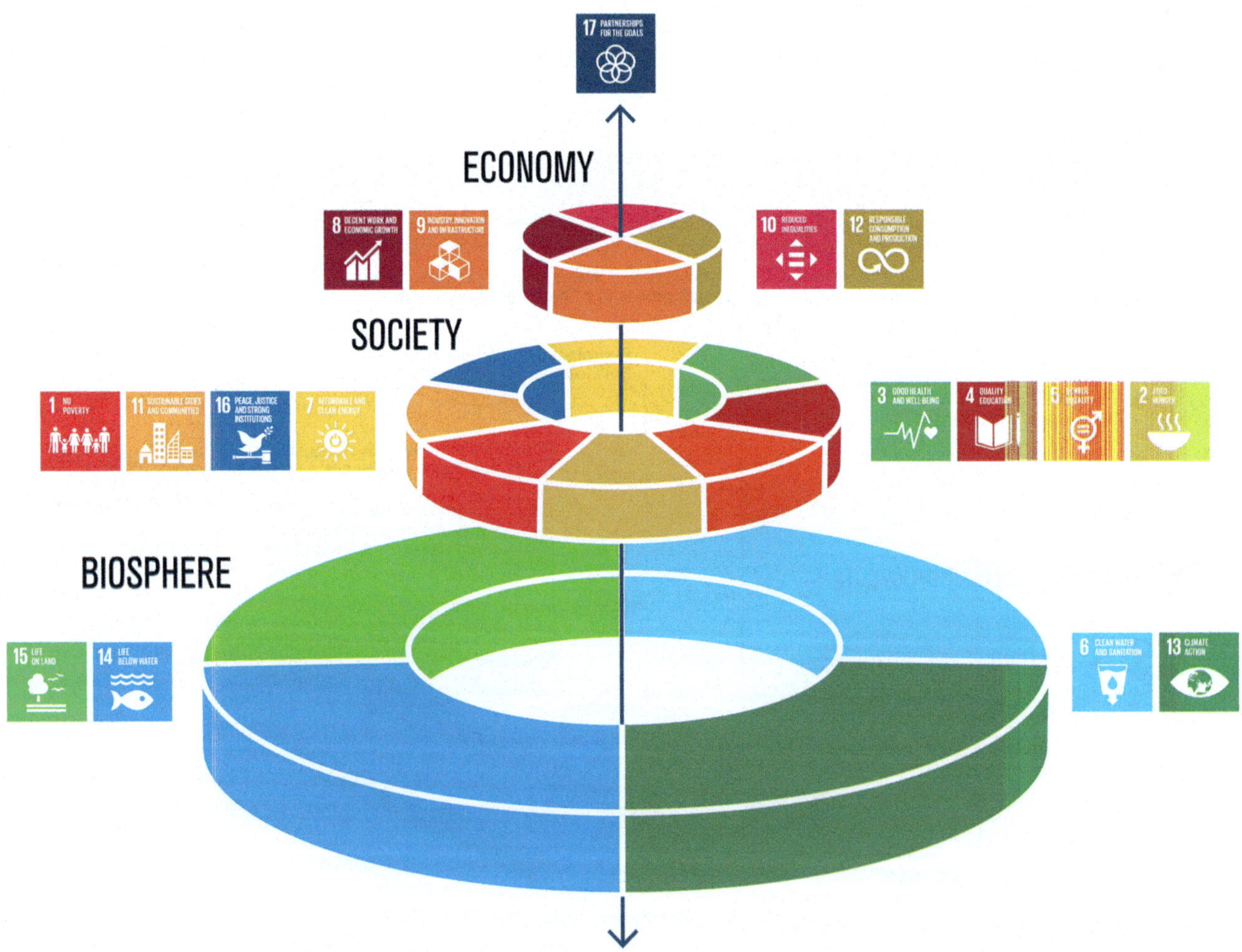

FIGURE 2 The Sustainable Development Goals (SDGs) wedding cake. The SDGs wedding cake diagram represents the relationship among different domains within the SDGs—biosphere, society, and economy. *Courtesy of Azote for Stockholm Resilience Center, Stockholm University CC BY-ND 3.0.*

Resilience, and Culture/Education) has been expanded to include the critical dimensions of law, policy, and governance (LPG). This reflects the recognition, particularly emphasized during the 2023 UN Water Conference, that holistic and transdisciplinary approaches are essential to accelerate SDG implementation by: (a) reducing fragmentation in water management by adopting systemic, holistic approaches; (b) implementing low-cost ecohydrological solutions that complement hydroengineering while reducing diffuse pollution; and (c) promoting culture and education, drawing on both scientific advances and traditional knowledge systems.

By emphasizing that ecohydrology principles, such as quantifying hydrological processes at the catchment scale, managing the biota to regulate water cycles, and employing ecological engineering solutions are critical for building resilient, sustainable societies.

Water, wisdom, and the way forward to an evolutionary–ecosystemic policy for a resilient biosphere

At the dawn of the Anthropocene, humanity stands at a pivotal juncture between ecological degradation and systemic renewal. The evidence is unequivocal: accelerating climate variability, water scarcity, biodiversity decline, and soil degradation are no longer isolated phenomena but symptoms of a deeper structural imbalance between societal development and the biosphere's life-supporting processes. In this context, *Transdisciplinary Ecohydrology: Advanced Nature Based Solutions for the enhancement of catchment sustainability potential* provides both a theoretical compass and a methodological framework for addressing this imbalance bridging science, policy, and practice through an integrative, process-based approach.

The book's central argument is that **water is the primary regulator of sustainability**, a unifying element that interlinks biological productivity, ecosystem resilience, and socioeconomic stability. By treating water not merely as a resource, but as a *driver of systemic regulation*, ecohydrology redefines the foundations of sustainable management. It calls for a transition from reactive, sectoral interventions toward proactive, *catchment-scale governance* rooted in the scientific understanding of hydrological–ecological feedback.

The **transdisciplinary ecohydrology (TEH)** paradigm moves decisively beyond disciplinary boundaries. It integrates hydrology, ecology, biotechnology, and social sciences into a single framework that supports *Advanced Nature-based Solutions (ANbS)* and *hybrid systems*. These systemic interventions built upon the principle of **dual regulation** seek to harmonize anthropogenic and natural processes to enhance the biosphere's carrying capacity. Within this paradigm, the four core sustainability pillars, **WBSR**, form the operational foundation for environmental policy and strategic planning.

From a governance perspective, the implications are profound. Achieving sustainability enhancement requires the integration of **CE** with **LPG**, creating an adaptive policy architecture capable of responding to dynamic environmental and societal conditions. Policies must evolve from the linear management of resources to a circular, feedback-oriented regulation that reflects the interdependence of hydrological, ecological, and socioeconomic systems.

The enhancement concept (Zalewski, 2014), as advanced by this work, represents a paradigm shift from conservation and restoration toward **expanding the Earth system sustainability potential (ESSP)**. Rather than attempting to return ecosystems to past states, often unattainable under contemporary pressures, enhancement focuses on *increasing system functionality, adaptability, and resilience*. This proactive orientation transforms sustainability from a goal of preservation into a process of continuous improvement and coevolution between human and natural systems.

The **policy translation of TEH** demands a multiscalar strategy. At the **local and regional levels**, catchment-based frameworks must guide land and water management, embedding ecohydrological indicators into decision-making processes. At the **national level**, environmental legislation should institutionalize transdisciplinary collaboration, incentivize ANbS deployment, and ensure long-term investment in the ecological infrastructure. At the **international level**, alignment with UNESCO's Intergovernmental Hydrological Program (IHP-IX) and SDGs reinforces a coherent global agenda for integrated water and ecosystem management.

In conclusion, *Transdisciplinary Ecohydrology: Advanced Nature Based Solutions for the enhancement of catchment sustainability potential* calls for a redefinition of progress itself. It positions sustainability not as an endpoint, but as a continuous negotiation between the ecological limits and societal aspirations, a dynamic equilibrium maintained through science-based policies, participatory governance, and adaptive management. The integration of empirical research with ethical and cultural dimensions ensures that water management evolves into a platform for both environmental enhancement and social cohesion.

In summary, the **evolutionary–ecosystemic paradigm** outlined in this book provides the intellectual and operational foundation for the 21st-century sustainability policy. It invites decision-makers, scientists, and citizens alike to view the biosphere not as a reservoir of resources, but as a living system whose resilience depends on our capacity for informed stewardship. The challenge ahead is clear: to institutionalize transdisciplinary thinking, operationalize ecohydrological principles, and accelerate the transition from exploitation to enhancement.

Only through such systemic transformation anchored in the science of water and the ethics of coexistence, humanity can safeguard the integrity of the biosphere and ensure that the foundations of life remain viable for generations to come.

References

Arduino, G., 2021. Ecohydrology: engineering harmony for a sustainable world, theme 5 of the intergovernmental hydrological Programme (IHP) phase VIII (2014–2021). Ecohydrol. Hydrobiol. 21 (3), 380–386. https://doi.org/10.1016/j.ecohyd.2021.08.009.

Zalewski, M., 2006. Ecohydrology – an interdisciplinary tool for integrated protection and management of water bodies. Large Rivers 16 (4), 613–622.

Zalewski, M., 2014. Ecohydrology and hydrologic engineering: Regulation of hydrology–biota interactions for sustainability. J. Hydrol. Eng. 20 (1), A4014012. https://doi.org/10.1061/(ASCE)HE.1943-5584.0000999.

Zalewski, M., Naiman, R.J., 1985. The regulation of riverine fish communities by a continuum of abiotic-biotic factors. In: Alabaster, J.S. (Ed.), Habitat Modification and Freshwater Fisheries. Butterworths for FAO/UN, London, pp. 3–9.

Zalewski, M., Brewińska-Zaraś, B., Frankiewicz, P., Kalinowski, S., 1990. The potential for biomanipulation using fry communities in a lowland reservoir: Concordance between water quality and optimal recruitment. Hydrobiologia 200/201, 549–556.

Glossary

Abiotic and Biotic Factors are drivers of the biosphere. 1. The abiotic, or non-living, portion of each ecosystem includes the structure - geomorphology, climate - and the concentrations of organic and inorganic substances in the environment. Additionally, they include all the functional processes related to energy flow, the water cycle and nutrients dynamics. 2. The second component is the biotic, or "living organism, portion, which includes producers, consumers, and decomposers". (Britannica Dictionary).

In the context of ecohydrological processes, taking place in water bodies such as river basins and lakes, abiotic and biotic components are interconnected, and their functions are interdependent. Their interactions focus on

1. **hydraulics and fluid dynamics**, which determines nutrients cycling in lakes and spiraling in rivers also influence the energy gain-to-expenditure ratio in organisms,
2. **temperature fluctuations,** which determine the rate of biota metabolism, including microbial communities, plants, invertebrates, and fish, ultimately affecting growth, reproductive performance, and biological productivity.

Abiotic Biotic Regulatory Continuum (ABRC) is a conceptual framework that describes changes in hierarchy of abiotic and biotic factors along river continuum in different climate regions (Zalewski and Naiman, 1985).

Absorption is the physical or chemical process by which atoms, molecules, or ions enter the bulk phase (volume) of a liquid or solid matter in an organism.

Advanced Nature-Based Solutions (ANbS) are nature-inspired approaches to water management and sustainability within transdisciplinary ecohydrology. They are based on a holistic framework that integrates hydrological, ecological, and biogeochemical processes, and incorporate disciplines such as ecohydraulics, molecular biology, biotechnology, and phytotechnology. ANbS are cost-effective, complementary to, and synergistic with conventional hydroengineering solutions.

Albedo – the proportion of incoming solar (shortwave) radiation reflected by a surface (IPCC, 2021).

Assimilation is the process by which nutrients are taken up and incorporated into the tissues of an organism; in plants, it refers to the uptake of nutrients from the environment.

Bioeconomy the share of the economy based on products, services, and processes derived from biological resources, encompassing sectors such as agriculture, forestry, fisheries, chemicals, textiles, energy, and biotechnology (Gallo, 2022).

Biodegradation, the gradual destruction of a material due to natural or artificially induced biological activity.

Biological monitoring, (biomonitoring) the use of a biological entity as a detector and its response to a measure to determine environmental conditions. Biosurveys and toxicity tests are common biomonitoring methods.

Biomanipulation involves shaping the biological structure of aquatic ecosystems by stocking predatory fish to reduce populations of planktivorous fish, thereby increasing zooplankton abundance and enhancing their filtration capacity, ultimately leading to the reduction of algal blooms.

Biotechnology is the use of organisms and biological systems to develop products and processes for specific applications in fields such as medicine, agriculture, industry, and environmental management.

Catchment (or drainage basin) is the most basic concept of hydrologic science – the area of land contributing water to a nominated point on the earth's surface. This concept works well where there is a well-defined stream and ridge system. Equivalent terms are watersheds (US) and drainage basins (US & UK) (Bren, 2015).

Collective intelligence – the emergent ability of a group to solve problems, make decisions, and perform a wide range of tasks more effectively than individuals acting alone through the interaction and integration of diverse knowledge, perspectives, and skills (Galesic et al., 2023; Woolley et al., 2010).

Cultural ecosystem services – the non-material benefits that people obtain from ecosystems through spiritual enrichment, cognitive development, reflection, recreation, aesthetic experiences, and the development of cultural identity and sense of place, thereby contributing to human well-being and quality of life (Sarukhán and Whyte, 2005; Romanazzi et al., 2023).

Carrying capacity – the maximum number of individuals of a species that an ecosystem can sustainably support over time without degrading its resources or ecological functions (Postel, 1994).

Cyanobacteria (also Cyanophytes or blue-green algae) a group of phytoplankton, some of which can produce toxins, regulate their depth using gas-vacuole buoyancy mechanism, and/or fix atmospheric nitrogen for use in growth. They often occur in eutrophic water as a bloom.

Data model is a formal method of organizing data to represent an observed environment, including its structure, relationships, and constraints.

Design Thinking – a human-centred, iterative methodology for solving complex problems through empathy with users, ideation, rapid prototyping, testing, and continuous refinement of solutions (Brown, 2009; Stanford d.school, 2020).

Dual Regulation as a term was first introduced to describe the dynamic interplay between water and biota, highlighting their mutual influence on ecosystem processes (Zalewski, 2006). It explains how the structure and dynamics of freshwater ecosystems are regulated by the seasonality, timing, and intensity of hydrological mesocycles. In turn, the hydrological cycle itself is influenced by the drainage basin's structural components (such as forests and land-water ecotones) and ecological processes (including phytotechnology, food chain interactions, biomanipulation, and hydrobiomanipulation). This principle describes the reciprocal relationships between the hydrological cycle and ecosystem performance, driven by feedback mechanisms that shape the catchments' capacity to adapt to changes in climate and anthropopressures and further enhance sustainability—captured in the framework of Water-Biodiversity-Ecosystem Services-Resilience (WBSR). Understanding these mechanisms is fundamental to transdisciplinary ecohydrology, which promotes the use of ecosystem properties as management tools (Zalewski, 2000) for Integrated Water Resources Management (IWRM).

Ecohydraulics is an interdisciplinary field that integrates aquatic ecology and hydraulic engineering to study how water flow affects aquatic life and how organisms alter their habitats.

Ecohydrology is a transdisciplinary and applied science, a sub-discipline of hydrology that seeks to understand ecological processes controlled by the hydrological cycle and vice versa (Zalewski, 2000; Zalewski and Robarts, 2003). It offers solution-oriented methods to reduce anthropogenic impacts, proposes "the use of ecosystem properties as management tools' (Zalewski, 2000), enhance ecosystem services, and strengthen human-nature connections for sustainable management.

Ecological footprint – a measure of human demand on the Earth's ecosystems, expressed as the amount of biologically productive land and water required to supply consumed resources and absorb generated waste, compared with the Earth's available biocapacity (Wackernagel and Kitzes, 2008).

Ecological integrity refers to the ability of an ecosystem to maintain its structure, function, and resilience over time, sustaining biodiversity and ecosystem processes without significant human-induced degradation. It implies that the ecosystem is healthy, self-sustaining, and capable of adapting to environmental changes.

Ecological succession is the gradual, natural process by which the species composition of a biological community changes over time, often following a disturbance.

Ecosystem is a functional unit or complex of relations in which living organisms (plants, animals, fungi, and microorganisms) interact with one another and with their physical environment, forming a dynamic yet broadly stable system. It may be of any size (https://www.britannica.com/science/ecosystem-paradigm).

Ecosystem services benefit that people receive from ecosystems, including supporting, provisioning, regulating, and cultural services (Chapin et al., 2010).

Ecosystemic approach is a form of environmental governance that places ecosystemic dynamics at the heart of environmental policy making. The ecosystemic approach grounds policy making in a scientific understanding of the environment, the ecosystem paradigm (https://www.britannica.com/science/ecosystemic-approach).

Ecoregion is a relatively homogeneous area defined by a similar similarity of climate, landform, soil, potential natural vegetation, hydrology, or other ecological relevant variables.

Ecotones, the transition zone between two different types of ecosystems, such as river and a meadow, characterized by very high biodiversity. Ecotones may play an important role as buffers, modifying and limiting flows of nutrients and pollutants between ecosystem components.

Enhancement, in the context of transdisciplinary ecohydrology, extends beyond conservation, restoration, and regeneration. It reflects the need to expand the carrying capacity of catchments and the biosphere through Dual Regulation in order to sustain Life Supporting Processes (LSPs) under increasing human and environmental pressures. As the global population continues to grow, resource consumption intensifies, and competition over limited natural resources increases, enhancement—together with the four determinants of sustainability embodied in the Water–Bioproductivity/

Biodiversity–Ecosystem Services–Resilience (WBSR) framework—emerges as a crucial strategy for achieving long-term sustainability (Zalewski, 2014).

Going a step further than regenerative ecology, transdisciplinary ecohydrology defines both the objective—what should be achieved (WBSR)—and the mechanism for achieving it through Dual Regulation and Advanced Nature-based Solutions (ANbS). Enhancement is therefore essential for integrating knowledge across disciplines to overcome the fragmentation of water management and sustainable development. In turn, strengthening the Water–Bioproductivity Biodiversity–Ecosystem Services–Resilience (WBSR) framework provides the foundation for avoiding the tragedy of the commons.

Evolution, changes in organisms as a result of genetic responses to past events (Chapin et al., 2010) Charles Darwin's theory of natural selection posits that evolution occurs through the differential survival and reproduction of individuals with heritable traits that are advantageous in their environment. In essence, organisms with traits better suited to their environment are more likely to survive and pass those traits on to their offspring, leading to a gradual shift in the characteristics of a population over time.

Evolutionary-ecosystemic paradigm (Zalewski, 2020) extends the sociocentric-mechanistic paradigm that leads towards biosphere exploitation and degradation by a new evolutionary-ecosystemic perspective that prioritizes the enhancement of life supporting processes (WBRS) by Dual Regulation towards "engineering harmony" between humankind and the biosphere.

Feyerabend's methodological pluralism and epistemological anarchism, articulated in *Against Method* Against Method, reject the idea of a universal scientific method, asserting that "anything goes" as a condition for progress (McGill and Feyerabend, 1976). Through historical examples such as Galileo Galilei, he shows that strict adherence to methodological rules can constrain innovation (Lakatos and Feyerabend, 1999). This perspective underpins Maciej Zalewski's concept of paradigm evolution in river ecology (Chapter 3). In this context, different perspectives—Thomas Kuhn's paradigm shifts and Karl Popper's falsifiability—offer complementary ways of understanding scientific change and knowledge.

Indicator, a measure of one aspect of a system used to communicate its state or direction of change (Chapin et al., 2010).

Integrated Water Resources Management (IWRM) is a process that promotes the coordinated development and management of water, land and related resources in order to maximize the resultant economic and social welfare in an equitable manner without compromising the sustainability of vital ecosystems (https://iwrmactionhub.org/about/iwrm-explained).

Life-sustaining Processes (LSPs), refer to the fundamental ecological and hydrological mechanisms that regulate and maintain the stability, resilience, and functionality of ecosystems. In the context of ecohydrology, LSPs encompass the dynamic interactions between water, soil, vegetation, and atmospheric processes that create and sustain vital ecosystems. These processes include, but are not limited to, water retention and filtration, nutrient cycling, energy flow, biogeochemical transformations, and habitat connectivity. By regulating hydrological regimes and ecological functions, LSPs ensure biodiversity, productivity, biodiversity and ecosystem services such as carbon sequestration, groundwater recharge, flood mitigation, and climate regulation. Understanding and enhancing LSPs through ecohydrological approaches is essential for the long-term sustainability potential and recovery of degraded landscapes.

Mesoscale this term was introduced by radar meteorologist Ligda (1951) to describe phenomena occurring at an intermediate scale, later elaborated upon by Emanuel (1986). In this context, the intermediate hydrological scale—mesocycle–refers to the linkage between small- and large-scale (global) hydrological processes. Recently term has been used for description of the water budget at the basin scale.

Hydrological cycle – the continuous circulation of water between the atmosphere, hydrosphere, lithosphere, and biosphere, driven by solar energy and gravity through evaporation, transpiration, condensation, atmospheric transport, precipitation, infiltration, groundwater recharge, streamflow, and runoff. At the mesoscale (catchment or basin scale), these interconnected processes combine to produce an integrated hydrological response, providing the basis for water resources assessment and management (Dingman, 2015; Uhlenbrook et al., 2004). The hydrological cycle provides the framework for identifying threats and optimizing strategies to enhance Water–Biodiversity–Ecosystem Services–Resilience (WBSR) (Principle I of TEH).

Above hydrological processes are quantified and modeled in meso-scale for basins, by in meso-scale because "in meso-scale basins, processes from smaller scales combine in a complex way to produce an integrated response" and intermediate-scale is crucial from the standpoint of water resource management. (Uhlenbrook at al., 2004), Such analysis expanded by precipitation patterns, geomorphology, point and non-source pollution distribution and urbanization, provide the framework for identification of threats and optimization of strategy for enhancement WBSR (I Principle TEH)

- in **micro-scale** catchments s (i.e. headwater catchments less than about 1 km^2),
- In **meso-scale** basins, processes from smaller scales combine in a complex way to produce an integrated response (approximately 10^1 – 10^3 km^2)
- the **macro-scale** (basins larger than 1000 km^2 (Uhlenbrook at al., 2004)

Hydrobiomanipulation refers to the regulation of water levels in reservoirs as an alternative or a complementary approach to predatory fish stocking, aimed at increasing zooplanktivorous fish fry and zooplankton density, thereby reducing algal blooms driven by excessive nutrient loading; term based on Zalewski (1990).

Constructed wetland, a constructed system permanently or periodically flooded that can act as a water purification system or nutrient sinks. Purification is enhanced by the activity of vegetation and variety of microbiological and biochemical processes taking place within the substrate of the wetland.

Drainage basin (see catchment) Factors descriptive of a drainage-basin as related to its hydrology may be classified broadly as:

- Morphologic—These factors depend only on the topography of the landforms of which the drainage-basin is composed and, on the form, and extent of the stream-system or drainage-net within it.
- Soil factors—This group includes factors descriptive of the materials forming the groundwork of the drainage-basin, including all those physical properties involved in the moisture-relations of soils.
- Geologic-structural factors—These factors relate to the depths and characteristics of the underlying rocks and the nature of the geologic structures in so far as they are related to ground-water conditions or otherwise to the hydrology of the drainage-basin.
- Vegetational factors—These are factors which depend wholly or in part on the vegetation, natural or cultivated, growing within the drainage-basin.
- **Climatic–hydrologic factors** – the climatic and hydrologic conditions that regulate the movement, distribution, and availability of water within a catchment. Climatic factors include temperature, precipitation, humidity, solar radiation, and evapotranspiration, whereas hydrologic factors include surface runoff, infiltration, streamflow, groundwater recharge, and other processes governed by the hydrologic cycle (Horton, 1932).

Ecological restoration – the process of assisting the recovery of an ecosystem that has been degraded, damaged, or destroyed, while accounting for environmental change and supporting the recovery of ecological integrity and ecosystem functionality (Gann et al., 2019).

Ecosystem restoration – the process of reversing ecosystem degradation to regain ecological functionality, improve biodiversity, and restore the capacity of ecosystems to provide goods, services, and other benefits to people and nature (IUCN, 2021).

Hyporheic zone – the saturated portion of streambeds, banks, and adjacent floodplains where stream water and groundwater mix and exchange. The extent and duration of hyporheic flow vary in space and time, influencing water quality, nutrient cycling, and ecosystem functioning (Woessner, 2017).

The Multidimensional Sustainability Potential of Catchments (WBSR) represents a cumulative and integrative goal that coordinates Integrated Water Resources Management and Transdisciplinary Ecohydrology (IWRM–TEH), operating in parallel across four key determinants critical for enhancing sustainability. It refers to a catchment's inherent capacity to adapt and improve under specific environmental conditions. This capacity can be scientifically defined and assessed through the three principles of transdisciplinary ecohydrology.

Nature-based solutions are inspired and supported by nature, cost-effective, and simultaneously provide environmental, social, and economic benefits. They help build resilience by bringing more, and more diverse, nature and natural features and processes into cities, landscapes, and seascapes through locally adapted, resource-efficient, and systemic interventions (European Commission, 2015).

N:P the Nitrogen to Phosphorus ratio is a critical factor regulating the growth, composition, and dominance of algal communities in aquatic ecosystems.

Nonpoint source pollution, nutrients or toxins scoured by runoff from agricultural or urban lands (Chapin et al., 2010).

Nutrients, chemical elements that are necessary for growth and development of vegetation. The main nutrients are phosphorus, nitrogen and carbon. Increased nutrients concentrations stimulate the process of eutrophication in aquatic ecosystem.

Nutrients cycling refers to the process of how nutrients move through an ecosystem, influencing productivity, consumption, and food web resilience.

Nutrient load, the amount of nutrient transported into a water body by rivers, sewage, discharges, etc. over a given period of time calculated as concentration multiplied by discharge.

Nutrient spiraling – the coupled process of nutrient uptake, transformation, recycling, and downstream transport in flowing waters, whereby nutrients are repeatedly assimilated by organisms, temporarily retained within the ecosystem, and subsequently released back into the water column as they move downstream (Webster, 1975; Newbold et al., 1981).

Phytoremediation is a sustainable, cost-effective, and environmentally friendly 'green' technology that uses plants and their associated microorganisms to remove, stabilise, or degrade contaminants in soil, water, and air.

Point source pollution, pollution entering water bodies from concentrated outflows, such as pipes transporting municipal and industrial sewage, water from purification plants, irrigation channels, etc.

The River Continuum Concept (RCC) is a generalized conceptual framework for characterizing pristine running-water ecosystems. Among its principal tenets are the concepts of biological analogues of energy equilibrium and entropy, maximization of energy use through continuous species replacement, the absence of ecological succession in stream ecosystems, and maximum biotic diversity in mid-reaches resulting from peak environmental heterogeneity (Vannote et al., 1980). The framework was subsequently critically evaluated by Statzner and Higler (1985), who argued that local hydraulic conditions, habitat heterogeneity, and spatial discontinuities are not adequately represented by the continuum model.

River metabolism, river ecosystem metabolism represents the collective metabolic activity of all producers and consumers within a river system. It is measured through gross primary production (GPP) and ecosystem respiration (ER), integrating inputs of organic carbon (OC) and inorganic carbon (IC) from terrestrial sources. These processes interact with gas fluxes and oxygen dynamics, shaping the overall energy balance and ecological function of the river (Battin et al., 2023).

Rehabilitation – a management action that aims to restore ecosystem functioning and the continued provision of ecosystem services in degraded ecosystems without necessarily returning them to their original natural condition (Gann et al., 2019).

Regenerative Sustainability emphasizes systems and practices that restore, renew, and revitalize their own sources of energy and materials. It aims to create processes that are not just sustainable but also beneficial to the environment, effectively contributing to the regeneration of ecosystems and communities. (EcoCart. *Regenerative Sustainability).*

Regeneration refers to the ability to exist or be recreated, derived from the Latin *regenerare* ("to create again"). In practice, regenerative systems go beyond sustainability by integrating principles of resilience, adaptability, and co-evolution, ensuring long-term ecological balance and human prosperity. Regenerative social-ecological systems—regions, economies, cities, businesses, communities, and so forth—can be defined as those that maintain positive reinforcing cycles of wellbeing, within and beyond themselves, especially between humans and wider nature (Buckton et al., 2023).

Regenerative design, a system of technologies and strategies based on an understanding of the inner working of ecosystems that generates designs to regenerate rather than deplete underlying life support systems and resources within socioecological wholes (Mang and Reed, 2012).

Regenerative development, a system of technologies and strategies for generating the patterned whole-system understanding of a place and developing the strategic systemic thinking capacities and the self-organizing and self-evolving stakeholder engagement/commitment required to ensure regenerative design processes achieve maximum systemic leverage and support (Mang and Reed, 2012).

Remediation – actions taken to remove, reduce, or contain pollution and other sources of environmental degradation to protect human health and ecosystems, including the cleanup of contaminated sites (Gann et al., 2019).

Renaturalisation – the process of restoring rivers, wetlands, or landscapes toward a more natural state by re-establishing natural hydrological, ecological, and geomorphological processes, native species, and ecosystem functions (Hobbs and Harris, 2001).

Restorative design – a design approach that restores degraded ecosystems and environmental quality while strengthening the relationship between people and nature. It seeks to recover ecological functions and enhance human well-being by integrating natural processes into the built environment (Mang and Reed, 2012).

Rewilding – the process of rebuilding, following major human disturbance, a natural ecosystem by restoring natural ecological processes and the complete food web at all trophic levels, using native species to create a self-sustaining and resilient ecosystem (Carver et al., 2021).

Runoff. Water that moves as overland flow or groundwater from a terrestrial to an aquatic system (Chapin et al., 2010).

Semi-intensive Agriculture is the agricultural approach that prioritises the use of moderate levels of the fertilizers, irrigation, and mechanisation, and is designed to increase productivity without the high resource demands of fully intensive systems. This approach is often used in transitional economies, mixed farming, or small-to- medium-scale operations, where sustainable production is pursued alongside environmental and economic considerations (FAO, 2001).

Sociocentric-mechanistic paradigm refers to a conventional approach to resource management that prioritizes human-centered needs and mechanistic control over natural systems. This perspective often leads to the overexploitation of natural resources, disrupting ecological balance and contributing to environmental degradation. By emphasizing short-term economic gain over long-term sustainability, this paradigm accelerates resource depletion and increases the risk of ecological collapse and biodiversity loss.

Strahler stream order system is a hierarchical classification used to define river networks. In this system, headwater streams (first-order) are those with no tributaries, while a second-order stream forms when two first-order streams converge. Higher-order streams emerge as lower-order streams join, progressively increasing in complexity and discharge capacity. This classification is fundamental in understanding riverine ecology, as hydrological and ecological processes change with increasing stream order.

Supporting services, fundamental ecological processes that sustain ecosystem functioning (maintenance of soil fertility, cycling of essential elements, biological diversity, and cycles of disturbance and renewal) (Chapin et al., 2010).

Sustainability, the use of the environment and resources to meet the needs of the present without compromising the ability of future generations to meet their own needs. Maintenance of the productive base (total capital) over time (Chapin et al., 2010).

Sustainable development seeks to improve human well-being, while at the same time sustaining the natural resource base and opportunities on which future generations depend (Chapin et al., 2010).

Sustainable management to sustain the functional properties of socioecological systems that are important to society (Chapin et al., 2010).

Synergy, ecosystem services or societal benefits that co-occur with other services and benefits (e.g., aesthetic value and carbon sequestration provided by natural forests) (Chapin et al., 2010).

TEH Systemic Solutions, these hybrid solutions grounded in Transdisciplinary Ecohydrology, combine Nature-basded Solutions, Advanced Nature-based Solutions, and hydroengineering, offering a comprehensive approach that merges ecological science, hydrology, ecohydraulics, biotechnology, and engineering for sustainable and adaptive water and ecosystem management.

Quantum Entanglement refers to the phenomenon in quantum physics where two or more particles become connected in such a way that the state of one particle cannot be described independently of the state of the other particles, regardless of the distance between them (https://www.sciencedirect.com/topics/physics-and-astronomy/quantum-entanglement). Transdisciplinary Ecohydrology expresses the growing complexity of relations between humans and the biosphere in the Anthropocene.

Tragedy of the commons – the degradation or depletion of a shared (common-pool) resource that can occur when individuals act in their own self-interest under conditions of open access or ineffective governance, leading to overuse of resources such as fisheries, forests, or groundwater (Hardin, 1968; Ostrom, 2008).

Transdisciplinary Ecohydrology expands the understanding of life-supporting processes in freshwater and coastal ecosystems by emphasizing the dynamic interplay and feedbacks between the hydrological cycle and ecosystem functioning—Dual Regulation. Its hydrocentric approach indicates how ecosystem properties can be used as management tools (Zalewski, 2000). By highlighting the importance of ecosystem metabolism in hydrological processes at the catchment scale, it provides an important step for the development of Advanced Nature-Based Solutions (ANbS), enhancing the sustainability potential of catchments and the biosphere. Using the Water–Biodiversity–Ecosystem Services–Resilience (WBSR) framework, Transdisciplinary Ecohydrology translates these insights, according to three principles, into hybrid solutions (ANbS + hydroengineering), that reinforce Sustainable Development Goals (SDGs), Integrated Water Resources Management (IWRM) and support the implementation of the Water Framework Directive (WFD).

Trophic cascade (trophic effect) changes in species composition driven by changes in top predators that affect lower trophic levels, primary producers, bacteria and nutrient cycles (Chapin et al., 2010).

Regenerate (verb) by Merriam Webster Dictionary"

- formed or created again,
- spiritually reborn or converted
- restored to a better, higher, or more worthy state

WBSR + CE + LPG factors, a central principle of Transdisciplinary Ecohydrology is that every Integrated Water Resources Management (IWRM) intervention within a catchment should simultaneously address and enhance four key factors: Water resources (W); Biocomplexity, Bioproductivity and Biodiversity (B); Ecosystem Services for society (S); and Resilience to impacts and climate change (R), together, these form the acronym WBSR. The effective implementation of this comprehensive approach relies on the active engagement of society across the entire catchment, achieved through fostering a culture of sustainability (C) and promoting education (E). Moreover, it is crucial to secure the involvement of decision-makers and to embed these principles within proactive law (L), policy (P), and governance (G) structures to ensure the durability and institutionalisation of sustainable practices.

References

Battin, T.J., Lauerwald, R., Bernhardt, E.S., Bertuzzo, E., Gómez-Gener, L., Hall Jr., R.O., Hotchkiss, E.R., et al., 2023. River ecosystem metabolism and carbon biogeochemistry in achanging world. Nature 613 (7944), 449–459. https://doi.org/10.1038/s41586-022-05500-8.

Bren, L., 2015. Forest Hydrology and Catchment Management: An Australian Perspective. Springer Netherlands. https://doi.org/10.1007/978-94-017-9337-7.

Brown, T., 2009. Change by Design: How Design Thinking Transforms Organizations and Inspires Innovation. HarperBusiness, New York.

Buckton, S.J., Fazey, I., Sharpe, B., Om, E.S., Doherty, B., Ball, P., Denby, K., Bryant, M., Lait, R., Bridle, S., Cain, M., Carmen, E., Collins, L., Nixon, N., Yap, C., Connolly, A., Fletcher, B., Frankowska, A., Gardner, G., James, A., Kendrick, I., Kluczkovski, A., Mair, S., Morris, B., Sinclair, M., 2023. The Regenerative Lens: A conceptual framework for regenerative social-ecological systems. One Earth 6 (7), 824–842. https://doi.org/10.1016/j.oneear.2023.06.006.

Carver, S., Convery, I., Beyers, R., Eagle, A., Kun, Z., Dyke, A., von Essen, E., Fisher, M., Glikman, J., Hayward, M.W., Jørgensen, D., Lorimer, J., Nogués-Bravo, D., Pettorelli, N., Ceausu, S., 2021. Guiding Principles for Rewildin. International Union for Conservation of Nature (IUCN), Gland, Switzerland. Commission on Ecosystem Management, Rewilding Thematic Group.

Chapin III, F.S., Carpenter, S.R., Kofinas, G.P., Folke, C., Abel, N., Clark, W.C., Olsson, P., Stafford Smith, D.M., Walker, B., Young, O.R., Berkes, F., Biggs, R., Grove, J.M., Naylor, R.L., Pinkerton, E., Steffen, W., Swanson, F.J., 2010. Ecosystem stewardship: sustainability strategies for a rapidly changing planet. Trends Ecol. Evol. 25 (4), 241–249. https://doi.org/10.1016/j.tree.2009.10.008.

Dingman, S.L., 2015. Physical Hydrology, third ed. Waveland Press, Long Grove, IL.

Emanuel, K.A., 1986. Overview and definition of mesoscale Meteorology. In: Ray, P.S. (Ed.), Mesoscale Meteorology and Forecasting. American Meteorological Society, Boston, MA, pp. 1–16.

European Commission, 2015. Towards an EU Research and Innovation Policy Agenda forNature-Based Solutions & Re-Naturing Cities: Final Report of the Horizon 2020 Expert Groupon Nature-Based Solutions and Re-Naturing Cities. Brussels: European Commission.

FAO, 2001. Mixed crop–livestock farming: A review of traditional technologies based on literature and field experience. Food and Agriculture Organization of the United Nations.

Galesic, M., Barkoczi, D., Katsikopoulos, K.V., et al., 2023. Beyond collective intelligence: collective adaptation. J. R. Soc. Interface 20 (199), 20220736. https://doi.org/10.1098/rsif.2022.0736.

Gallo, M.E., 2022. The bioeconomy: a Primer. Congressional Research Service, R46881. Library of Congress, Washington, DC. Available at: Congressional Research Service reportR46881.

Gann, G.D., McDonald, T., Walder, B., Aronson, J., Nelson, C.R., Jonson, J., Hallett, J.G., Eisenberg, C., Guariguata, M.R., Liu, J., Hua, F., Echeverría, C., Gonzales, E., Shaw, N., Decleer, K., Dixon, K.W., 2019. International Principles and Standards for the Practice of Ecological Restoration, second ed. Restor. Ecol. 27 (S1), S1–S46. https://doi.org/10.1111/rec.13035.

Hardin, G., 1968. The tragedy of the commons. Science 162 (3859), 1243–1248. https://doi.org/10.1126/science.162.3859.1243.

Hasso Plattner Institute of Design at Stanford, 2020. An Introduction to Design Thinking: Process Guide. Stanford University, Stanford, CA.

Hobbs, R.J., Harris, J.A., 2001. Restoration ecology: repairing the Earth's ecosystems in the new millennium. Restor. Ecol. 9 (2), 239–246. https://doi.org/10.1046/j.1526-00X.2001.009002239.x.

Horton, R.E., 1932. Drainage-basin characteristics. Trans. Am. Geophys. Union. 13, 350–361. https://doi.org/10.1029/TR013i001p00350.

IPCC, 2021. Climate Change 2021: The Physical Science Basis. Cambridge University Press, Cambridge.

Lakatos, I., Feyerabend, P., 1999. For and Against Method. University of Chicago Press.

Ligda, M.G.H., 1951. Radar storm observation. In: Malone, T.F. (Ed.), Compendium of Meteorology. American Meteorological Society, Boston, MA, pp. 1265–1282. https://doi.org/10.1007/978-1-940033-70-9_103.

Mang, P., Reed, B., 2012. Designing from place: a regenerative framework and methodology. Build. Res. Inform. 40 (1), 23–38. https://doi.org/10.1080/09613218.2012.621341.

McGill, V.J., Feyerabend, P., 1976. Against method: Outline of an anarchistic theory of knowledge. Philos. Phenomenol. Res. 37 (1), 129. https://doi.org/10.2307/2106382.

Newbold, J.D., Elwood, J.W., O'Neill, R.V., Van Winkle, W., 1981. Measuring nutrient spiralling in streams. Can. J. Fish Aquat. Sci. 38 (7), 860–863. https://doi.org/10.1139/f81-114.

Ostrom, E., 2008. The challenge of common-pool resources. Environment: Science andPolicy for Sustainable Development 50 (4), 8–21. https://doi.org/10.3200/ENVT.50.4.8-21.

Postel, S., 1994. Carrying capacity: Earth's bottom line. In: Brown, L.R., Flavin, C., French, H. (Eds.), State of the World 1994. W.W. Norton & Company, New York, pp. 3–21.
Romanazzi, G.R., Koto, R., De Boni, A., Ottomano Palmisano, G., Cioffi, M., Roma, R., 2023. Cultural ecosystem services: a review of methods and tools for economic evaluation. Environ. Sustain. Ind. 20, 100304. https://doi.org/10.1016/j.indic.2023.100304.
Sarukhán, J., Whyte, A. (Eds.), 2005. Millennium Ecosystem Assessment: Ecosystems and Human Well-being: Synthesis. Island Press, Washington, DC.
Statzner, B., Higler, B., 1985. Questions and comments on the river continuum concept. Can. J. Fish. Aquat. Sci. 42 (5), 1038–1044. https://doi.org/10.1139/f85-129.
Uhlenbrook, S., Roser, S., Tilch, N., 2004. Hydrological process representation at themeso-scale: The potential of distributed, conceptual catchment models. J. Hydrol. 291 (3–4), 278–296.
UNEP and FAO, 2021. In: Becoming #GenerationRestoration: Ecosystem Restoration for People, Nature and Climate. United Nations Environment Programme and Rome. Food and Agriculture Organization of the United Nations, Nairobi.
Vannote, R.L., Minshall, G.W., Cummins, K.W., Sedell, J.R., Cushing, C.E., 1980. The river continuum concept. Can. J. Fish Aquat. Sci. 37 (1), 130–137. https://doi.org/10.1139/f80-017.
Wackernagel, M., Kitzes, J., 2008. Ecological footprint. In: Jørgensen, S.E., Fath, B.D. (Eds.), Encyclopedia of Ecology. Elsevier, Oxford, pp. 1031–1037.
Webster, J.R., 1975. Analysis of Potassium and Calcium Dynamics in Stream Ecosystems on Three Southern Appalachian Watersheds of Contrasting Vegetation. University of Georgia. PhD thesis.
Woessner, W.W., 2017. Hyporheic Zone. In: Singhal, B.B.S., Gupta, R.P. (Eds.), Applied Hydrogeology of Fractured Rocks. Springer, Dordrecht.
Woolley, A.W., Chabris, C.F., Pentland, A., Hashmi, N., Malone, T.W., 2010. Evidence for acollective intelligence factor in the performance of human groups. Science 330 (6004), 686–688. https://doi.org/10.1126/science.1193147.
Zalewski, M., 2000. Ecohydrology – the scientific background to use ecosystem properties as management tools toward sustainability of water resources. Ecol. Eng. 16 (1), 1–8. https://doi.org/10.1016/S0925-8574(00)00071-9.
Zalewski, M., 2006. Ecohydrology – an interdisciplinary tool for integrated protection and management of water bodies. Arch. Hydrobiol. 158 (Suppl. 4), 613–622.
Zalewski, M., 2011. Ecohydrology for implementation of the EU water framework directive. Proc. Inst. Civil Eng. Water Manage. 164 (8), 375–385. https://doi.org/10.1680/wama.1000030.
Zalewski, M., 2014. Ecohydrology and hydrologic engineering: Regulation of hydrology–biota interactions for sustainability. J. Hydrol. Eng. 19 (5), 1080–1088. https://doi.org/10.1061/(ASCE)HE.1943-5584.0000999.
Zalewski, M. (Ed.), 2020. Ekohydrologia. Wydawnictwo Naukowe PWN, ISBN 978-83-01-20804-2.
Zalewski, M., 2021. Ecohydrology: An integrative sustainability science. In: Hromodka, T.V., Rao, P. (Eds.), Hydrology. IntechOpen. https://doi.org/10.5772/intechopen.94169.
Zalewski, M., Brewińska-Zaraś, B., Frankiewicz, P., Kalinowski, S., 1990. The potential forbiomanipulation using fry communities in a lowland reservoir: Concordance between water quality and optimal recruitment. Hydrobiologia 200 (1), 549–556. https://doi.org/10.1007/BF02530373.
Zalewski, M., Naiman, R.J., 1985. The regulation of riverine fish communities by a continuum of abiotic–biotic factors. In: Alabaster, J.S. (Ed.), Habitat Modification and Freshwater Fisheries. Butterworths Scientific, London.
Zalewski, M., Robarts, R.D., 2003. Ecohydrology – a new paradigm for integrated water resources management. SIL News 40, 1–5.

Recommended Key Books Shaping the Discourse on Ecohydrology

There is an existing collection of published books on ecohydrology that share valuable knowledge and expertise on the subject. These resources provide in-depth insights into the interactions between water and ecological systems, offering guidance for research and practical applications.

Alvarez Mieles, M.G., 2019. Ecological Modelling of River-Wetland Systems: A Case Study for the Abras de Mantequilla Wetland in Ecuador. Routledge.
Baird, A.J., Wilby, R.L. (Eds.), 1999. Eco-hydrology: Plants and Water in Terrestrial and Aquatic Environments. Routledge.
Belete, M., 2023. Ecohydrology-based Landscape Restoration: Theory and Practice. Routledge.
Brooks, K.N., Ffolliott, P.F., Gregersen, H.M., DeBano, L.F., 2012. Hydrology and the Management of Watersheds, 4th ed. Wiley-Blackwell.
D'Odorico, P. (Ed.), 2014. Dryland Ecohydrology. Springer.
Dodds, W.K., Whiles, M.R., 2019. Freshwater Ecology: Concepts and Environmental Applications of Limnology, 3rd ed. Academic Press.
Eagleson, P.S., 2005. Ecohydrology: Darwinian Expression of Vegetation Form and Function. Cambridge University Press.
Eamus, D., Hatton, T., Cook, P., Colvin, C., 2006. Ecohydrology: Vegetation Function, Water and Resource Management. CSIRO Publishing.
Eslamian, S. (Ed.), 2014. Handbook of Engineering Hydrology: Fundamentals and Applications. Routledge.
Harper, D.M., Zalewski, M., Pacini, N. (Eds.), 2007. Ecohydrology: Processes, Models and Case Studies. CABI.
Heathcote, I.W., 2009. Ecohydrology from Catchment to Coast. CRC Press.
Hewlett, J.D., 2003. Principles of Forest Hydrology. University of Georgia Press.
Lachavanne, J.B., Juge, R. (Eds.), 1997. Biodiversity in Land-inland Water Ecotones. UNESCO Publishing/Parthenon Publishing Group.
Mitsch, W., Gosselink, J., 2015. Wetlands. Wiley.

Rodríguez-Iturbe, I., Porporato, A., 2004. Ecohydrology of Water-controlled Ecosystems: Soil Moisture and Plant Dynamics. Cambridge University Press.

Rowiński, P., Radecki-Pawlik, A. (Eds.), 2015. Rivers – Physical, Fluvial and Environmental Processes. Springer.

Tundisi, J.G., 2008. Limnologia. Oficina de Textos.

Wolanski, E. (Ed.), 2007. Coastal Wetlands: An Integrated Ecosystem Approach. Elsevier.

Wolanski, E., 2007. Estuarine Ecohydrology. Elsevier.

Wolanski, E., Elliott, M., 2015. Estuarine Ecohydrology: An introduction, 2nd ed. Elsevier.

Wood, P.J., Hannah, D.M., Sadler, J.P. (Eds.), 2007. Hydroecology and Ecohydrology: Past, Present and Future. John Wiley & Sons, Ltd.

Further reading

Cole, R.J., 2012. Regenerative design and development: current theory and practice. Build. Res. Inf. 40 (1), 1–6. https://doi.org/10.1080/09613218.2012.617516.

Crowther, T.W., Glick, H.B., Covey, K.R., Bettigole, C., Maynard, D.S., Thomas, S.M., Smith, J.R., Hintler, G., Duguid, M.C., Amatulli, G., Tuanmu, M.-N., Jetz, W., Salas, C., Stam, C., Piotto, D., Tavani, R., Green, S., Bruce, G., Williams, S.J., Wiser, S.K., Huber, M.O., Hengeveld, G. M., Nabuurs, G.-J., Tikhonova, E., Borchardt, P., Li, C.-F., Powrie, L.W., Fischer, M., Hemp, A., Homeier, J., Cho, P., Vibrans, A.C., Umunay, P. M., Piao, S.L., Rowe, C.W., Ashton, M.S., Crane, P.R., Bradford, M.A., 2015. Mapping tree density at a global scale. Nature 525 (7568), 201–205. https://doi.org/10.1038/nature14967.

Hobbs, R.J., Higgs, E., Harris, J.A., 2009. Novel ecosystems: implications for conservation and restoration. Trends Ecol. Evol. 24 (11), 599–605. https://doi.org/10.1016/j.tree.2009.05.012.

Holland, M.M., 1988. SCOPE 43: Biological Diversity Biodiversity, pp. 43–70.

Hoicea-Milcu, A.-I., Leventon, J., Hanspach, J., Fischer, J., 2013. Disaggregated contributions of ecosystem services to human well-being: a case study from Eastern Europe. Reg. Environ. Change 13 (6), 1329–1343.

Hotchkiss, T.M., Pavelsky, T.M., Ran, L., Raymond, P., Rosentreter, J.A., Regnier, P., 2023. River ecosystem metabolism and carbon biogeochemistry in a changing world. Nature 613 (7944), 449–459. https://doi.org/10.1038/s41586-022-05500-8.

Ketterings, Q.M., Czymmek, K.J., Swink, S.N., 2013. Adaptive management approaches for timely impact. J. Agric. Sci. 152 (S1), 1–10. https://doi.org/10.1017/S0021859613000937.

Kolasa, J., Zalewski, M., 1995. Notes on ecotone attributes and functions. Hydrobiologia 303 (1–3), 1–7. https://doi.org/10.1007/BF00034039.

Makarieva, A.M., Gorshkov, V.G., 2007. Biotic pump of atmospheric moisture as driver of the hydrological cycle on land. Hydrol. Earth Syst. Sci. 11 (2), 1013–1033. https://doi.org/10.5194/hess-11-1013-2007.

Peterson, G.D., Cumming, G.S., Carpenter, S.R., 2003. Scenario planning: A tool for conservation in an uncertain world. Conserv. Biol. 17 (2), 358–366. https://doi.org/10.1046/j.1523-1739.2003.01491.x.

Schwella, E., 2014. Knowledge based governance, governance as learning: the leadership implications. Int. J. Leadersh. Public Serv. 10 (2), 84–90. https://doi.org/10.1108/ijlps-05-2014-0004.

Literature

Further reading

Achkeeva, M.V., Romanyuk, N.V., Frolova, E.A., Kondakov, D.F., Khomyakov, D.M., Danilov, V.P., 2015. Deicing properties of sodium, potassium, magnesium, and calcium chlorides, sodium formate and salt compositions on their basis. Theor. Found. Chem. Eng. 49, 481–484. https://doi.org/10.1134/S0040579515040028.

Ahmed, W., Hamilton, K., Toze, S., Cook, S., Page, D., 2019. A review on microbial contaminants in stormwater runoff and outfalls: potential health risks and mitigation strategies. Sci. Total Environ. 692, 1304–1321. https://doi.org/10.1016/j. scitotenv.2019.07.055.

Albarracín, M., Ramón, G., González, J., Iñiguez-Armijos, C., Zakaluk, T., Martos-Rosillo, S., 2021. The ecohydrological approach in water sowing and harvesting systems: the case of the Paltas Catacocha Ecohydrology Demonstration Site, Ecuador. Ecohydrol. Hydrobiol. 21 (3), 454–466.

Amare, A., 2005. Study of sediment yield from the watershed of angereb reservoir. Master's Thesis, Department of Agricultural Engineering, Alemaya University, Ethiopia.

Asres, M.T., Awulachew, S.B., 2010. SWAT based run off and sediment yield modeling: a case study of the Gumara watershed in the Blue Nile basin. Ecohydrol. Hydrobiol. 10 (2–4), 191–200.

Ansari, I., El-Kady, M.M., Mahmoud, A.E.D., Arora, C., Verma, A., Rajarathinam, R., Singh, P., Verma, D.K., Mittal, J., 2024. Persistent pesticides: accumulation, health risk assessment, management and remediation – an overview. Desalination Water Treat. 317, 100274.

Arvidsson, A.K., 2017. The Winter Model – a new way to calculate socio-economic costs depending on winter maintenance strategy. Cold Reg. Sci. Technol. 136, 26–30. https://doi.org/10.1016/j.coldregions.2017.01.005.

Aslund, M.L.W., Lunney, A.L., Rutter, A., Zeeb, B.A., 2010. Effects of amendments on the uptake and distribution of DDT in Cucurbita pepo ssp pepo plants. Environ. Pollut. 158 (2), 508–513. https://doi.org/10.1016/j.envpol.2009.08.030.

Baran, A., Cz, J., Tarnawski, M., 2010. Effect of bottom deposit supplement on trace element content in light soil. Ecol. Chem. Eng. 17 (12), 1553–1561.

Baran, A., Tarnawski, M., Urbaniak, M., 2019. An assessment of bottom sediment as a source of plant nutrients and an agent for improving. Environ. Eng. Manag. J. 18 (8), 1647–1656.

Barendregt, A., 2003. Ecological impact of flooding in a fresh water tidal area. In: International Conference "Towards Natural Flood Reduction Strategies".

Bayley, P.B., 1995. Understanding large river: floodplain ecosystems. Bioscience 45 (3), 153–158.

Bedford, B.L., Preston, E.M., 1988. Developing the scientific basis for assessing cumulative effects of wetland loss and degradation on landscape functions: status, perspectives, and prospects. Environ. Manag. 12 (5), 751–771. https://doi.org/10.1007/BF01867550.

Bednarek, A., Szklarek, S., Zalewski, M., 2014. Nitrogen pollution removal from areas of intensive farming—comparison of various denitrification biotechnologies. Ecohydrol. Hydrobiol. 14 (2), 132–141.

Beven, K., Germann, P., 1982. Macropores and water flow in soils. Water Resour. Res. 18, 1311–1325. https://doi.org/10.1029/WR018i005p01311.

Bianchi, T.S., 2020. The evolution of biogeochemistry: revisited. Biogeochemistry 149 (1), 1–20. https://doi.org/10.1007/s10533-020-00708-0.

Bosman, C., 2009. The hidden dragon: nitrate pollution from open-pit mines- a case study from the Limpopo Province, South Africa. In: Sustainable Solutions.

Bradshaw, K., Snyder, B.J., Oladeinde, A., Spidle, D., Berrang, M.E., et al., 2016. Characterizing relationships among fecal indicator bacteria, microbial source tracking markers, and associated waterborne pathogen occurrence in stream water and sediments in a mixed land use watershed. Water Res. 101, 498–509. https://doi.org/10.1016/j. watres.2016.05.014.

Briggs, G.G., Bromilow, R.H., Evans, A.A., 1982. Relationships between lipophilicity and root uptake and translocation of non-ionised chemicals by barley. Pestic. Sci. 13 (5), 495–504. https://doi.org/10.1002/ps.2780130506.

Buckton, S.J., Fazey, I., Sharpe, B., Om, E.S., Doherty, B., Ball, P., Denby, K., Bryant, M., Lait, R., Bridle, S., Cain, M., Carmen, E., Collins, L., Nixon, N., Yap, C., Connolly, A., Fletcher, B., Frankowska, A., Gardner, G., James, A., Kendrick, I., Kluczkovski, A., Mair, S., Morris, B., Sinclair, M., 2023. The regenerative lens: a conceptual framework for regenerative social-ecological systems. One Earth 6 (7), 824–842. https://doi.org/10.1016/j.oneear.2023.06.006.

Bunn, S., Arthington, A., 2002. Basic principles and ecological consequences of altered flow regimes for aquatic biodiversity. Environ. Manag. 30, 492–507. https://doi.org/10.1007/s00267-002-2737-0.

Burt, T.P., Bates, P.D., Stewarts, M.D., Claxton, A.J., Anderson, M.G., Price, D.A., 2002. Water table fluctuations within the floodplain of the river Severn, England. J. Hydrol. 262, 1–20.

Canet, R., Chaves, C., Pomares, F., Albiach, R., 2003. Agricultural use of sediments from the Albufera Lake (Eastern Spain). Agric. Ecosyst. Environ. 95 (1), 29–36. https://doi.org/10.1016/S0167-8809(02)00171-8.

CEQG, 2011. Canadian water quality guidelines for the protection of aquatic life. Can. Environ. Qual. Guidel. 1–9.

Chahal, M., Shi, Z., Flury, M., 2016. Nutrient leaching and copper speciation in compost-amended bioretention systems. Sci. Total Environ. 556, 302–309. https://doi.org/10.1016/j.scitotenv.2016.02.125.

Chakraborty, P., Acharyya, T., Raghunadh Babu, P.V., Bandyopadhyay, D., 2011. Impact of salinity and pH on phytoplankton communities in a tropical freshwater system: an investigation with pigment analysis by HPLC. J. Environ. Monit. 13, 614. https://doi.org/10.1039/c0em00333f.

Chief Inspectorate of Environmental Protection, 2018. Report on the State of the Environment 2018. Library Environmental Monitoring, Warsaw.

Clark, W.C., 2007. Sustainability science: a room of its own. Proc. Natl. Acad. Sci. 104 (6), 1737–1738. https://doi.org/10.1073/pnas.0611291104.

Clark, W.C., Harley, A.G., October 17, 2020. Sustainability science: toward a synthesis. Annu. Rev. Environ. Resour. 45 (1), 331–386. https://doi.org/10.1146/annurev-environ-012420-043621.

Codd, G.A., Meriluoto, J., Metcalf, J.S., 2017. Introduction: cyanobacteria, cyanotoxins, their human impact and risk management. In: Meriluoto, J., Spoof, L., Codd, G.A. (Eds.), Handbook of Cyanobacterial Monitoring and Cy-Anotoxin Analysis. John Wiler & Sons, Ltd, The Atrium, UK, Chichester, UK, pp. 3–8.

Coldsnow, K.D., Mattes, B.M., Hintz, W.D., Relyea, R.A., 2017. Rapid evolution of tolerance to road salt in zooplankton. Environ. Pollut. 222. https://doi.org/10.1016/j.envpol.2016.12.024.

Corsi, S.R., De Cicco, L.A., Lutz, M.A., Hirsch, R.M., 2015. River chloride trends in snow-affected urban watersheds: increasing concentrations outpace urban growth rate and are common among all seasons. Sci. Total Environ. 508, 488–497. https://doi.org/10.1016/j.scitotenv.2014.12.012.

Crawford, R.L., 2011. Biodegradation : Principles, Scope , and Technologies, [w:] Crawford R.L., (Red), Comprehensive Biotechnology. Elsevier B.V., pp. 3–13

Czekalski, N., Gascn Díez, E., Bürgmann, H., 2014. Wastewater as a point source of antibiotic-resistance genes in the sediment of a freshwater lake. ISME J. 8, 1381.

de Cárcer, D.A., Martin, M., Mackova, M., Macek, T., Karlson, U., Rivilla, R., 2007. The introduction of genetically modified microorganisms designed for rhizoremediation induces changes on native bacteria in the rhizosphere but not in the surrounding soil. ISME J. 1, 215–223. https://doi.org/10.1038/ismej.2007.27.

de Loë, R.C., Patterson, J.J., 2017. Rethinking water governance: moving beyond water-centric perspectives in a connected and changing world. Nat. Resour. J. 57 (1), 75–100.

Devol, A.H., 2015. Denitrification, anammox, and N(2) production in marine sediments. Ann. Rev. Mar. Sci 7, 403–423. https://doi.org/10.1146/annurev-marine-010213-135040.

DiPinto, L.M., Coull, B.C., Chandler, G., 1993. Lethal and sublethal effects of a sediment-associated PCB Aroclor 1254 on a meiobenthic copepod. Environ. Toxicol. Chem. 12, 1909–1918.

Doroszczyk, L., Długoszewski, B., Kanigowska, E., Wiśniewolski, W., Godlewska, M., 2013. Spatial and temporal variability of a vendace (Coregonus albula (L.)) population in Lake Pluszne, Poland, based on hydroacoustics and trawl catches. Adv. Limnol. 64, 171–187.

Duan, S., Kaushal, S.S., 2015. Salinization alters fluxes of bioreactive elements from stream ecosystems across land use. Biogeosciences 12, 7331–7347. https://doi.org/10.5194/bg-12-7331-2015.

Dugan, H.A., Bartlett, S.L., Burke, S.M., Doubek, J.P., Krivak-Tetley, F.E., Skaff, N.K., Summers, J.C., Farrell, K.J., McCullough, I.M., Morales-Williams, A.M., Roberts, D.C., Ouyang, Z., Scordo, F., Hanson, P.C., Weathers, K.C., 2017. Salting our freshwater lakes. Proc. Natl. Acad. Sci. USA 114 (17), 4453–4458. https://doi.org/10.1073/pnas.1620211114.

Ebbs, S., Talbott, J., Sankaran, R., 2006. Cultivation of garden vegetables in Peoria Pool sediments from the Illinois River: a case study in trace element accumulation and dietary exposures. Environ. Int. 32 (6), 766–774. https://doi.org/10.1016/j. envint.2006.03.013.

ECDC (European Centre for Disease Prevention and Control), 2018. EFSA (European Food Safety Authority), and EMA (European Medicines Agency) ECDC/EFSA/EMA second joint report on the integrated analysis of the consumption of antimicrobial agents and occurrence of antimicrobial resistance in bacteria from humans and food-producing animals – Joint Interagency Antimicrobial Consumption and Resistance Analysis (JIACRA) (2017), Report. EFSA J. 15, 4872–487135. https://doi.org/10.2903/j.efsa.2017.487.

Eagleson, P.S., 1982. Ecological optimality in water-limited natural soil-vegetation systems: 1. Theory and hypothesis. Water Resour. Res. 18 (2), 325–340.

Erisman, J.W., Galloway, J.N., Dise, N.B., Sutton, M.A., Bleeker, A., Grizzetti, B., Leach, A.M., De Vries, W., 2015. Nitrogen: Too Much of a Vital Resource. WWF Netherlands, Zeist, the Netherlands. https://doi.org/10.13140/RG.2.1.3664.8163.

European Commission, 1991. Council Directive 91/676/EEC of 12 December 1991 concerning the protection of waters against pollution caused by nitrates from agricultural sources. Official J. L 375.

European Commission, 2013. Green Infrastructure – Enhancing Europe's Natural Capital. Commission Communication COM (2013) 249. (Brussels).

Fakhrudin, M., 2018. Development of Ecohydrology Demonstration Site in Saguling Reservoir, APCE Best Learning of Water Resources Management. Sekolah Pascasarjana UGM Publishing House, pp. 77–101.

Feyerabend, P.K., 1975. Against Method: Outline of an Anarchistic Theory of Knowledge. New Left Books, London.

Fonseca, R., Barriga, F.J.A.S., Fyfe, W.S., 2019. Dam Reservoir Sediments as Fertilizers and Artificial Soils: Case Studies from Portugal and Brazil. http://home.uevora.pt./~rfonseca/fct_2009/Fonseca%20et%20al%202003.pdf.

Font, N.A., Serwecińska, L., Gęgała-Borowska, I., Jurczak, T., Mankiewicz-Boczek, J., 2017. The characterization of a novel bacterial strain capable of microcystin degradation from the Jeziorsko reservoir, Poland: a preliminary study. Biol. Sec. Cell. Mol. Biol. 72/12, 1394–1402.

Font, N.A., Serwecińska, L., Szklarek, S., Mankiewicz-Boczek, J., 2020. Characterization and comparison of microbial communities in sequential sedimentation-biofiltration systems for removal of nutrients in urban rivers. Ecol. Engine. 149. https://doi.org/10.1016/j.ecoleng.2020.105796.

Font Nájera, A., Serwecińska, L., Mankiewicz-Boczek, J., 2021a. Culturable nitrogen-transforming bacteria from sequential sedimentation biofiltration systems and their potential for nutrient removal in urban polluted rivers. Sci. Rep.

Font Nájera, A., Serwecińska, L., Szklarek, S., Mankiewicz-Boczek, J., 2021b. Seasonal and spatial changes of N-transforming microbial communities in sequential sedimentation-biofiltration systems – influence of system design and environmental conditions. Int. Biodeterior. Biodegradation 67, 218.

Forman, D., 2004. Commentary: nitrites, nitrates, and nitrosation as causes of brain cancer in children: epidemiological challenges. Int. J. Epidemiol. 33, 1216–1218.

Fovet, O., Ruiz, L., Hrachowitz, M., Faucheux, M., Gascuel-Odoux, C., 2015. Hydrological hysteresis and its value for assessing process consistency in catchment conceptual models. Hydrol. Earth Syst. Sci. 19 (1), 105–123.

Gallagher, M.T., Snodgrass, J.W., Ownby, D.R., Brand, A.B., Casey, R.E., Lev, S., 2011. Watershed-scale analysis of pollutant distributions in stormwater management ponds. Urban Ecosyst. 14, 469–484. https://doi.org/10.1007/s11252-011-0162-y.

Galloway, J.N., Dentener, F.J., Capone, D.G., Boyer, E.W., Howarth, R.W., Seitzinger, S.P., Asner, G.P., Cleveland, C., Green, P., Holland, E., Karl, A. D.M., Michaels, F., Porter, J.H., Townsend, A., Vőrősmarty, C., 2003. Nitrogen cycles: past, present and future. Biogeochemistry. 70, 153–226.

Gamble, T.N., Betlach, H.R., Tiedje, J.M., 1977. Numerically dominant denitrifying bacteria from world soils. Appl. Environ. Microbiol. 33, 926–939.

Garnier, J., Billen, G., Hannon, E., Fonbonne, S., Videnina, Y., Soulie, M., 2002. Modelling the transfer and retention of nutrients in the drainage network of the Danube River. Estuar. Coast Shelf Sci. 54 (3), 285–308.

Gęgała, I., Izydorczyk, K., Jurczak, T., Pawełczyk, J., Dziadek, J., Wojtal-Frankiewicz, J., Jóźwik, A., Jaskulska, A., Mankiewicz-Boczek, J., 2014. Role of environmental factors and toxic genotypes in the regulation of microcystins-producing cyanobacterial blooms. Microbal Ecology" nr 67, 465–479.

Gerhardt, et al., 2009. Phytoremediation and rhizoremediation of organic soil contaminants: potential and challenges. Plant Sci. 176 (1), 20–30.

Gilbert, E.S., Crowley, D.E., 1997. Plant compounds that induce polychlorinated biphenyl biodegradation by Arthrobacter sp. strain B1B. App. Environ Microb. 63, 1933–1938.

Godlewska, M., Balk, H., Kaczkowski, Z., Jurczak, T., Izydorczyk, K., Długoszewski, B., Jaskulska, A., Gęgała-Borowska, I., Mankiewicz-Boczek, J., 2018. Night fish avoidance of Microcystis bloom revealed by simultaneous hydroacoustic measurements of both organisms. Fish. Res. 207, 74–84.

Godlewska, M., Świerzowski, A., Winfield, I.J., 2004. Hydroacoustics as a tool for studies of fish and their habitat. Ecohydrol. Hydrobiol. 4, 417–427.

Green, S.M., Machin, R., Cresser, M.S., 2008. Effect of long-term changes in soil chemistry induced by road salt applications on N-transformations in roadside soils. Environ. Pollut. 152, 20–31. https://doi.org/10.1016/j.envpol.2007.06.005.

Green, S.M., Machin, R., Cresser, M.S., 2009. Does road salting induce or ameliorate DOC mobilisation from roadside soils to surface waters in the long term? Environ. Monit. Assess. 153, 435–448. https://doi.org/10.1007/s10661-008-0369-4.

Greenwood, S.J., Rutter, A., Zeeb, B.A., 2011. The absorption and translocation of polychlorinated biphenyl congeners by Cucurbita pepo ssp pepo. Environ. Sci. Technol. 45 (15), 6511–6516. https://doi.org/10.1021/es200598u.

Gregory, K.J., Davis, R.J., Downs, P.W., 1985. Evaluation of Large Woody Debris in Watercourses. Eden Rivers Trust, pp. 1–24.

Gregory, K.J., Davis, R.J., 1992. Coarse woody debris in stream channels in relation to river channel management in woodland areas. Regul. Rivers Res. Manag. 7 (2), 117–136.

Grodziński, W., Klekowski, R.Z., Duncan, A., 1975. Methods for ecological bioenergetics. In: Environmental Science Series. Pergamon Press.

Guesdon, G., de Santiago-Martín, A., Galvez-Cloutier, R., 2016. Phytodesalinization potential of Typha angustifolia, Juncus maritimus, and Eleocharis palustris for removal of de-icing salts from runoff water. Environ. Sci. Pollut. Res. 23, 19634–19644. https://doi.org/10.1007/s11356-016-7176-1.

Guo, J., Peng, Y., Wang, S., Ma, B., Ge, S., Wang, Z., 2013. Pathways and organisms involved in ammonia oxidation and nitrous oxide emission. Crit. Rev. Environ. Sci. Technol. 43, 2213–2296. https://doi.org/10.1080/10643389.2012.672072.

Gupta, S.K., Gupta, R.C., Gupta, A.B., Seth, A.K., Bassin, J.K., Gupta, A., 2000. Recurrent acute respiratory tract infections in areas with high nitrate concentrations in drinking water. Environ. Health Perspect. 108, 363–366.

Haichar, F. el Z., 2008. Plant host habitant and root exudates shape soil bacterial community structure. ISME J. 2 (12), 1221–1230.

Hrbáček, J., Dvořáková, M., Kořínek, V., Procházková, L., 1961. Demonstration of the Effect of the Fish Stock on the Species Composition of Zooplankton and the Intensity of Metabolism of the Whole Plankton Association.

Harmon, M.E., Franklin, J.F., Swanson, F.J., Sollins, P., Gregory, S.V., Lattin, J.D., Anderson, N.H., Cline, S.P., Aumen, N.G., Sedell, J.R., Lienkaemper, G.W., Cromack Jr., K., Cummins, K.W., 1986. Ecology of coarse woody debris in temperate ecosystems. Adv. Ecol. Res. 15, 133–302.

Hehanusa, P.E., Machbub, B., Sunsanto, S.S., 1994. Kebutuhan Riset dan Koordinasi Pengelolaan Sumberdaya Air di Indonesia. (The Research and Coordination Needs of Water Resource Management in Indonesia). National Research Council, Group II Natural Resources and Energy, Jakarta.

Harper, D.M., Pacini, N., Upton, C., Morrison, E.H.J., Fox, R., Kiminta, E., 2013. Water Cooperation or Sustainable Utilization: Lake Naivasha, Kenya, Free Flow – Reaching Water Security through Cooperation. UNESCO, Tudor Rose, pp. 256–259.

Hobbs, R.J., Harris, J.A., 2001. Restoration ecology: repairing the earth's ecosystems in the new millennium. Restor. Ecol. 9 (2), 239–246.

Holger, D., Lebedeva, E.V., Pjevac, P., Han, P., Herbold, C., Albertsen, M., Jehmlich, N., Palatinszky, M., Vier-heilig, J., Bulaev, A., Kirkegaard, R.H., von Bergen, M., Rattei, T., Bendinger, B., Nielsen, P.H., Wagner, M., 2015. Complete nitrification by Nitrospira bacteria. Nature 528, 504–509. https://doi.org/10.1038/nature16461.

Holmes, D.E., Dang, Y., Smith, J.A., 2019. Chapter Four - nitrogen cycling during wastewater treatment. Adv. Appl. Microbiol. 106, 113–192. https://doi.org/10.1016/bs.aambs.2018.10.003.

Hu, C., Zhan, Y., Tang, X., Luo, W., 2014. PCB biodegradation and bphA1 gene expression induced by salicylic acid and biphenyl with Pseudomonas fluorescence P2W and Ralstonia eutropha H850. Pol. J. Environ. Stud. 23, 1591–1598.

Hu, H.W., He, J.Z., 2017. Comammox - a newly discovered nitrification process in the terrestrial nitrogen cy-cle. Frontiers in soils and sediments 17, 2709–2717. https://doi.org/10.1007/s11368-017-1851-9.

Hulster, A., Mueller, J.F., Marschner, H., 1994. Soil-plant transfer of polychlorinated dibenzo-p-dioxins and dibenzofurans to vegetables of the cucumber family (Cucurbitaceae). Environ. Sci. Technol. 28 (6), 1110–1115. https://doi.org/10.1021/es00055a021.

Jakopanec, I., Borgen, K., Vold, L., Lund, H., Forseth, T., Hannula, R., Nygård, K., 2008. A large waterborne outbreak of campylobacteriosis in Norway: the need to focus on distribution system safety. BMC Infect. Dis. 8, 128. https://doi.org/10.1186/1471-2334-8-128.

Jang, J., Hur, H.G., Sadowsky, M.J., Byappanahalli, M.N., Yan, T., Ishii, S., 2017. Environmental *Escherichia coli*: ecology and public health implications-a review. J. Appl. Microbiol. 123, 570–581.

Jones, D.K., Mattes, B.M., Hintz, W.D., Schuler, M.S., Stoler, A.B., Lind, L.A., Cooper, R.O., Relyea, R.A., 2017. Investigation of road salts and biotic stressors on freshwater wetland communities. Environ. Pollut. 221, 159–167. https://doi.org/10.1016/j.envpol.2016.11.060.

Kartal, B., Keltjens, J.T., 2016. Anammox biochemistry: a tale of heme c proteins. Trends Biochem. Sci. 41 (12), 998–1011. https://doi.org/10.1016/j.tibs.2016.08.015.

Kentzer, A., Dembowska, E., Gizinski, A., Napiorkowski, P., 2010. Influence of the Włocławek Reservoir on hydrochemistry and plankton of a large, lowland river (the Lower Vistula River, Poland). Ecol. Eng. 36, 1747–1753.

Kiedrzyńska, E., Jóźwik, A., Kiedrzyński, M., Zalewski, M., 2014a. Hierarchy of factors exerting an impact on the nutrient load of the Baltic Sea and sustainable management of its drainage basin. Mar. Pollut. Bullet. 88, 62–173.

Kiedrzyńska, E., Kiedrzyński, M., Urbaniak, M., Magnuszewski, A., Skłodowski, M., Wyrwicka, A., Zalewski, M., 2014b. Point sources of nutrient pollution in the lowland river catchment in the context of the Baltic Sea eutrophication. Ecol. Eng. 70, 379–387. https://doi.org/10.1016/j.ecoleng.2014.06.010.

Kiedrzyńska, E., Kiedrzyński, M., Zalewski, M., 2015a. Sustainable floodplain management for flood prevention and water quality improvement. Nat. Hazards 76 (2), 955–977.

Kiedrzyńska, E., Urbaniak, M., Kiedrzyński, M., Jóźwik, A., Bednarek, A., Gęgała, I., Zalewski, M., 2017. The use of a hybrid Sequential Biofiltration System for the improvement of nutrient removal and PCB control in municipal wastewater. Sci. Rep. 7, 5477.

Kiedrzyńska, E., Wagner, I., Zalewski, M., 2008a. Quantification of phosphorus retention efficiency by floodplain vegetation and a management strategy for a eutrophic reservoir restoration. Ecol. Eng. 33, 15–25.

Kiedrzyńska, E., Kiedrzyński, M., Zalewski, M., 2008b. Flood sediment deposition and phosphorus retention in a lowland river floodplain: impact on water quality of a reservoir, Sulejów, Poland. Ecohydrol. Hydrobiol. 8, 281–289. https://doi.org/10.2478/v10104-009-0022-z.

Kiedrzyńska, E., Kiedrzyński, M., Zalewski, M., 2015b. Sustainable floodplain management for flood prevention and water quality improvement. Nat. Hazards 76, 955–977. https://doi.org/10.1007/s11069-014-1529-1.

Kiedrzyńska, E., Kiedrzyński, M., Urbaniak, M., Magnuszewski, A., Skłodowski, M., Wyrwicka, A., Zalewski, M., 2014c. Point sources of nutrient pollution in the lowland river catchment in the context of the Baltic Sea eutrophication. Ecol. Eng. 70, 337–348. https://doi.org/10.1016/j.ecoleng.2014.06.010.

Kim, S.Y., Koretsky, C., 2013. Effects of road salt deicers on sediment biogeochemistry. Biogeochemistry 112, 343–358. https://doi.org/10.1007/s10533-012-9728-x.

Kleiber, M., 1975. The Fire of Life: An Introduction to Animal Energetics. R. E. Krieger Publishing Company, Huntington, NY.

Klimiuk, E., Łebkowska, M., 2005. Biotechnologia W Ochronie Środowiska. Wydawnictwo Naukowe PWN, Warszawa.

Konarzewski, M., 2015. Na początku był głód.

Korajkic, A., Brian, R., McMinn, B.R., Harwood, V.J., 2018. Relationships between microbial indicators and pathogens in recreational water settings. Int. J. Environ. Res. Publ. Health 15, 2842.

Korzeniowska, J., Mikulski, A., 2024. The effect of bed load sediments on self-purification of river – experimental approach. Ecohydrol. Hydrobiol. 24 (2), 321–326.

Lair, C.J., Zehetner, F., Fiebig, M., Gerzbek, M.H., van Gestel, C.A.M., Hein, T., et al., 2009. How do long-term development and periodical changes of river-floodplain systems affect the fate of contaminants? Results from European rivers. Environ. Pollut. 157, 3336–3346. https://doi.org/10.1016/j.envpol.2009.06.004.

Lamouroux, N., Capra, H., Pouilly, M., Souchon, Y., 1999. Fish habitat preferences in large streams of southern France. Freshw. Biol. 42 (4), 673–687. https://doi.org/10.1046/j.1365-2427.1999.00521.x.

Lancaster, N.A., Bushey, J.T., Tobias, C.R., Song, B., Vadas, T.M., 2016. Impact of chloride on denitrification potential in roadside wetlands. Environ. Pollut. 212, 216–223. https://doi.org/10.1016/j.envpol.2016.01.068.

Laughlin, R.J., Stevens, R.J., 2002. Evidence for fungal dominance of denitrification and codenitrification in a grassland soil. Soil Sci. Soc. Am. J. 66, 1540–1548. https://doi.org/10.2136/sssaj2002.1540.

Lee, B.D., Choi, Y.S., Kim, Y.G., Kim, I.S., Yang, E.I., 2017. A comparison study of performance and environmental impacts of chloride-based deicers and eco-label certified deicers in South Korea. Cold Reg. Sci. Technol. 143, 43–51. https://doi.org/10.1016/j.coldregions.2017.08.010.

Leigh, M.B., Fletcher, J.S., Fu, X., Schmitz, F.J., 2002. Root turnover: an important source of microbial substrates in rhizosphere remediation of recalcitrant contaminants. Environ. Sci. Technol. 36, 1579–1583.

Lakatos, I., 1978. In: Worrall, J., Currie, G. (Eds.), The Methodology of Scientific Research Programmes: Philosophical Papers, vol 1. Cambridge University Press.

Lembcke, D., Thompson, B., Read, K., Betts, A., Singaraja, D., 2017. Reducing road salt application by considering winter maintenance needs in parking lot design. J. Green Build. 12, 1–12. https://doi.org/10.3992/1943-4618.12.2.1.

Levantesi, C., Bonadonna, L., Briancesco, R., Grohmann, E., Toze, S., et al., 2012. Salmonella in surface and drinking water: occurrence and water-mediated transmission. Food Res. Int. 45, 587–602.

Likens, G.E., 2021. The watershed-ecosystem approach. Hydrol. Process. 35 (6), e13977.

Lim, C.-H., 2021. Water-centric nexus approach for the agriculture and forest sectors in response to climate change in the Korean Peninsula. Agronomy 11 (8), 1657. https://doi.org/10.3390/agronomy11081657.

Li, Q., Yu, M., Lu, G., Cai, T., Bai, X., Xia, Z., 2011. Impacts of the gezhouba and three gorges reservoirs on the sediment regime in the Yangtze River, China. J. Hydrol. 403, 224–233.

Loeb, R., Lamers, L., 2003. The effects of river water quality on the development of wet floodplain vegetations in the Netherlands. In: International Conference "Towards Natural Flood Reduction Strategies". September 2003, Warsaw, Poland.

Machado, F., et al., 2018. The importance of plants to development and maintenance of soil structure, microbial communities and ecosystem functions. Soil Tillage Res. 175, 139–149.

Mamindy-Pajany, Y., Geret, F., Roméo, M., Hurel, C., Marmier, N., 2012. Ex situ remediation of contaminated sediments using mineral additives: assessment of pollutant bioavailability with the Microtox solid phase test. Chemosphere 86 (11), 1112–1116. https://doi.org/10.1016/j.chemosphere.2011.12.001.

Manassaram, D.M., Backer, L.C., Moll, D.M., 2005. A review of nitrates in drinking water: maternal exposure and adverse reproductive and developmental outcomes. Environ. Health Perspect. 114, 320–327.

Mankiewicz-Boczek, J., Bednarek, A., Gęgała-Borowska, I., Serwecińska, L., Zaborowski, A., Kolate, E., Pawełczyk, J., Zaczek, A., Dziadek, J., Zalewski, M., 2017. The removal of nitrogen compounds from farming wastewater- the effect of different carbon substrates and different carbon activators. Ecol. Engine. 105, 341–354.

Mankiewicz-Boczek, J., Jaskulska, A., Gęgała, I., Pawełczyk, J., Serwecińska, L., Dziadek, J., 2016. Cyanophages infection of Microcystis bloom in lowland dam reservoir of Sulejów, Poland. Microb. Ecol. 71 (2), 315–325.

Master, E.R., Mohn, W.W., 2001. Induction of bphA, encoding biphenyl dioxygenase, in two polychlorinated biphenyl-degrading bacteria, psychotolerant Pseudomonas strain Cam-1 and mesophilic Burkholderia strain LB400. Appl. Environ. Microbiol. 67, 2669–2676.

Matsuo, S., Yamazaki, K., Gion, K., Eun, H., Inui, H., 2011. Structure-selective accumulation of polychlorinated biphenyls in Cucurbita pepo. J. Pestic. Sci. 36 (3), 363–369. https://doi.org/10.1584/jpestics.G11-03.

Mayer, T., Rochfort, Q., Marsalek, J., Parrott, J., Servos, M., Baker, M., Mcinnis, R., Jurkovic, A., Scott, I., 2011. Environmental characterization of surface runoff from three highway sites in Southern Ontario, Canada: 2. Toxicology. Water Qual. Res. J. Can. 46, 121–136. https://doi.org/10.2166/wqrjc.2011.036.

Mierzejewska, E., Baran, A., Tankiewicz, M., Urbaniak, M., 2019a. Removal and ecotoxicity of 2,4-D and MCPA in microbial cultures enriched with structurally-similar plant secondary metabolites. Water 11 (7), 1451.

Mierzejewska, E., Baran, A., Urbaniak, M., 2009b. Biodegradation potential and ecotoxicity assessment in soil extracts amended with phenoxy acid herbicide (2,4-D) and a structurally-similar plant secondary metabolite (ferulic acid). Bull. Environ. Contam. Toxicol. 104 (2), 200–205. http://link.springer.com/10.1007/s00128-019-02760-5.

Mierzejewska, E., Urbaniak, M., 2020. Molecular methods as potential tools in ecohydrological studies on emerging contaminants in freshwater ecosystems. Water 12, 2962. https://doi.org/10.3390/w12112962.

Moatar F., Descy J.P., Rodrigues S., Souchon Y., Leitao M., Floury M., Wantzen K.M., The Loire Basin, w: the Rivers of Europe, Red. K. Tockner, C. Robinson, Elsevier, London.

Musilova, L., Ridl, J., Polivkova, M., Macek, T., Uhlik, O., 2016. Effects of secondary plant metabolites on microbial populations: changes in community structure and metabolic activity in contaminated environments. Int. J. Mol. Sci. 29 (8), 1–31.

Naisbitt, J., 1982. In: Megatrends: Ten New Directions Transforming Our Lives. Warner Books.

Negussie, Y.Z., Urbaniak, M., Szklarek, S., Lont, K., Gęgała, I., Zalewski, M., 2012. Efficiency analysis of two sequential biofiltration systems in Poland and Ethiopia – the pilot study. Ecohydrol. Hydrobiol. 12 (4), 271–285.

Nutile, S.A., Solan, M.E., 2019. Toxicity testing of "eco-friendly" de-icing formulations using Chironomus dilutus. Environ. Pollut. 246, 408–413. https://doi.org/10.1016/j.envpol.2018.12.033.

Offre, P., Spang, A., Schleper, C., 2013. Archaea in biogeochemical cycles. Annu. Rev. Microbiol. 67, 437–457.

Okruszko, T., Zalewski, M., 2018. Advancement in ecohydrology, biotechnology and engineering for sustainable development goals – EcoSummit2016. Ecohydrol. Hydrobiol. 18, 245–246.

Ostrom, E., 2000. Reformulating the commons. Swiss Polit. Sci. Rev. 6 (1), 29–52.

Pacini, N., Hesslerová, P., Pokorný, J., Mwinami, T., Morrison, E.H.J., Cook, A.A., Zhang, S., Harper, D.M., 2018. Papyrus as an ecohydrological tool for restoring ecosystem services in Afro- -tropical wetlands. Ecohydrol. Hydrobiol. 18 (2), 142–154.

Papaspyrou, S., Smith, C.J., Dong, L.F., Whitby, C., Dumbrell, A.J., Nedwell, D.B., 2014. Nitrate reduction functional genes and nitrate reduction potentials persist in deeper estuarine sediments. Why? PLoS One 9 (4), e94111. https://doi.org/10.1371/journal.pone.0094111.

Palus, J.D., 2015. Effects of N:P Ratio on the Occurrence of Harmful Algal Blooms (Honors Thesis). The Ohio State University.

Paller, M.H., 2005. The influence of biomanipulation on fish community development in a southeastern United States cooling reservoir. Hydrobiologia 539, 69–81.

Penczak, T., Zalewski, M., 1974. Distribution of fish numbers and biomass in the barbel region of the river and the adjoining old riverbeds. Ekol. Pol. 22, 107–119.

Philippot, L., Piutti, S., Martin-Laurent, F., Hallet, S., Germon, K.C., 2002. Molecular analysis of the nitrate-reducing community from unplanted and maize-planted soils. Appl. Environ. Microbiol. 68 (12), 6121–6128. https://doi.org/10.1128/AEM.68.12.6121-6128.2002.

Piégay, H., Gurnell, A.M., 1997. Large woody debris and river geomorphological pattern: examples from S.E. France and S. England. Geomorphology 19 (1–2), 99–116.

Piña-Ochoa, E., Høgslunda, S., Geslinb, E., Cedhagend, T., Revsbeche, N.P., Nielsene, L.P., Schweizerf, M., Jorissenb, F., Rysgaardg, S., Risgaard-Petersen, N., 2010. Widespread occurrence of nitrate storage and denitrification among Foraminifera and Gromiida. Proc. Natl. Acad. Sci. USA 107 (3), 1148–1153.

Prosser, J.I., 2005. Nitrogen in soils, nitrification. In: Encyclopedia of Soils in the Environment, pp. 31–39. https://doi.org/10.1016/B0-12-348530-4/00512-9.

Ran, L., Lu, X.X., Xin, Z., Yang, X., 2013. Cumulative sediment trapping by reservoirs in large river basins: a case study of the Yellow River basin. Global Planet. Change 100, 308–319.

Riehm, M., Nordin, L., 2012. Optimization of winter road maintenance energy costs in Sweden: a critique of site specific frost warning techniques. Meteorol. Appl. 19, 443–453. https://doi.org/10.1002/met.275.

Risgaard-Petersen, N., Langezaal, A., Ingvardsen, S., et al., 2006. Evidence for complete denitrification in a benthic foraminifer. Nature 443, 93–96. https://doi.org/10.1038/nature05070.

Rivett, M.O., Cuthbert, M.O., Gamble, R., Connon, L.E., Pearson, A., Shepley, M.G., Davis, J., 2016. Highway deicing salt dynamic runoff to surface water and subsequent infiltration to groundwater during severe UK winters. Sci. Total Environ. 565, 324–338. https://doi.org/10.1016/j.scitotenv.2016.04.095.

Rizzo, L., Manaia, C., Merlin, C., Schwartz, T., Dagot, C., Ploy, M.C., et al., 2013. Urban wastewater treatment plants as hotspots for antibiotic resistant bacteria and genes spread into the environment: a review. Sci. Total Environ. 447, 345–360. https://doi.org/10.1016/j.scitotenv.2013.01.032.

Robinson, H.K., Hasenmueller, E.A., Chambers, L.G., 2017. Soil as a reservoir for road salt retention leading to its gradual release to groundwater. Appl. Geochem. 83, 72–85. https://doi.org/10.1016/j.apgeochem.2017.01.018.

Roy, J.W., McInnis, R., Bickerton, G., Gillis, P.L., 2015. Assessing potential toxicity of chloride-affected groundwater discharging to an urban stream using juvenile freshwater mussels (Lampsilis siliquoidea). Sci. Total Environ. 532, 309–315. https://doi.org/10.1016/j.scitotenv.2015.06.023.

Ryall, B., Eydallin, G., Ferenci, T., 2012. Culture history and population heterogeneity as determinants of bacterial adaptation: the adaptomics of a single environmental transition. Microbiol. Mol. Biol. Rev. 76, 597–625.

Schiemer, F., Waidbacher, H., 1992. Strategies for conservation of a Danubian fish fauna. In: Boon, P.J., Calow, P., Petts, G.E. (Eds.), River Conservation and Management. John Wiley & Sons Ltd., pp. 363–382

Schuler, M.S., Relyea, R.A., 2018. A review of the combined threats of road salts and heavy metals to freshwater systems. Bioscience 68, 327–335. https://doi.org/10.1093/biosci/biy018.

SedNet, 2004. Contaminated Sediments in European River Basins. https://sednet.org/wpcontent/uploads/2016/03/Sednet_booklet_final_2.pdf.

Serwecińska, L., Bednarek, A., Gęgała-Borowska, I., Szklarek, S., Kupiec, J.M., Żeleźnik, W., Draczyński, Z., Krucińska, I., Mankiewicz-Boczek, J., 2017. Innovative method of improving water quality on the example of Gniezno stream and lake Jelonek. In: International Symposium "Ecohydrology for the Circular Economy, and Nature-Based Solutions towards Mitigation/Adaptation to Climate Change.

Serwecińska, L., Kiedrzyńska, E., Kiedrzyński, M., 2021. A cachment-scale assessment of the saniutary conditions of treated wastewater and river water based on fecal indicator bacteria and carbapenem-resistant Acinetobacter spp. Sci. Tot. Environ. 750, 142266. https://doi.org/10.1016/j.scitotenv.2020.142266.

Shackelford, N., McDougall, C., 2023. Ecosystem restoration, regeneration and rewilding. BMC Ecol. Evol. 23. https://doi.org/10.1186/s12862-023-02165-3.

Shelford, V.E., Eddy, S., 1929. Methods for the study of stream communities. Ecology 10 (4), 382–391. https://doi.org/10.2307/1931146.

Siciliano, S.D., Germida, J.J., Banks, K., Greer, C.W., 2003. Changes in microbial community composition and function during a polyaromatic hydrocarbon phytoremediation field trial. Appl. Environ. Microbiol. 69, 483–489.

Sidhu, J.P.S., Ahmed, W., Gernjak, W., Aryal, R., McCarthy, D., Palmer, A., Kolotelo, P., Toze, S., 2013. Sewage pollution in urban stormwater runoff as evident from the widespread presence of multiple microbial and chemical source tracking markers. Sci. Total Environ. 463, 488–496.

Siebielec, S., Siebielec, G., Urbaniak, M., Smreczak, B., Grządа, E., Wyrwicka, A., Kidd, P.S., 2019. Impact of rhizobacterial inoculants on plant growth and enzyme activities in soil treated with contaminated bottom sediments. Int. J. Phytoremediation 21 (4), 325–333. https://doi.org/10.1080/15226514.2018.1524833.

Simmonds, J., MacLennan, D., 2005. Fisheries Acoustics: Theory and Practice. Blackwell Science, Oxford.

Singer, A.C., Crowley, D.E., Thompson, I.P., 2003. Secondary plant metabolites in phytoremediation and biotransformation. Trends Biotechnol. 21, 123–130.

Skłodowski, M., Kiedrzyńska, E., Kiedrzyński, M., Urbaniak, M., Zielińska, K.M., Kurowski, J.K., Zalewski, M., 2014. The role of riparian willow communities in phosphorus accumulation and dioxin control for water quality improvement in a lowland river. Ecol. Eng. 70, 1–10.

Skłodowski, M., Kiedrzyńska, E., Kiedrzyński, M., Urbaniak, M., Zielińska, K.M., Kurowski, J.K., Zalewski, M., 2014b. The role of riparian willows in phosphorus accumulation and PCB control for lotic water quality improvement. Ecol. Eng. 70, 1–10. https://doi.org/10.1016/j.ecoleng.2014.03.088.

Slater, H., Gouin, T., Leigh, M.B., 2011. Assessing the potential for rhizoremediation of PCB contaminated soils in northern regions using native tree species. Chemosphere 84, 199–206. https://doi.org/10.1016/j.chemosphere.2011.04.058.

Snodgrass, J.W., Moore, J., Lev, S.M., Casey, R.E., Ownby, D.R., Flora, R.F., Izzo, G., 2017. Influence of modern stormwater management practices on transport of road salt to surface waters. Environ. Sci. Technol. 51, 4165–4172. https://doi.org/10.1021/acs.est.6b03107.

Søberg, L.C., Viklander, M., Blecken, G.-T., 2017. Do salt and low temperature impair metal treatment in stormwater bioretention cells with or without a submerged zone? Sci. Total Environ. 579, 1588–1599. https://doi.org/10.1016/j.scitotenv.2016.11.179.

Steele, J., Griffith, J., Noble, R., Schif, K., 2017. Tracking Human Fecal Sources in an Urban.

Stoler, A., Sudol, K., Mruzek, J., Relyea, R., 2018. Interactive effects of road salt and sediment disturbance on the productivity of seven common aquatic macrophytes. Freshw. Biol. 63, 709–720. https://doi.org/10.1111/fwb.13110.

Suaire, R., Durickovic, I., Framont-Terrasse, L., Leblain, J.Y., De Rouck, A.C., Simonnot, M.O., 2016. Phytoextraction of Na+ and Cl- by Atriplex halimus L. and Atriplex hortensis L.: a promising solution for remediation of road runoff contaminated with deicing salts. Ecol. Eng. 94, 182–189. https://doi.org/10.1016/j.ecoleng.2016.05.055.

Sutapa, I.D.A., 2013. Managing Water: From Local Wisdom to Modern Science, Free Flow, Reaching Water Security through Cooperation. UNESCO Publishing, Tudor Rose, pp. 188–191.

Sutapa, I.D.A., 2018. Ecohydrology as New Way to Manage Water Related Problems in Order to Improve Ecosystem Services, UNESCO Special Session in World Water Forum. Brasilia, Brazil, 18–23 March 2018.

Sutapa, I.D.A., Apip, 2017. Ecohydrology demonstration site to support sustainable water resources management. In: The 2nd Symposium on JASTIP Disaster Prevention International Cooperation Research in Kyoto, Japan, 23 March 2017.

Sutapa, I.D.A., Maftuah, E., Sunaryani, A., Pawitan, H., 2017. Comprehensive ecohydrology study to support agriculture and water resources management in Peatland Area – central kalimantan. J. Teknol. 40.

Szklarek, S., Stolarska, M., Wagner, I., Mankiewicz-Boczek, J., 2015. The microbiotest battery as an important component in the assessment of snowmelt toxicity in urban watercourses—preliminary studies. Environ. Monit. Assess. 187, 1–11. https://doi.org/10.1007/s10661-014-4252-1.

Szklarek, S., Wagner, I., Jurczak, T., Zalewski, M., 2018. Sequential Sedimentation-Biofiltration System for the purification of a small urban river (the Sokolowka, Lodz) supplied by stormwater. J. Environ. Manag. 205, 201–208. https://doi.org/10.1016/j.jenvman.2017.09.066.

Tadesse, H., Lakew, A., Getahun, A., 2025. Community structure and habitat preferences of stream fish in major tributaries of the upper Awash River, Ethiopia. J. Ecohydraulics 10 (2), 145–160.

Tarnawski, M., Baran, A., Koniarz, T., 2015. The effect of bottom sediment supplement on changes of soil properties and on the chemical composition of plants. Geol. Geophys. Environ. 41 (3), 285. https://doi.org/10.7494/geol.2015.41.3.285.

Terzaghi, E., Zanardini, E., Morosini, C., Raspa, G., Borin, S., Mapelli, F., Vergani, L., Di Guardo, A., 2018. Rhizoremediation half-lives of PCBs: role of congener composition, organic carbon forms, bioavailability, microbial activity, plant species and soil conditions, on the prediction of fate and persistence in soil. Sci. Total Environ. 612, 544–560. https://doi.org/10.1016/j.scitotenv.2017.08.189.

The Moorish Invention that Tamed Spain's Mountains. BBC.

Tiedje, J.M., 1988. Ecology of denitrification and dissimilatory nitrate reduction toammonium. In: Zehnder, A.J.B. (Ed.), Biology of Anaerobic Microorganisms. John Wiley & Sons, Inc., New York, pp. 179–244.

Tixier, G., Rochfort, Q., Grapentine, L., Marsalek, J., Lafont, M., 2012. Spatial and seasonal toxicity in a stormwater management facility: evidence obtained by adapting an integrated sediment quality assessment approach. Water Res. 46, 6671–6682. https://doi.org/10.1016/j.watres.2011.12.031.

Tonk, L., Bosch, K., Visser, P.M., Huisman, J., 2007. Salt tolerance of the harmful cyanobacterium Microcystis aeruginosa. Aquat. Microb. Ecol. 46, 117–123. https://doi.org/10.3354/ame046117.

The Wildlife Trusts and FreshwaterLife. (n.d.). Managing woody debris in rivers, streams & floodplains. FreshwaterLife. Retrieved from https://www.yumpu.com/en/document/view/20993575/managing-woody-debris-freshwaterlife.

Uhlenbrook, S., Roser, S., Tilch, N., 2004. Hydrological process representation at the meso-scale: the potential of a distributed, conceptual catchment model. J. Hydrol. 291 (3–4), 278–296.

Uhlik, O., Musilova, L., Ridl, J., Hroudova, M., Vlcek, C., Koubek, J., Heleckova, M., Mackova, M., Macek, M., 2013. Plant secondary metabolite-induced shifts in bacterial community structure and degradative ability in contaminated soil. Appl. Microbial. Biotechnol. 97, 9245–9256.

Urbaniak, M., Mierzejewska, E., Tankiewicz, M., 2019a. The stimulating role of syringic acid, a plant secondary metabolite, in the microbial degradation of structurally-related herbicide, MCPA. PeerJ 7, e6745.

Urbaniak, M., Wyrwicka, A., Zieliński, M., Mankiewicz-Boczek, J., 2016. Potential for phytoremediation of PCDD/PCDF-contaminated sludge and sediments using Cucurbitaceae plants: a pilot study. Bull. Environ. Contam. Toxicol. https://doi.org/10.1007/s00128-016-1868-6.

Urbaniak, M., 2007. Polychlorinated Biphenyls (PCBs): Sources, Distribution.

Urbaniak, M., 2013. Biodegradation of PCDDs/PCDFs and dl- PCBs. In: Chamy, R., Rosenkran, F. (Eds.), Biodegradation - Engineering and Technology. INTECH Publisher, pp. 73–100.

Urbaniak, M., Kiedrzyńska, E., Zalewski, M., 2012a. The role of a lowland reservoir in the transport of micropollutants, nutrients and the suspended particulate matter along the river continuum. Hydrol. Res. 43 (4), 400–411.

Urbaniak, M., Kiedrzyńska, E., Kiedrzyński, M., Mendra, M., Grochowalski, A., 2014. The impact of point sources of pollution on the transport of micropollutants along the river continuum. Hydrol. Res. 45 (3), 391–410.

Urbaniak, M., Kiedrzyńska, E., Wyrwicka, A., Zieliński, M., Mierzejewska, E., Kiedrzyński, M., Kannan, K., Zalewski, M., 2019b. An ecohydrological approach to the river contamination by PCDDs, PCDFs and dl-PCBs – concentrations, distribution and removal using phytoremediation techniques. Sci. Rep. 9, 19310. https://doi.org/10.1038/s41598-019-55973-3.

Urbaniak, M., Kiedrzyńska, E., Zalewski, M., 2012b. The role of a lowland reservoir in the transport of micropollutants, nutrients and the suspended particulate matter along the river continuum. Hydrol. Res. 43 (4), 400–411.

US Committee of Scientists, 1999. Sustaining the People's Lands: Recommendations for Stewardship of the National Forests and Grasslands into the Next Century. U.S. Department of Agriculture, Washington DC.

US EPA, 1986. Ambient Water Quality Criteria for Bacteria. USEPA, U.S.A, Washington, DC.

US EPA, 1988. Ambient aquatic life water quality criteria for chloride. United States Environ. Prot. Agency, 1–24.

US EPA, 2004. U.S. Environmental Protection Agency - Final Rule, Water Quality Standards.

US EPA, 2005. Microbial Source Tracking Guide Document.
van Peborgh, E., August 18, 2023. Reimagining our future: a regenerative approach to economy and ecology. Medium. https://ernesto-87727.medium.com/reimagining-our-future-a-regenerative-approach-to-economy-and-ecology-cfe70ceb9298.
Wahl, D.C., May 18, 2016. Sustainability is not enough: we need regenerative cultures. Medium. https://designforsustainability.medium.com/sustainability-is-not-enough-we-need-regenerative-cultures-4abb3c78e68b.
Wang, L., Kwok, J.S., Tsang, D.C., Poon, C.S., 2015. Mixture design and treatment methods for recycling contaminated sediment. J. Hazard. Mater. 283, 623–632. https://doi.org/10.1016/j.jhazmat.2014.09.056.
Wang, S., Lin, X., Yu, H., Wang, Z., Xia, H., An, J., Fan, G., 2017. Nitrogen removal from urban stormwater runoff by stepped bioretention systems. Ecol. Eng. 340–348. https://doi.org/10.1016/j.ecoleng.2017.05.055.
Wanjala, S.W., Mwinami, T., Harper, D.M., Morrison, E.H.J., Pacini, N., 2018. Ecohydrological tools for the preservation and enhancement of ecosystem services in the Naivasha Basin, Kenya. Ecohydrol. Hydrobiol. 18 (2), 155–173.
Wantzen, K.M., Blettler, M.C.M., Marchese, M.R., Amsler, M.L., Bacchi, M., Ezcurra de Drago, I.D., Drago, E., 2014. Sandy rivers: a review on general ecohydrological patterns of benthic invertebrate assemblages across continents. Int. J. River Basin Manag. 12, 163–174.
Watershed during Wet Weather. Southern California Coastal Water Research Project, Technical Report 1002.
White, J.C., 2010. Inheritace of p, p'-DDE phytoetraction ability in hybridized Cucurbita pep cultivars. Environ. Sci. Technol. 44 (33), 5165–5169.
White, J.C., 2002. Differential bioavaibility of field-weathered p,p0-DDE to plants of the Cucurbita and Cucumis genera. Chemosphere 49 (2), 143–152. https://doi.org/10.1016/S0045-6535(02)00277-1.
Wintenberger, C.L., Rodrigues, S., Bréhéret, J.-G., Villar, M., 2015. Fluvial islands: first stage of development from forced bars and vegetation interactions. Geomorphology 246, 305–320.
Wyrwicka, A., Urbaniak, M., 2016. The different physiological and antioxidative responses of zucchini and cucumber to sewage sludge application. PLoS One. https://doi.org/10.1371/journal.pone.0157782.
Wyrwicka, A., Steffani, S., Urbaniak, M., 2014. The effect of PCB-contaminated sewage sludge and sediment on metabolism of cucumber plants (Cucumis sativus L.). Ecohydrol. Hydrobiol. 14, 75–82.
Zalewski, M., 1982. An attempt to adopt an electrofishing method for sampling rivers with low conductivity. In: Annual Report for Matamak Research Program, 29. Woods Hole Oceanographic Institution, pp. 207–209.
Zalewski, M., 1986. Regulacja zespołów ryb w rzekach przez kontinuum czynników abiotycznych i biotycznych. Acta Universitatis Lodziensis 31, 31–68.
Zalewski, M., 1995. Freshwater habitat management and restoration in the face of global change: keynote lecture, Theme 1 – condition of the world's aquatic habitats. In: Voigtlander, C.W. (Ed.), Proceedings of the World Fisheries Congress. Oxford and IBH Publishing, New Delhi.
Zalewski, M., 1997. Fish diversity and ecotonal habitat. In: Lachavanne, J.-B., Juge, R. (Eds.), Biodiversity in Land-Inland Water Ecotones. UNESCO; The Parthenon Publishing Group, Paris, pp. 183–203.
Zalewski, M., 1999. Minimising the risk and amplifying the opportunities for restoration of shallow reservoirs. Hydrobiologia 395–396, 107–114.
Zalewski, M., 2001a. Ecohydrology: the scientific background to use ecosystem properties as management tools toward sustainability of water resources. Ecol. Eng. 16 (1), 1–8.
Zalewski, M., 2000a. Ecohydrology – the scientific background to use ecosystem properties as management tools toward sustainability of water resources. Ecol. Eng. 16, 1–8.
Zalewski, M., 2000b. Strategia wyprzedzania jako maksymalizacja szans rozwoju nauk ekologicznych w Polsce w warunkach postąpującej globalizacji. Kosmos 49 (1–2), 263–274.
Zalewski, M., 2001b. Ecohydrology—integrative science for sustainable water, environment and society. Ecohydrol. Hydrobiol. 2 (1), 3–10.
Zalewski, M., 2002a. Ecohydrology – the use of ecological and hydrological processes for sustainable management of water resources. Hydrol. Sci. J. 47 (5), 825–834.
Zalewski, M. (Ed.), 2002b. *Guidelines for the Integrated Management of the Watershed: Phytotechnology And Ecohydrology* (Freshwater Management Series No, vol 5. United Nations Environment Programme, International Environmental Technology Centre, p. 188.
Zalewski, M., 2005. Engineering harmony. Academia 1 (5), 4–7.
Zalewski, M., 2006a. Ecohydrology – an interdisciplinary tool for integrated protection and management of water bodies. Arch. Hydrobiol. 158 (Suppl. 4), 613–622.
Zalewski, M., 2006b. Flood pulses and river ecosystem robustness, in: Frontiers in Flood Research, red. In: Tchiguirinskaia, I., Thein, K.N.N., Hubert, P. (Eds.), Kovacs Colloquium, 305. IAHS Publication, p. 212.
Zalewski, M. (Ed.), 2008a. *Perspektywy zrównoważonego rozwoju regionu łódzkiego: Szanse i zagrożenia. LORIS Wizja – Regionalny Foresight Technologiczny*. Społeczna Wyższa Szkoła Przedsiąbiorczości i Zarzędzania w Łodzi.
Zalewski, M., 2008b. Rationale for the "Floodplain Declaration": from environmental conservation toward sustainability science. Ecohydrol. Hydrobiol. 8 (2–4), 107–113.
Zalewski, M., 2009. Ecohydrology for engineering harmony between environment and society. Danube News 19 (May), 11.
Zalewski, M., 2011a. Ecohydrology for implementation of the EU water framework directive. Water Manag. 164 (8), 375–385.
Zalewski, M., 2011b. Towards engineering harmony between water, ecosystem and society: editorial. Ecohydrol. Hydrobiol. 11 (3–4), 137–140.
Zalewski, M., 2012a. Ecohydrology – process oriented thinking for sustainability of river basins. Ecohydrol. Hydrobiol. 12 (2), 89–92.
Zalewski, M., 2012b. Blue-green city for compensating global climate change. In: The Parliament Magazine, 350, pp. 2–3.
Zalewski, M., 2013. Ecohydrology: process-oriented thinking towards sustainable river basins. Ecohydrol. Hydrobiol. 13 (2), 97–103.

Zalewski, M., 2014a. Ecohydrology and hydrologic engineering: regulation of hydrology-biota interactions for sustainability. J. Hydrol. Eng. 20 (1).
Zalewski, M., 2014b. Ecohydrology, biotechnology and engineering for cost efficiency in reaching the sustainability of biogeosphere. Ecohydrol. Hydrobiol. 14, 14–20.
Zalewski, M., 2017. Ecohydrology measures for sustainable catchment management: an outline. In: Radecki-Pawlik, A., Pagliara, S., Hradecky, J. (Eds.), Open Channel Hydraulics, River Hydraulic Structures and Fluvial Geomorphology: For Engineers, Geomorphologists and Physical Geographers. CRC Press.
Zalewski, M., Arduino, G., Bidoglio, G., Junk, W., Cullman, J., Uhlenbrook, S., Xia, J., Garcia de Leaniz, C., Rowiński, P.M., Vörösmarty, C., Chicharo, L., 2018a. Ecohydrol. Hydrobiol. 18, 309–310.
Zalewski, M., Bis, B., 1998. Ekologiczne podstawy ochrony i kształtowania środowiska [Ecological concepts for environmental protection and management]. In: Olaczek, R., Warcholińska, A. (Eds.), Ochrona I Kształtowanie Środowiska Przyrodniczego [Protection and Management of the Natural Environment]. Wydawnictwo Uniwersytetu Łódzkiego.
Zalewski, M., Bis, B., Łapińska, M., Frankiewicz, P., Puchalski, W., 1998. The importance of the riparian ecotone and river hydraulics for sustainable basin-scale restoration scenarios. Aquat. Conserv. Mar. Freshw. Ecosyst. 8 (2), 287–301.
Zalewski, M., Brewińska-Zaraś, B., Frankiewicz, P., 1990a. Fry communities as a biomanipulating tool in a temperate lowland reservoir. Archiv für Hydrobiologie – Beiheft: Ergeb. Limnol. 33, 763–774.
Zalewski, M., Brewińska-Zaraś, B., Frankiewicz, P., Kalinowski, S., 1990b. The potential for biomanipulation using fry communities in a lowland reservoir: concordance between water quality and optimal recruitment. Hydrobiologia 200/201, 549–556.
Zalewski, M., Frankiewicz, P., Brewinska-Zaras, B., 1986. The production of brown trout (*Salmo trutta* L.) introduced to streams of various orders in an upland watershed. Pol. Arch. Hydrobiol. 33 (3–4), 411–422.
Zalewski, M., Frankiewicz, P., Brewińska, B., 1985. The factors limiting growth and survival of brown trout, *Salmo trutta* m. fario L., introduced to different types of streams. J. Fish. Biol. 27 (A), 59–73.
Zalewski, M., Frankiewicz, P., Przybylski, M., Bańbura, J., Nowak, M., 1990c. Structure and dynamics of fish communities in temperate rivers in relation to the abiotic–biotic regulatory continuum concept. Pol. Arch. Hydrobiol. 37 (1–2), 151–176.
Zalewski, M., Harper, D., 2001. Rationale: ecohydrology – the use of ecosystem properties as enhancement of absorbing capacity of ecosystem against human impact (UNESCO IHP-V). Ecohydrol. Hydrobiol. 1 (1–2).
Zalewski, M., Harper, D.M., Demars, B., Jolankai, G., Crosa, G., Janauer, G.A., Pacini, N., 2008. Linking biological and physical processes at the river basin scale: the origins, scientific background and scope of ecohydrology. In: Harper, D.M., Zalewski, M., Pacini, N. (Eds.), Ecohydrology: Processes, Models and Case Studies. CAB International, pp. 1–17.
Zalewski, M., Janauer, G.A., Jolánkai, G., 1997a. Ecohydrology: A New Paradigm for the Sustainable Use of Aquatic Resources. Conceptual Background, Working Hypothesis, Rationale and Scientific Guidelines for the Implementation of the IHP-V Projects 2.3/2.4. UNESCO.
Zalewski, M., Janauer, G.A., Jolankai, G., 1997b. Ecohydrology: a new paradigm for the sustainable use of aquatic resources. In: UNESCO IHP Technical Documents in Hydrology, vol 7. UNESCO, Paris.
Zalewski, M., Jarosiewicz, P., Kiedrzyńska, E., Czerniawski, R., Izydorczyk, K., Mankiewicz- -Boczek, J., Kaczkowski, Z., Sługocki, Ł., Jurczak, T., Bednarek, A., Godlewska, M., Parasiewicz, P., 2019. Environmental Guidelines for the Application of Ecohydrology Solutions for Planning and Designing Investments Aimed at Improving the Navigability of the River Odra River.
Zalewski, M., Kiedrzyńska, E., 2010. System approach to sustainable management of inland floodplains – declaration on sustainable floodplain management. CAB Rev. 5 (056), 1–8.
Zalewski, M., Łapińska, M., Bayley, P., 2003a. Fish relationships with wood in large rivers. In: Proceedings of the American Fisheries Society Symposium.
Zalewski, M., Naiman, R.J., 1985. The regulation of riverine fish communities by a continuum of abiotic-biotic factors, [w:] Alabaster J.S. (red.), Habitat modification and freshwater fisheries. Butterworths Scientific. Londyn. Wielka Brytania.
Zalewski, M., Penczak, T., 1981. Characterization of the fish community of the Utarta River drainage basin and evaluation of the efficiency of catching methods. Pol. Arch. Hydrobiol. 28, 385–396.
Zalewski, M., Puchalski, W., Frankiewicz, P., Bis, B., 1994. Riparian ecotones and fish communities in rivers – intermediate complexity hypothesis. In: Cowx, I.G. (Ed.), Rehabilitation of Freshwater Fisheries. Fishing News Books.
Zalewski, M., Robarts, R., 2003. Ecohydrology – a new paradigm for integrated water resource management. SIL News 40, 1–5.
Zalewski, M., Santiago-Fandino, V., Neate, J., 2003b. Energy, water, plant interactions: "Green feedback" as a mechanism for environmental management and control through the application of phytotechnology and ecohydrology. Hydrol. Process. 17 (14), 2753–2767.
Zalewski, M., Thorpe, J.E., Gaudin, P. (Eds.), 1991. Fish and Land-Inland Water Ecotones. UNESCO-MAB, University of Łódź. Stirling–Lyon.
Zalewski, M., Thorpe, J.E., Naiman, R.J., 2001a. Fish and riparian ecotones – a hypothesis. Ecohydrol. Hydrobiol. 1 (1–2), 11–24.
Zalewski, M., Thorpe, J.E., Scheimer, F., 2001b. Guest editorial. Ecohydrol. Hydrobiol. 1, 5–7.
Zalewski, M., Urbaniak, M., Zerihun Negussie, Y., 2010. Ecohydrological systematic solutions for reduction of siltation, eutrophication and dioxin-induced toxicity. The pilot study of the Asella BioFarm Park Lake, Ethiopia. Ecohydrol. Hydrobiol. 10 (2–4), 363–368.
Zalewski, M., Wagner, I., 1998. Temperature and nutrient dynamics in eutrophic freshwater ecosystems. Geogr. Pol. 71, 79–92.
Zalewski, M., Wagner, I., 2004. Integrated Watershed Management: Ecohydrology and Phytotechnology – Manual. UNESCO/UNEP Publication.
Zalewski, M., Wagner, I., 2005. Ecohydrology – the use of water and ecosystem processes for healthy urban environments. In *Special issue: Aquatic habitats in integrated urban water management*. Ecohydrol. Hydrobiol. 5 (4), 263–268.

Zalewski, M., Wagner, I., Frętczak, W., Mankiewicz-Boczek, J., Parniewski, P., 2012. Blue-Green City compensating global climate change. The Parliament Magazine 6.

Zalewski, M., Wagner-Lutowska, I., Robarts, R.D. (Eds.), 2004. Integrated Watershed Management: Ecohydrology and Phytotechnology – Manual. UNESCO.

Zalewski, M., Wiśniewski, R.J. (Eds.), 1997. *Application of Ecosystem Biotechnologies to Improvement of the Water Body Quality*. Institute of Ecology. Polish Academy of Sciences Press.

Zalewski, M., Arduino, G., Bidoglio, G., Junk, W., Cullmann, J., Uhlenbrook, S., Xia, J., García de Leaniz, C., Rowinski, P.M., Vörösmarty, C.J., Chicharo, L., 2018b. Low cost, nature-based solutions for managing aquatic resources: integrating the principles of Ecohydrology and the Circular Economy. Ecohydrol. Hydrobiol. 18, 309–310. https://doi.org/10.1016/j.ecohyd.2018.12.001.

Zedler, J.B., 2000. Progress in wetland restoration ecology. Trends Ecol. Evol. 15 (10), 402–407.

Zhang, D.Q., Jinadasa, K.B.S.N., Gersberg, R.M., Liu, Y., Ng, W.J., Tan, S.K., 2014. Application of constructed wetlands for wastewater treatment in developing countries – a review of recent developments (2000–2013). J. Environ. Manag. 141, 116–131.

Zhang, Z., Lu, W.X., Zhao, Y., Song, W.B., 2014. Development tendency analysis and evaluation of the water ecological carrying capacity in the Siping area of Jilin Province in China based on system dynamics and analytic hierarchy process. Ecol. Model. 275, 9–21.

Zhao, M.S., Running, S.W., 2010. Drought-induced reduction in global terrestrial net primary production from 2000 through 2009. Science 329 (5994), 940–943.

Zrihun, N.Y., Urbaniak, M., Zalewski, M., 2011. Ecohydrology for a sustainable future in Africa – the cases of Ethiopia, Kenya and Tanzania. Ecohydrol. Hydrobiol. 11 (3–4), 229–244.

Żelazo, J., Popek, Z., 2002. Podstawy Renaturyzacji Rzek. Wydawnictwo SGGW.

Index

'*Note:* Page numbers followed by "f" indicate figures "t" indicate tables and "b" indicate boxes.'

D

E

F

G

T

U

V

W

X

Z

d, Croydon, CR0 4YY

3